Lecture Notes in Mathematics

Volume 2397

This series reports on new developments in all areas of mathematics and their applications - quickly, informally and at a high level. Mathematical texts analysing new developments in modelling and numerical simulation are welcome. The type of material considered for publication includes:

1. Research monographs
2. Lectures on a new field or presentations of a new angle in a classical field
3. Summer schools and intensive courses on topics of current research.

Texts which are out of print but still in demand may also be considered if they fall within these categories. The timeliness of a manuscript is sometimes more important than its form, which may be preliminary or tentative. Please visit the LNM Editorial Policy (https://drive.google.com/file/d/1MOg4TbwOSokRnFJ3ZR3ciEeKs9hOnNX_/view?usp=sharing)

Titles from this series are indexed by Scopus, Web of Science, Mathematical Reviews, and zbMATH.

Christophe Breuil · Zicheng Qian

Splitting and Making Explicit the de Rham Complex of the Drinfeld Space

Springer

Christophe Breuil
Département de Mathématiques
Faculté d'Orsay
CNRS
Université Paris-Saclay
Orsay, France

Zicheng Qian
Morningside Center of Mathematics
Academy of Mathematics and Systems Sciences
Chinese Academy of Sciences
Beijing, China

ISSN 0075-8434 ISSN 1617-9692 (electronic)
Lecture Notes in Mathematics
ISBN 978-3-032-20094-5 ISBN 978-3-032-20095-2 (eBook)
https://doi.org/10.1007/978-3-032-20095-2

Mathematics Subject Classification: 22E50, 14F10, 22E35, 14G22

This Springer imprint is published by the registered company Springer Nature Switzerland AG
The registered company address is: Gewerbestrasse 11, 6330 Cham, Switzerland

Preface

Let p be a prime number and K a finite field extension of the field of p-adic numbers Q_p. This monograph has to do with certain p-adic representations of $\mathrm{GL}_n(K)$. One of its main aims is to completely describe the complex of differential forms of the Drinfeld space of dimension $n-1$ as a complex of representations of $\mathrm{GL}_n(K)$, where $n \geq 2$. The Drinfeld spaces are very important p-adic spaces over K introduced by Drinfeld in the seventies, and there is one Drinfeld space (of dimension $n-1$) for each integer $n \geq 2$. The group $\mathrm{GL}_n(K)$ acts on the Drinfeld space of dimension $n-1$, hence on its complex of differential forms, yielding representations of $\mathrm{GL}_n(K)$ that mathematicians have started studying in the eighties. Understanding these representations was one of the main motivations for the development of the theory of locally analytic representations of $\mathrm{GL}_n(K)$ in the 2000's, which can be seen as a p-adic analogue of Harish-Chandra's (gl_n, K)-modules, where K here is not the above field K but a maximal compact subgroup of $\mathrm{GL}_n(\mathrm{R})$. One difference is that, whereas $\mathrm{GL}_n(\mathrm{R})$ does not act on (gl_n, K)-modules (though it acts on analytic globalizations), the whole group $\mathrm{GL}_n(K)$ acts on locally analytic representations. Though this monograph is largely self-contained and almost all of its proofs are given with full details, it is rather aimed at mathematicians with some elementary knowledge of smooth representations of p-adic groups, of modules over enveloping algebras and of p-adic functional analysis.

Paris, France
Beijing, China
August 2025

Christophe Breuil
Zicheng Qian

Acknowledgements This monograph is the culmination of a long-term work. At various stages of its gestation, we benefited from discussions with several people. In particular, we would like to thank Laurent Clozel, Jean-François Dat, Yiwen Ding, Gabriel Dospinescu, Guy Henniart, Valentin Hernandez, Florian Herzig, Joaquín Rodrigues Jacinto, Stefano Morra, Vincent Pilloni, and Benjamin Schraen. We also thank Benchao Su for pointing out an inaccuracy. Z. Q. is partially supported by National Key R&D Program of China 2025YFA 1018000.

Competing Interests The authors have no competing interests to declare that are relevant to the content of this manuscript.

Contents

Chapter 1
Introduction

Let p be a prime number and K a finite field extension of $\mathbb{Q}_p$, we review the history of the theory of locally analytic representations of $\mathrm{GL}_n(K)$ and its relations to the Drinfeld space. We then we explain the main results of this book.

1.1 Historical Background

We fix $n \geq 2$ an integer and we let $\mathbb{P}^{n-1}_{\mathrm{rig}/K}$ be the rigid analytic projective space of dimension $n-1$ over K. Fifty years ago, Drinfeld introduced in [24] what is now called the Drinfeld space $\mathbb{H}_{/K}$ over K of dimension $n-1$, which is the rigid analytic (admissible) open subspace of $\mathbb{P}^{n-1}_{\mathrm{rig}}$ defined as

$$\mathbb{H} \stackrel{\mathrm{def}}{=} \mathbb{P}^{n-1}_{\mathrm{rig}} \setminus \bigcup_{\mathcal{H}} \mathcal{H} \tag{1.1}$$

where $\mathcal{H}$ runs through the K-rational hyperplanes inside $\mathbb{P}^{n-1}_{\mathrm{rig}}$. It has become such a familiar space in p-adic arithmetic geometry that it is hard to imagine a time when it was not defined. In [25] Drinfeld gave a second definition of $\mathbb{H}$ as a moduli space of certain p-divisible groups, and used this definition to define a tower of étale coverings of $\mathbb{H}$. These coverings have now been vastly generalized into the Rapoport-Zink spaces [58], and more recently the local Shimura varieties [57, 74]. Their cohomology plays a fundamental role in the Langlands program.

In this work, we only use the definition of $\mathbb{H}$ given in (1.1). The group $G \stackrel{\mathrm{def}}{=} \mathrm{GL}_n(K)$ naturally acts on $\mathbb{P}^{n-1}_{\mathrm{rig}}$ and preserves the open subspace $\mathbb{H}$. By functoriality, it follows that any cohomology group of $\mathbb{H}$ is naturally endowed with a left action of G. More than thirty years ago, Schneider and Stuhler in [67] computed this action on any abstract cohomology theory satisfying certain axioms (see [67, Sect. 2] for

C. Breuil and Z. Qian, *Splitting and Making Explicit the de Rham Complex of the Drinfeld Space*, Lecture Notes in Mathematics 2397,
https://doi.org/10.1007/978-3-032-20095-2_1

more details, we won't need these axioms). The first example of such a cohomology theory is the ℓ-adic étale cohomology

$$H^\bullet_{\text{ét}}(\mathbb{H} \times_K \widehat{\overline{K}}, \mathbb{Q}_\ell)$$

where $\widehat{\overline{K}}$ is the p-adic completion of an algebraic closure $\overline{K}$ of K and ℓ is a prime number distinct from p[1]. The second example is the de Rham cohomology

$$H^\bullet_{\text{dR}}(\mathbb{H}).$$

We recall their result. For $P \subseteq G$ the K-points of a lower standard parabolic subgroup of GL_n, we write $\mathrm{Ind}_P^G(1)$ for the smooth parabolic induction of the trivial representation of P. For $k \in \{0, \dots, n-1\}$ we let $P_{[1,n-k-1]}$ be the K-points of the lower standard parabolic subgroup of G of Levi $\mathrm{GL}_{n-k} \times \mathrm{GL}_1 \times \cdots \times \mathrm{GL}_1$. Following our notation in the text (see (2.40) and (2.6)), we define:

$$V^\infty_{[1,n-k-1],\Delta} \stackrel{\text{def}}{=} (\mathrm{Ind}^G_{P_{[1,n-k-1]}} 1)^\infty \;/\; \sum_{P_{[1,n-k-1]} \subsetneq P} (\mathrm{Ind}^G_P 1)^\infty \tag{1.2}$$

(where $(\mathrm{Ind}_P^G(-))^\infty$ is the usual smooth parabolic induction and Δ the set of simple roots of GL_n) which is called a smooth generalized Steinberg representation of G. The representations $V^\infty_{[1,n-k-1],\Delta}$ are absolutely irreducible, and note that $V^\infty_{[1,n-1],\Delta} = 1$ (the trivial representation of G) while $V^\infty_{\emptyset,\Delta}$ is the smooth Steinberg representation of G, that we also denote by St_n^∞.

Theorem 1.1.1 ([67]) *We have $H^k_{\text{ét}}(\mathbb{H} \times_K \widehat{\overline{K}}, \mathbb{Q}_\ell) = H^k_{\text{dR}}(\mathbb{H}) = 0$ for $k \geq n$, and for $k \in \{0, \dots, n-1\}$ we have G-equivariant isomorphisms*

$$H^k_{\text{ét}}(\mathbb{H} \times_K \widehat{\overline{K}}, \mathbb{Q}_\ell) \cong (V^\infty_{[1,n-k-1],\Delta})^\vee \quad \text{and} \quad H^k_{\text{dR}}(\mathbb{H}) \cong (V^\infty_{[1,n-k-1],\Delta})^\vee$$

where the first $V^\infty_{[1,n-k-1],\Delta}$ is seen with $\mathbb{Q}_\ell$-coefficients, the second with K-coefficients, and where $(-)^\vee$ is the corresponding algebraic dual.

More precisely the above theorem follows from [67, Sect. 3 Theorem 1] (with $A = \mathbb{Q}_\ell$ or $A = K$) together with [67, Sect. 4 Lemma 1]. In particular we see that the ℓ-adic and de Rham cohomology groups of $\mathbb{H}$ are duals of *smooth* representations of G, which is not obvious *a priori*.

By [67, Sect. 1 Proposition 4] the rigid analytic space $\mathbb{H}$ is *quasi-Stein*, by which we mean that $\mathbb{H}$ admits an admissible covering given by the union of an ascending sequence $U_1 \subseteq U_2 \subseteq \cdots \subseteq U_n \subseteq \cdots$ of affinoid open subspaces such that the restriction maps of Banach spaces $\Gamma(U_n, \mathcal{O}_{U_n}) \to \Gamma(U_{n-1}, \mathcal{O}_{U_{n-1}})$ have dense image. It follows that $H^k(\mathbb{H}, \mathcal{F}) = 0$ for $k \geq 1$ and $\mathcal{F}$ any coherent sheaf on $\mathbb{H}$ ([38, Satz 2.4]). In particular the de Rham cohomology of $\mathbb{H}$ (which for an arbitrary

[1] The case $\ell = p$ was treated more recently in [17].

rigid space is defined as the hypercohomology of its de Rham complex) is just here the cohomology of the complex $\Omega^\bullet$ of its *global sections*, i.e.

$$H^k_{\mathrm{dR}}(\mathbb{H}) = H^k(\Omega^\bullet)$$

where

$$\Omega^\bullet \stackrel{\mathrm{def}}{=} [\Omega^0 \longrightarrow \Omega^1 \longrightarrow \cdots \longrightarrow \Omega^{n-1}]$$

with

$$\Omega^k \stackrel{\mathrm{def}}{=} \Gamma(\mathbb{H}, \Omega^k_{\mathbb{H}/K}). \tag{1.3}$$

By functoriality each Ω^k is also a representation of G over K and the complex $\Omega^\bullet$ is G-equivariant. Moreover, as we have seen in Theorem 1.1.1, $H^k(\Omega^\bullet)$ is the (algebraic) dual of a smooth representation of G. However, the individual representations Ω^i are far from being (duals of) smooth representations of G.

Investigating which type of G-action they carry was one of the main motivations for Schneider and Teitelbaum's theory of (admissible) locally analytic representations of G, and completely determining the representations Ω^k of G and the differential maps in $\Omega^\bullet$ is one of the main goals of this monograph. Indeed, motivated by the beginnings of the p-adic Langlands program, the first author has been fascinated during years (in the early 2000s) by the idea of fully understanding the internal structure of $\Omega^\bullet$ as a complex of representations of G, with the hope that it was maybe hiding "secrets". In the end, this internal structure is not that complicated and does not hide that many secrets, but completely unravelling $\Omega^\bullet$ remains an interesting and challenging (though sometimes technical) task.

Writing $\mathbb{H} = \cup_n U_n$ for affinoids $U_n \subseteq U_{n+1}$ as above, one has $\Omega^k \xrightarrow{\sim} \varprojlim_n \Gamma(U_n, \Omega^k_{\mathbb{H}/K})$. Since all $\Gamma(U_n, \Omega^k_{\mathbb{H}/K})$ are (p-adic) Banach spaces, the projective limit topology gives a natural (p-adic) Fréchet topology on each Ω^k (recall that a p-adic Fréchet space is a projective limit of countably many p-adic Banach spaces), and it is not very hard to check that the map $G \times \Omega^k \to \Omega^k$, $(g, v) \mapsto gv$ is continuous. In the eighties, Morita pioneered the study of the continuous G-representations Ω^0 and Ω^1 when $n = 2$ [45–47]. In particular in [46, Sect. 5] he proved the following theorem:

Theorem 1.1.2 ([46]) *Let $B \subseteq G = \mathrm{GL}_2(K)$ be the K-points of the lower Borel of* GL_2, *then one has G-equivariant topological isomorphisms*

$$\Omega^0/1 \cong \big((\mathrm{Ind}_B^G\, t^{-1} \boxtimes t)^{\mathrm{an}}\big)^\vee \quad and \quad \Omega^1 \cong \big((\mathrm{Ind}_B^G\, 1)^{\mathrm{an}}/1\big)^\vee.$$

Here $((\mathrm{Ind}_B^G \chi_1 \boxtimes \chi_2)^{\mathrm{an}})^\vee$ is the topological dual of the locally K-analytic principal series $(\mathrm{Ind}_B^G \chi_1 \boxtimes \chi_2)^{\mathrm{an}}$ where $\chi_i : K^\times \to K$ are locally K-analytic characters and t is the identity character $K^\times \to K$, $t \mapsto t$.

Recall that $(\mathrm{Ind}_B^G \chi_1 \boxtimes \chi_2)^{\mathrm{an}}$ is the K-vector space of locally K-analytic functions $f : G \longrightarrow K$ such that

$$f(gb) = b^{-1} \cdot f(g) = (\chi_1(t_1)\chi_2(t_2))^{-1} f(g), \quad g \in G,\ b = \left(\begin{smallmatrix} t_1 & 0 \\ * & t_2 \end{smallmatrix}\right) \in B$$

with the (left) action of G given by

$$(g(f))(g') \stackrel{\text{def}}{=} f(g^{-1}g'), \quad g, g' \in G. \tag{1.4}$$

It is naturally endowed with a topology which makes it a locally convex K-vector space of compact type for which the action of G is continuous (see [73, Sect. 6] and [70, Lemma 2.1]). The representation $(\mathrm{Ind}_B^G 1)^{\mathrm{an}}/1$ is called the locally K-analytic Steinberg representation of G. Note that Morita more generally proves Theorem 1.1.2 for (what is now called) Schneider's holomorphic discrete series ([64, Sect. 3]). In [47, Theorem 1] and [46, Theorem 2] (see also [46, Theorem 1] and [46, Theorem 2]), Morita moreover proves that $(\mathrm{Ind}_B^G t^{-1} \boxtimes t)^{\mathrm{an}}$ is (topologically) irreducible as a G-representation and that $(\mathrm{Ind}_B^G 1)^{\mathrm{an}}/1$ is an extension of $(\mathrm{Ind}_B^G t^{-1} \boxtimes t)^{\mathrm{an}}$ by the (irreducible) smooth Steinberg representation of G.[2]

For many years, there was no progress on and (maybe) no interest in these p-adic questions (except [64]) until Schneider and Teitelbaum, using Féaux de Lacroix's thesis [43], decided to start the theory from scratch in [69–73]. In particular in [72] they defined an important abelian category of admissible locally K-analytic representations of G (more generally of a locally K-analytic group) on locally convex K-vector spaces of compact type, or equivalently taking continuous duals an abelian category of (so called) coadmissible $D(G)$-modules. Here $D(G)$ is the K-algebra of locally K-analytic distributions on G, i.e. the continuous dual of

$$C^{\mathrm{an}}(G) \stackrel{\text{def}}{=} \{f : G \longrightarrow K, \quad f \text{ locally } K\text{-analytic}\}$$

endowed with its natural locally convex topology ([70, Sect. 2]). Shortly after, Emerton gave his own account of the theory in [27] (which was published years later). All this work ultimately lead, a decade or so later, to the beautiful theory of Orlik-Strauch representations in [52], which can be seen as one of the key achievements of the theory.

Orlik-Strauch representations $\mathcal{F}_P^G(M, \pi^\infty)$ are now widely used and are, essentially, the only admissible locally K-analytic representations of G which are so far well understood. We briefly review their main properties in the text (see Theorem 4.3.7). Suffice it here to say that, if P is the K-points of a (lower standard) parabolic subgroup of GL_n, M a $U(\mathfrak{g})$-module in Bernstein-Gelfand-Gelfand category's $\mathcal{O}^{\mathfrak{p}}$ ([33, Sect. 9.3]) such that all its weights are integral (here $\mathfrak{g}$, $\mathfrak{p}$ are the respective K-Lie algebras of G, P and $U(-)$ the enveloping algebra) and π^∞ a smooth admissible representation of the Levi factor L_P of P, then

$$\mathcal{F}_P^G(M, \pi^\infty) \stackrel{\text{def}}{=} \left((\mathrm{Ind}_P^G\, W^\vee \otimes \pi^\infty)^{\mathrm{an}} \right)^{\ker = 0} \tag{1.5}$$

---

[2] Morita's results are proven for $\mathrm{SL}_2(K)$ instead of $\mathrm{GL}_2(K)$. Moreover the proof of the irreducibility is actually flawed, see [70, pp. 443–444].

where W is any finite dimensional algebraic representation of P such that one has a surjection of $U(\mathfrak{g})$-modules $U(\mathfrak{g}) \otimes_{\mathfrak{p}} W \twoheadrightarrow M$ and where $(-)^{\ker=0}$ is the (closed) subrepresentation of the locally K-analytic parabolic induction $(\mathrm{Ind}_P^G W^\vee \otimes \pi^\infty)^{\mathrm{an}}$ of vectors killed by

$$\ker \stackrel{\mathrm{def}}{=} \ker\big(U(\mathfrak{g}) \otimes_{\mathfrak{p}} W \twoheadrightarrow M\big).$$

Here $(\mathrm{Ind}_P^G W^\vee \otimes \pi^\infty)^{\mathrm{an}}$ is defined similarly to $(\mathrm{Ind}_B^G \chi_1 \boxtimes \chi_2)^{\mathrm{an}}$ above with left action (1.4) of G. We refer the reader to [52] for more details on the admissible representations $\mathcal{F}_P^G(M, \pi^\infty)$. In particular, when π^∞ is of finite length, they are moreover (topologically) of finite length.

Going back to the G-representations Ω^i of (1.3), the first result after Morita's Theorem 1.1.2 came in [71] where the authors could describe explicitly the graded pieces of a filtration on Ω^{n-1} (for any $n \geq 2$). More precisely in [71, Theorem 8.6] they proved:

Theorem 1.1.3 ([71]) *The $D(G)$-module Ω^{n-1} admits a filtration by closed $D(G)$-submodules*

$$\Omega^{n-1} = \mathrm{Fil}^0(\Omega^{n-1}) \supsetneq \mathrm{Fil}^1(\Omega^{n-1}) \supsetneq \cdots \supsetneq \mathrm{Fil}^n(\Omega^{n-1}) = 0$$

such that for $0 \leq j \leq n-1$:

$$\mathrm{gr}^j(\Omega^{n-1}) \stackrel{\mathrm{def}}{=} \mathrm{Fil}^j(\Omega^{n-1})/\mathrm{Fil}^{j+1}(\Omega^{n-1}) \cong \mathcal{F}_{P_{\widehat{j}}}^G\big(L(s_j s_{j-1} \cdots s_1 \cdot 0), 1_{\mathrm{GL}_j} \boxtimes \mathrm{St}_{n-j}^\infty\big)^{\vee 3} \tag{1.6}$$

where $P_{\widehat{j}}$ is the K-points of the lower parabolic subgroup of GL_n of Levi factor $\mathrm{GL}_j \times \mathrm{GL}_{n-j}$, $s_1, \ldots, s_{n-1}$ are the simple reflections of GL_n and $L(s_j s_{j-1} \cdots s_1 \cdot 0)$ is the unique irreducible $U(\mathfrak{g})$-module in $\mathcal{O}^{\mathfrak{p}_{\widehat{j}}}$ of highest weight $s_j s_{j-1} \cdots s_1 \cdot 0$ (with $P_{\widehat{j}} = G$, $s_j s_{j-1} \cdots s_1 = 1$ when $j = 0$, and the dot action $\cdot$ being relative to the lower Borel subgroup of GL_n).

Since all graded pieces in Theorem 1.1.3 are finite length coadmissible $D(G)$-modules, and since coadmissibility and finite length are preserved under extensions (for coadmissibility see for instance the proof of [11, Lemme 2.1.1]), we see in particular that Ω^{n-1} is also a finite length coadmissible $D(G)$-module. To prove some of the results of this monograph (see Sect. 1.2 below), we use a weak variant of Theorem 1.1.3 as a *key* ingredient (see Theorem 5.4.14 in the text together with the comment before Theorem 5.4.16).

A few years later, Pohlkamp in [55] proved a result analogous to Theorem 1.1.3 but where Ω^{n-1} is replaced by the global sections Ω^0 of the structural sheaf of $\mathbb{H}$. Finally, a few more years later Orlik considerably generalized both statements to Ω^k for all k in [49] (see also [50] or Theorem 5.4.14 in the text). In particular all

---

[3] The reference [52] was not available at the time of [71] but they directly used the description on the right hand side of (1.5).

$D(G)$-modules Ω^k are coadmissible and (topologically) of finite length. Note that, when $0 < k < n-1$, Ω^k is more evolved because it has (essentially) twice as many irreducible constituents as Ω^{n-1} or Ω^0.

Although the above theorems describe the graded pieces of a filtration on Ω^k, and in particular the irreducible constituents of Ω^k, this does not give its full internal structure. For instance we do not know the extensions between the graded pieces (some extensions as subquotients could be split). Or it could be that there are several analogous finite length coadmissible $D(G)$-modules with the same graded pieces but with different extensions as subquotients. As an example, let us go back to the case $n = 2$ and Theorem 1.1.2, and denote $\mathrm{SP} \stackrel{\mathrm{def}}{=} (\mathrm{Ind}_B^G\, t^{-1} \boxtimes t)^{\mathrm{an}}$. Although Ω^1 is fully determined there, this is not the case of Ω^0: we only know it is an extension of SP by the trivial representation 1. To obtain the full structure of Ω^0, we need to know that it is the unique such non-split extension.[4] In particular for $n = 2$ we can make the complex $\Omega^\bullet = [\Omega^0 \longrightarrow \Omega^1]$ transparent by rewriting it

$$\Omega^\bullet = \Big[(1 \text{——} \mathrm{SP}^\vee) \longrightarrow (\mathrm{SP}^\vee \text{——} (\mathrm{St}_2^\infty)^\vee) \Big] \tag{1.7}$$

where as usual a line means a non-split extension between two irreducible constituents (with the sole on the left and the cosocle on the right) and where we use that $(\mathrm{Ind}_B^G 1)^{\mathrm{an}}/1$ is the (unique) non-split extension of SP by the smooth Steinberg St_2^∞. An analogous complete description of $\Omega^\bullet$ for $G = \mathrm{GL}_3(\mathbb{Q}_p)$ was given by Schraen in his thesis ([76, Sect. 6.4]).

The history of $\Omega^\bullet$ (so far) did not quite stop there. By another result of Schraen ([76, Theorem 6.1][5]) crucially based on results of Orlik ([48, Theorem 1]) and on a theorem of Dat ([18, Corollary A.1.3]) which itself is an elaboration of Deligne's splitting result ([19]), the complex $\Omega^\bullet$ splits in the bounded derived category of all (abstract) $D(G)$-modules, i.e. there exists an isomorphism in this derived category:

$$\Omega^\bullet \cong \bigoplus_{k=0}^{n-1} (V^\infty_{[1,n-k-1],\Delta})^\vee[-k]. \tag{1.8}$$

However, trying to unravel the abstract proofs of [18, Corollary A.1.3] and [76, Theorem 6.1] to produce an explicit such isomorphism, say in the bounded derived category of finite length coadmissible $D(G)$-modules (instead of all $D(G)$-modules), seems seriously challenging. Yet, when $n = 2$, one can easily produce an explicit such isomorphism as follows. Choosing a p-adic logarithm $\log : K^\times \to K$, one can "glue"

---

[4] Curiously the authors could not find this classical result explicitly and clearly stated in the literature (it follows for instance from the baby case $n = 2$, $k = 0$, $\mu_0 = (0, 0)$ of Theorem 5.4.16 in the text).

[5] Although the result is stated there only for $K = \mathbb{Q}_p$, its proof works for arbitrary K using Lemma 4.2.3.

Ω^0 and Ω^1 in (1.7) into one length 3 $D(G)$-module $1 — \mathrm{SP}^\vee — (\mathrm{St}_2^\infty)^\vee$ (see for instance [11, Sect. 3.2]). One then has an explicit section in the derived category $(\mathrm{St}_2^\infty)^\vee[-1] \dashrightarrow \Omega^\bullet$ to the canonical morphism of complexes $\Omega^\bullet \to H^1(\Omega^\bullet)[-1] \cong (\mathrm{St}_2^\infty)^\vee[-1]$ provided by (see [75, Sect. 5.1]):

$$(\mathrm{St}_2^\infty)^\vee[-1] \longleftarrow \Big[(1 — \mathrm{SP}^\vee) \to (1 — \mathrm{SP}^\vee — (\mathrm{St}_2^\infty)^\vee)\Big] \longrightarrow \Omega^\bullet \quad (1.9)$$

(where the morphisms of complexes are easily guessed and are quasi-isomorphisms). Adding up the (trivial) morphism of complexes $1[0] \to \Omega^\bullet$, we deduce an isomorphism as in (1.8) $1[0] \oplus (\mathrm{St}_2^\infty)^\vee[-1] \overset{\sim}{\dashrightarrow} \Omega^\bullet$. In fact, there exists a slightly better variant which, in the case $K = \mathbb{Q}_p$, is more directly related to the p-adic local Langlands correspondence for $\mathrm{GL}_2(\mathbb{Q}_p)$. Let $\varepsilon : K^\times \to K$ be the p-adic cyclotomic character (which factors as $K^\times \to \mathbb{Q}_p^\times \twoheadrightarrow \mathbb{Z}_p^\times \subset K$ where the first map is the norm and the second the projection sending p to 1) and let $\mathrm{SP}' \overset{\mathrm{def}}{=} (\mathrm{Ind}_B^G \varepsilon^{-1} \boxtimes \varepsilon)^{\mathrm{an}}$. Then one can add in a unique way the constituent $\mathrm{SP}'^\vee$ to each length 3 $D(G)$-module as above and obtain a length 4 $D(G)$-module

$$\mathrm{SP}'^\vee — 1 — \mathrm{SP}^\vee — (\mathrm{St}_2^\infty)^\vee .$$

The isomorphism classes of such $D(G)$-modules are in non-canonical bijection with $\mathbb{A}^1(K)$ (which corresponds to the choice of $\log(p) \in K$ in the p-adic logarithm). When $K = \mathbb{Q}_p$, such length 4 representations precisely correspond to 2-dimensional semi-stable non-crystalline representations of $\mathrm{Gal}(\overline{\mathbb{Q}_p}/\mathbb{Q}_p)$ (over $\mathbb{Q}_p$) of Hodge-Tate weights (0, 1) (see [16] or [11, Sect. 3.1], the value of $\log(p) \in K$ is related to Fontaine's $\mathcal{L}$-invariant on the Galois side). One then has a section (in the derived category) similar to (1.9)

$$(\mathrm{St}_2^\infty)^\vee[-1] \longleftarrow \Big[(\mathrm{SP}'^\vee — 1 — \mathrm{SP}^\vee) \to (\mathrm{SP}'^\vee — 1 — \mathrm{SP}^\vee — (\mathrm{St}_2^\infty)^\vee)\Big] \longrightarrow \Omega^\bullet. \quad (1.10)$$

Following the notation of this text, the constituent $\mathrm{SP}^\vee$ is denoted X_1 and the constituent $\mathrm{SP}'^\vee$ is denoted Y_1.

With an explicit description of $\Omega^\bullet$ at hand for $n \geq 2$ (as we will soon have), it becomes tempting to look for a generalization of (1.10) to $n \geq 3$. We provide such explicit sections in this work, which also apply to Schneider's holomorphic discrete series.

1.2 The Main Results

The first aim of this monograph is to finish the work started by Schneider-Teitelbaum and continued by Orlik (and Schraen) by giving a transparent description of $\Omega^\bullet$ for $n \geq 2$ analogous to (1.7) when $n = 2$.

We first need a bit of notation. For $\underline{j} = (j_0, j_1, j_2)$ such that

$$1 \leq j_0, j_1 \leq n-1, \ 1 \leq j_2 \leq n \text{ and } 0 \leq j_2 - j_1 \leq n-1 \tag{1.11}$$

we set using the notation in (1.5) (see (5.68))

$$C_{\underline{j}} = C_{(j_0,j_1,j_2)} \stackrel{\text{def}}{=} \mathcal{F}^G_{P_{\widehat{j_1}}}(L(w_{j_1,j_0} \cdot 0), \pi^\infty_{j_1,j_2}) \tag{1.12}$$

where $P_{\widehat{j_1}}$ is as in Theorem 1.1.3, $w_{j_1,j_0} \stackrel{\text{def}}{=} \begin{cases} s_{j_1}s_{j_1-1}\cdots s_{j_0} & \text{if } j_1 \geq j_0 \\ s_{j_1}s_{j_1+1}\cdots s_{j_0} & \text{if } j_1 \leq j_0 \end{cases}$, $L(w_{j_1,j_0} \cdot 0)$ is as in Theorem 1.1.3 and where $\pi^\infty_{j_1,j_2}$ is an explicit irreducible smooth representation of $\mathrm{GL}_{j_1}(K) \times \mathrm{GL}_{n-j_1}(K)$ defined in (2.71). The $C_{\underline{j}}$ are irreducible admissible locally analytic representations of G over K. For $k \in \{1, \dots, n-1\}$ one first proves that there exists a unique finite length coadmissible $D(G)$-module X_k of the following form (see (5.135) with Theorem 5.3.6 and recall $(-)^\vee$ means the continuous dual):

$$X_k = C^\vee_{(n-k,n-1,n-1)} \text{ ——— } C^\vee_{(n-k,n-2,n-2)} \text{ ——— } \cdots \text{ ——— } C^\vee_{(n-k,2,2)} \text{ ——— } C^\vee_{(n-k,1,1)} \tag{1.13}$$

(we say that X_k is uniserial).

Theorem 1.2.1 (Theorem 5.4.16, (ii) of Corollary 5.4.3) *For $k \in \{0, \dots, n-1\}$ Ω^k is the unique coadmissible $D(G)$-module of the form*

$$X_k \text{ ——— } (V^\infty_{[1,n-k-1],\Delta})^\vee \text{ ——— } X_{k+1}$$

with $X_0 = X_n \stackrel{def}{=} 0$. It is indecomposable multiplicity free with an irreducible socle and cosocle. Moreover the k-th differential map is the unique (up to non-zero scalar) non-zero map of $D(G)$-modules $\Omega^k \to \Omega^{k+1}$.

The first statement of Theorem 1.2.1 follows from Theorem 5.4.16 and (ii) of Corollary 5.4.3, while the second statement follows from (5.136) and (5.137). Note that the differential map in Theorem 1.2.1 obviously factors as

$$X_k - (V^\infty_{[1,n-k-1],\Delta})^\vee - X_{k+1} \twoheadrightarrow X_{k+1} \hookrightarrow X_{k+1} - (V^\infty_{[1,n-k-2],\Delta})^\vee - X_{k+2}\,.$$

In fact one exactly knows the form of $\Omega^k \to \Omega^{k+1}$, which looks like (for $2 \leq k \leq n-4$):

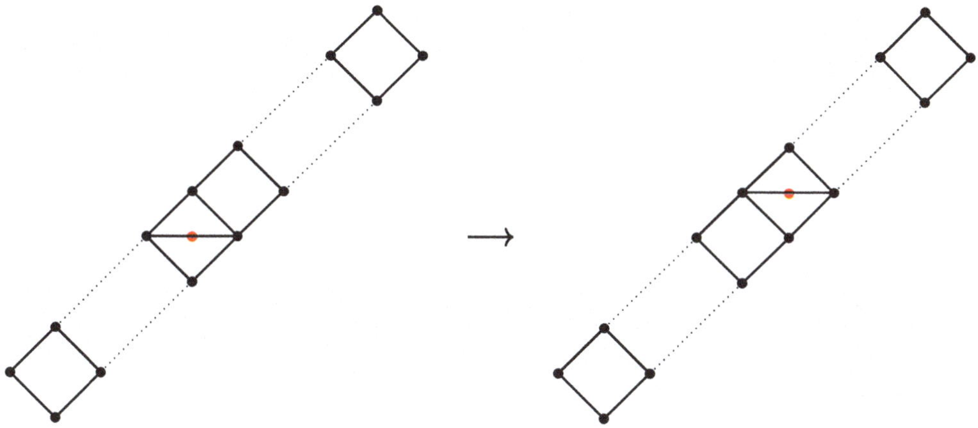

(1.14)

where each bullet is an irreducible constituent, where the 4 long diagonals are (from left to right) X_k, X_{k+1}, X_{k+1}, X_{k+2}, where the left (resp. right) red bullet is $(V^{\infty}_{[1,n-k-1],\Delta})^{\vee}$ (resp. $(V^{\infty}_{[1,n-k-2],\Delta})^{\vee}$) and where the socle (resp. cosocle) of Ω^k or Ω^{k+1} is the leftmost (resp. rightmost) bullet. All this follows from Theorem 5.4.16 and the description of D_k in (5.137). See also the full complex $\Omega^{\bullet}$ for $n = 4$ and $n = 5$ in Figure 1 and Figure 7 of Appendix B where we see that the dual of the smooth constituent (i.e. the red bullet) goes up and up when moving from left to right.

Theorem 1.2.1 is more generally proven for holomorphic discrete series ([64]), where for instance $(V^{\infty}_{[1,n-k-1],\Delta})^{\vee}$ is replaced by $(V^{\infty}_{[1,n-k-1],\Delta})^{\vee} \otimes_K L(\mu_0)$ for an arbitrary dominant weight μ_0 (with respect to the lower Borel).

Let $\mathrm{St}_n^{\mathrm{an}} \overset{\mathrm{def}}{=} (\mathrm{Ind}_B^G 1)^{\mathrm{an}} / \sum_{B \subsetneq P} (\mathrm{Ind}_P^G 1)^{\mathrm{an}}$ be the locally K-analytic Steinberg representation of G (where B is the K-points of the lower Borel of GL_n). Then Theorem 1.2.1 is used in [56] to show that the K-vector space of homomorphisms $(\mathrm{St}_n^{\mathrm{an}})^{\vee}[1-n] \dashrightarrow \Omega^{\bullet}$ in the derived category of $D(G)$-modules has a natural structure of an admissible filtered (φ, N)-module in the sense of Fontaine [28] corresponding to a certain explicit "universal" semi-stable non-crystalline p-adic representation of $\mathrm{Gal}(\overline{K}/K)$.

The second aim of this monograph is to generalize the section (1.10) to $n \geq 2$.

But there is a problem to fix. As soon as $n > 2$, it is *impossible* in general to "glue" consecutive Ω^k as was done to obtain (1.9). For instance, already for $n = 3$, one cannot "glue" Ω^0 and Ω^1 as it turns out that the $D(G)$-module (using the notation of (1.14) for $n = 3$)

$$1 \text{ --- } X_1 \text{ --- } (V^{\infty}_{\{1\},\Delta})^{\vee} \text{ --- } X_2 \simeq$$ 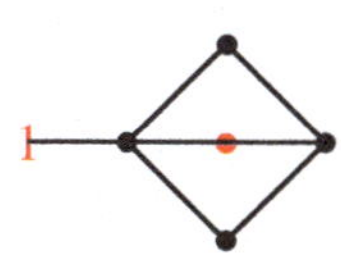

(where the red bullet is $(V^{\infty}_{\{1\},\Delta})^{\vee}$) does not exist. When $K = \mathbb{Q}_p$ and $n = 3$, the first author proved in 2019 that there exists a coadmissible length 2 $D(G)$-module $Y_2^{\flat}$ such that there exist a unique coadmissible $D(G)$-module of the form

$$\widetilde{\Omega}^{1\flat} \overset{\text{def}}{=} 1 \begin{array}{c} \diagup\; Y_2^{\flat} \;\diagdown \\ \\ \diagdown\; X_1 \;\diagup \end{array} (V^{\infty}_{\{1\},\Delta})^{\vee} \;\text{---}\; X_2, \tag{1.15}$$

and non-unique coadmissible $D(G)$-modules of the form

$$\widetilde{\Omega}^{2\flat} \overset{\text{def}}{=} Y_2{}^{\flat} \;\text{---}\; (V^{\infty}_{\{1\},\Delta})^{\vee} \;\text{---}\; X_2 \;\text{---}\; (\mathrm{St}_3^{\infty})^{\vee}$$

where the non-unicity is similar to the non-unicity of $1 \;\text{---}\; X_1 \;\text{---}\; (\mathrm{St}_2^{\infty})^{\vee}$ in (1.9). Hence, using $Y_2{}^{\flat}$ and remembering that $\Omega^0 \simeq 1 \;\text{---}\; X_1$, one still has a surjection of complexes $[\Omega^0 \to \widetilde{\Omega}^{1\flat} \to \widetilde{\Omega}^{2\flat}] \twoheadrightarrow (\mathrm{St}_3^{\infty})^{\vee}[-2]$ which is a quasi-isomorphism and a natural morphism of complexes $[\Omega^0 \to \widetilde{\Omega}^{1\flat} \to \widetilde{\Omega}^{2\flat}] \to \Omega^{\bullet}$ which give an explicit section $(\mathrm{St}_3^{\infty})^{\vee}[-2] \dashrightarrow \Omega^{\bullet}$ to the morphism $\Omega^{\bullet} \to H^2(\Omega^{\bullet})[-2] \cong (\mathrm{St}_3^{\infty})^{\vee}[-2]$ in the derived category of $D(G)$-modules:

$$(\mathrm{St}_3^{\infty})^{\vee}[-2] \longleftarrow [\Omega^0 \to \widetilde{\Omega}^{1\flat} \to \widetilde{\Omega}^{2\flat}] \longrightarrow \Omega^{\bullet}.$$

Just like (1.10) is better than (1.9) when $n = 2$, we can also define $\widetilde{\Omega}^0 \overset{\text{def}}{=} Y_1 \;\text{---}\; 1 \;\text{---}\; X_1$ where Y_1 is a certain finite length coadmissible $D(G)$-module Y_1, and modify accordingly $Y_2{}^{\flat}$ and $\widetilde{\Omega}^{1\flat}$, $\widetilde{\Omega}^{2\flat}$ into slightly larger (finite length coadmissible) $D(G)$-modules Y_2 and

$$\widetilde{\Omega}^{1} \overset{\text{def}}{=} Y_1 \;\text{---}\; 1 \begin{array}{c} \diagup\; Y_2 \;\diagdown \\ \\ \diagdown\; X_1 \;\diagup \end{array} (V^{\infty}_{\{1\},\Delta})^{\vee} \;\text{---}\; X_2\,,\quad \widetilde{\Omega}^{2} \overset{\text{def}}{=} Y_2 \;\text{---}\; (V^{\infty}_{\{1\},\Delta})^{\vee} \;\text{---}\; X_2 \;\text{---}\; (\mathrm{St}_3^{\infty})^{\vee}\,.$$

This has two advantages. The first is that, as for $n = 2$, the continuous dual of $\widetilde{\Omega}^2$ is then similar when $K = \mathbb{Q}_p$ to the representations of $\mathrm{GL}_3(\mathbb{Q}_p)$ appearing in the completed cohomology for 3-dimensional semi-stable non-crystalline representations of $\mathrm{Gal}(\overline{\mathbb{Q}_p}/\mathbb{Q}_p)$ with Hodge-Tate weights $(0, 1, 2)$ (for instance the dual of $\widetilde{\Omega}^2$ is exactly like the "upper branch" of the representation in [12, (1.1)]). The second (which was discovered in the present work) is that the $D(G)$-modules $\widetilde{\Omega}^0$, $\widetilde{\Omega}^1$, $\widetilde{\Omega}^2$ behave remarkably well with respect to wall-crossing functors (as will be explained in Sect. 1.3 below) contrary to the $D(G)$-modules Ω^0, $\widetilde{\Omega}^{1\flat}$, $\widetilde{\Omega}^{2\flat}$.

The coadmissible $D(G)$-modules Y_k and $\widetilde{\Omega}^k$ turn out to nicely generalize to any $n \geq 2$.

Theorem 1.2.2 *There exist indecomposable multiplicity free finite length coadmissible $D(G)$-modules $Y_1, \dots, Y_{n-1}$ with irreducible socle and cosocle satisfying the following properties.*

(i) For $k \in \{0, \dots, n-2\}$ there exists a unique coadmissible $D(G)$-module $\widetilde{\Omega}^k$ of the form

$$\begin{array}{ccccccccc} & & & & Y_{k+1} & & & & \\ & & & \nearrow & & \searrow & & & \\ Y_k & \text{———} & (V^{\infty}_{[1,n-k],\Delta})^{\vee} & & & & (V^{\infty}_{[1,n-k-1],\Delta})^{\vee} & \text{———} & X_{k+1} \\ & & & \searrow & & \nearrow & & & \\ & & & & X_k & & & & \end{array}$$

where $Y_0 = X_0 = (V^{\infty}_{[1,n],\Delta})^{\vee} \overset{def}{=} 0$. Moreover $\widetilde{\Omega}^k$ is indecomposable multiplicity free with irreducible socle and cosocle.

(ii) The set of isomorphism classes of coadmissible $D(G)$-modules $\widetilde{\Omega}^{n-1}$ of the form

$$Y_{n-1} \text{ ——— } (V^{\infty}_{\{1\},\Delta})^{\vee} \text{ ——— } X_{n-1} \text{ ——— } (\mathrm{St}^{\infty}_n)^{\vee}$$

is in non-canonical bijection with $\mathbb{A}^{n-1}(K)$.[6] Moreover any such $\widetilde{\Omega}^{n-1}$ is indecomposable multiplicity free with irreducible socle and cosocle.

Contrary to the $D(G)$-modules X_k in (1.13), the $D(G)$-modules Y_k are not uniserial, they look like "triangles", and contrary to X_k they contain duals of smooth irreducible constituents (when $k > 1$), see for instance Figs. 2 and 8 in Appendix B when $n = 4$ and $n = 5$. To represent the $D(G)$-modules $\widetilde{\Omega}^k$ one needs 3-dimensional drawings (at least when $k \in \{1, \dots, n-2\}$), see for instance Figs. 3–6 and 9–13 in Appendix B when $n = 4$ and $n = 5$ (where $\widetilde{\Omega}^k$ is denoted $\widetilde{D}_k$).

The proof of Theorem 1.2.2 is disseminated throughout the paper. The first statement about the Y_k follows from (5.135). The first statement of (i) of Theorem 1.2.2 follows from (iv) of Theorem 5.3.10 with (i), (iii) of Corollary 5.4.3, while the second statement follows from (5.137). The first statement of (ii) of Theorem 1.2.2 follows from (ii) of Theorem 5.3.11 with (ii) of Theorem 5.3.10 (for $k = n-1$) and the discussion around (5.148), while the second statement follows from (ii) of Lemma 5.2.2, the definition of X_{n-1} in (5.135) (or (1.13)) and the definition and unicity of Z_{n-1} in (5.135) and (i) of Corollary 5.4.3. Moreover, like Theorems 1.2.1, 1.2.2 is also proved in the setting of holomorphic discrete series.

As for $n = 2$ and $n = 3$ when $K = \mathbb{Q}_p$, one can expect that, for all $n \geq 2$ and all K, the duals of some of the $D(G)$-modules $\widetilde{\Omega}^{n-1}$ in (ii) of Theorem 1.2.2 (i.e. for some values of the parameter in $\mathbb{A}^{n-1}(K)$) occur in the completed cohomology for n-dimensional semi-stable non-crystalline representations of $\mathrm{Gal}(\overline{K}/K)$ with Hodge-Tate weights $(0, 1, \dots, n-1)$ in all directions. The values of the parameter should

---

[6] One has $\dim_K \mathrm{Ext}^1_{D(G)}\big((\mathrm{St}^{\mathrm{alg}}_n)^{\vee}, Y_{n-1} \text{ —— } (V^{\infty}_{\{1\},\Delta})^{\vee} \text{ —— } X_{n-1}\big) = n$, see (ii) of Theorem 5.3.11.

then be determined by the Hodge filtration on Fontaine's filtered (φ, N)-module associated to the $\mathrm{Gal}(\overline{K}/K)$-representation, see for instance [12, Theorem 1.1] or [13, Theorem 1.9] when $n = 3$ and $K = \mathbb{Q}_p$. More generally one can wonder if the duals of *all* the $D(G)$-modules $\widetilde{\Omega}^k$ for $k \in \{0, \dots, n-2\}$ in (i) of Theorem 1.2.2 do not also occur as *subquotients* in the completed cohomology for such Galois representations (for instance this is obvious for $n = 2$ and one can check it for $\widetilde{\Omega}^1$, and thus $\widetilde{\Omega}^0$, when $n = 3$ and $K = \mathbb{Q}_p$).

Using Theorem 1.2.2 (and Theorem 1.2.1), we immediately obtain a section to

$$\Omega^\bullet \longrightarrow H^{n-1}(\Omega^\bullet)[-(n-1)] \cong (\mathrm{St}_n^\infty)^\vee[-(n-1)]$$

as follows. There are unique (up to scalar) non-zero $D(G)$-equivariant morphisms $\widetilde{\Omega}^k \to \widetilde{\Omega}^{k+1}$ for $k \in \{0, \dots, n-2\}$. We can thus define the complex of finite length coadmissible $D(G)$-modules (fixing an arbitrary choice for $\widetilde{\Omega}^{n-1}$)

$$\widetilde{\Omega}^\bullet \stackrel{\text{def}}{=} [\widetilde{\Omega}^0 \to \widetilde{\Omega}^1 \to \cdots \to \widetilde{\Omega}^{n-2} \to \widetilde{\Omega}^{n-1}]$$

which is exact in degree $< n-1$ and has cohomology $(\mathrm{St}_n^\infty)^\vee$ in degree $n-1$. In particular there is a quasi-isomorphism $\widetilde{\Omega}^\bullet \to (\mathrm{St}_n^\infty)^\vee[-(n-1)]$. There are also unique (up to non-zero scalar) non-zero $D(G)$-equivariant morphisms $\widetilde{\Omega}^k \to \Omega^k$ for $k \in \{0, \dots, n-1\}$ (which are surjections) which give a morphism of complexes $\widetilde{\Omega}^\bullet \to \Omega^\bullet$.

Corollary 1.2.3 (Theorem 5.3.13, Corollary 5.4.19) *There is an explicit section*

$$(\mathrm{St}_n^\infty)^\vee[-(n-1)] \dashrightarrow \Omega^\bullet$$

in the derived category of finite length coadmissible $D(G)$-modules to the canonical morphism of complexes $\Omega^\bullet \to (\mathrm{St}_n^\infty)^\vee[-(n-1)]$ given by

$$(\mathrm{St}_n^\infty)^\vee[-(n-1)] \longleftarrow \widetilde{\Omega}^\bullet \longrightarrow \Omega^\bullet.$$

We see that this explicit section exists in the derived category of finite length coadmissible $D(G)$-modules with all irreducible constituents being (duals of) Orlik-Strauch representations (1.5). Corollary 1.2.3 and (1.8) raise the question of the existence of analogous explicit sections to the morphisms of complexes $\tau_{\le \ell}\Omega^\bullet \twoheadrightarrow H^\ell(\Omega^\bullet)[-\ell] \cong (V^\infty_{[1,n-\ell-1],\Delta})^\vee[-\ell]$ for $\ell \in \{0, \dots, n-2\}$. This is obvious for $\ell = 0$ and not too difficult for $\ell = 1$ (Proposition 5.3.14). In particular we obtain a full explicit splitting of $\Omega^\bullet$ when $n = 3$ (Corollary 5.3.15). For some time we thought we could use the $D(G)$-modules $\widetilde{\Omega}^k$ to also obtain sections for all $\ell \in \{2, \dots, n-2\}$ (in the derived category of finite length coadmissible $D(G)$-modules with Orlik-Strauch irreducible constituents), but this does not seem to work. Nevertheless, we do expect such sections to exist:

Conjecture 1.2.4 (Conjecture 5.3.16) For $\ell \in \{2, \dots, n-2\}$ the morphism of complexes

$$\tau_{\leq \ell}\Omega^{\bullet} \twoheadrightarrow H^{\ell}(\Omega^{\bullet})[-\ell]$$

admits a section in the derived category of finite length coadmissible $D(G)$-modules with Orlik-Strauch irreducible constituents.

We could prove the above conjecture for $n = 4$ and $\ell = 2$, but the complex we construct for that (analogous to $\widetilde{\Omega}^{\bullet}$ when $\ell = 3$ or to the complex in Proposition 5.3.14 when $\ell = 1$) looks much more complicated. In particular it is not clear to us if the above complexes $\widetilde{\Omega}^{\bullet}$ admit nice generalizations to $\tau_{\leq \ell}\Omega^{\bullet}$ and $(V^{\infty}_{[1,n-\ell-1],\Delta})^{\vee}[-\ell]$ for $\ell \in \{1, \ldots, n-2\}$.

1.3 Some Intermediate Results and Ideas of Proofs

If V_0, V_1 are admissible locally K-analytic representations of G over K, it is not difficult to check that $\mathrm{Ext}^i_G(V_0, V_1) \cong \mathrm{Ext}^i_{D(G)}(V_1^{\vee}, V_0^{\vee})$ for $i = 0, 1$ where the first Ext^i is computed in the abelian category of admissible locally K-analytic representations of G (à la Yoneda for $i = 1$) and the second Ext^i is computed in the category of all (abstract) $D(G)$-modules (see [72, Sect. 6] when $i = 0$, [11, Lemma 2.1.1] when $i = 1$). Hence, in order to prove Theorems 1.2.1 and 1.2.2, it is enough to control the dimensions of the K-vector spaces $\mathrm{Ext}^i_{D(G)}(V_1^{\vee}, V_0^{\vee})$ for $i = 0, 1$ and certain V_0, V_1. By dévissage, it is enough to control $\mathrm{Ext}^i_{D(G)}(V_1^{\vee}, V_0^{\vee})$ for $i = 0, 1, 2$ where V_0, V_1 are Orlik-Strauch representations (1.5) such that π^{∞} is of finite length. However, this is still a non-trivial task. For instance, when $i \neq 0$ it is not even clear that such K-vector spaces are finite dimensional.

To explain our method, we need some more notation. For $i = 0, 1$ write $V_i = \mathcal{F}^G_{P_{I_i}}(M_i, \pi_i^{\infty})$ where $I_i \subseteq \Delta$, P_{I_i} is the K-points of the associated lower parabolic subgroup of GL_n, L_{I_i} its Levi factor, M_i an object of $\mathcal{O}^{\mathfrak{p}_{I_i}}$ with integral weights and π_i^{∞} a smooth finite length representation of L_{I_i}. By dévissage, we can always reduce ourselves to the case where M_1 is a generalized Verma module, i.e. $M_1 \cong U(\mathfrak{g}) \otimes_{U(\mathfrak{p}_{I_1})} L^{I_1}(\mu)$ where μ is an (integral) dominant weight with respect to the lower Borel of L_{I_1} and $L^{I_1}(\mu)$ the irreducible finite dimensional algebraic representation of L_{I_1} of highest weights μ. In this case V_1 is the locally K-analytic parabolic induction $(\mathrm{Ind}^G_{P_{I_1}} L^{I_1}(\mu)^{\vee} \otimes \pi_1^{\infty})^{\mathrm{an}}$.

By a result of Schneider-Teitelbaum when $K = \mathbb{Q}_p$ completed by Schmidt when K is arbitrary (proof of [73, Lemma 6.3(ii)] replacing [73, Lemma 6.2] by [62, Proposition 2.6]), we have in this case isomorphisms for $i \geq 0$

$$\mathrm{Ext}^i_{D(G)}(V_1^{\vee}, V_0^{\vee}) \cong \mathrm{Ext}^i_{D(P_{I_1})}(L^{I_1}(\mu) \otimes_K (\pi_1^{\infty})^{\vee}, V_0^{\vee}) \tag{1.16}$$

where $D(P_{I_1})$ is the K-algebra of locally K-analytic distributions on P_{I_1} and $\mathrm{Ext}^i_{D(P_{I_1})}$ is computed in the category of $D(P_{I_1})$-modules. So we are reduced to computing the (hopefully finite) dimension of $\mathrm{Ext}^i_{D(P_{I_1})}(L^{I_1}(\mu) \otimes_K (\pi_1^\infty)^\vee, V_0^\vee)$. The method is then to use a spectral sequence, but there are two possible choices.

The first choice is to write $P_{I_1} = N_{I_1} L_{I_1}$, where N_{I_1} is the unipotent radical of P_{I_1}, and use the spectral sequence:

$$\mathrm{Ext}^i_{D(L_{I_1})}(L^{I_1}(\mu) \otimes_K (\pi_1^\infty)^\vee, \mathrm{Ext}^i_{D(N_{I_1})}(1, V_0^\vee)) \Longrightarrow \mathrm{Ext}^{i+j}_{D(P_{I_1})}(L^{I_1}(\mu) \otimes_K (\pi_1^\infty)^\vee, V_0^\vee).$$

This is for instance the method used in [76, Sect. 4] or in [11, Sect. 5] when $K = \mathbb{Q}_p$ and $n = 3$. Though one can maybe proceed this way, it is not the spectral sequence that we use in this work. One reason is that, already when $K = \mathbb{Q}_p$ and $n = 3$, using this spectral sequence proved quite laborious in *loc. cit.*

The second choice, which is ours, is to use the spectral sequence ([73, Sect. 3], see (4.85)):

$$\mathrm{Ext}^i_{D^\infty(P_{I_1})}\big((\pi_1^\infty)^\vee, \mathrm{Ext}^j_{U(\mathfrak{p}_{I_1})}(L^{I_1}(\mu), V_0^\vee)\big) \Longrightarrow \mathrm{Ext}^{i+j}_{D(P_{I_1})}\big(L^{I_1}(\mu) \otimes_K (\pi_1^\infty)^\vee, V_0^\vee\big) \tag{1.17}$$

where $D^\infty(P_{I_1})$ is the (algebraic) dual of the locally constant K-valued functions on P_{I_1}, $\mathrm{Ext}^i_{D^\infty(P_{I_1})}$ is computed in the category of $D^\infty(P_{I_1})$-modules and $\mathrm{Ext}^j_{U(\mathfrak{p}_{I_1})}$ in the category of $U(\mathfrak{p}_{I_1})$-modules. This spectral sequence seems more suited to our dimension calculations because (as we will see in (1.20) below) it turns out we can "separate" the smooth part and the Lie part on the left hand side of (1.17), so that we are essentially reduced to computing dimensions of extensions groups either in the world of smooth representations or in the world of modules over Lie algebras, which is much easier (and where there is no more topology).

We thus need to compute the left hand side of (1.17). However, we do not compute it directly. Rather we define a filtration on $V_0^\vee$ and first compute

$$\mathrm{Ext}^i_{D^\infty(P_{I_1})}\big((\pi_1^\infty)^\vee, \mathrm{Ext}^j_{U(\mathfrak{p}_{I_1})}(L^{I_1}(\mu), \text{graded pieces})\big).$$

Denote by $W^{I_0,I_1} \subseteq W(G)$ the subset of minimal length representatives of the double coset $W(L_{I_0})\backslash W(G)/W(L_{I_1})$ where $W(-)$ are the respective (finite) Weyl groups. The Bruhat decomposition $G = \bigsqcup_{w\in W^{I_0,I_1}} P_{I_1} w^{-1} P_{I_0}$ induces a filtration indexed by W^{I_0,I_1} on $V_0^\vee$ by closed $D(P_{I_1})$-submodules (see (4.71)), and we denote by $\mathrm{gr}_w(V_0^\vee)$ its graded pieces. We first prove the following key description of $\mathrm{Ext}^j_{U(\mathfrak{p}_{I_1})}(L^{I_1}(\mu), \mathrm{gr}_w(V_0^\vee))$.

Theorem 1.3.1 (Theorem 4.5.10 with Lemma 4.2.13) *For $w \in W^{I_0,I_1}$ and $j \geq 0$ there is a $D^\infty(P_{I_1})$-equivariant isomorphism of Fréchet spaces*

$$\mathrm{Ext}^j_{U(\mathfrak{p}_{I_1})}(L^{I_1}(\mu), \mathrm{gr}_w(V_0^\vee)) \cong \mathrm{Ext}^j_{U(\mathfrak{p}_{I_1})}(L^{I_1}(\mu), M_0^w) \otimes_K \big((\mathrm{ind}_{P_{I_1}\cap w^{-1}P_{I_0}w}^{P_{I_1}} \pi_0^{\infty,w})^\infty\big)^\vee \tag{1.18}$$

where the K-vector space $\mathrm{Ext}^j_{U(\mathfrak{p}_{I_1})}(L^{I_1}(\mu), M_0^w)$ *is finite dimensional and has trivial action of* $D^\infty(P_{I_1})$. *Here* $(-)^w$ *means that* $\mathfrak{x} \in U(\mathfrak{g})$ *(resp.* $x \in P_{I_1} \cap w^{-1} P_{I_0} w$*) acts by* $w\mathfrak{x}w^{-1}$ *(resp.* wxw^{-1}*) and* $(\mathrm{ind}_{P_{I_1} \cap w^{-1} P_{I_0} w}^{P_{I_1}} \pi_0^{\infty,w})^\infty$ *is the smooth induction with compact support.*

The first main ingredient for the proof of Theorem 1.3.1 is an explicit description of the continuous dual $\mathrm{gr}_w(V_0^\vee)$ (Proposition 4.4.3), where the *canonical Fréchet completion* $\mathcal{M}_0$ of M_0 defined in [63] (see also Proposition 4.3.1) shows up, or more precisely its twist $\mathcal{M}_0^w$. The second main ingredient is the following statement.

Theorem 1.3.2 (Lemmas 4.5.1 and 4.5.2) *For* $w \in W^{I_0,I_1}$ *and* $j \geq 0$ *the natural morphism*

$$\mathrm{Ext}^j_{U(\mathfrak{p}_{I_1})}(L^{I_1}(\mu), M_0^w) \longrightarrow \mathrm{Ext}^j_{U(\mathfrak{p}_{I_1})}(L^{I_1}(\mu), \mathcal{M}_0^w)$$

is an isomorphism of finite dimensional K-vector spaces. Moreover $\mathrm{Ext}^j_{U(\mathfrak{p}_{I_1})}(L^{I_1}(\mu), \mathcal{M}_0^w)$ *is separated for its natural topology (coming from the topology on* $\mathcal{M}_0^w$ *and from the Chevalley-Eilenberg complex).*

We remark that Theorem 1.3.2 is also proved in [9].

Using these two key ingredients, the proof of Theorem 1.3.1 then consists in carefully analysing the Chevalley-Eilenberg complex of $\mathrm{gr}_w(V_0^\vee)$. It is given in Sect. 4.5, and is quite long and tedious because we give all the (topological) technical details. Note that, had the topological vector space $\mathrm{Ext}^j_{U(\mathfrak{p}_{I_1})}(L^{I_1}(\mu), \mathcal{M}_0^w)$ been non-separated, the solid techniques of [60] would probably have been necessary.

Plugging in (1.18) inside the left hand side of (1.17) with $\mathrm{gr}_w(V_0^\vee)$ instead of $V_0^\vee$, we deduce canonical isomorphisms for $w \in W^{I_0,I_1}$ and $i, j \geq 0$ (see Corollary 4.5.11)

$$\begin{aligned}&\mathrm{Ext}^i_{D^\infty(P_{I_1})}\big((\pi_1^\infty)^\vee, \mathrm{Ext}^j_{U(\mathfrak{p}_{I_1})}(L^{I_1}(\mu), \mathrm{gr}_w(V_0^\vee))\big)\\ &\qquad\cong \mathrm{Ext}^j_{U(\mathfrak{p}_{I_1})}(L^{I_1}(\mu), M_0^w) \otimes_K \mathrm{Ext}^i_{L_{I_1}}\Big(\big((\mathrm{ind}_{P_{I_1} \cap w^{-1} P_{I_0} w}^{P_{I_1}} \pi_0^{\infty,w})^\infty\big)_{N_{I_1}}, \pi_1^\infty\Big)^\infty\end{aligned} \tag{1.19}$$

where $\mathrm{Ext}^i_{L_{I_1}}(-)^\infty$ means extensions in the category of smooth representations of L_{I_1}. Moreover the K-vector space on the right hand side of (1.19) is finite dimensional for all w, i, j (using Theorem 1.3.2 for $\mathrm{Ext}^j_{U(\mathfrak{p}_{I_1})}(L^{I_1}(\mu), M_0^w)$ and the finite length of π_0^∞, π_1^∞ for the other factor). Hence we obtain from the analogue of (1.17) with $\mathrm{gr}_w(V_0^\vee)$ instead of $V_0^\vee$ that the K-vector space $\mathrm{Ext}^i_{D(P_{I_1})}(L^{I_1}(\mu) \otimes_K (\pi_1^\infty)^\vee, \mathrm{gr}_w(V_0^\vee))$ is also finite dimensional for all $i \geq 0$. By dévissage on the filtration on $V_0^\vee$ and by (1.16), we deduce the following nice by-product result, which is new.

Theorem 1.3.3 (Theorem 4.5.16) *Let* Π_0, Π_1 *be finite length admissible locally K-analytic representations of G over a finite extension of K with all (topological) irreducible constituents being Orlik-Strauch representations (1.5). Then the K-vector space* $\mathrm{Ext}^i_{D(G)}(\Pi_0^\vee, \Pi_1^\vee)$ *is finite dimensional for* $i \geq 0$.

We mention here another aside result, purely on the Lie algebra side, that we need in the proofs and that we couldn't find in the literature (it is also proved in [9]):

Proposition 1.3.4 (Proposition 3.1.5) *Let I a subset of Δ. Denote by $\mathfrak{b}$ the K-Lie algebra of the lower Borel of G, $\mathfrak{n}_I$, $\mathfrak{l}_I$ the K-Lie algebras of N_I, L_I, and $\mathfrak{b}_I$ the K-Lie algebra of the lower Borel of L_I. Then for any M in $\mathcal{O}^{\mathfrak{b}}$ with integral weights and any $j \geq 0$ the Lie algebra cohomology group $H^j(\mathfrak{n}_I, M)$ is an object of the category $\mathcal{O}^{\mathfrak{b}_I}$ relative to $\mathfrak{l}_I$ (with integral weights).*

Going back to (1.19), in practice we only need to apply it with π_0^∞ such that the smooth L_{I_1}-representations $((\mathrm{ind}_{P_{I_1}\cap w^{-1}P_{I_0}w}^{P_{I_1}} \pi_0^{\infty,w})^\infty)_{N_{I_1}}$ lie in distinct Bernstein blocks when w varies in W^{I_0,I_1}. Assuming that π_1^∞ lies in a unique Bernstein block (which will be our case), we thus deduce that there is *at most one* $w \in W^{I_0,I_1}$ only depending on π_0^∞ and π_1^∞ such that

$$\mathrm{Ext}^i_{L_{I_1}}\Big(\big((\mathrm{ind}_{P_{I_1}\cap w^{-1}P_{I_0}w}^{P_{I_1}} \pi_0^{\infty,w})^\infty\big)_{N_{I_1}}, \pi_1^\infty\Big)^\infty \neq 0$$

and hence such that (1.19) is possibly non-zero. By an obvious dévissage on the filtration on $V_0^\vee$, we thus see that, in order to compute the left hand side of (1.17), it is sufficient (in our case) to compute (1.19) because there is at most one w for which (1.19) is non-zero! In particular by (1.16) the spectral sequence (1.17) becomes

$$\mathrm{Ext}^j_{U(\mathfrak{p}_{I_1})}(L^{I_1}(\mu), M_0^w) \otimes_K \mathrm{Ext}^i_{L_{I_1}}\Big(\big((\mathrm{ind}_{P_{I_1}\cap w^{-1}P_{I_0}w}^{P_{I_1}} \pi_0^{\infty,w})^\infty\big)_{N_{I_1}}, \pi_1^\infty\Big)^\infty \Longrightarrow \mathrm{Ext}^{i+j}_{D(G)}(V_1^\vee, V_0^\vee) \quad (1.20)$$

for this unique w (with possibly all vector spaces being 0).

Now, the big advantage of (1.20) is that the term on its left hand side is computable because $\mathrm{Ext}^j_{U(\mathfrak{p}_{I_1})}(L^{I_1}(\mu), M_0^w)$ is purely in the category of $U(\mathfrak{p}_{I_1})$-modules while $\mathrm{Ext}^i_{L_{I_1}}(((\mathrm{ind}_{P_{I_1}\cap w^{-1}P_{I_0}w}^{P_{I_1}} \pi_0^{\infty,w})^\infty)_{N_{I_1}}, \pi_1^\infty)^\infty$ is purely in the category of smooth representations of L_{I_1} (in other words we have "separated" the Lie part and the smooth part). We give in Sect. 2.1 all the material needed to compute the dimensions of the K-vector spaces $\mathrm{Ext}^\bullet_{L_{I_1}}(((\mathrm{ind}_{P_{I_1}\cap w^{-1}P_{I_0}w}^{P_{I_1}} \pi_0^{\infty,w})^\infty)_{N_{I_1}}, \pi_1^\infty)^\infty$, at least for the π_i^∞ we need, and we give in Sect. 3 all the material needed to compute the dimensions of $\mathrm{Ext}^\bullet_{U(\mathfrak{p}_{I_1})}(L^{I_1}(\mu), M_0^w)$. Very often, either all terms on the left hand side of (1.20) are 0, and we obtain the vanishing of $\mathrm{Ext}^i_{D(G)}(V_1^\vee, V_0^\vee)$, or only one of them is non-zero, and we obtain a useful description of $\mathrm{Ext}^i_{D(G)}(V_1^\vee, V_0^\vee)$. Combining this with some dévissage, we can compute the dimensions of lots of $\mathrm{Ext}^i_{D(G)}(V_1^\vee, V_0^\vee)$ (for M_1 in $\mathcal{O}^{\mathfrak{p}_{I_1}}$ not necessarily a generalized Verma module), see the various results in Sects. 5.1, 5.2. As a sample, let us mention the following statement:

Corollary 1.3.5 (Proposition 5.1.14) *Let $\underline{j} = (j_0, j_1, j_2)$ and $\underline{j}' = (j_0', j_1', j_2')$ as in (1.11) and assume $(j_0, j_1) \neq (j_0', j_1')$, there is an isomorphism of (finite dimensional) K-vector spaces*

$$\begin{aligned}&\mathrm{Ext}^1_{D(G)}(C_{\underline{j}}^{\vee}, C_{\underline{j}'}^{\vee})\\ &\cong \mathrm{Ext}^1_{U(\mathfrak{g})}(L(w_{j_1,j_0}\cdot 0), L(w_{j_1',j_0'}\cdot 0)) \otimes_K \mathrm{Hom}_{L_{\widehat{j}_1}}\Big(\big((\mathrm{ind}_{P_{\widehat{j}_1}\cap P_{\widehat{j}_1'}}^{P_{\widehat{j}_1}} \pi^{\infty}_{j_1',j_2'})^{\infty}\big)_{N_{\widehat{j}_1}}, \pi^{\infty}_{j_1,j_2}\Big).\end{aligned} \tag{1.21}$$

One then deduces $\dim_K \mathrm{Ext}^1_{D(G)}(C_{\underline{j}}^{\vee}, C_{\underline{j}'}^{\vee})$ by computing the dimension of the right hand side of (1.21), see for instance Lemma 5.2.1.

Pushing further this kind of arguments, we prove the existence and unicity of the $D(G)$-modules X_k in (1.13), of the $D(G)$-modules X_k —— $(V^{\infty}_{[1,n-k-1],\Delta})^{\vee}$ —— X_{k+1} in Theorem 1.2.1, and of all the $D(G)$-modules in Theorem 1.2.2. Note that, apart from the X_k, most of these $D(G)$-modules are not uniserial: they involve subquotients which look like "squares" or "cubes". We call them Ext-squares or Ext-cubes and we construct them in Sect. 5.2 (and Sect. 3.4 for the Lie counterpart). For instance the Ext-squares involved in the $D(G)$-modules (1.14) are computed in (ii) of Proposition 5.2.10, the Ext-squares involved in the $D(G)$-modules $\underline{Y_k}$ of Theorem 1.2.2 are computed in Proposition 5.2.18, while the $D(G)$-modules $\widetilde{\Omega}^k$ of Theorem 1.2.2 also involve Ext-cubes which are computed in Proposition 5.2.28. These computations can be long and technical, but since we give all details they should not be hard to follow by an interested reader.

(It is worth mentioning here that the smooth generalized Steinberg $V^{\infty}_{[1,n-k-1],\Delta}$ of (1.2) and the smooth representations $\pi^{\infty}_{j_1,j_2}$ in (1.12) are instances of more general smooth representations that we call *G-basic* (see Definition 2.1.4) and that we completely study in Sect. 2.1 (see Sect. 2.3 for the special case of the $\pi^{\infty}_{j_1,j_2}$). These G-basic representations are quite convenient because one can easily compute explicitly everything that is needed for this work: Jacquet functors, constituents of smooth parabolic inductions, smooth extension groups, etc.).

But this is still not the end of the proof of Theorem 1.2.1.

For $k \in \{0, \ldots, n-1\}$ denote by D_k the unique coadmissible $D(G)$-module X_k —— $(V^{\infty}_{[1,n-k-1],\Delta})^{\vee}$ —— X_{k+1} of Theorem 1.2.1. The $D(G)$-modules D_k and all the $D(G)$-modules of Theorem 1.2.2 are constructed (in Sect. 5.3) *independently* of the Drinfeld space. This is enough for Theorem 1.2.2 where there is no mention of the Drinfeld space. But in order to complete the proof of Theorem 1.2.1, we need to show $\Omega^k \cong D_k$. This is non-trivial and uses two more key ingredients that we explain now.

The first ingredient is a unicity statement which strengthens Schneider-Teitelbaum's Theorem 1.1.3.

Theorem 1.3.6

(i) The coadmissible $D(G)$-module $D_{n-1} = X_{n-1}$ —— $(\mathrm{St}_n^{\infty})^{\vee}$ *of Theorem 1.2.1 (for $k = n-1$) is the unique coadmissible $D(G)$-module D which admits a filtration by closed $D(G)$-submodules*

$$D = \mathrm{Fil}^0(D) \supsetneq \mathrm{Fil}^1(D) \supsetneq \cdots \supsetneq \mathrm{Fil}^n(D) = 0$$

satisfying (1.6) and such that $H^0(N_{\widehat{n-1}}, D) \cong s_{n-1}s_{n-2}\cdots s_1 \cdot 0$ *(recall* $N_{\widehat{n-1}}$ *is the unipotent radical of* $P_{\widehat{n-1}}$*).*

(ii) We have $H^0(N_{\widehat{n-1}}, \Omega^{n-1}) \cong s_{n-1}s_{n-2}\cdots s_1 \cdot 0$.

Part (i) is proven in Theorem 5.4.4. Its proof is a bit long but it is remarkable that the condition $H^0(N_{\widehat{n-1}}, D) \cong s_{n-1}s_{n-2}\cdots s_1 \cdot 0$ is enough to "rigidify everything" and ensure unicity. In particular all extensions between the graded pieces of the filtration $\mathrm{Fil}^\bullet(D)$ are then non-split. The easier part (ii) is proven in Lemma 5.4.15 for any $k \in \{0, \ldots, n-1\}$ (not just $k = n-1$). Note that both parts were already proven in [76] when $G = \mathrm{GL}_3(\mathbb{Q}_p)$ (see [76, Proposition 6.3] and [76, Sect. 6.4]), and the strategy in *loc. cit.* actually inspired Theorem 1.3.6.

From Theorems 1.1.3 and 1.3.6 we immediately obtain Theorem 1.2.1 for $k = n-1$. We now need to go from $k = n-1$ to smaller k, and this is where we use the second ingredient: translation functors.

Translation functors were first defined in the setting of representations of real Lie algebras and of $(\mathfrak{g}, K)$[7]-modules in the late seventies. Together with wall-crossing functors, they quickly became a major tool in the study of BGG's category $\mathcal{O}^{\mathfrak{p}}$, see for instance [33, Sect. 7]. Strangely, it is only recently that these functors have been introduced in the framework of locally analytic representations [35], and they are only beginning to be used [21]. Let us quickly recall their definition.

Let $Z(\mathfrak{g})$ be the center of $U(\mathfrak{g})$, a $D(G)$-module D is said to be $Z(\mathfrak{g})$-finite, if for any $v \in M$, the K-vector subspace $\langle Z(\mathfrak{g})v \rangle$ of D generated by elements in $Z(\mathfrak{g})v$ is finite dimensional. Let λ, μ be integral weights, $\xi_\lambda, \xi_\mu : Z(\mathfrak{g}) \to K$ the infinitesimal characters of the respective Verma modules $U(\mathfrak{g}) \otimes_{U(\mathfrak{b})} \lambda$, $U(\mathfrak{g}) \otimes_{U(\mathfrak{b})} \mu$, and $\overline{\lambda - \mu}$ the unique highest weight (relative to the lower Borel of GL_n) in the $W(G)$-orbit of $\lambda - \mu$ for the standard action of $W(G)$ (not the dot action). We denote by $L(\overline{\lambda - \mu})$ the unique finite dimensional algebraic representation of G over K with highest weight $\overline{\lambda - \mu}$. Then for any $Z(\mathfrak{g})$-finite $D(G)$-module D we define (following [35])

$$\mathcal{T}_\lambda^\mu(D) \overset{\text{def}}{=} \mathrm{pr}_\mu\big(L(\overline{\lambda - \mu}) \otimes_K \mathrm{pr}_\lambda(D)\big) \tag{1.22}$$

where $\mathrm{pr}_\lambda, \mathrm{pr}_\mu$ is the projection onto the generalized eigenspace for the infinitesimal character ξ_λ, ξ_μ (which is well-defined thanks to the $Z(\mathfrak{g})$-finiteness). The $D(G)$-module $\mathcal{T}_\lambda^\mu(D)$ is still $Z(\mathfrak{g})$-finite and the functor $D \to \mathcal{T}_\lambda^\mu(D)$ is exact on $Z(\mathfrak{g})$-finite $D(G)$-modules. It is called a translation functor. Moreover with the notation of (1.5) one has $\mathcal{T}_\lambda^\mu(\mathcal{F}_P^G(M, \pi^\infty)) \cong \mathcal{F}_P^G(T_\lambda^\mu M, \pi^\infty)$ where T_λ^μ is the usual translation functor on the category $\mathcal{O}^{\mathfrak{p}}$ ([33, Sect. 7.1], in fact T_λ^μ is defined exactly as in (1.22) replacing D by M).

Translation functors are already interesting. For instance one can use them to revisit the construction of [64] and show that Schneider's holomorphic discrete series are translations of $\Omega^\bullet$, see Lemma 5.4.11. However, there exist even more interesting functors.

[7] Here K is of course a maximal compact subgroup, not the field K!

Let ρ be half the sum of the roots of the lower Borel of GL_n and fix a simple reflection s_k of GL_n for $1 \leq k \leq n-1$. Let μ be an integral weight such that $\langle \mu + \rho, \alpha^\vee \rangle \geq 0$ for $\alpha \in \Delta$ and the stabilizer of μ in the Weyl group $W(G)$ for the dot action is $\{1, s_k\}$. Then we define the *wall-crossing* functor on $Z(\mathfrak{g})$-finite $D(G)$-modules as[8]

$$\Theta_\mu \stackrel{\text{def}}{=} \mathcal{T}^{\mu}_{w_0 \cdot 0} \circ \mathcal{T}^{w_0 \cdot 0}_{\lambda}$$

where w_0 is the longest element of $W(G)$ and $w_0 \cdot 0 = w_0(\rho) - \rho$. The terminology comes from that fact that the stabilizer of $w_0 \cdot 0$ in $W(G)$ (for the dot action) is trivial, which is not the case for μ, see [33, Sect. 7.15]. By the argument in [33, Sect. 7.2] one has canonical and functorial adjunction morphisms of ($Z(\mathfrak{g})$-finite) $D(G)$-modules $D \to \Theta_\mu(D)$ and $\Theta_\mu(D) \to D$ which are non-zero when D and $\Theta_\mu(D)$ are non-zero (see (5.107)).

Although their definition may look puzzling at first, wall-crossing functors have remarkable properties. For instance they behave very well on the $D(G)$-modules D_k and $\widetilde{\Omega}^k$:

Theorem 1.3.7 *Let $1 \leq k \leq n-1$ and let μ be an integral weight such that $\langle \mu + \rho, \alpha^\vee \rangle \geq 0$ for $\alpha \in \Delta$ and the stabilizer of μ in the Weyl group $W(G)$ for the dot action is $\{1, s_k\}$.*

(i) We have non-split short exact sequences of coadmissible $D(G)$-modules $0 \to D_k \to \Theta_\mu(D_k) \to D_{k-1} \to 0$ where $D_k \to \Theta_\mu(D_k)$ is the canonical adjunction map.
(ii) We have short exact sequences of coadmissible $D(G)$-modules $0 \to \Omega^k \to \Theta_\mu(\Omega^k) \to \Omega^{k-1} \to 0$.
(iii) If $k \neq n-1$ we have non-split short exact sequences of coadmissible $D(G)$-modules $0 \to \widetilde{\Omega}^k \to \Theta_\mu(\widetilde{\Omega}^k) \to \widetilde{\Omega}^{k-1} \to 0$ where $\widetilde{\Omega}^k \to \Theta_\mu(\widetilde{\Omega}^k)$ is the canonical adjunction map.

Part (i) of Theorem 1.3.7 is proven in (iii) of Theorem 5.4.1, part (ii) of Theorem 1.3.7 is proven in Lemma 5.4.12 and part (iii) of Theorem 1.3.7 is proven in (iv) of Theorem 5.4.1. When $k = n-1$, (iii) of Theorem 1.3.7 is not true anymore, but it is almost true, see (v) of Theorem 5.4.1. Note that, going back to $n = 3$ and Ω^0, $\widetilde{\Omega}^{1\flat}$ in (1.15), we do *not* have a short exact sequence $0 \to \widetilde{\Omega}^{1\flat} \to \Theta_\mu(\widetilde{\Omega}^{1\flat}) \to \Omega^0 \to 0$ (some constituents are missing in Ω^0).

Now, when $k = n-1$, we know that $\Omega^{n-1} \cong D_{n-1}$. Using (ii) of Theorem 1.3.7 and (i) of Theorem 1.3.7 for $k = n-1$, we can deduce $\Omega^{n-2} \cong D_{n-2}$. Applying again *loc. cit.* for $k = n-2$ we obtain $\Omega^{n-3} \cong D_{n-3}$ and so on, see Theorem 5.4.16. We finally obtain $\Omega^k \cong D_k$ for all k, which finishes the proof of Theorem 1.2.1. Note that a posteriori the exact sequence in (ii) of Theorem 1.3.7 is also non-split and the injection is also the adjunction map.

---

[8] Experts on wall-crossing functors are more used to the notation Θ_{s_k}. However we do not know if the functor Θ_μ defined on all $Z(\mathfrak{g})$-finite $D(G)$-modules only depends on s_k and not on the choice of μ (as it does when defined on the category $\mathcal{O}^\flat$), see Remark 5.2.11. This doesn't affect this work.

Other nice properties are satisfied. For instance, for $1 \le k \le n-1$, the composition $\Theta_\mu(\Omega^k) \twoheadrightarrow \Omega^{k-1} \to \Omega^k$, where the surjection is in (ii) of Theorem 1.3.7 and the second map is the differential map on $\Omega^\bullet$, is nothing other than the other adjunction morphism $\Theta_\mu(\Omega^k) \to \Omega^k$, and likewise with $\widetilde{\Omega}^k$, see Remark 5.4.2. Moreover the wall-crossing functors Θ_μ are important tools for constructing Ext-squares and Ext-cubes of $D(G)$-modules, see for instance (among many other statements) Proposition 3.4.5, Lemmas 3.4.7, 5.2.13, 5.2.20, Proposition 5.3.5, etc.

We finally briefly give the contents of each section. In Sect. 2.1 we introduce (smooth) G-basic representations, among which are the $V^\infty_{[1,n-k-1],\Delta}$ and the $\pi^\infty_{j_1,j_2}$, and we prove all the necessary material on smooth representations for the later sections. In Sect. 3 we prove all results on $U(\mathfrak{g})$-modules and Ext groups of $U(\mathfrak{g})$-modules needed in Sects. 4 and 5. In Sect. 4 we prove Theorem 1.3.1 and deduce many consequences (e.g. Theorem 1.3.3). Finally, in Sect. 5 we construct all the relevant finite length coadmissible $D(G)$-modules and we prove the main results of Sect. 1.2. Appendix A is devoted to technical combinatorial lemmas on certain elements of $W(G)$ while Appendix B gives pictures for most of the previous $D(G)$-modules when $G = \mathrm{GL}_4(K)$ and $G = \mathrm{GL}_5(K)$, which are fairly representative of the general case. At the very end, the reader can find an index of most of the notation used in the text.

1.4 Some General Notation

We end up this introduction with general notation which will be used throughout this work. More specialized notation will be gradually introduced within the text.

We fix E an arbitrary finite extension of K which can be K. If $x \in K$ we let $|x|_K \stackrel{\text{def}}{=} q^{-e\mathrm{val}(x)}$ where q is the cardinality of the residue field of K, e the ramification index of K and val is normalized by $\mathrm{val}(p) = 1$.

We fix an integer $n \ge 2$ and, *unless otherwise stated*, G is either the algebraic group GL_n/K or its K-points $\mathrm{GL}_n(K)$ (the context being clear). We denote by T the diagonal matrices in G, B the lower triangular matrices of G and U the lower unipotent matrices of B (so $B = UT$). We let $\mathfrak{g} = \mathfrak{gl}_n$, $\mathfrak{t}$ the Lie algebra of T, $\mathfrak{b}$ the one of B, $\mathfrak{u} \subseteq \mathfrak{b}$ its radical, $\mathfrak{b}^+$ the upper Borel, $\mathfrak{u}^+ \subseteq \mathfrak{b}^+$ its radical and $Z(\mathfrak{g})$ the center of the enveloping algebra $U(\mathfrak{g})$. By extension of scalars from K to E, all Lie algebras are considered as E-vector spaces.

We let $\Lambda \stackrel{\text{def}}{=} X(T) \simeq \mathbb{Z}^n$ be the integral weights, Λ^{dom} the set of (integral) dominant weights *with respect to* $\mathfrak{b}$, Φ^+ the set of positive roots in $\mathfrak{u}^+$ and $\Delta \subseteq \Phi^+$ the subset of positive simple roots. We fix a bijection $\{1, \cdots, n-1\} \cong \Delta$ sending j to $e_j - e_{j+1}$ and we use both notation α or $j \in \{1, \ldots, n-1\}$ for an element of Δ. We set $\widehat{j} \stackrel{\text{def}}{=} \Delta \setminus \{j\}$ for $j \in \Delta$.

For $I \subseteq \Delta$ a subset, we let P_I (resp. P_I^+) be the (K-points of) the lower (resp. upper) standard parabolic subgroup of GL_n associated to I, N_I the unipotent radical of P_I and $L_I \cong P_I/N_I$ its Levi factor (so $P_I = N_I L_I$). We denote by

$B_I = B \cap L_I$ the lower Borel of L_I and $U_I \subseteq B_I$ its unipotent radical. We let $\mathfrak{p}_I, \mathfrak{p}_I^+$, $\mathfrak{n}_I \subseteq \mathfrak{p}_I$, $\mathfrak{l}_I \cong \mathfrak{p}_I/\mathfrak{n}_I$, $\mathfrak{b}_I = \mathfrak{b} \cap \mathfrak{l}_I$ and $\mathfrak{u}_I \subseteq \mathfrak{b}_I$ be the respective Lie algebras of P_I, P_I^+, N_I, L_I, B_I and U_I. We let $Z(\mathfrak{l}_I)$ be the center of the enveloping algebra $U(\mathfrak{l}_I)$, $\mathfrak{n}_I^+$ the radical of $\mathfrak{p}_I^+$, $\mathfrak{b}_I^+ = \mathfrak{b}^+ \cap \mathfrak{l}_I$ the upper Borel of $\mathfrak{l}_I$ and $\mathfrak{u}_I^+ \subseteq \mathfrak{b}_I^+$ the radical of $\mathfrak{b}_I^+$. We denote by Λ_I^{dom} the set of (integral) dominant weights with respect to $\mathfrak{b}_I$ and $\Phi_I^+ \subseteq \Phi^+$ the roots of $\mathfrak{u}_I^+$.

We let $W(G)$ be Weyl group of G, $W(L_I)$ the Weyl group of the Levi L_I corresponding to I, $\ell(w) \in \mathbb{Z}_{\geq 0}$ the length of $w \in W(G)$ and $W^{I_0,I_1}(L_I)$ for $I_0, I_1 \subseteq I$ the set of minimal length representatives of $W(L_{I_0})\backslash W(L_I)/W(L_{I_1})$ (see [20, Lemma 5.4]). When $L_I = G$ we write W^{I_0,I_1}. We let ρ_I be half the sum of the roots in $\mathfrak{b}_I$ (so $\langle \rho_I, \alpha^\vee \rangle = -1$ for $\alpha \in I$ and ρ_I is dominant with respect to $\mathfrak{b}_I$) and we define the dot action $w \cdot \mu \overset{\mathrm{def}}{=} w(\mu + \rho_I) - \rho_I$ for $w \in W(L_I)$ and $\mu \in X(T)$. Note that, since $\rho - \rho_I$ is invariant under $W(L_I)$, the choice of ρ or ρ_I doesn't change the above dot action. When $I = \emptyset$, we forget the index I in the notation. We denote by w_I the longest element in $W(L_I)$ and $w_0 \overset{\mathrm{def}}{=} w_\Delta$. We endow $W(L_I)$ with the Bruhat order $<$ and for $w \in W(L_I)$, we set

$$D_L(w) \overset{\mathrm{def}}{=} \{\alpha \in I \mid s_\alpha w < w\} = \{\alpha \in I \mid -w^{-1}(\alpha) \in \Phi_I^+\} \tag{1.23}$$

$$D_R(w) \overset{\mathrm{def}}{=} \{\alpha \in I \mid w s_\alpha < w\} = \{\alpha \in I \mid -w(\alpha) \in \Phi_I^+\} \tag{1.24}$$

(so $D_L(1) = D_R(1) = \emptyset$). As $x < w$ if and only if $x^{-1} < w^{-1}$, we have $D_L(w) = D_R(w^{-1})$.

If $\mathcal{C}$ is an abelian category, we recall that a finite length object in $\mathcal{C}$ is an object M of $\mathcal{C}$ such that there exists finitely many subobjects $M_1 \subseteq M_2 \subseteq \cdots \subseteq M_m = M$ in $\mathcal{C}$ such that M_i/M_{i-1} is a simple object of $\mathcal{C}$. We then write $\mathrm{JH}_{\mathcal{C}}(M)$ for the (finite) set of isomorphism classes of the simple subquotients of M. If M is multiplicity free, we equip $\mathrm{JH}_{\mathcal{C}}(M)$ with the following partial order: given M_1, M_2 in $\mathrm{JH}_{\mathcal{C}}(M)$ we write $M_1 \leq M_2$ if M_1 is a subquotient of the unique subobject of M with cosocle isomorphic to M_2. If $\mathcal{C}$ is a full subcategory of the category of left A-modules (for A any unital associative ring) or of representations of G (for G any group), we write $\mathrm{JH}_A(M)$, $\mathrm{JH}_G(M)$ instead of $\mathrm{JH}_{\mathcal{C}}(M)$.

If A is a unital associative ring, we denote by Mod_A the (abelian) category of all abstract left A-modules. If M is a finite length object in Mod_A, we write

$$\cdots \subseteq \mathrm{Rad}^{k+1}(M) \subseteq \mathrm{Rad}^k(M) \subseteq \cdots \subseteq \mathrm{Rad}^1(M) \subseteq \mathrm{Rad}^0(M) = M$$

for the radical filtration of M, where $\mathrm{Rad}^{k+1}(M)$ is the minimal submodule of $\mathrm{Rad}^k(M)$ such that $\mathrm{Rad}_k(M) \overset{\mathrm{def}}{=} \mathrm{Rad}^k(M)/\mathrm{Rad}^{k+1}(M)$ is semi-simple. We also write $\mathrm{rad}(M) = \mathrm{Rad}^1(M)$ (the radical of M). The Loewy length of M, written $\ell\ell(M)$ is by definition the minimal integer $k \geq 0$ such that $\mathrm{Rad}^k(M) = 0$. Similarly we define the socle filtration $\cdots \supseteq \mathrm{Soc}^k(M) \supseteq \cdots \supseteq \mathrm{Soc}^0(M) = 0$ of M where $\mathrm{Soc}^{k+1}(M)$ is the maximal submodule of M containing $\mathrm{Soc}^k(M)$ such that

$\mathrm{Soc}^{k+1}(M)/\mathrm{Soc}^k(M)$ is semi-simple. Note that $\ell\ell(M)$ is also the minimal integer $k \geq 0$ such that $\mathrm{Soc}^{k+1}(M)/\mathrm{Soc}^k(M) = M$. The module M is called *rigid* if $\mathrm{Rad}^k(M) = \mathrm{Soc}^{\ell\ell(M)-k}(M)$ for $0 \leq k \leq \ell\ell(M)$.

If π is a representation of a subgroup H' of some group H and $h \in H$, we denote by π^h the representation of $h^{-1}H'h$ with same underlying space as π where $h' \in h^{-1}H'h$ acts by $hh'h^{-1}$. If H is a locally compact group, $\mathrm{Ext}^{\bullet}_H(-,-)^{\infty}$ means extensions in the category of smooth representations of H over E-vector spaces. If π is a smooth representation of L_I (for some $I \subseteq \Delta$), we denote by $\pi^{\sim}$ its smooth contregredient.

Chapter 2
Preliminaries on Smooth Representation

We prove all results on smooth representations of G needed in Sect. 4 and especially in Sect. 5. Most results are not really new, but we provide complete proofs. In particular, using Bernstein-Zelevinsky's theory, we define and study the convenient notion of G-basic smooth representation of L_I for some $I \subseteq \Delta$ and we give several results on certain G-basic representations which are crucially used afterwards.

2.1 G-Basic Representations and Bernstein-Zelevinsky's Theory

We define G-basic representations (Definition 2.1.4) and use Bernstein-Zelevinsky's geometric lemma and segment theory ([4, 79]) to prove several useful (and presumably well-known) results on them.

Let G be a locally compact topological group with left Haar measure μ_G, recall its *modulus character* $\delta_G : G \to \mathbb{R}_{>0}$ is the unique character satisfying $\mu_G(A \cdot x) = \delta_G(x)\mu_G(A)$ for any Borel subset $A \subseteq G$. When G is moreover p-adic analytic, δ_G is $\mathbb{Q}^\times$-valued and we will see δ_G as an $E^\times$-valued character.

Let G be a locally profinite group. We write $\mathrm{Rep}^\infty(G)$ for the abelian category of all smooth representations of G over E and $\mathrm{Rep}^\infty_{\mathrm{adm}}(G)$ for the full abelian subcategory of admissible ones. Let $H \subseteq G$ be a closed subgroup and $\pi^\infty \in \mathrm{Rep}^\infty(H)$. We define $(\mathrm{Ind}_H^G \pi^\infty)^\infty$ to be the E-vector space of uniformly locally constant functions $f : G \to \pi^\infty$ such that $f(xh) = h^{-1} \cdot f(x)$ for $x \in G$ and $h \in H$, which is naturally a (left) smooth G-representation via $(g(f))(x) \stackrel{\mathrm{def}}{=} f(g^{-1}x)$ $(g, x \in G$, $f \in (\mathrm{Ind}_H^G \pi^\infty)^\infty)$. We also consider the subspace $(\mathrm{ind}_H^G \pi^\infty)^\infty \subseteq (\mathrm{Ind}_H^G \pi^\infty)^\infty$ consisting of those f for which there exists a compact open subset C_f of G such that $f(x) = 0$ for $x \notin C_f H$. They give the so-called (unnormalized) induction and compact induction functors

C. Breuil and Z. Qian, *Splitting and Making Explicit the de Rham Complex of the Drinfeld Space*, Lecture Notes in Mathematics 2397,
https://doi.org/10.1007/978-3-032-20095-2_2

$$(\mathrm{Ind}_H^G)^\infty,\ (\mathrm{ind}_H^G)^\infty : \mathrm{Rep}^\infty(H) \to \mathrm{Rep}^\infty(G)$$

which are both exact. Note that they do not send $\mathrm{Rep}^\infty_{\mathrm{adm}}(H)$ to $\mathrm{Rep}^\infty_{\mathrm{adm}}(G)$ in general.

Remark 2.1.1 We add $(-)^\infty$ as exponent to avoid possible confusion with locally analytic inductions. Moreover our convention in the definition of $(\mathrm{Ind}_H^G)^\infty$ and $(\mathrm{ind}_H^G)^\infty$ (which is the one used in [51, 52, 72, 73], etc.) is different from the more commonly one used for instance in [14] or [59, Sect. III.2.2], where $f \in (\mathrm{Ind}_H^G \pi^\infty)^\infty$ satisfies $f(hx) = h \cdot f(x)$ and $g \in G$ acts on $(\mathrm{Ind}_H^G \pi^\infty)^\infty$ by $(g(f))(x) \stackrel{\mathrm{def}}{=} f(xg)$. But the isomorphism $f \mapsto [g \mapsto f(g^{-1})]$ gives an isomorphism between our $(\mathrm{Ind}_H^G \pi^\infty)^\infty$ and theirs, and we can freely use the results of [59].

We start with a general lemma (which will be used in Lemma 2.1.8 below).

Lemma 2.1.2 *Let P be a locally profinite group and $N \subseteq H \subseteq P$ be closed subgroups such that N is* normal *in P. Assume that there exists a continuous section $s : P/H \hookrightarrow P$ of the canonical surjection $P \twoheadrightarrow P/H$ which induces a homeomorphism of locally profinite topological spaces $P/H \times H \xrightarrow{\sim} P$. Then for each π^∞ in* $\mathrm{Rep}^\infty(H)$, *we have a canonical isomorphism in* $\mathrm{Rep}^\infty(P/N)$

$$\big((\mathrm{ind}_H^P \pi^\infty)^\infty\big)_N \xrightarrow{\sim} \big(\mathrm{ind}_{H/N}^{P/N} (\pi^\infty)_N\big)^\infty$$

where $(-)_N$ means the usual N-coinvariants.

Proof As N acts trivially on π^∞_N, we have a natural isomorphism $(\mathrm{ind}_{H/N}^{P/N} (\pi^\infty)_N)^\infty \cong (\mathrm{ind}_H^P (\pi^\infty)_N)^\infty$. For a E-vector space M equipped with a smooth N-action, let $V(M) \subseteq M$ be the subspace spanned by vectors $n \cdot v - v$ for $v \in M$ and $n \in N$. Then $(\pi^\infty)_N = \pi^\infty / V(\pi^\infty)$ by definition, and thus $(\mathrm{ind}_H^P (\pi^\infty)_N)^\infty = (\mathrm{ind}_H^P \pi^\infty)^\infty / (\mathrm{ind}_H^P V(\pi^\infty))^\infty$ by the exactness of $(\mathrm{ind}_H^P)^\infty$. As $((\mathrm{ind}_H^P \pi^\infty)^\infty)_N = (\mathrm{ind}_H^P \pi^\infty)^\infty / V((\mathrm{ind}_H^P \pi^\infty)^\infty)$ by definition, it suffices to show that we have

$$V((\mathrm{ind}_H^P \pi^\infty)^\infty) = (\mathrm{ind}_H^P V(\pi^\infty))^\infty \tag{2.1}$$

as subspaces of $(\mathrm{ind}_H^P \pi^\infty)^\infty$. Let $f \in (\mathrm{ind}_H^P \pi^\infty)^\infty$, then for $x \in P$ and $n \in N$, we have $x^{-1} n x \in N \subseteq H$ and

$$(n \cdot f)(x) = f(n^{-1} x) = f(x(x^{-1} n^{-1} x)) = x^{-1} n x \cdot f(x). \tag{2.2}$$

In particular, $(n \cdot f - f)(x) = x^{-1} n x \cdot f(x) - f(x) \in V(\pi^\infty)$, and thus we have an inclusion $V((\mathrm{ind}_H^P \pi^\infty)^\infty) \subseteq (\mathrm{ind}_H^P V(\pi^\infty))^\infty$. Let us prove that it is a surjection. Write $X \stackrel{\mathrm{def}}{=} s(P/H) \subseteq P$ and define $C_c^\infty(X, \pi^\infty)$ as the E-vector space of locally constant function $h : X \to \pi^\infty$ with compact support, and similarly with $C_c^\infty(X, V(\pi^\infty))$. By definition of $(\mathrm{ind}_H^P \pi^\infty)^\infty$, the map $f \mapsto f|_X$ induces an isomorphism of E-vector spaces

$$(\mathrm{ind}_H^P \pi^\infty)^\infty \xrightarrow{\sim} C_c^\infty(X, \pi^\infty), \tag{2.3}$$

which induces an isomorphism of E-vector spaces

$$(\mathrm{ind}_H^P V(\pi^\infty))^\infty \xrightarrow{\sim} C_c^\infty(X, V(\pi^\infty)). \tag{2.4}$$

Let $n \in N$, $v \in \pi^\infty$ and $x \in X$. As the N-action on π^∞ is smooth and the P-action on N by conjugation is continuous, there exists a compact open subset $C_x \subseteq X$ such that $x \in C_x$ and $(y^{-1}ny) \cdot v = (x^{-1}nx) \cdot v$ for each $y \in C_x$. Let $h_{x,v} \in C_c^\infty(X, \pi^\infty)$ be the function defined by $h_{x,v}(y) = v$ for $y \in C_x$ and $h_{x,v}(y) = 0$ for $y \in X \setminus C_x$, and let $f_{x,v} \in (\mathrm{ind}_H^P \pi^\infty)^\infty$ correspond to $h_{x,v}$ under (2.3). Similarly to (2.2), we have $(n \cdot f_{x,v})(y) = (y^{-1}ny) \cdot f_{x,v}(y) = (y^{-1}ny) \cdot h_{x,v}(y) = (y^{-1}ny) \cdot v = (x^{-1}nx) \cdot v$ for $y \in C_x$ and $(n \cdot f_{x,v})(y) = (y^{-1}ny) \cdot f_{x,v}(y) = 0$ for each $y \in X \setminus C_x$. Hence, we have $(n \cdot f_{x,v} - f_{x,v})(y) = (x^{-1}nx) \cdot v - v$ for $y \in C_x$ and $(n \cdot f_{x,v} - f_{x,v})(y) = 0$ for $y \in X \setminus C_x$. As $n \in N$, $v \in \pi^\infty$, $x \in X$ are arbitrary and as C_x can be an arbitrarily small neighborhood of x in X, elements of the form $(n \cdot f_{x,v} - f_{x,v})|_X$ span $C_c^\infty(X, V(\pi^\infty))$ over E, and thus elements of the form $n \cdot f_{x,v} - f_{x,v}$ span $(\mathrm{ind}_H^P V(\pi^\infty))^\infty$ over E by (2.4). Since we clearly have $n \cdot f_{x,v} - f_{x,v} \in V((\mathrm{ind}_H^P \pi^\infty)^\infty)$, we see that (2.1) holds. □

Let G be the K-points of a p-adic reductive algebraic group over K. For (the K-points of) a parabolic subgroup $P = L_P N_P \subseteq G$, the unnormalized parabolic induction functor $(\mathrm{Ind}_P^G)^\infty : \mathrm{Rep}^\infty(L_P) \to \mathrm{Rep}^\infty(G)$ restricts to a functor $(\mathrm{Ind}_P^G)^\infty : \mathrm{Rep}^\infty_{\mathrm{adm}}(L_P) \to \mathrm{Rep}^\infty_{\mathrm{adm}}(G)$ (see [59, Sect. III.2.3]). The functor $(\mathrm{Ind}_P^G)^\infty$ admits a left adjoint functor $J_{N_P} : \mathrm{Rep}^\infty(G) \to \mathrm{Rep}^\infty(L_P)$ which is exact and restricts to a functor $J_{N_P} : \mathrm{Rep}^\infty_{\mathrm{adm}}(G) \to \mathrm{Rep}^\infty_{\mathrm{adm}}(L_P)$ (cf. [59, Sect. VI.1.1, VI.6.1] but note that our J_{N_P} is the unnormalized Jacquet functor). Arguing as in the beginning of [59, Sect. VI.9.6], the functor $(\mathrm{Ind}_P^G)^\infty$ also admits a right adjoint $J'_{N_P} : \mathrm{Rep}^\infty(G) \to \mathrm{Rep}^\infty(L_P)$ which is also exact and restricts to $J'_{N_P} : \mathrm{Rep}^\infty_{\mathrm{adm}}(G) \to \mathrm{Rep}^\infty_{\mathrm{adm}}(L_P)$ (cf. [59, Sect. VI.9.6] taking care again that we are unnormalized).

We also need the *normalized parabolic induction functor* $i_P^G : \mathrm{Rep}^\infty(L_P) \to \mathrm{Rep}^\infty(G)$ defined by $i_P^G(-) \stackrel{\mathrm{def}}{=} (\mathrm{Ind}_P^G((-) \otimes_E \delta_P^{-1/2}))^\infty$, and the *normalized parabolic restriction functor* $r_P^G : \mathrm{Rep}^\infty(G) \to \mathrm{Rep}^\infty(L_P)$ defined by $r_P^G(-) \stackrel{\mathrm{def}}{=} J_{N_P}(-) \otimes_E \delta_P^{1/2}$ (cf. [59, Sect. VI.1.2] and recall that $\delta_P : P \to E^\times$ factors through L_P). Note that, here, we might need to extend scalars to $E' = E(\sqrt{q})$ so that $\delta_P^{1/2}$ is E'-valued. But these normalized functors will only play a minor intermediate role in that paper, and everything is ultimately K-rational (see for instance Remarks 2.1.3 and 2.1.10 below).

For parabolic subgroups $Q = L_Q N_Q \subseteq P = L_P N_P$, $Q \cap L_P$ is a parabolic subgroup of L_p with reductive quotient L_Q, and we have the formula

$$\delta_Q = \delta_{Q \cap L_P} \cdot (\delta_P|_{L_Q}) : L_Q \to E^\times. \tag{2.5}$$

From now on $G = \mathrm{GL}_n(K)$ as in Sect. 1.4. For $I \subseteq I_0 \subseteq \Delta$, we use the shortened notation

$$i_{I,I_0}^\infty \stackrel{\mathrm{def}}{=} \left(\mathrm{Ind}_{P_I \cap L_{I_0}}^{L_{I_0}}\right)^\infty, \ J_{I_0,I} \stackrel{\mathrm{def}}{=} J_{N_I \cap L_{I_0}}, \ J'_{I_0,I} \stackrel{\mathrm{def}}{=} J'_{N_I \cap L_{I_0}}. \tag{2.6}$$

Then for each $I' \subseteq I$, we clearly have

$$i^{\infty}_{I',I_0} \cong i^{\infty}_{I,I_0}(i^{\infty}_{I',I}),\ J_{I_0,I'} \cong J_{I,I'}(J_{I_0,I}),\ J'_{I_0,I'} \cong J'_{I,I'}(J'_{I_0,I}).$$

For π^∞ in $\mathrm{Rep}^{\infty}_{\mathrm{adm}}(L_I)$ and π_0^∞ in $\mathrm{Rep}^{\infty}_{\mathrm{adm}}(L_{I_0})$, we have canonical isomorphisms for $k \geq 0$:

$$\mathrm{Ext}^k_{L_{I_0}}(\pi_0^\infty, i^{\infty}_{I,I_0}(\pi^\infty))^\infty \cong \mathrm{Ext}^k_{L_I}(J_{I_0,I}(\pi_0^\infty), \pi^\infty)^\infty \tag{2.7}$$

and

$$\mathrm{Ext}^k_{L_{I_0}}(i^{\infty}_{I,I_0}(\pi^\infty), \pi_0^\infty)^\infty \cong \mathrm{Ext}^k_{L_I}(\pi^\infty, J'_{I_0,I}(\pi_0^\infty))^\infty. \tag{2.8}$$

Recall that, for $I, I' \subseteq I_0$, $W^{I',I}(L_{I_0}) \subseteq W(L_{I_0})$ is the set of minimal length representatives of the double coset $W(L_{I'})\backslash W(L_{I_0})/W(L_I)$. Let $w_{I_0,I}$ be the *longest* element inside $W^{\emptyset,I}(L_{I_0})$, we have $w_{I_0,I}(I) \subseteq I_0$ and

$$w_{I_0,I}^{-1}(\Phi^+_{I_0}) \cap \Phi^+_{I_0} = \Phi^+_I. \tag{2.9}$$

Note that $P_I^+ \cap L_{I_0}$ is the parabolic of L_{I_0} opposite to $P_I \cap L_{I_0}$. It thus follows from [59, (VI.9.6.1)], $w_{I_0,I}^{-1} L_{w_{I_0,I}(I)} w_{I_0,I} = L_I$ and $\delta_{P_I^+ \cap L_{I_0}} = \delta^{-1}_{P_I \cap L_{I_0}}$ that

$$\begin{aligned}
\mathrm{Hom}_{L_{I_0}}(i^{\infty}_{I,I_0}(\pi^\infty), \pi_0^\infty) &= \mathrm{Hom}_{L_{I_0}}(i^{L_{I_0}}_{P_I \cap L_{I_0}}(\pi^\infty \otimes_E \delta^{1/2}_{P_I \cap L_{I_0}}), \pi_0^\infty)\\
&\cong \mathrm{Hom}_{L_I}(\pi^\infty \otimes_E \delta^{1/2}_{P_I \cap L_{I_0}}, r^{L_{I_0}}_{P_I^+ \cap L_{I_0}}(\pi_0^\infty))\\
&\cong \mathrm{Hom}_{L_I}\left(\pi^\infty \otimes_E \delta^{1/2}_{P_I \cap L_{I_0}}, J_{N_I^+ \cap L_{I_0}}(\pi_0^\infty) \otimes_E \delta^{1/2}_{P_I^+ \cap L_{I_0}}\right)\\
&\cong \mathrm{Hom}_{L_I}\left(\pi^\infty, (J_{I_0, w_{I_0,I}(I)}(\pi_0^\infty))^{w_{I_0,I}} \otimes_E \delta^{1/2}_{P_I^+ \cap L_{I_0}} \otimes_E \delta^{-1/2}_{P_I \cap L_{I_0}}\right)\\
&\cong \mathrm{Hom}_{L_I}\left(\pi^\infty, (J_{I_0, w_{I_0,I}(I)}(\pi_0^\infty))^{w_{I_0,I}} \otimes_E \delta^{-1}_{P_I \cap L_{I_0}}\right).
\end{aligned}$$

By (2.8) for $k = 0$ and since this holds for arbitrary π^∞ in $\mathrm{Rep}^{\infty}_{\mathrm{adm}}(L_I)$ we deduce

$$J'_{I_0,I}(\pi_0^\infty) \cong (J_{I_0, w_{I_0,I}(I)}(\pi_0^\infty))^{w_{I_0,I}} \otimes_E \delta^{-1}_{P_I \cap L_{I_0}}. \tag{2.10}$$

We write $\widehat{T}^\infty$ for the set of smooth E-valued characters of T, which is naturally an abelian group under multiplication. We let $W(G)$ act on the left on $\widehat{T}^\infty$ via the following dot action

$$w \cdot \chi \stackrel{\mathrm{def}}{=} w(\chi \otimes_E \delta_B^{1/2}) \otimes_E \delta_B^{-1/2} = (\chi \otimes_E \delta_B^{1/2})(w^{-1} \cdot w) \otimes_E \delta_B^{-1/2} \tag{2.11}$$

and recall that δ_B sends $(t_1, \ldots, t_n) \in T$ to $\prod_{i=1}^n |t_i|_K^{(n+1)-2i}$, which can be rewritten as $|\sum_{\alpha \in \Phi^+} \alpha|_K = |-2\rho|_K$. If $w \in W(L_I)$ for some $I \subseteq \Delta$ one can check that $w \cdot \chi = w(\chi \otimes_E \delta_{B_I}^{1/2}) \otimes_E \delta_{B_I}^{-1/2}$ (recall $B_I = B \cap L_I$).

Remark 2.1.3 Note that the character δ_B is $K^\times$-valued but $\delta_B^{1/2}$ is not in general. Nevertheless, $w(\delta_B^{1/2}) \otimes_E \delta_B^{-1/2}$ is $K^\times$-valued for any $w \in W(G)$. Similarly, $w(\delta_{B_I}^{1/2}) \otimes_E \delta_{B_I}^{-1/2}$ is $K^\times$-valued for $I \subseteq \Delta$ and $w \in W(L_I)$. Consequently, $\chi \in \widehat{T}^\infty$ is $K^\times$-valued if and only if $w \cdot \chi = w(\chi) \otimes_E (w(\delta_B^{1/2}) \otimes_E \delta_B^{-1/2})$ is $K^\times$-valued for each $w \in W(G)$.

Definition 2.1.4 Let $I \subseteq \Delta$, π^∞ a finite length representation in $\mathrm{Rep}^\infty_{\mathrm{adm}}(L_I)$ and recall that $J_{I,\emptyset}(\pi^\infty)$ has finite length (cf. [59, Sect. VI.6.4]). We write $\mathcal{J}(\pi^\infty) \subseteq \widehat{T}^\infty$ for the subset of χ such that

$$\mathrm{Hom}_{L_I}(\pi^\infty, i^\infty_{\emptyset,I}(\chi)) \cong \mathrm{Hom}_T(J_{I,\emptyset}(\pi^\infty), \chi) \neq 0.$$

(i) A character $\chi \in \widehat{T}^\infty$ is called *G-regular* if $w \cdot \chi \neq \chi$ for $1 \neq w \in W(G)$.
(ii) If π^∞ is irreducible, it is called *G-regular* if $\mathcal{J}(\pi^\infty)$ contains a G-regular element. In general, π^∞ is called *G-basic* if there exists $I_1 \subseteq I$ and an irreducible G-regular π_1^∞ in $\mathrm{Rep}^\infty_{\mathrm{adm}}(L_{I_1})$ such that $\pi^\infty \cong i^\infty_{I_1,I}(\pi_1^\infty)$.
(iii) For $i = 0, 1$ let $I_i \subseteq I \subseteq \Delta$ and π_i^∞ in $\mathrm{Rep}^\infty_{\mathrm{adm}}(L_{I_i})$, we define the *$L_I$-distance from π_0^∞ to π_1^∞* as

$$d_I(\pi_0^\infty, \pi_1^\infty) \stackrel{\mathrm{def}}{=} \inf\{k \mid \mathrm{Ext}^k_{L_I}(i^\infty_{I_0,I}(\pi_0^\infty), i^\infty_{I_1,I}(\pi_1^\infty))^\infty \neq 0\} \leq \infty.$$

We write $d(\pi_0^\infty, \pi_1^\infty) \stackrel{\mathrm{def}}{=} d_\Delta(\pi_0^\infty, \pi_1^\infty)$.

Note that the trivial character $1_T \in \widehat{T}^\infty$ is G-regular. By (2.7) for $k = 0$ applied to π_1^∞ as in (ii) of Definition 2.1.4 and the left-exactness of induction we see that $\pi^\infty \cong i^\infty_{I_1,I}(\pi_1^\infty)$ embeds into $i^\infty_{\emptyset,I}(\chi) \cong i^\infty_{I_1,I}(i^\infty_{\emptyset,I_1}(\chi))$ for $\chi \in \mathcal{J}(\pi_1^\infty)$, and hence $\chi \in \mathcal{J}(\pi^\infty)$. In particular we have $\mathcal{J}(\pi_1^\infty) \subseteq \mathcal{J}(\pi^\infty)$. Note also that any G-basic representation of L_I is (admissible) of finite length (as so is $i^\infty_{I_1,I}(\pi_1^\infty)$, see [59, Sect. VI.6.2]) and that $i^\infty_{I,I'}(\pi^\infty) \in \mathrm{Rep}^\infty_{\mathrm{adm}}(L_{I'})$ is again G-basic for any $I \subseteq I' \subseteq \Delta$.

Remark 2.1.5 In Definition 2.1.4, as $J_{I,\emptyset}(\pi^\infty)$ is a finite length smooth representation of T, we have $J_{I,\emptyset}(\pi^\infty) \cong \bigoplus_{\chi \in \mathcal{J}(\pi^\infty)} J_{I,\emptyset}(\pi^\infty)_\chi$ with $J_{I,\emptyset}(\pi^\infty)_\chi$ having only χ as Jordan-Hölder factor. In particular, the set $\mathcal{J}(\pi^\infty)$ equals $\mathrm{JH}_T(J_{I,\emptyset}(\pi^\infty))$ as well as the set of χ such that

$$\mathrm{Hom}_T(\chi, J_{I,\emptyset}(\pi^\infty)) \neq 0.$$

Remark 2.1.6 For $I \subseteq I_0 \subseteq \Delta$, by an easy calculation using (2.9) one can check

$$\delta^{-1}_{P_I \cap L_{I_0}}|_T = \delta^{1/2}_{w^{-1}_{I_0,I} B_{I_0} w_{I_0,I}} \otimes_E \delta^{-1/2}_{B_{I_0}} = \delta^{1/2}_{B_{I_0}}(w_{I_0,I} \cdot w^{-1}_{I_0,I}) \otimes_E \delta^{-1/2}_{B_{I_0}}.$$

Using this and (2.10) we deduce isomorphisms of T-representations for π_0^∞ in $\mathrm{Rep}^\infty_{\mathrm{adm}}(L_{I_0})$

$$\begin{aligned}J_{I,\emptyset}(J'_{I_0,I}(\pi_0^\infty)) &\cong (J_{w_{I_0,I}(I),\emptyset}(J_{I_0,w_{I_0,I}(I)}(\pi_0^\infty)))^{w_{I_0,I}} \otimes_E \delta^{-1}_{P_I\cap L_{I_0}}|_T\\ &\cong J_{I_0,\emptyset}(\pi_0^\infty)^{w_{I_0,I}} \otimes_E \delta^{-1}_{P_I\cap L_{I_0}}|_T\\ &\cong (J_{I_0,\emptyset}(\pi_0^\infty)\otimes_E \delta_{B_{I_0}}^{1/2})^{w_{I_0,I}} \otimes_E \delta_{B_{I_0}}^{-1/2}.\end{aligned}$$

From $w_{I_0,I}^{-1}\cdot\chi = (\chi\otimes_E \delta_{B_{I_0}}^{1/2})(w_{I_0,I}\cdot w_{I_0,I}^{-1})\otimes_E \delta_{B_{I_0}}^{-1/2}$ we finally obtain

$$\mathcal{J}(J'_{I_0,I}(\pi_0^\infty)) = w_{I_0,I}^{-1}\cdot\mathcal{J}(\pi_0^\infty) \tag{2.12}$$

where the set $\mathcal{J}(-)$ is as in Definition 2.1.4.

Let $I\subseteq\Delta$. We refer to [59, Sect. VI.5.2] for the definition of equivalence classes of cuspidal data (for the group L_I). We consider here equivalence classes of cuspidal data $(T,\chi\otimes_E\delta_{B_I}^{1/2})$ for $\chi\in\widehat{T}^\infty$. Recall that it consists of the cuspidal data $(gTg^{-1},(\chi\otimes_E\delta_{B_I}^{1/2})(g^{-1}\cdot g)$ for $g\in L_I$. In particular for $\chi'\in W(L_I)\cdot\chi$ all cuspidal data $(T,\chi'\otimes_E\delta_{B_I}^{1/2})$ are equivalent. To a left $W(L_I)$-coset Σ (under the dot action (2.11)) we can thus associate an element in the set $\Omega(L_I)$ of all equivalence classes of cuspidal data.

For each finite subset $\Sigma\subseteq\widehat{T}^\infty$ which is stable under the left dot action of $W(L_I)$, we write $\mathcal{B}^I_\Sigma$ for the category of finite length representations π^∞ in $\mathrm{Rep}^\infty_{\mathrm{adm}}(L_I)$ with each $\sigma^\infty\in\mathrm{JH}_{L_I}(\pi^\infty)$ satisfying $\emptyset\neq\mathcal{J}(\sigma^\infty)\subseteq\Sigma$. By [59, Sect. VI.7.2] and Remark 2.1.7 below, we know that $\mathcal{B}^I_\Sigma$ is the direct sum of the $\mathcal{B}^I_{\Sigma'}$ for Σ' running through the $W(L_I)$-cosets contained in Σ. In particular, for each object π^∞ in $\mathcal{B}^I_\Sigma$, we have a canonical decomposition

$$\pi^\infty\cong\bigoplus_{\Sigma'}\pi^\infty_{\mathcal{B}^I_{\Sigma'}} \tag{2.13}$$

with $\pi^\infty_{\mathcal{B}^I_{\Sigma'}}\in\mathcal{B}^I_{\Sigma'}$ and Σ' running through $W(L_I)$-cosets contained in Σ. The exactness of $J_{I,\emptyset}$ implies that $\mathcal{J}(\pi^\infty_{\mathcal{B}^I_{\Sigma'}})=\mathcal{J}(\pi^\infty)\cap\Sigma'$. We say that Σ is *G-regular* if each $\chi\in\Sigma$ is G-regular.

Remark 2.1.7 The coefficient field in [59, Sect. VI.7.2] being algebraically closed, the decomposition (2.13) deserves some justification. Let $\overline{E}$ an algebraic closure of E and $\mathrm{Rep}^\infty_{\mathrm{adm}}(L_I,\overline{E})$ the abelian category of smooth admissible representations of L_I over $\overline{E}$-vector spaces. Let $I\subseteq\Delta$ and $\Sigma\subseteq\widehat{T}^\infty$ a finite subset which is stable under the left dot action of $W(L_I)$ (and that we also see inside the smooth $\overline{E}$-valued characters of T). We write $\mathcal{B}^I_{\Sigma,\overline{E}}$ for the category of finite length representations $\pi^\infty_{\overline{E}}$ in $\mathrm{Rep}^\infty_{\mathrm{adm}}(L_I,\overline{E})$ with each $\sigma^\infty_{\overline{E}}\in\mathrm{JH}_{L_I}(\pi^\infty_{\overline{E}})$ satisfying $\emptyset\neq\mathcal{J}(\sigma^\infty_{\overline{E}})\subseteq\Sigma$. Now let π^∞ in $\mathcal{B}^I_\Sigma\subseteq\mathrm{Rep}^\infty_{\mathrm{adm}}(L_I)$ and $\pi^\infty_{\overline{E}}\stackrel{\mathrm{def}}{=}\pi^\infty\otimes_E\overline{E}$. By [59, Sect. VI.7.2] we have a decomposition

$$\pi^\infty_{\overline{E}}\cong\bigoplus_{\Sigma'}(\pi^\infty_{\overline{E}})_{\mathcal{B}^I_{\Sigma',\overline{E}}} \tag{2.14}$$

where $(\pi^{\infty}_{\overline{E}})_{\mathcal{B}^I_{\Sigma',\overline{E}}} \in \mathcal{B}^I_{\Sigma',\overline{E}}$ and Σ' runs through the $W(L_I)$-cosets contained in Σ. There exists an obvious E-linear action of $\mathrm{Gal}(\overline{E}/E)$ on $\pi^{\infty}_{\overline{E}}$ such that $(\pi^{\infty}_{\overline{E}})^{\mathrm{Gal}(\overline{E}/E)} \cong \pi^{\infty}$, which induces by (2.14) an E-linear action of $\mathrm{Gal}(\overline{E}/E)$ on $\oplus_{\Sigma'}(\pi^{\infty}_{\overline{E}})_{\mathcal{B}^I_{\Sigma',\overline{E}}}$, and also by scalar extension from E to $\overline{E}$ an $\overline{E}$-linear action of $\mathrm{Gal}(\overline{E}/E)$ on

$$(\pi^{\infty}_{\overline{E}}) \otimes_E \overline{E} \cong \bigoplus_{\Sigma'} \big((\pi^{\infty}_{\overline{E}})_{\mathcal{B}^I_{\Sigma',\overline{E}}}\big) \otimes_E \overline{E}.$$

But the smooth reducible representations $\big((\pi^{\infty}_{\overline{E}})_{\mathcal{B}^I_{\Sigma',\overline{E}}}\big) \otimes_E \overline{E}$ do not share any irreducible constituent over $\overline{E}$ for *distinct* Σ' since the Σ' consist of E-valued characters, hence the action of $\mathrm{Gal}(\overline{E}/E)$ on $\pi^{\infty}_{\overline{E}}$ must stabilize each $(\pi^{\infty}_{\overline{E}})_{\mathcal{B}^I_{\Sigma',\overline{E}}}$. Let $\pi^{\infty}_{\mathcal{B}^I_{\Sigma'}} \stackrel{\mathrm{def}}{=} ((\pi^{\infty}_{\overline{E}})_{\mathcal{B}^I_{\Sigma',\overline{E}}})^{\mathrm{Gal}(\overline{E}/E)}$, we deduce from (2.14)

$$\pi^{\infty} \cong \bigoplus_{\Sigma'} \pi^{\infty}_{\mathcal{B}^I_{\Sigma'}}$$

and $\pi^{\infty}_{\mathcal{B}^I_{\Sigma'}} \otimes_E \overline{E} \cong (\pi^{\infty}_{\overline{E}})_{\mathcal{B}^I_{\Sigma',\overline{E}}}$ in $\mathrm{Rep}^{\infty}_{\mathrm{adm}}(L_I, \overline{E})$ (extending scalars to $\overline{E}$ again). This is the decomposition (2.13).

Let $I_0, I_1 \subseteq I \subseteq \Delta$ and recall $W^{I_0,I_1}(L_I) \subseteq W(L_I)$ is the subset of minimal length representatives of $W(L_{I_0})\backslash W(L_I)/W(L_{I_1})$. Recall also the Bruhat decomposition ([20, Lemma 5.5])

$$\bigsqcup_{w\in W^{I_0,I_1}(L_I)} (P_{I_0} \cap L_I)w(P_{I_1} \cap L_I) = L_I = \bigsqcup_{w\in W^{I_0,I_1}(L_I)} (P_{I_1} \cap L_I)w^{-1}(P_{I_0} \cap L_I).$$

For $w \in W^{I_0,I_1}(L_I)$ we have

$$\begin{aligned} B_{I_0} &= B \cap L_{I_0} = wBw^{-1} \cap L_{I_0} \subseteq wP_{I_1}w^{-1} \cap L_{I_0} \\ B_{I_1} &= B \cap L_{I_1} = w^{-1}Bw \cap L_{I_1} \subseteq w^{-1}P_{I_0}w \cap L_{I_1}. \end{aligned} \tag{2.15}$$

Hence $wP_{I_1}w^{-1} \cap L_{I_0}$ is a parabolic subgroup of L_{I_0} with Levi quotient $wL_{I_1}w^{-1} \cap L_{I_0} = L_{w(I_1)\cap I_0}$, and $w^{-1}P_{I_0}w \cap L_{I_1}$ is a parabolic subgroup of L_{I_1} with Levi quotient $w^{-1}L_{I_0}w \cap L_{I_1} = L_{w^{-1}(I_0)\cap I_1}$. By (2.15) and the equality $L_{w(I_1)\cap I_0} = wL_{w^{-1}(I_0)\cap I_1} w^{-1}$ we have

$$\begin{aligned} B_{w(I_1)\cap I_0} = B_{I_0} \cap L_{w(I_1)\cap I_0} = wBw^{-1} \cap L_{I_0} \cap L_{w(I_1)\cap I_0} &= w(B \cap L_{w^{-1}(I_0)\cap I_1})w^{-1} \\ &= wB_{w^{-1}(I_0)\cap I_1}w^{-1}. \end{aligned} \tag{2.16}$$

As $w(I_1) \cap I_0 \subseteq I_0$, we have $P_{w(I_1)\cap I_0} \subseteq P_{I_0}$ and thus we have the decomposition

$$P_{w(I_1)\cap I_0} \cap L_I = (N_{I_0} \cap L_I) \cdot (P_{w(I_1)\cap I_0} \cap L_{I_0}) = (N_{I_0} \cap L_I) \cdot (wP_{I_1}w^{-1} \cap L_{I_0}),$$

which together with (2.5) (taking $P = P_{I_0} \cap L_I$ and $Q = P_{w(I_1)\cap I_0} \cap L_I \subseteq P$) implies

$$\delta_{wP_{I_1}w^{-1}\cap L_{I_0}} \cdot (\delta_{P_{I_0}\cap L_I}|_{wL_{I_1}w^{-1}\cap L_{I_0}}) = \delta_{P_{w(I_1)\cap I_0}\cap L_I}. \tag{2.17}$$

Likewise, as $w^{-1}(I_0) \cap I_1 \subseteq I_1$, we have $P_{w^{-1}(I_0)\cap I_1} \subseteq P_{I_1}$ and thus the decomposition

$$P_{w^{-1}(I_0)\cap I_1} \cap L_I = (N_{I_1} \cap L_I) \cdot (P_{w^{-1}(I_0)\cap I_1} \cap L_{I_1}) = (N_{I_1} \cap L_I) \cdot (w^{-1}P_{I_0}w \cap L_{I_1}),$$

which together with (2.5) (taking $P = P_{I_1} \cap L_I$ and $Q = P_{w^{-1}(I_0)\cap I_1} \cap L_I \subseteq P$) implies

$$\delta_{w^{-1}P_{I_0}w\cap L_{I_1}} \cdot (\delta_{P_{I_1}\cap L_I}|_{L_{I_1}\cap w^{-1}L_{I_0}w}) = \delta_{P_{w^{-1}(I_0)\cap I_1}\cap L_I}. \tag{2.18}$$

We now consider the two functors (see (2.6) for the notation)

$$J_{I_0,I_1,w}(-) \stackrel{\text{def}}{=} (J_{I_0,w(I_1)\cap I_0}(-))^w \otimes_E \delta_{I_0,I_1,w} : \mathrm{Rep}^{\infty}_{\mathrm{adm}}(L_{I_0}) \longrightarrow \mathrm{Rep}^{\infty}_{\mathrm{adm}}(L_{w^{-1}(I_0)\cap I_1}) \tag{2.19}$$

where

$$\delta_{I_0,I_1,w} \stackrel{\text{def}}{=} \delta^{1/2}_{w^{-1}P_{w(I_1)\cap I_0}w\cap L_I} \cdot \delta^{-1/2}_{P_{w^{-1}(I_0)\cap I_1}\cap L_I}, \tag{2.20}$$

(note that $\delta_{I_0,I_1,1} = 1$) and

$$i^{\infty}_{I_0,I_1,w}(-) \stackrel{\text{def}}{=} i^{\infty}_{w^{-1}(I_0)\cap I_1,I_1}(-) : \mathrm{Rep}^{\infty}_{\mathrm{adm}}(L_{w^{-1}(I_0)\cap I_1}) \longrightarrow \mathrm{Rep}^{\infty}_{\mathrm{adm}}(L_{I_1}). \tag{2.21}$$

It follows from (2.16) (and (2.5)) that

$$\begin{aligned}\delta_{I_0,I_1,w}|_T &= \delta^{1/2}_{w^{-1}P_{w(I_1)\cap I_0}w\cap L_I}|_T \cdot \delta^{1/2}_{w^{-1}B_{w(I_1)\cap I_0}w} \cdot \delta^{-1/2}_{B_{w^{-1}(I_0)\cap I_1}} \cdot \delta^{-1/2}_{P_{w^{-1}(I_0)\cap I_1}\cap L_I}|_T \\ &= \delta^{1/2}_{w^{-1}B_Iw} \otimes_E \delta^{-1/2}_{B_I}.\end{aligned} \tag{2.22}$$

In particular $\delta_{I_0,I_1,w}$ is $K^{\times}$-valued (as in Remark 2.1.3). Moreover if $I_0 = \emptyset$ (so that $L_{I_0} = L_{w^{-1}(I_0)\cap I_1} = T$), we have for $\chi \in \widehat{T}^{\infty}$ and $w \in W^{\emptyset,I_1}(L_I)$

$$J_{\emptyset,I_1,w}(\chi) = (w^{-1}(\chi) \otimes_E \delta^{1/2}_{w^{-1}B_Iw}) \otimes_E \delta^{-1/2}_{B_I} = w^{-1} \cdot \chi. \tag{2.23}$$

Lemma 2.1.8 *Let I_0, $I_1 \subseteq I \subseteq \Delta$, $w \in W^{I_0,I_1}(L_I)$ and π_0^{∞} in $\mathrm{Rep}^{\infty}_{\mathrm{adm}}(L_{I_0})$. We have an isomorphism in $\mathrm{Rep}^{\infty}_{\mathrm{adm}}(L_{I_1} \cap w^{-1}L_{I_0}w)$*

$$J_{I_0,I_1,w}(\pi_0^{\infty}) \cong \left((\mathrm{ind}^{N_{I_1}}_{N_{I_1}\cap w^{-1}P_{I_0}w}\pi_0^{\infty,w})^{\infty}\right)_{N_{I_1}} \tag{2.24}$$

and an isomorphism in $\mathrm{Rep}^{\infty}_{\mathrm{adm}}(L_{I_1})$

$$i^{\infty}_{I_0,I_1,w}(J_{I_0,I_1,w}(\pi_0^{\infty})) \cong \left((\mathrm{ind}^{P_{I_1}}_{P_{I_1}\cap w^{-1}P_{I_0}w}\pi_0^{\infty,w})^{\infty}\right)_{N_{I_1}} \tag{2.25}$$

(where we view $\pi_0^{\infty,w} \in \mathrm{Rep}^{\infty}_{\mathrm{adm}}(w^{-1}L_{I_0}w) \subseteq \mathrm{Rep}^{\infty}_{\mathrm{adm}}(w^{-1}P_{I_0}w)$ *as a smooth representation of* $P_{I_1} \cap w^{-1}P_{I_0}w$ *by restriction).*

Proof The isomorphism (2.24) easily follows from [59, (VI.5.1.3)] (applied with $J = N = N_{I_1}$ and $H = N_{I_1} \cap w^{-1}P_{I_0}w$) and from $(J_{I_0,w(I_1)\cap I_0}(\pi_0^{\infty}))^w \cong J_{N_{I_1\cap w^{-1}(I_0)}\cap L_{w^{-1}(I_0)}}(\pi_0^{\infty,w})$, noting that the action of $N_{I_1} \cap w^{-1}P_{I_0}w$ on $\pi_0^{\infty,w}$ factors through $N_{I_1} \cap w^{-1}L_{I_0}w = N_{I_1\cap w^{-1}(I_0)} \cap L_{w^{-1}(I_0)}$. Note that N_{I_1} is a normal subgroup of P_{I_1} and thus $N_{I_1}(P_{I_1} \cap w^{-1}P_{I_0}w)$ is a subgroup of P_{I_1} satisfying

$$N_{I_1}(P_{I_1} \cap w^{-1}P_{I_0}w)/(P_{I_1} \cap w^{-1}P_{I_0}w) \cong N_{I_1}/(N_{I_1} \cap w^{-1}P_{I_0}w) \tag{2.26}$$

and

$$P_{I_1}/N_{I_1}(P_{I_1} \cap w^{-1}P_{I_0}w) \cong L_{I_1}/(L_{I_1} \cap w^{-1}P_{I_0}w). \tag{2.27}$$

The isomorphism (2.25) then follows from

$$\begin{aligned}
((\mathrm{ind}_{P_{I_1}\cap w^{-1}P_{I_0}w}^{P_{I_1}}\pi_0^{\infty,w})^{\infty})_{N_{I_1}} &\cong ((\mathrm{ind}_{N_{I_1}(P_{I_1}\cap w^{-1}P_{I_0}w)}^{P_{I_1}}(\mathrm{ind}_{P_{I_1}\cap w^{-1}P_{I_0}w}^{N_{I_1}(P_{I_1}\cap w^{-1}P_{I_0}w)}\pi_0^{\infty,w})^{\infty})^{\infty})_{N_{I_1}} \\
&\cong (\mathrm{ind}_{N_{I_1}(P_{I_1}\cap w^{-1}P_{I_0}w)}^{P_{I_1}}((\mathrm{ind}_{P_{I_1}\cap w^{-1}P_{I_0}w}^{N_{I_1}(P_{I_1}\cap w^{-1}P_{I_0}w)}\pi_0^{\infty,w})^{\infty})_{N_{I_1}})^{\infty} \\
&\cong (\mathrm{Ind}_{L_{I_1}\cap w^{-1}P_{I_0}w}^{L_{I_1}}((\mathrm{ind}_{N_{I_1}\cap w^{-1}P_{I_0}w}^{N_{I_1}}\pi_0^{\infty,w})^{\infty})_{N_{I_1}})^{\infty} \\
&\cong i_{I_0,I_1,w}^{\infty}(J_{I_0,I_1,w}(\pi_0^{\infty}))
\end{aligned}$$

where the second isomorphism follows from Lemma 2.1.2 applied with $P = P_{I_1}$, $H = N_{I_1}(P_{I_1} \cap w^{-1}P_{I_0}w)$, $N = N_{I_1}$ and $\pi^{\infty} = (\mathrm{ind}_{P_{I_1}\cap w^{-1}P_{I_0}w}^{N_{I_1}(P_{I_1}\cap w^{-1}P_{I_0}w)}\pi_0^{\infty,w})^{\infty}$ (using that $P_{I_1} \twoheadrightarrow P_{I_1}/N_{I_1}(P_{I_1} \cap w^{-1}P_{I_0}w)$ admits a continuous section), the third isomorphism follows from (2.26) and (2.27), and the last from (2.24). □

Now we recall the classical Bernstein-Zelevinsky geometric lemma.

Lemma 2.1.9 *Let* $I_0, I_1 \subseteq I \subseteq \Delta$ *and* π_0^{∞} *in* $\mathrm{Rep}^{\infty}_{\mathrm{adm}}(L_{I_0})$ *of finite length. Then* $J_{I,I_1}(i_{I_0,I}^{\infty}(\pi_0^{\infty})) \in \mathrm{Rep}^{\infty}_{\mathrm{adm}}(L_{I_1})$ *admits a canonical decreasing filtration with graded pieces* $i_{I_0,I_1,w}^{\infty}(J_{I_0,I_1,w}(\pi_0^{\infty}))$ *for* $w \in W^{I_0,I_1}(L_I)$.

Proof This is [4, Sect. 2.12] (see also [59, Sect. VI.5.1]), however since we use different normalizations, we need to make a translation. It follows from [59, Theorem VI.5.1] that

$$J_{I,I_1}(i_{I_0,I}^{\infty}(\pi_0^{\infty})) \cong r_{P_{I_1}\cap L_I}^{L_I}(i_{P_{I_0}\cap L_I}^{L_I}(\pi_0^{\infty} \otimes_E \delta_{P_{I_0}\cap L_I}^{1/2})) \otimes_E \delta_{P_{I_1}\cap L_I}^{-1/2}$$

admits a decreasing filtration with graded pieces ($w \in W^{I_0,I_1}(L_I)$):

$$
\begin{aligned}
&\Big(i^{L_{I_1}}_{w^{-1}P_{I_0}w\cap L_{I_1}} \circ w \circ r^{L_{I_0}}_{wP_{I_1}w^{-1}\cap L_{I_0}}(\pi_0^\infty \otimes_E \delta^{1/2}_{P_{I_0}\cap L_I})\Big) \otimes_E \delta^{-1/2}_{P_{I_1}\cap L_I}\\
&\cong i^\infty_{I_0,I_1,w}\Big(\big(J_{wP_{I_1}w^{-1}\cap L_{I_0}}(\pi_0^\infty \otimes_E \delta^{1/2}_{P_{I_0}\cap L_I}) \otimes_E \delta^{1/2}_{wP_{I_1}w^{-1}\cap L_{I_0}}\big)^w \otimes_E \delta^{-1/2}_{w^{-1}P_{I_0}w\cap L_{I_1}}\Big) \otimes_E \delta^{-1/2}_{P_{I_1}\cap L_I}\\
&\cong i^\infty_{I_0,I_1,w}\Big(\big(J_{wP_{I_1}w^{-1}\cap L_{I_0}}(\pi_0^\infty) \otimes_E \delta^{1/2}_{P_{w(I_1)\cap I_0}\cap L_I}\big)^w \otimes_E \delta^{-1/2}_{P_{w^{-1}(I_0)\cap I_1}\cap L_I}\Big)\\
&\cong i^\infty_{I_0,I_1,w}\Big(J_{wP_{I_1}w^{-1}\cap L_{I_0}}(\pi_0^\infty)^w \otimes_E \big(\delta^{1/2}_{w^{-1}P_{w(I_1)\cap I_0}w\cap L_I} \cdot \delta^{-1/2}_{P_{w^{-1}(I_0)\cap I_1}\cap L_I}\big)\Big)\\
&\cong i^\infty_{I_0,I_1,w}(J_{I_0,I_1,w}(\pi_0^\infty))
\end{aligned}
$$

where the second isomorphism follows from (2.17) and (2.18) and the last from (2.20) in the definition of $J_{I_0,I_1,w}$. □

Remark 2.1.10 We emphasize that the statement of Lemma 2.1.9 holds for $E = K$, though we cannot take $E = K$ in its proof because of characters such as $\delta^{1/2}_{P_{I_0}\cap L_I}$ and $\delta^{-1/2}_{P_{I_1}\cap L_I}$ which are not $K^\times$-valued in general. The usual formulation of the geometric lemma ([4, Sect. 2.12], see also [59, Sect. VI.5.1]) uses *normalized* induction and restriction functors. But if one reformulates the geometric lemma using *unnormalized* induction and restriction functors (as we do in Lemma 2.1.9), one sees that its proof works for any E, including $E = K$.

Lemma 2.1.11 *Let $I_0 \subseteq I \subseteq \Delta$ and π_0^∞ in $\mathrm{Rep}^\infty_{\mathrm{adm}}(L_{I_0})$ of finite length. Then we have*

$$
\mathcal{J}(i^\infty_{I_0,I}(\pi_0^\infty)) = \bigcup_{w\in W^{I_0,\emptyset}(L_I)} w^{-1}\cdot\mathcal{J}(\pi_0^\infty) \subseteq W(L_I)\cdot\mathcal{J}(\pi_0^\infty). \tag{2.28}
$$

Proof By Lemma 2.1.9, we know that $J_{I,\emptyset}(i^\infty_{I_0,I}(\pi_0^\infty))$ admits a canonical filtration with graded pieces $J_{I_0,\emptyset,w}(\pi_0^\infty)$ for $w \in W^{I_0,\emptyset}$, which implies that

$$
\mathcal{J}(i^\infty_{I_0,I}(\pi_0^\infty)) = \bigcup_{w\in W^{I_0,\emptyset}} \mathcal{J}(J_{I_0,\emptyset,w}(\pi_0^\infty)). \tag{2.29}
$$

It follows from (2.19) and (2.22) that

$$
\mathcal{J}(J_{I_0,\emptyset,w}(\pi_0^\infty)) = w^{-1}\cdot\mathcal{J}(J_{I_0,\emptyset}(\pi_0^\infty)) = w^{-1}\cdot\mathcal{J}(\pi_0^\infty),
$$

which together with (2.29) gives (2.28). □

Remark 2.1.12 By taking $I_0 = \emptyset$ and π_0^∞ to be some $\chi \in \widehat{T}^\infty$ in Lemma 2.1.11, we deduce that $\mathcal{J}(i^\infty_{\emptyset,I}(\chi)) = W(L_I)\cdot\chi$. Suppose $\pi^\infty \in \mathrm{Rep}^\infty_{\mathrm{adm}}(L_I)$ is irreducible with $\chi \in \mathcal{J}(\pi^\infty)$, then π^∞ embeds into $i^\infty_{\emptyset,I}(\chi)$ (using (2.7)) and $W(L_I)\cdot\mathcal{J}(\pi^\infty) = W(L_I)\cdot\chi$.

Lemma 2.1.13 *Let $I_0, I_1 \subseteq I \subseteq \Delta$, $w \in W^{I_0,I_1}(L_I)$ and π_0^∞ in $\mathrm{Rep}^\infty_{\mathrm{adm}}(L_{I_0})$ of finite length. Then we have*

$$
\mathcal{J}(J_{I_0,I_1,w}(\pi_0^\infty)) = w^{-1}\cdot\mathcal{J}(J_{I_0,w(I_1)\cap I_0}(\pi_0^\infty)) = w^{-1}\cdot\mathcal{J}(\pi_0^\infty) \subseteq w^{-1}W(L_{I_0})\cdot\mathcal{J}(\pi_0^\infty) \tag{2.30}
$$

and

$$\mathcal{J}(i^{\infty}_{I_0,I_1,w}(J_{I_0,I_1,w}(\pi_0^{\infty}))) \subseteq W(L_{I_1}) \cdot \mathcal{J}(J_{I_0,I_1,w}(\pi_0^{\infty})) \subseteq W(L_{I_1})w^{-1}W(L_{I_0}) \cdot \mathcal{J}(\pi_0^{\infty}). \tag{2.31}$$

Proof The first claim (2.30) follows directly from (2.19) and (2.22). The second claim (2.31) follows from (2.30) together with (2.28) (replacing I_0, I and π_0^{∞} there by $w^{-1}(I_0) \cap I_1$, I_1 and $J_{I_0,I_1,w}(\pi_0^{\infty})$). □

Lemma 2.1.14 *For a G-regular $\chi \in \widehat{T}^{\infty}$ (see (i) of Definition 2.1.4) and $I \subseteq \Delta$, we have a canonical isomorphism in* $\mathrm{Rep}^{\infty}_{\mathrm{adm}}(T)$

$$J_{I,\emptyset}(i^{\infty}_{\emptyset,I}(\chi)) \cong \bigoplus_{w \in W(L_I)} w^{-1} \cdot \chi. \tag{2.32}$$

Proof By (2.23) we have $i^{\infty}_{\emptyset,\emptyset,w}(J_{\emptyset,\emptyset,w}(\chi)) = J_{\emptyset,\emptyset,w}(\chi) \cong w^{-1} \cdot \chi$. Then the statement follows from Lemma 2.1.9 applied with $I_0 = I_1 = \emptyset$ and $\pi_0^{\infty} = \chi$, noting that, as χ is G-regular, the $w \cdot \chi$ are distinct for $w \in W(L_I)$, so the canonical filtration in Lemma 2.1.9 must split. □

Lemma 2.1.15 *Let $I \subseteq \Delta$.*

(i) Let π^{∞} in $\mathrm{Rep}^{\infty}_{\mathrm{adm}}(L_I)$ be G-basic, then the T-representation $J_{I,\emptyset}(\pi^{\infty})$ is semi-simple and multiplicity free and $W(L_I) \cdot \mathcal{J}(\pi^{\infty})$ is a single regular left $W(L_I)$-coset.

(ii) Let π^{∞} in $\mathrm{Rep}^{\infty}_{\mathrm{adm}}(L_I)$ be irreducible and G-regular, then the L_I-representation π^{∞} is uniquely determined by the set $\mathcal{J}(\pi^{\infty})$. Moreover non-isomorphic irreducible G-regular $\pi_0^{\infty}, \pi_1^{\infty} \in \mathrm{Rep}^{\infty}_{\mathrm{adm}}(L_I)$ satisfy $\mathcal{J}(\pi_0^{\infty}) \cap \mathcal{J}(\pi_1^{\infty}) = \emptyset$.

(iii) Let $\chi \in \widehat{T}^{\infty}$ be G-regular, then $i^{\infty}_{\emptyset,I}(\chi)$ is multiplicity free and any irreducible constituent π^{∞} of $i^{\infty}_{\emptyset,I}(\chi)$ is such that $\mathcal{J}(\pi^{\infty}) \neq 0$. Moreover for $w \in W(L_I)$ the semi-simplification of $i^{\infty}_{\emptyset,I}(w \cdot \chi)$ doesn't depend on w.

Proof We prove (i). For a G-regular χ, Lemma 2.1.14 implies that $J_{I,\emptyset}(i^{\infty}_{\emptyset,I}(\chi))$ is multiplicity free and

$$W(L_I) \cdot \chi = \mathcal{J}(i^{\infty}_{\emptyset,I}(\chi)) = \bigsqcup_{\pi^{\infty}} \mathcal{J}(\pi^{\infty}) \tag{2.33}$$

where π^{∞} runs through the Jordan-Hölder factors of $i^{\infty}_{\emptyset,I}(\chi)$. This implies (i) as each G-basic π^{∞} embeds into $i^{\infty}_{\emptyset,I}(\chi)$ for some G-regular χ by (ii) of Definition 2.1.4. The first half of (iii) follows from [79, Proposition 2.1(b),(c)]. We prove (ii). The first half of (iii) together with (2.33) easily imply the first half of (ii). For the second half of (ii), any $\pi_0^{\infty}, \pi_1^{\infty} \in \mathrm{Rep}^{\infty}_{\mathrm{adm}}(L_I)$ satisfying $\chi \in \mathcal{J}(\pi_0^{\infty}) \cap \mathcal{J}(\pi_1^{\infty})$ both inject into $i^{\infty}_{\emptyset,I}(\chi)$, again contradicting (2.33). Finally, the second half of (iii) follows from (ii) and the fact that $\mathcal{J}(i^{\infty}_{\emptyset,I}(w \cdot \chi)) = W(L_I) \cdot \chi$ (by (2.33)) is independent of the choice of $w \in W(L_I)$. □

In particular it follows from Lemma 2.1.15 that if π^∞ is a G-basic representation of L_I then π^∞ is in $\mathcal{B}^I_\Sigma$ for $\Sigma = W(L_I) \cdot \mathcal{J}(\pi^\infty)$. We give below several other useful remarks on G-basic representations.

Remark 2.1.16

(i) Let $I \subseteq \Delta$ and $\pi^\infty \in \mathrm{Rep}^\infty_{\mathrm{adm}}(L_I)$ be G-basic, then by (ii) of Definition 2.1.4 there exist $I_1 \subseteq I$ and an irreducible G-regular $\pi_1^\infty \in \mathrm{Rep}^\infty_{\mathrm{adm}}(L_{I_1})$ such that $\pi^\infty \cong i^\infty_{I_1,I}(\pi_1^\infty)$. As $J_{I_1,\emptyset}(\pi_1^\infty)$ is semi-simple and multiplicity free by (i) of Lemma 2.1.15, we deduce from (2.10) that $J'_{I_1,\emptyset}(\pi_1^\infty)$ is also semi-simple and multiplicity free with

$$\mathrm{JH}_T(J'_{I_1,\emptyset}(\pi_1^\infty)) = w_{I_1} \cdot \mathrm{JH}_T(J_{I_1,\emptyset}(\pi_1^\infty)) \tag{2.34}$$

where we recall from Sect. 1.4 that w_{I_1} is the longest element of $W(L_{I_1})$ (so that $w_{I_1}(\delta_{B_{I_1}}) = w_{I_1}^{-1}(\delta_{B_{I_1}}) = \delta_{B_{I_1}}^{-1}$) and $\mathrm{JH}_T(-)$ the set of irreducible constituents. Hence, by combining (2.34) with (2.7) and (2.8) for $k = 0$, we see that π_1^∞ embeds into $i^\infty_{\emptyset,I_1}(\chi)$ if and only if π_1^∞ is a quotient of $i^\infty_{\emptyset,I_1}(w_{I_1} \cdot \chi)$. In particular, we deduce that $\pi^\infty \cong i^\infty_{I_1,I}(\pi_1^\infty)$ is a quotient of $i^\infty_{\emptyset,I}(w_{I_1} \cdot \chi)$ for any $\chi \in \mathcal{J}(\pi_1^\infty)$.

(ii) Let $I \subseteq \Delta$, $\pi^\infty \in \mathrm{Rep}^\infty_{\mathrm{adm}}(L_I)$ be irreducible and G-regular, and $\chi \in \widehat{T}^\infty$ be G-regular. By (2.7) for $k = 0$ (resp. by (2.8) for $k = 0$ with (2.34)) π^∞ is in the socle (resp. cosocle) of $i^\infty_{\emptyset,I}(\chi)$ if and only if $\chi \in \mathcal{J}(\pi^\infty)$ (resp. $\chi \in w_I \cdot \mathcal{J}(\pi^\infty)$). The second half of (ii) of Lemma 2.1.15 then implies that $i^\infty_{\emptyset,I}(\chi)$ has simple socle and cosocle.

(iii) For G-regular $\chi, \chi' \in \widehat{T}^\infty$ such that $W(L_I) \cdot \chi = W(L_I) \cdot \chi'$, we have canonical isomorphisms by (2.7) (for $k = 0$) and Lemma 2.1.14:

$$\mathrm{Hom}_{L_I}(i^\infty_{\emptyset,I}(\chi'), i^\infty_{\emptyset,I}(\chi)) \cong \mathrm{Hom}_T(J_{I,\emptyset}(i^\infty_{\emptyset,I}(\chi')), \chi) \cong \mathrm{Hom}_T\Big(\bigoplus_{w \in W(L_I)} w^{-1} \cdot \chi', \chi\Big)$$

of one dimensional spaces. In particular, there exists a unique (up to scalar) non-zero map $i^\infty_{\emptyset,I}(\chi') \to i^\infty_{\emptyset,I}(\chi)$ for such χ, χ'.

(iv) Let $I \subseteq \Delta$ and $\pi^\infty \in \mathrm{Rep}^\infty_{\mathrm{adm}}(L_I)$ be G-basic. By (ii) of Definition 2.1.4 and (i) above there exist G-regular $\chi, \chi' \in \widehat{T}^\infty$ such that π^∞ is a subrepresentation of $i^\infty_{\emptyset,I}(\chi)$ and a quotient of $i^\infty_{\emptyset,I}(\chi')$. As both $i^\infty_{\emptyset,I}(\chi)$ and $i^\infty_{\emptyset,I}(\chi')$ are multiplicity free by (iii) of Lemma 2.1.15, we deduce from (iii) above that π^∞ is the image of the unique (up to scalar) non-zero map $i^\infty_{\emptyset,I}(\chi') \to i^\infty_{\emptyset,I}(\chi)$, is multiplicity free and (using (ii) above) has simple socle and cosocle.

Lemma 2.1.17 *Let $I \subseteq \Delta$, $w \in W(G)$ such that $w^{-1}(I) \subseteq \Delta$, $\delta : L_{w^{-1}(I)} \to E^\times$ a smooth character and $\pi^\infty \in \mathrm{Rep}^\infty_{\mathrm{adm}}(L_I)$ a G-basic representation. Assume that $\mathcal{J}((\pi^\infty)^w \otimes_E \delta) \subseteq W(G) \cdot \mathcal{J}(\pi^\infty)$. Then $(\pi^\infty)^w \otimes_E \delta \in \mathrm{Rep}^\infty_{\mathrm{adm}}(L_{w^{-1}(I)})$ is again G-basic.*

Proof By (ii) of Definition 2.1.4 there exist $I_1 \subseteq I$ and an irreducible G-regular $\pi_1^\infty \in \mathrm{Rep}^\infty_{\mathrm{adm}}(L_{I_1})$ such that $\pi^\infty \cong i^\infty_{I_1,I}(\pi_1^\infty)$, which implies that

$$(\pi^\infty)^w \otimes_E \delta \cong i^\infty_{w^{-1}(I_1),w^{-1}(I)}((\pi_1^\infty)^w \otimes_E \delta|_{L_{w^{-1}(I_1)}}). \tag{2.35}$$

As $W(L_I) \cdot \mathcal{J}(\pi^\infty)$ is a G-regular left $W(L_I)$-coset by (i) of Lemma 2.1.15, the sets

$$\mathcal{J}((\pi_1^\infty)^w \otimes_E \delta|_{L_{w^{-1}(I_1)}}) \subseteq \mathcal{J}((\pi^\infty)^w \otimes_E \delta) \subseteq W(G) \cdot \mathcal{J}(\pi^\infty)$$

consist of G-regular elements. This forces $(\pi_1^\infty)^w \otimes_E \delta|_{L_{I_1}}$ to be G-regular (irreducible) and thus (2.35) implies that $(\pi^\infty)^w \otimes_E \delta$ is G-basic. □

Lemma 2.1.18 *Let $I_0, I_1 \subseteq I \subseteq \Delta$, $\Sigma_0 \subseteq \widehat{T}^\infty$ be a left $W(L_{I_0})$-stable finite subset which is G-regular, and $\pi_0^\infty \in \mathcal{B}^{I_0}_{\Sigma_0}$.*

(i) We have a canonical decomposition in $\mathrm{Rep}^\infty_{\mathrm{adm}}(L_{I_1})$

$$J_{I,I_1}(i^\infty_{I_0,I}(\pi_0^\infty)) \cong \bigoplus_{w \in W^{I_0,I_1}(L_I)} i^\infty_{I_0,I_1,w}(J_{I_0,I_1,w}(\pi_0^\infty)). \tag{2.36}$$

(ii) If Σ_0 is a G-regular left $W(L_{I_0})$-coset, we have a canonical decomposition for $w \in W^{I_0,I_1}(L_I)$ induced by (2.13)

$$i^\infty_{I_0,I_1,w}(J_{I_0,I_1,w}(\pi_0^\infty)) \cong \bigoplus_{\Sigma} i^\infty_{I_0,I_1,w}(J_{I_0,I_1,w}(\pi_0^\infty))_{\mathcal{B}^{I_1}_{\Sigma}} \tag{2.37}$$

where Σ runs through the (G-regular) left $W(L_{I_1})$-cosets contained in

$$W(L_{I_1})w^{-1}W(L_{I_0}) \cdot \mathcal{J}(\pi_0^\infty).$$

In particular $J_{I,I_1}(i^\infty_{I_0,I}(\pi_0^\infty))_{\mathcal{B}^{I_1}_{\Sigma}} \cong i^\infty_{I_0,I_1,w}(J_{I_0,I_1,w}(\pi_0^\infty))_{\mathcal{B}^{I_1}_{\Sigma}}$.

Proof Using (2.13), it suffices to treat $(\pi_0^\infty)_{\mathcal{B}^{I_0}_{\Sigma'_0}}$ separately for each left $W(L_{I_0})$-coset Σ'_0 contained in Σ_0. Hence, we assume from now that Σ_0 is a single G-regular left $W(L_{I_0})$-coset, namely $\Sigma_0 = W(L_{I_0}) \cdot \mathcal{J}(\pi_0^\infty) = W(L_{I_0}) \cdot \chi$ for an arbitrary $\chi \in \mathcal{J}(\pi_0^\infty)$. As $\chi \in \mathcal{J}(\pi_0^\infty)$ is G-regular, we know that

$$\begin{aligned} W(L_I) \cdot \mathcal{J}(\pi_0^\infty) = W(L_I) \cdot \chi &= \bigsqcup_{w \in W^{I_0,I_1}(L_I)} W(L_{I_1})w^{-1}W(L_{I_0}) \cdot \chi \\ &= \bigsqcup_{w \in W^{I_0,I_1}(L_I)} W(L_{I_1})w^{-1}W(L_{I_0}) \cdot \mathcal{J}(\pi_0^\infty), \end{aligned}$$

which together with (2.31) and (2.13) forces the canonical filtration on $J_{I,I_1}(i^\infty_{I_0,I}(\pi_0^\infty))$ described in Lemma 2.1.9 to be split, and thus (i) follows. (ii) follows directly from (i), (2.31) and (2.13). □

Remark 2.1.19 If we take $I_0 = \emptyset$ and $\pi_0^\infty = \chi$ in Lemma 2.1.18, then by (2.23) we deduce a canonical isomorphism in $\mathrm{Rep}^\infty_{\mathrm{adm}}(L_{I_1})$

$$J_{I,I_1}(i^{\infty}_{\emptyset,I}(\chi)) \cong \bigoplus_{w \in W^{\emptyset,I_1}(L_I)} i^{\infty}_{\emptyset,I_1}(w^{-1} \cdot \chi).$$

Let $I \subseteq \Delta$ and $\Sigma \subseteq \widehat{T}^{\infty}$ be a G-regular left $W(L_I)$-coset. We now attach an undirected graph Γ_{Σ} to Σ following [79, Sect. 2.2]. We choose an arbitrary $\chi \in \Sigma$ and write

$$\chi \otimes_E \delta_B^{1/2} = \rho_1 \boxtimes \cdots \boxtimes \rho_n \tag{2.38}$$

where the ρ_k, $1 \le k \le n$, are smooth distinct characters of $K^{\times}$ (as χ is G-regular). We define the set of vertices of Γ_{Σ} as $V(\Gamma_{\Sigma}) \stackrel{\text{def}}{=} \{\rho_1, \dots, \rho_n\}$ and note that the set $V(\Gamma_{\Sigma})$ is independent of the choice of $\chi \in \Sigma$. Two vertices $\rho, \rho' \in V(\Gamma_{\Sigma})$ are connected by one edge if and only if $\rho'\rho^{-1} \in \{|\cdot|_K, |\cdot|_K^{-1}\}$ and $\rho = \rho_k, \rho' = \rho_{k'}$ for k, k' in the same Levi block of L_I. So each connected component of Γ_{Σ} (there are at least as many as the number of blocks in L_I) has its vertices of the form $\{\rho, |\cdot|_K\rho, \dots, |\cdot|_K^{\ell-1}\rho\}$ for some $\ell \ge 1$, which is called a segment $[\rho, |\cdot|_K^{\ell-1}\rho]$ of length ℓ. An *orientation* $\mathbf{\Gamma}_{\Sigma}$ (on Γ_{Σ}) is a directed graph whose underlying undirected graph is Γ_{Σ}. Each $\chi \in \Sigma$ as above determines an orientation denoted $\mathbf{\Gamma}_{\Sigma}(\chi)$ by requiring that an edge connecting ρ and ρ' has direction $\rho \to \rho'$ if and only if $\rho = \rho_k$ and $\rho' = \rho_{k'}$ for some $k < k'$.

Recall from the last statement in (iii) of Lemma 2.1.15 that the set of irreducible constituents of $i^{\infty}_{\emptyset,I}(\chi)$ is independent of $\chi \in \Sigma$, and we denote it by JH_{Σ}.

Theorem 2.1.20 ([79], Theorem 2.2) *There exists a unique bijection* $\mathbf{\Gamma}_{\Sigma} \mapsto \omega(\mathbf{\Gamma}_{\Sigma})$ *between the set of all orientations on* Γ_{Σ} *and the set* JH_{Σ} *such that*

$$\mathcal{J}(\omega(\mathbf{\Gamma}_{\Sigma})) = \{\chi \mid \mathbf{\Gamma}_{\Sigma}(\chi) = \mathbf{\Gamma}_{\Sigma}\}. \tag{2.39}$$

Remark 2.1.21 Note that (2.39) differs from the statement in [79, Theorem 2.2] by a twist $\delta_B^{1/2}$, and that the unicity of ω in Theorem 2.1.20 follows from (ii) of Lemma 2.1.15.

Remark 2.1.22 Let $I \subseteq \Delta$ and $\Sigma \subseteq \widehat{T}^{\infty}$ be a G-regular left $W(L_I)$-coset. We say that two elements of Σ are *equivalent* if they correspond to the same orientation on Γ_{Σ}. As $\delta_B^{1/2}$ might not be $K^{\times}$-valued, Γ_{Σ} might not be defined as in (2.38) in general. However, the equivalence relation on Σ discussed above is always well-defined regardless of whether $\delta_B^{1/2}$ is $K^{\times}$-valued or not. In particular, Theorem 2.1.20 still makes sense when $E = K$. We also observe that the equivalence relation on Σ remains unchanged if we view Σ as a set of $(E')^{\times}$-valued characters for some finite extension E' of E. This together with Theorem 2.1.20 implies that each constituents in JH_{Σ} is absolutely irreducible.

Given $I \subseteq \Delta$ and a multiplicity free finite length representation π^{∞} in $\mathrm{Rep}^{\infty}_{\mathrm{adm}}(L_I)$, recall that we have defined in Sect. 1.4 a partial order on the set $\mathrm{JH}_{L_I}(\pi^{\infty})$ of constituents of π^{∞}. We slightly reformulate [79, Theorem 2.8] as follows.

Theorem 2.1.23 ([79], Theorem 2.8) *Let* $\chi \in \Sigma$. *The partial order on the set of orientations on* Γ_{Σ} *given via the bijection in Theorem 2.1.20 by the partial order on*

$\mathrm{JH}_{L_I}(i^{\infty}_{\emptyset,I}(\chi))$ *is the following: two orientations* $\mathbf{\Gamma}_{\Sigma}$, $\mathbf{\Gamma}'_{\Sigma}$ *satisfy* $\mathbf{\Gamma}_{\Sigma} \leq \mathbf{\Gamma}'_{\Sigma}$ *if and only if each edge of* Γ_{Σ} *which has the same direction in* $\mathbf{\Gamma}'_{\Sigma}$ *and* $\mathbf{\Gamma}_{\Sigma}(\chi)$ *also has the same direction in* $\mathbf{\Gamma}_{\Sigma}$.

Concretely, $\mathbf{\Gamma}_{\Sigma}(\chi)$ is the orientation on Γ_{Σ} corresponding to the socle $\omega(\mathbf{\Gamma}_{\Sigma}(\chi))$ of $i^{\infty}_{\emptyset,I}(\chi)$ (which is irreducible by (iv) of Remark 2.1.16), and one gets the orientations corresponding to the Jordan-Hölder factors in higher layers by successively reversing (more and more) arrows in $\mathbf{\Gamma}_{\Sigma}(\chi)$, until all arrows of $\mathbf{\Gamma}_{\Sigma}(\chi)$ are reversed which gives the orientation corresponding to the (irreducible by *loc. cit.*) cosocle of $i^{\infty}_{\emptyset,I}(\chi)$.

For $I_1 \subseteq I \subseteq \Delta$ we define

$$V^{\infty}_{I_1,I} \stackrel{\text{def}}{=} i^{\infty}_{I_1,I}(1_{L_{I_1}}) / \sum_{I_1 \subsetneq I'_1} i^{\infty}_{I'_1,I}(1_{L_{I'_1}}) \tag{2.40}$$

which is an (absolutely irreducible) smooth generalized Steinberg representation of L_I (the smooth Steinberg being $V^{\infty}_{\emptyset,I}$). Note that it is G-regular (for instance by (iii) of Lemma 2.1.15 and (2.32)). The following corollary is classical, we provide a short proof for the reader's convenience.

Corollary 2.1.24 *Let* $I \subseteq \Delta$ *and* $w \in W(L_I)$.

(i) *For* $I_1 \subseteq I$ *we have*

$$\mathcal{J}(V^{\infty}_{I_1,I}) = \{x \cdot 1_T \mid x \in W(L_I), I_1 = I \setminus D_R(x)\}. \tag{2.41}$$

(ii) *The representation* $i^{\infty}_{\emptyset,I}(w \cdot 1_T)$ *is multiplicity free with socle* $V^{\infty}_{I \setminus D_R(w),I}$, *cosocle* $V^{\infty}_{I \cap D_R(w),I}$ *and constituents*

$$\mathrm{JH}_{L_I}(i^{\infty}_{\emptyset,I}(w \cdot 1_T)) = \{V^{\infty}_{I_1,I} \mid I_1 \subseteq I\}. \tag{2.42}$$

(iii) *The partial order* $\leq_w$ *on* $\{V^{\infty}_{I_1,I} \mid I_1 \subseteq I\}$ *induced from the one on* $\mathrm{JH}_{L_I}(i^{\infty}_{\emptyset,I}(w \cdot 1_T))$ *via (2.42) is:* $V^{\infty}_{I_1,I} \leq_w V^{\infty}_{I_2,I}$ *(with* $I_1, I_2 \subseteq I$*) if and only if* $I_2 \cap (I \setminus D_R(w)) \subseteq I_1 \subseteq I_2 \cup (I \setminus D_R(w))$.

Proof We only prove the case $I = \Delta$, the general case follows by treating each Levi block of L_I separately. The graph $\Gamma \stackrel{\text{def}}{=} \Gamma_{W(G)\cdot 1_T}$ attached to 1_T (or equivalently to any $w \cdot 1_T$ with $w \in W(G)$) has vertices $\{1, |\cdot|_K, \ldots, |\cdot|_K^{n-1}\}$. The set of orientations $\mathbf{\Gamma}$ is in natural bijection to the set of subsets of Δ by sending $\mathbf{\Gamma}$ to the subset $I_{\mathbf{\Gamma}}$ of $j \in \Delta$ such that $\mathbf{\Gamma}$ contains an arrow $|\cdot|_K^{n-j} \to |\cdot|_K^{n-j-1}$. For $w \in W(G)$, the attached orientation $\mathbf{\Gamma}(w) \stackrel{\text{def}}{=} \mathbf{\Gamma}(w \cdot 1_T)$ is characterized as follows: for $j \in \Delta$, there exists an arrow $|\cdot|_K^{n-j} \to |\cdot|_K^{n-j-1}$ if $j \notin D_R(w)$, and an arrow $|\cdot|_K^{n-j-1} \to |\cdot|_K^{n-j}$ if $j \in D_R(w)$, i.e. $I_{\mathbf{\Gamma}(w)} = \Delta \setminus D_R(w)$. This together with (2.39) and the discussion after Theorem 2.1.23 imply

$$\begin{aligned} \mathcal{J}(\mathrm{soc}_G(i^{\infty}_{\emptyset,\Delta}(w \cdot 1_T))) &= \{x \cdot 1_T \mid x \in W(G), D_R(x) = D_R(w)\} \\ \mathcal{J}(\mathrm{cosoc}_G(i^{\infty}_{\emptyset,\Delta}(w \cdot 1_T))) &= \{x \cdot 1_T \mid x \in W(G), D_R(x) = \Delta \setminus D_R(w)\}. \end{aligned} \tag{2.43}$$

We now prove (i). By (2.28) with I_0, I, π_0^∞ there being I_1, Δ, $1_{L_{I_1}}$, we have for $I_1 \subseteq \Delta$

$$\mathcal{J}(i_{I_1,\Delta}^\infty(1_{L_{I_1}})) = \{x^{-1} \cdot 1_T \mid x \in W^{I_1,\emptyset}\} = \{x \cdot 1_T \mid I_1 \subseteq \Delta \setminus D_R(x)\}.$$

Together with $V_{I_1,\Delta}^\infty = i_{I_1,\Delta}^\infty(1_{L_{I_1}})/\sum_{I_1 \subsetneq I_1'} i_{I_1',\Delta}^\infty(1_{L_{I_1'}})$ this implies

$$\begin{aligned}\mathcal{J}(V_{I_1,\Delta}^\infty) &= \{x \cdot 1_T \mid I_1 \subseteq \Delta \setminus D_R(x)\} \setminus \{x \cdot 1_T \mid I_1 \subsetneq \Delta \setminus D_R(x)\}\\ &= \{x \cdot 1_T \mid I_1 = \Delta \setminus D_R(x)\} \quad (2.44)\end{aligned}$$

which is (i). Then (ii) follows from (2.44), (2.43) and

$$\{x \cdot 1_T \mid x \in W(G)\} = \bigsqcup_{I_1 \subseteq \Delta} \{x \cdot 1_T \mid \Delta \setminus D_R(x) = I_1\}.$$

We prove (iii). We write $\mathbf{\Gamma}_i$ for the orientation on Γ with $I_{\mathbf{\Gamma}_i} = I_i$, $i = 1, 2$. It follows from Theorem 2.1.23 that $V_{I_1,\Delta}^\infty \leq_w V_{I_2,\Delta}^\infty$ if and only if each edge of Γ which has the same direction in $\mathbf{\Gamma}_2$ and $\mathbf{\Gamma}(w)$ also has this direction in $\mathbf{\Gamma}_1$. Using the above definition of $I_{\mathbf{\Gamma}_i}$ and $I_{\mathbf{\Gamma}(w)} = \Delta \setminus D_R(w)$, this can be easily translated to $I_2 \cap (\Delta \setminus D_R(w)) \subseteq I_1$ and $(\Delta \setminus I_2) \cap D_R(w) \subseteq \Delta \setminus I_1$, or equivalently $I_2 \cap (I \setminus D_R(w)) \subseteq I_1 \subseteq I_2 \cup (I \setminus D_R(w))$ which is (iii). □

Corollary 2.1.25 *Let $I \subseteq \Delta$, π^∞ in $\mathrm{Rep}_{\mathrm{adm}}^\infty(L_I)$ irreducible G-regular and Γ_Σ the undirected graph attached to $\Sigma \stackrel{def}{=} W(L_I) \cdot \mathcal{J}(\pi^\infty)$ above Theorem 2.1.20 (recall that Σ is a single regular left $W(L_I)$-coset by (i) of Lemma 2.1.15). If Γ_Σ has one connected component for each Levi block of L_I, then there exists $I_1 \subseteq I$ and a smooth character $\delta : L_I \to E^\times$ such that $\pi^\infty \cong V_{I_1,I}^\infty \otimes_E \delta$.*

Proof It is harmless to treat each Levi block of L_I separately, so we may assume $I = \Delta$ and Γ_Σ connected. By definition of Γ_Σ, it is connected if and only if its set of vertices has the form $\{\rho, \rho \otimes_E |\cdot|_K, \dots, \rho \otimes_E |\cdot|_K^{n-1}\}$ for some smooth character $\rho : K^\times \to E^\times$. Let $\delta \stackrel{\mathrm{def}}{=} \rho \circ \det$ where $\det : G \to E^\times$ is the determinant character. Then the graph $\Gamma_{\Sigma'}$ attached to $\Sigma' \stackrel{\mathrm{def}}{=} W(G) \cdot \mathcal{J}(\pi^\infty \otimes_E \delta^{-1})$ has vertices $\{1, |\cdot|_K, \dots, |\cdot|_K^{n-1}\}$, which forces $\pi^\infty \otimes_E \delta^{-1}$ to be a Jordan-Hölder factor of $i_{\emptyset,\Delta}^\infty(1_T)$, and thus to be $V_{I_1,\Delta}^\infty$ for some $I_1 \subseteq \Delta$ by (2.42). □

Corollary 2.1.26 *Let $I \subseteq \Delta$ and $\chi \in \widehat{T}^\infty$ G-regular. Let π^∞ be a subquotient of $i_{\emptyset,I}^\infty(\chi)$ with simple socle and cosocle. Then π^∞ is G-basic and there exist $I_0 \subseteq I$ and an irreducible G-regular $\sigma^\infty \in \mathrm{Rep}_{\mathrm{adm}}^\infty(L_{I_0})$ such that $\pi^\infty \cong i_{I_0,I}^\infty(\sigma^\infty)$.*

Proof Write σ_1^∞ (resp. σ_2^∞) for the socle (resp. cosocle) of π^∞. For $\chi' \in \mathcal{J}(\sigma_1^\infty)$, the image of the unique (up to scalar) non-zero map $i_{\emptyset,I}^\infty(\chi) \to i_{\emptyset,I}^\infty(\chi')$ between multiplicity free objects (see (iii) of Lemma 2.1.15 and (iii) of Remark 2.1.16) is the unique quotient of $i_{\emptyset,I}^\infty(\chi)$ with socle σ_1^∞, and thus contains π^∞ (as π^∞ has simple socle σ_1^∞). Upon replacing χ with χ', we assume from on that π^∞ is a subrepresentation of $i^\infty_{\emptyset,I}(\chi)$, hence σ_1^∞ is the socle of $i^\infty_{\emptyset,I}(\chi)$.

Let Γ_Σ be the undirected graph attached to $\Sigma \stackrel{\text{def}}{=} W(L_I)\cdot\chi$ (see before Theorem 2.1.20), $\mathbf{\Gamma}_{\Sigma,i} \stackrel{\text{def}}{=} \omega^{-1}(\sigma_i^\infty)$, $i=1,2$ (see (2.39)), and recall that $\mathbf{\Gamma}_{\Sigma,1}=\mathbf{\Gamma}_\Sigma(\chi)$. Write $\chi\otimes_E\delta_B^{1/2}=\rho_1\boxtimes\cdots\boxtimes\rho_n$, modifying χ within $W(L_I)\cdot\chi$ without changing $\mathbf{\Gamma}_\Sigma(\chi)$, we can assume that each pair of vertices ρ_k, $\rho_{k'}$ that are connected by an edge of Γ_Σ are such that $|k-k'|=1$. For such a χ, the set of edges of Γ_Σ having the same orientation in $\mathbf{\Gamma}_{\Sigma,1}$ and $\mathbf{\Gamma}_{\Sigma,2}$ naturally determines a subset $I_0\subseteq I$. Let Γ_{Σ_0} be the undirected graph attached to $\Sigma_0 \stackrel{\text{def}}{=} W(L_{I_0})\cdot\chi$, and note that Γ_{Σ_0} is obtained from Γ_Σ by exactly deleting the edges corresponding to elements *not* in I_0. Each orientation on Γ_Σ induces one on Γ_{Σ_0}, and by definition of I_0 $\mathbf{\Gamma}_{\Sigma,1}$ and $\mathbf{\Gamma}_{\Sigma,2}$ induce the same orientation $\mathbf{\Gamma}_{\Sigma_0}(\chi)$ on Γ_{Σ_0}, hence a well-defined Jordan-Hölder factor σ^∞ of $i^\infty_{\emptyset,I_0}(\chi)$ by Theorem 2.1.20, which is actually its socle. Note that $\chi\in\mathcal{J}(\sigma^\infty)$ from (2.7) (for $k=0$), hence σ^∞ is G-regular.

We claim that $i^\infty_{I_0,I}(\sigma^\infty)\cong\pi^\infty$. As $i^\infty_{I_0,I}(\sigma^\infty)$ is a subrepresentation of $i^\infty_{\emptyset,I}(\chi)$, it suffices to show that $i^\infty_{I_0,I}(\sigma^\infty)$ has cosocle σ_2^∞. Let w_{I,I_0} be the longest element inside $W^{\emptyset,I_0}(L_I)$, by (2.12) we have

$$\mathcal{J}(J'_{I,I_0}(\sigma_2^\infty)) = w_{I,I_0}^{-1}\cdot\mathcal{J}(\sigma_2^\infty). \tag{2.45}$$

The orientation on Γ_Σ associated to $w_{I,I_0}\cdot\chi$ is easily checked to be obtained from the one associated to χ by reversing the orientation of the edges of Γ_Σ corresponding to $I\setminus I_0$, which is $\mathbf{\Gamma}_{\Sigma,2}=\omega^{-1}(\sigma_2^\infty)$ by definition of I_0. In particular we have $w_{I,I_0}\cdot\chi\in\mathcal{J}(\sigma_2^\infty)$, which together with (2.45) implies $\chi\in\mathcal{J}(J'_{I,I_0}(\sigma_2^\infty))$. By (2.8) for $k=0$ we have for $\tau^\infty \stackrel{\text{def}}{=} \mathrm{cosoc}_{L_I}(i^\infty_{I_0,I}(\sigma^\infty))$

$$0\neq \mathrm{Hom}_{L_I}(i^\infty_{I_0,I}(\sigma^\infty),\tau^\infty)\cong\mathrm{Hom}_{L_{I_0}}(\sigma^\infty, J'_{I,I_0}(\tau^\infty)).$$

Thus the irreducible σ^∞ injects into $J'_{I,I_0}(\tau^\infty)$ and $\mathcal{J}(\sigma^\infty)\subseteq\mathcal{J}(J'_{I,I_0}(\tau^\infty))$, which implies $\chi\in\mathcal{J}(J'_{I,I_0}(\tau^\infty))$ since $\chi\in\mathcal{J}(\sigma^\infty)$. But it follows from (2.10) and Lemma 2.1.14, and the exactness of J'_{I,I_0} and $J_{w_{I,I_0}(I_0),\emptyset}$, that σ_2^∞ is the only Jordan-Hölder factor of $i^\infty_{\emptyset,I}(\chi)$ satisfying $\chi\in\mathcal{J}(J'_{I,I_0}(\sigma_2^\infty))$. Hence we must have $\sigma_2^\infty=\tau^\infty$. □

Corollary 2.1.27 *Let $I\subseteq\Delta$ and π^∞ in $\mathrm{Rep}^\infty_{\mathrm{adm}}(L_I)$ G-basic and* reducible. *Then there exists G-basic π_i^∞ in $\mathrm{Rep}^\infty_{\mathrm{adm}}(L_I)$ for $i=0,1$ such that π^∞ fits into $0\to\pi_1^\infty\to\pi^\infty\to\pi_0^\infty\to 0$ which is non-split.*

Proof We continue to use the notation from the proof of Corollary 2.1.26 and write σ_1^∞ (resp. σ_2^∞) for the socle (resp. cosocle) of π^∞, with $\mathbf{\Gamma}_{\Sigma,i}=\omega^{-1}(\sigma_i^\infty)$ for $i=1,2$. Since π^∞ is reducible, we have $\sigma_1^\infty\neq\sigma_2^\infty$ and thus $\mathbf{\Gamma}_{\Sigma,1}\neq\mathbf{\Gamma}_{\Sigma,2}$. We fix an arbitrary edge e of Γ_Σ on which $\mathbf{\Gamma}_{\Sigma,1}$ and $\mathbf{\Gamma}_{\Sigma,2}$ have opposite direction. Let τ_i^∞ be the Jordan-Hölder factor of π^∞ whose attached orientation on Γ_Σ differs from $\mathbf{\Gamma}_{\Sigma,i}$ by changing only the direction of the fixed edge e. We define π_1^∞ (resp. π_0^∞) as the unique sub (resp. quotient) of π^∞ with socle σ_1^∞ and cosocle τ_2^∞ (resp. with socle τ_1^∞ and cosocle

σ_2^∞). By applying Theorem 2.1.23 to $i^\infty_{\emptyset,I}(\chi)$ for some $\chi \in \mathcal{J}(\sigma_1^\infty)$ (with π^∞ being the unique sub of $i^\infty_{\emptyset,I}(\chi)$ with cosocle σ_2^∞ as in the proof of Corollary 2.1.26), we see that $\mathrm{JH}_{L_I}(\pi_1^\infty)$ (resp. $\mathrm{JH}_{L_I}(\pi_0^\infty)$) consists of exactly those Jordan-Hölder factors whose attached orientation on Γ_Σ have the same direction as $\mathbf{\Gamma_{\Sigma,1}}$ (resp. $\mathbf{\Gamma_{\Sigma,2}}$) on the fixed edge e, and in particular, π^∞ fits into a non-split short exact sequence $0 \to \pi_1^\infty \to \pi^\infty \to \pi_0^\infty \to 0$. □

Lemma 2.1.28 *Let $I \subseteq \Delta$ and π^∞ in $\mathrm{Rep}^\infty_{\mathrm{adm}}(L_I)$ G-basic. Then there exist $I_1 \subseteq I_0 \subseteq I$ and a smooth character $\delta : L_{I_0} \to E^\times$ such that $V^\infty_{I_1,I_0} \otimes_E \delta$ is irreducible G-regular and $\pi^\infty \cong i^\infty_{I_0,I}(V^\infty_{I_1,I_0} \otimes_E \delta)$.*

Proof By the definition of G-basic ((ii) of Definition 2.1.4), we can assume π^∞ irreducible G-regular. Recall that $\Sigma \overset{\text{def}}{=} W(L_I) \cdot \mathcal{J}(\pi^\infty)$ is a single left $W(L_I)$-coset ((i) of Lemma 2.1.15). As in the proof of Corollary 2.1.26, we fix an arbitrary G-regular $\chi \in \Sigma$ such that two vertices $\rho_k, \rho_{k'}$ of the graph Γ_Σ that are connected by an edge are adjacent, i.e. such that $|k - k'| = 1$. Then there exists $I_0 \subseteq I$ such that there is a bijection between the Levi blocks of L_{I_0} and the connected components of Γ_Σ, hence also a bijection between the set of orientations on Γ_{Σ_0} ($\overset{\text{def}}{=}$ the graph of $\Sigma_0 \overset{\text{def}}{=} W(L_{I_0}) \cdot \chi$) and on Γ_Σ. Using Theorem 2.1.20 applied to both Γ_Σ and Γ_{Σ_0}, we deduce a bijection between the set of constituents of $i^\infty_{\emptyset,I}(\chi)$ and of $i^\infty_{\emptyset,I_0}(\chi)$. In particular they have same length, and the exactness of $i^\infty_{I_0,I}(-)$ then implies that $\sigma_0^\infty \mapsto i^\infty_{I_0,I}(\sigma_0^\infty)$ induces a bijection between both sets of constituents. Using (iii) of Lemma 2.1.15, we see that π^∞ is a constituent of $i^\infty_{\emptyset,I}(\chi)$, hence $\pi^\infty \cong i^\infty_{I_0,I}(\pi_0^\infty)$ for a (unique) constituent π_0^∞ of $i^\infty_{\emptyset,I_0}(\chi)$. Applying Corollary 2.1.25 to π_0^∞, we deduce $\pi_0^\infty \cong V^\infty_{I_1,I_0} \otimes_E \delta$ for a smooth character $\delta : L_{I_0} \to E^\times$ and $I_1 \subseteq I_0$, and thus $\pi^\infty \cong i^\infty_{I_0,I}(V^\infty_{I_1,I_0} \otimes_E \delta)$. □

Lemma 2.1.29 *Let $I_0 \subseteq I \subseteq \Delta$ and π^∞ in $\mathrm{Rep}^\infty_{\mathrm{adm}}(L_I)$ G-basic. Then we have a canonical decomposition in $\mathrm{Rep}^\infty_{\mathrm{adm}}(L_{I_0})$ induced by (2.13)*

$$J_{I,I_0}(\pi^\infty) \cong \bigoplus_{\Sigma} J_{I,I_0}(\pi^\infty)_{\mathcal{B}^{I_0}_\Sigma} \tag{2.46}$$

where Σ runs through left $W(L_{I_0})$-cosets in $W(L_I) \cdot \mathcal{J}(\pi^\infty)$ and $J_{I,I_0}(\pi^\infty)_{\mathcal{B}^{I_0}_\Sigma}$ is G-basic if non-zero. Moreover, if π^∞ is simple, then $J_{I,I_0}(\pi^\infty)_{\mathcal{B}^{I_0}_\Sigma}$ is simple if non-zero.

Proof Note first that we can indeed apply (2.13) to $J_{I,I_0}(\pi^\infty)$ since $\mathcal{J}(J_{I,I_0}(\pi^\infty)) = \mathcal{J}(\pi^\infty)$ consists of G-regular weights by the last assertion in (i) of Lemma 2.1.15. By (iv) of Remark 2.1.16 there exist G-regular $\chi, \chi' \in \widehat{T}^\infty$ such that π^∞ is the image of the unique (up to scalar) non-zero map $i^\infty_{\emptyset,I}(\chi') \to i^\infty_{\emptyset,I}(\chi)$. By the exactness of $J_{I,I_0}(-)$ we deduce that $J_{I,I_0}(\pi^\infty)$ is the image of the induced map $J_{I,I_0}(i^\infty_{\emptyset,I}(\chi')) \to J_{I,I_0}(i^\infty_{\emptyset,I}(\chi))$, which by Remark 2.1.19 is the same as the map

$$\bigoplus_{w' \in W^{\emptyset,I_0}(L_I)} i^\infty_{\emptyset,I_0}((w')^{-1} \cdot \chi') \longrightarrow \bigoplus_{w \in W^{\emptyset,I_0}(L_I)} i^\infty_{\emptyset,I_0}(w^{-1} \cdot \chi).$$

Consequently, for any left $W(L_{I_0})$-coset Σ in $W(L_I)\cdot\mathcal{J}(\pi^\infty)$, $J_{I,I_0}(\pi^\infty)_{\mathcal{B}_\Sigma^{I_0}}$ is necessarily the image of a (possibly zero) map

$$i^\infty_{\emptyset,I_0}((w')^{-1}\cdot\chi')\longrightarrow i^\infty_{\emptyset,I_0}(w^{-1}\cdot\chi) \tag{2.47}$$

for some $w,w'\in W^{\emptyset,I_0}(L_I)$ such that $\Sigma=W(L_{I_0})(w')^{-1}\cdot\chi'=W(L_{I_0})w^{-1}\cdot\chi$. But the image of (2.47) is either zero or has simple socle and cosocle (since both representations in (2.47) have simple socle and cosocle by the last assertion in (ii) of Remark 2.1.16), hence is either zero or G-basic by Corollary 2.1.26. It follows that $J_{I,I_0}(\pi^\infty)_{\mathcal{B}_\Sigma^{I_0}}$ is either zero or G-basic.

Now we assume that π^∞ is simple and prove that $J_{I,I_0}(\pi^\infty)_{\mathcal{B}_\Sigma^{I_0}}$ is either zero or simple. We can assume $\sigma_0^\infty\overset{\text{def}}{=}J_{I,I_0}(\pi^\infty)_{\mathcal{B}_\Sigma^{I_0}}\neq 0$, and we write σ^∞ for its cosocle. The natural surjections $J_{I,I_0}(\pi^\infty)\twoheadrightarrow\sigma_0^\infty\twoheadrightarrow\sigma^\infty$ induce by (2.7) (for $k=0$) maps $\pi^\infty\to i^\infty_{I,I_0}(\sigma_0^\infty)\twoheadrightarrow i^\infty_{I,I_0}(\sigma^\infty)$ the composition of which is non-zero. Hence π^∞ (which is simple) appears in the socle of both $i^\infty_{I,I_0}(\sigma_0^\infty)$ and $i^\infty_{I,I_0}(\sigma^\infty)$. As σ_0^∞ is G-basic, so is $i^\infty_{I,I_0}(\sigma_0^\infty)$. Moreover $i^\infty_{I,I_0}(\sigma^\infty)$ is clearly G-basic. Hence by the end of (iv) of Remark 2.1.16 both $i^\infty_{I,I_0}(\sigma_0^\infty)$ and $i^\infty_{I,I_0}(\sigma^\infty)$ have simple socle and cosocle, and thus have same socle π^∞. As $i^\infty_{I,I_0}(\sigma^\infty)$ is a quotient of $i^\infty_{I,I_0}(\sigma_0^\infty)$ this forces $i^\infty_{I,I_0}(\sigma^\infty)=i^\infty_{I,I_0}(\sigma_0^\infty)$ and thus $\sigma^\infty=\sigma_0^\infty$. In particular $\sigma_0^\infty=J_{I,I_0}(\pi^\infty)_{\mathcal{B}_\Sigma^{I_0}}$ is simple. □

Remark 2.1.30 Let $I_0\subseteq I\subseteq\Delta$ and π^∞ in $\mathrm{Rep}^\infty_{\mathrm{adm}}(L_I)$ G-basic. As $J_{I,w_{I,I_0}(I_0)}(\pi^\infty)$ is a direct sum of G-basic objects in distinct Bernstein blocks by Lemma 2.1.29, we deduce from (2.10), (2.12) and Lemma 2.1.17 that an analogous statement as Lemma 2.1.29 holds for $J'_{I,I_0}(\pi^\infty)$. Similarly, for $I_0,I_1\subseteq\Delta$, $w\in W^{I_0,I_1}$ and π_0^∞ in $\mathrm{Rep}^\infty_{\mathrm{adm}}(L_{I_0})$ G-basic, as $J_{I_0,w(I_1)\cap I_0}(\pi_0^\infty)$ is a direct sum of G-basic objects in distinct Bernstein blocks by Lemma 2.1.29, we deduce from (2.19), (2.20), (2.22) and Lemma 2.1.17 that the statement of Lemma 2.1.29 holds for $J_{I_0,I_1,w}(\pi_0^\infty)$. All these statements remain true if we replace everywhere G-basic by irreducible G-regular.

Lemma 2.1.31 *Let $I_1\subseteq I\subseteq\Delta$, $w\in W(L_I)$ and $\pi^\infty\overset{def}{=}\mathrm{cosoc}_{L_{I_1}}(i^\infty_{\emptyset,I_1}(w\cdot 1_T))$. Then $i^\infty_{I_1,I}(\pi^\infty)$ is isomorphic to the unique quotient of $i^\infty_{\emptyset,I}(w\cdot 1_T)$ with socle $V^\infty_{I_2,I}$ (and cosocle $V^\infty_{D_R(w),I}$) where $I_2\overset{def}{=}I\setminus D_R(w_{I_1}w)$.*

Proof Note first that $i^\infty_{I_1,I}(\pi^\infty)$ is clearly a quotient of $i^\infty_{\emptyset,I}(w\cdot 1_T)$. As $i^\infty_{I_1,I}(\pi^\infty)$ is G-basic (recall π^∞ is irreducible by (ii) of Remark 2.1.16), it has simple socle by (iv) of Remark 2.1.16. By (2.42) this socle has the form $V^\infty_{I_2,I}$ for some $I_2\subseteq I$. It follows from (i) of Remark 2.1.16 that $\pi^\infty\cong\mathrm{soc}_{L_{I_1}}(i^\infty_{\emptyset,I_1}(w_{I_1}w\cdot 1_T))$, and hence $w_{I_1}w\cdot 1_T\in\mathcal{J}(\pi^\infty)$ ((2.7) with $k=0$). Since by *loc. cit.*

$$\mathrm{Hom}_{L_{I_1}}(J_{I,I_1}(V^\infty_{I_2,I}),\pi^\infty)\cong\mathrm{Hom}_{L_I}(V^\infty_{I_2,I},i^\infty_{I_1,I}(\pi^\infty))\neq 0$$

we see that $\mathcal{J}(\pi^\infty)\subseteq\mathcal{J}(J_{I,I_1}(V^\infty_{I_2,I}))=\mathcal{J}(V^\infty_{I_2,I})$ as π^∞ is irreducible and thus $w_{I_1}w\cdot 1_T\in\mathcal{J}(V^\infty_{I_2,I})$. In particular (2.41) implies $I_2=I\setminus D_R(w_{I_1}w)$. □

Lemma 2.1.32 *Let $I \subseteq \Delta$ and $\pi_0^\infty, \pi_1^\infty$ in $\mathrm{Rep}_{\mathrm{adm}}^\infty(L_I)$ both G-basic with $\mathrm{JH}_{L_I}(\pi_0^\infty) \cap \mathrm{JH}_{L_I}(\pi_1^\infty) = \emptyset$. Let π^∞ in $\mathrm{Rep}_{\mathrm{adm}}^\infty(L_I)$ which fits into a non-split short exact sequence*

$$0 \to \pi_1^\infty \to \pi^\infty \to \pi_0^\infty \to 0. \tag{2.48}$$

Then π^∞ admits a unique subquotient σ^∞ which is G-basic with simple socle $\mathrm{soc}_{L_I}(\pi_1^\infty)$ and simple cosocle $\mathrm{cosoc}_{L_I}(\pi_0^\infty)$.

Proof As π_i^∞ is G-basic, by (iv) of Remark 2.1.16 it is multiplicity free with simple socle and cosocle. In particular, π^∞ is multiplicity free and we write π_2^∞ for its unique subrepresentation with cosocle $\mathrm{cosoc}_{L_I}(\pi_0^\infty)$. In particular the composition of $\pi_2^\infty \to \pi^\infty \to \pi_0^\infty$ is a surjection. If $\pi_2^\infty \cap \pi_1^\infty = 0$ inside π^∞, then (2.48) induces an isomorphism $\pi_2^\infty \xrightarrow{\sim} \pi_0^\infty$ and thus (2.48) splits, a contradiction to our assumption. If $\pi_2^\infty \cap \pi_1^\infty \neq 0$, as π_1^∞ has simple socle, we must have $\mathrm{soc}_{L_I}(\pi_1^\infty) \subseteq \pi_2^\infty$, so we can define σ^∞ as the unique quotient of π_2^∞ with socle $\mathrm{soc}_{L_I}(\pi_1^\infty)$. Since π_0^∞ and π_1^∞ are G-basic with $\mathrm{JH}_{L_I}(\pi_0^\infty) \cap \mathrm{JH}_{L_I}(\pi_1^\infty) = \emptyset$, by (i) and (iii) of Lemma 2.1.15 we know that $J_{I,\emptyset}(\pi_0^\infty)$ and $J_{I,\emptyset}(\pi_1^\infty)$ are multiplicity free, semi-simple and share no common constituent. Since we have a short exact sequence $0 \to J_{I,\emptyset}(\pi_1^\infty) \to J_{I,\emptyset}(\pi^\infty) \to J_{I,\emptyset}(\pi_0^\infty) \to 0$ (as $J_{I,\emptyset}$ is exact), it follows that $J_{I,\emptyset}(\pi^\infty)$ is also semi-simple and multiplicity free, and thus so is $J_{I,\emptyset}(\sigma^\infty)$. Now we choose an arbitrary $\chi \in \mathcal{J}(\mathrm{soc}_{L_I}(\pi_1^\infty)) \subseteq \mathcal{J}(\sigma^\infty)$, which by (2.7) (with $k = 0$) gives a non-zero map $\sigma^\infty \to i_{\emptyset,I}^\infty(\chi)$. Since $i_{\emptyset,I}^\infty(\chi)$ is multiplicity free with socle $\mathrm{soc}_{L_I}(\pi_1^\infty) \cong \mathrm{soc}_{L_I}(\sigma^\infty)$ by (iii) of Lemma 2.1.15 and (ii) of Remark 2.1.16, the map $\sigma^\infty \to i_{\emptyset,I}^\infty(\chi)$ is an injection. Since σ^∞ has simple socle and cosocle by definition, Corollary 2.1.26 finally implies that σ^∞ is G-basic. □

2.2 Results on Smooth Ext Groups

We prove several useful results on smooth Ext groups of G-basic representations.

We start with some preliminaries. For $I_0, I_1 \subseteq \Delta$, we define

$$d(I_0, I_1) \stackrel{\mathrm{def}}{=} \#(I_0 \setminus I_1) + \#(I_1 \setminus I_0) \tag{2.49}$$

and $[I_0, I_1] \stackrel{\mathrm{def}}{=} \{I \subseteq \Delta \mid d(I_0, I_1) = d(I_0, I) + d(I, I_1)\}$. One easily checks $d(I_0, I_1) = d(I_0, I) + d(I, I_1)$ if and only if

$$(I_0 \setminus I_1) \sqcup (I_1 \setminus I_0) = (I_0 \setminus I) \sqcup (I \setminus I_1) \sqcup (I_1 \setminus I) \sqcup (I \setminus I_0),$$

if and only if $(I_0 \setminus I) \cap (I_1 \setminus I) = \emptyset$ and $(I \setminus I_0) \cap (I \setminus I_1) = \emptyset$, if and only if $I_0 \cap I_1 \subseteq I \subseteq I_0 \cup I_1$. In other words, we have

$$[I_0, I_1] = \{I \subseteq \Delta \mid I_0 \cap I_1 \subseteq I \subseteq I_0 \cup I_1\}. \tag{2.50}$$

Given another $I_1' \subseteq \Delta$, it is clear that we have

$$[I_0, I_1] \cap [I_0, I_1'] = [I_0, I_1''] \tag{2.51}$$

with $I_1'' \stackrel{\text{def}}{=} (I_0 \setminus (I_1 \cup I_1')) \sqcup ((I_1 \cap I_1') \setminus I_0)$.

Lemma 2.2.1 *Let $I \subseteq \Delta$.*

(i) For $I_0, I_1 \subseteq I$, there exists a unique G-basic $Q_I(I_0, I_1) \in \mathrm{Rep}^{\infty}_{\mathrm{adm}}(L_I)$ with socle $V^{\infty}_{I_0,I}$ and cosocle $V^{\infty}_{I_1,I}$, and it has set of Jordan-Hölder factors $\{V^{\infty}_{I',I} \mid I' \in [I_0, I_1]\}$.

(ii) For $I_0, I_1, I_0', I_1' \subseteq I$, there exists a non-zero map $Q_I(I_0', I_1') \to Q_I(I_0, I_1)$ if and only if $I_1' \in [I_0, I_1]$ and $I_0 \in [I_0', I_1']$, in which case the map is unique (up to scalar) with image isomorphic to $Q_I(I_0, I_1')$.

(iii) For $I_0, I_1 \subseteq I$ and $I_0', I_1' \in [I_0, I_1]$, $Q_I(I_0', I_1')$ is a subquotient of $Q_I(I_0, I_1)$ if and only if $I_0' \in [I_0, I_1']$ if and only if $I_1' \in [I_0', I_1]$.

Proof We prove (i). By (ii) and (iv) of Remark 2.1.16 any G-basic representation of L_I with socle $V^{\infty}_{I_0,I}$ and cosocle $V^{\infty}_{I_1,I}$ must be the image of the unique (up to scalar) non-zero map $i^{\infty}_{\emptyset,I}(\chi') \to i^{\infty}_{\emptyset,I}(\chi)$ for any $\chi \in \mathcal{J}(V^{\infty}_{I_0,I})$ and $\chi' \in w_I \cdot \mathcal{J}(V^{\infty}_{I_1,I})$. This implies the unicity of such a representation. For its existence, note that by (iii) of Remark 2.1.16 and (2.41) there is a non-zero map $i^{\infty}_{\emptyset,I}(\chi') \to i^{\infty}_{\emptyset,I}(\chi)$ for any $\chi \in \mathcal{J}(V^{\infty}_{I_0,I})$ and $\chi' \in w_I \cdot \mathcal{J}(V^{\infty}_{I_1,I})$. Its image, which has socle $V^{\infty}_{I_0,I}$ and cosocle $V^{\infty}_{I_1,I}$, is G-basic by Corollary 2.1.26. For the last statement of (i), choose $w \in W(L_I)$ such that $I_0 = I \setminus D_R(w)$. One can check that $Q_I(I_0, I \setminus I_0) \cong i^{\infty}_{\emptyset,I}(w \cdot 1_T)$ and (using (ii) of Corollary 2.1.24) that $Q_I(I_0, I_1)$ is the unique subrepresentation of $i^{\infty}_{\emptyset,I}(w \cdot 1_T)$ with cosocle $V^{\infty}_{I_1,I}$. By *loc. cit.* the constituents of $Q_I(I_0, I_1)$ are the $V^{\infty}_{I',I}$ such that $V^{\infty}_{I',I} \leq_w V^{\infty}_{I_1,I}$ (where $\leq_w$ is the partial order defined by $i^{\infty}_{\emptyset,I}(w \cdot 1_T)$). The last statement in (i) follows then from (iii) of Corollary 2.1.24 and (2.50).

We prove (ii). By (ii) and (iv) of Remark 2.1.16 again, we can choose $\chi \in \mathcal{J}(V^{\infty}_{I_0,I})$ and $\chi' \in w_I \cdot \mathcal{J}(V^{\infty}_{I_1',I})$ such that $Q_I(I_0, I_1) \hookrightarrow i^{\infty}_{\emptyset,I}(\chi)$ and $i^{\infty}_{\emptyset,I}(\chi') \twoheadrightarrow Q_I(I_0', I_1')$. Assume that there is a non-zero map $Q_I(I_0', I_1') \to Q_I(I_0, I_1)$, then the composition

$$i^{\infty}_{\emptyset,I}(\chi') \twoheadrightarrow Q_I(I_0', I_1') \to Q_I(I_0, I_1) \hookrightarrow i^{\infty}_{\emptyset,I}(\chi)$$

is also non-zero and has same image. By the existence part in (i) there is a unique (up to scalar) non-zero map $i^{\infty}_{\emptyset,I}(\chi') \to i^{\infty}_{\emptyset,I}(\chi)$ with image $Q_I(I_0, I_1')$. Hence $Q_I(I_0, I_1')$ must be a quotient of $Q_I(I_0', I_1')$ and a subrepresentation of $Q_I(I_0, I_1)$, forcing $I_1' \in [I_0, I_1]$ and $I_0 \in [I_0', I_1']$ by the last statement in (i). Conversely, if $I_1' \in [I_0, I_1]$, then $V^{\infty}_{I_1',I}$ shows up in $Q_I(I_0, I_1)$ by (i), hence $Q_I(I_0, I_1')$ is a subrepresentation of $Q_I(I_0, I_1)$ (by unicity of $Q_I(I_0, I_1')$). Similarly, $I_0 \in [I_0', I_1']$ implies that $Q_I(I_0, I_1')$ is a quotient of $Q_I(I_0', I_1')$. So if $I_1' \in [I_0, I_1]$ and $I_0 \in [I_0', I_1']$ there is a non-zero map $Q_I(I_0', I_1') \to Q_I(I_0, I_1)$ (with image $Q_I(I_0, I_1')$).

Finally, (iii) follows from (ii) and the observation that $Q_I(I_0', I_1')$ for $I_0', I_1' \in [I_0, I_1]$ is a subquotient of $Q_I(I_0, I_1)$ if and only if it is a subrepresentation of $Q_I(I_0', I_1)$ if and only it is a quotient of $Q_I(I_0, I_1')$. □

Remark 2.2.2 Let $I \subseteq \Delta$ and $w \in W(L_I)$. It follows from [59, Theorem III.2.7] (with W there being $(w \cdot 1_T)^\sim \cong \delta_{B_I}^{1/2} \cdot w(\delta_{B_I})^{-1/2}$ and recalling that $(-)^\sim$ is the smooth contragredient):

$$\begin{aligned} i^\infty_{\emptyset,I}(w \cdot 1_T) &\cong i^\infty_{\emptyset,I}((w \cdot 1_T)^\sim \cdot \delta_{B_I}^{-1})^\sim \cong i^\infty_{\emptyset,I}(w(\delta_{B_I})^{-1/2} \cdot \delta_{B_I}^{-1/2}) \\ &\cong i^\infty_{\emptyset,I}(w(\delta_{B_I}^{-1/2}) \cdot \delta_{B_I}^{-1/2}) \cong i^\infty_{\emptyset,I}(ww_I(\delta_{B_I}^{1/2}) \cdot \delta_{B_I}^{-1/2}) = i^\infty_{\emptyset,I}(ww_I \cdot 1_T). \end{aligned}$$

Together with (the proof of) (i) of Lemma 2.2.1 this implies for $I_0, I_1 \subseteq I$

$$Q_I(I_0, I_1)^\sim \cong Q_I(I_1, I_0). \tag{2.52}$$

By [77, Proposition 5] and (2.52) we have for $k \geq 0$

$$\begin{aligned} \mathrm{Ext}^k_{L_I}(Q_I(I_0, I_1), Q_I(I_0', I_1'))^\infty &\cong \mathrm{Ext}^k_{L_I}(Q_I(I_0', I_1')^\sim, Q_I(I_0, I_1)^\sim)^\infty \\ &\cong \mathrm{Ext}^k_{L_I}(Q_I(I_1', I_0'), Q_I(I_1, I_0))^\infty \end{aligned} \tag{2.53}$$

and in particular for $I_0, I_1, I_0', I_1' \subseteq I$

$$d_I(Q_I(I_0, I_1), Q_I(I_0', I_1')) = d_I(Q_I(I_1', I_0'), Q_I(I_1, I_0)).$$

We now recall the following classical result (see (iii) of Definition 2.1.4 for $d(-,-)$).

Lemma 2.2.3 *Let $I_0, I_1 \subseteq I \subseteq \Delta$. Then we have $d(V^\infty_{I_0,I}, V^\infty_{I_1,I}) = d(I_0, I_1)$ and*

$$\dim_E \mathrm{Ext}^{d(I_0,I_1)}_{L_I}(V^\infty_{I_0,I}, V^\infty_{I_1,I})^\infty = 1.$$

Proof This follows directly from [48, Corollary 2] or [18, Theorem 1.3]. □

Lemma 2.2.4 *Let $I_i \subseteq I \subseteq \Delta$ and π_i^∞ in $\mathrm{Rep}^\infty_{\mathrm{adm}}(L_{I_i})$ G-basic for $i = 0, 1$. Assume that $d_I(\pi_0^\infty, \pi_1^\infty) < \infty$. Then $d_{I'}(\pi_0^\infty, \pi_1^\infty) = d_I(\pi_0^\infty, \pi_1^\infty)$ for $I' \supseteq I$ and*

$$\mathrm{Ext}^{d_I(\pi_0^\infty, \pi_1^\infty)}_{L_I}(i^\infty_{I_0,I}(\pi_0^\infty), i^\infty_{I_1,I}(\pi_1^\infty))^\infty \tag{2.54}$$

is one dimensional.

Proof Let $\Sigma \overset{\text{def}}{=} W(L_I) \cdot \mathcal{J}(\pi_1^\infty) = W(L_I) \cdot \mathcal{J}(i^\infty_{I_1,I}(\pi_1^\infty))$ (see Lemma 2.1.11), then $d_I(\pi_0^\infty, \pi_1^\infty) < \infty$ implies in particular

$$i^\infty_{I_0,I}(\pi_0^\infty) \in \mathcal{B}^I_\Sigma.$$

Consequently, for $I' \supseteq I$, we have by the last assertion in (ii) of Lemma 2.1.18 applied with π_0^∞ there being $i^\infty_{I_0,I}(\pi_0^\infty)$ and with $w = 1$

$$J_{I',I}(i^\infty_{I,I'}(i^\infty_{I_0,I}(\pi_0^\infty)))_{\mathcal{B}^I_\Sigma} \cong i^\infty_{I,I,1}(J_{I,I,1}(i^\infty_{I_0,I}(\pi_0^\infty))) \cong i^\infty_{I_0,I}(\pi_0^\infty).$$

Together with (2.7) this implies canonical isomorphisms for $k \geq 0$

$$\begin{aligned}\mathrm{Ext}^k_{L_{I'}}(i^\infty_{I_0,I'}(\pi^\infty_0), i^\infty_{I_1,I'}(\pi^\infty_1))^\infty &\cong \mathrm{Ext}^k_{L_I}(J_{I',I}(i^\infty_{I_0,I'}(\pi^\infty_0)), i^\infty_{I_1,I}(\pi^\infty_1))^\infty\\ &\cong \mathrm{Ext}^k_{L_I}(i^\infty_{I_0,I}(\pi^\infty_0), i^\infty_{I_1,I}(\pi^\infty_1))^\infty,\end{aligned}$$

and thus $d_{I'}(\pi^\infty_0, \pi^\infty_1) = d_I(\pi^\infty_0, \pi^\infty_1)$ for $I' \supseteq I$.

Now we prove the second assertion by induction on I. We assume inductively that

$$\dim_E \mathrm{Ext}^{d_{I'}(\sigma^\infty_0,\sigma^\infty_1)}_{L_{I'}}(i^\infty_{I'_0,I'}(\sigma^\infty_0), i^\infty_{I'_1,I'}(\sigma^\infty_1))^\infty = 1$$

for any $I'_0, I'_1 \subseteq I' \subsetneq I$ and G-basic σ^∞_i in $\mathrm{Rep}^\infty_{\mathrm{adm}}(L_{I'_i})$, $i = 0, 1$, such that $d_{I'}(\sigma^\infty_0, \sigma^\infty_1) < \infty$. Note that the induction hypothesis trivially holds when $I = I_0 = I_1 = \emptyset$ since we only have smooth characters then.

Case 1: If there exist $I_2 \subsetneq I$ and an irreducible G-regular π^∞_2 in $\mathrm{Rep}^\infty_{\mathrm{adm}}(L_{I_2})$ such that $i^\infty_{I_1,I}(\pi^\infty_1) \cong i^\infty_{I_2,I}(\pi^\infty_2)$, then by (2.7) we have isomorphisms for $k \geq 0$

$$\begin{aligned}\mathrm{Ext}^k_{L_I}(i^\infty_{I_0,I}(\pi^\infty_0), i^\infty_{I_1,I}(\pi^\infty_1))^\infty &\cong \mathrm{Ext}^k_{L_I}(i^\infty_{I_0,I}(\pi^\infty_0), i^\infty_{I_2,I}(\pi^\infty_2))^\infty\\ &\cong \mathrm{Ext}^k_{L_{I_2}}(J_{I,I_2}(i^\infty_{I_0,I}(\pi^\infty_0)), \pi^\infty_2)^\infty \cong \mathrm{Ext}^k_{L_{I_2}}(\pi^\infty_3, \pi^\infty_2)^\infty\end{aligned} \tag{2.55}$$

where $\Sigma \overset{\text{def}}{=} W(L_{I_2}) \cdot \mathcal{J}(\pi^\infty_2)$ and $\pi^\infty_3 \overset{\text{def}}{=} J_{I,I_2}(i^\infty_{I_0,I}(\pi^\infty_0))_{\mathcal{B}^{I_2}_\Sigma}$. By Lemma 2.1.29 π^∞_3 is either zero or G-basic. The assumption $d_I(\pi^\infty_0, \pi^\infty_1) < \infty$ forces $\pi^\infty_3 \neq 0$, and thus π^∞_3 is G-basic with $d_{I_2}(\pi^\infty_3, \pi^\infty_2) = d_I(\pi^\infty_0, \pi^\infty_1)$ by (2.55). Hence, (2.54) is isomorphic to $\mathrm{Ext}^{d_{I_2}(\pi^\infty_3,\pi^\infty_2)}_{L_{I_2}}(\pi^\infty_3, \pi^\infty_2)^\infty$ which is one dimensional by our induction hypothesis as $I_2 \subsetneq I$.

Case 2: If there exist $I_2 \subsetneq I$ and an irreducible G-regular π^∞_2 in $\mathrm{Rep}^\infty_{\mathrm{adm}}(L_{I_2})$ such that $i^\infty_{I_0,I}(\pi^\infty_0) \cong i^\infty_{I_2,I}(\pi^\infty_2)$, then by (2.8) we have isomorphisms for $k \geq 0$

$$\begin{aligned}\mathrm{Ext}^k_{L_I}(i^\infty_{I_0,I}(\pi^\infty_0), i^\infty_{I_1,I}(\pi^\infty_1))^\infty &\cong \mathrm{Ext}^k_{L_I}(i^\infty_{I_2,I}(\pi^\infty_2), i^\infty_{I_1,I}(\pi^\infty_1))^\infty\\ &\cong \mathrm{Ext}^k_{L_{I_2}}(\pi^\infty_2, J'_{I,I_2}(i^\infty_{I_1,I}(\pi^\infty_1)))^\infty \cong \mathrm{Ext}^k_{L_{I_2}}(\pi^\infty_2, \pi^\infty_3)^\infty\end{aligned}$$

where $\Sigma \overset{\text{def}}{=} W(L_{I_2}) \cdot \mathcal{J}(\pi^\infty_2)$ and $\pi^\infty_3 \overset{\text{def}}{=} J'_{I,I_2}(i^\infty_{I_1,I}(\pi^\infty_1))_{\mathcal{B}^{I_2}_\Sigma}$. We deduce from Remark 2.1.30 that π^∞_3 is either zero or G-basic. We conclude by induction as in Case 1 since $I_2 \subsetneq I$.

Case 3: If we are in none of the above two cases, then we must have $I_0 = I_1 = I$ and, using Lemma 2.1.28, subsets $I_2, I_3 \subseteq I$ and smooth characters $\delta_2, \delta_3 : L_I \to E^\times$ such that $\pi^\infty_0 \cong V^\infty_{I_2,I} \otimes_E \delta_2$ and $\pi^\infty_1 \cong V^\infty_{I_3,I} \otimes_E \delta_3$. Our assumption $d_I(\pi^\infty_0, \pi^\infty_1) < \infty$ forces π^∞_0 and π^∞_1 to lie in the same Bernstein block and thus

$$W(L_I)\cdot\delta_2|_T=W(L_I)\cdot\mathcal{J}(V^{\infty}_{I_2,I}\otimes_E\delta_2)=W(L_I)\cdot\mathcal{J}(V^{\infty}_{I_3,I}\otimes_E\delta_3)=W(L_I)\cdot\delta_3|_T,$$

which implies $\delta_2=\delta_3$ as $\delta_2|_T$ is the unique element in $W(L_I)\cdot\delta_2|_T$ which extends to a smooth character $L_I\to E^\times$. Consequently, we obtain isomorphisms $\mathrm{Ext}^k_{L_I}(\pi_0^\infty,\pi_1^\infty)^\infty\cong\mathrm{Ext}^k_{L_I}(V^{\infty}_{I_2,I},V^{\infty}_{I_3,I})$ for $k\geq 0$ which by Lemma 2.2.3 forces (2.54) to be one dimensional. □

Recall that $\widehat{j}=\Delta\setminus\{j\}$ for $j\in\Delta$.

Lemma 2.2.5 *Let $I_1,I_2\subseteq\Delta$ with $d(I_1,I_2)=1$, then there exists a unique non-split extension $0\to V^{\infty}_{I_1,\Delta}\to\pi^\infty\to V^{\infty}_{I_2,\Delta}\to 0$ in $\mathrm{Rep}^{\infty}_{\mathrm{adm}}(G)$. More precisely $\pi^\infty\cong i^{\infty}_{\widehat{j},\Delta}(\tau^\infty)$ (hence π^∞ is G-basic) for $j\in\Delta$ and $\tau^\infty\in\mathrm{Rep}^{\infty}_{\mathrm{adm}}(L_{\widehat{j}})$ irreducible G-regular which are as follows:*

(i) if $I_2=I_1\setminus\{j_1\}$, then $j=j_1$ and $\tau^\infty=V^{\infty}_{I_2,\widehat{j}}$;

(ii) if $I_2=I_1\sqcup\{j_2\}$, then $j=n-j_2$ and $\tau^\infty=V^{\infty}_{w^{\widehat{j}}(I_1),\widehat{j}}\otimes_E\delta$ where $w^{\widehat{j}}$ is the longest element of $W^{\widehat{j},\emptyset}$ and $\delta: L_{\widehat{j}}\to E^\times$ is the unique character such that $\delta|_T\cong w^{\widehat{j}}\cdot 1_T$ (see (2.11)).

Proof As $d(I_0,I_1)=1$, by Lemma 2.2.3 there exists a unique (up to isomorphism) length two π^∞ in $\mathrm{Rep}^{\infty}_{\mathrm{adm}}(G)$ with socle $V^{\infty}_{I_1,I}$ and cosocle $V^{\infty}_{I_2,I}$.

Case 1: If $I_2=I_1\setminus\{j_1\}$, set $j\overset{\mathrm{def}}{=}j_1$ and $\tau^\infty\overset{\mathrm{def}}{=}V^{\infty}_{I_2,\widehat{j}}$. It follows from (ii) of Corollary 2.1.24 that $\tau^\infty\cong\mathrm{cosoc}_{L_{\widehat{j}}}(i_{\emptyset,\widehat{j}}(w_1\cdot 1_T)^\infty)$ for any $w_1\in W(L_{\widehat{j}})$ such that $D_R(w_1)=I_1\setminus\{j\}=I_2$. Let $w_{\widehat{j}}\in W(L_{\widehat{j}})$ be the longest element, we have $D_R(w_{\widehat{j}}w_1)=\widehat{j}\setminus D_R(w_1)=\Delta\setminus I_1$, hence $\Delta\setminus D_R(w_{\widehat{j}}w_1)=I_1$. Lemma 2.1.31 then implies that $i^{\infty}_{\widehat{j},\Delta}(\tau^\infty)$ is the unique quotient of $i^{\infty}_{\emptyset,\Delta}(w_1\cdot 1_T)$ with socle $V^{\infty}_{I_1,\Delta}$ and cosocle $V^{\infty}_{I_2,\Delta}$, which has length 2 by (iii) of Corollary 2.1.24 and is thus isomorphic to π^∞.

Case 2: If $I_2=I_1\sqcup\{j_2\}$, set $j\overset{\mathrm{def}}{=}n-j_2$ and $w^{\widehat{j}}\overset{\mathrm{def}}{=}w_{\widehat{j}}w_0=w_0w_{\widehat{j_2}}$. Then $w^{\widehat{j}}$ is the longest element in $W^{\emptyset,\widehat{j_2}}$ and we have $w^{\widehat{j}}(\Delta\setminus\{j_2\})=\Delta\setminus\{j\}$, or equivalently $w^{\widehat{j}}L_{\widehat{j_2}}{w^{\widehat{j}}}^{-1}=L_{\widehat{j}}$ (note then that $w^{\widehat{j}}$ is also the longest element in $W^{\widehat{j},\emptyset}$). In particular $w^{\widehat{j}}(I_1)\subseteq\Delta\setminus\{j\}$ since $j_2\notin I_1$ and there exists $w_2\in W(L_{\widehat{j}})$ such that $D_R(w_2)=w^{\widehat{j}}(I_1)$. We choose any such w_2 and define $\tau^\infty\overset{\mathrm{def}}{=}\mathrm{cosoc}_{L_{\widehat{j}}}(i_{\emptyset,\widehat{j}}(w_2w^{\widehat{j}}\cdot 1_T)^\infty)\cong V^{\infty}_{w^{\widehat{j}}(I_1),\widehat{j}}\otimes_E\delta$ (using (ii) of Corollary 2.1.24). By Lemma 2.1.31 $i^{\infty}_{\widehat{j},\Delta}(\tau^\infty)$ is the unique quotient of $i^{\infty}_{\emptyset,\Delta}(w_2w^{\widehat{j}}\cdot 1_T)$ with socle $V^{\infty}_{\Delta\setminus D_R(w_{\widehat{j}}w_2w^{\widehat{j}}),\Delta}$ and cosocle $V^{\infty}_{D_R(w_2w^{\widehat{j}}),\Delta}$. Let us first compute $\Delta\setminus D_R(w_{\widehat{j}}w_2w^{\widehat{j}})$. Let $w_2'\overset{\mathrm{def}}{=}(w^{\widehat{j}})^{-1}w_2w^{\widehat{j}}\in W(L_{\widehat{j_2}})$, we have $D_R(w_2')=(w^{\widehat{j}})^{-1}(D_R(w_2))=(w^{\widehat{j}})^{-1}w^{\widehat{j}}(I_1)=I_1$ and hence $D_R(w_0w_2')=\Delta\setminus I_1$ by (1.24). This implies $\Delta\setminus D_R(w_{\widehat{j}}w_2w^{\widehat{j}})=\Delta\setminus D_R(w_0w_2')=I_1$. Let us now compute $D_R(w_2w^{\widehat{j}})$. Let $j'\in\Delta\setminus\{j_2\}$, then $w^{\widehat{j}}(j')\in\Delta\setminus\{j\}$ and we have $\ell(w_2w^{\widehat{j}}s_{j'})=\ell(w_2s_{w^{\widehat{j}}(j')}w^{\widehat{j}})=\ell(w_2s_{w^{\widehat{j}}(j')})+\ell(w^{\widehat{j}})$ as $w^{\widehat{j}}\in W^{\widehat{j},\emptyset}$ and

$w_2 s_{w^{\widehat{j}}(j')} \in W(L_{\widehat{j}})$. By (1.24) we see that $j' \in D_R(w_2 w^{\widehat{j}})$ if and only if $w^{\widehat{j}}(j') \in D_R(w_2)$. If $j' = j_2$, then as $w^{\widehat{j}} > w^{\widehat{j}} s_{j_2}$ (using that $w^{\widehat{j}} s_{j_2} \in W^{\widehat{j},\emptyset}$ and that $w^{\widehat{j}} \in W^{\widehat{j},\emptyset}$ has the longest possible length) we have $\ell(w_2 w^{\widehat{j}} s_{j'}) = \ell(w_2) + \ell(w^{\widehat{j}} s_{j'}) = \ell(w_2 w^{\widehat{j}}) - 1$, hence $j' \in D_R(w_2 w^{\widehat{j}})$. Thus $D_R(w_2 w^{\widehat{j}}) = (w^{\widehat{j}})^{-1}(D_R(w_2)) \sqcup \{j_2\} = I_1 \sqcup \{j_2\} = I_2$. We deduce that $i^{\infty}_{\widehat{j},\Delta}(\tau^{\infty})$ is the (unique) quotient of $i^{\infty}_{\emptyset,\Delta}(w_2 w^{\widehat{j}} \cdot 1_T)$ with socle $V^{\infty}_{I_1,\Delta}$ and cosocle $V^{\infty}_{I_2,\Delta}$. It has length 2 by (iii) of Corollary 2.1.24 and is thus isomorphic to π^{∞}. □

Lemma 2.2.6 *Let $I \subseteq I' \subseteq \Delta$ and $I_0, I_1 \subseteq I$. Then we have the following isomorphism in* $\mathrm{Rep}^{\infty}_{\mathrm{adm}}(L_{I'})$

$$i^{\infty}_{I,I'}(Q_I(I_0, I_1)) \cong Q_{I'}(I_0 \sqcup (I' \setminus I), I_1).$$

Proof By choosing a sequence of subsets $I' = J_0 \supsetneq J_1 \supsetneq \cdots \supsetneq J_t = I$ with $t = \# I' \setminus I$ (and thus $\# J_{t'-1} \setminus J_{t'} = 1$ for $1 \leq t' \leq t$), we easily reduce to the case when $I' \setminus I = \{j\}$ for some $j \in \Delta$. As $Q_I(I_0, I_1)$ is G-basic, so is $i^{\infty}_{I,I'}(Q_I(I_0, I_1))$. By (iv) of Remark 2.1.16 we know that $i^{\infty}_{I,I'}(Q_I(I_0, I_1))$ is multiplicity free with simple socle and cosocle. As $i^{\infty}_{I,I'}(-)$ is exact and $Q_I(I_0, I_1)$ has socle $V^{\infty}_{I_0,I}$ and cosocle $V^{\infty}_{I_1,I}$, by (i) of Lemma 2.2.5 we know that $V^{\infty}_{I_0 \sqcup \{j\},I'}$ (resp. $V^{\infty}_{I_1,I'}$) occurs in the socle (resp. cosocle) of $i^{\infty}_{I,I'}(Q_I(I_0, I_1))$, which forces the latter to have socle $V^{\infty}_{I_0 \sqcup \{j\},I'}$ (resp. cosocle $V^{\infty}_{I_1,I'}$). By (i) of Lemma 2.2.1 we conclude that $i^{\infty}_{I,I'}(Q_I(I_0, I_1)) \cong Q_{I'}(I_0 \sqcup (I' \setminus I), I_1')$. □

Lemma 2.2.7 *Let $I_0, I_1 \subseteq I \subseteq \Delta$ with $I_0 \neq I_1$, and $I_2 \subseteq I$ such that $d(I_0, I_2) = d(I_0, I_1) - 1$ and $d(I_2, I_1) = 1$. Let π^{∞} in $\mathrm{Rep}^{\infty}_{\mathrm{adm}}(L_I)$ the unique non-split extension $0 \to V^{\infty}_{I_1,I} \to \pi^{\infty} \to V^{\infty}_{I_2,I} \to 0$ (Lemma 2.2.3). We have $d_I(V^{\infty}_{I_0,I}, V^{\infty}_{I_2,I}) = d_I(V^{\infty}_{I_0,I}, V^{\infty}_{I_1,I}) - 1$ and $d_I(V^{\infty}_{I_0,I}, \pi^{\infty}) = \infty$.*

Proof Note first that $d_I(V^{\infty}_{I_0,I}, \pi^{\infty}) = \infty$ and the short exact sequence $0 \to V^{\infty}_{I_1,I} \to \pi^{\infty} \to V^{\infty}_{I_2,I} \to 0$ imply $\mathrm{Ext}^{k-1}_{L_I}(V^{\infty}_{I_0,I}, V^{\infty}_{I_2,I})^{\infty} \cong \mathrm{Ext}^{k}_{L_I}(V^{\infty}_{I_0,I}, V^{\infty}_{I_1,I})^{\infty}$ for $k \geq 1$, and in particular $d_I(V^{\infty}_{I_0,I}, V^{\infty}_{I_2,I}) = d_I(V^{\infty}_{I_0,I}, V^{\infty}_{I_1,I}) - 1$. Hence it is enough to prove $d_I(V^{\infty}_{I_0,I}, \pi^{\infty}) = \infty$, i.e. $\mathrm{Ext}^{k}_{L_I}(V^{\infty}_{I_0,I}, \pi^{\infty})^{\infty} = 0$ for $k \geq 0$.

Case 1: If $I_2 = I_1 \setminus \{j_1\}$ for some $j_1 \in I_1 \setminus I_0$, then we have $\pi^{\infty} \cong i^{\infty}_{I \setminus \{j_1\},I}(V^{\infty}_{I_2,I\setminus\{j_1\}})$ by (i) of Lemma 2.2.5 (which extends verbatim with Δ replaced by I and G by L_I when $I_0, I_1 \subseteq I$). Let $I^- \stackrel{\mathrm{def}}{=} I \setminus \{j_1\}$, $\Sigma \stackrel{\mathrm{def}}{=} W(L_{I^-}) \cdot 1_T$ and recall that $V^{\infty}_{I_2,I^-} \in \mathcal{B}^{I^-}_{\Sigma}$ by (2.41). By (2.7) and (2.46) we deduce for $k \geq 0$

$$\mathrm{Ext}^{k}_{L_I}(V^{\infty}_{I_0,I}, \pi^{\infty})^{\infty} \cong \mathrm{Ext}^{k}_{L_{I^-}}(J_{I,I^-}(V^{\infty}_{I_0,I})_{\mathcal{B}^{I^-}_{\Sigma}}, V^{\infty}_{I_2,I^-})^{\infty}.$$

Hence, to prove $d_I(V^{\infty}_{I_0,I}, \pi^{\infty}) = \infty$, it suffices to show that $J_{I,I^-}(V^{\infty}_{I_0,I})_{\mathcal{B}^{I^-}_{\Sigma}} = 0$, or equivalently $\mathcal{J}(V^{\infty}_{I_0,I}) \cap \Sigma = \emptyset$ (see the discussion below (2.13)). Let $w \cdot 1_T \in \mathcal{J}(V^{\infty}_{I_0,I}) \cap \Sigma$ (using (2.41)), then $w \in W(L_{I^-})$ by definition of Σ and $I_0 = I \setminus D_R(w)$ by (2.41). However, $w \in W(L_{I^-})$ implies $w(j_1) \in \Phi^+$ and thus $j_1 \notin D_R(w)$ by (1.24). Hence $j_1 \in I \setminus D_R(w) = I_0$ which contradicts $j_1 \in I_1 \setminus I_0$. Consequently we have $\mathcal{J}(V^{\infty}_{I_0,I}) \cap \Sigma = \emptyset$.

Case 2: If $I_2 = I_1 \sqcup \{j_2\}$ for some $j_2 \in I_0 \setminus I_1$, then we deduce from (ii) of Lemma 2.2.5 (applied with L_I instead of G) that $\pi^\infty \cong i^\infty_{I\setminus\{j\},I}(V^\infty_{w^{J^-}(I_1),I\setminus\{j\}} \otimes_E \delta)$ where $J \subseteq I$ is the subset corresponding to the Levi block of L_I containing j_2, j is the unique element of J such that, for $J^- \overset{\text{def}}{=} J \setminus \{j\}$ and w^{J^-} the longest element of $W^{J^-,\emptyset}(L_J)$, we have $j_2 \in D_R(w^{J^-})$, and $\delta : L_I \to E^\times$ is the unique smooth character such that $\delta|_T \cong w^{J^-} \cdot 1_T$. Let $I^- \overset{\text{def}}{=} I \setminus \{j\}$, $\Sigma \overset{\text{def}}{=} W(L_{I^-})w^{J^-} \cdot 1_T$, and note that $W^{J^-,\emptyset}(L_J) \subseteq W^{I^-,\emptyset}(L_I)$. As in Case 1 we have $V^\infty_{w^{J^-}(I_1),I^-} \otimes_E \delta \in \mathcal{B}^{I^-}_\Sigma$ and for $k \geq 0$:

$$\mathrm{Ext}^k_{L_I}(V^\infty_{I_0,I}, \pi^\infty)^\infty \cong \mathrm{Ext}^k_{L_{I^-}}(J_{I,I^-}(V^\infty_{I_0,I})_{\mathcal{B}^{I^-}_\Sigma}, V^\infty_{w^{J^-}(I_1),I^-} \otimes_E \delta)^\infty.$$

Again it is enough to prove $\mathcal{J}(V^\infty_{I_0,I}) \cap \Sigma = \emptyset$. Let $w \cdot 1_T \in \mathcal{J}(V^\infty_{I_0,I}) \cap \Sigma$, then $w \in W(L_{I^-})w^{J^-}$ and $I_0 = I \setminus D_R(w)$ as in Case 1. However, since $w \in W(L_{I^-})w^{J^-}$ and $w^{J^-} \in W^{I^-,\emptyset}(L_I)$ we have $D_R(w^{J^-}) \subseteq D_R(w)$. As $j_2 \in D_R(w^{J^-})$, we have $j_2 \in D_R(w)$ and thus $j_2 \notin I \setminus D_R(w) = I_0$, which contradicts $j_2 \in I_0 \setminus I_1$. Hence, we must have $\mathcal{J}(V^\infty_{I_0,I}) \cap \Sigma = \emptyset$. □

For $I_0, I_1, I_2 \subseteq \Delta$, we define $d(I_2, [I_0, I_1]) \overset{\text{def}}{=} \min\{d(I_2, I'_1) \mid I'_1 \in [I_0, I_1]\}$.

Lemma 2.2.8 *For $I_0, I_1, I_2 \subseteq \Delta$, there exists a unique $I_3 \in [I_0, I_1]$ such that*

$$d(I_2, I_3) = d(I_2, [I_0, I_1]). \tag{2.56}$$

Proof Let us first prove that

$$I_3 \overset{\text{def}}{=} (I_0 \cap I_1) \cup (I_2 \cap (I_0 \cup I_1)) \tag{2.57}$$

satisfies (2.56). In fact, we prove that it satisfies the stronger statement: for all $I'_1 \in [I_0, I_1]$

$$d(I_2, I_3) = d(I_2, I'_1) - d(I_3, I'_1). \tag{2.58}$$

Recall from (2.50) that $I'_1 \in [I_0, I_1]$ if and only if

$$I_0 \cap I_1 \subseteq I'_1 \subseteq I_0 \cup I_1. \tag{2.59}$$

It follows from (2.59) and (2.57) that, for any $I'_1 \in [I_0, I_1]$, $I_3 \setminus I_2 = (I_0 \cap I_1) \setminus I_2 \subseteq I'_1 \setminus I_2$ and $I'_1 \setminus I_3 \subseteq I'_1 \setminus I_2$, which implies $I'_1 \setminus I_2 = I'_1 \setminus I_3 \sqcup I_3 \setminus I_2$. Likewise we have $I_2 \setminus I_3 = I_2 \setminus (I_0 \cup I_1) \subseteq I_2 \setminus I'_1$ and $I_3 \setminus I'_1 \subseteq I_2 \setminus I'_1$, hence $I_2 \setminus I'_1 = I_2 \setminus I_3 \sqcup I_3 \setminus I'_1$. This implies $d(I_2, I'_1) = d(I_2, I_3) + d(I_3, I'_1)$, i.e. (2.58), and therefore (2.56).

Now we prove unicity. Let $I'_3 \in [I_0, I_1]$ satisfying (2.56), which implies $d(I_2, I'_3) = d(I_2, I_3)$. By the previous paragraph we have $I_3 \setminus I_2 \subseteq I'_3 \setminus I_2$ and $I_2 \setminus I_3 \subseteq I_2 \setminus I'_3$, hence we must have $I_3 \setminus I_2 = I'_3 \setminus I_2$ and $I_2 \setminus I_3 = I_2 \setminus I'_3$. Assume there is $x \in I'_3 \setminus I_3$, then necessarily $x \notin I_2$ by (2.57), hence $x \in I'_3 \setminus I_2$ but $x \notin I_3 \setminus I_2$, which is a contradiction. Assume there is $x \in I_3 \setminus I'_3$, then necessarily $x \notin I_0 \cap I_1$, hence

$x \in I_2$. Thus $x \in I_2 \setminus I_3'$ but $x \notin I_2 \setminus I_3$, which is again a contradiction. We therefore must have $I_3 = I_3'$. □

Lemma 2.2.9 *Let $I \subseteq \Delta$ and $I_0, I_1, I_2 \subseteq I$. We have $d_I(Q_I(I_0, I_1), V_{I_2,I}^{\infty}) < \infty$ if and only if*

$$d(I_2, I_1) = d(I_2, [I_0, I_1]), \tag{2.60}$$

in which case $d_I(Q_I(I_0, I_1), V_{I_2,I}^{\infty}) = d(I_2, I_1)$.

Proof Let I_3 as in Lemma 2.2.8, so (2.60) is equivalent to $I_3 = I_1$. If $I_3 = I_1$, then by (2.58) we have $d(I_2, I_1') = d(I_2, I_1) + d(I_1, I_1') > d(I_2, I_1)$ for $I_1' \in [I_0, I_1] \setminus \{I_1\}$. By (i) of Lemmas 2.2.1, 2.2.3 and an obvious dévissage we have $\mathrm{Ext}^k_{L_I}(\ker(Q_I(I_0, I_1) \to V_{I_1,I}^{\infty}), V_{I_2,I}^{\infty})^{\infty} = 0$ for $k \le d(I_2, I_1)$. Hence, the surjection $Q_I(I_0, I_1) \twoheadrightarrow V_{I_1,I}^{\infty}$ induces an isomorphism for $k \le d(I_2, I_1)$

$$\mathrm{Ext}^k_{L_I}(V_{I_1,I}^{\infty}, V_{I_2,I}^{\infty})^{\infty} \xrightarrow{\sim} \mathrm{Ext}^k_{L_I}(Q_I(I_0, I_1), V_{I_2,I}^{\infty})^{\infty}.$$

As $d_I(V_{I_1,I}^{\infty}, V_{I_2,I}^{\infty}) = d(I_1, I_2) = d(I_2, I_1)$ (Lemma 2.2.3) we deduce $d_I(Q_I(I_0, I_1), V_{I_2,I}^{\infty}) = d(I_2, I_1)$.

We assume from now $I_3 \neq I_1$ and prove that $d_I(Q_I(I_0, I_1), V_{I_2,I}^{\infty}) = \infty$. As $I_3 \neq I_1$, one easily sees that there exists $I_4 \in [I_3, I_1]$ such that $d(I_3, I_4) = 1$ and $d(I_3, I_1) = d(I_4, I_1) + 1$ (I_4 is obtained by either adding to I_3 an element of $I_1 \setminus I_3$ or withdrawing from I_3 an element of $I_3 \setminus I_1$ (which is hence in $I_3 \cap I_0$)). By applying (2.58) (with I_4 replacing I_1' there) we deduce $d(I_4, I_2) = d(I_3, I_2) + 1$, and thus we have either $\emptyset \neq I_3 \setminus I_4 \subseteq I_2$, or $I_4 \setminus I_3 \neq \emptyset$ with $(I_4 \setminus I_3) \cap I_2 = \emptyset$. Now consider any $I_3', I_4' \in [I_0, I_1]$ such that $I_3 \setminus I_4 = I_3' \setminus I_4'$ and $I_4 \setminus I_3 = I_4' \setminus I_3'$. This implies $d(I_3', I_4') = 1$ and one easily checks that one still has $d(I_3', I_1) = d(I_4', I_1) + 1$, and that we have either $\emptyset \neq I_4' \setminus I_3' \subseteq I_1$, or $I_3' \setminus I_4' \neq \emptyset$ with $(I_3' \setminus I_4') \cap I_1 = \emptyset$. In all cases $I_4' \in [I_3', I_1]$ and by (iii) of Lemma 2.2.1 we deduce that $Q_I(I_3', I_4')$ is a subquotient of $Q_I(I_0, I_1)$. Moreover we also have either $\emptyset \neq I_3' \setminus I_4' \subseteq I_2$, or $I_4' \setminus I_3' \neq \emptyset$ with $(I_4' \setminus I_3') \cap I_2 = \emptyset$, and in both cases $d(I_2, I_4') = d(I_2, I_3') + 1$. We then deduce by (2.53) for $k \ge 0$:

$$\mathrm{Ext}^k_{L_I}(Q_I(I_3', I_4'), V_{I_2,I}^{\infty})^{\infty} \cong \mathrm{Ext}^k_{L_I}((V_{I_2,I}^{\infty})^{\sim}, Q_I(I_3', I_4')^{\sim})^{\infty} \cong \mathrm{Ext}^k_{L_I}(V_{I_2,I}^{\infty}, Q_I(I_4', I_3'))^{\infty} = 0 \tag{2.61}$$

where the last equality follows from the last assertion in Lemma 2.2.7 together with (i) of Lemma 2.2.1. Now, for a given $j \in I_1 \setminus I_0$ (resp. $j \in I_0 \setminus I_1$), one can write $[I_0, I_1]$ as a disjoint union of $[I_0', I_1']$ for $I_0', I_1' \in [I_0, I_1]$ such that $I_1' = I_0' \sqcup \{j\}$ (resp. $I_0' = I_1' \sqcup \{j\}$). It follows that $[I_0, I_1]$ is the disjoint union of $[I_3', I_4']$ for some $I_3', I_4' \in [I_0, I_1]$ as discussed above, and hence that $Q_I(I_0, I_1)$ admits a filtration whose graded pieces are the $Q_I(I_3', I_4')$ (using (iii) of Lemma 2.2.1). By (2.61) and an obvious dévissage we deduce $\mathrm{Ext}^k_{L_I}(Q_I(I_0, I_1), V_{I_2,I}^{\infty})^{\infty} = 0$ for $k \ge 0$, i.e. $d_I(Q_I(I_0, I_1), V_{I_2,I}^{\infty}) = \infty$. □

Lemma 2.2.10 *Let $I \subseteq \Delta$ and $I_0, I_1, I_0', I_1' \subseteq I$ with $[I_0, I_1] \cap [I_0', I_1'] = \emptyset$. Then we have*

$$\mathrm{Ext}^1_{L_I}(Q_I(I'_0, I'_1), Q_I(I_0, I_1))^\infty \neq 0 \tag{2.62}$$

if and only if $[I_0, I'_1] \subseteq [I_0, I_1] \sqcup [I'_0, I'_1]$*, in which case (2.62) is one dimensional.*

Proof It follows from (2.51) that we have $[I_0, I'_1] \cap [I_0, I_1] = [I_0, I''_1]$ with $I''_1 \overset{\text{def}}{=} (I_0 \cap (I_1 \cup I'_1)) \sqcup ((I_1 \cap I'_1) \setminus I_0)$, and similarly $[I_0, I'_1] \cap [I'_0, I'_1] = [I''_0, I'_1]$ with $I''_0 \overset{\text{def}}{=} (I'_1 \cap (I_0 \cup I'_0)) \sqcup ((I_0 \cap I'_0) \setminus I'_1)$.

Assume first that $[I_0, I'_1] \subseteq [I_0, I_1] \sqcup [I'_0, I'_1]$, which together with $[I_0, I_1] \cap [I'_0, I'_1] = \emptyset$ implies that

$$[I_0, I'_1] = ([I_0, I'_1] \cap [I_0, I_1]) \sqcup ([I_0, I'_1] \cap [I'_0, I'_1]) = [I_0, I''_1] \sqcup [I''_0, I'_1]. \tag{2.63}$$

By (i) and (ii) of Lemma 2.2.1 and since $I''_0 \in [I_0, I'_1]$ (resp. $I''_1 \in [I_0, I'_1]$), there exists a (unique up to scalar) surjection $Q_I(I_0, I'_1) \twoheadrightarrow Q_I(I''_0, I'_1)$ (resp. injection $Q_I(I_0, I''_1) \hookrightarrow Q_I(I_0, I'_1)$). By (i) of Lemma 2.2.1 and (2.63), we deduce a short exact sequence

$$0 \to Q_I(I_0, I''_1) \to Q_I(I_0, I'_1) \to Q_I(I''_0, I'_1) \to 0$$

which has to be non-split as $Q_I(I_0, I'_1)$ has simple socle and cosocle. In particular, we have

$$\mathrm{Ext}^1_G(Q_I(I''_0, I'_1), Q_I(I_0, I''_1)) \neq 0. \tag{2.64}$$

Our assumption $[I_0, I_1] \cap [I'_0, I'_1] = \emptyset$ implies that $Q_I(I'_0, I'_1)$ and $Q_I(I_0, I_1)$ have no common Jordan-Hölder factor. Thus the surjection $Q_I(I'_0, I'_1) \twoheadrightarrow Q_I(I''_0, I'_1)$ and the injection $Q_I(I_0, I''_1) \hookrightarrow Q_I(I_0, I_1))$ from (ii) of Lemma 2.2.1 induce an injection

$$\mathrm{Ext}^1_G(Q_I(I''_0, I'_1), Q_I(I_0, I''_1)) \hookrightarrow \mathrm{Ext}^1_G(Q_I(I'_0, I'_1), Q_I(I_0, I_1)),$$

which together with (2.64) gives (2.62). It is clear from $[I_0, I_1] \cap [I'_0, I'_1] = \emptyset$ and (2.62) that $d(Q_I(I'_0, I'_1), Q_I(I_0, I_1)) = 1$, which by Lemma 2.2.4 implies that (2.62) is one dimensional.

Now we assume that (2.62) holds and let V be a representationt which fits into a non-split extension

$$0 \to Q_I(I_0, I_1) \to V \to Q_I(I'_0, I'_1) \to 0.$$

By Lemma 2.1.32 we know that V admits unique subquotient V' which is G-basic with socle $\mathrm{soc}_{L_I}(Q_I(I_0, I_1)) \cong V^\infty_{I_0,I}$ and cosocle $\mathrm{cosoc}_{L_I}(Q_I(I'_0, I'_1)) \cong V^\infty_{I'_1,I}$. By (i) of Lemma 2.2.1 we must have $V' \cong Q_I(I_0, I'_1)$ and in particular

$$\mathrm{JH}_{L_I}(Q_I(I_0, I'_1)) \subseteq \mathrm{JH}_{L_I}(V) = \mathrm{JH}_{L_I}(Q_I(I_0, I_1)) \sqcup \mathrm{JH}_{L_I}(Q_I(I'_0, I'_1))$$

which (again by *loc. cit.*) gives $[I_0, I'_1] \subseteq [I_0, I_1] \sqcup [I'_0, I'_1]$. $\square$

Lemma 2.2.11 *For $i=0,1$ let $I_i \subseteq \Delta$ and π_i^∞ in $\mathrm{Rep}^\infty_{\mathrm{adm}}(L_{I_i})$ irreducible G-regular with $\Sigma_i \overset{def}{=} W(L_{I_i}) \cdot \mathcal{J}(\pi_i^\infty)$. Let $w \in W^{I_0,I_1}$.*

(i) We have

$$\mathrm{Hom}_{L_{I_1}}(i^\infty_{I_0,I_1,w}(J_{I_0,I_1,w}(\pi_0^\infty)), \pi_1^\infty) \neq 0 \tag{2.65}$$

if and only if we have an isomorphism of irreducible G-regular representations

$$J_{I_0,I_1,w}(\pi_0^\infty)_{\mathcal{B}^{w^{-1}(I_0)\cap I_1}_{w^{-1}\cdot\Sigma_0\cap\Sigma_1}} \cong J'_{I_1,w^{-1}(I_0)\cap I_1}(\pi_1^\infty)_{\mathcal{B}^{w^{-1}(I_0)\cap I_1}_{w^{-1}\cdot\Sigma_0\cap\Sigma_1}}, \tag{2.66}$$

in which case (2.65) is one dimensional. If we denote by π^∞ the representation in (2.66), we have $\pi_1^\infty \cong \mathrm{cosoc}_{L_{I_1}}(i^\infty_{w^{-1}(I_0)\cap I_1,I_1}(\pi^\infty))$ and

$$\begin{aligned}\pi_0^\infty &\cong \mathrm{soc}_{L_{I_0}}(i^\infty_{w^{-1}(I_0)\cap I_1,w^{-1}(I_0)}(\pi^\infty \otimes_E \delta^{-1}_{I_0,I_1,w})^{w^{-1}})\\ &\cong \mathrm{soc}_{L_{I_0}}(i^\infty_{I_0\cap w(I_1),I_0}((\pi^\infty \otimes_E \delta^{-1}_{I_0,I_1,w})^{w^{-1}})).\end{aligned}$$

(ii) Assume (2.65) and let σ_0^∞ be an irreducible constituent of $i^\infty_{I_0\cap w(I_1),I_0}((\pi^\infty \otimes_E \delta^{-1}_{I_0,I_1,w})^{w^{-1}})$ and σ_1^∞ an irreducible constituent of $i^\infty_{w^{-1}(I_0)\cap I_1,I_1}(\pi^\infty)$. Then we have

$$J_{I_0,I_1,w}(\sigma_0^\infty)_{\mathcal{B}^{w^{-1}(I_0)\cap I_1}_{w^{-1}\cdot\Sigma_0\cap\Sigma_1}} = 0 \text{ if } \sigma_0^\infty \neq \pi_0^\infty,\quad J'_{I_1,w^{-1}(I_0)\cap I_1}(\sigma_1^\infty)_{\mathcal{B}^{w^{-1}(I_0)\cap I_1}_{w^{-1}\cdot\Sigma_0\cap\Sigma_1}} = 0 \text{ if } \sigma_1^\infty \neq \pi_1^\infty,$$

and we also have

$$\mathrm{Ext}^k_{L_{I_1}}(i^\infty_{I_0,I_1,w}(J_{I_0,I_1,w}(\sigma_0^\infty)), \sigma_1^\infty) \neq 0 \text{ for some } k \geq 0 \tag{2.67}$$

if and only if $\sigma_0^\infty = \pi_0^\infty$ and $\sigma_1^\infty = \pi_1^\infty$.

Proof Recall that by (2.8) and (2.21) we have canonical isomorphisms for $k \geq 0$

$$\mathrm{Ext}^k_{L_{I_1}}(J_{I_0,I_1,w}(\pi_0^\infty), J'_{I_1,w^{-1}(I_0)\cap I_1}(\pi_1^\infty)) \cong \mathrm{Ext}^k_{L_{I_1}}(i^\infty_{I_0,I_1,w}(J_{I_0,I_1,w}(\pi_0^\infty)), \pi_1^\infty) \tag{2.68}$$

and that any G-regular left $W(L_{w^{-1}(I_0)\cap I_1})$-coset $\Sigma \subseteq \widehat{T}^\infty$ gives a block $\mathcal{B}_\Sigma^{w^{-1}(I_0)\cap I_1}$ (see the discussion around (2.13)). We write $I \overset{def}{=} w^{-1}(I_0)\cap I_1$ and $\delta \overset{def}{=} \delta_{I_0,I_1,w}$.

We prove (i). Assume that (2.65) holds, thus the vector spaces in (2.68) are non-zero for $k=0$. Then by Lemma 2.1.29 and Remark 2.1.30 there exists a left $W(L_I)$-coset Σ such that

$$J_{I_0,I_1,w}(\pi_0^\infty)_{\mathcal{B}^I_\Sigma} \neq 0 \neq J'_{I_1,I}(\pi_1^\infty)_{\mathcal{B}^I_\Sigma}. \tag{2.69}$$

Recall that $\Sigma_i = W(L_{I_i}) \cdot \mathcal{J}(\pi_i^\infty)$ are single left regular $W(L_{I_i})$-cosets for $i=0,1$ by (the last statement in) (i) of Lemma 2.1.15. An easy exercise using that all characters here are G-regular shows that $w^{-1}\cdot\Sigma_0\cap\Sigma_1$ is again a single (regular) left $W(L_{w^{-1}(I_0)\cap I_1})$-coset. As we have $\mathcal{J}(J_{I_0,I_1,w}(\pi_0^\infty)) \subseteq w^{-1}\cdot\Sigma_0$ and $\mathcal{J}(J'_{I_1,I}(\pi_1^\infty)) \subseteq$

Σ_1, we necessarily have $\Sigma = w^{-1} \cdot \Sigma_0 \cap \Sigma_1$. In other words, Σ as in (2.69) is uniquely determined by π_0^∞, π_1^∞ and w, and thus we have a canonical isomorphism

$$0 \neq \mathrm{Hom}_{L_{I_1}}(J_{I_0,I_1,w}(\pi_0^\infty), J'_{I_1,I}(\pi_1^\infty)) \cong \mathrm{Hom}_{L_{I_1}}(J_{I_0,I_1,w}(\pi_0^\infty)_{\mathcal{B}^I_\Sigma}, J'_{I_1,I}(\pi_1^\infty)_{\mathcal{B}^I_\Sigma}). \tag{2.70}$$

As π_0^∞ and π_1^∞ are irreducible G-regular, both $J_{I_0,I_1,w}(\pi_0^\infty)_{\mathcal{B}^I_\Sigma}$ and $J'_{I_1,I}(\pi_1^\infty)_{\mathcal{B}^I_\Sigma}$ are irreducible G-regular by (the last statement in) Lemma 2.1.29 and Remark 2.1.30. Hence they must be isomorphic by (2.70) and we denote them by π^∞. Note in particular that (2.70) is one dimensional and thus (2.65) is one dimensional. We have by (2.8)

$$\mathrm{Hom}_{L_{I_1}}(i^\infty_{I,I_1}(\pi^\infty), \pi_1^\infty) \cong \mathrm{Hom}_{L_I}(\pi^\infty, J'_{I_1,I}(\pi_1^\infty)) \neq 0$$

and by (2.7) (together with (2.19))

$$\mathrm{Hom}_{L_{I_0}}((\pi_0^\infty)^w, i^\infty_{I,w^{-1}(I_0)}(\pi^\infty \otimes_E \delta^{-1})) \cong \mathrm{Hom}_{L_I}(J_{I_0,I_1,w}(\pi_0^\infty), \pi^\infty) \neq 0.$$

As π_0^∞ and π_1^∞ are irreducible G-regular, we deduce $\pi_0^\infty \cong \mathrm{soc}_{L_{I_0}}(i^\infty_{w(I),I_0}(\pi^\infty \otimes_E \delta^{-1})^{w^{-1}})$ and $\pi_1^\infty \cong \mathrm{cosoc}_{L_{I_1}}(i^\infty_{I,I_1}(\pi^\infty))$.

We prove (ii). We borrow Σ and π^∞ from (i) and we fix $\sigma_0^\infty \in \mathrm{JH}_{L_{I_0}}(i^\infty_{w(I),I_0}((\pi^\infty \otimes_E \delta^{-1})^{w^{-1}}))$ and $\sigma_1^\infty \in \mathrm{JH}_{L_{I_1}}(i^\infty_{I,I_1}(\pi^\infty))$. It follows from Lemma 2.1.18 that the natural injection

$$\begin{aligned} J_{I_0,I_1,w}(\pi_0^\infty)_{\mathcal{B}^I_\Sigma} &\cong \pi^\infty \cong i^\infty_{I,I,1}(J_{I,I,1}(\pi^\infty \otimes_E \delta^{-1})) \otimes_E \delta \\ &\hookrightarrow (J_{w^{-1}(I_0),I}(i^\infty_{I,w^{-1}(I_0)}(\pi^\infty \otimes_E \delta^{-1})) \otimes_E \delta)_{\mathcal{B}^I_\Sigma} \end{aligned}$$

is an isomorphism, and hence that $J_{I_0,I_1,w}(i^\infty_{I,w^{-1}(I_0)}(\pi^\infty \otimes_E \delta^{-1})^{w^{-1}}/\pi_0^\infty)_{\mathcal{B}^I_\Sigma} = 0$, or equivalently $J_{I_0,I_1,w}(\sigma_0^\infty)_{\mathcal{B}^I_\Sigma} = 0$, for $\sigma_0^\infty \neq \pi_0^\infty$. Similarly, for w_1 the longest element of $W^{\emptyset,I}(L_{I_1})$, it follows from (2.10) and Lemma 2.1.18 that

$$\begin{aligned} J'_{I_1,I}(i^\infty_{I,I_1}(\pi^\infty))_{\mathcal{B}^I_\Sigma} &\cong (J_{I_1,w_1(I)}(i^\infty_{I,I_1}(\pi^\infty))^{w_1} \otimes_E \delta^{-1}_{P_I \cap L_{I_1}})_{\mathcal{B}^I_\Sigma} \\ &\cong (i^\infty_{I,w_1(I),w_1}(J_{I,w_1(I),w_1}(\pi^\infty))^{w_1} \otimes_E \delta^{-1}_{P_I \cap L_{I_1}})_{\mathcal{B}^I_\Sigma} \\ &\cong ((\pi^\infty)^{w_1^{-1}} \otimes_E \delta_{P_{w_1(I)} \cap L_{I_1}})^{w_1} \otimes_E \delta^{-1}_{P_I \cap L_{I_1}} \\ &\cong \pi^\infty \cong J'_{I_1,I}(\pi_1^\infty)_{\mathcal{B}^I_\Sigma} \end{aligned}$$

which implies $J'_{I_1,I}(\sigma_1^\infty)_{\mathcal{B}^I_\Sigma} = 0$ for $\sigma_1^\infty \neq \pi_1^\infty$. Furthermore, arguing as in (i) with σ_i^∞ instead of π_i^∞ (and $\mathrm{Ext}^k_{L_I}$ instead of Hom_{L_I}), we have seen that (2.67) forces

$$J_{I_0,I_1,w}(\sigma_0^\infty)_{\mathcal{B}^I_{\Sigma'}} \neq 0 \neq J'_{I_1,I}(\sigma_1^\infty)_{\mathcal{B}^I_{\Sigma'}}$$

where $\Sigma' \stackrel{\mathrm{def}}{=} w^{-1} \cdot W(L_{I_0}) \cdot \mathcal{J}(\sigma_0^\infty) \cap W(L_{I_1}) \cdot \mathcal{J}(\sigma_1^\infty)$. But it easily follows from (i) of Lemma 2.1.15 that

$$W(L_{I_1})\cdot\mathcal{J}(\sigma_1^\infty)=W(L_{I_1})\cdot\mathcal{J}(i^\infty_{I,I_1}(\pi^\infty))=W(L_{I_1})\cdot\mathcal{J}(\pi_1^\infty)=\Sigma_1$$

and similarly $W(L_{I_0})\cdot\mathcal{J}(\sigma_0^\infty)=\Sigma_0$, which forces $\Sigma'=\Sigma$. By what we have proven before on $J_{I_0,I_1,w}(\sigma_0^\infty)_{\mathcal{B}^I_\Sigma}$ and $J'_{I_1,I}(\sigma_1^\infty)_{\mathcal{B}^I_\Sigma}$, we deduce that (2.67) forces $\sigma_0^\infty=\pi_0^\infty$ and $\sigma_1^\infty=\pi_1^\infty$. □

2.3 Examples of G-Basic Representations

We study specific G-basic representations: the representations $\pi^\infty_{j_1,j_2}$ below. We prove several technical lemmas on these representations which will be mainly used in Sect. 5.2 below.

Given $j,j'\in\mathbb{Z}$, we write $[j,j']\overset{\text{def}}{=}\{j''\in\mathbb{Z}\mid j\le j''\le j'\}$ (hence $[j,j']=\emptyset$ if $j'<j$). Recall that $\widehat{j}=\Delta\setminus\{j\}$ for $j\in\Delta$. We also use the convenient notation $\widehat{n}\overset{\text{def}}{=}\Delta$. We define

$$\mathbf{J}^\infty\overset{\text{def}}{=}\{(j_1,j_2)\mid 1\le j_1\le j_2\le n,\ \ j_1\le n-1\}$$

equipped with the partial order $(j_1,j_2)\le(j_1',j_2')$ if and only if $j_2\le j_2'$ and $j_2-j_1\le j_2'-j_1'$. If $j\ge 1$, we write St_j^∞ (resp. 1_j) for the smooth Steinberg (resp. the trivial representation) of $\mathrm{GL}_j(K)$. For $(j_1,j_2)\in\mathbf{J}^\infty$, we set

$$\sigma^\infty_{j_1,j_2}\overset{\text{def}}{=}|\det{}_{j_1}|_K^{j_1-j_2}\boxtimes_E\left(|\det{}_{j_2-j_1}|_K^{j_1}\otimes_E\mathrm{St}^\infty_{j_2-j_1}\right)\boxtimes_E\mathrm{St}^\infty_{n-j_2}\in\mathrm{Rep}^\infty_{\mathrm{adm}}(L_{\widehat{j_1}\cap\widehat{j_2}}),$$

which is irreducible G-regular, and we define the G-basic representation

$$\begin{cases}\pi^\infty_{j_1,j_2}\overset{\text{def}}{=}i^\infty_{\widehat{j_1}\cap\widehat{j_2},\widehat{j_1}}(\sigma^\infty_{j_1,j_2}) \text{ if } 1\le j_1<j_2\le n-1\\ \pi^\infty_{j_1,j_2}\overset{\text{def}}{=}\sigma^\infty_{j_1,j_2}\ \text{ if } j_1=j_2 \text{ or } j_2=n.\end{cases}\tag{2.71}$$

In particular, we have $\pi^\infty_{j_1,j_1}=\sigma^\infty_{j_1,j_1}=1_{j_1}\boxtimes_E\mathrm{St}^\infty_{n-j_1}=V^\infty_{[1,j_1-1],\widehat{j_1}}$ when $j_2=j_1$, and $\pi^\infty_{j_1,n}=\sigma^\infty_{j_1,n}=|\det{}_{j_1}|_K^{j_1-n}\boxtimes_E\left(|\det{}_{n-j_1}|_K^{j_1}\otimes_E\mathrm{St}^\infty_{n-j_1}\right)$ when $j_2=n$. Recall that

$$\Sigma_{j_1,j_2}\overset{\text{def}}{=}W(L_{\widehat{j_1}})\cdot\mathcal{J}(\sigma^\infty_{j_1,j_2})=W(L_{\widehat{j_1}})\cdot\mathcal{J}(\pi^\infty_{j_1,j_2})\tag{2.72}$$

and $\Sigma'_{j_1,j_2}\overset{\text{def}}{=}W(L_{\widehat{j_1}\cap\widehat{j_2}})\cdot\mathcal{J}(\sigma^\infty_{j_1,j_2})$ are single (regular) left cosets by (i) of Lemma 2.1.15. Let $\Gamma_{\Sigma_{j_1,j_2}}$ and $\Gamma_{\Sigma'_{j_1,j_2}}$ be their respective associated undirected graphs (see above Theorem 2.1.20). They have the same vertices but $\Gamma_{\Sigma_{j_1,j_2}}$ possibly has more edges. However, when $j_1<j_2<n$ or equivalently when $L_{\widehat{j_1}\cap\widehat{j_2}}$ has 3 Levi blocks, one can explicitly check that both $\Gamma_{\Sigma'_{j_1,j_2}}$ and $\Gamma_{\Sigma_{j_1,j_2}}$ have 3 connected components, which implies $\Gamma_{\Sigma_{j_1,j_2}}=\Gamma_{\Sigma'_{j_1,j_2}}$. Thus any orientation on $\Gamma_{\Sigma'_{j_1,j_2}}$ defines a unique orientation on $\Gamma_{\Sigma_{j_1,j_2}}$, which means by Theorem 2.1.20 that, for any $\chi\in\mathcal{J}(\sigma^\infty_{j_1,j_2})$, the two

principal series $i^{\infty}_{\emptyset,\widehat{j_1}\cap\widehat{j_2}}(\chi)$ and $i^{\infty}_{\emptyset,\widehat{j_1}}(\chi)$ have the same number of constituents. This implies in particular that $\pi^{\infty}_{j_1,j_2} = i^{\infty}_{\widehat{j_1}\cap\widehat{j_2},\widehat{j_1}}(\sigma^{\infty}_{j_1,j_2})$ remains irreducible (G-regular).

For $(j_1, j_2) \in \mathbf{J}^{\infty}$, let x_{j_1,j_2} be the longest element in

$$\{x \in W(G) \mid D_L(x) = \{j_1\},\ \mathrm{Supp}(x) \subseteq [1, j_2 - 1]\} \tag{2.73}$$

with $x_{j_1,j_2} \stackrel{\mathrm{def}}{=} 1$ if $j_1 = j_2$. Recall that the condition $D_L(x) = \{j_1\}$ is equivalent to $x \in W^{j_1,\emptyset}$ (use for instance (1.23)). The following lemma gives the structure of $i^{\infty}_{\widehat{j_1}\cap\widehat{j_2},\Delta}(\sigma^{\infty}_{j_1,j_2})$.

Lemma 2.3.1 *For $(j_1, j_2) \in \mathbf{J}^{\infty}$, we have $\Sigma_{j_1,j_2} = W(L_{\widehat{j_1}})x_{j_1,j_2} \cdot 1_T$ and the explicit structure of $i^{\infty}_{j_1,\Delta}(\pi^{\infty}_{j_1,j_2}) \cong i^{\infty}_{\widehat{j_1}\cap\widehat{j_2},\Delta}(\sigma^{\infty}_{j_1,j_2})$ is given as follows.*

(i) If $1 \le j_1 < j_2 \le n-1$, then $i^{\infty}_{\widehat{j_1}\cap\widehat{j_2},\Delta}(\sigma^{\infty}_{j_1,j_2})$ has Loewy length 3, with socle $V^{\infty}_{[j_2-j_1+1,j_2],\Delta}$, cosocle $V^{\infty}_{[j_2-j_1,j_2-1],\Delta}$, and middle layer $V^{\infty}_{[j_2-j_1+1,j_2-1],\Delta} \oplus V^{\infty}_{[j_2-j_1,j_2],\Delta}$.

(ii) If $1 \le j_1 = j_2 \le n-1$, then $i^{\infty}_{\widehat{j_1}\cap\widehat{j_2},\Delta}(\sigma^{\infty}_{j_1,j_2})$ has Loewy length 2, with socle $V^{\infty}_{[1,j_1],\Delta}$ and cosocle $V^{\infty}_{[1,j_1-1],\Delta}$.

(iii) If $1 \le j_1 < j_2 = n$, then $i^{\infty}_{\widehat{j_1}\cap\widehat{j_2},\Delta}(\sigma^{\infty}_{j_1,j_2})$ has Loewy length 2, with socle $V^{\infty}_{[n-j_1+1,n-1],\Delta}$ and cosocle $V^{\infty}_{[n-j_1,n-1],\Delta}$.

Proof Let $z'_{j_1,j_2} \in W(L_{\widehat{j_1}\cap\widehat{j_2}}) \subseteq W(G)$ such that $z'_{j_1,j_2} \cdot 1_T \in \mathcal{J}(\sigma^{\infty}_{j_1,j_2})$, $w_{\widehat{j_1}\cap\widehat{j_2}} \in W(L_{\widehat{j_1}\cap\widehat{j_2}})$ the longest element and $z_{j_1,j_2} \stackrel{\mathrm{def}}{=} w_{\widehat{j_1}\cap\widehat{j_2}} z'_{j_1,j_2}$, then by (ii) of Remark 2.1.16 we have

$$\sigma^{\infty}_{j_1,j_2} \cong \mathrm{soc}_{L_{\widehat{j_1}\cap\widehat{j_2}}}(i^{\infty}_{\emptyset,\widehat{j_1}\cap\widehat{j_2}}(z'_{j_1,j_2} \cdot 1_T)) \cong \mathrm{cosoc}_{L_{\widehat{j_1}\cap\widehat{j_2}}}(i^{\infty}_{\emptyset,\widehat{j_1}\cap\widehat{j_2}}(z_{j_1,j_2} \cdot 1_T)).$$

By Lemma 2.1.31 this implies that $i^{\infty}_{\widehat{j_1}\cap\widehat{j_2},\Delta}(\sigma^{\infty}_{j_1,j_2})$ is the unique quotient of $i^{\infty}_{\emptyset,\Delta}(z_{j_1,j_2} \cdot 1_T)$ with socle $V^{\infty}_{\Delta\setminus D_R(z'_{j_1,j_2}),\Delta}$ (and cosocle $V^{\infty}_{D_R(z_{j_1,j_2}),\Delta}$). Using (iii) of Corollary 2.1.24, we only need to find an explicit z'_{j_1,j_2} as above and compute $D_R(z_{j_1,j_2})$ and $\Delta \setminus D_R(z'_{j_1,j_2})$.

Let y_{j_1,j_2} be the longest element in $\{x \in W(G) \mid \mathrm{Supp}(x) \subseteq [1, j_1 - 1]\}$. Note that $y_{j_1,j_2} \in W(L_{[1,j_1-1]}) \cong W(\mathrm{GL}_{j_1}) \subseteq W(L_{\widehat{j_1}\cap\widehat{j_2}})$ and that $D_R(y_{j_1,j_2}) = [1, j_1 - 1]$. We set

$$\delta_{j_1,j_2} \stackrel{\mathrm{def}}{=} |\det{}_{j_1}|_K^{j_1-j_2} \boxtimes_E |\det{}_{j_2-j_1}|_K^{j_1} \boxtimes_E 1_{\mathrm{GL}_{n-j_2}} : L_{\widehat{j_1}\cap\widehat{j_2}} \to E^{\times}.$$

We observe that $x_{j_1,j_2} \cdot 1_T = \delta_{j_1,j_2}|_T$ and that $y_{j_1,j_2}x_{j_1,j_2} \cdot 1_T = (y_{j_1,j_2} \cdot 1_T) \otimes_E (\delta_{j_1,j_2}|_T)$ (use $y_{j_1,j_2} \in W(L_{\widehat{j_1}\cap\widehat{j_2}})$). This implies

$$i^{\infty}_{\emptyset,\widehat{j_1}\cap\widehat{j_2}}(y_{j_1,j_2}x_{j_1,j_2} \cdot 1_T) \cong i^{\infty}_{\emptyset,\widehat{j_1}\cap\widehat{j_2}}((y_{j_1,j_2} \cdot 1_T) \otimes_E \delta_{j_1,j_2}|_T) \cong i^{\infty}_{\emptyset,\widehat{j_1}\cap\widehat{j_2}}(y_{j_1,j_2} \cdot 1_T) \otimes_E \delta_{j_1,j_2}. \tag{2.74}$$

It follows from (ii) of Lemma 2.1.24 that

$$\mathrm{cosoc}_{L_{\widehat{j_1}\cap\widehat{j_2}}}(i^{\infty}_{\emptyset,\widehat{j_1}\cap\widehat{j_2}}(y_{j_1,j_2}\cdot 1_T))\cong 1_{\mathrm{GL}_{j_1}}\boxtimes_E \mathrm{St}^{\infty}_{j_2-j_1}\boxtimes_E \mathrm{St}^{\infty}_{n-j_2},$$

which together with (2.74) implies

$$\mathrm{cosoc}_{L_{\widehat{j_1}\cap\widehat{j_2}}}(i^{\infty}_{\emptyset,\widehat{j_1}\cap\widehat{j_2}}(y_{j_1,j_2}x_{j_1,j_2}\cdot 1_T))\cong \sigma^{\infty}_{\widehat{j_1,j_2}}. \tag{2.75}$$

Hence we may take $z_{j_1,j_2}=y_{j_1,j_2}x_{j_1,j_2}$. Note that (2.75) and (i) of Lemma 2.1.15 imply

$$\Sigma_{j_1,j_2}=W(L_{\widehat{j_1}})\cdot\mathcal{J}(\sigma^{\infty}_{\widehat{j_1,j_2}})=W(L_{\widehat{j_1}})y_{j_1,j_2}x_{j_1,j_2}\cdot 1_T=W(L_{\widehat{j_1}})x_{j_1,j_2}\cdot 1_T.$$

Let $y'_{j_1,j_2}\stackrel{\mathrm{def}}{=}w_{\widehat{j_1}\cap\widehat{j_2}}y_{j_1,j_2}$, it is the longest element of $W(\mathrm{GL}_{j_2-j_1}\times\mathrm{GL}_{n-j_2})$ (with $\mathrm{GL}_{j_2-j_1}\times\mathrm{GL}_{n-j_1}$ seen as Levi blocks of $L_{\widehat{j_1}\cap\widehat{j_2}}$), and we have $D_R(y'_{j_1,j_2})=[j_1+1,j_2-1]\sqcup[j_2+1,n]$. We have the following two cases.

Case 1: If $j_1=j_2$, then $z_{j_1,j_2}=y_{j_1,j_2}$ and we have $D_R(z_{j_1,j_2})=D_R(y_{j_1,j_2})=[1,j_1-1]$ and $\Delta\setminus D_R(z'_{j_1,j_2})=\Delta\setminus D_R(y'_{j_1,j_2})=\Delta\setminus[j_1+1,n]=[1,j_1]$. This gives (ii).

Case 2: We assume $j_2>j_1$ and write x, y, y', $z=yx$, $z'=y'x$ for x_{j_1,j_2}, y_{j_1,j_2}, y'_{j_1,j_2}, z_{j_1,j_2}, z'_{j_1,j_2}. As $x\in W^{\widehat{j_1},\emptyset}$, we have $\ell(ux)=\ell(u)+\ell(x)$ for each $u\in W(L_{\widehat{j_1}})$. In particular, as $y,y'\in W(L_{\widehat{j_1}})$, we have $\ell(z)=\ell(y)+\ell(x)$ and $\ell(z')=\ell(y')+\ell(x)$. We also have $D_R(z)\subseteq\mathrm{Supp}(yx)\subseteq[1,j_2-1]$ and $D_R(z')\subseteq\mathrm{Supp}(y'x)\subseteq\widehat{j_2}$. If $1\leq j\leq j_2-j_1-1$, we have $j+j_1>j_1$ and thus $ys_{j+j_1},y's_{j+j_1}\in W(L_{\widehat{j_1}})$. This together with $ys_{j+j_1}>y$ and $y's_{j+j_1}<y$ gives $yxs_j=ys_{j+j_1}x>yx$ and $y'xs_j=y's_{j+j_1}x<yx$, and thus $j\in D_R(z')\setminus D_R(z)$. If $j_2-j_1+1\leq j\leq j_2-1$, we have $j-(j_2-j_1)<j_1$ and thus $ys_{j-(j_2-j_1)},y's_{j-(j_2-j_1)}\in W(L_{\widehat{j_1}})$. This together with $ys_{j-(j_2-j_1)}<y$ and $y's_{j-(j_2-j_1)}>y'$ gives $yxs_j=ys_{j-(j_2-j_1)}x<yx$ and $y'xs_j=y's_{j-(j_2-j_1)}x>y'x$, and thus $j\in D_R(z)\setminus D_R(z')$. If $j_2+1\leq j\leq n-1$, we have $y'xs_j=y's_jx<y'x$ and $j\notin\mathrm{Supp}(z)$, and thus $j\in D_R(z')\setminus D_R(z)$. To finish, it suffices to prove that $j_2-j_1\in D_R(x)$, which implies $j_2-j_1\in D_R(z)\cap D_R(z')$. Let $I\stackrel{\mathrm{def}}{=}[1,j_2-1]$, $I'\stackrel{\mathrm{def}}{=}[1,j_2-1]\setminus\{j_1\}$ and recall that w_I, $w_{I'}$ is the longest element of respectively $W(L_I)$, $W(L_{I'})$. By definition of x we have $x=w_{I'}w_I$ (with $D_R(x)\subseteq\mathrm{Supp}(x)\subseteq I$) and thus $xs_{j'}=w_{I'}s_{j_2-j'}w_I$ for $j'\in I$. Note that $u<u'$ if and only if $uw_I>u'w_I$ for each $u,u'\in W(L_I)$. It follows that $j'\in D_R(x)\subseteq I$ if and only if $xs_{j'}<x$ (see (1.24)) if and only if $w_{I'}s_{n-j'}>w_{I'}$ if and only if $j'\in I\setminus w_I(D_R(w_{I'}))=I\setminus w_I(I')=\{j_2-j_1\}$.

To sum up, we have proven $D_R(z)=[j_2-j_1,j_2-1]$, $D_R(z')=[1,j_2-j_1]\sqcup[j_2+1,n-1]$ and $\Delta\setminus D_R(z')=[j_2-j_1+1,j_2]\cap\Delta$. This gives (i) and (iii). □

For $(j_1,j_2)\in\mathbf{J}^{\infty}$, we define subsets $I^+_{j_1,j_2},I^-_{j_1,j_2}\subseteq\Delta$ by

$$\mathrm{soc}_G(i^{\infty}_{\widehat{j_1},\Delta}(\pi^{\infty}_{j_1,j_2}))\cong V^{\infty}_{I^+_{j_1,j_2},\Delta},\quad \mathrm{cosoc}_G(i^{\infty}_{\widehat{j_1},\Delta}(\pi^{\infty}_{j_1,j_2}))\cong V^{\infty}_{I^-_{j_1,j_2},\Delta} \tag{2.76}$$

and we note that $I^+_{j_1,j_2}$, $I^-_{j_1,j_2}$ are explicitly given in Lemma 2.3.1. We have $i^\infty_{\widehat{j_1},\Delta}(\pi^\infty_{j_1,j_2}) \cong Q_\Delta(I^+_{j_1,j_2}, I^-_{j_1,j_2})$ by (i) of Lemma 2.2.1. The following lemma will be used in Sect. 5.2 to study extensions between (certain) non-locally algebraic Orlik-Strauch representations.

Lemma 2.3.2 *Let $(j_1, j_2), (j_1', j_2') \in \mathbf{J}^\infty$ with $(j_1, j_2) < (j_1', j_2')$. Then we have*

(i) $d(\pi^\infty_{j_1',j_2'}, \pi^\infty_{j_1,j_2}) = 0$ if and only if $(j_1', j_2') \in \{(j_1+1, j_2+1), (j_1-1, j_2), (j_1, j_2+1)\}$;
(ii) $d(\pi^\infty_{j_1',j_2'}, \pi^\infty_{j_1,j_2}) = 1$ if and only if $(j_1', j_2') \in \{(j_1+2, j_2+2), (j_1-2, j_2)\}$;
(iii) $\mathcal{J}(i^\infty_{\widehat{j_1'},\Delta}(\pi^\infty_{j_1',j_2'})) \cap \mathcal{J}(\pi^\infty_{j_1,j_2}) = \emptyset$ if $(j_1', j_2') \notin \{(j_1+1, j_2+1), (j_1-1, j_2), (j_1, j_2+1)\}$.

Proof We prove (i). By (i) of Lemma 2.2.1, it suffices to check the conditions $I^-_{j_1',j_2'} \in [I^+_{j_1,j_2}, I^-_{j_1,j_2}]$ and $I^+_{j_1,j_2} \in [I^+_{j_1',j_2'}, I^-_{j_1',j_2'}]$. This is a straightforward check using Lemma 2.3.1:

- If $(j_1', j_2') \notin \{(j_1+1, j_2+1), (j_1-1, j_2), (j_1, j_2+1)\}$, then one can check that $[I^+_{j_1,j_2}, I^-_{j_1,j_2}] \cap [I^+_{j_1',j_2'}, I^-_{j_1',j_2'}] \neq \emptyset$ only when $j_1 = j_1' = 1$ and $j_2' > j_2+1$, in which case $[I^+_{j_1,j_2}, I^-_{j_1,j_2}] \cap [I^+_{j_1',j_2'}, I^-_{j_1',j_2'}] = \{\emptyset\}$ (i.e. the empty set is the only subset of Δ in this intersection). Since $I^+_{j_1,j_2}, I^-_{j_1',j_2'} \neq \emptyset$, the conditions $I^-_{j_1',j_2'} \in [I^+_{j_1,j_2}, I^-_{j_1,j_2}]$ and $I^+_{j_1,j_2} \in [I^+_{j_1',j_2'}, I^-_{j_1',j_2'}]$ are never satisfied.
- If $(j_1', j_2') \in \{(j_1+1, j_2+1), (j_1-1, j_2), (j_1, j_2+1)\}$, then one can check that $[I^+_{j_1,j_2}, I^-_{j_1,j_2}] \cap [I^+_{j_1',j_2'}, I^-_{j_1',j_2'}] = [I^+_{j_1,j_2}, I^-_{j_1',j_2'}]$, more precisely $[I^+_{j_1,j_2}, I^-_{j_1,j_2}] \cap [I^+_{j_1',j_2'}, I^-_{j_1',j_2'}] = \{[j_2-j_1+1, j_2], [j_2-j_1, j_2]\}$ if $(j_1', j_2') = (j_1+1, j_2+1)$, $[I^+_{j_1,j_2}, I^-_{j_1,j_2}] \cap [I^+_{j_1',j_2'}, I^-_{j_1',j_2'}] = \{[j_2-j_1+1, j_2], [j_2-j_1+1, j_2-1]\}$ if $(j_1', j_2') = (j_1-1, j_2)$, and $I^+_{j_1,j_2} = I^-_{j_1',j_2'} = [j_2-j_1+1, j_2]$ if $(j_1', j_2') = (j_1, j_2+1)$.

We prove (ii). By (i) it suffices to find all $(j_1, j_2), (j_1', j_2') \in \mathbf{J}^\infty$ satisfying $(j_1, j_2) < (j_1', j_2')$, $(j_1', j_2') \notin \{(j_1+1, j_2+1), (j_1-1, j_2), (j_1, j_2+1)\}$ and

$$\mathrm{Ext}^1_G(i^\infty_{\widehat{j_1'},\Delta}(\pi^\infty_{j_1',j_2'}), i^\infty_{\widehat{j_1},\Delta}(\pi^\infty_{j_1,j_2}))^\infty \tag{2.77}$$

is non-zero. We have two cases.

Case 1: If $[I^+_{j_1,j_2}, I^-_{j_1,j_2}] \cap [I^+_{j_1',j_2'}, I^-_{j_1',j_2'}] \neq \emptyset$, as $(j_1', j_2') \notin \{(j_1+1, j_2+1), (j_1-1, j_2), (j_1, j_2+1)\}$, we must have $j_1 = j_1' = 1$, $j_2' > j_2+1$ and $[I^+_{j_1,j_2}, I^-_{j_1,j_2}] \cap [I^+_{j_1',j_2'}, I^-_{j_1',j_2'}] = \{\emptyset\}$ (see the proof of (i)). Then $i^\infty_{\widehat{j_1},\Delta}(\pi^\infty_{j_1,j_2})$ contains a length two subrepresentation σ^∞ with socle $V^\infty_{\{j_2\},\Delta}$ and socle $\mathrm{St}^\infty_n = V^\infty_{\emptyset,\Delta}$. It follows from Lemma 2.2.7 that $\mathrm{Ext}^1_G(V^\infty_{I,\Delta}, \sigma^\infty)^\infty = 0$ for each $I \subseteq \Delta$ satisfying $d(I, \{j_2\}) = d(I, \emptyset) + 1$, or equivalently $j_2 \notin I$. As any $I \in [I^+_{j_1',j_2'}, I^-_{j_1',j_2'}]$ satisfies $j_2 \notin I$, we deduce by dévissage

$$\mathrm{Ext}^1_G(i^\infty_{\widehat{j_1'},\Delta}(\pi^\infty_{j_1',j_2'}), \sigma^\infty)^\infty = 0. \tag{2.78}$$

If $i^{\infty}_{\widehat{j_1},\Delta}(\pi^{\infty}_{j_1,j_2})/\sigma^{\infty} \neq 0$, then it has length two with socle $V^{\infty}_{[j_2-1,j_2],\Delta}$ and cosocle $V^{\infty}_{\{j_2-1\},\Delta}$, and it follows from Lemma 2.2.7 that

$$\mathrm{Ext}^1_G(V^{\infty}_{\emptyset,\Delta}, i^{\infty}_{\widehat{j_1},\Delta}(\pi^{\infty}_{j_1,j_2})/\sigma^{\infty})^{\infty} = 0. \tag{2.79}$$

As any $\emptyset \neq I \in [I^+_{j_1',j_2'}, I^-_{j_1',j_2'}]$ satisfies $d(I, [j_2-1, j_2]) \geq 3$ and $d(I, \{j_2-1\}) \geq 2$, we deduce from the first statement of Lemma 2.2.3 that $\mathrm{Ext}^1_G(V^{\infty}_{I,\Delta}, i^{\infty}_{\widehat{j_1},\Delta}(\pi^{\infty}_{j_1,j_2})/\sigma^{\infty})^{\infty} = 0$ for such I, which together with (2.79) implies $\mathrm{Ext}^1_G(i^{\infty}_{\widehat{j_1'},\Delta}(\pi^{\infty}_{j_1',j_2'}), i^{\infty}_{\widehat{j_1},\Delta}(\pi^{\infty}_{j_1,j_2})/\sigma^{\infty})^{\infty} = 0$. With (2.78) we deduce that (2.77) is zero.

Case 2: We assume $[I^+_{j_1,j_2}, I^-_{j_1,j_2}] \cap [I^+_{j_1',j_2'}, I^-_{j_1',j_2'}] = \emptyset$, then we must have $\max\{j_1, j_1'\} \geq 2$. If (2.77) is non-zero, then $i^{\infty}_{\widehat{j_1},\Delta}(\pi^{\infty}_{j_1,j_2})$ (resp. $i^{\infty}_{\widehat{j_1'},\Delta}(\pi^{\infty}_{j_1',j_2'})$) must contain a Jordan-Hölder factor $V^{\infty}_{I_0,\Delta}$ (resp. $V^{\infty}_{I_1,\Delta}$) such that $I_0 \neq I_1$ and

$$\mathrm{Ext}^1_G(V^{\infty}_{I_0,\Delta}, V^{\infty}_{I_1,\Delta})^{\infty} \neq 0, \tag{2.80}$$

which by Lemma 2.2.3 implies $d(I_0, I_1) = 1$. We have the possibilities.

- If $(j_1', j_2') \in \{(j_1+2, j_2+2), (j_1-2, j_2)\}$, then one can check that $[I^+_{j_1,j_2}, I^-_{j_1,j_2}] \sqcup [I^+_{j_1',j_2'}, I^-_{j_1',j_2'}] = [I^+_{j_1,j_2}, I^-_{j_1',j_2'}]$, and thus $Q_{\Delta}(I^+_{j_1,j_2}, I^-_{j_1',j_2'})$ gives a non-zero element of (2.77).
- If $(j_1', j_2') \in \{(j_1+1, j_2+2), (j_1-1, j_2+1)\}$, then the only possible choice of I_0, I_1 for (2.80) to hold is $I_0 = I^+_{j_1',j_2'}$ and $I_1 = I^-_{j_1,j_2}$, and there always exists $I_2 \in [I^+_{j_1,j_2}, I^-_{j_1,j_2}] \setminus \{I_1\}$ such that $d(I_1, I_2) = 1$ and $d(I_0, I_2) = 2$. Write π^{∞} for the unique length two quotient of $i^{\infty}_{\widehat{j_1},\Delta}(\pi^{\infty}_{j_1,j_2})$ with socle $V^{\infty}_{I_2,\Delta}$ and cosocle $V^{\infty}_{I_1,\Delta}$, then it follows from Lemma 2.2.7 that $\mathrm{Ext}^1_G(V^{\infty}_{I_0,\Delta}, \pi^{\infty}) = 0$. By a dévissage on both $i^{\infty}_{\widehat{j_1'},\Delta}(\pi^{\infty}_{j_1',j_2'})$ and $i^{\infty}_{\widehat{j_1},\Delta}(\pi^{\infty}_{j_1,j_2})$ (using a filtration on $i^{\infty}_{\widehat{j_1'},\Delta}(\pi^{\infty}_{j_1',j_2'})$ with simple graded pieces and a filtration on $i^{\infty}_{\widehat{j_1},\Delta}(\pi^{\infty}_{j_1,j_2})$ with π^{∞} being the unique non-simple graded piece), we deduce that (2.77) is zero.
- If $(j_1', j_2') \notin \{(j_1+2, j_2+2), (j_1-2, j_2), (j_1+1, j_2+2), (j_1-1, j_2+1)\}$, then a pair I_0, I_1 as in (2.80) does not exist since we have $d(I_0, I_1) \geq 2$ for any $I_0 \in [I^+_{j_1,j_2}, I^-_{j_1,j_2}]$ and $I_1 \in [I^+_{j_1',j_2'}, I^-_{j_1',j_2'}]$. Hence (2.77) is zero by dévissage.

Finally we prove (iii). By (2.7) we have

$$\mathrm{Hom}_{L_{\widehat{j_1}}}(J_{\Delta,\widehat{j_1}}(V^{\infty}_{I^+_{j_1,j_2},\Delta}), \pi^{\infty}_{j_1,j_2}) \cong \mathrm{Hom}_G(V^{\infty}_{I^+_{j_1,j_2},\Delta}, i^{\infty}_{\widehat{j_1},\Delta}(\pi^{\infty}_{j_1,j_2})) \neq 0$$

and thus $\mathcal{J}(\pi^{\infty}_{j_1,j_2}) \subseteq \mathcal{J}(J_{\Delta,\widehat{j_1}}(V^{\infty}_{I^+_{j_1,j_2},\Delta})) = \mathcal{J}(V^{\infty}_{I^+_{j_1,j_2},\Delta})$ since $\pi^{\infty}_{j_1,j_2}$ is irreducible. Hence, to prove (iii), it suffices to show that $\mathcal{J}(i^{\infty}_{\widehat{j_1'},\Delta}(\pi^{\infty}_{j_1',j_2'})) \cap \mathcal{J}(V^{\infty}_{I^+_{j_1,j_2},\Delta}) = \emptyset$, or equivalently by (ii) of Lemma 2.1.15 that $V^{\infty}_{I^+_{j_1,j_2},\Delta}$ is not a constituent of $i^{\infty}_{\widehat{j_1'},\Delta}(\pi^{\infty}_{j_1',j_2'}) \cong Q_{\Delta}(I^+_{j_1',j_2'}, I^-_{j_1',j_2'})$, i.e. $I^+_{j_1,j_2} \notin [I^+_{j_1',j_2'}, I^-_{j_1',j_2'}]$ by (i) of Lemma 2.2.1. By the proof of (i), we already see that $(j_1, j_2) < (j_1', j_2')$, $(j_1', j_2') \notin \{(j_1+1, j_2+1), (j_1-1, j_2), (j_1, j_2+1)\}$ and $[I^+_{j_1,j_2}, I^-_{j_1,j_2}] \cap [I^+_{j_1',j_2'}, I^-_{j_1',j_2'}] \neq \emptyset$ happens only when $j_1 = j_1' = 1$ and $j_2' > j_2 + 1$, in which case $[I^+_{j_1,j_2}, I^-_{j_1,j_2}] \cap [I^+_{j_1',j_2'}, I^-_{j_1',j_2'}] = \{\emptyset\}$. But we never have $I^+_{j_1,j_2} = \emptyset$, which proves the statement. □

The following lemma will be used in Sect. 5.2 to study extensions between (certain) non-locally algebraic Orlik-Strauch representations and (certain) locally algebraic representations.

Lemma 2.3.3 *Let $(j_1, j_2) \in \mathbf{J}^\infty$ and $I \subseteq \Delta$.*

(i) We have $d(\pi^\infty_{j_1,j_2}, V^\infty_{I,\Delta}) = 0$ if and only if $I = I^-_{j_1,j_2}$, and $d(V^\infty_{I,\Delta}, \pi^\infty_{j_1,j_2}) = 0$ if and only if $I = I^+_{j_1,j_2}$.

(ii) If $I \neq I^-_{j_1,j_2}$, then we have $d(\pi^\infty_{j_1,j_2}, V^\infty_{I,\Delta}) = 1$ if and only if $I \notin [I^+_{j_1,j_2}, I^-_{j_1,j_2}]$ and $d(I^-_{j_1,j_2}, I) = 1$. Similarly, if $I \neq I^+_{j_1,j_2}$, then $d(V^\infty_{I,\Delta}, \pi^\infty_{j_1,j_2}) = 1$ if and only if $I \notin [I^+_{j_1,j_2}, I^-_{j_1,j_2}]$ and $d(I, I^+_{j_1,j_2}) = 1$.

Proof (i) follows from $i^\infty_{\widehat{j_1},\Delta}(\pi^\infty_{j_1,j_2}) \cong Q_\Delta(I^+_{j_1,j_2}, I^-_{j_1,j_2})$. (ii) follows from this and Lemma 2.2.9, using moreover the isomorphisms

$$\mathrm{Ext}^1_G(i^\infty_{\widehat{j_1}}(\pi^\infty_{j_1,j_2}), V^\infty_{I,\Delta})^\infty \cong \mathrm{Ext}^1_G(V^\infty_{I,\Delta}, i^\infty_{\widehat{j_1}}(\pi^\infty_{j_1,j_2})^\sim)^\infty \cong \mathrm{Ext}^1_G(V^\infty_{I,\Delta}, Q_\Delta(I^-_{j_1,j_2}, I^+_{j_1,j_2}))^\infty$$

for the second statement (see Remark 2.2.2). □

Recall that Σ_{j_1,j_2} is defined in (2.72). The following lemma will be used in Lemma 2.3.5 below (and in Sect. 5.2).

Lemma 2.3.4 *Let $(j_1, j_2), (j_1', j_2') \in \mathbf{J}^\infty$ with $(j_1, j_2) < (j_1', j_2')$.*

(i) We have $\Sigma_{j_1,j_2} \cap \Sigma_{j_1',j_2'} \neq \emptyset$ if and only if either $j_2' = j_2$ or $j_2' - j_1' = j_2 - j_1$.

(ii) If $j_1' = j_1$ then $\Sigma_{j_1,j_2} \cap (w^{-1} \cdot \Sigma_{j_1,j_2'}) \neq \emptyset$ for some $w \in W^{\widehat{j_1},\widehat{j_1}}$ if and only if w is the representative of $w_{[1,j_2'-1]} w^{-1}_{[1,j_2-1]}$, i.e. $w_{[1,j_2'-1]} w^{-1}_{[1,j_2-1]} \in W(L_{\widehat{j_1}}) w W(L_{\widehat{j_1}})$. Moreover $\Sigma_{j_1,j_2} \cap (s_{j_1} \cdot \Sigma_{j_1',j_2'}) \neq \emptyset$ (i.e. $w = s_{j_1}$) if and only if either $j_1 = 1$ or $j_2' = j_2 + 1$.

Proof We prove (i). Recall that $\Sigma_{j_1,j_2} = W(L_{\widehat{j_1}}) x_{j_1,j_2} \cdot 1_T$, $\Sigma_{j_1',j_2'} = W(L_{\widehat{j_1'}}) x_{j_1',j_2'} \cdot 1_T$ (see (2.73) and Lemma 2.3.1). We need to find all pairs of $(j_1, j_2) < (j_1', j_2')$ such that $\Sigma_{j_1,j_2} \cap \Sigma_{j_1',j_2'} \neq \emptyset$. As in the proof of (i) of Lemma 2.2.11, one checks that $\Sigma_{j_1,j_2} \cap \Sigma_{j_1',j_2'}$ is a single left $W(L_{\widehat{j_1} \cap \widehat{j_1'}})$-coset if non-empty, and thus contains a unique element $x' \cdot 1_T$ with $x' \in W(L_{\widehat{j_1}}) x_{j_1,j_2} \cap W(L_{\widehat{j_1'}}) x_{j_1',j_2'}$ of miminal length. We write $x' = x'_{j_1,j_2} x_{j_1,j_2}$ and $x' = x'_{j_1',j_2'} x_{j_1',j_2'}$ with $x'_{j_1,j_2} \in W(L_{\widehat{j_1}})$, $x'_{j_1',j_2'} \in W(L_{\widehat{j_1'}})$ and both expressions reduced. Then it is clear that $x'_{j_1,j_2} \in W(L_{\widehat{j_1}})$ is minimal in $W(L_{\widehat{j_1} \cap \widehat{j_1'}}) x'_{j_1,j_2}$, and that $x'_{j_1',j_2'} \in W(L_{\widehat{j_1'}})$ is minimal in $W(L_{\widehat{j_1} \cap \widehat{j_1'}}) x'_{j_1',j_2'}$. We have the following cases.

- If $j_1' = j_1$, then Σ_{j_1,j_2} and $\Sigma_{j_1,j_2'}$ are both $W(L_{\widehat{j_1}})$-cosets, and thus $\Sigma_{j_1,j_2} \cap \Sigma_{j_1',j_2'} \neq \emptyset$ if and only if $\Sigma_{j_1,j_2} = \Sigma_{j_1',j_2'}$ if and only if $x_{j_1,j_2} = x_{j_1',j_2'}$, which contradicts $(j_1, j_2) < (j_1', j_2')$. Thus $\Sigma_{j_1,j_2} \cap \Sigma_{j_1',j_2'} = \emptyset$.
- If $j_1' < j_1$ and $j_2' = j_2$, then $W(L_{\widehat{j_1}})x_{j_1,j_2}$ and $W(L_{\widehat{j_1'}})x_{j_1',j_2'} = W(L_{\widehat{j_1'}})x_{j_1',j_2}$ both contain the (unique) maximal length element in the set $\{x \in W(G) \mid \mathrm{Supp}(x) \subseteq [1, j_2 - 1]\}$, and thus $\Sigma_{j_1,j_2} \cap \Sigma_{j_1',j_2'} \neq \emptyset$.
- If $j_1' < j_1$ and $j_2' > j_2$, assume $\Sigma_{j_1,j_2} \cap \Sigma_{j_1',j_2'} \neq \emptyset$. The minimal length element $x' \in W(L_{\widehat{j_1}})x_{j_1,j_2} \cap W(L_{\widehat{j_1'}})x_{j_1',j_2'}$ satisfies $x' = x'_{j_1',j_2'}x_{j_1',j_2'} \geq x_{j_1',j_2'}$. Thus $j_2' - 1 \in [1, j_2' - 1] = \mathrm{Supp}(x_{j_1',j_2'}) \subseteq \mathrm{Supp}(x')$, which together with $x' = x'_{j_1,j_2}x_{j_1,j_2}$ and $j_2' - 1 \notin [1, j_2 - 1] = \mathrm{Supp}(x_{j_1,j_2})$ forces $j_2' - 1 \in \mathrm{Supp}(x'_{j_1,j_2})$. As $j_1' < j_1$ and $x'_{j_1,j_2} \in W(L_{\widehat{j_1}})$ is minimal in $W(L_{\widehat{j_1} \cap \widehat{j_1'}})x'_{j_1,j_2}$, we deduce $\mathrm{Supp}(x'_{j_1,j_2}) \subseteq [1, j_1 - 1]$ (use that $L_{\widehat{j_1}}$ and $L_{\widehat{j_1} \cap \widehat{j_1'}}$ have the lower right block GL_{n-j_1} in common when $j_1' < j_1$). But $[1, j_1 - 1]$ doesn't contain $j_2' - 1$ as $j_1 < j_2'$, a contradiction. Thus $\Sigma_{j_1,j_2} \cap \Sigma_{j_1',j_2'} = \emptyset$.
- If $j_1' > j_1$ and $j_2' - j_1' = j_2 - j_1$, then one checks that $x_{j_1',j_2'} \in W(L_{\widehat{j_1}})x_{j_1,j_2}$ and thus $\Sigma_{j_1,j_2} \cap \Sigma_{j_1',j_2'} \neq \emptyset$.
- If $j_1' > j_1$ and $j_2' - j_1' > j_2 - j_1$, assume $\Sigma_{j_1,j_2} \cap \Sigma_{j_1',j_2'} \neq \emptyset$. The minimal length element $x' \in W(L_{\widehat{j_1}})x_{j_1,j_2} \cap W(L_{\widehat{j_1'}})x_{j_1',j_2'}$ satisfies $x' = x'_{j_1',j_2'}x_{j_1',j_2'} \geq x_{j_1',j_2'}$. On the other hand, we have $x' \leq x''_{j_1,j_2}x_{j_1,j_2}$ where x''_{j_1,j_2} is the element of maximal length in $W^{\widehat{j_1} \cap \widehat{j_1'}, \emptyset}(L_{\widehat{j_1}})$, thus $x_{j_1',j_2'} \leq x''_{j_1,j_2}x_{j_1,j_2}$. But it follows from (i) of Lemma A.16 (together with $j_1' > j_1$ and $j_2' - j_1' > j_2 - j_1$) that $x_{j_1',j_2'} \nleq x''_{j_1,j_2}x_{j_1,j_2}$, a contradiction. Thus $\Sigma_{j_1,j_2} \cap \Sigma_{j_1',j_2'} = \emptyset$.

We have shown that $\Sigma_{j_1,j_2} \cap \Sigma_{j_1',j_2'} \neq \emptyset$ for $(j_1, j_2) < (j_1', j_2')$ if and only if either $j_2 = j_2'$ or $j_2' - j_1' = j_2 - j_1$, which is (i).

We prove (ii) and assume from now on $j_1 = j_1'$. Note first that $\Sigma_{j_1,j_2} \cap (w^{-1} \cdot \Sigma_{j_1',j_2'}) \neq \emptyset$ is equivalent to $\Sigma_{j_1,j_2} \subseteq W(L_{\widehat{j_1}})w^{-1}\Sigma_{j_1',j_2'}$, and hence can hold for at most one $w \in W^{\widehat{j_1},\widehat{j_1}}$ (by the double coset decomposition, see for instance the end of the proof of Lemma 2.1.18). From the definitions we have $w_{[1,j_2-1]} = w'x_{j_1,j_2}$ where w' is the longest element in $W(L_{[1,j_2-1]\setminus\{j_1\}})$, in particular $w_{[1,j_2-1]} \in W(L_{\widehat{j_1}})x_{j_1,j_2}$. Likewise we have $w_{[1,j_2'-1]} \in W(L_{\widehat{j_1}})x_{j_1,j_2'}$. Let $w \in W^{\widehat{j_1},\widehat{j_1}}$ such that $w_{[1,j_2'-1]}w_{[1,j_2-1]}^{-1} \in W(L_{\widehat{j_1}})wW(L_{\widehat{j_1}})$, i.e. $w_{[1,j_2'-1]}w_{[1,j_2-1]}^{-1} = y'wy$ for some $y, y' \in W(L_{\widehat{j_1}})$. Then $yw_{[1,j_2-1]} = w^{-1}(y')^{-1}w_{[1,j_2'-1]}$ with

$$
\begin{aligned}
yw_{[1,j_2-1]} &\in W(L_{\widehat{j_1}})w_{[1,j_2-1]} = W(L_{\widehat{j_1}})x_{j_1,j_2} \\
w^{-1}(y')^{-1}w_{[1,j_2'-1]} &\in w^{-1}W(L_{\widehat{j_1}})w_{[1,j_2'-1]} = w^{-1}W(L_{\widehat{j_1}})x_{j_1,j_2'},
\end{aligned}
$$

and thus $\Sigma_{j_1,j_2} \cap (w^{-1} \cdot \Sigma_{j_1,j_2'}) \neq \emptyset$. Finally, it follows from (ii) of Lemma A.16 that $w_{[1,j_2'-1]}w_{[1,j_2-1]}^{-1} \in W(L_{\widehat{j_1}})s_{j_1}W(L_{\widehat{j_1}})$ if and only if either $j_1 = 1$ or $j_2' = j_2 + 1$. □

Let $(j_1, j_2), (j_1', j_2') \in \mathbf{J}^\infty$ with $(j_1, j_2) < (j_1', j_2')$. From (2.7) and (2.36) we have canonical isomorphisms for $k \geq 0$

$$\mathrm{Ext}^k_G(i^\infty_{\widehat{j'_1},\Delta}(\pi^\infty_{j'_1,j'_2}), i^\infty_{\widehat{j_1},\Delta}(\pi^\infty_{j_1,j_2}))^\infty \cong \bigoplus_{w\in W^{\widehat{j'_1},\widehat{j_1}}} \mathrm{Ext}^k_{L_{\widehat{j_1}}}(i^\infty_{\widehat{j'_1},\widehat{j_1},w}(J_{\widehat{j'_1},\widehat{j_1},w}(\pi^\infty_{j'_1,j'_2})), \pi^\infty_{j_1,j_2}). \tag{2.81}$$

Using (2.37) we see that (2.81) is non-zero for some $k \geq 0$ if only if there exists $w \in W^{\widehat{j'_1},\widehat{j_1}}$ (necessarily unique) such that $\Sigma_{j_1,j_2} \subseteq W(L_{\widehat{j_1}})w^{-1}\Sigma_{j'_1,j'_2}$ (equivalently $\Sigma_{j_1,j_2} \cap (w^{-1}\cdot\Sigma_{j'_1,j'_2}) \neq \emptyset$) and such that 2.81 induces an isomorphism

$$0 \neq \mathrm{Ext}^k_G(i^\infty_{\widehat{j'_1},\Delta}(\pi^\infty_{j'_1,j'_2}), i^\infty_{\widehat{j_1},\Delta}(\pi^\infty_{j_1,j_2}))^\infty \cong \mathrm{Ext}^k_{L_{\widehat{j_1}}}(i^\infty_{\widehat{j'_1},\widehat{j_1},w}(J_{\widehat{j'_1},\widehat{j_1},w}(\pi^\infty_{j'_1,j'_2}))_{\mathcal{B}^{\widehat{j_1}}_{\Sigma_{j_1,j_2}}}, \pi^\infty_{j_1,j_2}). \tag{2.82}$$

Lemma 2.3.5 *Let $(j_1, j_2), (j'_1, j'_2) \in \mathbf{J}^\infty$ with $(j_1, j_2) < (j'_1, j'_2)$.*

(i) We have

$$\mathrm{Hom}_{L_{\widehat{j_1}}}\big(i^\infty_{\widehat{j_1}\cap\widehat{j'_1},\widehat{j_1}}(J_{\widehat{j'_1},\widehat{j_1}\cap\widehat{j'_1}}(\pi^\infty_{j'_1,j'_2})), \pi^\infty_{j_1,j_2}\big) \neq 0 \tag{2.83}$$

if and only if $(j'_1, j'_2) \in \{(j_1+1, j_2+1), (j_1-1, j_2)\}$.

(ii) When $j'_1 = j_1$, we have

$$\mathrm{Hom}_{L_{\widehat{j_1}}}\big(i^\infty_{\widehat{j_1},\widehat{j_1},s_{j_1}}(J_{\widehat{j_1},\widehat{j_1},s_{j_1}}(\pi^\infty_{j_1,j'_2})), \pi^\infty_{j_1,j_2}\big) \neq 0$$

if and only if $j'_2 = j_2 + 1$.

Proof We prove (i). By definition (see (2.19) and (2.21)) we have

$$i^\infty_{\widehat{j_1}\cap\widehat{j'_1},\widehat{j_1}}(J_{\widehat{j'_1},\widehat{j_1}\cap\widehat{j'_1}}(\pi^\infty_{j'_1,j'_2})) = i^\infty_{\widehat{j'_1},\widehat{j_1},1}(J_{\widehat{j'_1},\widehat{j_1},1}(\pi^\infty_{j'_1,j'_2})),$$

hence by (2.82) (and the line above it) we see that (2.83) holds if and only if $\Sigma_{j_1,j_2} \cap \Sigma_{j'_1,j'_2} \neq \emptyset$ and $d(\pi^\infty_{j'_1,j'_2}, \pi^\infty_{j_1,j_2}) = 0$. We thus deduce (i) from (i) of Lemma 2.3.4 and (i) of Lemma 2.3.2.

We prove (ii). Likewise, (2.83) holds if and only if $\Sigma_{j_1,j_2} \cap s_{j_1}\Sigma_{j_1,j'_2} \neq \emptyset$ and $d(\pi^\infty_{j_1,j'_2}, \pi^\infty_{j_1,j_2}) = 0$. Thus (ii) follows from the last statement in (ii) of Lemma 2.3.4 and (i) of Lemma 2.3.2. □

Let $(j_1, j_2) \in \mathbf{J}^\infty$ with $1 < j_1 < n-1$ and $j_2 < n$ and thus $(j_1+1, j_2+1), (j_1-1, j_2), (j_1, j_2+1) \in \mathbf{J}^\infty$. We write

$$I_+ \stackrel{\mathrm{def}}{=} \Delta\setminus\{j_1, j_1+1\},\quad I_- \stackrel{\mathrm{def}}{=} \Delta\setminus\{j_1, j_1-1\},\quad I_\pm \stackrel{\mathrm{def}}{=} \Delta\setminus\{j_1-1, j_1, j_1+1\} \tag{2.84}$$

and we set

$$\begin{cases} \Sigma_{+,0} \stackrel{\mathrm{def}}{=} \Sigma_{j_1+1,j_2+1} \cap \Sigma_{j_1,j_2+1} & \Sigma_{-,0} \stackrel{\mathrm{def}}{=} \Sigma_{j_1-1,j_2} \cap \Sigma_{j_1,j_2+1} \\ \Sigma_{+,1} \stackrel{\mathrm{def}}{=} \Sigma_{j_1+1,j_2+1} \cap \Sigma_{j_1,j_2} & \Sigma_{-,1} \stackrel{\mathrm{def}}{=} \Sigma_{j_1-1,j_2} \cap \Sigma_{j_1,j_2} \\ \Sigma_\pm \stackrel{\mathrm{def}}{=} \Sigma_{j_1,j_1} \cap s_{j_1}\cdot\Sigma_{j_1,j_2+1}. \end{cases} \tag{2.85}$$

It follows from Lemma 2.3.4 that $\Sigma_{*,0}$ and $\Sigma_{*,1}$ are single left $W(L_{I_*})$-cosets for $* \in \{+,-\}$, and $\Sigma_\pm$ is a single left $W(L_{I_\pm})$-coset. Note that $\pi^\infty_{j_1,j_2}$, $\pi^\infty_{j_1+1,j_2+1}$, $\pi^\infty_{j_1-1,j_2}$ and $\pi^\infty_{j_1,j_2+1}$ are all irreducible G-regular. In all cases (i) to (v) below, (both parts of) Lemma 2.1.18 together with (2.7) (for $k=0$) and (i) of Lemma 2.3.2 imply that (2.65) holds, which gives us by the first statement in (i) of Lemma 2.2.11 isomorphisms of irreducible G-regular representations:

(i) $\pi^\infty_{+,0} \overset{\text{def}}{=} J_{\widehat{j_1},I_+}(\pi^\infty_{j_1,j_2+1})_{\mathcal{B}^{I_+}_{\Sigma_{+,0}}} \cong J'_{\Delta\setminus\{j_1+1\},I_+}(\pi^\infty_{j_1+1,j_2+1})_{\mathcal{B}^{I_+}_{\Sigma_{+,0}}}$;

(ii) $\pi^\infty_{-,0} \overset{\text{def}}{=} J_{\widehat{j_1},I_-}(\pi^\infty_{j_1,j_2+1})_{\mathcal{B}^{I_-}_{\Sigma_{-,0}}} \cong J'_{\Delta\setminus\{j_1-1\},I_-}(\pi^\infty_{j_1-1,j_2})_{\mathcal{B}^{I_-}_{\Sigma_{-,0}}}$;

(iii) $\pi^\infty_{+,1} \overset{\text{def}}{=} J_{\Delta\setminus\{j_1+1\},I_+}(\pi^\infty_{j_1+1,j_2+1})_{\mathcal{B}^{I_+}_{\Sigma_{+,1}}} \cong J'_{\widehat{j_1},I_+}(\pi^\infty_{j_1,j_2})_{\mathcal{B}^{I_+}_{\Sigma_{+,1}}}$;

(iv) $\pi^\infty_{-,1} \overset{\text{def}}{=} J_{\Delta\setminus\{j_1-1\},I_-}(\pi^\infty_{j_1-1,j_2})_{\mathcal{B}^{I_-}_{\Sigma_{-,1}}} \cong J'_{\widehat{j_1},I_-}(\pi^\infty_{j_1,j_2})_{\mathcal{B}^{I_-}_{\Sigma_{-,1}}}$;

(v) $\pi^\infty_{\pm} \overset{\text{def}}{=} J_{\widehat{j_1},\widehat{j_1},s_{j_1}}(\pi^\infty_{j_1,j_2+1})_{\mathcal{B}^{I_\pm}_{\Sigma_\pm}} \cong J'_{\widehat{j_1},I_\pm}(\pi^\infty_{j_1,j_2})_{\mathcal{B}^{I_\pm}_{\Sigma_\pm}}$.

Using the irreducibility of $\pi^\infty_{j_1,j_2}$, $\pi^\infty_{j_1,j_2+1}$, $\pi^\infty_{j_1+1,j_2+1}$ and $\pi^\infty_{j_1-1,j_2}$ as well as (2.7) and (2.8), we observe that $\pi^\infty_{j_1,j_2} \cong \mathrm{cosoc}_{L_{\widehat{j_1}}}(i^\infty_{I_*,\widehat{j_1}}(\pi^\infty_{*,1}))$ and $\pi^\infty_{j_1,j_2+1} \cong \mathrm{soc}_{L_{\widehat{j_1}}}(i^\infty_{I_*,\widehat{j_1}}(\pi^\infty_{*,0}))$ for $* \in \{+,-\}$. Similarly, we have $\mathrm{soc}_{L_{\widehat{j_1}}}(i^\infty_{I_+,\Delta\setminus\{j_1+1\}}(\pi^\infty_{+,1})) \cong \pi^\infty_{j_1+1,j_2+1} \cong \mathrm{cosoc}_{L_{\widehat{j_1}}}(i^\infty_{I_+,\Delta\setminus\{j_1+1\}}(\pi^\infty_{+,0}))$ and $\mathrm{soc}_{L_{\widehat{j_1}}}(i^\infty_{I_-,\Delta\setminus\{j_1-1\}}(\pi^\infty_{-,1})) \cong \pi^\infty_{j_1-1,j_2} \cong \mathrm{cosoc}_{L_{\widehat{j_1}}}(i^\infty_{I_-,\Delta\setminus\{j_1-1\}}(\pi^\infty_{-,0}))$. As $\pi^\infty_\pm$ is irreducible G-regular, for each $I \supseteq I_\pm$ the representation $i^\infty_{I_\pm,I}(\pi^\infty_\pm)$ is G-basic and thus multiplicity free with simple socle and cosocle by the last statement of (iv) of Remark 2.1.16.

The last lemma below will be used in Lemmas 5.2.16 and 5.2.17, themselves used in the important Proposition 5.2.18.

Lemma 2.3.6 *Let* $(j_1,j_2) \in \mathbf{J}^\infty$ *with* $1 < j_1 < n-1$ *and* $j_2 < n$.

(i) We have $s_{j_1} \cdot \Sigma_{j_1+1,j_2+1} = \Sigma_{j_1+1,j_2+1}$ *and* $s_{j_1} \cdot \Sigma_{j_1-1,j_2} = \Sigma_{j_1-1,j_2}$, *which induce the following equalities for* $* \in \{+,-\}$

$$\Sigma_\pm = \Sigma_{j_1,j_2} \cap s_{j_1} \cdot \Sigma_{*,0} = \Sigma_{*,1} \cap s_{j_1} \cdot \Sigma_{j_1,j_2+1}. \tag{2.86}$$

(ii) We have

$$\pi^\infty_\pm \cong J'_{I_*,I_\pm}(\pi^\infty_{*,1})_{\mathcal{B}^{I_\pm}_{\Sigma_\pm}} \cong J_{I_*,\widehat{j_1},s_{j_1}}(\pi^\infty_{*,0})_{\mathcal{B}^{I_\pm}_{\Sigma_\pm}}$$

and $\pi^\infty_{*,1} \cong \mathrm{cosoc}_{L_{I_*}}(i^\infty_{I_\pm,I_*}(\pi^\infty_\pm))$ *for* $* \in \{+,-\}$. *We also have*

$$\pi^\infty_{j_1+1,j_2+1} \in \mathrm{JH}_{L_{\Delta\setminus\{j_1+1\}}}(i^\infty_{I_\pm,\Delta\setminus\{j_1+1\}}(\pi^\infty_\pm)) \text{ and } \pi^\infty_{j_1-1,j_2} \in \mathrm{JH}_{L_{\Delta\setminus\{j_1-1\}}}(i^\infty_{I_\pm,\Delta\setminus\{j_1-1\}}(\pi^\infty_\pm)).$$

(iii) We have $J'_{\Delta\setminus\{j_1+1\},I_+}(\tau^\infty)_{\mathcal{B}^{I_+}_{\Sigma_{+,0}}} = 0$ *for each* $\tau^\infty \in \mathrm{JH}_{L_{\Delta\setminus\{j_1+1\}}}(i^\infty_{I_\pm,\Delta\setminus\{j_1+1\}}(\pi^\infty_\pm))$ *satisfying* $\tau^\infty < \pi^\infty_{j_1+1,j_2+1}$, *and similarly* $J'_{\Delta\setminus\{j_1-1\},I_-}(\tau^\infty)_{\mathcal{B}^{I_-}_{\Sigma_{-,0}}} = 0$ *for each* $\tau^\infty \in \mathrm{JH}_{L_{\Delta\setminus\{j_1-1\}}}(i^\infty_{I_\pm,\Delta\setminus\{j_1-1\}}(\pi^\infty_\pm))$ *satisfying* $\tau^\infty < \pi^\infty_{j_1-1,j_2}$.

(iv) We have $J_{\widehat{j_1},\widehat{j_1},s_{j_1}}(\tau^\infty)_{\mathcal{B}^{I_\pm}_{\Sigma_\pm}} = 0$ *for each* $\tau^\infty \in \mathrm{JH}_{L_{\widehat{j_1}}}(i^\infty_{I_*,\widehat{j_1}}(\pi^\infty_{*,0})/\pi^\infty_{j_1,j_2+1})$ *and* $* \in \{+,-\}$.

Proof We prove (i). It follows from $s_{j_1} \in W(L_{\Delta\setminus\{j_1+1\}}) \cap W(L_{\Delta\setminus\{j_1-1\}})$ that we have $s_{j_1} \cdot \Sigma_{j_1+1,j_2+1} = \Sigma_{j_1+1,j_2+1}$ and $s_{j_1} \cdot \Sigma_{j_1-1,j_2} = \Sigma_{j_1-1,j_2}$, which together with (2.85) give $s_{j_1} \cdot \Sigma_{+,0} = \Sigma_{j_1+1,j_2+1} \cap s_{j_1} \cdot \Sigma_{j_1,j_2+1}$ and $s_{j_1} \cdot \Sigma_{-,0} = \Sigma_{j_1-1,j_2} \cap s_{j_1} \cdot \Sigma_{j_1,j_2+1}$. Hence, in order to prove (2.86), it suffices to prove that

$$\Sigma_{\pm} \subseteq \Sigma_{j_1+1,j_2+1} \cap \Sigma_{j_1-1,j_2}. \tag{2.87}$$

Following the proof of $\Sigma_{j_1,j_2} \cap s_{j_1} \Sigma_{j_1,j_2+1} \neq \emptyset$ in (ii) of Lemma 2.3.4, we write $w_{[1,j_2]} w_{[1,j_2-1]}^{-1} = y' s_{j_1} y$ for some $y, y' \in W(L_{\widehat{j_1}})$. We have

$$y w_{[1,j_2-1]} = s_{j_1} (y')^{-1} w_{[1,j_2]} \in W(L_{\widehat{j_1}}) x_{j_1,j_2} \cap s_{j_1} W(L_{\widehat{j_1}}) x_{j_1,j_2+1}.$$

More precisely, as $w_{[1,j_2]} w_{[1,j_2-1]}^{-1} = s_1 \cdots s_{j_2}$, we have $y = s_{j_1+1} \cdots s_{j_2} \in W(L_{\Delta\setminus\{j_1-1\}})$ (resp. $s_{j_1}(y')^{-1} = s_{j_1} s_{j_1-1} \cdots s_1 \in W(L_{\Delta\setminus\{j_1+1\}})$), and since $w_{[1,j_2-1]} \in W(L_{\Delta\setminus\{j_1-1\}}) x_{j_1-1,j_2}$ (resp. $w_{[1,j_2]} \in W(L_{\Delta\setminus\{j_1+1\}}) x_{j_1+1,j_2+1}$) this implies $y w_{[1,j_2-1]} \in W(L_{\Delta\setminus\{j_1-1\}}) x_{j_1-1,j_2}$ (resp. $s_{j_1}(y')^{-1} w_{[1,j_2]} \in W(L_{\Delta\setminus\{j_1+1\}}) x_{j_1+1,j_2+1}$). In other words, we have shown

$$y w_{[1,j_2-1]} \cdot 1_T = s_{j_1} (y')^{-1} w_{[1,j_2]} \cdot 1_T \in \Sigma_{j_1,j_2} \cap s_{j_1} \Sigma_{j_1,j_2+1} \cap \Sigma_{j_1-1,j_2} \cap \Sigma_{j_1+1,j_2+1}.$$

In particular, $\Sigma_{\pm}$ (resp. Σ_{j_1+1,j_2+1}, resp. Σ_{j_1-1,j_2}) is the unique $W(L_{I_{\pm}})$-coset (resp. $W(L_{\Delta\setminus\{j_1+1\}})$-coset, resp. $W(L_{\Delta\setminus\{j_1-1\}})$-coset) containing $y w_{[1,j_2-1]} \cdot 1_T$, which forces (2.87) and thus (2.86). This finishes the proof of (i).

We prove (ii). On one hand, we have $J'_{\widehat{j_1},I_{\pm}}(-) \cong J'_{I_*,I_{\pm}}(J'_{\widehat{j_1},I_*}(-))$ which together with $\Sigma_{\pm} \subseteq \Sigma_{*,1}$ (and the definition of $\pi^{\infty}_{\pm}$ and $\pi^{\infty}_{*,1}$) gives the isomorphisms

$$\begin{aligned}\pi^{\infty}_{\pm} \cong J'_{\widehat{j_1},I_{\pm}}(\pi^{\infty}_{\widehat{j_1},j_2})_{\mathcal{B}^{I_{\pm}}_{\Sigma_{\pm}}} &\cong J'_{I_*,I_{\pm}}(J'_{\widehat{j_1},I_*}(\pi^{\infty}_{\widehat{j_1},j_2}))_{\mathcal{B}^{I_{\pm}}_{\Sigma_{\pm}}} \\ &\cong J'_{I_*,I_{\pm}}(J'_{\widehat{j_1},I_*}(\pi^{\infty}_{\widehat{j_1},j_2})_{\mathcal{B}^{I_*}_{\Sigma_{*,1}}})_{\mathcal{B}^{I_{\pm}}_{\Sigma_{\pm}}} \cong J'_{I_*,I_{+,-}}(\pi^{\infty}_{*,1})_{\mathcal{B}^{I_{\pm}}_{\Sigma_{\pm}}},\end{aligned}$$

which together with (2.8) (and the irreducibility of $\pi^{\infty}_{\pm}$) give $\pi^{\infty}_{*,1} \cong \mathrm{cosoc}_{L_{I_*}}(i^{\infty}_{I_{\pm},I_*}(\pi^{\infty}_{\pm}))$ for $* \in \{+,-\}$. On the other hand, we have the equalities

$$\widehat{j_1} \cap s_{j_1}(\widehat{j_1}) = I_{\pm} = I_* \cap s_{j_1}(\widehat{j_1}), \tag{2.88}$$

which together with (2.19), (2.20) and $\Sigma_{\pm} \subseteq s_{j_1} \cdot \Sigma_{*,0}$ (and the definitions of $\pi^{\infty}_{\pm}$, $\pi^{\infty}_{*,0}$) give isomorphisms

$$\begin{aligned}\pi^{\infty}_{\pm} \cong J_{\widehat{j_1},\widehat{j_1},s_{j_1}}(\pi^{\infty}_{\widehat{j_1},j_2+1})_{\mathcal{B}^{I_{\pm}}_{\Sigma_{\pm}}} &\cong J_{I_*,\widehat{j_1},s_{j_1}}(J_{\widehat{j_1},I_*}(\pi^{\infty}_{\widehat{j_1},j_2+1}))_{\mathcal{B}^{I_{\pm}}_{\Sigma_{\pm}}} \\ &\cong J_{I_*,\widehat{j_1},s_{j_1}}(J_{\widehat{j_1},I_*}(\pi^{\infty}_{\widehat{j_1},j_2+1})_{\mathcal{B}^{I_*}_{\Sigma_{*,0}}})_{\mathcal{B}^{I_{\pm}}_{\Sigma_{\pm}}} \cong J_{I_*,\widehat{j_1},s_{j_1}}(\pi^{\infty}_{*,0})_{\mathcal{B}^{I_{\pm}}_{\Sigma_{\pm}}}.\end{aligned}$$

Finally, recall from the discussion right before this lemma that

$$\pi^\infty_{j_1+1,j_2+1} \cong \mathrm{soc}_{L_{\widehat{j_1}}}(i^\infty_{I_+,\Delta\setminus\{j_1+1\}}(\pi^\infty_{+,1})) \text{ and } \pi^\infty_{j_1-1,j_2} \cong \mathrm{soc}_{L_{\widehat{j_1}}}(i^\infty_{I_-,\Delta\setminus\{j_1-1\}}(\pi^\infty_{-,1})),$$

which together with $\pi^\infty_{*,1} \cong \mathrm{cosoc}_{L_{I_*}}(i^\infty_{I_\pm,I_*}(\pi^\infty_\pm))$ for $* \in \{+,-\}$ implies that $\pi^\infty_{j_1+1,j_2+1} \in \mathrm{JH}_{L_{\Delta\setminus\{j_1+1\}}}(i^\infty_{I_\pm,\Delta\setminus\{j_1+1\}}(\pi^\infty_\pm))$ and $\pi^\infty_{j_1-1,j_2} \in \mathrm{JH}_{L_{\Delta\setminus\{j_1-1\}}}(i^\infty_{I_\pm,\Delta\setminus\{j_1-1\}}(\pi^\infty_\pm))$. This finishes the proof of (ii).

We prove the first half of (iii) and leave the second half, which is similar, to the reader. We write $\delta \overset{\text{def}}{=} \delta_{\widehat{j_1},\widehat{j_1},s_{j_1}} : L_{I_\pm} \to E^\times$ (see (2.20)) and note that $\delta = \delta_{I_*,\widehat{j_1},s_{j_1}}$ for $* \in \{+,-\}$ (using (2.88)). Recall from the discussion right before this lemma that $i^\infty_{I_\pm,\Delta\setminus\{j_1+1\}}(\pi^\infty_\pm)$ is G-basic and multiplicity free with simple socle and cosocle, and from (ii) that $\pi^\infty_{j_1+1,j_2+1} \in \mathrm{JH}_{L_{\Delta\setminus\{j_1+1\}}}(i^\infty_{I_\pm,\Delta\setminus\{j_1+1\}}(\pi^\infty_\pm))$. We write $\pi^\infty_{j_1+1,j_2+1,-}$ for the unique subrepresentation of $i^\infty_{I_\pm,\Delta\setminus\{j_1+1\}}(\pi^\infty_\pm)$ with cosocle $\pi^\infty_{j_1+1,j_2+1}$. Note that $\pi^\infty_{j_1+1,j_2+1,-}$ is G-basic by Corollary 2.1.26, and that we have an injection $q_1 : \pi^\infty_{j_1+1,j_2+1,-} \hookrightarrow i^\infty_{I_\pm,\Delta\setminus\{j_1+1\}}(\pi^\infty_\pm)$ and a surjection $q_2 : \pi^\infty_{j_1+1,j_2+1,-} \twoheadrightarrow \pi^\infty_{j_1+1,j_2+1}$. We write $J_+(-) \overset{\text{def}}{=} J'_{\Delta\setminus\{j_1+1\},I_+}(-)_{\mathcal{B}^{I_+}_{\Sigma_{+,0}}}$ and recall that $J_+(-)$ is an exact functor. By the exactness of $J_+(-)$, the injection q_1 induces an injection $q'_1 : J_+(\pi^\infty_{j_1+1,j_2+1,-}) \hookrightarrow J_+(i^\infty_{I_\pm,\Delta\setminus\{j_1+1\}}(\pi^\infty_\pm))$, and the surjection q_2 induces a surjection $q'_2 : J_+(\pi^\infty_{j_1+1,j_2+1,-}) \twoheadrightarrow J_+(\pi^\infty_{j_1+1,j_2+1})$. For any G-basic representation σ^∞ in $\mathrm{Rep}^\infty_{\mathrm{adm}}(L_{\Delta\setminus\{j_1+1\}})$, it follows from Lemma 2.1.29 and Remark 2.1.30 that $J_+(\sigma^\infty)$ is either zero or G-basic. As $J_+(\pi^\infty_{j_1+1,j_2+1}) = \pi^\infty_{+,0}$ by definition of $\pi^\infty_{+,0}$, we have $J_+(\pi^\infty_{j_1+1,j_2+1,-}) \neq 0$, hence $J_+(\pi^\infty_{j_1+1,j_2+1,-})$ and $J_+(i^\infty_{I_\pm,\Delta\setminus\{j_1+1\}}(\pi^\infty_\pm))$ are G-basic, and thus multiplicity free with simple socle and cosocle (last statement in (iv) of Remark 2.1.16). In particular, the surjection q'_2 implies that $J_+(\pi^\infty_{j_1+1,j_2+1,-})$ has cosocle $\pi^\infty_{+,0}$. It follows from (2.8) and then (2.7) that

$$\begin{aligned}
&\mathrm{Hom}_{L_{I_+}}(\pi^\infty_{+,0}, J_+(i^\infty_{I_\pm,\Delta\setminus\{j_1+1\}}(\pi^\infty_\pm)))\\
&\quad\cong \mathrm{Hom}_{L_{\Delta\setminus\{j_1+1\}}}(i^\infty_{I_+,\Delta\setminus\{j_1+1\}}(\pi^\infty_{+,0}), i^\infty_{I_\pm,\Delta\setminus\{j_1+1\}}(\pi^\infty_\pm))\\
&\qquad\cong \mathrm{Hom}_{L_{I_\pm}}(J_{\Delta\setminus\{j_1+1\},I_\pm}(i^\infty_{I_+,\Delta\setminus\{j_1+1\}}(\pi^\infty_{+,0}))_{\mathcal{B}^{I_\pm}_{\Sigma_\pm}}, \pi^\infty_\pm). \qquad (2.89)
\end{aligned}$$

Applying (ii) of Lemma 2.1.18 with $I_0 = I_+$ and $I_1 = I_\pm$, we see that

$$J_{\Delta\setminus\{j_1+1\},I_\pm}(i^\infty_{I_+,\Delta\setminus\{j_1+1\}}(\pi^\infty_{+,0}))_{\mathcal{B}^{I_\pm}_{\Sigma_\pm}} \cong J_{I_+,I_\pm,s_{j_1}}(\pi^\infty_{+,0})_{\mathcal{B}^{I_\pm}_{\Sigma_\pm}}.$$

Since $J_{I_+,I_\pm,s_{j_1}}(-) = J_{I_+,\widehat{j_1},s_{j_1}}(-)$ from (2.19) and $\pi^\infty_\pm \cong J_{I_+,\widehat{j_1},s_{j_1}}(\pi^\infty_{+,0})_{\mathcal{B}^{I_\pm}_{\Sigma_\pm}}$ from (ii), this implies by (2.89)

$$\mathrm{Hom}_{L_{I_+}}(\pi^\infty_{+,0}, J_+(i^\infty_{I_\pm,\Delta\setminus\{j_1+1\}}(\pi^\infty_\pm))) \neq 0. \qquad (2.90)$$

As $J_+(i^\infty_{I_\pm,\Delta\setminus\{j_1+1\}}(\pi^\infty_\pm))$ has simple socle by the previous discussion, by (2.90) it must have socle $\pi^\infty_{+,0}$, hence the same holds for its subrepresentation $J_+(\pi^\infty_{j_1+1,j_2+1,-})$. But recall that $J_+(\pi^\infty_{j_1+1,j_2+1,-})$ is multiplicity free with cosocle $\pi^\infty_{+,0}$ by the previous discussion, hence we have $J_+(\pi^\infty_{j_1+1,j_2+1,-}) \cong \pi^\infty_{+,0}$ and in particular $J_+(\ker(q_2)) = 0$ by the exactness of $J_+(-)$. As $\tau^\infty \in \mathrm{JH}_{L_{\Delta\setminus\{j_1+1\}}}(\ker(q_2))$ if and only if $\tau^\infty \in$

$\mathrm{JH}_{L_{\Delta\backslash\{j_1+1\}}}(i^{\infty}_{I_\pm,\Delta\backslash\{j_1+1\}}(\pi^{\infty}_{\pm}))$ and $\tau^{\infty} < \pi^{\infty}_{j_1+1,j_2+1}$ (from the definition of the partial order on $\mathrm{JH}_{L_{\Delta\backslash\{j_1+1\}}}(i^{\infty}_{I_\pm,\Delta\backslash\{j_1+1\}}(\pi^{\infty}_{\pm}))$ in Sect. 1.4), we obtain the first half of (iii) using $J_{+}(\ker(q_2)) = 0$.

We prove (iv). As $\pi^{\infty}_{\pm} \cong J_{I_*,\widehat{j_1},s_{j_1}}(\pi^{\infty}_{*,0})_{B^{I_\pm}_{\Sigma_\pm}}$ by (ii), we have $\pi^{\infty}_{*,0} \cong \mathrm{soc}_{L_{I_*}}(i^{\infty}_{I_\pm,s_{j_1}(I_*)}(\pi^{\infty}_{\pm} \otimes_E \delta^{-1})^{s_{j_1}})$ for $* \in \{+,-\}$ by the last statement in (i) of Lemma 2.2.11 applied with $I_0 = I_*$, $I_1 = \widehat{j_1}$ and $w = s_{j_1}$, and thus $i^{\infty}_{I_*,\widehat{j_1}}(\pi^{\infty}_{*,0})$ is a subrepresentation of

$$i^{\infty}_{I_*,\widehat{j_1}}(i^{\infty}_{I_\pm,s_{j_1}(I_*)}(\pi^{\infty}_{\pm} \otimes_E \delta^{-1})^{s_{j_1}}) \cong i^{\infty}_{I_\pm,\widehat{j_1}}((\pi^{\infty}_{\pm} \otimes_E \delta^{-1})^{s_{j_1}}).$$

By the first statement of (ii) of Lemma 2.2.11 (which uses the last statement of (i) of Lemma 2.2.11) applied with $I_0 = I_1 = \widehat{j_1}$, $w = s_{j_1}$, $\pi^{\infty} = \pi^{\infty}_{\pm}$ and $\sigma^{\infty}_0 = \tau^{\infty}$, we have $J_{\widehat{j_1},\widehat{j_1},s_{j_1}}(\tau^{\infty})_{B^{I_\pm}_{\Sigma_\pm}} = 0$ for $\tau^{\infty} \in \mathrm{JH}_{L_{\widehat{j_1}}}(i^{\infty}_{I_\pm,\widehat{j_1}}((\pi^{\infty}_{\pm} \otimes_E \delta^{-1})^{s_{j_1}})/\pi^{\infty}_{j_1,j_2+1})$, and in particular for $\tau^{\infty} \in \mathrm{JH}_{L_{\widehat{j_1}}}(i^{\infty}_{I_*,\widehat{j_1}}(\pi^{\infty}_{*,0})/\pi^{\infty}_{j_1,j_2+1})$ as $i^{\infty}_{I_*,\widehat{j_1}}(\pi^{\infty}_{*,0}) \subseteq i^{\infty}_{I_\pm,\widehat{j_1}}((\pi^{\infty}_{\pm} \otimes_E \delta^{-1})^{s_{j_1}})$. □

Chapter 3
Results on Lie Algebra Cohomology Groups

We prove all results on $U(\mathfrak{g})$-modules needed in Sect. 4 and especially in Sect. 5. In particular we prove many statements on unipotent cohomology groups and on various Ext groups of $U(\mathfrak{g})$-modules, and we construct important explicit finite length $U(\mathfrak{g})$-modules. We use the notation in Sect. 1.4 and fix throughout a weight $\mu_0 \in \Lambda^{\mathrm{dom}}$. For standard facts on the Bernstein-Gelfand-Gelfand category and on Kazhdan-Lusztig theory, we use [33] (the reader can find in *loc. cit.* the original references where the results we use are actually proven).

3.1 Categories of $U(\mathfrak{g})$-Modules

We introduce various abelian categories of $U(\mathfrak{g})$-modules and prove several basic results on unipotent cohomology groups and Ext groups of $U(\mathfrak{g})$-modules in these categories, and on the relations between the two.

For M in $\mathrm{Mod}_{U(\mathfrak{t})}$ and $\mu \in \Lambda = X(T)$, we define $M_\mu \subseteq M$ as the maximal $U(\mathfrak{t})$-submodule of M on which $t - \mu(t)$ acts *nilpotently* for each $t \in \mathfrak{t}$. So M_μ is the generalized weight space attached to the weight μ. Hence, we always have a $U(\mathfrak{t})$-equivariant embedding

$$\bigoplus_{\mu\in\Lambda} M_\mu \hookrightarrow M. \tag{3.1}$$

We define $\mathcal{C}_{\mathrm{alg}}$ as the full subcategory of $\mathrm{Mod}_{U(\mathfrak{g})}$ of those M such that the embedding (3.1) is an isomorphism, and $\mathcal{C}_{\mathrm{alg}}^{\mathrm{fin}} \subset \mathcal{C}_{\mathrm{alg}}$ as the full subcategory of M satisfying moreover $\dim_E M_\mu < \infty$ for each $\mu \in \Lambda$. In particular each object of $\mathcal{C}_{\mathrm{alg}}^{\mathrm{fin}}$ has countable dimension as E-vector space. We define $\widetilde{\mathcal{O}}_{\mathrm{alg}}^{\mathfrak{b}} \subset \mathcal{C}_{\mathrm{alg}}$ as the full subcategory of M which are *locally* $\mathfrak{b}$*-finite*, i.e. M is the union of its finite dimensional $U(\mathfrak{b})$-submodules. Note that the full subcategory of the category $\mathcal{O}$ of [33] of objects with

C. Breuil and Z. Qian, *Splitting and Making Explicit the de Rham Complex of the Drinfeld Space*, Lecture Notes in Mathematics 2397,
https://doi.org/10.1007/978-3-032-20095-2_3

integral (equivalently algebraic) weights is the full subcategory $\mathcal{O}^{\mathfrak{b}}_{\mathrm{alg}} \subset \widetilde{\mathcal{O}}^{\mathfrak{b}}_{\mathrm{alg}} \cap \mathcal{C}^{\mathrm{fin}}_{\mathrm{alg}}$ consisting of those M which are moreover $U(\mathfrak{t})$-semi-simple and finitely generated as $U(\mathfrak{g})$-modules [33, Sect. 1.1]. We also write $\mathcal{O}^{\mathfrak{b},\infty}_{\mathrm{alg}} \subseteq \widetilde{\mathcal{O}}^{\mathfrak{b}}_{\mathrm{alg}}$ for the full abelian subcategory consisting of finite length objects.

For each $\mu \in \Lambda$, we have a Verma module $M(\mu) \stackrel{\mathrm{def}}{=} U(\mathfrak{g}) \otimes_{U(\mathfrak{b})} \mu \in \mathcal{O}^{\mathfrak{b}}_{\mathrm{alg}}$, which has an irreducible cosocle denoted by $L(\mu)$ [33, Theorem 1.2(f)]. Moreover recall that each simple object of $\mathcal{O}^{\mathfrak{b}}_{\mathrm{alg}}$ has the form $L(\mu)$ for some $\mu \in \Lambda$ [33, Sect. 1.3] and that each object of $\mathcal{O}^{\mathfrak{b}}_{\mathrm{alg}}$ has finite length [33, Sect. 1.11], and thus $\mathcal{O}^{\mathfrak{b}}_{\mathrm{alg}}$ is the full subcategory of $\mathcal{O}^{\mathfrak{b},\infty}_{\mathrm{alg}}$ consisting of those M which are moreover $U(\mathfrak{t})$-semi-simple. As each M in $\widetilde{\mathcal{O}}^{\mathfrak{b}}_{\mathrm{alg}}$ contains at least one $\mathfrak{b}$-stable E-line with weight $\mu \in \Lambda$ (since M is locally $\mathfrak{b}$-finite), there exists a non-zero map $M(\mu) \to M$. It follows that $\widetilde{\mathcal{O}}^{\mathfrak{b}}_{\mathrm{alg}}$, $\mathcal{O}^{\mathfrak{b},\infty}_{\mathrm{alg}}$ and $\mathcal{O}^{\mathfrak{b}}_{\mathrm{alg}}$ all share the same simple objects, namely the $L(\mu)$ for $\mu \in \Lambda$. (It is thus clear that $\mathcal{O}^{\mathfrak{b},\infty}_{\mathrm{alg}} \subseteq \mathcal{C}^{\mathrm{fin}}_{\mathrm{alg}}$.) In the sequel, we write $N(\mu)$ for the kernel of the surjection $M(\mu) \twoheadrightarrow L(\mu)$.

We say that a full subcategory $\mathcal{C} \subseteq \mathrm{Mod}_{U(\mathfrak{g})}$ is *stable under extensions* if for each short exact sequence $0 \to M_1 \to M \to M_2 \to 0$ in $\mathrm{Mod}_{U(\mathfrak{g})}$ with M_1, M_2 in $\mathcal{C}$, we necessarily have M in $\mathcal{C}$. We observe that both $\mathcal{C}_{\mathrm{alg}}$ and $\mathcal{C}^{\mathrm{fin}}_{\mathrm{alg}}$ are stable under extensions. It is not difficult to check that $\widetilde{\mathcal{O}}^{\mathfrak{b}}_{\mathrm{alg}}$ and $\mathcal{O}^{\mathfrak{b},\infty}_{\mathrm{alg}}$ are also stable under extensions (by pull-back one can replace M_2 by a finite dimensional $U(\mathfrak{b})$-submodule $M_2' \subseteq M_2$, and then by induction on $\dim_E M_2'$ one can reduce to $M_2' = E$ by taking a $\mathfrak{b}$-stable E-line in M_2' and twisting, and then use that $\mathrm{Ext}^1_{U(\mathfrak{b})}(E, M_1) = H^1(\mathfrak{b}, M_1) = \varinjlim H^1(\mathfrak{b}, M_1')$ where the limit runs along the finite dimensional $U(\mathfrak{b})$-submodules M_1' of M_1). Recall however that the extension in $\mathrm{Mod}_{U(\mathfrak{g})}$ of two objects of $\mathcal{O}^{\mathfrak{b}}_{\mathrm{alg}}$ is not an object of $\mathcal{O}^{\mathfrak{b}}_{\mathrm{alg}}$ in general.

For each object M of $\mathcal{C}^{\mathrm{fin}}_{\mathrm{alg}}$, we set

$$M^{\tau} \stackrel{\mathrm{def}}{=} \bigoplus_{\mu \in \Lambda} M_{\mu}^{*} \tag{3.2}$$

with $M_{\mu}^{*} \stackrel{\mathrm{def}}{=} \mathrm{Hom}_E(M_{\mu}, E)$ and we make $\mathfrak{g}$ act on M^{τ} by (see [33, Sect. 3.2])

$$(x \cdot f)(v) \stackrel{\mathrm{def}}{=} f(\tau(x) \cdot v)$$

where $x \in \mathfrak{g}$ and $\tau : \mathfrak{g} \to \mathfrak{g}$ is Chevalley's anti-involution introduced at the end of [33, Sect. 0.5] (for GL_n is it induced by the transpose map). Then $M \mapsto M^{\tau}$ defines a (contravariant) endo-functor of $\mathcal{C}^{\mathrm{fin}}_{\mathrm{alg}}$ which is an exact involution, and thus a self-equivalence. As $\mathcal{C}^{\mathrm{fin}}_{\mathrm{alg}}$ is stable under extensions, the functor τ induces an isomorphism for M_1, M_2 in $\mathcal{C}^{\mathrm{fin}}_{\mathrm{alg}}$

$$\mathrm{Ext}^1_{U(\mathfrak{g})}(M_1, M_2) \cong \mathrm{Ext}^1_{U(\mathfrak{g})}(M_2^{\tau}, M_1^{\tau}). \tag{3.3}$$

Moreover, τ restricts to an exact involution of $\mathcal{O}^{\mathfrak{b}}_{\mathrm{alg}}$ which satisfies

$$L(\mu)^{\tau} \cong L(\mu) \tag{3.4}$$

for each $\mu \in \Lambda$ [33, Theorem 3.2].

For $I \subseteq \Delta$, we consider the full subcategory $\widetilde{\mathcal{O}}^{\mathfrak{p}_I}_{\mathrm{alg}} \subseteq \widetilde{\mathcal{O}}^{\mathfrak{b}}_{\mathrm{alg}}$ of those M which are locally $\mathfrak{p}_I$-finite, i.e. equal to the union of their finite dimensional $U(\mathfrak{p}_I)$-submodules. As $\mathfrak{p}_I = \mathfrak{l}_I \oplus \mathfrak{n}_I$ and the category of finite dimensional $U(\mathfrak{l}_I)$-modules is semi-simple, $M \in \mathcal{O}^{\mathfrak{b}}_{\mathrm{alg}}$ is locally $\mathfrak{p}_I$-finite if and only if the (underlying) $U(\mathfrak{l}_I)$-module M is a direct sum of (simple) finite dimensional $U(\mathfrak{l}_I)$-modules. We also define $\mathcal{O}^{\mathfrak{p}_I,\infty}_{\mathrm{alg}} \stackrel{\mathrm{def}}{=} \widetilde{\mathcal{O}}^{\mathfrak{p}_I}_{\mathrm{alg}} \cap \mathcal{O}^{\mathfrak{b},\infty}_{\mathrm{alg}}$ and $\mathcal{O}^{\mathfrak{p}_I}_{\mathrm{alg}} \stackrel{\mathrm{def}}{=} \widetilde{\mathcal{O}}^{\mathfrak{p}_I}_{\mathrm{alg}} \cap \mathcal{O}^{\mathfrak{b}}_{\mathrm{alg}}$. Replacing $\mathfrak{g}$ with $\mathfrak{l}_I$, we can define analogous full subcategories $\mathcal{C}_{\mathfrak{l}_I,\mathrm{alg}}$, $\mathcal{C}^{\mathrm{fin}}_{\mathfrak{l}_I,\mathrm{alg}}$, $\widetilde{\mathcal{O}}^{\mathfrak{b}_I}_{\mathfrak{l}_I,\mathrm{alg}}$, $\mathcal{O}^{\mathfrak{b}_I,\infty}_{\mathfrak{l}_I,\mathrm{alg}}$, $\mathcal{O}^{\mathfrak{b}_I}_{\mathfrak{l}_I,\mathrm{alg}}$ of $\mathrm{Mod}_{U(\mathfrak{l}_I)}$ with fully faithful embeddings $\mathcal{O}^{\mathfrak{b}_I}_{\mathfrak{l}_I,\mathrm{alg}} \hookrightarrow \mathcal{C}^{\mathrm{fin}}_{\mathfrak{l}_I,\mathrm{alg}} \hookrightarrow \mathcal{C}_{\mathfrak{l}_I,\mathrm{alg}}$ and $\mathcal{O}^{\mathfrak{b}_I}_{\mathfrak{l}_I,\mathrm{alg}} \hookrightarrow \mathcal{O}^{\mathfrak{b}_I,\infty}_{\mathfrak{l}_I,\mathrm{alg}} \hookrightarrow \widetilde{\mathcal{O}}^{\mathfrak{b}_I}_{\mathfrak{l}_I,\mathrm{alg}}$. We also define $\widetilde{\mathcal{O}}^{\mathfrak{l}_I\cap\mathfrak{p}_{I'}}_{\mathfrak{l}_I,\mathrm{alg}} \hookrightarrow \widetilde{\mathcal{O}}^{\mathfrak{b}_I}_{\mathfrak{l}_I,\mathrm{alg}}$, $\mathcal{O}^{\mathfrak{l}_I\cap\mathfrak{p}_{I'},\infty}_{\mathfrak{l}_I,\mathrm{alg}}$ and $\mathcal{O}^{\mathfrak{l}_I\cap\mathfrak{p}_{I'}}_{\mathfrak{l}_I,\mathrm{alg}}$ for each $I' \subseteq I$ (note that $\mathfrak{l}_I \cap \mathfrak{p}_{I'}$ is a lower parabolic in $\mathfrak{l}_I$). We write $L^I(\mu) \in \mathcal{O}^{\mathfrak{b}_I}_{\mathfrak{l}_I,\mathrm{alg}}$ for the unique simple quotient of $U(\mathfrak{l}_I) \otimes_{U(\mathfrak{b}_I)} \mu$, and set

$$M^I(\mu) \stackrel{\mathrm{def}}{=} U(\mathfrak{g}) \otimes_{U(\mathfrak{p}_I)} L^I(\mu). \tag{3.5}$$

Be careful that we allow $M^I(\mu)$ and $L^I(\mu)$ to be defined for any $\mu \in \Lambda$, in particular $L^I(\mu)$ can be infinite dimensional. As $M^I(\mu)$ is a quotient of

$$M(\mu) = U(\mathfrak{g}) \otimes_{U(\mathfrak{b})} \mu \cong U(\mathfrak{g}) \otimes_{U(\mathfrak{p}_I)} (U(\mathfrak{l}_I) \otimes_{U(\mathfrak{b}_I)} \mu),$$

we see that $M^I(\mu)$ is in $\mathcal{O}^{\mathfrak{b}}_{\mathrm{alg}}$ and has $L(\mu)$ as unique simple quotient. We write $N^I(\mu)$ for the kernel of the surjection $M^I(\mu) \twoheadrightarrow L(\mu)$. Moreover it follows from [33, Proposition 9.3(e)] and [33, Theorem 9.4] that $M^I(\mu)$ is in $\mathcal{O}^{\mathfrak{p}_I}_{\mathrm{alg}}$ if and only if $L(\mu)$ is in $\mathcal{O}^{\mathfrak{p}_I}_{\mathrm{alg}}$ if and only if $\mu \in \Lambda_I^{\mathrm{dom}}$ if and only if $L^I(\mu)$ is finite dimensional.

For $w \in W(G)$ and $I \subseteq \Delta$ we set

$$\begin{aligned} M(w) &\stackrel{\mathrm{def}}{=} M(w\cdot\mu_0), & L(w) &\stackrel{\mathrm{def}}{=} L(w\cdot\mu_0), & N(w) &\stackrel{\mathrm{def}}{=} N(w\cdot\mu_0) \\ M^I(w) &\stackrel{\mathrm{def}}{=} M^I(w\cdot\mu_0), & L^I(w) &\stackrel{\mathrm{def}}{=} L^I(w\cdot\mu_0), & N^I(w) &\stackrel{\mathrm{def}}{=} N^I(w\cdot\mu_0). \end{aligned}$$

Note that all Jordan-Hölder factors of $N(w)$ have the form $L(w')$ for some $w' > w$ [33, Sect. 5.1, 5.2] and [33, Sect. 8.3(a)]. For $w \in W(G)$, recall the sets $D_L(w)$, $D_R(w) \subseteq \Delta$ from (1.23) and (1.24).

Lemma 3.1.1 *Let $w \in W(G)$ be an element and $I \subseteq \Delta$ be a subset. Then $L(w)$ is in $\mathcal{O}^{\mathfrak{p}_I}_{\mathrm{alg}}$ if and only if $I \cap D_L(w) = \emptyset$.*

Proof We have $L(w) \in \mathcal{O}_{\mathrm{alg}}^{\mathfrak{p}_I}$ if and only if $w \cdot \mu_0 \in \Lambda_I^{\mathrm{dom}}$ if and only if $\langle w \cdot \mu_0, \alpha^\vee \rangle \leq 0$ for $\alpha \in I$ if and only if $\langle w(\mu_0 + \rho), \alpha^\vee \rangle \leq \langle \rho, \alpha^\vee \rangle = -1$ for $\alpha \in I$ if and only if $\langle \mu_0 + \rho, w^{-1}(\alpha)^\vee \rangle < 0$ for $\alpha \in I$ if and only if $w^{-1}(\alpha) \in \Phi^+$ for $\alpha \in I$ (as $\mu_0 \in \Lambda^{\mathrm{dom}}$ and hence $\mu_0 + \rho \in \Lambda^{\mathrm{dom}}$) if and only if $s_\alpha w > w$ for $\alpha \in I$ [33, Sect. 0.3(4)] if and only if $I \cap D_L(w) = \emptyset$. □

For M_1, M_2 in $\mathrm{Mod}_{U(\mathfrak{g})}$ we write $\mathrm{Ext}^k_{U(\mathfrak{g})}(M_1, M_2)$ for the extension groups computed in the category $\mathrm{Mod}_{U(\mathfrak{g})}$. When M_1, M_2 are in $\mathcal{O}^{\mathfrak{b}}_{\mathrm{alg}}$, we write $\mathrm{Ext}^k_{\mathcal{O}^{\mathfrak{b}}_{\mathrm{alg}}}(M_1, M_2)$ for the extension groups computed in $\mathcal{O}^{\mathfrak{b}}_{\mathrm{alg}}$ (which still has enough projective and injective objects by [33, Theorem 3.8]). Given two objects M_1, M_2 in $\mathcal{O}^{\mathfrak{b}}_{\mathrm{alg}}$, the fully faithful embedding $\mathcal{O}^{\mathfrak{b}}_{\mathrm{alg}} \hookrightarrow \mathrm{Mod}_{U(\mathfrak{g})}$ induces an injection

$$\mathrm{Ext}^1_{\mathcal{O}^{\mathfrak{b}}_{\mathrm{alg}}}(M_1, M_2) \hookrightarrow \mathrm{Ext}^1_{U(\mathfrak{g})}(M_1, M_2),$$

but the comparison between $\mathrm{Ext}^k_{\mathcal{O}^{\mathfrak{b}}_{\mathrm{alg}}}(M_1, M_2)$ and $\mathrm{Ext}^k_{U(\mathfrak{g})}(M_1, M_2)$ for $k \geq 2$ is more complicated in general. Since the dual functor $\tau : \mathcal{C}^{\mathrm{fin}}_{\mathrm{alg}} \to \mathcal{C}^{\mathrm{fin}}_{\mathrm{alg}}$ in (3.2) restricts to an exact involution of $\mathcal{O}^{\mathfrak{b}}_{\mathrm{alg}}$, we have a canonical isomorphism for M_1, M_2 in $\mathcal{O}^{\mathfrak{b}}_{\mathrm{alg}}$ and $k \geq 0$

$$\mathrm{Ext}^k_{\mathcal{O}^{\mathfrak{b}}_{\mathrm{alg}}}(M_1, M_2) \cong \mathrm{Ext}^k_{\mathcal{O}^{\mathfrak{b}}_{\mathrm{alg}}}(M_2^\tau, M_1^\tau). \tag{3.6}$$

For M in $\mathrm{Mod}_{U(\mathfrak{g})}$ and $I \subseteq \Delta$, we consider the Chevalley-Eilenberg complex (see for instance [73, Sect. 3])

$$M \to \mathrm{Hom}_E(\mathfrak{n}_I, M) \to \cdots \to \mathrm{Hom}_E(\wedge^k \mathfrak{n}_I, M) \to \cdots \tag{3.7}$$

with M in degree zero, and we define $H^k(\mathfrak{n}_I, M)$ as the cohomology group of this complex in degree $k \geq 0$. As the complex (3.7) is $U(\mathfrak{l}_I)$-equivariant, $H^k(\mathfrak{n}_I, M)$ is naturally a $U(\mathfrak{l}_I)$-module, and thus in particular a $U(\mathfrak{t})$-module. We will use the following lemma.

Lemma 3.1.2 *Let $\mu, \mu' \in \Lambda$, $I \subseteq \Delta$ and $k \geq 0$. Assume that*

$$H^k(\mathfrak{n}_I, L(\mu))_{\mu'} \neq 0.$$

Then there exists k distinct roots $\alpha_1, \ldots, \alpha_k \in \Phi^+ \setminus \Phi_I^+$ (no roots if $k = 0$) such that

$$\mu' - \mu - \sum_{\ell=1}^{k} \alpha_\ell \in \mathbb{Z}_{\geq 0}\Phi^+. \tag{3.8}$$

Proof The statement is obvious for $k = 0$, hence we can assume $k \geq 1$. Let $\mu'' \in \Lambda$ be an (integral) weight. We observe that $(\mathfrak{n}_I)_{\mu''} \neq 0$ if and only if $\mu'' = -\alpha$ for some $\alpha \in \Phi^+ \setminus \Phi_I^+$, and thus $(\wedge^k \mathfrak{n}_I)_{\mu''} \neq 0$ if and only if $\mu'' = -\sum_{\ell=1}^k \alpha_\ell$ for k distinct roots $\alpha_1, \ldots, \alpha_k$ in $\Phi^+ \setminus \Phi_I^+$. Consequently, $\mathrm{Hom}_E(\wedge^k \mathfrak{n}_I, L(\mu))_{\mu'} \neq 0$ if and only if there exists $\mu'' \in \Lambda$ such that $(\wedge^k \mathfrak{n}_I)_{\mu''} \neq 0$ and $L(\mu)_{\mu'+\mu''} \neq 0$. Note that $L(\mu)_{\mu'+\mu''} \neq 0$ implies $\mu' + \mu'' - \mu \in \mathbb{Z}_{\geq 0}\Phi^+$. Hence, $\mathrm{Hom}_E(\wedge^k \mathfrak{n}_I, L(\mu))_{\mu'} \neq 0$ implies the existence of k distinct roots $\alpha_1, \ldots, \alpha_k \in \Phi^+ \setminus \Phi_I^+$ such that (3.8) holds. As (3.7) is $U(\mathfrak{t})$-equivariant and $H^k(\mathfrak{n}_I, L(\mu))$ is a subquotient of $\mathrm{Hom}_E(\wedge^k \mathfrak{n}_I, L(\mu))$ as $U(\mathfrak{t})$-module, we obtain the statement. □

It is not difficult to check that, if M is in $\mathcal{C}_{\mathrm{alg}}$ (resp. $\mathcal{C}^{\mathrm{fin}}_{\mathrm{alg}}$, resp. $\widetilde{\mathcal{O}}^{\mathfrak{b}}_{\mathrm{alg}}$), then the $U(\mathfrak{l}_I)$-module $H^k(\mathfrak{n}_I, M)$ is also in $\mathcal{C}_{\mathfrak{l}_I,\mathrm{alg}}$ (resp. $\mathcal{C}^{\mathrm{fin}}_{\mathfrak{l}_I,\mathrm{alg}}$, resp. $\widetilde{\mathcal{O}}^{\mathfrak{b}_I}_{\mathfrak{l}_I,\mathrm{alg}}$, for the latter recall that $\mathfrak{n}_I \subseteq \mathfrak{b}$). If M is in $\mathcal{O}^{\mathfrak{b}}_{\mathrm{alg}}$, we will prove in Proposition 3.1.5 below that $H^k(\mathfrak{n}_I, M) \in \mathcal{O}^{\mathfrak{b}_I}_{\mathfrak{l}_I,\mathrm{alg}}$.

For $I, I' \subseteq \Delta$, we have in $\mathfrak{g}$

$$\mathfrak{u}_I \cap \mathfrak{n}_{I'} = \mathfrak{u}_I \cap \mathfrak{n}_{I\cap I'} = \mathfrak{l}_I \cap \mathfrak{n}_{I\cap I'} = \mathfrak{l}_I \cap \mathfrak{n}_{I'},$$

which is the nilpotent radical of $\mathfrak{l}_I \cap \mathfrak{p}_{I'} = \mathfrak{l}_I \cap \mathfrak{p}_{I\cap I'}$, and $H^k(\mathfrak{l}_I \cap \mathfrak{n}_{I'}, M_I)$ is naturally a $U(\mathfrak{l}_{I\cap I'})$-module for M_I in $\mathrm{Mod}_{U(\mathfrak{l}_I)}$. Since $\mathfrak{n}_I$ is an ideal in $\mathfrak{n}_{I\cap I'}$ with quotient naturally identified with $\mathfrak{l}_I \cap \mathfrak{n}_{I'}$, we have a $U(\mathfrak{l}_{I\cap I'})$-equivariant spectral sequence for M in $\mathrm{Mod}_{U(\mathfrak{g})}$ (see for instance [78, Sect. 7.5])

$$H^{\ell_1}(\mathfrak{l}_I \cap \mathfrak{n}_{I'}, H^{\ell_2}(\mathfrak{n}_I, M)) \Longrightarrow H^{\ell_1+\ell_2}(\mathfrak{n}_{I\cap I'}, M). \tag{3.9}$$

When $I' = \emptyset$ and $I = \{j\}$ for some $j \in \Delta$, $\mathfrak{l}_I \cap \mathfrak{n}_{I'} = \mathfrak{l}_{\{j\}} \cap \mathfrak{n} = \mathfrak{u}_{\{j\}}$ is a 1-dimensional Lie algebra, and (3.9) induces the short exact sequence for $k \geq 1$ (see e.g. [78, Exercise 5.2.1])

$$0 \to H^1(\mathfrak{u}_{\{j\}}, H^{k-1}(\mathfrak{n}_{\{j\}}, M)) \to H^k(\mathfrak{u}, M) \to H^0(\mathfrak{u}_{\{j\}}, H^k(\mathfrak{n}_{\{j\}}, M)) \to 0. \tag{3.10}$$

For $I \subseteq \Delta$, M_I an $U(\mathfrak{l}_I)$-module and $\xi : Z(\mathfrak{l}_I) \to E$ a character (i.e. an E-algebras homomorphism), we write $M_{I,\xi}$ for the maximal $U(\mathfrak{l}_I)$-submodule of M_I on which $z - \xi(z)$ acts nilpotently for each $z \in Z(\mathfrak{l}_I)$. For a simple object $L^I(\mu)$ of $\widetilde{\mathcal{O}}^{\mathfrak{b}_I}_{\mathfrak{l}_I,\mathrm{alg}}$, we denote by $\xi_\mu : Z(\mathfrak{l}_I) \to E$ the unique character such that $L^I(\mu)_{\xi_\mu} \neq 0$ [33, Sect. 1.7]. By Harish-Chandra's theorem [33, Theorem 1.10] $\xi_\mu = \xi_{\mu'}$ if and only if there exists $w \in W(L_I)$ such that $\mu' = w \cdot \mu$. We consider the endo-functor

$$\mathrm{pr}_\xi : \mathcal{C}^{\mathrm{fin}}_{\mathfrak{l}_I,\mathrm{alg}} \to \mathcal{C}^{\mathrm{fin}}_{\mathfrak{l}_I,\mathrm{alg}},\ M_I \mapsto M_{I,\xi}.$$

Lemma 3.1.3 *For $\xi : Z(\mathfrak{l}_I) \to E$ the endo-functor pr_ξ has the following properties.*

(i) The functor pr_ξ is exact and an idempotent, and we have a natural transformation

$$\mathrm{id} \cong \bigoplus_\xi \mathrm{pr}_\xi.$$

(ii) *Let* $M_{I,1}, M_{I,2}$ *in* $\mathcal{C}^{\mathrm{fin}}_{\mathfrak{l}_I,\mathrm{alg}}$ *such that* $\mathrm{Ext}^k_{U(\mathfrak{l}_I)}(M_{I,1}, M_{I,2}) \neq 0$ *for some* $k \geq 0$. *Then there exists* ξ *such that* $M_{I,1,\xi} \neq 0$ *and* $M_{I,2,\xi} \neq 0$. *The same statement holds for* $M_{I,1}, M_{I,2}$ *in* $\mathcal{O}^{\mathfrak{b}_I}_{\mathfrak{l}_I,\mathrm{alg}}$ *when* $\mathrm{Ext}^k_{\mathcal{O}^{\mathfrak{b}_I}_{\mathfrak{l}_I,\mathrm{alg}}}(M_{I,1}, M_{I,2}) \neq 0$.

(iii) *For each indecomposable object* M_I *in* $\mathcal{C}^{\mathrm{fin}}_{\mathfrak{l}_I,\mathrm{alg}}$, *there exists a unique* $\xi : Z(\mathfrak{l}_I) \to E$ *such that* $M_I = M_{I,\xi}$.

Proof It suffices to prove $M_I \cong \bigoplus_\xi M_{I,\xi}$ for M_I in $\mathcal{C}^{\mathrm{fin}}_{\mathfrak{l}_I,\mathrm{alg}}$ and the rest is abstract non-sense. For $\mu \in \Lambda$, the action of $Z(\mathfrak{l}_I)$ on M_I stabilizes $M_{I,\mu}$. As $M_{I,\mu}$ is finite dimensional by assumption, we deduce $M_{I,\mu} \cong \bigoplus_\xi M_{I,\xi,\mu}$. As $M_I \cong \bigoplus_{\mu\in\Lambda} M_{I,\mu}$ by assumption, we obtain $M_I \cong \bigoplus_\xi M_{I,\xi}$. □

Recall that there is an E-linear projection $\mathrm{pr} : U(\mathfrak{g}) \twoheadrightarrow U(\mathfrak{t})$ obtained by sending to 0 all monomials of $U(\mathfrak{g})$ (in a standard Poincaré-Birkhoff-Witt basis associated to the decomposition $\mathfrak{g} = \mathfrak{u}^+ \oplus \mathfrak{t} \oplus \mathfrak{u}$) containing factors which are not in $\mathfrak{t}$. Using the decompositions $\mathfrak{g} = \mathfrak{n}_I^+ \oplus \mathfrak{l}_I \oplus \mathfrak{n}_I$ and $\mathfrak{l}_I = \mathfrak{u}_I^+ \oplus \mathfrak{t} \oplus \mathfrak{u}_I$ for $I \subseteq \Delta$, we see that pr uniquely factors as $U(\mathfrak{g}) \twoheadrightarrow U(\mathfrak{l}_I) \twoheadrightarrow U(\mathfrak{t})$. It then follows from Harish-Chandra's theory (see [33, Sects. 1.7, 1.9, 1.10] and from the fact that $\rho - \rho_I$ is invariant under $W(L_I)$ that the above surjections restrict to injective morphisms of commutative E-algebras $Z(\mathfrak{g}) \hookrightarrow Z(\mathfrak{l}_I) \hookrightarrow Z(\mathfrak{t}) = U(\mathfrak{t})$. We denote by $\psi_I : Z(\mathfrak{g}) \hookrightarrow Z(\mathfrak{l}_I)$ the first injection.

We recall the following version of a classical theorem by Casselman-Osborne [15, Theorem 2.6].

Lemma 3.1.4 *Let* M *in* $\mathcal{C}^{\mathrm{fin}}_{\mathrm{alg}}$, $I \subseteq \Delta$ *and* $\xi : Z(\mathfrak{l}_I) \to E$. *Then we have for* $k \geq 0$

$$H^k(\mathfrak{n}_I, M)_\xi \subseteq H^k(\mathfrak{n}_I, M_{\xi\circ\psi_I}). \tag{3.11}$$

Proof Recall first that $H^k(\mathfrak{n}_I, M_{\xi\circ\psi_I})$ is an $U(\mathfrak{l}_I)$-submodule of $H^k(\mathfrak{n}_I, M)$ by (i) of Lemma 3.1.3 (applied to $I = \emptyset$ and M). Making explicit the action of $U(\mathfrak{p}_I)$ on the Chevalley-Eilenberg complex (3.7) (using $[\mathfrak{p}_I, \mathfrak{n}_I] \subseteq \mathfrak{n}_I$) and using that $\mathfrak{n}_I$ acts trivially on $H^k(\mathfrak{n}_I, M_{\xi\circ\psi_I})$, it is not difficult to check that the characters $\xi' : Z(\mathfrak{l}_I) \to E$ such that $H^k(\mathfrak{n}_I, M_{\xi\circ\psi_I})_{\xi'} \neq 0$ satisfy $\xi' \circ \psi_I = \xi \circ \psi_I$. As any character of $Z(\mathfrak{g})$ can be written $\xi'' \circ \psi_I$ for some character ξ'' of $Z(\mathfrak{l}_I)$, decomposing M using (i) of Lemma 3.1.3 and applying $H^k(\mathfrak{n}_I, -)_\xi$ implies the inclusion (3.11). □

The following important proposition is probably known, but we couldn't find it in the published literature (it is also proved in [9]).

Proposition 3.1.5 *Let* M *in* $\mathcal{O}^{\mathfrak{b}}_{\mathrm{alg}}$ *and* $I \subseteq \Delta$. *Then we have for* $k \geq 0$.

$$H^k(\mathfrak{n}_I, M) \in \mathcal{O}^{\mathfrak{b}_I}_{\mathfrak{l}_I,\mathrm{alg}}.$$

Proof As M is in $\mathcal{O}^{\mathfrak{b}}_{\mathrm{alg}}$, M is locally $\mathfrak{b}_I$-finite and $U(\mathfrak{t})$-semi-simple with finite dimensional weight spaces. Since $\wedge^k\mathfrak{n}_I$ is a finite dimensional $U(\mathfrak{l}_I)$-module which is semi-simple as $U(\mathfrak{t})$-module, $\mathrm{Hom}_E(\wedge^k \mathfrak{n}_I, M)$ is still locally $\mathfrak{b}_I$-finite and $\mathfrak{t}$-semi-simple with finite dimensional weight spaces. Hence so is $H^k(\mathfrak{n}_I, M)$ for $k \geq 0$ by $U(\mathfrak{l}_I)$-equivariance of the Chevalley-Eilenberg complex (3.7). It follows that any finitely generated $U(\mathfrak{l}_I)$-submodule of $H^k(\mathfrak{n}_I, M)$ is necessarily in $\mathcal{O}^{\mathfrak{b}_I}_{\mathfrak{l}_I,\mathrm{alg}}$, and thus has finite length by [33, Sect. 1.11].

It remains to show that $H^k(\mathfrak{n}_I, M)$ itself has finite length. Since M is in $\mathcal{O}^{\mathfrak{b}}_{\mathrm{alg}}$, M has finite length and hence by (i) of Lemma 3.1.3 there exists a *finite* set Ω of characters $\xi : Z(\mathfrak{g}) \to E$ such that $M \cong \bigoplus_{\xi \in \Omega} M_\xi$ wich $M_\xi \neq 0$. Let $\xi_I : Z(\mathfrak{l}_I) \to E$ be a character such that $H^k(\mathfrak{n}_I, M)_{\xi_I} \neq 0$. By Lemma 3.1.4 we know that $M_{\xi_I \circ \psi_I} \neq 0$ and thus $\xi_I \circ \psi_I \in \Omega$. By (i) of Lemma 3.1.3 applied to the object $H^k(\mathfrak{n}_I, M)$ of $\mathcal{C}^{\mathrm{fin}}_{\mathfrak{l}_I,\mathrm{alg}}$ we deduce an isomorphism of $U(\mathfrak{l}_I)$-modules where $\Omega_I \stackrel{\mathrm{def}}{=} \{\xi_I : Z(\mathfrak{l}_I) \to E \mid \xi_I \circ \psi_I \in \Omega\}$:

$$H^k(\mathfrak{n}_I, M) \cong \bigoplus_{\xi_I \in \Omega_I} H^k(\mathfrak{n}_I, M)_{\xi_I}.$$

Write $U(\mathfrak{t})$ as a polynomial algebra $E[t_1, \ldots, t_n]$, then there are constants $\rho_1, \ldots, \rho_n \in E$ such that, for any $\mu : U(\mathfrak{t}) \to E$, the polynomial $\prod_{i=1}^n (X - \rho_i - \mu(t_i)) \in E[X]$ only depends on $\mu|_{Z(\mathfrak{g})}$ by [33, Theorem 1.10(a)] (the ρ_i being related to the shift by ρ in *loc. cit.*). Since it has finitely many roots, we deduce that there is only a finite number of characters $\mu : U(\mathfrak{t}) \to E$ with a given $\mu|_{Z(\mathfrak{g})}$. Since any character of $Z(\mathfrak{l}_I)$ is the restriction of a character of $U(\mathfrak{t})$, it follows that the set $\{\xi_I \mid \xi_I \circ \psi_I = \xi\}$ is *a fortiori* finite for each $\xi \in \Omega$, hence Ω_I is again finite (as Ω is). Now, assume on the contrary that $H^k(\mathfrak{n}_I, M)$ has infinite length. Then there exists $\xi_I \in \Omega_I$ such that $H^k(\mathfrak{n}_I, M)_{\xi_I}$ has infinite length. Using [33, Theorem 1.10(b)], we deduce that there exists at least one $\mu \in \Lambda$ such that $L^I(\mu)$ appears infinitely many times as a subquotient of (finitely generated $U(\mathfrak{l}_I)$-submodules of) $H^k(\mathfrak{n}_I, M)_{\xi_I}$. But $H^k(\mathfrak{n}_I, M)_{\xi_I}$ is $\mathfrak{t}$-semi-simple, so the infinite multiplicity of $L^I(\mu)$ in $H^k(\mathfrak{n}_I, M)_{\xi_I}$ forces the μ-weight space of $H^k(\mathfrak{n}_I, M)_{\xi_I}$ to have infinite dimension, a contradiction. Consequently $H^k(\mathfrak{n}_I, M)$ has finite length and thus lies in $\mathcal{O}^{\mathfrak{b}_I}_{\mathfrak{l}_I,\mathrm{alg}}$. □

For M in $\mathrm{Mod}_{U(\mathfrak{g})}$ and M_I in $\mathrm{Mod}_{U(\mathfrak{l}_I)}$, recall the Hochschild-Serre spectral sequence [78, Sect. 7.5]

$$\mathrm{Ext}^{\ell_1}_{U(\mathfrak{l}_I)}(M_I, H^{\ell_2}(\mathfrak{n}_I, M)) \implies \mathrm{Ext}^{\ell_1+\ell_2}_{U(\mathfrak{p}_I)}(M_I, M) \cong \mathrm{Ext}^{\ell_1+\ell_2}_{U(\mathfrak{g})}(U(\mathfrak{g}) \otimes_{U(\mathfrak{p}_I)} M_I, M) \tag{3.12}$$

where the last isomorphism is Shapiro's lemma for Lie algebra cohomology. In particular, we have a canonical isomorphism

$$\mathrm{Hom}_{U(\mathfrak{l}_I)}(M_I, H^0(\mathfrak{n}_I, M)) \cong \mathrm{Hom}_{U(\mathfrak{g})}(U(\mathfrak{g}) \otimes_{U(\mathfrak{p}_I)} M_I, M), \tag{3.13}$$

and an exact sequence

$$\begin{aligned} 0 \to \mathrm{Ext}^1_{U(\mathfrak{l}_I)}(M_I, H^0(\mathfrak{n}_I, M)) &\to \mathrm{Ext}^1_{U(\mathfrak{g})}(U(\mathfrak{g}) \otimes_{U(\mathfrak{p}_I)} M_I, M) \\ &\to \mathrm{Hom}_{U(\mathfrak{l}_I)}(M_I, H^1(\mathfrak{n}_I, M)) \to \mathrm{Ext}^2_{U(\mathfrak{l}_I)}(M_I, H^0(\mathfrak{n}_I, M)). \end{aligned} \tag{3.14}$$

Lemma 3.1.6 *Let M_I in $\mathcal{O}^{\mathfrak{b}_I}_{\mathfrak{l}_I,\mathrm{alg}}$ and M in $\mathcal{O}^{\mathfrak{b}}_{\mathrm{alg}}$. Then we have a spectral sequence*

$$\mathrm{Ext}^{\ell_1}_{\mathcal{O}^{\mathfrak{b}_I}_{\mathfrak{l}_I,\mathrm{alg}}}(M_I, H^{\ell_2}(\mathfrak{n}_I, M)) \Longrightarrow \mathrm{Ext}^{\ell_1+\ell_2}_{\mathcal{O}^{\mathfrak{b}}_{\mathrm{alg}}}(U(\mathfrak{g}) \otimes_{U(\mathfrak{p}_I)} M_I, M). \tag{3.15}$$

Proof Note first that $H^{\ell_2}(\mathfrak{n}_I, M)$ is indeed in $\mathcal{O}^{\mathfrak{b}_I}_{\mathfrak{l}_I,\mathrm{alg}}$ by Lemma 3.1.5. For $I \subseteq \Delta$ we use in this proof the extension groups $\mathrm{Ext}^k_{\mathfrak{l}_I,\mathfrak{t}}(-,-)$ computed in the abelian category of $(\mathfrak{l}_I, \mathfrak{t})$-modules as defined in [7, Sect. I.2], i.e. the full subcategory of $\mathrm{Mod}_{U(\mathfrak{l}_I)}$ of $U(\mathfrak{l}_I)$-modules which are semi-simple as $U(\mathfrak{t})$-modules (i.e. such that (3.1) is an isomorphism with all generalized weight spaces M_μ being weight spaces). We have analogous groups replacing $(\mathfrak{l}_I, \mathfrak{t})$ by $(\mathfrak{p}_I, \mathfrak{t})$. We first need to recall a little background.

Let $H^k(\mathfrak{l}_I, \mathfrak{t}, -) \stackrel{\mathrm{def}}{=} \mathrm{Ext}^k_{\mathfrak{l}_I,\mathfrak{t}}(E, -)$ and $H^k(\mathfrak{p}_I, \mathfrak{t}, -) \stackrel{\mathrm{def}}{=} \mathrm{Ext}^k_{\mathfrak{p}_I,\mathfrak{t}}(E, -)$. For M, N in $\mathrm{Mod}_{U(\mathfrak{l}_I)}$, we endow $\mathrm{Hom}_E(N, M)$ with the unique structure of left $U(\mathfrak{l}_I)$-module such that $(\mathfrak{x} \cdot f)(n) \stackrel{\mathrm{def}}{=} \mathfrak{x}(f(n)) - f(\mathfrak{x}(n))$ for $\mathfrak{x} \in \mathfrak{l}_I$, $f \in \mathrm{Hom}_E(N, M)$ and $n \in N$. If M is in $\mathrm{Mod}_{U(\mathfrak{p}_I)}$, by the same formula $\mathrm{Hom}_E(N, M)$ is naturally a left $U(\mathfrak{p}_I)$-module (with $U(\mathfrak{p}_I)$ acting on N via $U(\mathfrak{p}_I) \twoheadrightarrow U(\mathfrak{l}_I)$). If M and N are moreover semi-simple as $U(\mathfrak{t})$-modules (in the above sense), then by standard homological arguments we have canonical isomorphisms $H^\ell(\mathfrak{l}_I, \mathfrak{t}, \mathrm{Hom}_E(N, M)) \cong \mathrm{Ext}^\ell_{\mathfrak{l}_I,\mathfrak{t}}(N, M)$ (or $H^\ell(\mathfrak{p}_I, \mathfrak{t}, \mathrm{Hom}_E(N, M)) \cong \mathrm{Ext}^\ell_{\mathfrak{p}_I,\mathfrak{t}}(N, M)$ if M is in $\mathrm{Mod}_{U(\mathfrak{p}_I)}$) for $\ell \geq 0$.

Let M_I in $\mathrm{Mod}_{U(\mathfrak{l}_I)}$, M in $\mathrm{Mod}_{U(\mathfrak{g})}$ and assume that M and M_I are semi-simple as $U(\mathfrak{t})$-modules. As $\mathfrak{n}_I$ is an ideal in $\mathfrak{p}_I$ (with $\mathfrak{l}_I \cong \mathfrak{p}_I/\mathfrak{n}_I$) and $\mathfrak{n}_I \cap \mathfrak{t} = 0$, it then follows from [7, Theorem I.6.5] and [7, Remark I.6.7] that there is a Hochschild-Serre type spectral sequence

$$H^{\ell_1}(\mathfrak{l}_I, \mathfrak{t}, H^{\ell_2}(\mathfrak{n}_I, \mathrm{Hom}_E(M_I, M))) \Longrightarrow H^{\ell_1+\ell_2}(\mathfrak{p}_I, \mathfrak{t}, \mathrm{Hom}_E(M_I, M)). \tag{3.16}$$

As M_I is semi-simple as $U(\mathfrak{t})$-module, so is $U(\mathfrak{g}) \otimes_{U(\mathfrak{p}_I)} M_I$, and we have canonical isomorphisms (the second being the usual Shapiro's lemma)

$$\begin{aligned} H^{\ell_1+\ell_2}(\mathfrak{p}_I, \mathfrak{t}, \mathrm{Hom}_E(M_I, M)) &\cong \mathrm{Ext}^{\ell_1+\ell_2}_{\mathfrak{p}_I,\mathfrak{t}}(M_I, M) \\ &\cong \mathrm{Ext}^{\ell_1+\ell_2}_{\mathfrak{g},\mathfrak{t}}(U(\mathfrak{g}) \otimes_{U(\mathfrak{p}_I)} M_I, M). \end{aligned} \tag{3.17}$$

As $\mathfrak{n}_I$ acts trivially on M_I, we have a $U(\mathfrak{l}_I)$-equivariant isomorphism

$$H^{\ell_2}(\mathfrak{n}_I, \mathrm{Hom}_E(M_I, M)) \cong \mathrm{Hom}_E(M_I, H^{\ell_2}(\mathfrak{n}_I, M)),$$

which together with (3.17) implies that (3.16) can be rewritten

$$\mathrm{Ext}^{\ell_1}_{\mathfrak{l}_I,\mathfrak{t}}(M_I, H^{\ell_2}(\mathfrak{n}_I, M)) \Longrightarrow \mathrm{Ext}^{\ell_1+\ell_2}_{\mathfrak{g},\mathfrak{t}}(U(\mathfrak{g}) \otimes_{U(\mathfrak{p}_I)} M_I, M). \tag{3.18}$$

Now take M_I in $\mathcal{O}^{\mathfrak{b}_I}_{\mathfrak{l}_I,\mathrm{alg}}$, M in $\mathcal{O}^{\mathfrak{b}}_{\mathrm{alg}}$. By Delorme's theorem ([33, Theorem 6.15]) we have in that case for $k \geq 0$

$$\mathrm{Ext}^k_{\mathcal{O}^{\mathfrak{b}}_{\mathrm{alg}}}(U(\mathfrak{g}) \otimes_{U(\mathfrak{p}_I)} M_I, M) \cong \mathrm{Ext}^k_{\mathfrak{g},\mathfrak{t}}(U(\mathfrak{g}) \otimes_{U(\mathfrak{p}_I)} M_I, M),$$

and for $\ell_1, \ell_2 \geq 0$

$$\mathrm{Ext}^{\ell_1}_{\mathcal{O}^{\mathfrak{b}_I}_{\mathfrak{l}_I,\mathrm{alg}}}(M_I, H^{\ell_2}(\mathfrak{n}_I, M)) \cong \mathrm{Ext}^{\ell_1}_{\mathfrak{l}_I,\mathfrak{t}}(M_I, H^{\ell_2}(\mathfrak{n}_I, M)).$$

Then (3.18) gives the spectral sequence (3.15). □

Lemma 3.1.7 *Let $w \in W(G)$, $I \subseteq \Delta$ and M in $\mathcal{O}^{\mathfrak{b}_I}_{\mathfrak{l}_I,\mathrm{alg}}$. Assume there is $k \geq 0$ such that*

$$\mathrm{Ext}^k_{U(\mathfrak{l}_I)}(L^I(w), M) \neq 0 \tag{3.19}$$

or

$$\mathrm{Ext}^k_{\mathcal{O}^{\mathfrak{b}_I}_{\mathfrak{l}_I,\mathrm{alg}}}(L^I(w), M) \neq 0 \tag{3.20}$$

or

$$\mathrm{Hom}_{U(\mathfrak{t})}(w \cdot \mu_0, H^k(\mathfrak{u}_I, M)) \neq 0. \tag{3.21}$$

Then there exists $x \in W(L_I)w$ such that

$$\mathrm{Hom}_{U(\mathfrak{l}_I)}(L^I(x), M) \neq 0.$$

Proof Let $\xi : Z(\mathfrak{l}_I) \to E$ be the unique homomorphism such that $L^I(w)_\xi \neq 0$.

We first prove that the three hypothesis all imply $M_\xi \neq 0$. If either (3.19) or (3.20) holds, this follows from (ii) of Lemma 3.1.3. Replacing $(\mathfrak{g}, \mathfrak{l}_I, \mathfrak{t})$ by $(\mathfrak{l}_I, \mathfrak{t}, \mathfrak{t})$, we have a spectral sequence analogous to (3.15). But since $\mathrm{Ext}^k_{\mathcal{O}^{\mathfrak{t}}_{\mathfrak{t},\mathrm{alg}}} = 0$ if $k > 0$, it gives isomorphisms for $k \geq 0$

$$\mathrm{Hom}_{U(\mathfrak{t})}(w \cdot \mu_0, H^k(\mathfrak{u}_I, M)) \cong \mathrm{Ext}^k_{\mathcal{O}^{\mathfrak{b}_I}_{\mathfrak{l}_I,\mathrm{alg}}}(U(\mathfrak{l}_I) \otimes_{U(\mathfrak{b}_I)} w \cdot \mu_0, M).$$

If (3.21) holds, we thus obtain $\mathrm{Ext}^k_{\mathcal{O}^{\mathfrak{b}_I}_{\mathfrak{l}_I,\mathrm{alg}}}(L^I(w'), M) \neq 0$ for some Jordan-Hölder factor $L^I(w')$ of $U(\mathfrak{l}_I) \otimes_{U(\mathfrak{b}_I)} w \cdot \mu_0$ (with $w' \in W(L_I)w$ and $L^I(w')_\xi \neq 0$ since all constituents of $U(\mathfrak{l}_I) \otimes_{U(\mathfrak{b}_I)} w \cdot \mu_0$ have the same infinitesimal character). As above with (3.20), this implies $M_\xi \neq 0$.

We now prove the statement. Take $\mu \in \Lambda$ such that $\mathrm{Hom}_{U(\mathfrak{l}_I)}(L^I(\mu), M_\xi) \neq 0$, thus in particular $L^I(\mu)_\xi \neq 0$. From Harish-Chandra's theorem ([33, Theorem 1.10]) we deduce $\mu = x \cdot \mu_0$ for some $x \in W(L_I)$. This finishes the proof. □

For $\mu, \mu' \in \Lambda$, (3.13) applied with $\mathfrak{l}_I = \mathfrak{t}$ gives an isomorphism

$$\mathrm{Hom}_{U(\mathfrak{t})}(\mu', H^0(\mathfrak{u}, L(\mu))) \cong \mathrm{Hom}_{U(\mathfrak{g})}(M(\mu'), L(\mu)).$$

As $H^0(\mathfrak{u}, L(\mu))$ is a semi-simple $U(\mathfrak{t})$-module and $\mathrm{Hom}_{U(\mathfrak{g})}(M(\mu'), L(\mu)) \neq 0$ if $\mu' \neq \mu$ and has dimension 1 if $\mu' = \mu$, we deduce a $U(\mathfrak{t})$-equivariant isomorphism $H^0(\mathfrak{u}, L(\mu)) \cong \mu$. Similarly, we have a $U(\mathfrak{t})$-equivariant isomorphism $H^0(\mathfrak{u}_I, L^I(\mu)) \cong \mu$ for each $I \subseteq \Delta$.

Lemma 3.1.8 *Let $I \subseteq \Delta$.*

(i) For $I' \subseteq \Delta$, M in $\mathcal{O}_{\mathrm{alg}}^{\mathfrak{p}_{I'}}$ and $k \geq 0$, the $U(\mathfrak{l}_I)$-module $H^k(\mathfrak{n}_I, M)$ is locally $\mathfrak{l}_I \cap \mathfrak{p}_{I'}$-finite.
(ii) For $\mu \in \Lambda$ we have $H^0(\mathfrak{n}_I, L(\mu)) \cong L^I(\mu)$.
(iii) For $\mu \in \Lambda$, the unique (by (ii)) $\xi : Z(\mathfrak{l}_I) \to E$ such that $H^0(\mathfrak{n}_I, L(\mu))_\xi \neq 0$ is such that $H^k(\mathfrak{n}_I, L(\mu))_\xi = 0$ for $k \geq 1$.

Proof As M is in $\mathcal{O}_{\mathrm{alg}}^{\mathfrak{p}_{I'}}$, M is locally $\mathfrak{p}_{I'}$-finite and *a fortiori* locally $\mathfrak{l}_I \cap \mathfrak{p}_{I'}$-finite. Hence so is $\mathrm{Hom}_E(\wedge^k \mathfrak{n}_I, M)$ for $k \geq 0$ (as $\wedge^k \mathfrak{n}_I$ is finite dimensional), and (i) follows by the $U(\mathfrak{l}_I)$-equivariance of (3.7).

We prove (ii). Note first that for $\mu' \in \Lambda$

$$\begin{aligned}\mathrm{Hom}_{U(\mathfrak{l}_I)}(U(\mathfrak{l}_I) \otimes_{U(\mathfrak{b}_I)} \mu', H^0(\mathfrak{n}_I, L(\mu))) &\cong \mathrm{Hom}_{U(\mathfrak{t})}(\mu', H^0(\mathfrak{u}_I, H^0(\mathfrak{n}_I, L(\mu))))\\ &\cong \mathrm{Hom}_{U(\mathfrak{t})}(\mu', H^0(\mathfrak{u}, L(\mu))) \cong \mathrm{Hom}_{U(\mathfrak{t})}(\mu', \mu)\end{aligned}$$

is non-zero if and only if $\mu' = \mu$. This implies that $H^0(\mathfrak{n}_I, L(\mu))$ has simple socle $L^I(\mu)$. It is thus enough to prove that $H^0(\mathfrak{n}_I, L(\mu))$ is a highest weight $U(\mathfrak{l}_I)$-module of weight μ. The following argument is due to Florian Herzig (note that we know that $H^0(\mathfrak{n}_I, L(\mu))$ is in $\mathcal{O}_{\mathfrak{l}_I,\mathrm{alg}}^{\mathfrak{b}_I}$ by Proposition 3.1.5, but we need the above more precise statement). Recall that $-\Phi^+$ (resp. $-\Phi_I^+$) are the roots of $\mathfrak{u}$ (resp. $\mathfrak{u}_I$) and $L(\mu) = \sum_{\lambda \in \mathbb{Z}_{\geq 0}\Phi^+} L(\mu)_{\mu+\lambda}$. Consider the following $U(\mathfrak{t})$-submodules of $L(\mu)$

$$L(\mu)' \stackrel{\mathrm{def}}{=} \sum_{\lambda \in \mathbb{Z}_{\geq 0}\Phi_I^+} L(\mu)_{\mu+\lambda} \text{ and } L(\mu)'' \stackrel{\mathrm{def}}{=} \sum_{\lambda \notin \mathbb{Z}_{\geq 0}\Phi_I^+} L(\mu)_{\mu+\lambda}.$$

We have $L(\mu) = L(\mu)' \oplus L(\mu)''$, $L(\mu)_\mu = L(\mu)'_\mu$ and $L(\mu)' = U(\mathfrak{u}_I^+) \cdot L(\mu)_\mu$. Since the action of $\mathfrak{l}_I$ modifies a weight by a character in $\mathbb{Z}\Phi_I^+$ and since $\mathbb{Z}_{\geq 0}\Phi^+ \cap \mathbb{Z}\Phi_I^+ = \mathbb{Z}_{\geq 0}\Phi_I^+$, we see that $L(\mu')$ and $L(\mu'')$ are $U(\mathfrak{l}_I)$-submodules of $L(\mu)$. Since the action of $\mathfrak{n}_I$ modifies a weight by a character in $-\mathbb{Z}_{\geq 0}(\Phi^+ \setminus \Phi_I^+)$, we see that $\mathfrak{n}_I$ necessarily acts by 0 on $L(\mu)_{\mu+\lambda}$ for $\lambda \in \mathbb{Z}_{\geq 0}\Phi_I^+$, i.e. $L(\mu)' \subseteq H^0(\mathfrak{n}_I, L(\mu))$. Assume that $L(\mu)' \subsetneq H^0(\mathfrak{n}_I, L(\mu))$, or equivalently $L(\mu)'' \cap H^0(\mathfrak{n}_I, L(\mu)) \neq 0$. The action of $\mathfrak{b}$ on $L(\mu)'' \cap H^0(\mathfrak{n}_I, L(\mu))$ factors through $\mathfrak{b}_I$ and since $L(\mu)$ is locally $\mathfrak{b}$-finite, then $L(\mu)'' \cap H^0(\mathfrak{n}_I, L(\mu))$ is locally $\mathfrak{b}_I$-finite. In particular it contains a non-zero maximal vector for the action of $\mathfrak{u}_I$, or equivalently $\mathfrak{u}$, i.e. $H^0(\mathfrak{u}, L(\mu)'') \neq 0$. However $H^0(\mathfrak{u}, L(\mu)'') \subseteq H^0(\mathfrak{u}, L(\mu)) = L(\mu)_\mu = L(\mu)'_\mu$ which contradicts $L(\mu)' \cap L(\mu)'' = 0$. It follows that $L(\mu)' = H^0(\mathfrak{n}_I, L(\mu))$, hence $H^0(\mathfrak{n}_I, L(\mu)) = U(\mathfrak{u}_I^+) \cdot L(\mu)_\mu$ is a highest weight $U(\mathfrak{l}_I)$-module of weight μ, which finishes the proof of (ii).

We prove (iii). Recall first that $H^k(\mathfrak{n}_I, L(\mu))$ is in $\mathcal{O}_{\mathfrak{l}_I,\mathrm{alg}}^{\mathfrak{b}_I}$ by Proposition 3.1.5. Let $\xi : Z(\mathfrak{l}_I) \to E$ be the unique homomorphism such that $L^I(\mu)_\xi \neq 0$ and assume that there exists $k \geq 1$ and a Jordan-Hölder factor $L^I(\mu')$ of $H^k(\mathfrak{n}_I, L(\mu))$ such that $L^I(\mu')_\xi \neq 0$. Then we have $\mu' = w \cdot \mu$ for some $w \in W(L_I)$ by Harish-Chandra's theorem. As the weight space $H^k(\mathfrak{n}_I, L(\mu))_{\mu'}$ is non-zero, by Lemma 3.1.2 there exists distinct $\alpha_1, \ldots, \alpha_k \in \Phi^+ \setminus \Phi_I^+$ such that $\mu' - \mu - \sum_{\ell=1}^k \alpha_\ell \in \mathbb{Z}_{\geq 0}\Phi^+$. However, $\mu' = w \cdot \mu$ (with $w \in W(L_I)$) implies $\mu' - \mu \in \mathbb{Z}\Phi_I^+$, which is a contradiction as

$$\left(\left(\sum_{\ell=1}^k \alpha_\ell\right) + \mathbb{Z}_{\geq 0}\Phi^+\right) \cap \mathbb{Z}\Phi_I^+ = \emptyset.$$

It follows that $H^k(\mathfrak{n}_I, L(\mu))_\xi = 0$ for $k \geq 1$. □

Lemma 3.1.9 *Let $\mu \in \Lambda$, $I \subseteq \Delta$ and M_I in $\mathcal{C}^{\mathrm{fin}}_{\mathfrak{l}_I,\mathrm{alg}}$ such that $M_I = M_{I,\xi}$ for some $\xi : Z(\mathfrak{l}_I) \to E$.*

(i) The map

$$d_2^{k,1} : \mathrm{Ext}^k_{U(\mathfrak{l}_I)}(M_I, H^1(\mathfrak{n}_I, L(\mu))) \to \mathrm{Ext}^{k+2}_{U(\mathfrak{l}_I)}(M_I, H^0(\mathfrak{n}_I, L(\mu)))$$

in (3.12) is zero for $k \geq 0$. In fact either the source or the target of $d_2^{k,1}$ is 0.

(ii) If $H^0(\mathfrak{n}_I, L(\mu)))_\xi \neq 0$ then $\mathrm{Ext}^k_{U(\mathfrak{l}_I)}(M_I, H^\ell(\mathfrak{n}_I, L(\mu))) = 0$ for $k \geq 0$, $\ell \geq 1$ and we have isomorphisms for $k \geq 0$

$$\mathrm{Ext}^k_{U(\mathfrak{l}_I)}(M_I, H^0(\mathfrak{n}_I, L(\mu))) \cong \mathrm{Ext}^k_{U(\mathfrak{g})}(U(\mathfrak{g}) \otimes_{U(\mathfrak{p}_I)} M_I, L(\mu)).$$

(iii) If $H^0(\mathfrak{n}_I, L(\mu))_\xi = 0$ then $\mathrm{Ext}^k_{U(\mathfrak{l}_I)}(M_I, H^0(\mathfrak{n}_I, L(\mu))) = 0$ for $k \geq 0$ and we have an isomorphism

$$\mathrm{Hom}_{U(\mathfrak{l}_I)}(M_I, H^1(\mathfrak{n}_I, L(\mu))) \cong \mathrm{Ext}^1_{U(\mathfrak{g})}(U(\mathfrak{g}) \otimes_{U(\mathfrak{p}_I)} M_I, L(\mu))$$

and an injection

$$\mathrm{Ext}^1_{U(\mathfrak{l}_I)}(M_I, H^1(\mathfrak{n}_I, L(\mu))) \hookrightarrow \mathrm{Ext}^2_{U(\mathfrak{g})}(U(\mathfrak{g}) \otimes_{U(\mathfrak{p}_I)} M_I, L(\mu)). \quad (3.22)$$

(iv) Analogous statements as (i), (ii), (iii) hold when M_I is in $\mathcal{O}^{\mathfrak{b}_I}_{\mathfrak{l}_I,\mathrm{alg}}$ and replacing $\mathrm{Ext}^k_{U(\mathfrak{l}_I)}$ and $\mathrm{Ext}^k_{U(\mathfrak{g})}$ by respectively $\mathrm{Ext}^k_{\mathcal{O}^{\mathfrak{b}_I}_{\mathfrak{l}_I,\mathrm{alg}}}$ and $\mathrm{Ext}^k_{\mathcal{O}^{\mathfrak{b}}_{\mathrm{alg}}}$.

Proof Combining (ii) and (iii) of Lemma 3.1.8 with (ii) of Lemma 3.1.3 and using (3.12), (3.14) and (3.15), we obtain (i), (ii) and the first two statements in (iii) in both cases $M_I \in \mathcal{C}^{\mathrm{fin}}_{\mathfrak{l}_I,\mathrm{alg}}$ and $M_I \in \mathcal{O}^{\mathfrak{b}_I}_{\mathfrak{l}_I,\mathrm{alg}}$. For (3.22), note that the bottom line of the E_2-terms in the spectral sequence (3.12) is identically 0. This implies that $E_2^{1,1} = E_\infty^{1,1} = \mathrm{Ext}^1_{U(\mathfrak{l}_I)}(M_I, H^1(\mathfrak{n}_I, L(\mu)))$ is the first non-zero graded piece of the abutment filtration on $E_\infty^2 = \mathrm{Ext}^2_{U(\mathfrak{g})}(U(\mathfrak{g}) \otimes_{U(\mathfrak{p}_I)} M_I, L(\mu))$, whence (3.22). When M_I is in $\mathcal{O}^{\mathfrak{b}_I}_{\mathfrak{l}_I,\mathrm{alg}}$ the proof is analogous using (3.15) instead. □

3.2 Results on Ext^1 Groups in the Category $\mathcal{O}^{\mathfrak{b}}_{\mathrm{alg}}$

We prove results on Ext groups (mainly Ext^1 groups) in the category $\mathcal{O}^{\mathfrak{b}}_{\mathrm{alg}}$, in the category of all $U(\mathfrak{g})$-modules, and on the comparison between the two.

We denote by $w_0 \in W(G)$ the element of maximal length. For $x, w \in W(G)$ such that $x \leq w$ we let $P_{x,w}(q) \in \mathbb{Z}[q]$ be the associated Kazhdan-Lusztig polynomial [37] and let $P_{x,w}(q) \stackrel{\mathrm{def}}{=} 0$ if $x \nleq w$. When $x \leq w$ recall that $\deg P_{x,w}(q) \leq \frac{1}{2}(\ell(w) - \ell(x) - 1)$ if $x \neq w$ and $P_{x,w} = 1$ if $\ell(w) \leq \ell(x) + 2$. Moreover the Kazhdan-Lusztig conjectures (independently proved in the 1980s by Beilinson-Bernstein and Brylinski-Kashiwara, and reproved by several people ever since) imply the following formula in the Grothendieck group of $\mathcal{O}^{\mathfrak{b}}_{\mathrm{alg}}$ (with obvious notation)

$$[L(w)] = \sum_{w \leq x} (-1)^{\ell(x)-\ell(w)} P_{xw_0, ww_0}(1)[M(x)].$$

Following [37, Definition 1.2], we write $x \prec w$ if $x < w$ and $\deg P_{x,w}(q) = \frac{1}{2}(\ell(w) - \ell(x) - 1)$ (so $x \prec w$ implies that $\ell(w) - \ell(x)$ is odd) and we define $\mu(x, w)$ as the leading coefficient of $P_{x,w}$ if $x \prec w$, and $\mu(x, w) \stackrel{\mathrm{def}}{=} 0$ otherwise. For instance if $x < w$ and $\ell(w) = \ell(x) + 1$, we have $x \prec w$ (and $\mu(x, w) = 1$).

Lemma 3.2.1 *Let $x, w \in W(G)$ and $\mu \in \Lambda$.*

(i) We have an isomorphism for $k \geq 0$

$$\mathrm{Ext}^k_{\mathcal{O}^{\mathfrak{b}}_{\mathrm{alg}}}(M(\mu), L(w)) \cong \mathrm{Hom}_{U(\mathfrak{t})}(\mu, H^k(\mathfrak{u}, L(w))). \tag{3.23}$$

Moreover, if (3.23) is non-zero then there exists $x' \geq w$ in $W(G)$ such that $\mu = x' \cdot \mu_0$.

(ii) The dimension of $\mathrm{Ext}^k_{\mathcal{O}^{\mathfrak{b}}_{\mathrm{alg}}}(M(x), L(w))$ (for $k \geq 0$) is equal to the coefficient of the monomial of degree $\frac{1}{2}(\ell(x) - \ell(w) - k)$ in P_{xw_0, ww_0}.

(iii) If $k \geq 0$ is the minimal integer such that (3.23) is non-zero and if x' is as in (i), then we have for $k' \leq k$

$$\mathrm{Ext}^{k'}_{U(\mathfrak{g})}(M(x'), L(w)) \cong \mathrm{Ext}^{k'}_{\mathcal{O}^{\mathfrak{b}}_{\mathrm{alg}}}(M(x'), L(w)).$$

Proof We prove (i). The isomorphism (3.23) is simply [33, Theorem 6.15(b)]. If (3.23) is non-zero, then by (ii) of Lemma 3.1.3 $Z(\mathfrak{g})$ acts on $M(\mu)$ and $L(w)$ by the same character, which together with Harish-Chandra's theorem gives some $x' \in W(G)$ such that $\mu = x' \cdot \mu_0$. Moreover the non-vanishing of (3.23) forces $x' \geq w$ by [33, Theorem 6.11(a)]. (ii) follows from (the second statement in) (i) and [33, Theorem 8.11(b), (c)]. We prove (iii). By assumption we have

$$H^{k'}(\mathfrak{u}, L(w))_{x'\cdot\mu_0} \cong \mathrm{Hom}_{U(\mathfrak{t})}(x' \cdot \mu_0, H^{k'}(\mathfrak{u}, L(w))) = 0 \text{ for } k' < k.$$

Hence, from (ii) of Lemma 3.1.3 applied with $I = \emptyset$ (recall from Proposition 3.1.5 that $H^{k'}(\mathfrak{u}, L(w))$ is in $\mathcal{O}^{\mathfrak{t}}_{\mathfrak{t},\mathrm{alg}} \subset \mathcal{C}^{\mathrm{fin}}_{\mathfrak{t},\mathrm{alg}}$) we obtain $\mathrm{Ext}^{k''}_{U(\mathfrak{t})}(x' \cdot \mu_0, H^{k'}(\mathfrak{u}, L(w))) = 0$ for $k'' \geq 0$ and $k' < k$. By (3.12) (applied with $I = \emptyset$) we deduce for $k' \leq k$

$$\mathrm{Ext}^{k'}_{U(\mathfrak{g})}(M(x'), L(w)) \cong \mathrm{Hom}_{U(\mathfrak{t})}(x' \cdot \mu_0, H^{k'}(\mathfrak{u}, L(w)))$$

(which is zero if $k' < k$). □

We collect the following standard results on $\mathrm{Ext}^1_{\mathcal{O}^{\mathfrak{b}}_{\mathrm{alg}}}$.

Lemma 3.2.2 *Let $x, w \in W(G)$.*

(i) If $\mathrm{Ext}^1_{\mathcal{O}^{\mathfrak{b}}_{\mathrm{alg}}}(L(x), L(w)) \neq 0$, then either $x < w$ or $w < x$.

(ii) If $x < w$, then $\mathrm{Ext}^1_{\mathcal{O}^{\mathfrak{b}}_{\mathrm{alg}}}(L(x), L(w)) \neq 0$ if and only if $x \prec w$, and is $\mu(x, w)$-dimensional in that case.

(iii) If $x < w$ and $\ell(w) = \ell(x) + 1$, then $\dim_E \mathrm{Ext}^1_{\mathcal{O}^{\mathfrak{b}}_{\mathrm{alg}}}(L(x), L(w)) = 1$.

Proof We start with (i). It follows from (3.6) and $L(x)^{\tau} \cong L(x)$, $L(w)^{\tau} \cong L(w)$ that

$$\mathrm{Ext}^1_{\mathcal{O}^{\mathfrak{b}}_{\mathrm{alg}}}(L(x), L(w)) \cong \mathrm{Ext}^1_{\mathcal{O}^{\mathfrak{b}}_{\mathrm{alg}}}(L(w), L(x)). \tag{3.24}$$

So upon exchanging x and w, it suffices to treat the case $x \not\geq w$. Note that $x \not\geq w$ is equivalent to $x \cdot \mu_0 - w \cdot \mu_0 \notin \mathbb{Z}_{\geq 0}\Phi^+$. Hence, by [33, Proposition 3.1(a)] (and our conventions)

$$\mathrm{Ext}^1_{\mathcal{O}^{\mathfrak{b}}_{\mathrm{alg}}}(M(x), L(w)) = 0. \tag{3.25}$$

As $L(x)$ is the cosocle of $M(x)$, we obviously have

$$\mathrm{Hom}_{\mathcal{O}^{\mathfrak{b}}_{\mathrm{alg}}}(L(x), L(w)) \cong \mathrm{Hom}_{\mathcal{O}^{\mathfrak{b}}_{\mathrm{alg}}}(M(x), L(w)). \tag{3.26}$$

The short exact sequence $0 \to N(x) \to M(x) \to L(x) \to 0$ together with (3.25) and (3.26) induce an isomorphism

$$\mathrm{Hom}_{\mathcal{O}^{\mathfrak{b}}_{\mathrm{alg}}}(N(x), L(w)) \cong \mathrm{Ext}^1_{\mathcal{O}^{\mathfrak{b}}_{\mathrm{alg}}}(L(x), L(w)) \neq 0 \tag{3.27}$$

which in particular implies $w > x$. We prove (ii). If $x < w$ we have by (3.24), [33, Theorem 8.15(c)] and [10, p. 9]

$$\dim_E \mathrm{Ext}^1_{\mathcal{O}^{\mathfrak{b}}_{\mathrm{alg}}}(L(x), L(w)) = \mu(ww_0, xw_0) = \mu(x, w), \tag{3.28}$$

which implies (ii). We prove (iii). If $x < w$ and $\ell(w) = \ell(x) + 1$, then $P_{x,w} = 1$ and $\deg P_{x,w} = \frac{1}{2}(\ell(w) - \ell(x) - 1) = 0$, so we have $x \prec w$ and $\mu(x, w) = 1$, and thus $\dim_E \mathrm{Ext}^1_{\mathcal{O}^{\mathfrak{b}}_{\mathrm{alg}}}(L(x), L(w)) = 1$ by (3.28). □

Lemma 3.2.3 *Let $x \in W(G)$ and M in $\mathcal{O}^{\mathfrak{b}}_{\mathrm{alg}}$ with all irreducible constituents isomorphic to $L(x')$ for some $x' \nleq x$. Then we have a canonical isomorphism*

$$\mathrm{Ext}^1_{\mathcal{O}^{\mathfrak{b}}_{\mathrm{alg}}}(L(x), M) \xrightarrow{\sim} \mathrm{Ext}^1_{U(\mathfrak{g})}(L(x), M). \tag{3.29}$$

Proof As (3.29) is clearly injective, it is enough to show that it is also surjective. Given a $U(\mathfrak{g})$-module M' representing a non-split extension of M by $L(x)$, it is enough to show that M' lies in $\mathcal{O}^{\mathfrak{b}}_{\mathrm{alg}}$ (note that it lies in $\mathcal{C}^{\mathrm{fin}}_{\mathrm{alg}}$). For $x' \nleq x$ and $\mu \in \mathbb{Z}_{\geq 0}\Phi^+$, we have $L(x')_{x\cdot\mu_0-\mu} = 0$. Indeed, $x' \nleq x$ if and only if $x\cdot\mu_0 - x'\cdot\mu_0 \notin \mathbb{Z}_{\geq 0}\Phi^+$ and $L(x')_{\mu'} = 0$ if and only if $\mu' - x'\cdot\mu_0 \notin \mathbb{Z}_{\geq 0}\Phi^+$. By the assumption and an obvious dévissage, we deduce $M_{x\cdot\mu_0-\mu} = 0$ and thus $M'_{x\cdot\mu_0-\mu} \xrightarrow{\sim} L(x)_{x\cdot\mu_0-\mu}$ for $\mu \in \mathbb{Z}_{\geq 0}\Phi^+$. In particular, $\dim_E M'_{x\cdot\mu_0} = 1$ and $M'_{x\cdot\mu_0-\alpha} = 0$ for $\alpha \in \Phi^+$, forcing $M'[\mathfrak{u}_\alpha]_{x\cdot\mu_0} \xrightarrow{\sim} M'_{x\cdot\mu_0}$ for $\alpha \in \Phi^+$ ($\mathfrak{u}_\alpha$ being the one dimensional Lie subalgebra of $\mathfrak{u}$ corresponding to $\alpha \in \Delta$) and thus $M'[\mathfrak{u}]_{x\cdot\mu_0} \xrightarrow{\sim} M'_{x\cdot\mu_0}$. We deduce isomorphisms of one dimensional vector spaces

$$\mathrm{Hom}_{U(\mathfrak{t})}(x\cdot\mu_0, M'[\mathfrak{u}]) \cong \mathrm{Hom}_{U(\mathfrak{b})}(x\cdot\mu_0, M') \xrightarrow{\sim} \mathrm{Hom}_{U(\mathfrak{b})}(x\cdot\mu_0, L(x)),$$

giving a non-zero map $M(x) \to M'$ whose image M'' is necessarily distinct from M. In particular M' is the amalgamate sum of M and M'' over $M \cap M'' \subseteq M$. But M'' is in $\mathcal{O}^{\mathfrak{b}}_{\mathrm{alg}}$ as it is a quotient of $M(x)$ and M is in $\mathcal{O}^{\mathfrak{b}}_{\mathrm{alg}}$ by assumption. It follows that the quotient M' of $M \oplus M''$ is in also in $\mathcal{O}^{\mathfrak{b}}_{\mathrm{alg}}$. □

We will extensively use the following consequences of Lemmas 3.2.2 and 3.2.3:

Lemma 3.2.4 *Let $x, w \in W(G)$.*

(i) We have $\mathrm{Ext}^1_{\mathcal{O}^{\mathfrak{b}}_{\mathrm{alg}}}(L(x), L(w)) \neq 0$ if and only if $x \prec w$ or $w \prec x$, in which case it has dimension $\mu(x, w)$ or $\mu(w, x)$ respectively and $|\ell(x) - \ell(w)|$ is odd.
(ii) Assume $x \neq w$, then $\mathrm{Ext}^1_{\mathcal{O}^{\mathfrak{b}}_{\mathrm{alg}}}(L(x), L(w)) \cong \mathrm{Ext}^1_{U(\mathfrak{g})}(L(x), L(w))$. In particular for $x \neq w$ we have $\mathrm{Ext}^1_{U(\mathfrak{g})}(L(x), L(w)) \neq 0$ if and only if $x \prec w$ or $w \prec x$, in which case it has dimension $\mu(x, w)$ or $\mu(w, x)$ respectively.

Proof (i) follows from parts (i) and (ii) of Lemma 3.2.2 together with (3.24). (ii) follows from Lemma 3.2.3 together with (3.3) and (3.6), and from (i). □

Lemma 3.2.5 *If $x \prec w$ and $\ell(w) > \ell(x) + 1$, then we have $D_L(w) \subseteq D_L(x)$ and $D_R(w) \subseteq D_R(x)$.*

Proof The inclusion $D_R(w) \subseteq D_R(x)$ is [6, Proposition 5.1.9], and the inclusion $D_L(w) \subseteq D_L(x)$ is its symmetric version which follows from $P_{x,w} = P_{x^{-1},w^{-1}}$ [10, p. 9]. □

Recall that $W^{I,\emptyset}$ is the set of minimal length representatives of $W(L_I)\backslash W(G)$.

Lemma 3.2.6 *Let $I \subseteq \Delta$, $y \in W^{I,\emptyset}$, $x, w \in W(L_I)$ with $x < w$ and $P^I_{x,w}$ the Kazhdan-Lusztig polynomial for the Coxeter group $W(L_I)$. Then we have $P^I_{x,w} = P_{x,w} = P_{xy,wy}$.*

Proof Recall that the multiplication on the right by y gives a bijection preserving the Bruhat order between the interval $[x, w]$ in $W(G)$ and the interval $[xy, wy]$ in $W(G)$, and that $\ell(x'y) = \ell(x') + \ell(y)$ for $x' \in W(L_I)$ (one reference for such facts is [6, Sect. 2]). Using this, the equality $P_{x,w} = P_{xy,wy}$ easily follows by induction on $\ell(y)$ from [10, Theorem 2(iii)] and [10, Theorem 4(iv)] (arguing as in the last paragraph of [10, p. 8]). By the same combinatorics, the equality $P^I_{x,w} = P_{x,w}$ comes from the fact that the interval $[x, w]$ in $W(G)$ is the same as the interval $[x, w]$ in $W(L_I)$. □

Lemma 3.2.7 *Let $x, w \in W(G)$ and $I \subseteq \Delta$.*

(i) If $w \in W(L_I)x$, then we have canonical isomorphisms for $k \geq 0$

$$\mathrm{Ext}^k_{\mathcal{O}^{\mathfrak{b}_I}_{\mathfrak{l}_I,\mathrm{alg}}}(L^I(w), H^0(\mathfrak{n}_I, L(x))) \cong \mathrm{Ext}^k_{\mathcal{O}^{\mathfrak{b}}_{\mathrm{alg}}}(M^I(w), L(x)).$$

(ii) The E-vector space $\mathrm{Hom}_{U(\mathfrak{l}_I)}(L^I(w), H^1(\mathfrak{n}_I, L(x)))$ is non-zero if and only if $x \prec w$ and $w \notin W(L_I)x$, and has dimension $\mu(x, w)$ if non-zero.

(iii) For $k \geq 0$ the canonical map

$$H^k(\mathfrak{u}_I, H^1(\mathfrak{n}_I, L(x))) \to H^{k+2}(\mathfrak{u}_I, H^0(\mathfrak{n}_I, L(x))) \tag{3.30}$$

induced from the spectral sequence (3.9) (with $I' = \emptyset$) is zero.

(iv) The inclusion $\mathrm{soc}_{U(\mathfrak{l}_I)}(H^1(\mathfrak{n}_I, L(x))) \subseteq H^1(\mathfrak{n}_I, L(x))$ induces a $U(\mathfrak{t})$-equivariant isomorphism

$$H^0\big(\mathfrak{u}_I, \mathrm{soc}_{U(\mathfrak{l}_I)}(H^1(\mathfrak{n}_I, L(x)))\big) \xrightarrow{\sim} H^0(\mathfrak{u}_I, H^1(\mathfrak{n}_I, L(x))) \tag{3.31}$$

with both $U(\mathfrak{t})$-modules in (3.31) isomorphic to $\bigoplus_w (w \cdot \mu_0)^{\oplus \mu(x,w)}$ where w runs through those $w \in W(G)$ such that $w \notin W(L_I)x$ and $x \prec w$.

Proof (i) follows from (ii) of Lemma 3.1.8 and (iv) of Lemma 3.1.9 (together with Harish-Chandra's theorem).

We prove (ii). Let $\mu \in \Lambda$, by Lemma 3.1.2 if $H^1(\mathfrak{n}_I, L(x))_\mu \neq 0$ then $\mu - x \cdot \mu_0 - \alpha \in \mathbb{Z}_{\geq 0}\Phi^+$ for some $\alpha \in \Phi^+ \setminus \Phi^+_I$. In particular

$$\mathrm{Hom}_{U(\mathfrak{l}_I)}(L^I(w), H^1(\mathfrak{n}_I, L(x))) \neq 0 \implies w \cdot \mu_0 - x \cdot \mu_0 \in \alpha + \mathbb{Z}_{\geq 0}\Phi^+ \implies w > x. \tag{3.32}$$

On the other hand, each Jordan-Hölder factor $L(w')$ of $N(w)$ satisfies $w' > w$. So for $w > x$ the surjections $M(w) \twoheadrightarrow M^I(w) \twoheadrightarrow L(w)$ induce embeddings

$$\mathrm{Ext}^1_{\mathcal{O}^{\mathfrak{b}}_{\mathrm{alg}}}(L(w), L(x)) \hookrightarrow \mathrm{Ext}^1_{\mathcal{O}^{\mathfrak{b}}_{\mathrm{alg}}}(M^I(w), L(x)) \hookrightarrow \mathrm{Ext}^1_{\mathcal{O}^{\mathfrak{b}}_{\mathrm{alg}}}(M(w), L(x)). \tag{3.33}$$

But since the first and third vector spaces both have dimension $\mu(x, w)$ by [10, p.9], [33, Theorem 8.15(c)] and [33, Theorem 8.11(b)], we deduce that the embeddings in (3.33) are all isomorphisms. In particular, when $x < w$ each E-vector space in (3.33) has dimension $\mu(x, w)$ and is non-zero if and only if $x \prec w$.

Now, from (ii), (iii) and (iv) of Lemma 3.1.9 we have $\mathrm{Hom}_{U(\mathfrak{l}_I)}(L^I(w), H^1(\mathfrak{n}_I, L(x))) \neq 0$ if and only if $\mathrm{Ext}^1_{\mathcal{O}^{\mathfrak{b}}_{\mathrm{alg}}}(M^I(w), L(x)) \neq 0$ and $H^0(\mathfrak{n}_I, L(x))_\xi = 0$ where ξ is the infinitesimal character of $L^I(w)$. By Harish-Chandra's theorem and (ii) of Lemma 3.1.8, $H^0(\mathfrak{n}_I, L(x))_\xi = 0$ if and only if $w \notin W(L_I)x$. By the previous paragraph $x \prec w$ if and only if $x < w$ and $\mathrm{Ext}^1_{\mathcal{O}^{\mathfrak{b}}_{\mathrm{alg}}}(M^I(w), L(x)) \neq 0$. Using (3.32), we get (ii).

We prove (iii). Since $H^0(\mathfrak{n}_I, L(x)) \cong L^I(x)$ by (ii) of Lemma 3.1.8, for each $\mu \in \Lambda$ satisfying $H^{k+2}(\mathfrak{u}_I, H^0(\mathfrak{n}_I, L(x)))_\mu \neq 0$, we deduce in particular from Lemma 3.1.2 (applied with $\mathfrak{l}_I$ instead of $\mathfrak{g}$) that

$$\mu - x \cdot \mu_0 \in \mathbb{Z}_{\geq 0}\Phi_I^+. \tag{3.34}$$

Let $\mu' \in \Lambda$ such that $H^k(\mathfrak{u}_I, H^1(\mathfrak{n}_I, L(x)))_{\mu'} \neq 0$. By a dévissage on $H^1(\mathfrak{n}_I, L(x))$ (using Proposition 3.1.5) and by applying Lemma 3.1.2 twice, first to $H^1(\mathfrak{n}_I, L(x))$ and then to $H^k(\mathfrak{u}_I, L^I(\mu''))$ for each Jordan-Hölder factor $L^I(\mu'')$ of $H^1(\mathfrak{n}_I, L(x))$, we deduce the existence of $\alpha \in \Phi^+ \setminus \Phi_I^+$ such that

$$\mu' - x \cdot \mu_0 - \alpha \in \mathbb{Z}_{\geq 0}\Phi^+. \tag{3.35}$$

If the map (3.30) is non-zero, then there must exists $\mu = \mu'$ satisfying both (3.34) and (3.35), which is impossible as

$$\mathbb{Z}_{\geq 0}\Phi_I^+ \cap (\alpha + \mathbb{Z}_{\geq 0}\Phi^+) = \emptyset.$$

We prove (iv). Note first that (iii) together with (3.9) imply a short exact sequence

$$0 \to H^1(\mathfrak{u}_I, H^0(\mathfrak{n}_I, L(x))) \to H^1(\mathfrak{u}, L(x)) \to H^0(\mathfrak{u}_I, H^1(\mathfrak{n}_I, L(x))) \to 0. \tag{3.36}$$

Secondly, from (i) and (ii) of Lemma 3.2.1 (and $\mu(ww_0, xw_0) = \mu(x, w)$) we have

$$H^1(\mathfrak{u}, L(x)) \cong \bigoplus_{x \prec w} (w \cdot \mu_0)^{\oplus \mu(x,w)}. \tag{3.37}$$

Thirdly, write $x = x_I x^I$ where $x_I \in W(L_I)$ and $x^I \in W^{I,\emptyset}$ and note that $x^I \cdot \mu_0 \in \Lambda_I^{\mathrm{dom}}$, so that, when dealing with the reductive group L_I, we can replace μ_0 by $x^I \cdot \mu_0$. Then a proof analogous to the proof of (i) and (ii) of Lemma 3.2.1 replacing G by L_I and using Lemma 3.2.6 gives

$$H^1(\mathfrak{u}_I, L^I(x)) \cong \bigoplus_{x \prec w, w \in W(L_I)x} (w \cdot \mu_0)^{\oplus \mu(x,w)}. \tag{3.38}$$

As $H^0(\mathfrak{n}_I, L(x)) \cong L^I(x)$ by (ii) of Lemma 3.1.8, we deduce from (3.36), (3.37) and (3.38)

$$H^0(\mathfrak{u}_I, H^1(\mathfrak{n}_I, L(x))) \cong \bigoplus_{x \prec w, w \notin W(L_I)x} (w \cdot \mu_0)^{\oplus \mu(x,w)}.$$

But by (ii) we also have

$$\mathrm{soc}_{U(\mathfrak{l}_I)}(H^1(\mathfrak{n}_I, L(x))) \cong \bigoplus_{x \prec w, w \notin W(L_I)x} L^I(w)^{\oplus \mu(x,w)}$$

which clearly implies (3.31). □

Remark 3.2.8 Let $x, w \in W(G)$ with $x \neq w$ and $I \stackrel{\mathrm{def}}{=} \Delta \setminus D_L(w)$. The short exact sequence $0 \to N^I(w) \to M^I(w) \to L(w) \to 0$ together with $x \neq w$ give an exact sequence

$$0 \to \mathrm{Hom}_{U(\mathfrak{g})}(N^I(w), L(x)) \to \mathrm{Ext}^1_{U(\mathfrak{g})}(L(w), L(x)) \to \mathrm{Ext}^1_{U(\mathfrak{g})}(M^I(w), L(x)). \tag{3.39}$$

By [33, Theorem 9.4(c)] and Lemma 3.1.1 we have $\mathrm{Hom}_{U(\mathfrak{g})}(N^I(w), L(x)) \neq 0$ if and only if $\mathrm{Hom}_{U(\mathfrak{g})}(N(w), L(x)) \neq 0$ and $D_L(x) \subseteq D_L(w)$, and by (3.27) and (ii) of Lemma 3.2.2 that $\mathrm{Hom}_{U(\mathfrak{g})}(N(w), L(x)) \neq 0$ if and only if $w \prec x$. Since $w \prec x$ and $D_L(x) \subseteq D_L(w)$ if and only if $w \prec x$ and $x \notin W(L_I)w$ (using Lemma 3.2.5 to deal with the implication $\Leftarrow$ when $\ell(x) > \ell(w) + 1$), we finally deduce (with (ii) of Lemma 3.2.2) that $\mathrm{Hom}_{U(\mathfrak{g})}(N^I(w), L(x)) \neq 0$ if and only if $w \prec x$ and $x \notin W(L_I)w$, in which case it has dimension $\mu(w, x)$.

Moreover it follows from (ii), (iii) of Lemma 3.1.9 (with Harish-Chandra's theorem) and (ii) of Lemma 3.1.8 that one has isomorphisms

$$\mathrm{Ext}^1_{U(\mathfrak{g})}(M^I(w), L(x)) \cong \mathrm{Ext}^1_{U(\mathfrak{l}_I)}(L^I(w), H^0(\mathfrak{n}_I, L(x))) \cong \mathrm{Ext}^1_{U(\mathfrak{l}_I)}(L^I(w), L^I(x)) \tag{3.40}$$

if $w \in W(L_I)x$, and

$$\mathrm{Ext}^1_{U(\mathfrak{g})}(M^I(w), L(x)) \cong \mathrm{Hom}_{U(\mathfrak{l}_I)}(L^I(w), H^1(\mathfrak{n}_I, L(x))) \tag{3.41}$$

if $w \notin W(L_I)x$. When $w \in W(L_I)x$, it then follows from Lemma 3.2.6 (upon writing $x = x'y$, $w = w'y$ for $w' < x'$ in $W(L_I)$ and $y \in W^{I,\emptyset}$) and (ii) of Lemma 3.2.4 (applied with $\mathfrak{l}_I$ instead of $\mathfrak{g}$) that (3.40) has dimension $\mu(x, w)$ (resp. $\mu(w, x)$) when $x < w$ (resp. when $w < x$), which is also $\dim_E \mathrm{Ext}^1_{U(\mathfrak{g})}(L(w), L(x))$ by (ii) of Lemma 3.2.4. When $w \notin W(L_I)x$, it then follows from (ii) of Lemma 3.2.7 that (3.41) is non-zero if and only if $x \prec w$ (and $w \notin W(L_I)x$) and has dimension $\mu(x, w) = \dim_E \mathrm{Ext}^1_{U(\mathfrak{g})}(L(w), L(x))$ in that case.

By (3.39) and the above discussion on $\mathrm{Hom}_{U(\mathfrak{g})}(N^I(w), L(x))$ and $\mathrm{Ext}^1_{U(\mathfrak{g})}(M^I(w), L(x))$, we see that $\mathrm{Ext}^1_{U(\mathfrak{g})}(L(w), L(x)) \neq 0$ if and only if exactly one of the following holds:

- $w \prec x, x \notin W(L_I)(w), \mathrm{Ext}^1_{U(\mathfrak{g})}(M^I(w), L(x)) = 0$ and (3.39) induces an isomorphism
$$\mathrm{Hom}_{U(\mathfrak{g})}(N^I(w), L(x)) \xrightarrow{\sim} \mathrm{Ext}^1_{U(\mathfrak{g})}(L(w), L(x));$$
- $\mathrm{Hom}_{U(\mathfrak{g})}(N^I(w), L(x)) = 0$ and (3.39) induces an isomorphism
$$\mathrm{Ext}^1_{U(\mathfrak{g})}(L(w), L(x)) \xrightarrow{\sim} \mathrm{Ext}^1_{U(\mathfrak{g})}(M^I(w), L(x))$$
(as it is then an embedding between two vector spaces of the same dimension).

Lemma 3.2.9 *Let $w \in W(G)$. We have canonical isomorphisms for $\ell \leq \ell(w)$ induced by $M(w) \twoheadrightarrow L(w)$*

$$\mathrm{Ext}^{\ell}_{\mathcal{O}^{\mathfrak{b}}_{\mathrm{alg}}}(L(w), L(1)) \xrightarrow{\sim} \mathrm{Ext}^{\ell}_{\mathcal{O}^{\mathfrak{b}}_{\mathrm{alg}}}(M(w), L(1)) \cong \mathrm{Hom}_{U(\mathfrak{t})}(w \cdot \mu_0, H^{\ell}(\mathfrak{u}, L(1))) \tag{3.42}$$

where all E-vector spaces in (3.42) are 0 if $\ell < \ell(w)$ and 1-dimensional if $\ell = \ell(w)$.

Proof Recall that by definition $L(1) = L(\mu_0)$. By Bott's formula (see [33, Sect. 6.6]) there is a $U(\mathfrak{t})$-equivariant isomorphism for $\ell \geq 0$

$$H^{\ell}(\mathfrak{u}, L(1)) \cong \bigoplus_{\ell(w)=\ell} w \cdot \mu_0. \tag{3.43}$$

We deduce from (3.15) applied with $I = \emptyset$ (using $\mathrm{Ext}^{\ell_1}_{\mathcal{O}^{\mathfrak{t}}_{\mathfrak{t},\mathrm{alg}}} = 0$ if $\ell_1 > 0$) that for $\ell \geq 0$

$$\mathrm{Ext}^{\ell}_{\mathcal{O}^{\mathfrak{b}}_{\mathrm{alg}}}(M(w), L(1)) \cong \mathrm{Hom}_{U(\mathfrak{t})}\big(w \cdot \mu_0, H^{\ell}(\mathfrak{u}, L(1))\big),$$

which is 1-dimensional if $\ell = \ell(w)$ and 0 otherwise by (3.43). This gives the second isomorphism in (3.42). We prove the first isomorphism in (3.42) when $\ell \leq \ell(w)$ by decreasing induction on $\ell(w)$. If $w = w_0$, then $L(w_0) = M(w_0)$ and there is nothing more to prove. If $w < w_0$, we have by induction $\mathrm{Ext}^{\ell}_{\mathcal{O}^{\mathfrak{b}}_{\mathrm{alg}}}(L(w'), L(1)) = 0$ for $\ell \leq \ell(w)$ and $w < w'$, which implies $\mathrm{Ext}^{\ell}_{\mathcal{O}^{\mathfrak{b}}_{\mathrm{alg}}}(N(w), L(1)) = 0$ for $\ell \leq \ell(w)$ by an obvious dévissage on $N(w)$. The short exact sequence $0 \to N(w) \to M(w) \to L(w) \to 0$ then induces an isomorphism

$$\mathrm{Ext}^{\ell}_{\mathcal{O}^{\mathfrak{b}}_{\mathrm{alg}}}(L(w), L(1)) \xrightarrow{\sim} \mathrm{Ext}^{\ell}_{\mathcal{O}^{\mathfrak{b}}_{\mathrm{alg}}}(M(w), L(1))$$

for $\ell \leq \ell(w)$. □

Lemma 3.2.10 *Let M in $\mathcal{O}^{\mathfrak{b}}_{\mathrm{alg}}$. Assume that the inclusion $\mathrm{soc}_{U(\mathfrak{g})}(M) \subseteq M$ induces an isomorphism*

$$H^0(\mathfrak{u}, \mathrm{soc}_{U(\mathfrak{g})}(M)) \xrightarrow{\sim} H^0(\mathfrak{u}, M) \tag{3.44}$$

and that

$$\mathrm{Hom}_{U(\mathfrak{t})}(\mu_0, H^0(\mathfrak{u}, M)) = 0, \tag{3.45}$$

then it also induces an isomorphism

$$\mathrm{Ext}^1_{U(\mathfrak{g})}(L(1), \mathrm{soc}_{U(\mathfrak{g})}(M)) \xrightarrow{\sim} \mathrm{Ext}^1_{U(\mathfrak{g})}(L(1), M). \tag{3.46}$$

Proof We have the decomposition $M = \bigoplus_\xi M_\xi$ from (i) of Lemma 3.1.3. Since both vector spaces in (3.46) are 0 when M is replaced by M_ξ with $\xi \neq \xi_{\mu_0}$ by (ii) of Lemma 3.1.3, we can assume $M = M_{\xi_{\mu_0}}$, i.e. that all Jordan-Hölder factors of M are of the form $L(x)$ for some $x \in W(G)$.

For $\alpha \in \Phi^+$, we have $L(x)_{\mu_0-\alpha} = 0$ for each $x \in W(G)$ and thus $M_{\mu_0-\alpha} = 0$. Consequently, any $v \in M_{\mu_0}$ must be killed by $\mathfrak{u}$, i.e. $M_{\mu_0} \subseteq H^0(\mathfrak{u}, M)$. Then (3.45) forces $M_{\mu_0} = 0$, i.e. $L(1)$ is not a constituent of M. Likewise, for $\alpha \in \Phi^+$ and $x, x' \in W(G)$ with $x' \neq 1$ and $\ell(x) = 1$, we have $L(x')_{x\cdot\mu_0-\alpha} = 0$. Consequently $M_{x\cdot\mu_0-\alpha} = 0$ for $\alpha \in \Phi^+$ and $x \in W(G)$ with $\ell(x) = 1$ (using that $L(1)$ does not appear in M) and thus $M_{x\cdot\mu_0} \subseteq H^0(\mathfrak{u}, M)$. From (3.44) we then obtain $M_{x\cdot\mu_0} \subseteq H^0(\mathfrak{u}, \mathrm{soc}_{U(\mathfrak{g})}(M)) \subseteq \mathrm{soc}_{U(\mathfrak{g})}(M)$. Since any constituent $L(x)$ of M contributes to $M_{x\cdot\mu_0}$, we deduce that all constituents $L(x)$ of M with $\ell(x) = 1$ can only appear in $\mathrm{soc}_{U(\mathfrak{g})}(M)$. Since, by (ii) of Lemma 3.2.4, for $x \neq 1$ we have $\mathrm{Ext}^1_{U(\mathfrak{g})}(L(1), L(x)) \neq 0$ if and only if $\ell(x) = 1$, an easy dévissage implies (3.46). □

3.3 $W(G)$-Conjugates of Objects of $\mathcal{O}^{\mathfrak{b}}_{\rm alg}$

We study unipotent cohomology groups and Ext^1, Ext^2 groups of $U(\mathfrak{g})$-modules which are conjugates of $U(\mathfrak{g})$-modules in the category $\mathcal{O}^{\mathfrak{b}}_{\rm alg}$ by elements of $W(G)$.

Recall we defined full subcategories $\mathcal{C}_{\rm alg}$, $\mathcal{C}^{\rm fin}_{\rm alg}$, $\widetilde{\mathcal{O}}^{\mathfrak{b}}_{\rm alg}$ and $\mathcal{O}^{\mathfrak{b}}_{\rm alg}$ of $\mathrm{Mod}_{U(\mathfrak{g})}$ in Sect. 3.1. For M in $\mathrm{Mod}_{U(\mathfrak{g})}$ and $g \in G$, we define M^g in $\mathrm{Mod}_{U(\mathfrak{g})}$ as the same underlying $U(\mathfrak{g})$-module as M but where $x \in U(\mathfrak{g})$ acts by $\mathrm{ad}(g)(x) = gxg^{-1}$. For $g_1, g_2 \in G$ we have $(M^{g_1})^{g_2} \cong M^{g_1 g_2}$. For M_1, M_2 in $\mathrm{Mod}_{U(\mathfrak{g})}$ and $g \in G$, we (clearly) have isomorphisms for $k \geq 0$

$$\mathrm{Ext}^k_{U(\mathfrak{g})}(M_1, M_2) \cong \mathrm{Ext}^k_{U(\mathfrak{g})}(M_1^g, M_2^g). \tag{3.47}$$

Recall that an algebraic action of $\mathfrak{t}$, resp. of $\mathfrak{b}$, on a finite dimensional E-vector space lifts uniquely to an (algebraic) action of T, resp. of B, see for instance [52, Lemma 3.2]. It follows that, for M in $\mathcal{C}^{\rm fin}_{\rm alg}$ and $t \in T$, the resulting action of t on M induces an isomorphism $M \xrightarrow{\sim} M^t$ in $\mathcal{C}^{\rm fin}_{\rm alg}$. Likewise, for M in $\widetilde{\mathcal{O}}^{\mathfrak{b}}_{\rm alg}$ and $b \in B$, we have $M \xrightarrow{\sim} M^b$ in $\widetilde{\mathcal{O}}^{\mathfrak{b}}_{\rm alg}$. In particular, for M in $\mathcal{C}^{\rm fin}_{\rm alg}$, the isomorphism class of M^g is independent of the choice of $g \in N_G(T)$ lifting $w \in N_G(T)/T$ and we denote it by M^w. For $I \subseteq \Delta$ and $M \in \mathcal{O}^{\mathfrak{p}_I}_{\rm alg}$, by [52, Lemma 3.2] the action of $U(\mathfrak{p}_I)$ on M lifts uniquely to an action of P_I on M, and thus as above $M \xrightarrow{\sim} M^g$ as $U(\mathfrak{g})$-modules when $g \in P_I$. In particular, $M^w \cong M$ for each $w \in W(L_I)$.

Lemma 3.3.1 *Let $I, I' \subseteq \Delta$, M in $\mathcal{O}^{\mathfrak{p}_{I'}}_{\mathrm{alg}}$ and $w \in W(G)$. We write $w = w_1 w_2 w_3$ for (unique) $w_1 \in W(L_{I'})$, $w_3 \in W(L_I)$ and $w_2 \in W^{I',I}$.*

(i) For $k \geq 0$ the $U(\mathfrak{l}_I)$-module $H^k(\mathfrak{n}_I, M^{w_2})$ lies in $\mathcal{O}^{\mathfrak{b}_I}_{\mathfrak{l}_I,\mathrm{alg}}$ and we have $H^k(\mathfrak{n}_I, M^w) \cong (H^k(\mathfrak{n}_I, M^{w_2}))^{w_3}$ in $\mathrm{Mod}_{U(\mathfrak{l}_I)}$.

(ii) For $\mu \in \Lambda$ such that I' is maximal for the condition $L(\mu) \in \mathcal{O}^{\mathfrak{p}_{I'}}_{\mathrm{alg}}$, we have $H^0(\mathfrak{n}_I, L(\mu)^w) \neq 0$ if and only if $w_2 = 1$.

(iii) For $\mu \in \Lambda$ such that I' is maximal for the condition $L(\mu) \in \mathcal{O}^{\mathfrak{p}_{I'}}_{\mathrm{alg}}$, we have $L(\mu)^w \in \mathcal{O}^{\mathfrak{b}}_{\mathrm{alg}}$ if and only if $L(\mu)^w \cong L(\mu)$ if and only if $w \in W(L_{I'})$.

Proof We prove the second statement in (i). Since any lift of $w_1 \in W(L_{I'})$ in $N_G(T)$ lies in $L_{I'} \subseteq P_{I'}$, we have $M^{w_1} \cong M$ as $U(\mathfrak{g})$-modules by the sentence just before Lemma 3.3.1. Since L_I normalizes N_I, we have $w_3^{-1}\mathfrak{n}_I w_3 \cong \mathfrak{n}_I$ and thus (using (3.7) for the last isomorphism)

$$H^k(\mathfrak{n}_I, M_1^{w_3}) \cong H^k(w_3^{-1}\mathfrak{n}_I w_3, M_1^{w_3}) \cong H^k(\mathfrak{n}_I, M_1)^{w_3} \tag{3.48}$$

for any $U(\mathfrak{g})$-module M_1 and $k \geq 0$. We apply this to $M_1 = M^{w_1 w_2} \cong M^{w_2}$, which gives the second statement. We prove the first statement in (i). The minimal length assumption on w_2 implies in particular $\mathfrak{b}_I = \mathfrak{b} \cap \mathfrak{l}_I \subseteq w_2^{-1}\mathfrak{b} w_2$, and thus M^{w_2} is locally $\mathfrak{b}_I$-finite, which together with the $U(\mathfrak{l}_I)$-equivariance of the Chevalley-Eilenberg complex (3.7) (and the fact $\dim_E \mathfrak{n}_I < +\infty$) implies that the $U(\mathfrak{l}_I)$-module $H^k(\mathfrak{n}_I, M^{w_2})$ is locally $\mathfrak{b}_I$-finite for $k \geq 0$. A similar argument shows that $H^k(\mathfrak{n}_I, M^{w_2})$ is also $\mathfrak{t}$-semi-simple. Then the argument to show that $H^k(\mathfrak{n}_I, M^{w_2})$ has finite length (hence is in $\mathcal{O}^{\mathfrak{b}_I}_{\mathfrak{l}_I,\mathrm{alg}}$) is parallel to the one for the proof of Proposition 3.1.5, using as (crucial) inputs Lemma 3.1.4 and $\dim_E H^k(\mathfrak{n}_I, M^{w_2})_\mu < \infty$ for $\mu \in \Lambda$. Note that M^{w_2} itself is not in general in $\mathcal{O}^{\mathfrak{b}}_{\mathrm{alg}}$ as it might not be locally $\mathfrak{b}$-finite

We prove (ii). By the second statement in (i), it suffices to treat the case when $w = w_2 \in W^{I',I}$. If $w_2 = 1$, we have $H^0(\mathfrak{n}_I, L(\mu)) \cong L^I(\mu) \neq 0$ from (ii) of Lemma 3.1.8. We assume from now on $w_2 \neq 1$ and $H^0(\mathfrak{n}_I, L(\mu)^{w_2}) \neq 0$ and seek a contradiction. Let M_I be any non-zero simple $U(\mathfrak{l}_I)$-module that embeds into $H^0(\mathfrak{n}_I, L(\mu)^{w_2})$ (M_I exists and belongs to $\mathcal{O}^{\mathfrak{b}_I}_{\mathfrak{l}_I,\mathrm{alg}}$ by the first statement in (i)). Then the injection $M_I \hookrightarrow H^0(\mathfrak{n}_I, L(\mu)^{w_2})$ induces a non-zero map $U(\mathfrak{g}) \otimes_{U(\mathfrak{p}_I)} M_I \to L(\mu)^{w_2}$ by (3.13) which has to be a surjection as $L(\mu)^{w_2}$ is irreducible. As $U(\mathfrak{g}) \otimes_{U(\mathfrak{p}_I)} M_I$ is in $\mathcal{O}^{\mathfrak{b}}_{\mathrm{alg}}$ (see below (3.5)), $L(\mu)^{w_2}$ is also an object of $\mathcal{O}^{\mathfrak{b}}_{\mathrm{alg}}$, and in particular $L(\mu)$ is locally $w_2\mathfrak{b}w_2^{-1}$-finite. As $1 \neq w_2$, there exists $\alpha \in \Delta$ (recall it is a positive simple root for $\mathfrak{b}^+$) such that $w_2^{-1}(\alpha) \in \mathfrak{b}$, or equivalently $\mathfrak{u}_\alpha \subseteq w_2\mathfrak{b}w_2^{-1}$ (where $\mathfrak{u}_\alpha \subseteq \mathfrak{u} \subset \mathfrak{b}^+$ is the one dimensional root subspace corresponding to α), which implies that $L(\mu)$ is locally $\mathfrak{u}_\alpha$-finite. Note that $\alpha \notin I'$ because $w_2^{-1}(I') \subseteq \Phi^+$. Let $v \in M(\mu)_\mu$ be a highest weight vector of $M(\mu)$ and $0 \neq x_\alpha \in \mathfrak{u}_\alpha$. As $L(\mu)$ is locally $\mathfrak{u}_\alpha$-finite, there exists $N \geq 1$ such that $x_\alpha^N \cdot v \in N(\mu)_{\mu+N\alpha} \subseteq M(\mu)_{\mu+N\alpha}$, which implies $N(\mu) \neq 0$ as $M(\mu) \cong U(\mathfrak{u}^+)$. As each Jordan-Hölder factor of $N(\mu)$

has highest weight $w'\cdot\mu$ for some $w'\in W(G)$ such that $w'\cdot\mu-\mu\in\mathbb{Z}_{>0}\Phi^+$ ([33, Theorem 5.1], in particular $w'\neq 1$), we deduce $\mu+N\alpha-w'\cdot\mu\in\mathbb{Z}_{\geq 0}\Phi^+$ for such a w'. Since $w'\neq 1$, there also exists $\beta\in\Delta$ such that $w'\cdot\mu-\mu=(w'\cdot\mu-s_\beta\cdot\mu)+(s_\beta\cdot\mu-\mu)\in\beta+\mathbb{Z}_{\geq 0}\Phi^+$. We can choose $\beta\neq\alpha$, as otherwise this would mean $w'\cdot\mu-\mu\in\mathbb{Z}_{>0}\alpha$, hence $w'=s_\alpha$, but this is impossible since $\alpha\notin I'$ and I' is maximal such that $L(\mu)\in\mathcal{O}^{\mathfrak{p}_{I'}}_{\mathrm{alg}}$, equivalently I' is maximal such that μ is dominant with respect to $\mathfrak{b}_{I'}$, which implies $s_\alpha\cdot\mu-\mu\in\mathbb{Z}_{\leq 0}\Phi^+$. It follows that $(\mu+N\alpha-w'\cdot\mu)+(w'\cdot\mu-\mu)=N\alpha\in\beta+\mathbb{Z}_{\geq 0}\Phi^+$ which is impossible as $\beta\neq\alpha$.

Finally we prove (iii). As in (i) above, $w\in W(L_{I'})$ implies $L(\mu)^w\cong L(\mu)$ which implies $L(\mu)^w\in\mathcal{O}^{\mathfrak{b}}_{\mathrm{alg}}$. Conversely, assume $L(\mu)^w\in\mathcal{O}^{\mathfrak{b}}_{\mathrm{alg}}$, which implies $H^0(\mathfrak{u},L(\mu)^w)\neq 0$. By (ii) applied with $I=\emptyset$ we get $w\in W(L_{I'})W(L_I)=W(L_{I'})$. $\square$

The following consequence of Lemma 3.3.1 will be used later.

Lemma 3.3.2 *For $k\geq 0$, M, M' in $\mathcal{O}^{\mathfrak{b}}_{\mathrm{alg}}$ and $w\in W(G)$, the E-vector space $\mathrm{Ext}^k_{U(\mathfrak{g})}(M',M^w)$ is finite dimensional.*

Proof It follows from the first statement in (i) of Lemma 3.3.1 applied with $I=\emptyset$ that $H^\ell(\mathfrak{u},L(\mu)^w)$ is a finite dimensional semi-simple $U(\mathfrak{t})$-module for $\ell\geq 0$, $\mu\in\Lambda$ and $w\in W(G)$, and thus $\mathrm{Ext}^k_{U(\mathfrak{t})}(\mu',H^\ell(\mathfrak{u},L(\mu)^w))$ is finite dimensional for $k,\ell\geq 0$, $\mu,\mu'\in\Lambda$. By (3.12) (applied with $I=\emptyset$) we deduce

$$\dim_E\mathrm{Ext}^k_{U(\mathfrak{g})}(M(\mu'),L(\mu)^w)<+\infty \tag{3.49}$$

for $k\geq 0$, $\mu,\mu'\in\Lambda$ and $w\in W(G)$. Now, let $\mu'\in\Lambda$, if there is no $\mu''\neq\mu'$ such that $\mu''\uparrow\mu'$ (where $\uparrow$ is the strong linkage relation from [33, Sect. 5.1]) then $M(\mu')\xrightarrow{\sim}L(\mu')$ by [33, Theorem 5.1(b)] and hence $\mathrm{Ext}^k_{U(\mathfrak{g})}(L(\mu'),L(\mu)^w)$ is finite dimensional for $k\geq 0$ by (3.49). Assume by induction that $\mathrm{Ext}^k_{U(\mathfrak{g})}(L(\mu''),L(\mu)^w)$ is finite dimensional for $k\geq 0$ and any $\mu''\uparrow\mu'$, $\mu''\neq\mu'$. Then $\mathrm{Ext}^k_{U(\mathfrak{g})}(N(\mu'),L(\mu)^w)$ and $\mathrm{Ext}^k_{U(\mathfrak{g})}(M(\mu'),L(\mu)^w)$ are both finite dimensional, the first by dévissage and induction using [33, Theorem 5.1(a),(b)], and the second by (3.49). We deduce $\dim_E\mathrm{Ext}^k_{U(\mathfrak{g})}(L(\mu'),L(\mu)^w)<+\infty$ for $k\geq 0$ by an obvious dévissage. The statement of the lemma then follows by (another) obvious dévissage on the constituents of M' and M^w. $\square$

Recall from Sect. 3.1 that, for $x\in W(G)$ and $I\subseteq\Delta$, $L^I(x)$ is the unique simple quotient of $U(\mathfrak{l}_I)\otimes_{U(\mathfrak{b}_I)}x\cdot\mu_0$ (and lies in $\mathcal{O}^{\mathfrak{b}_I}_{\mathfrak{l}_I,\mathrm{alg}}$).

Lemma 3.3.3 *Let $x\in W(G)$ and $j\in\Delta$.*

(i) We have $L^{\{j\}}(x)\cong L^{\{j\}}(x)^{s_j}$ if and only if $j\notin D_L(x)$, in which case we have $U(\mathfrak{t})$-equivariant isomorphisms $H^0(\mathfrak{u}_{\{j\}},L^{\{j\}}(x))\cong x\cdot\mu_0$ and $H^1(\mathfrak{u}_{\{j\}},L^{\{j\}}(x))\cong s_jx\cdot\mu_0$.

(ii) *If* $j \in D_L(x)$, *we have* $U(\mathfrak{t})$-*equivariant isomorphisms* $H^0(\mathfrak{u}_{\{j\}}, L^{\{j\}}(x)) \cong x \cdot \mu_0$, $H^1(\mathfrak{u}_{\{j\}}, L^{\{j\}}(x)) = 0$, $H^0(\mathfrak{u}_{\{j\}}, L^{\{j\}}(x)^{s_j}) = 0$ *and* $H^1(\mathfrak{u}_{\{j\}}, L^{\{j\}}(x)^{s_j}) \cong s_j x \cdot \mu_0$.

Proof We prove (i). The isomorphism $H^0(\mathfrak{u}_{\{j\}}, L^{\{j\}}(x)) \cong x \cdot \mu_0$ always holds by (ii) of Lemma 3.1.8. By (ii) of Lemma 3.2.7 (applied with the reductive group $L_{\{j\}}$ instead of G and with $I = \emptyset$) and the fact that $H^1(\mathfrak{u}_{\{j\}}, L^{\{j\}}(x))$ is a semi-simple $U(\mathfrak{t})$-module (cf.Proposition 3.1.5) we deduce that $H^1(\mathfrak{u}_{\{j\}}, L^{\{j\}}(x)) \cong s_j x \cdot \mu_0$ if $x < s_j x$ (equivalently if $j \notin D_L(x)$), and $H^1(\mathfrak{u}_{\{j\}}, L^{\{j\}}(x)) = 0$ if $s_j x < x$ (equivalently if $j \in D_L(x)$). By (iii) of Lemma 3.3.1 (applied with $L_{\{j\}}$ and $I = \emptyset$), we have $L^{\{j\}}(x) \cong L^{\{j\}}(x)^{s_j}$ if and only if $j \notin D_L(x)$ (noting that $I' \neq \emptyset$ in *loc. cit.* if and only if $I' = \{j\}$ if and only if $j \notin D_L(x)$ by Lemma 3.1.1).

We prove (ii). By (ii) of Lemma 3.3.1 (applied with $L_{\{j\}}$ and $I = \emptyset$) if $j \in D_L(x)$ we have $I' = \emptyset$ and $H^0(\mathfrak{u}_{\{j\}}, L^{\{j\}}(x)^{s_j}) = 0$. In view of what was proven before, it remains to show that $H^1(\mathfrak{u}_{\{j\}}, L^{\{j\}}(x)^{s_j}) \cong s_j x \cdot \mu_0$ when $j \in D_L(x)$. Assume $j \in D_L(x)$, then $L^{\{j\}}(x) \cong U(\mathfrak{l}_{\{j\}}) \otimes_{U(\mathfrak{b}_{\{j\}})} x \cdot \mu_0$ is a free $U(\mathfrak{u}^+_{\{j\}})$-module with a generator $0 \neq v \in L^{\{j\}}(x)_{x \cdot \mu_0}$. So $L^{\{j\}}(x)^{s_j}$ is a free $U(\mathfrak{u}_{\{j\}}) = U((\mathfrak{u}^+_{\{j\}})^{s_j})$-module with a generator $0 \neq v^{s_j} \in (L^{\{j\}}(x)^{s_j})_{s_j(x \cdot \mu_0)}$. Recall that $H^1(\mathfrak{u}_{\{j\}}, L^{\{j\}}(x)^{s_j})$ is by definition the cokernel of the map of $U(\mathfrak{t})$-modules

$$L^{\{j\}}(x)^{s_j} \to L^{\{j\}}(x)^{s_j} \otimes_E \mathfrak{u}^\vee_{\{j\}}, \ m \mapsto u(m) \otimes u^\vee$$

where $\mathfrak{u}^\vee_{\{j\}}$ is the E-vector space dual to $\mathfrak{u}_{\{j\}}$, $u \neq 0$ a fixed element of $\mathfrak{u}_{\{j\}}$ and $u^\vee$ the dual basis (recall $\dim_E \mathfrak{u}_{\{j\}} = 1$). By an easy computation (we are with GL_2) we have that $v^{s_j} \otimes u^\vee$ spans this cokernel as E-vector space. In particular as $U(\mathfrak{t})$-module the cokernel is isomorphic to $s_j(x \cdot \mu_0) + \alpha_j = s_j \cdot (x \cdot \mu_0) = s_j x \cdot \mu_0$ where $\alpha_j = e_j - e_{j+1} \in \Delta$ is the positive simple root also denoted j. This finishes the proof of (ii). □

Lemma 3.3.4 *Let* $I, I' \subseteq \Delta$, $1 \neq w_1 \in W^{I',I}$ *and* $w \in W(G)$ *such that* $I' = \Delta \setminus D_L(w)$. *Let* $x \in W(G)$, *then the following statements are equivalent.*

(i) $\mathrm{Hom}_{U(\mathfrak{l}_I)}(L^I(x), H^1(\mathfrak{n}_I, L(w)^{w_1})) \neq 0$;
(ii) $\mathrm{Ext}^1_{U(\mathfrak{g})}(L(x), L(w)^{w_1}) \neq 0$;
(iii) $\mathrm{Ext}^1_{U(\mathfrak{g})}(L(x), L(w)) \neq 0$ *and* $w_1 \in W(L_{I'})W(L_{\Delta \setminus D_L(x)})$.

Moreover, if these statements hold, we have a canonical isomorphism of E-*vector spaces of dimension* $\dim_E \mathrm{Ext}^1_{U(\mathfrak{g})}(L(x), L(w))$ *induced by applying the functor* $H^0(\mathfrak{u}_I, -)$:

$$\mathrm{Hom}_{U(\mathfrak{l}_I)}(L^I(x), H^1(\mathfrak{n}_I, L(w)^{w_1})) \xrightarrow{\sim} \mathrm{Hom}_{U(\mathfrak{t})}\big(x \cdot \mu_0, H^0(\mathfrak{u}_I, H^1(\mathfrak{n}_I, L(w)^{w_1}))\big). \quad (3.50)$$

Proof By Lemma 3.1.1 if $w \in W(G)$ and $I' = \Delta \setminus D_L(w)$ then $L(w)$ is in $\mathcal{O}^{\mathfrak{p}_{I'}}_{\mathrm{alg}}$ and I' is maximal for that condition. In particular $L(w)^{w'} \cong L(w)$ for $w' \in W(L_{I'})$ (see just before Lemma 3.3.1) and, if $w_1 = w_2 w_3$ for some $w_2 \in W(L_{\Delta \setminus D_L(w)})$ and

$w_3 \in W(L_{\Delta\setminus D_L(x)})$, then $L(w)^{w_2} \cong L(w)$ and $L(x)^{w_3^{-1}} \cong L(x)$. Hence, we deduce from (3.47) (with $M_1 = L(x)$, $M_2 = L(w)^{w_1}$ and $g = w_3^{-1}$):

$$\mathrm{Ext}^1_{U(\mathfrak{g})}(L(x), L(w)^{w_1}) \cong \mathrm{Ext}^1_{U(\mathfrak{g})}(L(x)^{w_3^{-1}}, L(w)^{w_2}) \cong \mathrm{Ext}^1_{U(\mathfrak{g})}(L(x), L(w)). \tag{3.51}$$

In particular, (iii) implies (ii).

Since $w_1 \neq 1$ by (ii) of Lemma 3.3.1 we have $H^0(\mathfrak{n}_I, L(w)^{w_1}) = 0$, which together with (3.14) (applied with $M_I = L^I(x)$ and $M = L(w)^{w_1}$) gives

$$\mathrm{Ext}^1_{U(\mathfrak{g})}(M^I(x), L(w)^{w_1}) \cong \mathrm{Hom}_{U(\mathfrak{l}_I)}(L^I(x), H^1(\mathfrak{n}_I, L(w)^{w_1})). \tag{3.52}$$

By a parallel argument, we have $H^0(\mathfrak{u}, L(w)^{w_1}) = 0$ and a canonical isomorphism

$$\mathrm{Ext}^1_{U(\mathfrak{g})}(M(x), L(w)^{w_1}) \cong \mathrm{Hom}_{U(\mathfrak{t})}(x \cdot \mu_0, H^1(\mathfrak{u}, L(w)^{w_1})). \tag{3.53}$$

The vanishing $H^0(\mathfrak{n}_I, L(w)^{w_1}) = 0$ together with (3.9) (applied with $I' = \emptyset$) also implies

$$H^1(\mathfrak{u}, L(w)^{w_1}) \cong H^0(\mathfrak{u}_I, H^1(\mathfrak{n}_I, L(w)^{w_1})),$$

hence we have

$$\mathrm{Hom}_{U(\mathfrak{t})}\big(x \cdot \mu_0, H^0(\mathfrak{u}_I, H^1(\mathfrak{n}_I, L(w)^{w_1}))\big) \cong \mathrm{Hom}_{U(\mathfrak{t})}(x \cdot \mu_0, H^1(\mathfrak{u}, L(w)^{w_1})). \tag{3.54}$$

By (iii) of Lemma 3.3.1 we have $L(w)^{w_1} \notin \mathcal{O}^{\mathfrak{b}}_{\mathrm{alg}}$, and thus $\mathrm{Hom}_{U(\mathfrak{g})}(L(x'), L(w)^{w_1}) = 0$ for any $L(x') \in \mathcal{O}^{\mathfrak{b}}_{\mathrm{alg}}$. Hence, the surjections $M(x) \twoheadrightarrow M^I(x) \twoheadrightarrow L(x)$ induce injections

$$\mathrm{Ext}^1_{U(\mathfrak{g})}(L(x), L(w)^{w_1}) \hookrightarrow \mathrm{Ext}^1_{U(\mathfrak{g})}(M^I(x), L(w)^{w_1}) \overset{q_1}{\hookrightarrow} \mathrm{Ext}^1_{U(\mathfrak{g})}(M(x), L(w)^{w_1}). \tag{3.55}$$

It follows that (ii) implies $\mathrm{Ext}^1_{U(\mathfrak{g})}(M^I(x), L(w)^{w_1}) \neq 0$ which is equivalent to (i) by (3.52). Thus (ii) implies (i).

The injection q_1 corresponds under (3.52), (3.53) and (3.54) to the injection induced by applying the functor $H^0(\mathfrak{u}_I, -)$

$$\mathrm{Hom}_{U(\mathfrak{l}_I)}(L^I(x), H^1(\mathfrak{n}_I, L(w)^{w_1})) \overset{q_2}{\hookrightarrow} \mathrm{Hom}_{U(\mathfrak{t})}\big(x \cdot \mu_0, H^0(\mathfrak{u}_I, H^1(\mathfrak{n}_I, L(w)^{w_1}))\big). \tag{3.56}$$

Note that by (3.51), (3.52), (3.53) and (3.55), in order to prove that q_2 (i.e. (3.50)) is an isomorphism it is enough to prove that (3.54) has dimension $\dim_E \mathrm{Ext}^1_{U(\mathfrak{g})}(L(x), L(w))$. Moreover, if (i) holds, then (3.54) must be non-zero by (3.56). Consequently, in order to prove that (i) implies (iii) and the last statement of the lemma (and hence the lemma), it suffices to prove:

Sublemma 3.3.5 *Assume that (3.54) is non-zero, then (iii) holds and (3.54) has dimension* $\dim_E \mathrm{Ext}^1_{U(\mathfrak{g})}(L(x), L(w))$.

We now prove Sublemma 3.3.5 by an increasing induction on $\ell(w_1) \geq 1$. Note first that, if we take $k = 1$ and $M = L(w)^{w_1}$ in (3.10), we have the short exact sequence for any $j \in \Delta$

$$0 \to H^1(\mathfrak{u}_{\{j\}}, H^0(\mathfrak{n}_{\{j\}}, L(w)^{w_1})) \to H^1(\mathfrak{u}, L(w)^{w_1}) \\ \to H^0(\mathfrak{u}_{\{j\}}, H^1(\mathfrak{n}_{\{j\}}, L(w)^{w_1})) \to 0. \quad (3.57)$$

We first prove the case $\ell(w_1) = 1$, i.e. $w_1 = s_j$ for some $j \in \Delta$. As $w_1 = s_j \in W^{I',I}$ we have $j \notin I \cup I'$ and thus $j \in D_L(w)$. We also have $s_j \mathfrak{n}_{\{j\}} s_j = \mathfrak{n}_{\{j\}}$ and thus (arguing as in (3.48))

$$H^k(\mathfrak{n}_{\{j\}}, L(w)^{s_j}) \cong H^k(s_j \mathfrak{n}_{\{j\}} s_j, L(w)^{s_j}) \cong H^k(\mathfrak{n}_{\{j\}}, L(w))^{s_j} \quad (3.58)$$

for $k \geq 0$. By (ii) of Lemma 3.1.8 this implies $H^0(\mathfrak{n}_{\{j\}}, L(w)^{s_j}) \cong L^{\{j\}}(w)^{s_j}$. Using (3.57), the non-vanishing of $\mathrm{Hom}_{U(\mathfrak{t})}(x \cdot \mu_0, H^1(\mathfrak{u}, L(w)^{w_1}))$ forces either

$$\mathrm{Hom}_{U(\mathfrak{t})}(x \cdot \mu_0, H^1(\mathfrak{u}_{\{j\}}, L^{\{j\}}(w)^{s_j})) \neq 0 \quad (3.59)$$

or

$$\mathrm{Hom}_{U(\mathfrak{t})}(x \cdot \mu_0, H^0(\mathfrak{u}_{\{j\}}, H^1(\mathfrak{n}_{\{j\}}, L(w)^{s_j}))) \neq 0. \quad (3.60)$$

We treat these two possibilities separately.

- Since $j \in D_L(w)$, by (ii) of Lemma 3.3.3 (applied with $x = w$!) we see that (3.59) is non-zero if and only if $x = s_j w (< w)$, in which case (3.59) is one dimensional.
- Now we consider (3.60). We deduce from (ii) of Lemma 3.2.7 (applied with $I = \{j\}$) that $\mathrm{soc}_{U(\mathfrak{l}_{\{j\}})}(H^1(\mathfrak{n}_{\{j\}}, L(w))) \cong \bigoplus_{x'} L^{\{j\}}(x')^{\oplus \mu(w,x')}$ and therefore

$$\mathrm{soc}_{U(\mathfrak{l}_{\{j\}})}(H^1(\mathfrak{n}_{\{j\}}, L(w))^{s_j}) \cong \bigoplus_{x'} (L^{\{j\}}(x')^{s_j})^{\oplus \mu(w,x')} \quad (3.61)$$

where x' runs through elements of $W(G)$ such that $w \prec x'$ (note that, as $W(L_{\{j\}}) = \{1, s_j\}$ and $j \in D_L(w)$, $w \prec x'$ also ensures $x' \notin W(L_{\{j\}})w$). It follows from (ii) of Lemma 3.3.1 (applied with $L_{\{j\}}$ and $I = \emptyset$) that $H^0(\mathfrak{u}_{\{j\}}, L^{\{j\}}(x')^{s_j})$ is non-zero if and only if $s_j \in W(L_{\Delta \setminus D_L(x')})$ if and only if $j \notin D_L(x')$, in which case it is isomorphic to $x' \cdot \mu_0$ by (i) of Lemma 3.3.3. Using Proposition 3.1.5, each indecomposable direct summand of the $U(\mathfrak{l}_{\{j\}})$-module $H^1(\mathfrak{n}_{\{j\}}, L(w))$ has one of the five forms described in [33, Proposition 3.12]. But we also have

$$H^0\big(\mathfrak{u}_{\{j\}}, \mathrm{soc}_{U(\mathfrak{l}_{\{j\}})}(H^1(\mathfrak{n}_{\{j\}}, L(w)))\big) \cong H^0(\mathfrak{u}_{\{j\}}, H^1(\mathfrak{n}_{\{j\}}, L(w)))$$

by (iv) of Lemma 3.2.7 (applied with $I = \{j\}$). It is then easy to check that an indecomposable direct summand of $H^1(\mathfrak{n}_{\{j\}}, L(w))$ is either irreducible of the form $L^{\{j\}}(x')$ (for $w \prec x'$) or uniserial of length two with socle $L^{\{j\}}(x')$ and cosocle

$L^{\{j\}}(s_jx')$ (for $w \prec x'$ and $j \notin D_L(x')$). Since $H^0(\mathfrak{u}_{\{j\}}, L^{\{j\}}(s_jx')^{s_j}) = 0$ if $j \notin D_L(x')$ (as seen above), this in turns implies using (3.58) (for $k = 1$)

$$H^0\big(\mathfrak{u}_{\{j\}}, \mathrm{soc}_{U(\mathfrak{l}_{\{j\}})}(H^1(\mathfrak{n}_{\{j\}}, L(w)^{s_j}))\big) \cong H^0(\mathfrak{u}_{\{j\}}, H^1(\mathfrak{n}_{\{j\}}, L(w)^{s_j})).$$

By (3.61) and the discussion just after it, we deduce that (3.60) is non-zero if and only if $w \prec x$ and $j \notin D_L(x)$, in which case (3.60) is $\mu(w, x)$-dimensional.

Note that in both cases $w_1 = s_j \in W(L_{\Delta\setminus D_L(x)})$ and $x \neq w$. Moreover when either of (3.59), (3.60) is non-zero, then the other is zero. Using (ii) of Lemma 3.2.4 (together with the $U(\mathfrak{t})$-semi-simplicity of $H^1(\mathfrak{u}, L(w)^{w_1})$) we deduce that (3.54) always has dimension $\dim_E \mathrm{Ext}^1_{U(\mathfrak{g})}(L(x), L(w))$. This finishes the proof of Sublemma 3.3.5 when $\ell(w_1) = 1$.

We now assume $\ell(w_1) > 1$ and prove the induction step. We choose an arbitrary $j \in D_R(w_1)$ and set $w_2 \stackrel{\mathrm{def}}{=} w_1 s_j < w_1$. As $s_j \mathfrak{n}_{\{j\}} s_j = \mathfrak{n}_{\{j\}}$, we have as in (3.58)

$$H^k(\mathfrak{n}_{\{j\}}, L(w)^{w_1}) \cong H^k(s_j \mathfrak{n}_{\{j\}} s_j, L(w)^{w_2 s_j}) \cong H^k(\mathfrak{n}_{\{j\}}, L(w)^{w_2})^{s_j}. \qquad (3.62)$$

As $w_1 \in W^{I',I} \subseteq W^{I',\emptyset}$ and $w_1 = w_2 s_j > w_2$, we have $w_2 \in W^{I',\emptyset}$, which together with $j \in D_R(w_1) \setminus D_R(w_2)$ implies $w_2 \in W^{I',\{j\}}$. Note that $w_2 \neq 1$ since $\ell(w_1) > 1$, and therefore $H^0(\mathfrak{n}_{\{j\}}, L(w)^{w_1}) \cong H^0(\mathfrak{n}_{\{j\}}, L(w)^{w_2})^{s_j} = 0$ by (3.62) and (ii) of Lemma 3.3.1. By (3.57) we deduce an isomorphism (using again (3.62)) $H^1(\mathfrak{u}, L(w)^{w_1}) \cong H^0(\mathfrak{u}_{\{j\}}, H^1(\mathfrak{n}_{\{j\}}, L(w)^{w_2})^{s_j})$. In particular, $\mathrm{Hom}_{U(\mathfrak{t})}(x \cdot \mu_0, H^1(\mathfrak{u}, L(w)^{w_1}))$ is non-zero if and only if

$$\mathrm{Hom}_{U(\mathfrak{t})}\big(x \cdot \mu_0, H^0(\mathfrak{u}_{\{j\}}, H^1(\mathfrak{n}_{\{j\}}, L(w)^{w_2})^{s_j})\big) \neq 0. \qquad (3.63)$$

Moreover, as $\ell(w_2) = \ell(w_1) - 1$ and $w_2 \in W^{I',\{j\}}$, the induction assumption and the discussion before Sublemma 3.3.5 imply that

$$\mathrm{Hom}_{U(\mathfrak{l}_{\{j\}})}(L^{\{j\}}(x'), H^1(\mathfrak{n}_{\{j\}}, L(w)^{w_2})) \neq 0 \qquad (3.64)$$

if and only if $\mathrm{Ext}^1_{U(\mathfrak{g})}(L(x'), L(w)) \neq 0$ and $w_2 \in W(L_{I'})W(L_{\Delta\setminus D_L(x')})$, in which case (3.64) has the same dimension as $\mathrm{Ext}^1_{U(\mathfrak{g})}(L(x'), L(w))$.

Now let $x \in W(G)$ satisfying (3.63), we have the following two cases.

- Assume $j \in D_L(x)$. Then $L^{\{j\}}(x)^{s_j}$ is not locally $\mathfrak{b}_{\{j\}}$-finite ((iii) of Lemma 3.3.1 and (i) of Lemma 3.3.3). Let $M_{\{j\}} \stackrel{\mathrm{def}}{=} U(\mathfrak{l}_{\{j\}}) \otimes_{U(\mathfrak{b}_{\{j\}})} x \cdot \mu_0$ which is isomorphic to $L^{\{j\}}(x)$ since $j \in D_L(x)$. By (3.63) there exists a non-zero map $M_{\{j\}} \to H^1(\mathfrak{n}_{\{j\}}, L(w)^{w_2})^{s_j}$, and thus a non-zero map $M_{\{j\}}^{s_j} \to H^1(\mathfrak{n}_{\{j\}}, L(w)^{w_2})$. This is impossible as $H^1(\mathfrak{n}_{\{j\}}, L(w)^{w_2})$ is locally $\mathfrak{b}_{\{j\}}$-finite (use $w_2 \in W^{I',\{j\}}$ and (i) of Lemma 3.3.1) but $M_{\{j\}}^{s_j} \cong L^{\{j\}}(x)^{s_j}$ is not. Hence this case can't happen.
- Assume $j \notin D_L(x)$. Consider the natural embedding

$$\begin{aligned}\operatorname{Hom}_{U(\mathfrak{t})}\big(x\cdot\mu_0, H^0(\mathfrak{u}_{\{j\}}, \operatorname{soc}_{U(\mathfrak{l}_{\{j\}})}(H^1(\mathfrak{n}_{\{j\}}, L(w)^{w_2})^{s_j}))\big)\\ \hookrightarrow \operatorname{Hom}_{U(\mathfrak{t})}(x\cdot\mu_0, H^0(\mathfrak{u}_{\{j\}}, H^1(\mathfrak{n}_{\{j\}}, L(w)^{w_2})^{s_j})).\end{aligned} \tag{3.65}$$

Assume that (3.65) is not an isomorphism. Then $H^1(\mathfrak{n}_{\{j\}}, L(w)^{w_2})^{s_j}$ must contain $M_{\{j\}} = U(\mathfrak{l}_{\{j\}}) \otimes_{U(\mathfrak{b}_{\{j\}})} x\cdot\mu_0$, and thus $H^1(\mathfrak{n}_{\{j\}}, L(w)^{w_2})$ must contain $M_{\{j\}}^{s_j}$. This is impossible as $H^1(\mathfrak{n}_{\{j\}}, L(w)^{w_2})$ is locally $\mathfrak{b}_{\{j\}}$-finite but $\operatorname{soc}_{U(\mathfrak{l}_{\{j\}})}(M_{\{j\}}^{s_j}) \cong L^{\{j\}}(s_jx)^{s_j}$ is not (same argument as in the previous case). Hence (3.65) is an isomorphism, and in particular its left hand side is non-zero by (3.63). But since $j \notin D_L(x)$, we have $s_j \in W(L_{\Delta\setminus D_L(x)})$ and thus $L^{\{j\}}(x)^{s_j} \cong L^{\{j\}}(x)$ ((iii) of Lemma 3.3.1). Hence the two (non-zero) $U(\mathfrak{l}_{\{j\}})$-modules in (3.65), once "untwisted" by s_j, become isomorphic to $\operatorname{Hom}_{U(\mathfrak{l}_{\{j\}})}(L^{\{j\}}(x), H^1(\mathfrak{n}_{\{j\}}, L(w)^{w_2}))$, which is thus also non-zero and has the same dimension as $\operatorname{Ext}^1_{U(\mathfrak{g})}(L(x), L(w))$ by what follows (3.64) applied to $x' = x$.

We have shown that (3.63) implies $\operatorname{Ext}^1_{U(\mathfrak{g})}(L(x), L(w)) \neq 0$, $w_2 \in W(L_{I'})W(L_{\Delta\setminus D_L(x)})$ (using the induction) and $j \notin D_L(x)$, which implies $w_1 = w_2s_j \in W(L_{I'})W(L_{\Delta\setminus D_L(x)})$. Moreover, we have also seen that (3.63) has the same dimension as $\operatorname{Ext}^1_{U(\mathfrak{g})}(L(x), L(w))$. The proof of the induction step is thus finished. □

Remark 3.3.6 Let $I \subseteq \Delta$, $w \in W(G)$, $I' = \Delta \setminus D_L(w)$ and $1 \neq w_1 \in W^{I',I}$.

(i) Let $I \subseteq \Delta$, $w \in W(G)$, $I' \stackrel{\text{def}}{=} \Delta \setminus D_L(w)$ and $1 \neq w_1 \in W^{I',I}$, it follows from (3.50), the fact that $L(w)$ and $L(w)^{w_1}$ have the same infinitesimal character (which is the one of $L(\mu_0) \cong L(\mu_0)^{w_1}$) and Lemma 3.1.4 (together with Harish-Chandra's theorem) that the inclusion $\operatorname{soc}_{U(\mathfrak{l}_I)}(H^1(\mathfrak{n}_I, L(w)^{w_1})) \subseteq H^1(\mathfrak{n}_I, L(w)^{w_1})$ induces an isomorphism

$$H^0(\mathfrak{u}_I, \operatorname{soc}_{U(\mathfrak{l}_I)}(H^1(\mathfrak{n}_I, L(w)^{w_1}))) \xrightarrow{\sim} H^0(\mathfrak{u}_I, H^1(\mathfrak{n}_I, L(w)^{w_1})).$$

(ii) We deduce from Lemma 3.3.4 and (3.52) that, for $x \in W(G)$ such that $I = \Delta \setminus D_L(x)$, we must have

$$\begin{aligned}\operatorname{Ext}^1_{U(\mathfrak{g})}(L(x), L(w)^{w_1}) &= \operatorname{Ext}^1_{U(\mathfrak{g})}(M^I(x), L(w)^{w_1})\\ &= \operatorname{Hom}_{U(\mathfrak{l}_I)}(L^I(x), H^1(\mathfrak{n}_I, L(w)^{w_1})) = 0.\end{aligned}$$

Lemma 3.3.7 *Let $j \in \Delta$ and $x, w \in W(G)$ with $D_L(x) = D_L(w) = \{j\}$. Then we have*

$$\operatorname{Hom}_{U(\mathfrak{t})}(x\cdot\mu_0, H^2(\mathfrak{u}, L(w)^{s_j})) = 0. \tag{3.66}$$

Proof We deduce from (3.10) the following exact sequence

$$0 \to H^1(\mathfrak{u}_{\{j\}}, H^1(\mathfrak{n}_{\{j\}}, L(w)^{s_j})) \to H^2(\mathfrak{u}, L(w)^{s_j}) \to H^0(\mathfrak{u}_{\{j\}}, H^2(\mathfrak{n}_{\{j\}}, L(w)^{s_j})) \to 0 \tag{3.67}$$

and recall that $s_j\mathfrak{n}_{\{j\}}s_j = \mathfrak{n}_{\{j\}}$ implies $H^k(\mathfrak{n}_{\{j\}}, L(w)^{s_j}) \cong H^k(\mathfrak{n}_{\{j\}}, L(w))^{s_j}$ for $k \geq 0$ (see (3.48)). By Lemma 3.3.3, for $x' \in W(G)$, we have $H^0(\mathfrak{u}_{\{j\}}, L^{\{j\}}(x')^{s_j}) \neq 0$ if and only if $j \notin D_L(x')$, in which case $H^0(\mathfrak{u}_{\{j\}}, L^{\{j\}}(x')^{s_j}) \cong H^0(\mathfrak{u}_{\{j\}}, L^{\{j\}}(x')) \cong x' \cdot \mu_0$. As $j \in D_L(x)$, $H^2(\mathfrak{n}_{\{j\}}, L(w)^{s_j}) \cong H^2(\mathfrak{n}_{\{j\}}, L(w))^{s_j}$ and $H^2(\mathfrak{n}_{\{j\}}, L(w) \in \mathcal{O}^{\mathfrak{b}_{\{j\}}}_{\mathfrak{l}_{\{j\}},\mathrm{alg}}$ by Proposition 3.1.5, we deduce by dévissage on the constituents of $H^2(\mathfrak{n}_{\{j\}}, L(w)^{s_j})$

$$f\mathrm{Hom}_{U(\mathfrak{t})}(x \cdot \mu_0, H^0(\mathfrak{u}_{\{j\}}, H^2(\mathfrak{n}_{\{j\}}, L(w)^{s_j}))) = 0.$$

Assume as a contradiction that (3.66) fails. Then we deduce by 3.67

$$\mathrm{Hom}_{U(\mathfrak{t})}(x \cdot \mu_0, H^1(\mathfrak{u}_{\{j\}}, H^1(\mathfrak{n}_{\{j\}}, L(w)^{s_j}))) \neq 0. \tag{3.68}$$

Using [33, Theorem 1.10] and a standard argument, one sees that any irreducible constituent of $H^1(\mathfrak{n}_{\{j\}}, L(w))$ has the form $L^{\{j\}}(x'')$ with $x'' \in \{x', s_jx'\}$ for some $L^{\{j\}}(x')$ showing up in the socle of $H^1(\mathfrak{n}_{\{j\}}, L(w))$. By (iv) of Lemma 3.2.7 we have

$$H^0\big(\mathfrak{u}_{\{j\}}, \mathrm{soc}_{U(\mathfrak{l}_{\{j\}})}(H^1(\mathfrak{n}_{\{j\}}, L(w)))\big) \cong H^0(\mathfrak{u}_{\{j\}}, H^1(\mathfrak{n}_{\{j\}}, L(w)))$$

and by Proposition 3.1.5 $H^1(\mathfrak{n}_{\{j\}}, L(w))$ is in $\mathcal{O}^{\mathfrak{b}_{\{j\}}}_{\mathfrak{l}_{\{j\}},\mathrm{alg}}$. Using the explicit list of all indecomposable objects in $\mathcal{O}^{\mathfrak{b}_{\{j\}}}_{\mathfrak{l}_{\{j\}},\mathrm{alg}}$ (see [33, Sect. 3.12]) we easily deduce that each indecomposable direct summand of $H^1(\mathfrak{n}_{\{j\}}, L(w))$ is either irreducible of the form $L^{\{j\}}(x')$, or uniserial of length two with socle $L^{\{j\}}(x')$, cosocle $L^{\{j\}}(s_jx')$ and $j \notin D_L(x')$. Let $M_{\{j\}}$ be such an indecomposable direct summand. It follows from Lemma 3.3.3 (and a straightforward dévissage using $H^2(\mathfrak{u}_{\{j\}}, -) = 0$) that $H^1(\mathfrak{u}_{\{j\}}, M_{\{j\}}^{s_j}) \cong s_jx' \cdot \mu_0$ if $M_{\{j\}} = L^{\{j\}}(x')$, and $H^1(\mathfrak{u}_{\{j\}}, M_{\{j\}}^{s_j}) \cong s_jx' \cdot \mu_0 \oplus x' \cdot \mu_0$ if $M_{\{j\}}$ has length two with socle $L^{\{j\}}(x')$ (and $j \notin D_L(x')$). Since $j \in D_L(x)$, it follows that (3.68) forces $x = s_jx' > x'$ for some x' such that $L^{\{j\}}(x')$ shows up in the socle of $H^1(\mathfrak{n}_{\{j\}}, L(w))$ and $j \notin D_L(x')$ (so $x' \prec x$). By (ii) of Lemma 3.2.7 such an x' satisfies $w \prec x'$. The existence of such a triple $w \prec x' \prec x$ with $D_L(x) = D_L(w) = \{j\}$ and $j \notin D_L(x')$ contradicts Lemma A.7. □

Lemma 3.3.8 *Let $j \in \Delta$, $x, w \in W(G)$ with $D_L(x) = D_L(w) = \{j\}$ and $I \overset{def}{=} \Delta \setminus \{j\}$. Then we have*

$$\mathrm{Hom}_{U(\mathfrak{l}_I)}(L^I(x), H^2(\mathfrak{n}_I, L(w)^{s_j})) = 0. \tag{3.69}$$

Proof As $j \in D_L(w)$, it follows from (i), (ii) of Lemma 3.3.1 (applied with $I' = \Delta \setminus D_L(w) = I$) that $H^k(\mathfrak{n}_I, L(w)^{s_j}) \in \mathcal{O}^{\mathfrak{b}_I}_{\mathfrak{l}_I,\mathrm{alg}}$ for $k \geq 1$ and that $H^0(\mathfrak{n}_I, L(w)^{s_j}) = 0$. As $I \cap D_L(x) = \emptyset$ and thus x has minimal length in $W(L_I)x$, we deduce from Lemma 3.1.2 (applied with $\mathfrak{l}_I$ instead of $\mathfrak{g}$ and with I there being $\emptyset$) that $H^\ell(\mathfrak{u}_I, L^I(w'))_{x\cdot\mu_0} = 0$ for $\ell \geq 1$ and any $w' \in W(G)$. By a dévissage on the constituents of $H^k(\mathfrak{n}_I, L(w)^{s_j})$ we obtain for $\ell \geq 1$ and $k \geq 0$

$$\mathrm{Hom}_{U(\mathfrak{t})}\big(x \cdot \mu_0, H^\ell(\mathfrak{u}_I, H^k(\mathfrak{n}_I, L(w)^{s_j}))\big) = 0. \tag{3.70}$$

By (3.9) (applied with $I' = \emptyset$) we have a spectral sequence of semi-simple $U(\mathfrak{t})$-modules

$$H^{\ell}(\mathfrak{u}_I, H^k(\mathfrak{n}_I, L(w)^{s_j})) \implies H^{\ell+k}(\mathfrak{u}, L(w)^{s_j}).$$

Applying the exact functor $\mathrm{Hom}_{U(\mathfrak{t})}(x \cdot \mu_0, -)$ to this spectral sequence, we deduce in particular from (3.70) and Lemma 3.3.7:

$$\mathrm{Hom}_{U(\mathfrak{t})}(x \cdot \mu_0, H^0(\mathfrak{u}_I, H^2(\mathfrak{n}_I, L(w)^{s_j}))) \cong \mathrm{Hom}_{U(\mathfrak{t})}(x \cdot \mu_0, H^2(\mathfrak{u}, L(w)^{s_j})) = 0. \tag{3.71}$$

Now, from the surjection $U(\mathfrak{l}_I) \otimes_{U(\mathfrak{b}_I)} x \cdot \mu_0 \twoheadrightarrow L^I(x)$ and (3.13) we have

$$\begin{aligned}\mathrm{Hom}_{U(\mathfrak{l}_I)}(L^I(x), H^2(\mathfrak{n}_I, L(w)^{s_j})) &\hookrightarrow \mathrm{Hom}_{U(\mathfrak{l}_I)}(U(\mathfrak{l}_I) \otimes_{U(\mathfrak{b}_I)} x \cdot \mu_0, H^2(\mathfrak{n}_I, L(w)^{s_j}))\\ &\cong \mathrm{Hom}_{U(\mathfrak{t})}(x \cdot \mu_0, H^0(\mathfrak{u}_I, H^2(\mathfrak{n}_I, L(w)^{s_j}))),\end{aligned}$$

which together with (3.71) gives (3.69). □

Proposition 3.3.9 *Let $j \in \Delta$, $w \in W(G)$ such that $D_L(w) = \{j\}$ and $I \stackrel{def}{=} \Delta \setminus \{j\}$. Let $S_0 \stackrel{def}{=} \{x' \mid x' \in W(L_I)w, \ell(x') = \ell(w) + 1, j \notin D_L(x')\}$. Then we have*

$$\dim_E \mathrm{Ext}^2_{U(\mathfrak{g})}(M^I(w), L(w)^{s_j}) = \#S_0$$

and

$$\dim_E \mathrm{Ext}^2_{U(\mathfrak{g})}(M^I(x), L(w)^{s_j}) = 0$$

for each $x \neq w$ satisfying $D_L(x) = \{j\}$.

Proof Let $x \in W(G)$ such that $D_L(x) = \{j\}$ (allowing $x = w$). As $j \in D_L(w)$, it follows from (ii) of Lemma 3.3.1 that $H^0(\mathfrak{n}_I, L(w)^{s_j}) = 0$, which together with Lemma 3.3.8 and (3.12) give an isomorphism

$$\mathrm{Ext}^2_{U(\mathfrak{g})}(M^I(x), L(w)^{s_j}) \cong \mathrm{Ext}^1_{U(\mathfrak{l}_I)}(L^I(x), H^1(\mathfrak{n}_I, L(w)^{s_j})). \tag{3.72}$$

For any $x' \in W(G)$ we have by Lemma 3.3.4 that

$$\mathrm{Hom}_{U(\mathfrak{l}_I)}(L^I(x'), \mathrm{soc}_{U(\mathfrak{l}_I)} H^1(\mathfrak{n}_I, L(w)^{s_j})) = \mathrm{Hom}_{U(\mathfrak{l}_I)}(L^I(x'), H^1(\mathfrak{n}_I, L(w)^{s_j})) \neq 0$$

if and only if $\mathrm{Ext}^1_{U(\mathfrak{g})}(L(x'), L(w)) \neq 0$ and $j \notin D_L(x')$, in which case both dimensions are $\dim_E \mathrm{Ext}^1_{U(\mathfrak{g})}(L(x'), L(w))$. As $j \in D_L(x)$, the vector space in (3.50) is zero, which together with (i) of Remark 3.3.6 and Lemma 3.2.10 (which can then be applied to $\mathfrak{l}_I$ instead of $\mathfrak{g}$, noting that $L^I(x)$ is a twist of $L^I(\mu_0) = L^I(1)$ since $D_L(x) = \{j\}$) give an isomorphism by (3.72)

$$\mathrm{Ext}^2_{U(\mathfrak{g})}(M^I(x), L(w)^{s_j}) \cong \mathrm{Ext}^1_{U(\mathfrak{l}_I)}(L^I(x), \mathrm{soc}_{U(\mathfrak{l}_I)}(H^1(\mathfrak{n}_I, L(w)^{s_j}))). \tag{3.73}$$

By the discussion following (3.72), we see that the dimension of (3.73) is

$$\sum_{x'} \dim_E \mathrm{Ext}^1_{U(\mathfrak{l}_I)}(L^I(x), L^I(x'))\dim_E \mathrm{Ext}^1_{U(\mathfrak{g})}(L(x'), L(w)) \tag{3.74}$$

where x' runs through the elements of $W(G)$ such that $j \notin D_L(x')$ and

$$\mathrm{Ext}^1_{U(\mathfrak{l}_I)}(L^I(x), L^I(x')) \neq 0 \neq \mathrm{Ext}^1_{U(\mathfrak{g})}(L(x'), L(w)).$$

Let $x' \in W(G)$ such that $x' \neq x$ and $x' \neq w$. As x is minimal in $W(L_I)x$, by (ii) of Lemma 3.2.4 (applied to $\mathfrak{l}_I$ instead of $\mathfrak{g}$) and (ii) of Lemma 3.1.3 (with Harish-Chandra's theorem), we have $\mathrm{Ext}^1_{U(\mathfrak{l}_I)}(L^I(x), L^I(x')) \neq 0$ if and only if $x' \in W(L_I)x$ and $\ell(x') = \ell(x) + 1$, in which case it has dimension $\mu(x, x') = 1$. Likewise by (ii) of Lemma 3.2.4 $\mathrm{Ext}^1_{U(\mathfrak{g})}(L(x'), L(w)) \neq 0$ if and only if either $x' \prec w$ or $w \prec x'$, in which case it has dimension $\mu(x', w)$ or $\mu(w, x')$. So in (3.74) we sum up over those $x' \in W(G)$ such that $j \notin D_L(x')$, $x' \in W(L_I)x$ with $\ell(x') = \ell(x) + 1$ (which implies $x \prec x'$), and either $x' \prec w$ or $w \prec x'$.

Since $x \prec x'$, we can't have $x' \prec w$ because this would contradict Lemma A.7. Hence we have $w \prec x'$. Using Lemma 3.2.5, $D_L(w) = \{j\}$ and $j \notin D_L(x')$ then force $x' \in W(L_I)w$ and $\ell(x') = \ell(w) + 1$ (hence $\mu(w, x') = 1$), which together with $x' \in W(L_I)x$ and $D_L(x) = D_L(w) = \{j\}$ force $x = w$. Consequently, we conclude that (3.73) is non-zero only when $x = w$, in which case it has dimension $\#S_0$. □

Lemma 3.3.10 *Let $x, w \in W(G)$ and $I \subseteq \Delta$ such that $x \in W(L_I)w$, $x > w$ and $\ell(x) = \ell(w) + 1$. Let $M \in \mathcal{O}^{\mathfrak{b}}_{\mathrm{alg}}$ be the (unique) length two object with socle $L(x)$ and cosocle $L(w)$ ((ii) of Lemma 3.2.4). Then $H^0(\mathfrak{n}_I, M)$ is the unique length two object in $\mathcal{O}^{\mathfrak{b}_I}_{\mathfrak{l}_I,\mathrm{alg}}$ with socle $L^I(x)$ and cosocle $L^I(w)$.*

Proof Using (ii) of Lemma 3.1.8 the short exact sequence $0 \to L(x) \to M \to L(w) \to 0$ induces an exact sequence

$$0 \to L^I(x) \to H^0(\mathfrak{n}_I, M) \xrightarrow{q} L^I(w) \to H^1(\mathfrak{n}_I, L(x)).$$

Let $\xi : Z(\mathfrak{l}_I) \to E$ be the unique character such that $L^I(w)_\xi \neq 0$, or equivalently $L^I(x)_\xi \neq 0$. Then by (iii) of Lemma 3.1.8 $H^1(\mathfrak{n}_I, L(x))_\xi = 0$, which implies that the map $L^I(w) = L^I(w)_\xi \to H^1(\mathfrak{n}_I, L(x))$ is 0 and hence that q is surjective.

It remains to prove that the short exact sequence $0 \to L^I(x) \to H^0(\mathfrak{n}_I, M) \xrightarrow{q} L^I(w) \to 0$ is non-split. For M' in $\mathcal{O}^{\mathfrak{b}}_{\mathrm{alg}}$ write $[M' : L(x)] \in \mathbb{Z}_{\geq 0}$ for the multiplicity of $L(x)$ in M'. Since $M(w) \cong U(\mathfrak{g}) \otimes_{U(\mathfrak{p}_I)} (U(\mathfrak{l}_I) \otimes_{U(\mathfrak{b}_I)} w \cdot \mu_0)$ we have the following obvious equality (see (3.5) for $M^I(x')$)

$$[M(w) : L(x)] = \sum_{x' \in W(G)} e_I(w, x')[M^I(x') : L(x)] \tag{3.75}$$

where $e_I(w, x') \in \mathbb{Z}_{\geq 0}$ is the multiplicity of $L^I(x')$ in $U(\mathfrak{l}_I) \otimes_{U(\mathfrak{b}_I)} w \cdot \mu_0$. As $x \in W(L_I)w$, $x > w$ and $\ell(x) = \ell(w) + 1$, we have $e_I(w, x) = 1$. As $x > w$ and $\ell(x) = \ell(w) + 1$ we also have $[M(w) : L(x)] = 1$. As $[M^I(x) : L(x)] = 1$ and

$e_I(w, w) = 1$, we see that (3.75) and $[M(w) : L(x)] = 1$ force $[M^I(w) : L(x)] = 0$. Now, assume on the contrary that the above short exact sequence splits. Then $\mathrm{Hom}_{U(\mathfrak{g})}(M^I(w), M) \cong \mathrm{Hom}_{U(\mathfrak{l}_I)}(L^I(w), H^0(\mathfrak{n}_I, M)) \neq 0$, and in particular $L(x)$ must be a constituent of $M^I(w)$ which contradicts $[M^I(w) : L(x)] = 0$. □

The following lemma is essentially proved in [76, (4.71)], we provide a self-contained proof for the reader's convenience.

Lemma 3.3.11 *Let $w \in W(G)$ and $j_1, j_2 \in \Delta$ such that $|j_1 - j_2| = 1$, $s_{j_1}w > w$ and $D_L(w) \cap \{j_1, j_2\} = \emptyset$. Let M be the unique length two $U(\mathfrak{l}_{\{j_1,j_2\}})$-module with socle $L^{\{j_1,j_2\}}(s_{j_1}w)$ and cosocle $L^{\{j_1,j_2\}}(w)$ ((ii) of Lemma 3.2.4). Then we have*

(i) the $U(\mathfrak{l}_{\{j_2\}})$-module $H^1(\mathfrak{l}_{\{j_1,j_2\}} \cap \mathfrak{n}_{\{j_2\}}, M^{s_{j_1}})$ has length two with socle $L^{\{j_2\}}(s_{j_2}s_{j_1}w)$ and cosocle $L^{\{j_2\}}(s_{j_1}w)$;
(ii) the $U(\mathfrak{l}_{\{j_2\}})$-module $H^2(\mathfrak{l}_{\{j_1,j_2\}} \cap \mathfrak{n}_{\{j_2\}}, M^{s_{j_1}})$ has length two with socle $L^{\{j_2\}}(s_{j_2}s_{j_1}s_{j_2}w)$ and cosocle $L^{\{j_2\}}(s_{j_1}s_{j_2}w)$.

Proof Since $D_L(w) \cap \{j_1, j_2\} = \emptyset$, $L^{\{j_1,j_2\}}(w)$ is a twist of $L^{\{j_1,j_2\}}(\mu_0) = L^{\{j_1,j_2\}}(1)$, so that "untwisting everything", we can reduce to the case $\mathfrak{g} = \mathfrak{gl}_3$, $\Delta = \{j_1, j_2\}$, $w = 1$ and $x = s_{j_1}$, in which case we have natural $U(\mathfrak{l}_{\{j_2\}})$-equivariant isomorphisms

$$M = M^{\{j_2\}}(1) = U(\mathfrak{g}) \otimes_{U(\mathfrak{p}_{\{j_2\}})} L^{\{j_2\}}(1) \cong U(\mathfrak{n}^+_{\{j_2\}}) \otimes_E L^{\{j_2\}}(1).$$

Write α_s for the positive simple root corresponding to j_s for $s = 1, 2$, then $\mathfrak{n}_{\{j_2\}} \cong \mathfrak{n}_{-\alpha_1} \oplus \mathfrak{n}_{-\alpha_1-\alpha_2}$ where $\mathfrak{n}_\alpha \subseteq \mathfrak{g}$ is the one dimensional subspace corresponding to a root $\alpha \in \Phi$. Then $M^{s_{j_1}} \cong U(s_{j_1}\mathfrak{n}^+_{\{j_2\}}s_{j_1}) \otimes_E L^{\{j_2\}}(1)^{s_{j_1}} \cong U(\mathfrak{n}_{-\alpha_1}) \otimes_E U(\mathfrak{n}_{\alpha_2}) \otimes_E L^{\{j_2\}}(1)^{s_{j_1}}$ is a free $U(\mathfrak{n}_{-\alpha_1})$-module (of infinite rank). We have $H^0(\mathfrak{n}_{-\alpha_1}, U(\mathfrak{n}_{-\alpha_1})) = 0$ and a $U(\mathfrak{t})$-equivariant isomorphism $H^1(\mathfrak{n}_{-\alpha_1}, U(\mathfrak{n}_{-\alpha_1})) \cong \alpha_1$ (use $U(\mathfrak{n}_{-\alpha_1}) \cong E[X]$). As $\mathfrak{n}^+_{\{j_2\}}$ and hence $s_{j_1}\mathfrak{n}^+_{\{j_2\}}s_{j_1}$ are commutative Lie algebras, we deduce $H^0(\mathfrak{n}_{-\alpha_1}, M^{s_{j_1}}) = 0$. From the (analogue of) (3.10) for the spectral sequence $H^{\ell_1}(\mathfrak{n}_{-\alpha_1-\alpha_2}, H^{\ell_2}(\mathfrak{n}_{-\alpha_1}, M^{s_{j_1}})) \implies H^{\ell_1+\ell_2}(\mathfrak{n}_{\{j_2\}}, M^{s_{j_1}})$ we deduce then $H^k(\mathfrak{n}_{\{j_2\}}, M^{s_{j_1}}) = 0$ for $k \notin \{1, 2\}$ and $U(\mathfrak{b}^+_{\{j_2\}})$-equivariant isomorphisms for $k \in \{1, 2\}$:

$$\begin{aligned}
H^k(\mathfrak{n}_{\{j_2\}}, M^{s_{j_1}}) &\cong H^{k-1}(\mathfrak{n}_{-\alpha_1-\alpha_2}, H^1(\mathfrak{n}_{-\alpha_1}, M^{s_{j_1}})) \\
&\cong H^{k-1}\big(\mathfrak{n}_{-\alpha_1-\alpha_2}, U(\mathfrak{n}_{\alpha_2}) \otimes_E L^{\{j_2\}}(1)^{s_{j_1}} \otimes_E H^1(\mathfrak{n}_{-\alpha_1}, U(\mathfrak{n}_{-\alpha_1}))\big) \\
&\overset{q}{\cong} U(\mathfrak{n}_{\alpha_2}) \otimes_E \big(H^{k-1}(\mathfrak{n}_{-\alpha_1-\alpha_2}, L^{\{j_2\}}(1)^{s_{j_1}}) \otimes_E \alpha_1\big) \\
&\cong U(\mathfrak{n}_{\alpha_2}) \otimes_E \big(H^{k-1}(\mathfrak{n}_{-\alpha_2}, L^{\{j_2\}}(1))^{s_{j_1}} \otimes_E \alpha_1\big) \\
&\cong U(\mathfrak{n}_{\alpha_2}) \otimes_E s_{j_1} \cdot \mu_k = U(\mathfrak{u}^+_{\{j_2\}}) \otimes_E s_{j_1} \cdot \mu_k \\
&\cong U(\mathfrak{l}_{\{j_2\}}) \otimes_{U(\mathfrak{b}_{\{j_2\}})} s_{j_1} \cdot \mu_k
\end{aligned}$$

where $\mu_k \overset{\text{def}}{=} H^{k-1}(\mathfrak{n}_{-\alpha_2}, L^{\{j_2\}}(1))$, where by definition $\mathfrak{u}^+_{\{j_2\}} = \mathfrak{n}_{\alpha_2}$, and where the isomorphism q above is checked to be $U(\mathfrak{b}^+_{\{j_2\}})$-equivariant using that $\mathfrak{n}_{-\alpha_1}$ acts trivially on $L^{\{j_2\}}(1)^{s_{j_1}} \otimes_E H^1(\mathfrak{n}_{-\alpha_1}, U(\mathfrak{n}_{-\alpha_1}))$. By (i) of Lemma 3.3.3 we have $\mu_1 = \mu_0$ and $\mu_2 = s_{j_2} \cdot \mu_0$, and thus $U(\mathfrak{l}_{\{j_2\}})$-equivariant isomorphisms

$$H^1(\mathfrak{n}_{\{j_2\}}, M^{s_{j_1}}) \cong U(\mathfrak{l}_{\{j_2\}}) \otimes_{U(\mathfrak{b}_{\{j_2\}})} s_{j_1} \cdot \mu_0, \ H^2(\mathfrak{n}_{\{j_2\}}, M^{s_{j_1}}) \cong U(\mathfrak{l}_{\{j_2\}}) \otimes_{U(\mathfrak{b}_{\{j_2\}})} s_{j_1} s_{j_2} \cdot \mu_0,$$

where we use the standard fact that a $U(\mathfrak{l}_{\{j_2\}})$-module which is isomorphic to $U(\mathfrak{l}_{\{j_2\}}) \otimes_{U(\mathfrak{b}_{\{j_2\}})} \mu$ as $U(\mathfrak{b}^+_{\{j_2\}})$-module must in fact be isomorphic to $U(\mathfrak{l}_{\{j_2\}}) \otimes_{U(\mathfrak{b}_{\{j_2\}})} \mu$ as $U(\mathfrak{l}_{\{j_2\}})$-module. This finishes the proof. □

Lemma 3.3.12 *Let $w \in W(G)$, $j \in D_L(w)$ and $I \stackrel{def}{=} \Delta \setminus D_L(w)$. Let $M \in \mathcal{O}^{\mathfrak{b}}_{\mathrm{alg}}$ be the unique length two $U(\mathfrak{g})$-module with socle $L(w)$ and cosocle $L(s_j w)$. The $U(\mathfrak{l}_I)$-module $L^I(w)$ appears with multiplicity one in $H^1(\mathfrak{n}_I, M^{s_j}) \in \mathcal{O}^{\mathfrak{b}_I}_{\mathfrak{l}_I,\mathrm{alg}}$, and $H^1(\mathfrak{n}_I, M^{s_j})$ contains a (unique) $U(\mathfrak{l}_I)$-submodule M^1_I with cosocle $L^I(w)$ which fits into a short exact sequence:*

$$0 \longrightarrow \bigoplus_{\substack{x \in W(L_I)w \\ \ell(x)=\ell(w)+1 \\ j \notin D_L(x)}} L^I(x) \longrightarrow M^1_I \longrightarrow L^I(w) \longrightarrow 0. \tag{3.76}$$

Proof Note first that the existence of M is clear from (iii) of Lemma 3.2.2 and that $H^1(\mathfrak{n}_I, M^{s_j})$ is in $\mathcal{O}^{\mathfrak{b}_I}_{\mathfrak{l}_I,\mathrm{alg}}$ by (i) of Lemma 3.3.1 applied with $I' = \emptyset$.

As $j \notin D_L(s_j w)$, by Lemma 3.1.1 and (iii) of Lemma 3.3.1 we have $L(s_j w)^{s_j} \cong L(s_j w) \in \mathcal{O}^{\mathfrak{b}}_{\mathrm{alg}}$ and $L(w)^{s_j} \notin \mathcal{O}^{\mathfrak{b}}_{\mathrm{alg}}$. Since $L(w)^{s_j}$ is the socle of M^{s_j}, we deduce

$$H^0(\mathfrak{n}_I, M^{s_j}) = 0, \tag{3.77}$$

otherwise M^{s_j} would contain a non-zero $U(\mathfrak{g})$-submodule in $\mathcal{O}^{\mathfrak{b}}_{\mathrm{alg}}$ using (3.13). By (ii) of Lemma 3.1.8 we have $H^0(\mathfrak{n}_I, L(s_j w)^{s_j}) \cong L^I(s_j w)$. The short exact sequence $0 \to L(w)^{s_j} \to M^{s_j} \to L(s_j w)^{s_j} \to 0$ then induces the long exact sequence of $U(\mathfrak{l}_I)$-modules

$$\begin{aligned} 0 \to H^1(\mathfrak{n}_I, L(w)^{s_j})/L^I(s_j w) &\to H^1(\mathfrak{n}_I, M^{s_j}) \to H^1(\mathfrak{n}_I, L(s_j w)) \\ &\to H^2(\mathfrak{n}_I, L(w)^{s_j}) \to H^2(\mathfrak{n}_I, M^{s_j}) \to H^2(\mathfrak{n}_I, L(s_j w)). \end{aligned} \tag{3.78}$$

If $L^I(w)$ appears with multiplicity one in $H^1(\mathfrak{n}_I, M^{s_j})$, we define M^1_I to be the unique $U(\mathfrak{l}_I)$-submodule of $H^1(\mathfrak{n}_I, M^{s_j})$ with cosocle $L^I(w)$.

We first prove that $L^I(w)$ appears with multiplicity one in $H^1(\mathfrak{n}_I, L(s_j w))$. It follows from (iv) of Lemma 3.2.7 (applied with $x = s_j w$) that $L^I(w)$ appears with multiplicity one in the socle of $H^1(\mathfrak{n}_I, L(s_j w))$. As w is minimal in $W(L_I)w$ (since $I = \Delta \setminus D_L(w)$) and $H^1(\mathfrak{n}_I, L(s_j w)) \in \mathcal{O}^{\mathfrak{b}_I}_{\mathfrak{l}_I,\mathrm{alg}}$ is semi-simple as $U(\mathfrak{t})$-module (Proposition 3.1.5), we must have the inclusion

$$H^1(\mathfrak{n}_I, L(s_j w))_{w \cdot \mu_0} \subseteq H^0(\mathfrak{u}_I, H^1(\mathfrak{n}_I, L(s_j w)))$$

(otherwise, apply $\mathfrak{u}_I$ to any vector in $H^1(\mathfrak{n}_I, L(s_j w))_{w\cdot\mu_0}$). Together with (3.31), this implies that any copy of $L^I(w)$ that appears in $H^1(\mathfrak{n}_I, L(s_j w))$ must appear in its socle, and thus $L^I(w)$ has multiplicity one in $H^1(\mathfrak{n}_I, L(s_j w))$.

We now prove that $L^I(w)$ is not a Jordan-Hölder factor of $H^1(\mathfrak{n}_I, L(w)^{s_j})$. By (i) of Remark 3.3.6 we have an isomorphism

$$H^0(\mathfrak{u}_I, \mathrm{soc}_{U(\mathfrak{l}_I)}(H^1(\mathfrak{n}_I, L(w)^{s_j}))) \xrightarrow{\sim} H^0(\mathfrak{u}_I, H^1(\mathfrak{n}_I, L(w)^{s_j})).$$

Hence, by the same argument as in the previous paragraph, it is enough to prove that $L^I(w)$ is not in the socle of $H^1(\mathfrak{n}_I, L(w)^{s_j})$. But this follows from the first part of Lemma 3.3.4 (as $w_1 = s_j \notin W(L_I)$). Note that we can then apply Lemma 3.2.10 (with $\mathfrak{l}_I$ instead of $\mathfrak{g}$ and noting that $I = \Delta \setminus D_L(w)$ implies $w \cdot \mu_0 \in \Lambda_I^{\mathrm{dom}}$ hence $L^I(w) \cong L^I(1)$ up to twist) and deduce

$$\mathrm{Ext}^1_{U(\mathfrak{l}_I)}(L^I(w), \mathrm{soc}_{U(\mathfrak{l}_I)}(H^1(\mathfrak{n}_I, L(w)^{s_j}))) \xrightarrow{\sim} \mathrm{Ext}^1_{U(\mathfrak{l}_I)}(L^I(w), H^1(\mathfrak{n}_I, L(w)^{s_j})). \tag{3.79}$$

Assuming that M_I^1 exists (i.e. that $L^I(w)$ "survives" in $H^1(\mathfrak{n}_I, M^{s_j})$), we prove that the radical $\mathrm{rad}(M_I^1)$ (see Sect. 1.4) is contained in the kernel of the short exact sequence (3.76). Note first that, since $s_j \notin W(L_I)$, by Lemma 3.1.7 we have $\mathrm{Ext}^k_{U(\mathfrak{l}_I)}(L^I(w), L^I(s_j w)) = 0$ for $k \geq 0$ and thus

$$\mathrm{Ext}^1_{U(\mathfrak{l}_I)}(L^I(w), H^1(\mathfrak{n}_I, L(w)^{s_j})) \xrightarrow{\sim} \mathrm{Ext}^1_{U(\mathfrak{l}_I)}(L^I(w), H^1(\mathfrak{n}_I, L(w)^{s_j})/L^I(s_j w)).$$

Hence we can forget about the quotient by $L^I(s_j w)$ in (3.78). By the above results, in particular (3.79) and the fact $L^I(w)$ appears in $\mathrm{soc}_{U(\mathfrak{l}_I)} H^1(\mathfrak{n}_I, L(s_j w))$, we must have

$$\mathrm{rad}(M_I^1) \subseteq \mathrm{soc}_{U(\mathfrak{l}_I)}(H^1(\mathfrak{n}_I, L(w)^{s_j})), \tag{3.80}$$

so that $\mathrm{rad}(M_I^1)$ is semi-simple and its constituents $L^I(x)$ satisfy $\mathrm{Ext}^1_{U(\mathfrak{l}_I)}(L^I(w), L^I(x)) \neq 0$ with $x \neq w$. Since w is minimal in $W(L_I)w$, any $x \in W(L_I)w$ is such that $w \leq x$, and if moreover $w \prec x$ it follows from Lemma 3.2.5 that $\ell(x) = \ell(w) + 1$. Then by Lemma 3.1.7 and (ii) of Lemma 3.2.4 we see that, for $x \in W(G)$ with $x \neq w$, $\mathrm{Ext}^1_{U(\mathfrak{l}_I)}(L^I(w), L^I(x))$ is non-zero if and only if $x \in W(L_I)w$ and $\ell(x) = \ell(w) + 1$. Combining this with the equivalence between conditions (i) and (iii) in Lemma 3.3.4, we obtain from (3.80)

$$\mathrm{rad}(M_I^1) \subseteq \bigoplus_{\substack{x \in W(L_I)w \\ \ell(x)=\ell(w)+1 \\ j \notin D_L(x)}} L^I(x). \tag{3.81}$$

Finally we prove that $L^I(w)$ is a constituent of $H^1(\mathfrak{n}_I, M^{s_j})$ and that (3.81) is an isomorphism. It is enough to prove that for any $x \in W(L_I)w$ with $\ell(x) = \ell(w) + 1$ and $j \notin D_L(x)$, $H^1(\mathfrak{n}_I, M^{s_j})$ contains a subquotient with socle $L^I(x)$ and cosocle

$L^I(w)$. As $x \in W(L_I)w$ and $\ell(x) = \ell(w) + 1$, there exists $j_1 \in I$ such that $x = s_{j_1}w$. As $j \in D_L(w) \setminus D_L(x)$, we must have $j_1 \in \{j-1, j+1\}$. As $H^0(\mathfrak{n}_I, M^{s_j}) = 0$ by (3.77), we obtain the following isomorphism from the spectral sequence (3.9) (applied with $I' \stackrel{\text{def}}{=} \{j, j_1\}$ and thus $I \cap I' = \{j_1\}$)

$$H^0(\mathfrak{l}_I \cap \mathfrak{n}_{I'}, H^1(\mathfrak{n}_I, M^{s_j})) \cong H^1(\mathfrak{n}_{I\cap I'}, M^{s_j}). \tag{3.82}$$

Switching the roles of I and I', another application of (3.9) gives an injection

$$H^1(\mathfrak{l}_{I'} \cap \mathfrak{n}_I, H^0(\mathfrak{n}_{I'}, M^{s_j})) \hookrightarrow H^1(\mathfrak{n}_{I\cap I'}, M^{s_j}). \tag{3.83}$$

It follows from Lemma 3.3.10 (applied with $x = w$, $w = s_j w$ and $I = I'$ there) and $s_j \mathfrak{n}_{I'} s_j \cong \mathfrak{n}_{I'}$ that $H^0(\mathfrak{n}_{I'}, M^{s_j}) \cong H^0(\mathfrak{n}_{I'}, M)^{s_j}$ is a length two $U(\mathfrak{l}_{I'})$-module with socle $L^{I'}(w)^{s_j}$ and cosocle $L^{I'}(s_j w)^{s_j}$ (and thus isomorphic to $M_{\{j_1\}}^{s_j}$ where $M_{\{j_1\}} \stackrel{\text{def}}{=} U(\mathfrak{l}_{I'}) \otimes_{U(\mathfrak{l}_{I'} \cap \mathfrak{p}_{\{j_1\}})} L^{\{j_1\}}(s_j w)$). Now by (i) of Lemma 3.3.11 applied with M there being the $U(\mathfrak{l}_{I'})$-module $H^0(\mathfrak{n}_{I'}, M)$ we have

$$H^1(\mathfrak{l}_{I'} \cap \mathfrak{n}_I, H^0(\mathfrak{n}_{I'}, M^{s_j})) \cong U(\mathfrak{l}_{\{j_1\}}) \otimes_{U(\mathfrak{b}_{\{j_1\}})} w \cdot \mu_0,$$

which is a $U(\mathfrak{l}_{\{j_1\}})$-module of length two with socle $L^{\{j_1\}}(s_{j_1} w)$ and cosocle $L^{\{j_1\}}(w)$. By (3.83) and (3.82) we deduce that $U(\mathfrak{l}_{\{j_1\}}) \otimes_{U(\mathfrak{b}_{\{j_1\}})} w \cdot \mu_0$ *embeds into* $H^0(\mathfrak{l}_I \cap \mathfrak{n}_{I'}, H^1(\mathfrak{n}_I, M^{s_j}))$. By (3.13) (applied with $\mathfrak{l}_I$ instead of $\mathfrak{g}$ and $\mathfrak{l}_I \cap \mathfrak{n}_{I'}$ instead of $\mathfrak{n}_I$) this first forces the constituent $L^I(w)$ (the cosocle of $U(\mathfrak{l}_I) \otimes_{U(\mathfrak{b}_I)} w \cdot \mu_0 \cong U(\mathfrak{l}_I) \otimes_{U(\mathfrak{l}_I \cap \mathfrak{p}_{I'})} (U(\mathfrak{l}_{\{j_1\}}) \otimes_{U(\mathfrak{b}_{\{j_1\}})} w \cdot \mu_0)$) to show up in $H^1(\mathfrak{n}_I, M^{s_j})$. Then together with (3.81) this also forces $H^1(\mathfrak{n}_I, M^{s_j})$ to have a length two subquotient with socle $L^I(s_{j_1} w) = L^I(x)$ and cosocle $L^I(w)$. □

3.4 Ext-Squares of $U(\mathfrak{g})$-Modules

We use all previous results of Sect. 3 to construct important finite length $U(\mathfrak{g})$-modules which are uniserial (Lemma 3.4.14) or "squares" (Proposition 3.4.9, Lemma 3.4.11).

Definition 3.4.1 Let $w, x, w_1 \in W(G)$, $I \stackrel{\text{def}}{=} \Delta \setminus D_L(w)$ and $I' \stackrel{\text{def}}{=} \Delta \setminus D_L(x)$.

(i) A $U(\mathfrak{g})$-module $Q_{w_1}(x, w)$ is an Ext-*hypercube* if the following properties hold

- $w_1 \in W^{I,I'}$;
- $Q_{w_1}(x, w)$ is semi-simple as $U(\mathfrak{t})$-module;
- $Q_{w_1}(x, w)$ is finite length multiplicity free and rigid as $U(\mathfrak{g})$-module;
- $Q_{w_1}(x, w)$ has socle $L(w)^{w_1}$ and cosocle $L(x)$.

(ii) An Ext-hypercube is an Ext-*square* if it has Loewy length three, and an Ext-*cube* if it has Loewy length four.

(iii) An Ext-hypercube $Q_{w_1}(x, w)$ is *minimal* if for any $U(\mathfrak{g})$-submodules $M_4 \subsetneq M_3 \subseteq M_2 \subsetneq M_1 \subseteq Q_{w_1}(x, w)$, we have $\mathrm{Ext}^1_{U(\mathfrak{g})}(M_1/M_2, M_3/M_4) \neq 0$ if and only if $M_2 = M_3$, in which case $\mathrm{Ext}^1_{U(\mathfrak{g})}(M_1/M_2, M_3/M_4)$ has dimension 1 and the sequence $0 \to M_3/M_4 = M_2/M_4 \to M_1/M_4 \to M_1/M_2 \to 0$ is non-split.

(iv) When $w_1 = 1$, an Ext-hypercube $Q_1(x, w)$ is *minimal in* $\mathcal{O}^{\mathfrak{b}}_{\mathrm{alg}}$ if $Q_1(x, w)$ lies in $\mathcal{O}^{\mathfrak{b}}_{\mathrm{alg}}$ and if (iii) holds with $\mathrm{Ext}^1_{\mathcal{O}^{\mathfrak{b}}_{\mathrm{alg}}}$ instead of $\mathrm{Ext}^1_{U(\mathfrak{g})}$.

If an Ext-hypercube $Q_{w_1}(x, w)$ exists, then $Q_{w_1^{-1}}(w, x)$ exists for the triple (w, x, w_1^{-1}) by setting $Q_{w_1^{-1}}(w, x) \stackrel{\mathrm{def}}{=} (Q_{w_1}(x, w)^{\tau})^{w_1^{-1}}$.

Remark 3.4.2 It is not true that $Q_{w_1}(x, w)$ is uniquely determined by the triple (x, w, w_1). For example, if $\mathfrak{g} = \mathfrak{gl}_3$, $x = s_1 s_2$, $w = s_2 s_1$ and $w_1 = 1$, one can check that there are three different choices of $Q_{w_1}(x, w)$ with minimal possible length, with middle layer being respectively $L(s_1) \oplus L(s_2)$, $L(s_1) \oplus L(s_1 s_2 s_1)$ and $L(s_2) \oplus L(s_1 s_2 s_1)$.

Recall that Rad^k and Rad_k for $k \geq 0$ are defined in Sect. 1.4.

Lemma 3.4.3 *An* Ext-*square* $Q_{w_1}(x, w)$ *is minimal if and only if it satisfies the following conditions:*

(i) $\mathrm{Ext}^1_{U(\mathfrak{g})}(L(x), L(w)^{w_1}) = 0$;

(ii) $\dim_E \mathrm{Ext}^1_{U(\mathfrak{g})}(C, L(w)^{w_1}) = 1 = \dim_E \mathrm{Ext}^1_{U(\mathfrak{g})}(L(x), C)$ *for any irreducible constituent* C *of* $\mathrm{Rad}_1(Q_{w_1}(x, w))$;

(iii) there exists an irreducible constituent C *of* $\mathrm{Rad}_1(Q_{w_1}(x, w))$ *such that*

$$\mathrm{Ext}^1_{U(\mathfrak{g})}(L(x), \mathrm{Rad}^1(Q_{w_1}(x, w))^C) = 0$$

where $\mathrm{Rad}^1(Q_{w_1}(x, w))^C \subset \mathrm{Rad}^1(Q_{w_1}(x, w))$ *is the unique subobject not containing* C *(in its cosocle).*

Moreover, if $w_1 = 1$, $Q_1(x, w)$ *is minimal in* $\mathcal{O}^{\mathfrak{b}}_{\mathrm{alg}}$ *if and only if the same conditions as above hold replacing everywhere* $\mathrm{Ext}^1_{U(\mathfrak{g})}$ *by* $\mathrm{Ext}^1_{\mathcal{O}^{\mathfrak{b}}_{\mathrm{alg}}}$.

Proof Note that, contrary to what the terminology "square" may suggest, $\mathrm{Rad}_1(Q_{w_1}(x, w))$ can contain more than 2 constituents. We prove the $U(\mathfrak{g})$-module case, the proof for $\mathcal{O}^{\mathfrak{b}}_{\mathrm{alg}}$ being the same. Note first that (i) and (ii) are contained in (iii) of Definition 3.4.1. Let C be an irreducible constituent of $\mathrm{Rad}_1(Q_{w_1}(x, w))$ and assume that (i) and (ii) hold. Since $\dim_E \mathrm{Ext}^1_{U(\mathfrak{g})}(L(x), C) = 1$ and $Q_{w_1}(x, w)$ is an Ext-square, we have a short exact sequence

$$0 \longrightarrow \mathrm{Ext}^1_{U(\mathfrak{g})}(L(x), \mathrm{Rad}^1(Q_{w_1}(x, w))^C) \longrightarrow \mathrm{Ext}^1_{U(\mathfrak{g})}(L(x), \mathrm{Rad}^1(Q_{w_1}(x, w))) \longrightarrow \mathrm{Ext}^1_{U(\mathfrak{g})}(L(x), C) \longrightarrow 0.$$

So we see that (iii) holds if and only if $\dim_E \mathrm{Ext}^1_{U(\mathfrak{g})}(L(x), \mathrm{Rad}^1(Q_{w_1}(x,w))) = 1$ if and only if (iii) holds for all constituents C if and only if $\mathrm{Ext}^1_{U(\mathfrak{g})}(L(x), M) = 0$ for any $M \subsetneq \mathrm{Rad}^1(Q_{w_1}(x,w))$ if and only if there is no Ext-square with socle $L(w)^{w_1}$, cosocle $L(x)$ and middle layer strictly contained in $\mathrm{Rad}_1(Q_{w_1}(x,w))$ if and only if (iii) of Definition 3.4.1 holds. □

Remark 3.4.4 The proof of Lemma 3.4.3 shows that, in the presence of (i) and (ii) of Lemma 3.4.3, condition (iii) of *loc. cit.* is equivalent to: there exists an irreducible constituent C of $\mathrm{Rad}_1(Q_{w_1}(x,w))$ such that $\mathrm{Ext}^1_{U(\mathfrak{g})}((Q_{w_1}(x,w)/L(w)^{w_1})/C, L(w)^{w_1}) = 0$. Likewise with $\mathcal{O}^{\mathfrak{b}}_{\mathrm{alg}}$ instead of $U(\mathfrak{g})$-modules.

In the rest of this section, we construct several minimal Ext-squares. Our main tool to do that are *wall-crossing functors*.

For $\lambda, \mu \in \Lambda$ we first have an exact translation functor (see for instance [33, Sect. 7.1])

$$T^{\mu}_{\lambda} : \mathcal{O}^{\mathfrak{b}}_{\mathrm{alg}} \to \mathcal{O}^{\mathfrak{b}}_{\mathrm{alg}} : M \mapsto \mathrm{pr}_{\mu}(L \otimes_E \mathrm{pr}_{\lambda}(M)) \tag{3.84}$$

where L is the unique finite dimensional $U(\mathfrak{g})$-module with highest weight in the $W(G)$-orbit of $\mu - \lambda$ (for the naive action), and pr_{μ}, pr_{λ} is the projection onto the generalized eigenspace for the infinitesimal character associated with $L(\mu)$, $L(\lambda)$ respectively. Let $j \in \{1, \dots, n-1\}$ and $\mu \in \Lambda$ such that $\langle \mu + \rho, \alpha^\vee \rangle \geq 0$ for all $\alpha \in \Phi^+$ and the stabilizer of μ in $W(G)$ for the dot action is $\{1, s_j\}$. We define $\Theta_{s_j} \overset{\mathrm{def}}{=} T^{w_0 \cdot \mu_0}_{\mu} \circ T^{\mu}_{w_0 \cdot \mu_0} : \mathcal{O}^{\mathfrak{b}}_{\mathrm{alg}} \to \mathcal{O}^{\mathfrak{b}}_{\mathrm{alg}}$ which doesn't depend on the choice of μ as above [33, Example 10.8] and is called a wall-crossing functor ([33, Sect. 7.15], the w_0 comes from the conventions of *loc. cit.* which uses antidominant weights). For any M in $\mathcal{O}^{\mathfrak{b}}_{\mathrm{alg}}$ there are two canonical adjunction maps $\Theta_{s_j}(M) \to M$ and $M \to \Theta_{s_j}(M)$ which are non-zero as soon as both M and $\Theta_{s_j}(M)$ are non-zero [33, Proposition 7.2(a)].

Proposition 3.4.5 *Let $w \in W(G)$ and $j \in \Delta$. Then we have $\Theta_{w_0 s_j w_0}(L(w)) \neq 0$ if and only if $j \in D_R(w)$, in which case $\Theta_{w_0 s_j w_0}(L(w))$ has Loewy length three with both socle and cosocle isomorphic to $L(w)$ and middle layer isomorphic to*

$$L(ws_j) \oplus \bigoplus_{x \in S} L(x)^{\mu(x,w)} \tag{3.85}$$

where $S \overset{def}{=} \{x \mid w \prec x, j \notin D_R(x)\}$.

Proof It follows from [33, Theorem 7.14(c), (f), (g)] together with [33, Theorem 7.9] that $\Theta_{w_0 s_j w_0}(L(w)) \neq 0$ if and only if $j \in D_R(w)$ (i.e. $ws_j < w$), in which case $L(w)$ is both the socle and cosocle of $\Theta_{w_0 s_j w_0}(L(w))$, that each $L(x) \in \mathrm{JH}_{U(\mathfrak{g})}(\Theta_{w_0 s_j w_0}(L(w)))$ with $x \neq w$ satisfies $j \notin D_R(x)$ (i.e. $x < xs_j$), and that if $j \notin D_R(x)$ we have an isomorphism

$$\mathrm{Hom}_{\mathcal{O}^{\mathfrak{b}}_{\mathrm{alg}}}(\mathrm{rad}(\Theta_{w_0 s_j w_0}(L(w))), L(x)) \cong \mathrm{Ext}^1_{\mathcal{O}^{\mathfrak{b}}_{\mathrm{alg}}}(L(w), L(x)). \tag{3.86}$$

By Vogan's Conjecture (which follows from the proof of the Kazhdan-Lusztig Conjecture) $\mathrm{rad}(\Theta_{w_0 s_j w_0}(L(w)))/\mathrm{soc}(\Theta_{w_0 s_j w_0}(L(w)))$ is semi-simple (cf. [33, Sects. 7.15, 8.10]), which together with (i) of Lemma 3.2.4 and (3.86) gives (3.85). Here we use the fact that the only x satisfying $x \prec w$ and $j \in D_R(w) \setminus D_R(x)$ is $x = ws_j$ (see Lemma 3.2.5). □

For $j_0, j_1 \in \Delta$ we define

$$w_{j_1,j_0} \overset{\mathrm{def}}{=} s_{j_1}s_{j_1-1}\cdots s_{j_0} \in W(G) \text{ if } j_1 \geq j_0, \quad w_{j_1,j_0} \overset{\mathrm{def}}{=} s_{j_1}s_{j_1+1}\cdots s_{j_0} \in W(G) \text{ if } j_1 \leq j_0 \tag{3.87}$$

(with $w_{j_0,j_0} = s_{j_0}$). It is clear that $D_L(w_{j_1,j_0}) = \{j_1\}$ and $D_R(w_{j_1,j_0}) = \{j_0\}$, and one can check that w_{j_1,j_0} is the unique partial-Coxeter element satisfying these two properties.

Remark 3.4.6 Let $j_0, j_1 \in \Delta$, $w = w_{j_1,j_0}$ and S as in Proposition 3.4.5, we deduce from (iv) of Lemma A.11 that

$$S = \{w_{j_1,j_0'} \mid j_0' \in \Delta,\ |j_0' - j_0| = 1,\ w_{j_1,j_0'} > w_{j_1,j_0}\}$$

and that $\mu(w,x) = 1$ for $x \in S$. More precisely, $w_{j_1,j_0+1} \in S$ if and only if $j_1 \leq j_0 < n-1$, and $w_{j_1,j_0-1} \in S$ if and only if $j_1 \geq j_0 > 1$.

Lemma 3.4.7 *Let $j_0, j_1, j_0', j_1' \in \Delta$ with $|j_0 - j_0'| = 1$ and $|j_1 - j_1'| = 1$. Let M_0 be the unique length 2 object in $\mathcal{O}^{\mathfrak{b}}_{\mathrm{alg}}$ with socle $L(w_{j_1',j_0})$ and cosocle $L(w_{j_1,j_0})$ (see (i) of Lemmas 3.2.4 and A.11). Then $L(w_{j_1',j_0'})$ occurs with multiplicity one in $\Theta_{w_0 s_{j_0} w_0}(M_0)$, and the unique quotient of $\Theta_{w_0 s_{j_0} w_0}(M_0)$ with socle $L(w_{j_1',j_0'})$ is an* Ext*-square $Q_1(w_{j_1,j_0}, w_{j_1',j_0'})$ in $\mathcal{O}^{\mathfrak{b}}_{\mathrm{alg}}$ with socle $L(w_{j_1',j_0'})$, cosocle $L(w_{j_1,j_0})$ and middle layer contained in $L(w_{j_1,j_0'}) \oplus L(w_{j_1',j_0}) \oplus L(1)$ if $j_0 = j_1$ and $j_0' = j_1'$, and contained in $L(w_{j_1,j_0'}) \oplus L(w_{j_1',j_0})$ otherwise.*

Proof We write $S \overset{\mathrm{def}}{=} \{x \mid w_{j_1,j_0} \prec x,\ j_0 \notin D_R(x)\}$ and $S' \overset{\mathrm{def}}{=} \{x \mid w_{j_1',j_0} \prec x,\ j_0 \notin D_R(x)\}$. By Lemma 3.2.5 we have $\ell(x) = \ell(w_{j_1,j_0}) + 1$ and thus $\mu(w_{j_1,j_0}, x) = 1$ (resp. $\ell(x) = \ell(w_{j_1',j_0}) + 1$ and thus $\mu(w_{j_1',j_0}, x) = 1$) for each $x \in S$ (resp. for each $x \in S'$). We write $L_S \overset{\mathrm{def}}{=} \bigoplus_{x \in S} L(x)$ and similarly for $L_{S'}$. By Proposition 3.4.5 and Lemma A.9 we know that $\Theta_{w_0 s_{j_0} w_0}(L(w_{j_1,j_0}))$ (resp. $\Theta_{w_0 s_{j_0} w_0}(L(w_{j_1',j_0}))$) has Loewy length three with both socle and cosocle $L(w_{j_1,j_0})$ (resp. $L(w_{j_1',j_0})$) and with middle layer $L(w_{j_1,j_0}s_{j_0}) \oplus L_S$ (resp. $L(w_{j_1',j_0}s_{j_0}) \oplus L_{S'}$). Recall that for any non-zero M in $\mathcal{O}^{\mathfrak{b}}_{\mathrm{alg}}$ we have adjunction maps $\Theta_{w_0 s_{j_0} w_0}(M) \to M$, $M \to \Theta_{w_0 s_{j_0} w_0}(M)$ which are non-zero if $\Theta_{w_0 s_{j_0} w_0}(M) \neq 0$. Together with the previous discussion and the exactness of $\Theta_{w_0 s_{j_0} w_0}$, we easily deduce that $\Theta_{w_0 s_{j_0} w_0}(M_0) \to M_0$ is surjective and $M_0 \to \Theta_{w_0 s_{j_0} w_0}(M_0)$ is injective. As $w_{j_1',j_0'} \in S' \setminus S$ (see Remark A.10) and $w_{j_1',j_0'} \neq w_{j_1,j_0}, w_{j_1',j_0}$, we see that $L(w_{j_1',j_0'})$ appears with multiplicity one in $\Theta_{w_0 s_{j_0} w_0}(L(w_{j_1',j_0}))$ and $\Theta_{w_0 s_{j_0} w_0}(M_0)$. We define M (resp. Q) as the unique quotient of $\Theta_{w_0 s_{j_0} w_0}(L(w_{j_1',j_0}))$ (resp. $\Theta_{w_0 s_{j_0} w_0}(M_0)$) with socle $L(w_{j_1',j_0'})$ and cosocle $L(w_{j_1',j_0})$. Then M has length 2 and the (non-zero) composition

$$\Theta_{w_0 s_{j_0} w_0}(L(w_{j_1',j_0})) \hookrightarrow \Theta_{w_0 s_{j_0} w_0}(M_0) \twoheadrightarrow Q$$

must have image M. In particular Q/M is a quotient of $\Theta_{w_0 s_{j_0} w_0}(L(w_{j_1,j_0}))$. Since $L(w_{j_1',j_0'})$ does not occur in M_0 the composition

$$M_0 \hookrightarrow \Theta_{w_0 s_{j_0} w_0}(M_0) \twoheadrightarrow Q$$

must be zero, and thus Q is a quotient of $\Theta_{w_0 s_{j_0} w_0}(M_0)/M_0$. As $L(w_{j_1',j_0}) = \mathrm{soc}(M_0)$ does not occur in $\Theta_{w_0 s_{j_0} w_0}(L(w_{j_1,j_0}))$ and $L(w_{j_1,j_0}) = \mathrm{cosoc}(M_0)$ does not occur in $\Theta_{w_0 s_{j_0} w_0}(L(w_{j_1',j_0}))$ (using the above description), the composition

$$M_0 \hookrightarrow \Theta_{w_0 s_{j_0} w_0}(M_0) \twoheadrightarrow \Theta_{w_0 s_{j_0} w_0}(L(w_{j_1,j_0}))$$

factors through $M_0 \twoheadrightarrow L(w_{j_1,j_0}) \hookrightarrow \Theta_{w_0 s_{j_0} w_0}(L(w_{j_1,j_0}))$. It follows that Q/M is a quotient of $\Theta_{w_0 s_{j_0} w_0}(L(w_{j_1,j_0}))/L(w_{j_1,j_0})$. Consequently, Q is an Ext-square (see Definition 3.4.1) with socle $L(w_{j_1',j_0'})$, cosocle $L(w_{j_1,j_0})$ and middle layer

$$\mathrm{rad}_1(Q) \subseteq L(w_{j_1',j_0}) \oplus L(w_{j_1,j_0} s_{j_0}) \oplus \bigoplus_{x \in S''} L(x) \tag{3.88}$$

where $S'' = \{x \in S \mid \mathrm{Ext}^1_{U(\mathfrak{g})}(L(x), L(w_{j_1',j_0'})) \neq 0\}$.

By (iv) of Lemma A.11 we know that $x \in S$ if and only if $x = w_{j_1,j_0-1}$ with $j_1 \geq j_0$ or $x = w_{j_1,j_0+1}$ with $j_1 \leq j_0$. In both cases $D_L(x) = \{j_1\}$ (and $j_1 \neq j_1'$), hence by (iii) of Lemma A.11 and (3.3) we deduce that $x \in S''$ if and only if $x = w_{j_1,j_0'}$ (we already know $j_1 = j_1' \pm 1$). Since $w_{j_1,j_0} s_{j_0}$ is obviously not in S, we also see that the right hand side of (3.88) is multiplicity free.

Assume that $L(w_{j_1,j_0} s_{j_0})$ occurs in $\mathrm{rad}_1(Q)$. Then $\mathrm{Ext}^1_{U(\mathfrak{g})}(L(w_{j_1,j_0} s_{j_0}), L(w_{j_1',j_0'})) \neq 0$. Assume first $j_0 \neq j_1$ so that $D_L(w_{j_1,j_0} s_{j_0}) = \{j_1\}$, then by (iii) of Lemma A.11 again (and (3.3)) we have $w_{j_1,j_0} s_{j_0} = w_{j_1,j_0'}$. Assume now $j_0 = j_1$, then $w_{j_1,j_0} s_{j_0} = 1 \neq w_{j_1',j_0'}$ and (ii) of Lemma 3.2.4 implies $1 \prec w_{j_1',j_0'}$ which forces $j_1' = j_0'$ by Lemma 3.2.5.

Summing up, we have shown that $\mathrm{rad}_1(Q)$ is contained in $L(w_{j_1,j_0'}) \oplus L(w_{j_1',j_0}) \oplus L(1)$ if $j_0 = j_1$ and $j_0' = j_1'$, and is contained in $L(w_{j_1,j_0'}) \oplus L(w_{j_1',j_0})$ otherwise. □

Lemma 3.4.8 *Let $j_0, j_1, j_0', j_1' \in \Delta$ with $|j_0 - j_0'| = 1$ and $|j_1 - j_1'| = 1$. Let $Q_1(w_{j_1,j_0}, w_{j_1',j_0'})$ be an* Ext*-square with socle $L(w_{j_1',j_0'})$, cosocle $L(w_{j_1,j_0})$ and middle layer contained in $L(w_{j_1,j_0'}) \oplus L(w_{j_1',j_0}) \oplus L(1)$ if $j_0 = j_1$ and $j_0' = j_1'$, and contained in $L(w_{j_1,j_0'}) \oplus L(w_{j_1',j_0})$ otherwise. Then $Q_1(w_{j_1,j_0}, w_{j_1',j_0'})$ is minimal, and in particular unique up to isomorphism, and is in $\mathcal{O}^{\mathrm{b}}_{\mathrm{alg}}$.*

Proof We write $Q \stackrel{\mathrm{def}}{=} Q_1(w_{j_1,j_0}, w_{j_1',j_0'})$ for short and note that the very last statement follows from unicity and Lemma 3.4.7. Moreover unicity follows from (iii) and (ii) of Lemma 3.4.3, hence we only need to prove minimality.

By (iii) of Lemma A.11, we see that (i) and (ii) of Lemma 3.4.3 hold. Hence it suffices to prove (iii) of Lemma 3.4.3. More precisely it suffices to show $\mathrm{Ext}^1_{U(\mathfrak{g})}$

$(L(w_{j_1,j_0}), M) = 0$ where $M \subseteq Q$ is the maximal $U(\mathfrak{g})$-submodule such that $L(w_{j_1,j_0}), L(w_{j_1',j_0}) \notin \mathrm{JH}_{U(\mathfrak{g})}(M)$. Note that the vanishing $\mathrm{Ext}^1_{U(\mathfrak{g})}(L(w_{j_1,j_0}), M) = 0$ actually forces $L(w_{j_1',j_0}) \in \mathrm{JH}_{U(\mathfrak{g})}(\mathrm{rad}_1(Q))$.

Assume on the contrary $\mathrm{Ext}^1_{U(\mathfrak{g})}(L(w_{j_1,j_0}), M) \neq 0$, then an arbitrary object M^+ that fits into a non-split extension $0 \to M \to M^+ \to L(w_{j_1,j_0}) \to 0$ contains a unique submodule Q^- with cosocle $L(w_{j_1,j_0})$. Since $L(w_{j_1,j_0}), L(w_{j_1',j_0}) \notin \mathrm{JH}_{U(\mathfrak{g})}(M)$, any constituent $L(x)$ of Q^- distinct from its cosocle is such that $j_0 \notin D_R(x)$, hence is killed by $\Theta_{w_0 s_{j_0} w_0}$ by the first statement of Proposition 3.4.5. Thus the surjection $Q^- \twoheadrightarrow L(w_{j_1,j_0})$ induces an isomorphism

$$\Theta_{w_0 s_{j_0} w_0}(Q^-) \xrightarrow{\sim} \Theta_{w_0 s_{j_0} w_0}(L(w_{j_1,j_0})) \tag{3.89}$$

and in particular $\Theta_{w_0 s_{j_0} w_0}(Q^-)$ has cosocle $L(w_{j_1,j_0})$ by the second statement of Proposition 3.4.5. Since Q^- is multiplicity free (as Q is) with cosocle $L(w_{j_1,j_0})$, the canonical (non-zero) adjunction map $\Theta_{w_0 s_{j_0} w_0}(Q^-) \to Q^-$ must be a surjection. Note that we have $L(w_{j_1',j_0'}) \notin \mathrm{JH}_{U(\mathfrak{g})}(\Theta_{w_0 s_{j_0} w_0}(L(w_{j_1,j_0})))$ by Proposition 3.4.5 and Remark 3.4.6, and thus $L(w_{j_1',j_0'}) \notin \mathrm{JH}_{U(\mathfrak{g})}(\Theta_{w_0 s_{j_0} w_0}(Q^-))$ by (3.89). However $L(w_{j_1',j_0'})$ is the socle of M and hence occurs in Q^-. This contradicts the surjection $\Theta_{w_0 s_{j_0} w_0}(Q^-) \twoheadrightarrow Q^-$. □

Proposition 3.4.9 *Let $j_0, j_1, j_0', j_1' \in \Delta$ with $|j_0 - j_0'| = 1$ and $|j_1 - j_1'| = 1$. Then there exists a unique minimal* Ext*-square $Q_1(w_{j_1,j_0}, w_{j_1',j_0'})$ with socle $L(w_{j_1',j_0'})$, cosocle $L(w_{j_1,j_0})$ and middle layer $L(w_{j_1,j_0'}) \oplus L(w_{j_1',j_0}) \oplus L(1)$ if $j_0 = j_1$ and $j_0' = j_1'$, and $L(w_{j_1,j_0'}) \oplus L(w_{j_1',j_0})$ otherwise. Moreover $Q_1(w_{j_1,j_0}, w_{j_1',j_0'})$ is in $\mathcal{O}^{\mathfrak{b}}_{\mathrm{alg}}$.*

Proof By Lemma 3.4.8 we know that Q is minimal, unique up to isomorphism and is in $\mathcal{O}^{\mathfrak{b}}_{\mathrm{alg}}$. Switching w_{j_1,j_0} and $w_{j_1',j_0'}$, we have similar statements for $Q_1(w_{j_1',j_0'}, w_{j_1,j_0})$, which implies $Q^\tau \cong Q_1(w_{j_1',j_0'}, w_{j_1,j_0})$ by unicity. Moreover, the proof of Lemma 3.4.8 implies $L(w_{j_1',j_0}) \in \mathrm{JH}_{U(\mathfrak{g})}(\mathrm{rad}_1(Q))$, and similarly $L(w_{j_1,j_0'}) \in \mathrm{JH}_{U(\mathfrak{g})}(\mathrm{rad}_1(Q^\tau))$. Since $\mathrm{JH}_{U(\mathfrak{g})}(\mathrm{rad}_1(Q^\tau)) = \mathrm{JH}_{U(\mathfrak{g})}(\mathrm{rad}_1(Q))$, we deduce $L(w_{j_1,j_0'}) \oplus L(w_{j_1',j_0}) \subseteq \mathrm{rad}_1(Q)$. It remains to prove $L(1) \subseteq \mathrm{rad}_1(Q)$ when $j_0 = j_1$ and $j_0' = j_1'$. Assume on the contrary that $L(1) \notin \mathrm{JH}_{U(\mathfrak{g})}(\mathrm{rad}_1(Q))$ and recall (from our convention) that if a weight in $w \cdot \mu_0 - \mathbb{Z}_{\geq 0}\Phi^+$ occurs in $L(w') = L(w' \cdot \mu_0)$ then $w' \leq w$. It follows that none of $L(s_{j_0'})$, $L(w_{j_0,j_0'})$, $L(w_{j_0',j_0})$ contain weights in $s_{j_0} \cdot \mu_0 - \mathbb{Z}_{>0}\Phi^+$ and hence that $Q_{s_{j_0} \cdot \mu_0}$ is one dimensional and $Q_{s_{j_0} \cdot \mu_0 - \nu}$ is 0 for $\nu \in \mathbb{Z}_{>0}\Phi^+$. Since the action of a non-zero element of $\mathfrak{u}$ modifies a weight by a character in $-\mathbb{Z}_{>0}\Phi^+$, this forces $0 \neq Q_{s_{j_0} \cdot \mu_0} \subseteq H^0(\mathfrak{u}, Q)_{s_{j_0} \cdot \mu_0}$, which together with (3.13) (for $I = \emptyset$) gives a non-zero map $M(s_{j_0}) \to Q$. As Q is multiplicity free with cosocle $L(s_{j_0})$, it has to be a surjection. But this contradicts the fact that $L(s_{j_0'}) \in \mathrm{JH}_{U(\mathfrak{g})}(Q) \setminus \mathrm{JH}_{U(\mathfrak{g})}(M(s_{j_0}))$. Hence, we must have $\mathrm{rad}_1(Q) \cong L(w_{j_1,j_0'}) \oplus L(w_{j_1',j_0}) \oplus L(1)$ when $j_0 = j_1$ and $j_0' = j_1'$. □

Remark 3.4.10 By a similar argument as in Lemmas 3.4.7, 3.4.8 and 3.4.9, one can prove that, for $j_0, j_1 \in \Delta$ with $|j_0 - j_1| = 1$, there exists a unique minimal Ext-square $Q_1(1, w_{j_1,j_0})$ (resp. $Q_1(w_{j_1,j_0}, 1)$) with middle layer $L(s_{j_0}) \oplus L(s_{j_1})$, and that $Q_1(1, w_{j_1,j_0})$, $Q_1(w_{j_1,j_0}, 1)$ are in $\mathcal{O}^{\mathfrak{b}}_{\mathrm{alg}}$.

For $S \subseteq W(G)$, we write $L_S \stackrel{\text{def}}{=} \bigoplus_{x\in S} L(x)$.

Lemma 3.4.11 *Let $j \in \Delta$, $I = \Delta \setminus \{j\}$ and $w \in W(G)$ such that $D_L(w) = \{j\}$. Assume $S_0 \stackrel{def}{=} \{x' \mid x' \in W(L_I)w, \ell(x') = \ell(w) + 1, j \notin D_L(x')\} \neq \emptyset$. Then there exists a unique minimal* Ext*-square $Q_{s_j}(w, w)$ such that*

$$\mathrm{Rad}_1(Q_{s_j}(w, w)) \cong L(s_j w) \oplus L_{S_0}.$$

Proof There exists a unique $U(\mathfrak{g})$-module M_0 with socle $L(w)$ and cosocle $L(s_j w) \oplus L_{S_0}$. Since $L(x')^{s_j} \cong L(x')$ if $j \notin D_L(x')$ ((iii) of Lemma 3.3.1), $M_0^{s_j}$ has socle $L(w)^{s_j}$ and cosocle $L(s_j w) \oplus L_{S_0}$. By (ii) of Lemma 3.2.4 we have

$$\dim_E \mathrm{Ext}^1_{U(\mathfrak{g})}(L(w), M_0^{s_j}/L(w)^{s_j}) = \dim_E \mathrm{Ext}^1_{U(\mathfrak{g})}(L(w), L(s_j w) \oplus L_{S_0}) = 1 + \#S_0. \quad (3.90)$$

By the equivalence between conditions (ii) and (iii) in Lemma 3.3.4 (applied with $I = I' = \Delta \setminus \{j\}$ and $w_1 = s_j$) we have $\mathrm{Ext}^1_{U(\mathfrak{g})}(L(x), L(w)^{s_j}) = 0$ for $x \in W(G)$ such that $D_L(x) = \{j\}$, and in particular for each irreducible constituent $L(x)$ of $M^I(w)$. This together with Proposition 3.3.9 and a dévissage using $0 \to N^I(w) \to M^I(w) \to L(w) \to 0$ implies

$$\dim_E \mathrm{Ext}^2_{U(\mathfrak{g})}(L(w), L(w)^{s_j}) \leq \#S_0. \quad (3.91)$$

The short exact sequence $0 \to L(w)^{s_j} \to M_0^{s_j} \to M_0^{s_j}/L(w)^{s_j} \to 0$ yields the exact sequence

$$0 \to \mathrm{Ext}^1_{U(\mathfrak{g})}(L(w), M_0^{s_j}) \to \mathrm{Ext}^1_{U(\mathfrak{g})}(L(w), M_0^{s_j}/L(w)^{s_j}) \to \mathrm{Ext}^2_{U(\mathfrak{g})}(L(w), L(w)^{s_j})$$

which together with (3.90) and (3.91) implies

$$\dim_E \mathrm{Ext}^1_{U(\mathfrak{g})}(L(w), M_0^{s_j}) \geq 1.$$

In particular, there exists an Ext-square $Q_{s_j}(w, w)$ with socle $L(w)^{s_j}$, cosocle $L(w)$ and middle layer

$$\mathrm{Rad}_1(Q_{s_j}(w, w)) \subseteq M_0^{s_j}/L(w)^{s_j} = L(s_j w) \oplus L_{S_0}.$$

We now prove $L(s_j w) \oplus L_{S_0} \subseteq \mathrm{Rad}_1(Q_{s_j}(w, w))$ and the minimality of $Q_{s_j}(w, w)$.

Step 1: Let M be any $U(\mathfrak{g})$-module with socle $L(w)$ and M' a $U(\mathfrak{g})$-module of finite length with all irreducible constituents in $\mathcal{O}^{\mathfrak{b}}_{\mathrm{alg}}$. As $L(w)^{s_j}$ is not in $\mathcal{O}^{\mathfrak{b}}_{\mathrm{alg}}$ (by (iii) of Lemmas 3.3.1 and 3.1.1), we have $\mathrm{Hom}_{U(\mathfrak{g})}(M', M^{s_j}) = 0$.

Step 2: We show that $Q_{s_j}(w, w)$ is minimal and is unique such that $\mathrm{Rad}_1(Q_{s_j}(w, w)) \subseteq L(s_j w) \oplus L_{S_0}$.
Using again (ii) of Lemma 3.2.4, there exists a unique $U(\mathfrak{g})$-module M_1 with

cosocle $L(w)$ and socle $L(s_j w) \oplus L_{S_0}$. By unicity of M_1 there is a surjection $M_1 \twoheadrightarrow Q_{s_j}(w,w)/L(w)^{s_j}$. Moreover it follows from (i) of Lemma 3.2.4 and (3.27) (applied with $x = w$ and $w = x' \in S_0$) that $M_1/L(s_j w)$ is a quotient of $M(w)$. As $\mathrm{Hom}_{U(\mathfrak{g})}(\ker(M(w) \twoheadrightarrow M_1/L(s_j w)), L(w)^{s_j}) = 0$ since $L(w)^{s_j}$ is not in $\mathcal{O}^{\mathfrak{b}}_{\mathrm{alg}}$, the surjection $M(w) \twoheadrightarrow M_1/L(s_j w)$ induces an embedding $\mathrm{Ext}^1_{U(\mathfrak{g})}(M_1/L(s_j w), L(w)^{s_j}) \hookrightarrow \mathrm{Ext}^1_{U(\mathfrak{g})}(M(w), L(w)^{s_j})$, which using (ii) of Remark 3.3.6 (applied with $I = \emptyset$, $x = w$ and $w_1 = s_j$) implies

$$\mathrm{Ext}^1_{U(\mathfrak{g})}(M_1/L(s_j w), L(w)^{s_j}) = 0. \tag{3.92}$$

Then from $0 \to L(s_j w) \to M_1 \to M_1/L(s_j w) \to 0$ we deduce

$$\mathrm{Ext}^1_{U(\mathfrak{g})}(M_1, L(w)^{s_j}) \hookrightarrow \mathrm{Ext}^1_{U(\mathfrak{g})}(L(s_j w), L(w)^{s_j}). \tag{3.93}$$

But $\dim_E \mathrm{Ext}^1_{U(\mathfrak{g})}(L(s_j w), L(w)^{s_j}) = 1$ by condition (iii) in Lemma 3.3.4 with (3.51) (applied with $I = I' = \Delta \setminus D_L(w)$ and $w_1 = s_j$) and (ii) of Lemma 3.2.4. Since we have $\mathrm{Ext}^1_{U(\mathfrak{g})}(Q_{s_j}(w,w)/L(w)^{s_j}, L(w)^{s_j}) \neq 0$ (by definition of $Q_{s_j}(w,w)$) and

$$\mathrm{Ext}^1_{U(\mathfrak{g})}(Q_{s_j}(w,w)/L(w)^{s_j}, L(w)^{s_j}) \hookrightarrow \mathrm{Ext}^1_{U(\mathfrak{g})}(M_1, L(w)^{s_j}) \tag{3.94}$$

(by Step 1 applied with $M' = \ker(M_1 \twoheadrightarrow Q_{s_j}(w,w)/L(w)^{s_j})$ and $M = L(w)$), we deduce from (3.93) and (3.94) isomorphisms of 1-dimensional vector spaces

$$\mathrm{Ext}^1_{U(\mathfrak{g})}(M_1, L(w)^{s_j}) \xrightarrow{\sim} \mathrm{Ext}^1_{U(\mathfrak{g})}(L(s_j w), L(w)^{s_j})$$
$$\mathrm{Ext}^1_{U(\mathfrak{g})}(Q_{s_j}(w,w)/L(w)^{s_j}, L(w)^{s_j}) \xrightarrow{\sim} \mathrm{Ext}^1_{U(\mathfrak{g})}(M_1, L(w)^{s_j}).$$

The composition gives an isomorphism of 1-dimensional E-vector spaces

$$\mathrm{Ext}^1_{U(\mathfrak{g})}(Q_{s_j}(w,w)/L(w)^{s_j}, L(w)^{s_j}) \xrightarrow{\sim} \mathrm{Ext}^1_{U(\mathfrak{g})}(L(s_j w), L(w)^{s_j})$$

which implies that $L(s_j w)$ appears in $\mathrm{Rad}_1(Q_{s_j}(w,w))$. We now prove minimality. Using again Step 1, we deduce from (3.92)

$$\mathrm{Ext}^1_{U(\mathfrak{g})}((Q_{s_j}(w,w)/L(w)^{s_j})/L(s_j w), L(w)^{s_j}) = 0.$$

By Lemma 3.4.3 and Remark 3.4.4 this implies that $Q_{s_j}(w,w)$ is minimal, and is actually the unique Ext-square with socle $L(w)^{s_j}$, cosocle $L(w)$ and middle layer $\mathrm{Rad}_1(Q_{s_j}(w,w)) \subseteq L(s_j w) \oplus L_{S_0}$.

Step 3: We finally show that $L_{S_0} \subseteq \mathrm{Rad}_1(Q_{s_j}(w,w))$.
Let M_1' be the unique $U(\mathfrak{g})$-module with socle $L(w)$ and cosocle $L(s_j w) \oplus L_{S_0}$, which is in $\mathcal{O}^{\mathfrak{b}}_{\mathrm{alg}}$ (use once more (ii) of Lemma 3.2.4). Then $Q_{s_j}(w,w)$ contains a unique maximal $U(\mathfrak{g})$-submodule Q_1 with socle $L(w)^{s_j}$ and cosocle contained

in $L(s_jw) \oplus L_{S_0}$ and containing $L(s_jw)$. We have $Q_1 \subseteq (M_1')^{s_j}$, $Q_1/L(w)^{s_j} \in \mathcal{O}^{\mathfrak{b}}_{\mathrm{alg}}$ and $\mathrm{Ext}^1_{U(\mathfrak{g})}(L(w), Q_1) \neq 0$. In order to prove the statement, it suffices to show $Q_1 = (M_1')^{s_j}$. Assume on the contrary $Q_1 \subsetneq (M_1')^{s_j}$ and let $S_0' \subsetneq S_0$ such that Q_1 has cosocle $L(s_jw) \oplus L_{S_0'}$ (and socle $L(w)^{s_j}$). The surjection $M^I(w) \twoheadrightarrow L(w)$ together with $\mathrm{Hom}_{U(\mathfrak{g})}(\ker(M^I(w) \to L(w)), Q_1) = 0$ gives an embedding $\mathrm{Ext}^1_{U(\mathfrak{g})}(L(w), Q_1) \hookrightarrow \mathrm{Ext}^1_{U(\mathfrak{g})}(M^I(w), Q_1)$, which forces

$$\mathrm{Ext}^1_{U(\mathfrak{g})}(M^I(w), Q_1) \neq 0. \tag{3.95}$$

By (ii) of Lemma 3.3.1 (applied with $I = I'$ and $w = w_2 = s_j$) we have $H^0(\mathfrak{n}_I, L(w)^{s_j}) = 0$, from which we deduce by dévissage (and (ii) of Lemma 3.1.8) that $H^0(\mathfrak{n}_I, Q_1)$ is in $\mathcal{O}^{\mathfrak{p}_I}_{\mathrm{alg}}$. Since the socle $L(w)^{s_j}$ of Q_1 is not in $\mathcal{O}^{\mathfrak{b}}_{\mathrm{alg}}$, (3.13) implies $H^0(\mathfrak{n}_I, Q_1) = 0$. Hence (3.14) (applied with $M = Q_1$ and $M_I = L^I(w)$) and (3.95) give

$$\mathrm{Hom}_{U(\mathfrak{l}_I)}(L^I(w), H^1(\mathfrak{n}_I, Q_1) \neq 0. \tag{3.96}$$

Let M_2 be the unique $U(\mathfrak{g})$-module with socle $L(w)$ and cosocle $L(s_jw)$. The short exact sequence $0 \to M_2^{s_j} \to Q_1 \to Q_1/M_2^{s_j} \to 0$ and $H^0(\mathfrak{n}_I, Q_1) = 0$ give an exact sequence

$$0 \to H^0(\mathfrak{n}_I, Q_1/M_2^{s_j}) \to H^1(\mathfrak{n}_I, M_2^{s_j}) \to H^1(\mathfrak{n}_I, Q_1) \to H^1(\mathfrak{n}_I, Q_1/M_2^{s_j}). \tag{3.97}$$

Since $Q_1/M_2^{s_j} \cong L_{S_0'}^{s_j} \cong L_{S_0'}$, we have $H^0(\mathfrak{n}_I, Q_1/M_2^{s_j}) \cong L^I_{S_0'}$ by (ii) of Lemma 3.1.8. Let $\xi : Z(\mathfrak{l}_I) \to E$ such that $L^I(w)_\xi \neq 0$. Since $x' \in W(L_I)w$, we have $L^I(x')_\xi \neq 0$, and thus $H^1(\mathfrak{n}_I, L(x'))_\xi = 0$ by (iii) of Lemma 3.1.8. In particular $H^1(\mathfrak{n}_I, Q_1/M_2^{s_j}) \cong \oplus_{x' \in S_0'} H^1(\mathfrak{n}_I, L(x'))$ does not have $L^I(w)$ as Jordan-Hölder factor. By Lemma 3.3.12 $L^I(w)$ has multiplicity one in $H^1(\mathfrak{n}_I, M_2^{s_j})$ and $H^1(\mathfrak{n}_I, M_2^{s_j})$ contains a $U(\mathfrak{l}_I)$-submodule M_I with socle $L^I_{S_0}$ and cosocle $L^I(w)$. From (3.97) we deduce that $L^I(w)$ appears with multiplicity one in $H^1(\mathfrak{n}_I, Q_1)$ and that $H^1(\mathfrak{n}_I, Q_1)$ contains a submodule with socle $L^I_{S_0 \setminus S_0'}$ and cosocle $L^I(w)$ (the image of M_I in $H^1(\mathfrak{n}_I, Q_1)$). Since by assumption $S_0 \setminus S_0' \neq \emptyset$, we must have $\mathrm{Hom}_{U(\mathfrak{l}_I)}(L^I(w), H^1(\mathfrak{n}_I, Q_1)) = 0$, which contradicts (3.96). It follows that $S_0' = S_0$ and $Q_1 = (M_1')^{s_j}$, which finishes the proof. □

Remark 3.4.12 Keep the notation of Lemma 3.4.11 and let $M \subsetneq M_1'$ where M_1' is the unique $U(\mathfrak{g})$-module with socle $L(w)$ and cosocle $L(s_jw) \oplus L_{S_0}$, and assume that $L(s_jw)$ appears in (the cosocle of) M. Then we proved in Step 3 of the above proof that $H^0(\mathfrak{n}_I, M^{s_j}) = 0$ and $\mathrm{Hom}_{U(\mathfrak{l}_I)}(L^I(w), H^1(\mathfrak{n}_I, M^{s_j})) = 0$. By (3.14) (applied with M there being M^{s_j}) it follows that we have for $\ell \leq 1$:

$$\mathrm{Ext}^\ell_{U(\mathfrak{g})}(M^I(w), M^{s_j}) = 0.$$

Example 3.4.13 The following special cases of Lemma 3.4.11 will be useful. Let $j, j' \in \Delta$ and $w \overset{\mathrm{def}}{=} w_{j,j'}$ (so $D_L(w) = \{j\}$). If $j > j'$, then we have $S_0 = \{w_{j+1,j'}\}$

when $j < n - 1$ and $S_0 = \emptyset$ when $j = n - 1$. If $j < j'$, then we have $S_0 = \{w_{j-1,j'}\}$ when $j > 1$ and $S_0 = \emptyset$ when $j = 1$. If $j = j'$, then we have $S_0 = \{w_{j'',j'} \mid j'' \in \Delta, |j - j''| = 1\}$.

Lemma 3.4.14 *Let $j \in \Delta$. There exists a $\mathfrak{z}$-semi-simple uniserial $U(\mathfrak{g})$-module of length 3 with both socle and cosocle $L(1)$ and middle layer $L(s_j)$.*

Proof Upon applying $T_{w_0 \cdot 0}^{w_0 \cdot \mu_0}$ (and using [33, Theorem 7.8]), it is harmless to assume in the rest of the proof that $\mu_0 = 0$. We write $I \overset{\text{def}}{=} \widehat{j}$, Z (resp. Z_I) for the center of G (resp. L_I) and $\mathfrak{z}$ (resp. $\mathfrak{z}_I$) their associated Lie algebras. We write 1_I for the trivial object of $\mathrm{Mod}_{U(\mathfrak{z}_I/\mathfrak{z})}$. Recall we have (for instance using (3.7))

$$\mathrm{Ext}^1_{U(\mathfrak{z}_I/\mathfrak{z})}(1_I, 1_I) \cong \mathrm{Hom}_E(\mathfrak{z}_I/\mathfrak{z}, E) \neq 0.$$

Choose an arbitrary non-split extension $0 \to 1_I \to \widetilde{1}_I \to 1_I \to 0$ in $\mathrm{Mod}_{U(\mathfrak{z}_I/\mathfrak{z})}$. We define $\widetilde{L}^I(1) \overset{\text{def}}{=} L^I(1) \otimes_E \widetilde{1}_I$, which we see in $\mathrm{Mod}_{U(\mathfrak{l}_I/\mathfrak{z})}$ (and thus in $\mathrm{Mod}_{U(\mathfrak{l}_I)}$) writing $\mathfrak{l}_I/\mathfrak{z} \cong \mathfrak{l}_I/\mathfrak{z}_I \times \mathfrak{z}_I/\mathfrak{z}$ and noting that $L^I(1)$ is in $\mathrm{Mod}_{U(\mathfrak{l}_I/\mathfrak{z}_I)}$. It is a non-split extension

$$0 \longrightarrow L^I(1) \longrightarrow \widetilde{L}^I(1) \longrightarrow L^I(1) \longrightarrow 0, \tag{3.98}$$

and $\widetilde{M}^I(1) \overset{\text{def}}{=} U(\mathfrak{g}) \otimes_{U(\mathfrak{p}_I)} \widetilde{L}^I(1)$, which is in the category $\mathcal{O}^{\mathfrak{p}_I,\infty}_{\mathrm{alg}}$ (see the beginning of Sect. 3.1) and is an extension of $U(\mathfrak{g})$-modules

$$0 \longrightarrow M^I(1) \longrightarrow \widetilde{M}^I(1) \longrightarrow M^I(1) \longrightarrow 0. \tag{3.99}$$

We denote by M^+ the pushforward of $\widetilde{M}^I(1)$ along the surjection $M^I(1) \twoheadrightarrow L(1)$ (on the left). Then M^+ is a quotient of $\widetilde{M}^I(1)$ which fits into an exact sequence $0 \to L(1) \to M^+ \to M^I(1) \to 0$. Recall that by [33, Theorem 9.4(c)] and Lemma 3.1.1 $M^I(1)$ is the maximal quotient of $M(1)$ with constituents $L(y)$ such that $D_L(y) \subseteq \Delta \setminus I = \{j\}$. By (iii) of Lemma 3.2.2 and (3.27) $M(1)$ admits a (unique) length 2 quotient M^- with socle $L(s_j)$ and cosocle $L(1)$. Hence M^- is also a quotient of $M^I(1)$. Since the two conditions $D_L(y) \subseteq \{j\}$ and $\ell(y) \leq 1$ force $y \in \{1, s_j\}$, and since $L(1)$, $L(s_j)$ occur with multiplicity 1 in $M(1)$, it follows that any constituent $L(y)$ of $\ker(M^I(1) \twoheadrightarrow M^-)$ satisfies $\ell(y) \geq 2$. As $\mathrm{Ext}^1_{U(\mathfrak{g})}(L(y), L(1)) = 0$ by (ii) of Lemma 3.2.4, a dévissage yields $\mathrm{Ext}^1_{U(\mathfrak{g})}(M^-, L(1)) \xrightarrow{\sim} \mathrm{Ext}^1_{U(\mathfrak{g})}(M^I(1), L(1))$. In particular M^+ admits a unique length 3 quotient M that fits into an exact sequence of $U(\mathfrak{g})$-modules

$$0 \longrightarrow L(1) \longrightarrow M \longrightarrow M^- \longrightarrow 0. \tag{3.100}$$

Let $\xi : Z(\mathfrak{l}_I) \to E$ be the unique infinitesimal character such that $L^I(1)_\xi \neq 0$. By (ii) and (iii) of Lemma 3.1.8 we have $H^0(\mathfrak{n}_I, L(1))_\xi \cong L^I(1)_\xi = L^I(1) \neq 0$ and $H^k(\mathfrak{n}_I, L(1))_\xi = 0$ for $k \geq 1$. Given $L(y) \in \mathrm{JH}_{U(\mathfrak{g})}(M^I(1)) \setminus \{L(1)\}$, we have $y \neq 1$ and $D_L(y) \subseteq \{j\}$ (Lemma 3.1.1), hence $D_L(y) = \{j\}$ and thus $y \notin W(L_I)$. Let $k \geq 0$ and assume $H^k(\mathfrak{n}_I, L(y))_\xi \neq 0$. Let $L^I(z)$ be a constituent of $H^k(\mathfrak{n}_I, L(y))_\xi$

for some $z \in W(L_I)$ (using Proposition 3.1.5 and [33, Theorem 1.10] for L_I). As $L^I(z)_{z\cdot\mu_0} \neq 0$, we have $H^k(\mathfrak{n}_I, L(y))_{z\cdot\mu_0} \neq 0$ (weight spaces). By Lemma 3.1.2 this implies $z \cdot \mu_0 - y \cdot \mu_0 \in \mathbb{Z}_{\geq 0}\Phi^+$, which implies $z \geq y$ in $W(G)$ (in view of our conventions), i.e. y is a subword of z (in $W(G)$). But since $y \notin W(L_I)$ this contradicts $z \in W(L_I)$. Hence we have $H^k(\mathfrak{n}_I, L(y))_\xi = 0$ for $k \geq 0$, and by dévissage we obtain in particular $H^1(\mathfrak{n}_I, M^I(1))_\xi = 0$. It follows from all this that the surjection $M^1(1) \twoheadrightarrow L(1)$ induces an isomorphism $H^0(\mathfrak{n}_I, M^I(1))_\xi \xrightarrow{\sim} H^0(\mathfrak{n}_I, L(1))_\xi \cong L^I(1)$, and that (3.99) induces a short exact sequence

$$0 \to H^0(\mathfrak{n}_I, M^I(1))_\xi \to H^0(\mathfrak{n}_I, \widetilde{M}^I(1))_\xi \to H^0(\mathfrak{n}_I, M^I(1))_\xi \to 0. \quad (3.101)$$

By (3.13) (and (i) of Lemma 3.1.3) we have canonical isomorphisms

$$\begin{aligned} 0 \neq \mathrm{Hom}_{U(\mathfrak{g})}(\widetilde{M}^I(1), \widetilde{M}^I(1)) &\cong \mathrm{Hom}_{U(\mathfrak{l}_I)}(\widetilde{L}^I(1), H^0(\mathfrak{n}_I, \widetilde{M}^I(1))) \\ &\cong \mathrm{Hom}_{U(\mathfrak{l}_I)}(\widetilde{L}^I(1), H^0(\mathfrak{n}_I, \widetilde{M}^I(1))_\xi). \end{aligned}$$

As the identity map on $\widetilde{M}^I(1)$ does not factor through $\widetilde{M}^I(1) \twoheadrightarrow M^I(1)$, the corresponding map $\widetilde{L}^I(1) \to H^0(\mathfrak{n}_I, \widetilde{M}^I(1))_\xi$ (via (3.13)) does not factor through $\widetilde{L}^I(1) \twoheadrightarrow L^I(1)$. Hence it is injective and thus must be an isomorphism using (3.101) and $H^0(\mathfrak{n}_I, M^I(1))_\xi \cong L^I(1)$. As $H^k(\mathfrak{n}_I, L(y))_\xi = 0$ for $k \geq 0$ (in fact $k = 0, 1$ is enough) and $L(y) \in \mathrm{JH}_{U(\mathfrak{g})}(M^I(1)) \setminus \{L(1)\}$, the surjection $\widetilde{\widetilde{M}}^I(1) \twoheadrightarrow M$ induces an isomorphism $\widetilde{L}^I(1) \cong H^0(\mathfrak{n}_I, \widetilde{M}^I(1))_\xi \xrightarrow{\sim} H^0(\mathfrak{n}_I, M)_\xi$. In particular, the short exact sequence (3.100) is non-split, otherwise we would have $\widetilde{L}^I(1) \cong H^0(\mathfrak{n}_I, M)_\xi \cong L^I(1) \oplus H^0(\mathfrak{n}_I, M^-)_\xi \cong L^I(1) \oplus L^I(1)$ which contradicts (3.98). Let us prove that M is actually uniserial (of length 3). If it were not uniserial, it would contain as subquotient a non-split extension of $L(1)$ by $L(1)$. As $U(\mathfrak{z})$ acts on $\widetilde{L}^I(1)$ by scalars (in E), it also acts on $\widetilde{M}^I(1)$ and M by (the same) scalars. Since $\mathrm{Ext}^1_{U(\mathfrak{g}/\mathfrak{z})}(L(1), L(1)) = 0$ (cf. for instance [76, (3.27)]), this would yield a contradiction. So M must be uniserial of length 3, with both socle and cosocle $L(1)$ and middle layer $L(s_j)$. In particular, the unique length 2 $U(\mathfrak{g})$-submodule of M is isomorphic to $(M^-)^\tau$ (see (3.2 for the notation) which is the unique length 2 $U(\mathfrak{g})$-module with socle $L(1)$ and cosocle $L(s_j)$ by (ii) of Lemma 3.2.4. □

Lemma 3.4.15 *Let* $I \overset{def}{=} \Delta \setminus \{n-1\}$.

(i) We have $M^I(w_{n-1,1}) \cong L(w_{n-1,1})$*, and* $M^I(w_{n-1,n-k})$ *is the unique length 2* $U(\mathfrak{g})$*-module with socle* $L(w_{n-1,n-k-1})$ *and cosocle* $L(w_{n-1,n-k})$ *for each* $0 \leq k \leq n-2$ *(with the convention* $w_{n-1,n} = 1$*).*

(ii) Let $1 \leq k \leq n-1$*. Then* $\Theta_{s_k}(M^I(w_{n-1,n-k}))$ *admits a subquotient (resp. quotient) of the form* $M^I(\mu)$ *for some* $\mu \in \Lambda_J^{\mathrm{dom}}$ *if and only if* $\mu \in \{w_{n-1,n-k} \cdot \mu_0, w_{n-1,n-k+1} \cdot \mu_0\}$ *(resp. if and only if* $\mu = w_{n-1,n-k} \cdot \mu_0$*). Moreover,* $\Theta_{s_k}(M^I(w_{n-1,n-k}))$ *fits into the following short exact sequence*

$$0 \to M^I(w_{n-1,n-k+1}) \to \Theta_{s_k}(M^I(w_{n-1,n-k})) \to M^I(w_{n-1,n-k}) \to 0. \quad (3.102)$$

Proof We have $W^{I,\emptyset} = \{x \in W(G) \mid D_L(x) \subseteq \{n-1\}\} = \{1\} \sqcup \{w_{n-1,n-k} \mid 1 \leq k \leq n-1\}$ where the second equality follows from Lemma A.5 and Remark A.6. We fix $w \in W^{I,\emptyset}$, hence we have $w = w_{n-1,n-k}$ (with $w_{n-1,n} = 1$) for some $k \in \{0, \dots, n-1\}$.

We prove (i). By [33, Theorem 9.4(c)] $M^I(w)$ is the maximal length quotient of $M(w)$ which belongs to $\mathcal{O}^{\mathfrak{p}_I}_{\mathrm{alg}}$. We consider an arbitrary constituent $L(x)$ of $M^I(w)$ with $x > w$. By [33, Theorem 9.4(c)] $L(x) \in \mathcal{O}^{\mathfrak{p}_I}_{\mathrm{alg}}$ and thus by Lemma 3.1.1 $x \in W^{I,\emptyset} = \{1\} \sqcup \{w_{n-1,n-k} \mid 1 \leq k \leq n-1\}$. As x is partial-Coxeter (see above Lemma A.2), by (i) of Lemma A.12 $L(x)$ has multiplicity one in $M(x') \subseteq M(w)$ for each x' satisfying $w < x' < x$ (using [33, Theorem 5.1(a)]), so the quotient $M^I(w)$ of $M(w)$ which admits $L(x)$ as a constituent must also admit $L(x')$ as a constituent (use that the composition $M(x') \hookrightarrow M(w) \twoheadrightarrow M^I(w)$ is non-zero as its image contains the constituent $L(x)$). If $\ell(x) > \ell(w) + 1$, we may always choose $w < x' < x$ such that $D_L(x') \not\subseteq \{n-1\}$, but $L(x') \in \mathrm{JH}_{U(\mathfrak{g})}(M^I(w))$ forces $D_L(x') \subseteq \{n-1\}$ by [33, Theorem 9.4(c)] (and Lemma 3.1.1), a contradiction. Hence, we must have $\ell(x) = \ell(w) + 1$, which together with $w = w_{n-1,n-k}$ forces $x = w_{n-1,n-k-1}$ (and thus $k < n-1$). When $0 \leq k < n-1$, by (3.27) we know that a length 2 quotient of $M(w)$ with socle $L(x)$ and cosocle $L(w)$ exists, and is unique by (ii) of Lemma 3.2.4. This finishes the proof of (i).

We prove (ii). Assume now $1 \leq k \leq n-1$. By the description of $M^I(w)$ in (i), Proposition 3.4.5 and Remark 3.4.6 (and the exactness of Θ_{s_k}) we deduce that $\Theta_{s_k}(M^I(w)) \cong \Theta_{s_k}(L(w))$ has Loewy length 3 with socle and cosocle $L(w)$ and middle layer $L(w_{n-1,n-k+1}) \oplus L(w_{n-1,n-k-1})$ if $k < n-1$ (resp. $L(w_{n-1,n-k+1})$ if $k = n-1$). Any $M^I(\mu)$ that appears as a subquotient of $\Theta_{s_k}(L(w))$ satisfies $\mathrm{JH}_{U(\mathfrak{g})}(M^I(\mu)) \subseteq \mathrm{JH}_{U(\mathfrak{g})}(\Theta_{s_k}(L(w)))$, which by the previous description of $\Theta_{s_k}(L(w))$ and (i) forces $\mu \in \{w \cdot \mu_0, w_{n-1,n-k+1} \cdot \mu_0\}$. Moreover, as $\Theta_{s_k}(L(w))$ has cosocle $L(w)$ and $M^I(\mu)$ has cosocle $L(\mu)$, we see that $M^I(\mu)$ can be a quotient of $\Theta_{s_k}(L(w))$ only if $\mu = w \cdot \mu_0$. Finally (3.102) follows from the above explicit structure of $\Theta_{s_k}(M^I(w)) \cong \Theta_{s_k}(L(w))$ and from (i). □

Chapter 4
Computing Ext Groups of Locally Analytic Representations

We prove an important result (Corollary 4.5.11, which follows from Theorem 4.5.10) which enables us to compute the Ext groups (4.14) of certain locally analytic representations of G by a computation of Ext groups purely on the smooth side and purely on the Lie algebra side.

4.1 Fréchet Spaces with $U(\mathfrak{t})$-Action

We define and study certain (left) $U(\mathfrak{t})$-modules over E and canonical Fréchet completions of them. This section has an intersection with [63], but our treatment is self-contained.

Definition 4.1.1 A (semi-simple) $U(\mathfrak{t})$-module M over E is *small* if the following two conditions hold.

- There is a $U(\mathfrak{t})$-equivariant isomorphism $M \cong \bigoplus_{\mu\in\Lambda} M_\mu$ where M_μ is the eigenspace attached to the weight $\mu \in \Lambda = X(T)$, i.e. M is semi-simple with *integral* weights.
- There exist finitely many $\mu_1, \ldots, \mu_k \in \Lambda$ such that $M_\mu \neq 0$ only if $\mu - \mu_{k'} \in \mathbb{Z}_{\geq 0}\Phi^+$ for some $1 \leq k' \leq k$, and M_μ is always a finite dimensional E-vector space.

Note that by [63, Lemma 3.6.1] the second condition in Definition 4.1.1 implies that the set of weights of M is relatively compact ([63, Sect. 2]). For each $\mu \in \Lambda$, we write $\langle\mu\rangle$ for the one dimensional $U(\mathfrak{t})$-module such that $\langle\mu\rangle_\mu \neq 0$. If M is a small $U(\mathfrak{t})$-module M, then so is $\langle\mu\rangle \otimes_E M$ for each $\mu \in \Lambda$. When $\mu \in \mathbb{Z}\Phi^+$, we write $\mu = \sum_{\alpha\in\Delta} \mu_\alpha \alpha$ for some $\mu_\alpha \in \mathbb{Z}$ and set $|\mu| \overset{\text{def}}{=} \sum_{\alpha\in\Delta} \mu_\alpha \in \mathbb{Z}$.

C. Breuil and Z. Qian, *Splitting and Making Explicit the de Rham Complex of the Drinfeld Space*, Lecture Notes in Mathematics 2397,
https://doi.org/10.1007/978-3-032-20095-2_4

Remark 4.1.2 (i) For $\mu, \mu' \in \Lambda$ such that $\mu - \mu' \in \mathbb{Z}\Phi^+$, there always exists $\mu'' \in \Lambda$ such that $(\mu + \mathbb{Z}_{\geq 0}\Phi^+) \cup (\mu' + \mathbb{Z}_{\geq 0}\Phi^+) \subseteq \mu'' + \mathbb{Z}_{\geq 0}\Phi^+$. Putting "together" the weights in $\mu_i + \mathbb{Z}_{\geq 0}\Phi^+$ and $\mu_j + \mathbb{Z}_{\geq 0}\Phi^+$ for all i, j such that $\mu_i - \mu_j \in \mathbb{Z}\Phi^+$, it follows that a small $U(\mathfrak{t})$-module admits a *canonical* decomposition into *small* direct summands indexed by finitely many cosets of $\Lambda/\mathbb{Z}\Phi^+$. By considering each direct summand (with weights contained in some $\mu'' + \mathbb{Z}_{\geq 0}\Phi^+$) separately and twisting it by $\langle -\mu'' \rangle$, we will often reduce ourselves to small $U(\mathfrak{t})$-modules with weights in $\mathbb{Z}_{\geq 0}\Phi^+$ in the sequel.

(ii) Note that the Verma module $M(\mu) = U(\mathfrak{g}) \otimes_{U(\mathfrak{b})} \mu$, which is isomorphic to the twist $\langle \mu \rangle \otimes_E U(\mathfrak{u}^+)$ as $U(\mathfrak{t})$-module, is a small $U(\mathfrak{t})$-module, and thus so is any object in $\mathcal{O}^{\mathfrak{b}}_{\mathrm{alg}}$. Conversely, any $U(\mathfrak{g})$-module which is small as a $U(\mathfrak{t})$-module is necessarily such that $\mathfrak{u}^-$ acts nilpotently, hence is the union of its finitely generated $U(\mathfrak{g})$-submodules, which are all in $\mathcal{O}^{\mathfrak{b}}_{\mathrm{alg}}$. An instructive example of such an M which is not in $\mathcal{O}^{\mathfrak{b}}_{\mathrm{alg}}$ is an infinite direct sum of Verma modules $M(\mu)$ where $\mu \in -\Lambda^+$ and $|\mu|$ tends to $+\infty$ (because for a given weight, only a finite number of such $M(\mu)$ will contribute).

Now we explicitly define a Fréchet topology on a small $U(\mathfrak{t})$-module M as follows. By (i) of Remark 4.1.2, we choose $\mu_1, \dots, \mu_k$ in Λ which are distinct in $\Lambda/\mathbb{Z}\Phi^+$ and such that each weight μ of M is in $\mu_{i(\mu)} + \mathbb{Z}_{\geq 0}\Phi^+$ for a unique $i(\mu) \in \{1, \dots, k\}$. For each weight μ of M, we fix a choice of a p-adic norm $|\cdot|_\mu$ on M_μ, which is equivalent to the choice of an $\mathcal{O}_E$-lattice $\{x \in M_\mu \mid |x|_\mu \leq 1\}$ in the finite dimensional E-vector space M_μ. We write $\mathcal{L} = \{|\cdot|_\mu\}_\mu$ for this collection of norms. Then we define a semi-norm $|\cdot|_{\mathcal{L},r}$ on M for $r \in \mathbb{Q}_{>0}$ by

$$|x|_{\mathcal{L},r} \overset{\mathrm{def}}{=} \max_{\mu} |x_\mu|_\mu r^{|\mu - \mu_{i(\mu)}|} \tag{4.1}$$

where $x = \sum_\mu x_\mu \in M$. It is clear that $|x|_{\mathcal{L},r} \leq |x|_{\mathcal{L},r'}$ for $r' \geq r > 0$. If we modify the μ_i by elements in $\mathbb{Z}\Phi^+$, we get an equivalent semi-norm using that, for each $r \in \mathbb{Q}_{>0}$ and $\nu \in \mathbb{Z}\Phi^+$, there exist $C_{\nu,r}, C'_{\nu,r} \in \mathbb{Q}_{>0}$ such that $C'_{\nu,r} r^{|\mu'-\nu|} \leq r^{|\mu'|} \leq C_{\nu,r} r^{|\mu'-\nu|}$ for $\mu' \in \mathbb{Z}_{\geq 0}\Phi^+$. The countable family of semi-norms $\{|\cdot|_{\mathcal{L},r}\}_{r \in \mathbb{Q}_{>0}}$ (or equivalently the countable family $\{|\cdot|_{\mathcal{L},r}\}_{r \in p^{\mathbb{Q}}}$) defines a Fréchet topology on M which we denote by $\mathcal{T}_{\mathcal{L}}$. We write $\widehat{M}_{\mathcal{L}}$ for the completion of M under $\mathcal{T}_{\mathcal{L}}$. For each $U(\mathfrak{t})$-submodule $M' \subseteq M$, the induced subspace topology on M' can be defined by the family of semi-norms $\{|\cdot|_{\mathcal{L}',r}\}_{r \in \mathbb{Q}_{>0}}$ associated with the collection $\mathcal{L}' = \{|\cdot|'_\mu\}_\mu$ where $|\cdot|'_\mu$ is the restriction of $|\cdot|_\mu$ to M'_μ.

For later convenience, we introduce the following definition.

Definition 4.1.3 Given a small $U(\mathfrak{t})$-module M, a semi-norm $|\cdot|$ on M is called *standard* if, up to equivalence of semi-norms, it satisfies $|x| = \max_\mu |x_\mu|$ for $x = \sum_\mu x_\mu \in M$.

Fort instance the semi-norm (4.1) is standard for any $r \in \mathbb{Q}_{>0}$.

Remark 4.1.4 Let M be a small $U(\mathfrak{t})$-module with a Fréchet topology $\mathcal{T_L}$ as above. We set $M_{\Lambda'} \overset{\text{def}}{=} \bigoplus_{\mu\in\Lambda'} M_\mu$ for any subset $\Lambda' \subseteq \Lambda$, i.e. we only keep in $M_{\Lambda'}$ the weights of M that are in Λ'. We clearly have $M \cong M_{\Lambda'} \oplus M_{\Lambda\backslash\Lambda'}$. We equip $M_{\Lambda'}$ and $M_{\Lambda\backslash\Lambda'}$ with the subspace topology (again denoted by $\mathcal{T_L}$) induced from M. It is then clear from (4.1) that the original Fréchet topology $\mathcal{T_L}$ on M is equivalent to the direct sum topology on $M_{\Lambda'} \oplus M_{\Lambda\backslash\Lambda'}$, and thus we have a $U(\mathfrak{t})$-equivariant topological isomorphism $\widehat{M}_{\mathcal{L}} \cong \widehat{M}_{\Lambda',\mathcal{L}} \oplus \widehat{M}_{\Lambda\backslash\Lambda',\mathcal{L}}$.

Lemma 4.1.5 *Let M be a small $U(\mathfrak{t})$-module and let $\mathcal{L} = \{|\cdot|_\mu\}_\mu$ be as above. Let $q \in \mathrm{End}_{U(\mathfrak{t})}(M)$ with $q|_{M_\mu} = a_\mu \in \mathbb{Z}$ for each $\mu \in \Lambda$ (with $a_\mu \overset{\text{def}}{=} 0$ if $M_\mu = 0$). Then q is continuous for the topology on M given by the semi-norms (4.1) and extends to a continuous endomorphism $\widehat{q} \in \mathrm{End}_{U(\mathfrak{t})}\widehat{M}_{\mathcal{L}}$. Assume moreover that there exists $C \in \mathbb{Z}_{\geq 0}$ such that $|a_\mu|_\infty \leq |\mu - \mu_{i(\mu)}|^C$ for each weight μ of M (with the notation of (4.1)), then $\widehat{q}$ is strict, and is invertible if and only if $a_\mu \neq 0$ for each μ such that $M_\mu \neq 0$.*

Proof The continuity of q is clear from the definition of $|\cdot|_{\mathcal{L},r}$ as well as the fact that $|a_\mu|_p \leq 1$.

Now we assume that there exists $C \in \mathbb{Z}_{\geq 0}$ such that $|a_\mu|_\infty \leq |\mu - \mu_{i(\mu)}|^C$ for any μ such that $M_\mu \neq 0$. We set $\Lambda_0 \overset{\text{def}}{=} \{\mu \in \Lambda \mid M_\mu \neq 0, a_\mu = 0\}$, $\Lambda_1 \overset{\text{def}}{=} \{\mu \in \Lambda \mid M_\mu \neq 0, a_\mu \neq 0\}$ and $M_i \overset{\text{def}}{=} \bigoplus_{\mu\in\Lambda_i} M_\mu$ for $i = 0, 1$. We have a $U(\mathfrak{t})$-equivariant topological isomorphism $\widehat{M}_{\mathcal{L}} \cong \widehat{M}_{0,\mathcal{L}} \oplus \widehat{M}_{1,\mathcal{L}}$ by Remark 4.1.4. As $\widehat{q}(x) = \sum_{\mu\in\Lambda} a_\mu x_\mu$ for each convergent sum $x = \sum_{\mu\in\Lambda} x_\mu \in \widehat{M}_{\mathcal{L}}$, it is clear that $\ker(\widehat{q}) = \widehat{M}_{0,\mathcal{L}}$ and thus we have a $U(\mathfrak{t})$-equivariant topological isomorphism $\widehat{M}_{\mathcal{L}}/\ker(\widehat{q}) \cong \widehat{M}_{1,\mathcal{L}}$. It suffices to show that $q' : M_1 \to M_1, \sum_{\mu\in\Lambda_1} x_\mu \mapsto \sum_{\mu\in\Lambda_1} a_\mu^{-1} x_\mu$ extends continuously to $\widehat{M}_{1,\mathcal{L}}$ (this will necessarily give the inverse of $\widehat{q}|_{\widehat{M}_{1,\mathcal{L}}}$). For each $r' > r > 0$ in $\mathbb{Q}$, our assumption $|a_\mu|_\infty \leq |\mu - \mu_{i(\mu)}|^C$ (together with $a_\mu \in \mathbb{Z}$ and $a_\mu \neq 0$ for $\mu \in \Lambda_1$) implies the existence of $C_{r,r'} \in \mathbb{Q}_{>0}$ such that $|a_\mu|_p^{-1}(r/r')^{|\mu-\mu_{i(\mu)}|} \leq C_{r,r'}$ for each $\mu \in \Lambda_1$, which together with (4.1) gives the continuity of q' on M_1. □

Remark 4.1.6 It follows from Lemma 4.1.5 that the $U(\mathfrak{t})$-action on M extends to a continuous $U(\mathfrak{t})$-action on $\widehat{M}_{\mathcal{L}}$ (i.e. any element of $U(\mathfrak{t})$ acts continuously): an arbitrary element $x \in \widehat{M}_{\mathcal{L}}$ can be expressed as a convergent infinite sum $\sum_{\mu\in\Lambda} x_\mu$ with $x_\mu \in M_\mu$ for $\mu \in \Lambda$, and $t \cdot x = \sum_{\mu\in\Lambda} \mu(t) x_\mu$. In particular, we have $x \in (\widehat{M}_{\mathcal{L}})_\mu$, i.e. $t \cdot x = \mu(t)x$ for $t \in \mathfrak{t}$, if and only if $x = x_\mu$. In other words, the embedding $M \hookrightarrow \widehat{M}_{\mathcal{L}}$ induces an equality $M_\mu = (\widehat{M}_{\mathcal{L}})_\mu$ for each $\mu \in \Lambda$.

It is convenient to introduce the following definitions.

Definition 4.1.7 (i) Let V be a Fréchet space over E. We say that V is a *Fréchet $U(\mathfrak{t})$-module* if it is equipped with a continuous $U(\mathfrak{t})$-action.

(ii) We say that a Fréchet $U(\mathfrak{t})$-module V is *small* if there exists a (semi-simple) small $U(\mathfrak{t})$-module M (see Definition 4.1.1) and a collection of norms $\mathcal{L} = \{|\cdot|_\mu\}_\mu$ such that we have a $U(\mathfrak{t})$-equivariant topological isomorphism $V \cong \widehat{M}_{\mathcal{L}}$.

Remark 4.1.8 Given a small Fréchet $U(\mathfrak{t})$-module $V \cong \widehat{M}_{\mathcal{L}}$, it is clear that the Fréchet $U(\mathfrak{t})$-module $\langle \mu_1 \rangle \otimes_E V$ is small for each $\mu_1 \in \Lambda$.

We refer to [65, Sect. 17.B] for the definition and properties of the semi-norm tensor product of two semi-norms.

Lemma 4.1.9 *Let $V \cong \widehat{M}_{\mathcal{L}}$ be a small Fréchet $U(\mathfrak{t})$-module and N a finite dimensional $U(\mathfrak{t})$-module (with its canonical Banach topology). We fix an arbitrary norm $|\cdot|_{N,\nu}$ on N_ν for each $\nu \in \Lambda$ such that $N_\nu \neq 0$.*

(i) For $\mu \in \Lambda$ and $\sum_\nu y_\nu \otimes_E x_{\mu-\nu} \in (N \otimes_E M)_\mu$ with $y_\nu \in N_\nu \neq 0$ and $x_{\mu-\nu} \in M_{\mu-\nu}$ we define

$$\left|\sum_\nu y_\nu \otimes_E x_{\mu-\nu}\right|_{N\otimes_E M,\mu} \stackrel{def}{=} \max_\nu |y_\nu|_{N,\nu} |x_{\mu-\nu}|_{\mu-\nu}. \tag{4.2}$$

Then $N \otimes_E V \cong (\widehat{N \otimes_E M})_{\mathcal{L}'}$ where the collection of norms $\mathcal{L}'$ on the weight spaces $(N \otimes_E M)_\mu$ is uniquely determined by (4.2). In particular $N \otimes_E V$ is a small Fréchet $U(\mathfrak{t})$-module.

(ii) For any standard semi-norm $|\cdot|$ on M (Definition 4.1.3) the semi-norm $(\max_\nu |\cdot|_{N,\nu}) \otimes_E |\cdot|$ on $N \otimes_E M$ is standard.

Proof We prove (i). Note first that $N \otimes_E M$ is a small $U(\mathfrak{t})$-module and that the Fréchet topology on $N \otimes_E V$ is defined by the family of semi-norms $\{(\max_\nu |\cdot|_{N,\nu}) \otimes_E |\cdot|_{\mathcal{L},r}\}_{r\in\mathbb{Q}_{>0}}$. Let $z = \bigoplus_{\mu\in\Lambda} z_\mu \in N \otimes_E M$ with $z_\mu = \sum_\nu y_\nu \otimes_E x_{\mu-\nu}$ for some y_ν and $x_{\mu-\nu}$ as in (4.2). We have by (4.1)

$$|z|_{\mathcal{L}',r} = \max_{\mu,\nu} |y_\nu|_{N,\nu} |x_{\mu-\nu}|_{\mu-\nu} r^{|\mu-\mu_{i(\mu)}|}.$$

The evaluation of the semi-norm $(\max_\nu |\cdot|_{N,\nu}) \otimes_E |\cdot|_{\mathcal{L},r}$ on z is

$$\max_{\mu,\nu} |y_\nu|_{N,\nu} |x_{\mu-\nu}|_{\mu-\nu} r^{|\mu-\nu-\mu_{i(\mu-\nu)}|}. \tag{4.3}$$

As there exists only finitely many ν such that $N_\nu \neq 0$, we conclude from (4.3) that for $r \in \mathbb{Q}_{>0}$ the semi-norms $|\cdot|_{\mathcal{L}',r}$ and $(\max_\nu |\cdot|_{N,\nu}) \otimes_E |\cdot|_{\mathcal{L},r}$ are equivalent on $N \otimes_E M$. It follows that $N \otimes_E V \cong (\widehat{N \otimes_E M})_{\mathcal{L}'}$.

The proof of (ii) is an easy exercise that is left to the reader. □

The following result is contained in [43, Sect. 1.3.1], we reproduce a proof for the reader's convenience.

Lemma 4.1.10 *Let V_0 be a small Fréchet $U(\mathfrak{t})$-module and $V \subseteq V_0$ a closed Fréchet $U(\mathfrak{t})$-submodule. Then V is again small. In particular $\bigoplus_{\mu\in\Lambda} V_\mu$ is a small $U(\mathfrak{t})$-module which is dense in V.*

Proof We fix throughout this proof a small $U(\mathfrak{t})$-module M and a collection of norms $\mathcal{L} = \{|\cdot|_\mu\}_\mu$ such that we have a $U(\mathfrak{t})$-equivariant topological isomorphism $V_0 \cong \widehat{M}_{\mathcal{L}}$. We may write $x = \sum_{\mu\in\Lambda} x_\mu$ as a convergent infinite sum with $x_\mu \in M_\mu$ for each $\mu \in \Lambda$. Note that $V_\mu \subseteq (V_0)_\mu = M_\mu$ is finite dimensional for each $\mu \in \Lambda$. Hence to prove that V is small, it suffices to show that $\bigoplus_{\mu\in\Lambda} V_\mu$ is dense in V. In other words, for each convergent infinite sum $x = \sum_{\mu\in\Lambda} x_\mu \in V \subseteq V_0$ with $x_\mu \in M_\mu$, we need to show that

$$x_\mu \in V. \tag{4.4}$$

Let $e \in \mathfrak{t}$ be an element such that $\mu(e) \in \mathbb{Z}$ for all $\mu \in \Lambda$ and $\alpha(e) \in \mathbb{Z}_{>0}$ for all $\alpha \in \Delta$. Replacing V_0 by $\langle -\mu'' \rangle \otimes_E V_0$ for some $\mu'' \in \Lambda$, we can moreover assume that $\mu(e) \in \mathbb{Z}_{\geq 0}$ for all the weights μ of V_0. For each $N \in \mathbb{Z}_{\geq 0}$, we set

$$\Lambda_N \stackrel{\text{def}}{=} \{\mu \text{ weight of } V_0 \mid \mu(e) = N\}$$

and similarly for $\Lambda_{\leq N}$, $\Lambda_{<N}$ and $\Lambda_{>N}$. Writing $\mu = \mu_{i(\mu)} + \sum_{\alpha\in\Delta} \mu_\alpha \alpha$ with $\mu_{i(\mu)}$ as in (4.1) and $\mu_\alpha \in \mathbb{Z}_{\geq 0}$, we have $\mu(e) = \mu_{i(\mu)}(e) + \sum_{\alpha\in\Delta} \mu_\alpha \alpha(e)$. Since $\alpha(e) \in \mathbb{Z}_{>0}$ for all α and since there is a finite number of μ_i, there is only a finite number (possibly 0) of tuples $(\mu_\alpha)_{\alpha\in\Delta}$ in $\mathbb{Z}_{\geq 0}^{\Delta}$ such that $\sum_{\alpha\in\Delta} \mu_\alpha \alpha(e) = N - \mu_i(e)$ for some i, i.e. Λ_N is always a finite set. Therefore, to prove (4.4), it suffices to prove by an increasing induction on $N \in \mathbb{Z}_{\geq 0}$ that

$$\sum_{\mu\in\Lambda_N} x_\mu \in V \tag{4.5}$$

for each $N \in \mathbb{Z}_{\geq 0}$ (use the action of U(t) on $\sum_{\mu\in\Lambda_N} x_\mu$ and the fact the x_μ are weight vectors with distinct weights to deduce that each x_μ is then also in V). We fix $N \in \mathbb{Z}_{\geq 0}$, assume inductively that (4.5) holds if N is replaced by any $N' < N$, and prove the statement for N. Upon replacing x by $x - \sum_{N'<N} \sum_{\mu\in\Lambda_{N'}} x_\mu = x - \sum_{\mu\in\Lambda_{<N}} x_\mu$, we can assume $x_\mu = 0$ for $\mu \in \Lambda_{<N}$.

For each $N_1 \in \mathbb{Z}_{>0}$, we set

$$\Theta_e^{N,N_1} \stackrel{\text{def}}{=} \frac{\prod_{k=1}^{N_1}(N+k-e)}{N_1!} \in U(\mathfrak{t}).$$

Then x_μ for μ a weight of V_0 is an eigenvector for Θ_e^{N,N_1} with eigenvalue

$$C_{\mu,N,N_1} \stackrel{\text{def}}{=} \frac{\prod_{k=1}^{N_1}(N-\mu(e)+k)}{N_1!} \in \mathbb{Z}.$$

In particular, we have $C_{\mu,N,N_1} = 1$ if $\mu \in \Lambda_N$, and $C_{\mu,N,N_1} = 0$ if and only if $\mu \in \Lambda_{N'}$ for some $N < N' \leq N + N_1$. As $\Theta_e^{N,N_1} \in U(\mathfrak{t})$ and V is $U(\mathfrak{t})$-stable, we have $\Theta_e^{N,N_1}(x) \in V$. Since V is closed in V_0, to prove (4.5), it suffices to show that the sequence $\{\Theta_e^{N,N_1}(x)\}_{N_1\in\mathbb{Z}_{\geq 1}}$ of vectors of V converges to $\Theta_e^{N,N_1}(\sum_{\mu\in\Lambda_N} x_\mu) = \sum_{\mu\in\Lambda_N} x_\mu$ inside V_0 when $N_1 \to +\infty$, or equivalently that $\{\Theta_e^{N,N_1}(x - \sum_{\mu\in\Lambda_N} x_\mu)\}_{N_1\in\mathbb{Z}_{\geq 1}}$

converges to zero in V_0 when $N_1 \to +\infty$. As $C_{\mu,N,N_1} \in \mathbb{Z}$, we have $|C_{\mu,N,N_1}|_p \leq 1$ and thus for $r \in \mathbb{Q}_{>0}$:

$$\left| \Theta_e^{N,N_1}\Big(x - \sum_{\mu \in \Lambda_N} x_\mu\Big) \right|_{\mathcal{L},r} = \left| \sum_{\mu \in \Lambda_{>N+N_1}} C_{\mu,N,N_1} x_\mu \right|_{\mathcal{L},r} \leq \left| \sum_{\mu \in \Lambda_{>N+N_1}} x_\mu \right|_{\mathcal{L},r}.$$

Since $|\mu - \mu_{i(\mu)}| = \sum_{\alpha \in \Delta} \mu_\alpha$ tends to $+\infty$ if and only if $\mu(e) = \mu_{i(\mu)}(e) + \sum_{\alpha \in \Delta} \mu_\alpha \alpha(e)$ tends to $+\infty$, we see that the convergence of the infinite sum $x = \sum_{\mu \in \Lambda} x_\mu$ in V_0 implies $\varinjlim_{N_1 \to +\infty} |\sum_{\mu \in \Lambda_{>N+N_1}} x_\mu|_{\mathcal{L},r} = 0$ for $r \in \mathbb{Q}_{>0}$, which gives the desired result. □

Remark 4.1.11 By Lemma 4.1.10, there exists a natural bijection between $U(\mathfrak{t})$-submodules of M and closed Fréchet $U(\mathfrak{t})$-submodules of $\widehat{M}_{\mathcal{L}}$, given by sending $M' \subseteq M$ to its closure in $\widehat{M}_{\mathcal{L}}$, with inverse given by sending a closed Fréchet $U(\mathfrak{t})$-submodule $V \subseteq \widehat{M}_{\mathcal{L}}$ to its dense $U(\mathfrak{t})$-submodule

$$\bigoplus_{\mu \in \Lambda} V_\mu = \bigoplus_{\mu \in \Lambda} V \cap M_\mu = V \cap \Big(\bigoplus_{\mu \in \Lambda} M_\mu\Big) = V \cap M.$$

This is special case of [43, Satz 1.3.19], see also [63, Proposition 2.0.1].

Remark 4.1.12 Let M be a small $U(\mathfrak{t})$-module, $|\cdot|$ a standard semi-norm on M (see Definition 4.1.3) and $\widehat{M}$ the corresponding completion. The proof of Lemma 4.1.10 actually shows that, for any closed $U(\mathfrak{t})$-stable Banach subspace $V \subseteq \widehat{M}$, the subspace $V \cap M$ is dense in V. In particular, there exists a natural bijection between $U(\mathfrak{t})$-submodules of M and closed $U(\mathfrak{t})$-stable Banach subspaces of $\widehat{M}$ which sends $M' \subseteq M$ to its closure $\widehat{M}'$ in $\widehat{M}$. Note that the induced semi-norm on the submodule M' (resp. on the quotient M/M') of M is again standard, with $\widehat{M}'$ (resp. $\widehat{M}/\widehat{M}'$) being the corresponding completion.

Lemma 4.1.13 *Let V_0 be a small Fréchet $U(\mathfrak{t})$-module and $V \subseteq V_0$ a closed Fréchet $U(\mathfrak{t})$-submodule. Then there exists another closed Fréchet $U(\mathfrak{t})$-submodule $V' \subseteq V_0$ such that the natural map $V \oplus V' \to V_0$ is a $U(\mathfrak{t})$-equivariant topological isomorphism.*

Proof We fix a small $U(\mathfrak{t})$-module M and a collection of norms $\mathcal{L} = \{|\cdot|_\mu\}_\mu$ such that $V_0 \cong \widehat{M}_{\mathcal{L}}$. By Remark 4.1.11, the choice of V' is equivalent to the choice of a subspace $V'_\mu \subseteq M_\mu$ for each $\mu \in \Lambda$. From the definition of the semi-norms $|\cdot|_{\mathcal{L},r}$ for $r \in \mathbb{Q}_{>0}$ in (4.1), we see that it suffices to construct $V'_\mu \subseteq M_\mu$ such that $V_\mu \oplus V'_\mu \cong M_\mu$ and

$$|x|_\mu = \max\{|y_\mu|_\mu, |z_\mu|_\mu\} \tag{4.6}$$

for $x = y + z \in M_\mu$ with $y \in V_\mu$ and $z \in V'_\mu$. If we set $M_\mu^\circ \overset{\text{def}}{=} \{x \in M_\mu \mid |x|_\mu \leq 1\}$, which is an $\mathcal{O}_E$-lattice in the E-vector space M_μ, then (4.6) is equivalent to the equality (inside M_μ)

$$M_\mu^\circ = (V_\mu \cap M_\mu^\circ) + (V'_\mu \cap M_\mu^\circ).$$

But such $V'_\mu \subseteq M_\mu$ exists for each given V_μ (and our fixed $M_\mu^\circ \subseteq M_\mu$) because we can extend an arbitrary $\mathcal{O}_E$-basis of $V_\mu \cap M_\mu^\circ$ into one of M_μ°, and define V'_μ as the E-span of the new basis elements we added. This finishes the proof. □

Remark 4.1.14 We consider a ($U(\mathfrak{t})$-equivariant) strict exact sequence of Fréchet $U(\mathfrak{t})$-modules

$$0 \longrightarrow V_1 \longrightarrow V_2 \longrightarrow V_3 \longrightarrow 0 \tag{4.7}$$

with V_2 small (and thus V_1 is also small by Lemma 4.1.10). By Lemma 4.1.13 there exists another (small) Fréchet $U(\mathfrak{t})$-submodule $V'_3 \subseteq V_2$ such that the natural map $V_1 \oplus V'_3 \to V_2$ is a $U(\mathfrak{t})$-equivariant topological isomorphism, which together with our assumption $V_2/V_1 \xrightarrow{\sim} V_3$ gives a $U(\mathfrak{t})$-equivariant topological isomorphism $V'_3 \xrightarrow{\sim} V_3$, and thus a $U(\mathfrak{t})$-equivariant topological isomorphism $V_2 \cong V_1 \oplus V_3$. In other words, the short exact sequence (4.7) always splits (non-canonically). Moreover, since V'_3 is small so is V_3, and we have canonical isomorphisms of small $U(\mathfrak{t})$-modules

$$\Big(\bigoplus_{\mu\in\Lambda} V_{2,\mu}\Big)\Big/\Big(\bigoplus_{\mu\in\Lambda} V_{1,\mu}\Big) \cong \bigoplus_{\mu\in\Lambda} V_{2,\mu}/V_{1,\mu} \cong \bigoplus_{\mu\in\Lambda} V_{3,\mu}. \tag{4.8}$$

Lemma 4.1.15 *Let $e \in \mathfrak{t}$ such that $\alpha(e) \in \mathbb{Z}_{>0}$ for each $\alpha \in \Delta$ and let V be a small Fréchet $U(\mathfrak{t})$-module. Then the action of $e - N$ on*

$$V\Big/\Big(\bigoplus_{\mu\in\Lambda} V_\mu\Big)$$

is invertible for any $N \in \mathbb{Z}$.

Proof By (i) of Remark 4.1.2 we can assume $V_\mu \neq 0$ only if $\mu \in \mathbb{Z}_{\geq 0}\Phi^+$. As in the proof of Lemma 4.1.10, the set $\Lambda_N = \{\mu \in \mathbb{Z}_{\geq 0}\Phi^+ \mid \mu(e) = N\}$ is finite and we write $\Lambda_{\neq N}$ for its complement in $\mathbb{Z}_{\geq 0}\Phi^+$. Let $V_{\neq N}$ be the closure of $\bigoplus_{\mu\in\Lambda_{\neq N}} V_\mu$ in V, then by Remark 4.1.4 we have a canonical isomorphism of Fréchet $U(\mathfrak{t})$-modules

$$V \cong V_{\neq N} \oplus \Big(\bigoplus_{\mu\in\Lambda_N} V_\mu\Big), \tag{4.9}$$

with $\bigoplus_{\mu\in\Lambda_N} V_\mu$ being (finite dimensional and) exactly the kernel of $e - N$ in V. The isomorphism (4.9) induces the following isomorphism of $U(\mathfrak{t})$-modules

$$V\Big/\Big(\bigoplus_{\mu\in\Lambda} V_\mu\Big) \cong V_{\neq N}\Big/\Big(\bigoplus_{\mu\in\Lambda} (V_{\neq N})_\mu\Big) = V_{\neq N}\Big/\Big(\bigoplus_{\mu\in\Lambda_{\neq N}} V_\mu\Big). \tag{4.10}$$

By Lemma 4.1.5 applied to $M = \bigoplus_{\mu\in\Lambda} (V_{\neq N})_\mu$ and $q = e - N$, we know that $e - N$ has invertible action on $V_{\neq N}$, and thus on *any* $U(\mathfrak{t})$-equivariant quotient of $V_{\neq N}$. In particular, $e - N$ has invertible action on (4.10). □

Remark 4.1.16 The results in this section can be generalized to certain $U(\mathfrak{t})$-modules M which are not necessarily semi-simple as follows. In the first part of Definition 4.1.1, we can replace "where M_μ is the eigenspace attached to the weight μ" by "where M_μ is the generalized eigenspace attached to the weight μ such that there is a fixed $N \in \mathbb{Z}_{\geq 1}$ satisfying $(t - \mu(t))^N = 0$ on M_μ for all μ and all $t \in \mathfrak{t}$" (the second part of Definition 4.1.1 being unchanged). Then such an M contains a sequence of $U(\mathfrak{t})$-submodules $M = M_N \supseteq M_{N-1} \supseteq \cdots \supseteq M_1 \supseteq M_0 = 0$ such that M_k/M_{k-1} is the maximal semi-simple $U(\mathfrak{t})$-submodule of M/M_{k-1} for $1 \leq k \leq N$. We can also fix a collection of norms $\mathcal{L} = \{|\cdot|_\mu\}_\mu$ on each M_μ and take the completion $V \stackrel{\text{def}}{=} \widehat{M}_{\mathcal{L}}$ for the same semi-norms (4.1). For $0 \leq k \leq N$ we define V_k as the closure of M_k in V, and we easily check that V_k/V_{k-1} for $1 \leq k \leq N$ is the completion of M_k/M_{k-1} for the quotient topology (which is given by (4.1) for the quotient norms of the norms $|\cdot|_\mu$).

Let $V' \subseteq V$ be a closed Fréchet $U(\mathfrak{t})$-submodule. For $0 \leq k \leq N$ we have a closed Fréchet $U(\mathfrak{t})$-submodule $V'_k \stackrel{\text{def}}{=} V' \cap V_k \subseteq V_k$. As M_k/M_{k-1} is small (semi-simple) with completion V_k/V_{k-1}, by Lemma 4.1.10 the closed Fréchet $U(\mathfrak{t})$-submodule $V'_k/V'_{k-1} \subseteq V_k/V_{k-1}$ is also small with

$$\bigoplus_{\mu \in \Lambda} (V'_k/V'_{k-1})_\mu = (V'_k/V'_{k-1}) \cap (M_k/M_{k-1}) \subseteq M_k/M_{k-1}.$$

By an increasing induction on $k \in \{1, \ldots, N\}$ one easily checks that $V' \cap M_k$ is dense in V'_k with $V' \cap M_k = \bigoplus_{\mu \in \Lambda} (V'_k)_\mu$, and hence satisfies the generalized Definition 4.1.1 (as defined above).

For such non-semi-simple M as above, Lemma 4.1.13 clearly fails in general, and likewise a strict exact sequence as in (4.7) does not split in general. However, one can prove (using the above filtration $V = V_N \supseteq \cdots \supseteq V_0$) that (4.8) still holds. Lemma 4.1.15 also admits a straightforward generalization to such non-semi-simple M.

4.2 Preliminaries on Locally Analytic Distributions

We prove several important (technical) results on locally analytic and locally constant distribution algebras with prescribed support. All these results are used in the next sections.

For a paracompact locally K-analytic manifold M of finite dimension ([66, Sect. 8]), we write $C^{\text{an}}(M)$ for the space of E-valued locally K-analytic functions on M equipped with its usual locally convex topology ([66, Sect. 12]). We write $D(M) \stackrel{\text{def}}{=} C^{\text{an}}(M)_b^\vee$ for its continuous E-dual equipped with the strong topology ([65, Sect. 9], [70, Sect. 2]). For a closed subset $C \subseteq M$, we let $D(M)_C \subseteq D(M)$ be the closed subspace of distributions supported on C, which is the strong dual of the quotient $C^{\mathrm{an}}(M)/C^{\mathrm{an}}(M)_{M\setminus C}$ where $C^{\mathrm{an}}(M)_{M\setminus C}$ is the closed subspace of $C^{\mathrm{an}}(M)$ of functions with support contained in the open $M \setminus C$ ([40, Sect. 1.2]).

Similarly, we write $C^{\infty}(M)$ for the space of E-valued locally constant functions on M equipped with its usual locally convex topology and $D^{\infty}(M) \stackrel{\text{def}}{=} C^{\infty}(M)_b^{\vee}$ for its strong continuous E-dual. We have a closed embedding $C^{\infty}(M) \hookrightarrow C^{\mathrm{an}}(M)$ (use that $C^{\infty}(M)$ is the kernel of the continuous map $C^{\mathrm{an}}(M) \to C^{\mathrm{an}}(TM)$ where TM is the tangent bundle, see [66, Sect. 9], [66, Definition 12.4.i] and [66, Remark 6.2]) which induces a strict continuous surjection $D(M) \twoheadrightarrow D^{\infty}(M)$ ([69, Sect. 2]). We write $D^{\infty}(M)_C$ for the image of $D(M)_C$ under $D(M) \twoheadrightarrow D^{\infty}(M)$.

Lemma 4.2.1 *The subspace $D^{\infty}(M)_C$ is closed in $D^{\infty}(M)$.*

Proof Define $\widetilde{D}^{\infty}(M)_C$ as the closed subspace of $D^{\infty}(M)$ which is the strong dual of $C^{\infty}(M)/C^{\infty}(M)_{M\setminus C}$ where $C^{\infty}(M)_{M\setminus C}$ is the closed subspace of $C^{\infty}(M)$ of functions with support contained in $M \setminus C$. The composition $D(M)_C \hookrightarrow D(M) \twoheadrightarrow D^{\infty}(M)$ factors through a continuous map $D(M)_C \to \widetilde{D}^{\infty}(M)_C$, and it is enough to prove that this map is surjective. Arguing exactly as in the first part of the proof of [13, Lemme 3.12] for both $C^{\infty}(M)$ and $C^{\mathrm{an}}(M)$, we can assume that the closed subset C is compact. Then using [13, (3.3)] for both $\widetilde{D}^{\infty}(M)_C$ and $D(M)_C$, it is enough to prove that the natural injection of locally convex spaces of compact type $\varinjlim_U C^{\infty}(U) \hookrightarrow \varinjlim_U C^{\mathrm{an}}(U)$ is a closed embedding, where U runs among the compact open subsets of M containing C and the transition maps are the restrictions. But this easily follows from the short exact sequence

$$0 \longrightarrow \varinjlim_U C^{\infty}(U) \longrightarrow \varinjlim_U C^{\mathrm{an}}(U) \longrightarrow \varinjlim_U C^{\mathrm{an}}(TU)$$

where TU is the tangent bundle of U (see [66, Sect. 9]). □

If $M \cong M_1 \times M_2$ and the closed subset $C \subseteq M$ has the form $C_1 \times C_2$ with $C_i \subseteq M_i$ for $i = 1, 2$, then we have a canonical topological isomorphism ([13, Lemme 3.12])

$$D(M)_C \cong D(M_1)_{C_1} \widehat{\otimes}_{E,\iota} D(M_2)_{C_2} \tag{4.11}$$

and by the same proof as for *loc. cit.*

$$D^{\infty}(M)_C \cong D^{\infty}(M_1)_{C_1} \widehat{\otimes}_{E,\iota} D^{\infty}(M_2)_{C_2}. \tag{4.12}$$

Lemma 4.2.2 *Let M be a paracompact locally K-analytic manifold and $C \subseteq M$ a closed paracompact locally K-analytic submanifold. The embedding $C \hookrightarrow M$ induces a closed topological embedding $D^{\infty}(C) \hookrightarrow D^{\infty}(M)$, which induces a topological isomorphism*

$$D^{\infty}(C) \xrightarrow{\sim} D^{\infty}(M)_C. \tag{4.13}$$

Proof The first statement follows from [40, Proposition 1.1.2] (and its proof). It suffices to show that (4.13) is surjective. If $(M_i)_{i\in I}$ is a covering of M by compact open disjoint subsets, then $(C_i)_{i\in I} \stackrel{\text{def}}{=} (M_i \cap C)_{i\in I}$ is a covering of C by compact open disjoint subsets, and using $D^\infty(C) \cong \bigoplus_i D^\infty(C_i)$, $D(M)_C \cong \bigoplus_i D(M_i)_{C_i}$ (see e.g. the proof of [13, Lemme 3.12]), we have $D^\infty(M)_C \cong \bigoplus_i D^\infty(M_i)_{C_i}$ which shows that we can assume M and C compact. Let $\delta \in D^\infty(M)_C$, we define below an element $\overline{\delta} \in D^\infty(C)$ which maps to δ. For each $f \in C^\infty(C)$, there exists a finite partition $\mathcal{P}$ of C into compact open subsets U such that $f|_U$ is a constant for each $U \in \mathcal{P}$. We can moreover choose an arbitrary partition $\widetilde{\mathcal{P}}$ of M into compact open subsets $\widetilde{U}$ such that we have either $\widetilde{U} \cap C \in \mathcal{P}$ or $\widetilde{U} \cap C = \emptyset$ for $\widetilde{U} \in \widetilde{\mathcal{P}}$. Then we define $\widetilde{f} \in C^\infty(M)$ by requiring that $\widetilde{f}$ is a constant on each $\widetilde{U} \in \widetilde{\mathcal{P}}$ and that $\widetilde{f}|_C = f$. Now we define $\overline{\delta}(f) \stackrel{\text{def}}{=} \delta(\widetilde{f}) \in E$ and observe that $\delta(\widetilde{f})$ is independent of the choice of $\widetilde{f}$ (and the partitions $\mathcal{P}, \widetilde{\mathcal{P}}$) as above. Indeed, if $\widetilde{f}'$ is another choice, then by construction $\widetilde{f} - \widetilde{f}'$ is zero in an open neighborhood of C and thus $\delta(\widetilde{f} - \widetilde{f}') = 0$ since $\delta \in D^\infty(M)_C$. It is then clear that $\overline{\delta} \in \mathrm{Hom}_E(C^\infty(M), E) \cong D^\infty(C)$ has image δ under $D^\infty(C) \hookrightarrow D^\infty(M)$. □

If $M = G$ is a locally K-analytic group (automatically paracompact by [66, Corollary 18.8]), then $D(G)$ is a unital associative algebra (with multiplication being the convolution, see [70, Proposition 2.3]). For a locally K-analytic closed subgroup $H \subseteq G$, the closed subspace $D(G)_H$ is closed under convolution ([40, Corollary 1.2.6]), making $D(G)_H$ a unital associative subalgebra of $D(G)$. Similarly, $D^\infty(G)$ is also a unital associative algebra isomorphic to the quotient $D(G) \otimes_{U(\mathfrak{g})} E$ of $D(G)$ by the closed two-sided ideal generated by the Lie algebra $\mathfrak{g}$ of G (cf. [66, Sect. 13] and [73, Remark 1.1(iii)]). The surjective algebra homomorphism $D(G) \twoheadrightarrow D^\infty(G)$ induces a surjective homomorphism $D(G)_H \twoheadrightarrow D^\infty(G)_H$ showing that $D^\infty(G)_H$ is a closed unital associative subalgebra of $D^\infty(G)$ isomorphic to $D^\infty(H)$ (using Lemmas 4.2.1 and 4.2.2). When H is the trivial group, we write $D(G)_1$ and $D^\infty(G)_1$ instead of $D(G)_{\{1\}}$, $D^\infty(G)_{\{1\}}$.

For A a Fréchet-Stein algebra we let $\mathcal{C}_A$ be the abelian category of coadmissible left A-modules (see [72, Sect. 3]), which is a full subcategory of Mod_A ([72, Corollary 3.5]). By the discussion before and after [72, Lemma 3.6], each $M \in \mathcal{C}_A$ carries a canonical topology as a E-Fréchet space and any A-linear map in $\mathcal{C}_A$ is continuous and strict.

If G is a compact locally K-analytic group, by [72, Theorem 5.1] the algebra $D(G)$ is Fréchet-Stein, and so is its quotient $D^\infty(G)$ by [72, Proposition 3.7]. Moreover the continuous surjection $D(G) \twoheadrightarrow D^\infty(G)$ induces a fully faithful embedding $\mathcal{C}_{D^\infty(G)} \hookrightarrow \mathcal{C}_{D(G)}$. When G is not necessarily compact, one defines the category $\mathcal{C}_{D(G)}$ of coadmissible $D(G)$-modules over E as the full subcategory of $\mathrm{Mod}_{D(G)}$ of $D(G)$-modules which are coadmissible as $D(H)$-module for any—equivalently one—compact open subgroup H of G (see [72, Sect. 6] and the references there). Replacing $\mathrm{Mod}_{D(G)}$ by $\mathrm{Mod}_{D^\infty(G)}$, one defines in a similar way the full subcategory $\mathcal{C}_{D^\infty(G)}$ of $\mathcal{C}_{D(G)}$.

Let G be an arbitrary locally K-analytic group, then an admissible locally K-analytic - or just locally analytic—representation of G over E is a locally K-analytic G-representation on a E-vector space of compact type V such that the strong dual $V_b^\vee$ is in $\mathcal{C}_{D(G)}$. Here the left $D(G)$-action on $V^\vee$ is given by $(\delta_g \cdot \delta)(x) \stackrel{\text{def}}{=} \delta(g^{-1} \cdot x)$

for $g \in G$, $x \in V$ and $\delta \in V^\vee$. We write $\mathrm{Rep}^{\mathrm{an}}_{\mathrm{adm}}(G)$ for the abelian category of admissible locally analytic representations of G over E. The strong dual gives an anti-equivalence $\mathrm{Rep}^{\mathrm{an}}_{\mathrm{adm}}(G) \xrightarrow{\sim} \mathcal{C}_{D(G)}$ ([72, Theorem 6.3]). Via this equivalence, the "inverse image" of the full subcategory $\mathcal{C}_{D^\infty(G)}$ recovers the abelian category $\mathrm{Rep}^{\infty}_{\mathrm{adm}}(G)$ (see §2.1) of admissible smooth representations of G over E ([72, Theorem 6.6]).

Given a locally K-analytic group G, we write $\mathrm{Ext}^\bullet_{D(G)}(-,-)$ for the extension groups computed in the category $\mathrm{Mod}_{D(G)}$. For V_0, V_1 in $\mathrm{Rep}^{\mathrm{an}}_{\mathrm{adm}}(G)$, we use the notation

$$\mathrm{Ext}^\bullet_G(V_0, V_1) \stackrel{\mathrm{def}}{=} \mathrm{Ext}^\bullet_{D(G)}(V_1^\vee, V_0^\vee). \tag{4.14}$$

We will need the following lemmas.

Lemma 4.2.3 *Let G be a locally K-analytic group and V_0, V_1 in $\mathrm{Rep}^\infty(G)$. Assume either that both V_0, V_1 are admissible, or that the map $V_1 \to (V_1^\vee)^\vee$ is an isomorphism (i.e. V_1 is finite dimensional). Then we have isomorphisms $\mathrm{Ext}^\bullet_{D^\infty(G)}(V_1^\vee, V_0^\vee) \cong \mathrm{Ext}^\bullet_G(V_0, V_1)^\infty$.*

Proof The first case is proven in Corollary 0.2 of the erratum to [76] (http://math.univ-lyon1.fr/homes-www/schraen/Erratum_GL3.pdf), we prove the second. First recall that for any M_0, M_1 in $\mathrm{Mod}_{D^\infty(G)}$ one has a functorial isomorphism

$$\mathrm{Hom}_{D^\infty(G)}(M_1, M_0^\vee) \cong \mathrm{Hom}_{D^\infty(G)}(M_0, M_1^\vee) \tag{4.15}$$

where $M^\vee = \mathrm{Hom}_E(M, E)$ is the algebraic dual of M with $\delta \in D^\infty(G)$ acting on $f \in M^\vee$ by $(\delta \cdot f)(m) \stackrel{\mathrm{def}}{=} f(\tilde{\delta} \cdot m)$ where $m \in M$ and $\tilde{}$ is the unique anti-involution on $D^\infty(G)$ extending $g \mapsto g^{-1}$ on G. In particular for V_0, V_1 in $\mathrm{Rep}^\infty(G)$ we have a functorial isomorphism $\mathrm{Hom}_{D^\infty(G)}(V_1^\vee, V_0^\vee) \cong \mathrm{Hom}_{D^\infty(G)}(V_0, (V_1^\vee)^\vee)$ (recall that any smooth representation of G over E is also a module over $D^\infty(G)$, see [73, p. 300]). Under the assumption on V_1, we thus have a functorial isomorphism in V_0:

$$\mathrm{Hom}_{D^\infty(G)}(V_1^\vee, V_0^\vee) \cong \mathrm{Hom}_{D^\infty(G)}(V_0, V_1) = \mathrm{Hom}_G(V_0, V_1). \tag{4.16}$$

Take any projective resolution $P^\cdot$ of V_0 in $\mathrm{Rep}^\infty(G)$ (which exists by [3, Sect. 2]). Using (4.15) together with the fact that the functor from $\mathrm{Mod}_{D^\infty(G)}$ to $\mathrm{Rep}^\infty(G)$ sending a module M to the subspace of its smooth vectors (under the action of G) is exact (as follows from [73, Lemma 1.3]), one easily deduces that the algebraic dual $P^{\cdot\vee}$ is an injective resolution of $V_0^\vee$ in $\mathrm{Mod}_{D^\infty(G)}$. It then follows from (4.16) that $\mathrm{Ext}^i_{D^\infty(G)}(V_1^\vee, V_0^\vee) \cong \mathrm{Ext}^i_G(V_0, V_1)^\infty$ for $i \geq 0$. □

Lemma 4.2.4 *Let I be a countable directed index set and $(0 \to A_i \to B_i \to C_i \to 0)_{i\in I}$ a inverse system of strict exact sequences of Banach E-spaces with continuous transition maps. Assume that the image of $A_{i'}$ is dense in A_i for each pair of indices $i \leq i'$. Then the inverse limit*

$$0 \to \varprojlim_{i\in I} A_i \to \varprojlim_{i\in I} B_i \to \varprojlim_{i\in I} C_i \to 0$$

is a strict exact sequence of Fréchet E-spaces.

Proof Continuous maps between E-Banach spaces are uniformly continuous. So this is a special case of [8, Chap. II, Sect. 3.5, Theorem 1] (see also [31, Remark 13.2.4(i)]). □

We now introduce some notation largely following [52, Sect. 5.5]. Let $\mathbf{G}_0$ be a split reductive algebraic group scheme over $\mathcal{O}_K$ and $\mathbf{T}_0 \subseteq \mathbf{G}_0$ a maximal split torus. Let $\mathbf{P}_0 \subseteq \mathbf{G}_0$ be a parabolic subgroup scheme that contains $\mathbf{T}_0$, and $\mathbf{P}_0^-$ the opposite parabolic with unipotent radical $\mathbf{N}_0^-$. We write $G_0 \stackrel{\text{def}}{=} \mathbf{G}_0(\mathcal{O}_K)$, $G \stackrel{\text{def}}{=} \mathbf{G}_0(K)$, $\mathfrak{g}_0 \stackrel{\text{def}}{=} \mathrm{Lie}(G_0)$, $\mathfrak{g} \stackrel{\text{def}}{=} \mathrm{Lie}(G)$ and use similar notation for the other subgroup schemes (e.g. P_0, P, $\mathfrak{p}_0$, $\mathfrak{p}$, T_0, etc.). Note that $\mathfrak{g}_0$ is an $\mathcal{O}_K$-lattice in the K-vector space $\mathfrak{g}$. We fix an integer $m_0 \geq 1$ if $p > 2$, $m_0 \geq 2$ if $p = 2$, and set $\kappa \stackrel{\text{def}}{=} 1$ if $p > 2$, $\kappa \stackrel{\text{def}}{=} 2$ if $p = 2$. By [22, Sect. 9.4], the $\mathcal{O}_K$-lattices $p^{m_0}\mathfrak{g}_0 \subseteq \mathfrak{g}$, $p^{m_0}\mathfrak{p}_0 \subseteq \mathfrak{p}$ and $p^{m_0}\mathfrak{n}_0^-$ are powerful $\mathbb{Z}_p$-Lie algebras, and thus the exponential map $\exp_G : \mathfrak{g} \dashrightarrow G$ converges on these $\mathcal{O}_K$-lattices. Hence we may define $G_1 \stackrel{\text{def}}{=} \exp_G(p^{m_0}\mathfrak{g}_0)$, $P_1 \stackrel{\text{def}}{=} \exp_G(p^{m_0}\mathfrak{p}_0)$ and $N_1^- \stackrel{\text{def}}{=} \exp_G(p^{m_0}\mathfrak{n}_0^-)$ which are uniform pro-p groups by [22, Theorem 9.10]. Since the adjoint action of G_0 leaves $\mathfrak{g}_0$ invariant, G_1 is normal in G_0. Moreover using coordinates "of the second kind" ([66, Sect. 34], [22, Theorem 4.9]) one checks that $G_1 = N_1^- P_1$, $G_1 \cap P_0 = P_1$ and $G_1 \cap N_0^- = N_1^-$. In particular the embedding $N_1^- \hookrightarrow G_1$ gives a (locally K-analytic) section $G_1/P_1 \cong N_1^- \hookrightarrow G_1$ of the natural surjection $\mathrm{pr}_1 : G_1 \twoheadrightarrow G_1/P_1$. Using $G_0/P_0 = \bigsqcup_{g \in G_0/G_1P_0} gG_1/P_1$ and choosing a system of representatives of G_0/G_1P_0, we extend it to a section $s : G_0/P_0 \hookrightarrow G_0$ of $\mathrm{pr}_0 : G_0 \twoheadrightarrow G_0/P_0$. As $G_0/P_0 = G/P$ (by the Iwasawa decomposition of Bruhat-Tits, see e.g. [32, Lemma 3.4]), s determines a section (again denoted by) $s : G/P \hookrightarrow G$ of the surjection $\mathrm{pr} : G \twoheadrightarrow G/P$.

We let $\mathcal{I} \subseteq]0, 1[\cap p^{\mathbb{Q}}$ be the subset of r satisfying [52, (5.5.3)] and $|\cdot|_r$ the norm on $D(G_1)$ associated to $r \in \mathcal{I}$ and the canonical p-valuation on the uniform pro-p group G_1, see [51, Sects. 2.2.3, 2.2.6] (with our G_1 and $|\cdot|_r$ denoted by H and $\overline{q}_r$ there, see [51, Sect. 2.2.6 Step 3]). We write $D(G_1)_r$ for the completion of $D(G_1)$ with respect to $|\cdot|_r$. Then $D(G_1)_r$ is an E-Banach algebra and $D(G_1) \cong \varprojlim_{r \in \mathcal{I}} D(G_1)_r$ gives the Fréchet-Stein structure on $D(G_1)$ (the projective limit is for $r \to 1$ in $\mathcal{I}$). As in [52, (5.5.4), (5.5.5)], we extend the norm $|\cdot|_r$ on $D(G_1)$ to a maximum norm (still denoted) $|\cdot|_r$ on $D(G_0) \cong \bigoplus_{g \in G_0/G_1} \delta_g D(G_1)$. Then $D(G_0)_r \stackrel{\text{def}}{=} D(G_0) \otimes_{D(G_1)} D(G_1)_r$ is isomorphic to the completion of $D(G_0)$ with respect to $|\cdot|_r$ and $D(G_0) \cong \varprojlim_{r \in \mathcal{I}} D(G_0)_r$ is a Fréchet-Stein structure on $D(G_0)$. Similarly, we equip $D(P_1)$ (resp. $D(N_1^-)$) with the norm $|\cdot|_r$ attached to the canonical p-valuation on P_1 (resp. N_1^-), extend it to a norm $|\cdot|_r$ on $D(P_0)$ (resp. $D(N_0^-)$) and write $D(P_0)_r$ (resp. $D(N_0^-)_r$) for the corresponding completion. Since the canonical p-valuation of G_1 restricts to the one of P_1 and N_1^-, the norm $|\cdot|_r$ on G_1 restricts to the norm $|\cdot|_r$ on P_1 and N_1^-, and thus the norm $|\cdot|_r$ on G_0 restricts to the norm $|\cdot|_r$ on P_0 and N_0^- (cf. the discussion before [52, (5.5.8)]). Finally, using $D^\infty(G_0) \cong D(G_0) \otimes_{U(\mathfrak{g})} E$ (cf. the discussion below Lemma 4.2.2), we deduce from [72, Proposition 3.7] (and its proof) that $D^\infty(G_0)$ admits the Fréchet-Stein structure $D^\infty(G_0) \cong \varprojlim_{r\in\mathcal{I}} D^\infty(G_0)_r$ where $D^\infty(G_0)_r \stackrel{\text{def}}{=} D(G_0)_r \otimes_{U(\mathfrak{g})} E \cong E \otimes_{U(\mathfrak{g})} D(G_0)_r$.

Let $(P_m(G_1))_{m\geq 1}$ be the lower p-series of G_1 (cf. [22, Definition 1.15]). Following the notation of [52], we write $G_1^m \stackrel{\text{def}}{=} P_{m+1}(G_1)$. By the proof of [22, Theorem 4.2] we have that G_1^m is a uniform pro-p group and using [22, Theorem 3.6(iii)] we have that its $\mathbb{Z}_p$-Lie algebra is $p^m\mathrm{Lie}_{\mathbb{Z}_p}(G_1)$ (in fact $G_1^m \simeq G_1^{p^m} \stackrel{\text{def}}{=} \{x^{p^m}, x \in G_1\}$). Similarly, we define $P_1^m \stackrel{\text{def}}{=} P_{m+1}(P_1)$ for $m \geq 0$ and as before we have $P_1^m = G_1^m \cap P_1 = G_1^m \cap P_0$. Let $s = r^{p^m}$ with $s > \frac{1}{p}$ and $s^\kappa < p^{-1/(p-1)}$. Following [61, Sect. 6], we write $|\cdot|_s^{(m)}$ for the norm on $D(G_1^m)$ attached to s and the canonical p-valuation on G_1^m, and $D(G_1^m)_s$ for the corresponding completion. As before $|\cdot|_s^{(m)}$ restricts to the norm on $D(P_1^m)$ defined similarly using the canonical p-valuation on P_1^m. We have the following result from [61] (refining results by [29, 40]).

Lemma 4.2.5 *For r, m, s as above the restriction of the norm $|\cdot|_r$ of $D(G_1)$ to $D(G_1^m)$ is equivalent to $|\cdot|_s^{(m)}$, and $D(G_1)_r$ is a finite free right $D(G_1^m)_s$-module with a basis given by any set of coset representatives for G_1/G_1^m.*

Proof This is [61, Proposition 6.2, Corollary 6.4]. □

For $r \in \mathcal{I}, i = 0, 1$ and a closed subset $C \subseteq G_i$, we write $D(G_i)_{C,r}$ (resp. $D^\infty(G_i)_{C,r}$) for the closure of $D(G_i)_C$ in $D(G_i)_r$ (resp. of $D^\infty(G_i)_C$ in $D^\infty(G_i)_r$). As $U(\mathfrak{g})$ is dense in $D(G_1)_1$ ([40, Proposition 1.2.8]), in the particular case $C = \{1\}$ we see that $D(G_1)_{1,r}$ is also the closure of $U(\mathfrak{g})$ in $D(G_1)_r$.

Lemma 4.2.6 *If $r^\kappa < p^{-1/(p-1)}$, then $U(\mathfrak{g})$ is dense in $D(G_1)_r$. In particular $D(G_1)_{1,r} = D(G_1)_r$ and $D(G_1)_{1,r}\mathfrak{g} = \mathfrak{g}D(G_1)_{1,r} = \mathfrak{g}D(G_1)_r = D(G_1)_r\mathfrak{g}$ is the (two-sided) augmentation ideal of $D(G_1)_r$.*

Proof This is [52, (5.5.6)], which follows from [61, Proposition 5.6]. □

The following lemma will be used many times in the sequel.

Lemma 4.2.7 *Let $r \in \mathcal{I}$ and $m \geq 0$ such that $s = r^{p^m}$ satisfies $s > p^{-1}$ and $s^\kappa < p^{-1/(p-1)}$.*

(i) For each closed subset $C \subseteq G_0$, $D(G_0)_{C,r}$ (resp. $D^\infty(G_0)_{C,r}$) is a finite free right $D(G_0)_{1,r}$-module (resp. a finite dimensional E-vector space) with a basis given by any set of coset representatives of CG_1^m/G_1^m.

(ii) Let $H_0 \subseteq G_0$ be a closed subgroup, for each closed subset $C \subseteq G_0$ such that $CH_0 = C$ in G_0, $D(G_0)_{C,r}$ (resp. $D^\infty(G_0)_{C,r}$) is a finite free right $D(G_0)_{H_0,r}$-module (resp. a finite free right $D^\infty(G_0)_{H_0,r}$-module) with a basis given by any set of coset representatives of $CG_1^m/H_0G_1^m$.

Proof We fix r, m and $s = r^{p^m}$ as in the statement. As $s^\kappa < p^{-1/(p-1)}$, by Lemma 4.2.6 $D(G_1^m)_{1,s} = D(G_1^m)_s$. As G_1^m is compact open in G_1, we have $D(G_1^m) \xrightarrow{\sim} D(G_1)_{G_1^m}$ and $D(G_1^m)_1 \xrightarrow{\sim} D(G_1)_1$ and likewise with $D(G_0)$ instead of $D(G_1)$. Together with

Lemma 4.2.5 this implies $D(G_1)_{G_1^m,r} = D(G_1)_{1,r} = D(G_1^m)_{1,s} = D(G_1^m)_s$ (likewise with $D(G_0)$ instead of $D(G_1)$) and

$$D(G_1)_r \cong \bigoplus_{g\in G_1/G_1^m} \delta_g D(G_1)_{G_1^m,r} \cong \bigoplus_{g\in G_1/G_1^m} D(G_1)_{gG_1^m,r},$$

which together with $D(G_0)_r \cong \bigoplus_{g\in G_0/G_1} \delta_g D(G_1)_r$ implies

$$D(G_0)_r \cong \bigoplus_{g\in G_0/G_1^m} \delta_g D(G_0)_{G_1^m,r} \cong \bigoplus_{g\in G_0/G_1^m} D(G_0)_{gG_1^m,r}. \tag{4.17}$$

By Lemma 4.2.6 (and the above discussion) $D(G_1^m)_{1,s}\mathfrak{g} = D(G_1^m)_s\mathfrak{g} = D(G_0)_{G_1^m,r}\mathfrak{g}$ is the augmentation ideal of $D(G_1^m)_{1,s} = D(G_1^m)_s = D(G_0)_{G_1^m,r}$, which together with the first equality in (4.17) implies

$$D^\infty(G_0)_r = D(G_0)_r/D(G_0)_r\mathfrak{g} \cong E[G_0/G_1^m]. \tag{4.18}$$

Now we fix a closed subset $C \subseteq G_0$ and note that $D(G_0) \cong \bigoplus_{g\in G_0/G_1^m} \delta_g D(G_1^m) \cong \bigoplus_{g\in G_0/G_1^m} D(G_0)_{gG_1^m}$ implies $D(G_0)_C \cong \bigoplus_{g\in G_0/G_1^m} D(G_0)_{C\cap gG_1^m}$ (see e.g. the proof of [40, Lemma 1.2.5]) which in turn implies

$$D(G_0)_{C,r} = \bigoplus_{g\in G_0/G_1^m} D(G_0)_{C\cap gG_1^m,r}. \tag{4.19}$$

Step 1: We prove $D(G_0)_{C\cap gG_1^m,r} = D(G_0)_{gG_1^m,r}$ if $g \in CG_1^m$ and $D(G_0)_{C\cap gG_1^m,r} = 0$ otherwise.
Note that $g \in CG_1^m$ if and only if $C \cap gG_1^m \neq \emptyset$. If $C \cap gG_1^m = \emptyset$, we obviously have $D(G_0)_{C\cap gG_1^m,r} = 0$. If $C \cap gG_1^m \neq \emptyset$, for any $h \in C \cap gG_1^m$ we have closed embeddings of E-Banach spaces

$$D(G_0)_{gG_1^m,r} \supseteq D(G_0)_{C\cap gG_1^m,r} \supseteq D(G_0)_{\{h\},r}.$$

Writing $D(G_0)_{\{h\},r} = \delta_h D(G_0)_{1,r} = \delta_h D(G_0)_{G_1^m,r} = D(G_0)_{hG_1^m,r} = D(G_0)_{gG_1^m,r}$ (where the last equality follows from $hG_1^m = gG_1^m$), we deduce $D(G_0)_{\{h\},r} = D(G_0)_{C\cap gG_1^m,r} = D(G_0)_{gG_1^m,r}$.

Step 2: We prove (i) and (ii).
We combine Step 1 with (4.19) and first deduce

$$D(G_0)_{C,r} = \bigoplus_{g\in CG_1^m/G_1^m} D(G_0)_{C\cap gG_1^m,r} = \bigoplus_{g\in CG_1^m/G_1^m} \delta_g D(G_0)_{G_1^m,r} = \bigoplus_{g\in CG_1^m/G_1^m} \delta_g D(G_0)_{1,r} \tag{4.20}$$

which proves the first statement in (i). Define $\widetilde{D}^\infty(G_0)_{C,r} \stackrel{\text{def}}{=} D(G_0)_{C,r} \otimes_{U(\mathfrak{g})} E$ (noting that $D(G_0)_{C,r}$ is a right $D(G_0)_{1,r}$-module) and note that $D^\infty(G_0)_{C,r}$ is the closure of the image of $\widetilde{D}^{\infty}(G_0)_{C,r}$ in $D^{\infty}(G_0)_r$. Applying $(-) \otimes_{U(\mathfrak{g})} E$ to (4.20), we deduce

$$\widetilde{D}^{\infty}(G_0)_{C,r} = E[CG_1^m/G_1^m],$$

which together with (4.18) shows that $\widetilde{D}^{\infty}(G_0)_{C,r}$ is already closed in $D^{\infty}(G_0)_r$. This implies $\widetilde{D}^{\infty}(G_0)_{C,r} \xrightarrow{\sim} D^{\infty}(G_0)_{C,r}$ and also gives the second statement of (i). Note that G_1^m is normal in G_0 and thus $G_1^m H_0 = H_0 G_1^m$ is a compact open subgroup of G_0. If furthermore $CH_0 = C$, then we may rewrite (4.20) as

$$\begin{aligned} D(G_0)_{C,r} &= \bigoplus_{g \in CG_1^m/G_1^m} \delta_g D(G_0)_{G_1^m,r} = \bigoplus_{g \in CG_1^m/H_0G_1^m, h \in H_0G_1^m/G_1^m} \delta_g \delta_h D(G_0)_{G_1^m,r} \\ &= \bigoplus_{g \in CG_1^m/H_0G_1^m} \delta_g D(G_0)_{H_0G_1^m,r} = \bigoplus_{g \in CG_1^m/H_0G_1^m} \delta_g D(G_0)_{H_0,r} \end{aligned}$$

where the third equality follows from (4.20) applied with $C = H_0 G_1^m$ and the last equality follows from (i) applied with $C = H_0$ and $C = H_0 G_1^m$ (noting that $H_0 G_1^m / G_1^m = (H_0 G_1^m) G_1^m / G_1^m$). Applying $(-) \otimes_{U(\mathfrak{g})} E$ and using $\widetilde{D}^{\infty}(G_0)_{C,r} \xrightarrow{\sim} D^{\infty}(G_0)_{C,r}$ (for both C and H_0, see Step 1) we deduce

$$D^{\infty}(G_0)_{C,r} = \bigoplus_{g \in CG_1^m/H_0G_1^m} \delta_g D^{\infty}(G_0)_{H_0,r}$$

which finishes the proof of (ii). □

The following result is well-known, we provide a proof for lack of a precise reference.

Lemma 4.2.8 *Let $H_0 \subseteq G_0$ be a closed subgroup. The closed subalgebra $D(G_0)_{H_0}$ of $D(G_0)$ is Fréchet-Stein via $D(G_0)_{H_0} \cong \varprojlim_{r \in \mathcal{I}} D(G_0)_{H_0,r}$.*

Proof We consider $r, r' \in \mathcal{I}$ with $r \leq r'$. We note that the image of $D(G_0)_{H_0,r'}$ is dense in $D(G_0)_{H_0,r}$ (both contain as a dense subspace the image of $D(G_0)_{H_0}$). It suffices to check that $D(G_0)_{H_0,r}$ is flat over $D(G_0)_{H_0,r'}$. By (ii) of Lemma 4.2.7 (applied with $C = G_0$), we know that $D(G_0)_r$ (resp. $D(G_0)_{r'}$) is finite free as a $D(G_0)_{H_0,r}$-module (resp. $D(G_0)_{H_0,r'}$-module) with $D(G_0)_{H_0,r}$ being a direct summand. Since $D(G_0)_r$ is flat over $D(G_0)_{r'}$ (cf. [72, Theorem 5.1] which uses [72, Theorem 4.10, Proposition 3.7]) and $D(G_0)_{r'}$ is free as a $D(G_0)_{H_0,r'}$-module, we see that $D(G_0)_r$ is flat over $D(G_0)_{H_0,r'}$. Since $D(G_0)_{H_0,r}$ is a direct summand of $D(G_0)_r$ as $D(G_0)_{H_0,r'}$-modules, this implies that $D(G_0)_{H_0,r}$ is also flat over $D(G_0)_{H_0,r'}$. □

Let $H \subseteq G$ be a closed subgroup, we say that a left $D(G)_H$-module is coadmissible if it is coadmissible as a $D(G_0)_{H \cap G_0}$-module with the Fréchet-Stein structure given by Lemma 4.2.8. If $G_0' \subseteq G_0$ is another compact open subgroup, then $D(G_0)_{H \cap G_0}$ is free of finite rank over $D(G_0')_{H \cap G_0'}$ (see the argument for [40, (1.7)]) and arguing as in [73, Sect. 6] we see that being a coadmissible $D(G)_H$-module doesn't depend on the compact open subgroup G_0. We let $\mathcal{C}_{D(G)_H}$ be the abelian category of coadmissible (left) $D(G)_H$-modules. Recall that each coadmissible $D(G)_H$-module carries a canonical Fréchet topology, and any $D(G)_H$-linear map between coadmissible $D(G)_H$-modules is continuous and strict. See [2, Sect. 7.7] for a similar discussion (also) based on [40, Corollary 1.4.3].

Let $H_0 \subseteq G_0$ be a closed subgroup, as $D(G_0)_{H_0}$ is a Fréchet-Stein $U(\mathfrak{g})$-module, it follows from [72, Proposition 3.7] and its proof that $D(G_0)_{H_0} \otimes_{U(\mathfrak{g})} E$ is also Fréchet-Stein with $D(G_0)_{H_0} \otimes_{U(\mathfrak{g})} E \xrightarrow{\sim} \varprojlim_{r\in\mathcal{I}}(D(G_0)_{H_0,r} \otimes_{U(\mathfrak{g})} E)$. Since $D(G_0)_{H_0,r} \otimes_{U(\mathfrak{g})} E \xrightarrow{\sim} D^\infty(G_0)_{H_0,r} \subseteq D^\infty(G_0)_r$ (see Step 2 in the proof of Lemma 4.2.7), we have $D(G_0)_{H_0} \otimes_{U(\mathfrak{g})} E \xrightarrow{\sim} D^\infty(G_0)_{H_0} \subseteq D^\infty(G_0)$ (recall $D^\infty(G_0)_{H_0}$ is by definition the image of $D(G_0)_{H_0}$ in $D^\infty(G_0)$). This statement can in fact be generalized to any closed subset C in G_0.

Lemma 4.2.9 *For any closed subset $C \subseteq G_0$ we have a topological isomorphism*

$$D(G_0)_C \otimes_{U(\mathfrak{g})} E \xrightarrow{\sim} D(G_0)_C \otimes_{D(G_0)_1} E \xrightarrow{\sim} D^\infty(G_0)_C. \tag{4.21}$$

Proof From the definitions each map is surjective and continuous, hence it is enough to prove injectivity of the composition. Since $D^\infty(G_0)_C$ is closed in $D^\infty(G_0)$ (Lemma 4.2.1), we have $D^\infty(G_0)_C \cong \varprojlim_{r\in\mathcal{I}} D^\infty(G_0)_{C,r} \subseteq \varprojlim_{r\in\mathcal{I}} D^\infty(G_0)_r \cong D^\infty(G_0)$. Since $D(G_0)_{C,r} \otimes_{U(\mathfrak{g})} E \xrightarrow{\sim} D^\infty(G_0)_{C,r}$ (Step 2 in the proof of Lemma 4.2.7) we have exact sequences $0 \to D(G_0)_{C,r}\mathfrak{g} \to D(G_0)_{C,r} \to D^\infty(G_0)_{C,r} \to 0$ and since the maps $D(G_0)_{C,r'}\mathfrak{g} \to D(G_0)_{C,r}\mathfrak{g}$ for $r \leq r'$ have dense image (as the image of $D(G_0)_C\mathfrak{g}$ is dense everywhere by construction), Lemma 4.2.4 gives a short exact sequence of Fréchet E-spaces

$$0 \longrightarrow \varprojlim_{r\in\mathcal{I}}(D(G_0)_{C,r}\mathfrak{g}) \longrightarrow D(G_0)_C \longrightarrow D^\infty(G_0)_C \longrightarrow 0. \tag{4.22}$$

Since $D(G_0)_1 \xrightarrow{\sim} \varprojlim_{r\in\mathcal{I}} D(G_0)_{1,r}$ is a Fréchet-Stein algebra (e.g. by Lemma 4.2.8), both $D(G_0)_1 \otimes_E \mathfrak{g}$ and $D(G_0)_1\mathfrak{g}$ are coadmissible $D(G_0)_1$-modules (by [72, Corollary 3.4.iv] for the latter). By [72, Corollary 3.4.ii] we have in particular short exact sequences of finitely generated $D(G_0)_{1,r}$-modules $0 \to M_r \to D(G_0)_{1,r} \otimes_E \mathfrak{g} \to D(G_0)_{1,r}\mathfrak{g} \to 0$ where the maps $M_{r'} \to M_r$ have dense image. By (i) of Lemma 4.2.7 $D(G_0)_{C,r}$ is a free $D(G_0)_{1,r}$-module of finite rank (this rank growing when r tends to 1), hence tensoring by $D(G_0)_{C,r}$ over $D(G_0)_{1,r}$ again gives short exact sequences of Banach spaces

$$0 \longrightarrow D(G_0)_{C,r} \otimes_{D(G_0)_{1,r}} M_r \longrightarrow D(G_0)_{C,r} \otimes_E \mathfrak{g} \longrightarrow D(G_0)_{C,r}\mathfrak{g} \longrightarrow 0$$

where the maps $D(G_0)_{C,r'} \otimes_{D(G_0)_{1,r'}} M_{r'} \to D(G_0)_{C,r} \otimes_{D(G_0)_{1,r}} M_r$ have dense image. By Lemma 4.2.4 (and since $\mathfrak{g}$ is finite dimensional over E) we deduce in particular a surjection $D(G_0)_C \otimes_E \mathfrak{g} \twoheadrightarrow \varprojlim_{r\in\mathcal{I}}(D(G_0)_{C,r}\mathfrak{g})$. By (4.22) this implies $D(G_0)_C \otimes_{U(\mathfrak{g})} E \cong D(G_0)_C/D(G_0)_C\mathfrak{g} \xrightarrow{\sim} D^\infty(G_0)_C$. □

Recall the surjections $\mathrm{pr}_0 : G_0 \twoheadrightarrow G_0/P_0$ and $\mathrm{pr} : G \twoheadrightarrow G/P$. For each subset $C \subseteq G_0$ (resp. $C \subseteq G$), we have $CP_0 = \mathrm{pr}_0^{-1}(\mathrm{pr}_0(C))$ (resp. $CP = \mathrm{pr}^{-1}(\mathrm{pr}(C))$). As the inclusion $G_0 \subseteq G$ induces an isomorphism $G_0/P_0 \xrightarrow{\sim} G/P$, we have $C \cap G_0 = \mathrm{pr}_0^{-1}\mathrm{pr}(C)$ for each subset $C \subseteq G$ such that $CP = C$. The fixed section $s : G_0/P_0 \hookrightarrow G_0$ of pr_0 induces an isomorphism $G_0/P_0 \times P_0 \xrightarrow{\sim} G_0$ and thus an isomorphism $C_0/P_0 \times P_0 \xrightarrow{\sim} C_0$ for each closed subset $C_0 \subseteq G_0$ such that $C_0 P_0 = C_0$. We deduce a topological isomorphism for such closed subsets (by applying (4.11) with $M_1 = G_0/P_0$, $C_1 = C_0/P_0$ and $M_2 = C_2 = P_0$)

$$D(G_0/P_0)_{C_0/P_0} \widehat{\otimes}_E D(P_0) \cong D(G_0)_{C_0}. \tag{4.23}$$

Similarly, using (4.12) s also induces a topological isomorphism

$$D^\infty(G_0/P_0)_{C_0/P_0} \widehat{\otimes}_E D^\infty(P_0) \cong D^\infty(G_0)_{C_0}. \tag{4.24}$$

We now fix a locally closed subset $X \subseteq G$ (i.e. X is the intersection of a closed subset and an open subset of G, equivalently X is open in its closure $\overline{X} \subseteq G$) such that $XP = X$ and we set $X_0 \overset{\text{def}}{=} X \cap G_0$ and $Y \overset{\text{def}}{=} \overline{X}$. Note that X_0 is also open in its closure $Y_0 \overset{\text{def}}{=} \overline{X_0} \cong Y \cap G_0$ and that $X_0 = X_0 P_0$. A compact open subset of X_0 is the same thing as a compact open subset of Y_0 which is contained in X_0, in particular it is a compact subset of G_0 of the form $Y_0 \cap U_0$ where U_0 is a compact open subset of G_0 (of which there are only countably many). For each compact open subset C_0 of X_0, recall that $D(G_0)_{C_0}$ is a Fréchet space equipped with a separately continuous right action of $D(G_0)_{P_0}$. If $C_0 \subseteq C_0'$ are two compact open subsets of X_0, one can easily find compact open subsets $U_0 \subseteq U_0'$ of G_0 such that $C_0 = X_0 \cap U_0 = Y_0 \cap U_0$ and $C_0' = X_0 \cap U_0' = Y_0 \cap U_0'$, in particular $C_0 = C_0' \cap U_0$, $D(G_0)_{C_0} = D(U_0)_{C_0}$ and $D(G_0)_{C_0'} = D(U_0')_{C_0'}$. Writing $U_0 = U_0 \amalg U_0' \backslash U_0$ and noting that $U_0' \backslash U_0$ is also a compact open subset of G_0, we have $D(U_0')_{C_0'} \cong D(U_0)_{C_0} \oplus D(U_0' \backslash U_0)_{C_0' \cap U_0' \backslash U_0}$, and hence deduce a canonical projection of Fréchet spaces $D(G_0)_{C_0'} \twoheadrightarrow D(G_0)_{C_0}$. We then consider the projective limit

$$\widehat{D}(G_0)_{X_0} \overset{\text{def}}{=} \varprojlim_{C_0} D(G_0)_{C_0} \tag{4.25}$$

over the compact open subsets C_0 of X_0 such that $C_0 P_0 = C_0$ with transition maps given by the above projections for $C_0 \subseteq C_0'$. This is still a Fréchet space equipped with a separately continuous right $D(G_0)_{P_0}$-action and we have

$$\widehat{D}(G_0)_{X_0} \cong \varprojlim_{C_0} \big(D(G_0/P_0)_{C_0/P_0} \widehat{\otimes}_E D(P_0)\big) \cong \big(\varprojlim_{C_0} D(G_0/P_0)_{C_0/P_0}\big) \widehat{\otimes}_E D(P_0) \tag{4.26}$$

where the first isomorphism comes from (4.23) and the second from [27, Proposition 1.1.29]. We also deduce a right $D(G)_P$-action on

$$\widehat{D}(G)_X \stackrel{\text{def}}{=} \widehat{D}(G_0)_{X_0} \otimes_{D(G_0)_{P_0}} D(G)_P \cong \widehat{D}(G_0)_{X_0} \otimes_{D(P_0)} D(P) \\ \cong (\varprojlim_{C_0} D(G_0/P_0)_{C_0/P_0})\widehat{\otimes}_{E,\iota} D(P) \quad (4.27)$$

where the first isomorphism follows from $D(G)_P \cong D(G_0)_{P_0} \otimes_{D(P_0)} D(P)$ (see the argument for [40, (1.7)]) and the second from (4.26) and the same argument as at the end of the proof of [73, Proposition A.3]. Similarly, we consider the right $D^\infty(G_0)_{P_0} \cong D^\infty(P_0)$-module (see Lemma 4.2.2 for the latter isomorphism)

$$\widehat{D}^\infty(G_0)_{X_0} \stackrel{\text{def}}{=} \varprojlim_{C_0} D^\infty(G_0)_{C_0} \cong (\varprojlim_{C_0} D^\infty(G_0/P_0)_{C_0/P_0})\widehat{\otimes}_E D^\infty(P_0)$$

where we have used (4.24) for the last isomorphism, and the right $D^\infty(G)_P \cong D^\infty(P)$-module

$$\widehat{D}^\infty(G)_X \stackrel{\text{def}}{=} \widehat{D}^\infty(G_0)_{X_0} \otimes_{D^\infty(P_0)} D^\infty(P) \cong (\varprojlim_{C_0} D^\infty(G_0/P_0)_{C_0/P_0})\widehat{\otimes}_{E,\iota} D^\infty(P) \quad (4.28)$$

(note that $D^\infty(P) \cong D^\infty(P_0) \otimes_{D(P_0)} D(P)$).

Lemma 4.2.10 *For any compact open subset C_0 of X_0 such that $C_0 = C_0 P_0$ we have topological isomorphisms*

$$\widehat{D}(G_0)_{X_0} \otimes_{U(\mathfrak{g})} E \xrightarrow{\sim} \widehat{D}(G_0)_{X_0} \otimes_{D(G_0)_1} E \xrightarrow{\sim} \widehat{D}^\infty(G_0)_{X_0} \quad (4.29)$$

$$\widehat{D}(G)_X \otimes_{U(\mathfrak{g})} E \xrightarrow{\sim} \widehat{D}(G)_X \otimes_{D(G)_1} E \xrightarrow{\sim} \widehat{D}^\infty(G)_X. \quad (4.30)$$

Proof The proof of (4.29) is analogous to the proof of Lemma 4.2.9 using twice Lemma 4.2.4 (first with $\varprojlim_r$ for a given C_0 as in the proof of *loc. cit.*, then with $\varprojlim_{C_0}$) and noting that all transition maps always have dense image. Writing $D(P) = \bigoplus_{h\in P_0\backslash P} D(P_0)\delta_h$, we have and by (4.27) and (4.28)

$$\widehat{D}(G)_X \cong \bigoplus_{h\in P_0\backslash P} \widehat{D}(G_0)_{X_0}\delta_h \cong \widehat{D}(G_0)_{X_0} \otimes_{D(P_0)} D(P)$$
$$\widehat{D}^\infty(G)_X \cong \bigoplus_{h\in P_0\backslash P} \widehat{D}^\infty(G_0)_{X_0}\delta_h \cong \widehat{D}^\infty(G_0)_{X_0} \otimes_{D(P_0)} D(P) \quad (4.31)$$

from which we easily deduce (4.30) using (4.29). □

As the morphisms in (4.21) are right $D(G_0)_{P_0}$-equivariant, so are the morphisms in (4.29), and thus the morphisms (4.30) are also right $D(G)_P$-equivariant by definition. Note that the topological isomorphisms $G_0/P_0 \times P_0 \xrightarrow{\sim} G_0$, $G/P \times P \to G$, $G_0/P_0 \xrightarrow{\sim} G/P$ (given by the section s and the Iwasawa decomposition) induce topological isomorphisms $X_0/P_0 \times P_0 \xrightarrow{\sim} X_0$, $X/P \times P \to X$, $X_0/P_0 \xrightarrow{\sim} X/P$. If X is moreover such that $QX = X$ for some locally K-analytic closed subgroup $Q \subseteq G$, we then have a natural separately continuous left $D(Q)$-action on $\varprojlim_C D(G/P)_C$ and $\varprojlim_C D^\infty(G/P)_C$, where C runs among the compact open subsets of $X/P = X_0/P_0$ and the transition maps are defined as previously, which is uniquely determined by ($h \in Q$, $(\delta_C)_C \in \varprojlim D(G/P)_C$ or $\varprojlim D^\infty(G/P)_C$):

$$\delta_h \cdot (\delta_C)_C = (\delta_h \cdot \delta_C)_{hC}.$$

Via the last isomorphism in (4.27) (resp. (4.28)), we deduce a separately continuous left $D(Q)$-action on $\widehat{D}(G)_X$ (resp. $\widehat{D}^\infty(G)_X$), which makes (4.29) and (4.30) left $D(Q)$-equivariant.

Lemma 4.2.11 *Let $X \subseteq G$ be a locally closed subset such that $XP = X$. We have canonical right topological $D(G)_P$-equivariant isomorphisms*

$$D(G)_{\overline{X}}/D(G)_{\overline{X}\setminus X} \xrightarrow{\sim} \widehat{D}(G)_X \tag{4.32}$$

$$D^\infty(G)_{\overline{X}}/D^\infty(G)_{\overline{X}\setminus X} \xrightarrow{\sim} \widehat{D}^\infty(G)_X. \tag{4.33}$$

If moreover $QX = X$ for some locally K-analytic subgroup $Q \subseteq G$, then (4.32) and (4.33) are $D(Q)$-equivariant for the natural left $D(Q)$-action on both sides.

Proof We only give the proofs for (4.32), leaving the case of (4.33) to the reader (arguing as in the proof of (4.30) above). As before we note $Y = \overline{X}$, $X_0 = X \cap G_0$, $Y_0 = Y \cap G_0$, and we also define $Z \stackrel{\text{def}}{=} \overline{X} \setminus X$ and $Z_0 \stackrel{\text{def}}{=} Z \cap G_0$. Hence $Z_0 \subseteq Y_0$ are closed subspaces of the compact group G_0 with $X_0 = Y_0 \setminus Z_0$ and we have $YP = Y$, $ZP = Z$, $Y_0P_0 = Y_0$, $Z_0P_0 = Z_0$. Each compact open subset $C_0 \subseteq X_0$ is compact open in Y_0, and writing $Y_0 = C_0 \amalg Y_0 \setminus C_0$ we have as in (4.25) a surjection of Fréchet spaces $D(G_0)_{Y_0} \twoheadrightarrow D(G_0)_{C_0}$ with kernel $D(G_0)_{Y_0\setminus C_0}$ containing $D(G_0)_{Z_0}$ as closed subspace. Taking the projective limit over those C_0 such that $C_0P_0 = C_0$ we deduce a canonical morphism of Fréchet spaces

$$D(G_0)_{Y_0}/D(G_0)_{Z_0} \longrightarrow \varprojlim_{C_0} D(G_0)_{C_0} = \widehat{D}(G_0)_{X_0}. \tag{4.34}$$

Step 1: We prove that (4.34) is a topological isomorphism.
By (ii) of Lemma 4.2.7, $D(G_0)_{Z_0,r}$, $D(G_0)_{Y_0,r}$ and $D(G_0)_{Y_0,r}/D(G_0)_{Z_0,r}$ are finite free right $D(G_0)_{P_0,r}$-modules with a basis given by $\{\delta_g\}$ with g running through coset representatives of $Z_0G_1^m/P_0G_1^m$, $Y_0G_1^m/P_0G_1^m$ and $(Y_0G_1^m \setminus Z_0G_1^m)/P_0G_1^m$ respectively (using the notation of *loc. cit.*). Since $(Y_0G_1^m \setminus Z_0G_1^m)/P_0G_1^m = (Y_0 \setminus (Z_0G_1^m))G_1^m/P_0G_1^m$ (recall $G_1^mP_0 = P_0G_1^m$) it follows from *loc. cit.* again that we have short exact sequence of E-Banach spaces (which are finite free right $D(G_0)_{P_0,r}$-modules)

$$0 \to D(G_0)_{Z_0,r} \to D(G_0)_{Y_0,r} \to D(G_0)_{Y_0\setminus(Z_0G_1^m),r} \to 0. \tag{4.35}$$

Applying Lemma 4.2.4, we obtain a short exact sequence of E-Fréchet spaces

$$0 \to D(G_0)_{Z_0} \to D(G_0)_{Y_0} \to \varprojlim_{r \in \mathcal{I}} D(G_0)_{Y_0 \setminus (Z_0 G_1^m), r} \to 0. \tag{4.36}$$

As Z_0 is compact, the $(G_1^m)_m$ form a system of neighborhood of 1 in G_0 when $r \to 1$ in $\mathcal{I}$ and as $G_1^m P_0 = P_0 G_1^m$, it is easy to check that $(Y_0 \setminus (Z_0 G_1^m))_m$ is cofinal among compact open subsets of $X_0 = Y_0 \setminus Z_0$ stable under right multiplication by P_0, and thus

$$\varprojlim_{r} D(G_0)_{Y_0 \setminus (Z_0 G_1^m), r} \cong \varprojlim_{C_0} \varprojlim_{r} D(G_0)_{C_0, r} \cong \varprojlim_{C_0} D(G_0)_{C_0} \cong \widehat{D}(G_0)_{X_0}. \tag{4.37}$$

Step 2: We prove that (4.34) induces a topological isomorphism (4.32) which is $D(G)_P$-equivariant.
Writing $D(P) = \bigoplus_{h \in P_0 \setminus P} D(P_0)\delta_h$, we have by the argument at the end of the proof of [40, Lemma 1.2.5] for $S \in \{Z, Y\}$

$$D(G)_S \cong \bigoplus_{h \in P_0 \setminus P} D(G_0)_{S_0} \delta_h \cong D(G_0)_{S_0} \otimes_{D(P_0)} D(P). \tag{4.38}$$

Together with the first line in (4.31), it follows that (4.34) induces a topological isomorphism (4.32). As the projection $D(G_0)_{Y_0} \twoheadrightarrow D(G_0)_{C_0}$ is right $D(G_0)_{P_0}$-equivariant for each compact open $C_0 \subseteq X_0$ such that $C_0 P_0 = C_0$, we deduce that (4.34) is right $D(G_0)_{P_0}$-equivariant, and therefore (4.32) is right $D(G)_P \cong D(G_0)_{P_0} \otimes_{D(P_0)} D(P)$-equivariant.

Step 3: If furthermore $QX = X$, we prove that (4.32) is left $D(Q)$-equivariant. As X is left Q-stable, Y and Z are also left Q-stable, so we have natural left $D(Q)$-action on $D(G)_Y$ and $D(G)_Z$. Let $C_0 \subseteq X_0$ be a compact open subset such that $C_0 P_0 = C_0$, the above short exact sequence $0 \to D(G_0)_{Y_0 \setminus C_0} \to D(G_0)_{Y_0} \to D(G_0)_{C_0} \to 0$ induces a short exact sequence

$$\begin{aligned} 0 \longrightarrow D(G_0)_{Y_0 \setminus C_0} \otimes_{D(P_0)} D(P) \longrightarrow D(G)_Y \cong D(G_0)_{Y_0} \otimes_{D(P_0)} D(P) \\ \longrightarrow D(G_0)_{C_0} \otimes_{D(P_0)} D(P) \cong D(G)_{C_0 P} \longrightarrow 0 \end{aligned} \tag{4.39}$$

where the kernel contains the closed subspace $D(G)_Z \cong D(G_0)_{Z_0} \otimes_{D(P_0)} D(P)$ of $D(G)_Y$ and where the last isomorphism follows again from the proof of [40, Lemma 1.2.5]. We thus have a continuous morphism $D(G)_Y / D(G)_Z \to D(G)_{C_0 P}$. Using $Z_0 = \bigcap_{C_0} Y_0 \setminus C_0$, and hence $D(G_0)_{Z_0} = \cap_{C_0} D(G_0)_{Y_0 \setminus C_0} \cong \varprojlim_{C_0} D(G_0)_{Y_0 \setminus C_0}$, we deduce from (4.39) an embedding

$$D(G)_Y / D(G)_Z \hookrightarrow \varprojlim_{C_0} D(G)_{C_0 P}. \tag{4.40}$$

Note that $D(G)_{C_0 P}$ does not have a left action of $D(Q)$. However, for each $\delta \in D(Q)$ and $\delta_{C_0 P} \in D(G)_{C_0 P}$, $\mathrm{Supp}(\delta\delta') \subseteq Q C_0 P \subseteq X$ is compact ([40, Remark 1.2.3]), so there always exists a compact open subset $C'_0 \subseteq X_0$ such that $C'_0 P_0 = C'_0$ and $\mathrm{Supp}(\delta\delta') \subseteq C'_0 P$ (use $X/P \cong X_0/P_0$). Using this, we can equip $\varprojlim_{C_0} D(G)_{C_0 P}$ with a natural left $D(Q)$-action so that (4.40) is $D(Q)$-equivariant. From the first isomorphism in (4.27) we have a natural map

$$\widehat{D}(G)_X \to D(G_0)_{C_0} \otimes_{D(P_0)} D(P) \cong D(G)_{C_0 P}$$

for each compact open $C_0 \subseteq X_0$ such that $C_0 P_0 = C_0$, hence a map

$$\widehat{D}(G)_X \to \varprojlim_{C_0} D(G)_{C_0 P} \tag{4.41}$$

which is $D(Q)$-equivariant by the definition of the $D(Q)$-actions on both sides. We see that the $D(Q)$-equivariant embedding (4.40) factors as the composition of the isomorphism (4.32) with the $D(Q)$-equivariant map (4.41). This forces (4.32) to be also $D(Q)$-equivariant. □

If A is a locally convex E-vector space endowed with a structure of a separately continuous algebra and V (resp. W) is a locally convex E-vector space endowed with a structure of a separately continuous right (resp. left) A-module, we define $V \widehat{\otimes}_A W$ as the quotient of $V \widehat{\otimes}_{E,\iota} W$ by the closure of the E-vector subspace generated by elements $va \otimes w - v \otimes aw$, $(a, v, w) \in A \times V \times W$ with the quotient topology. Let $V \otimes_A W$ be the obvious quotient of $V \otimes_{E,\iota} W$ (without completing) endowed with the quotient topology, it is then not difficult to check using the various universal properties that $V \widehat{\otimes}_A W$ is also the universal Hausdorff completion of $V \otimes_A W$ ([65, Sect. 7]). The following lemma will be very useful.

Lemma 4.2.12 *Let $V = \varprojlim_r V_r$ and $W = \varprojlim_r W_r$ be Fréchet spaces written as countable projective limits of E-Banach spaces V_r, W_r and assume that the transition maps have dense image. Let $A = \varprojlim_r A_r$ be a Fréchet algebra which is a countable projective limit of Noetherian E-Banach algebras A_r with transition maps having dense image. Assume that V and W admit a separately continuous A-action induced by a separately continuous action of A_r on V_r and W_r. Assume finally that V_r is a finitely generated A_r-module for each r. Then we have a canonical isomorphism of Fréchet spaces $V \widehat{\otimes}_A W \xrightarrow{\sim} \varprojlim_r (V_r \otimes_{A_r} W_r)$.*

Proof By [27, Proposition 1.1.29] we have an isomorphism $V \widehat{\otimes}_E W \xrightarrow{\sim} \varprojlim_r (V_r \widehat{\otimes}_E W_r)$ and by [8, Chap. II, § 3.5, Th. 1] the image of V, W, A in respectively V_r, W_r, A_r has dense image. Let $C \subseteq V \widehat{\otimes}_E W$ (resp. $C_r \subseteq V_r \widehat{\otimes}_E W_r$) be the closure of the E-vector subspace generated by elements $va \otimes w - v \otimes aw$ for $(v, w) \in V \times W$ and $a \in A$ (resp. $(v, w) \in V_r \times W_r$ and $a \in A_r$), then C_r is also the closure of the image of C in $V_r \widehat{\otimes}_E W_r$ and hence $C \xrightarrow{\sim} \varprojlim_r C_r$. Applying Lemma 4.2.4 to the short exact sequences $0 \to C_r \to V_r \widehat{\otimes}_E W_r \to V_r \widehat{\otimes}_{A_r} W_r \to 0$ we deduce an isomorphism $V \widehat{\otimes}_A W \cong \varprojlim_r (V_r \widehat{\otimes}_{A_r} W_r)$. But since V_r is finitely generated over the

Noetherian algebra A_r, hence finitely presented, using for instance [72, Proposition 2.1.iii] we see that $V_r \otimes_{A_r} W_r$ is already complete, hence $V_r \widehat{\otimes}_A W_r \cong V_r \otimes_{A_r} W_r$. □

Recall that a locally closed locally K-analytic submanifold of G is a closed locally K-analytic submanifold of an open subset of G (with its induced structure of locally K-analytic manifold). We need to slightly generalize the definition of smooth compact induction given before Lemma 2.1.2. We fix $X \subseteq G$ a locally closed locally K-analytic submanifold of G such that $XP = X$. We also fix a representation π^∞ of P in $\mathrm{Rep}^\infty_{\mathrm{adm}}(P)$ and recall that $(\pi^\infty)^\vee = \mathrm{Hom}_E(\pi^\infty, E)$ is in $\mathcal{C}_{D^\infty(P)}$. We define $(\mathrm{ind}_P^X \pi^\infty)^\infty$ to be the set of locally constant functions $f : X \to \pi^\infty$ such that

- $f(xh) = h^{-1} \cdot f(x)$ for $x \in X$ and $h \in P$;
- there exists a compact open subset C_f of X such that $f(x) = 0$ for $x \notin C_f P$.

If moreover $QX = X$ for some locally K-analytic closed subgroup $Q \subseteq G$, then $(\mathrm{ind}_P^X \pi^\infty)^\infty$ is naturally a (left) smooth Q-representation via $(h'(f))(x) \overset{\mathrm{def}}{=} f((h')^{-1}x)$ $(h' \in Q, x \in X, f \in (\mathrm{ind}_P^X \pi^\infty)^\infty)$. As $D(G)_P \otimes_{U(\mathfrak{g})} E \cong (D(G_0)_{P_0} \otimes_{D(P_0)} D(P)) \otimes_{U(\mathfrak{g})} E \cong D^\infty(G_0)_{P_0} \otimes_{D^\infty(P_0)} D^\infty(P) \cong D^\infty(P)$ (Lemma 4.2.2) we see that $(\pi^\infty)^\vee$ is in $\mathcal{C}_{D(G)_P}$ by [72, Proposition 3.7] and [72, Lemma 3.8].

Lemma 4.2.13 *With the above notation* $(\mathrm{ind}_P^X \pi^\infty)^\infty$ *is a locally convex E-vector space of compact type and we have a canonical isomorphism of Fréchet E-spaces*

$$\widehat{D}(G)_X \widehat{\otimes}_{D(G)_P} (\pi^\infty)^\vee \cong \widehat{D}^\infty(G)_X \widehat{\otimes}_{D^\infty(G)_P} (\pi^\infty)^\vee \cong \big((\mathrm{ind}_P^X \pi^\infty)^\infty\big)^\vee. \tag{4.42}$$

If moreover $QX = X$ for some locally K-analytic closed subgroup $Q \subseteq G$, then the isomorphisms in (4.42) are left $D(Q)$-equivariant with the $D(Q)$-actions factoring through $D^\infty(Q)$.

Proof It follows from (4.28) and (4.30) that

$$\widehat{D}(G)_X \otimes_{D(G)_1} E \cong \widehat{D}^\infty(G)_X \cong (\varprojlim D^\infty(G_0)_{C_0}) \otimes_{D^\infty(P_0)} D^\infty(P) \tag{4.43}$$

where C_0 runs through the compact open subsets of $X_0 \overset{\mathrm{def}}{=} X \cap G_0$ such that $C_0 P_0 = C_0$. As the left $D(G)_P$-action on $(\pi^\infty)^\vee$ factors through $D^\infty(G)_P$ and as $D^\infty(G)_P \cong D^\infty(P)$ (Lemma 4.2.2), by (4.43) we have topological isomorphisms

$$\widehat{D}(G)_X \otimes_{D(G)_P} (\pi^\infty)^\vee \cong \widehat{D}(G)_X \otimes_{D^\infty(G)_P} (\pi^\infty)^\vee \cong (\varprojlim_{C_0} D^\infty(G_0)_{C_0}) \otimes_{D^\infty(P_0)} (\pi^\infty)^\vee.$$

Taking universal Hausdorff completion, we obtain topological isomorphisms

$$\widehat{D}(G)_X \widehat{\otimes}_{D(G)_P} (\pi^\infty)^\vee \cong \widehat{D}^\infty(G)_X \widehat{\otimes}_{D^\infty(G)_P} (\pi^\infty)^\vee \cong \big(\varprojlim_{C_0} D^\infty(G_0)_{C_0}\big) \widehat{\otimes}_{D^\infty(P_0)} (\pi^\infty)^\vee$$

where the last space is a Fréchet space since both $\varprojlim_{C_0} D^{\infty}(G_0)_{C_0}$ and $(\pi^{\infty})^{\vee}$ are. We now prove the last isomorphism in (4.42). We have topological isomorphisms of Fréchet spaces

$$\begin{aligned}
(\varprojlim_{C_0} D^{\infty}(G_0)_{C_0})\widehat{\otimes}_{D^{\infty}(P_0)}(\pi^{\infty})^{\vee} &\cong \Big(\varprojlim_{C_0}(D^{\infty}(G_0/P_0)_{C_0/P_0}\widehat{\otimes}_E D^{\infty}(P_0))\Big)\widehat{\otimes}_{D^{\infty}(P_0)}(\pi^{\infty})^{\vee}\\
\cong \Big((\varprojlim_{C_0} D^{\infty}(G_0/P_0)_{C_0/P_0})\widehat{\otimes}_E D^{\infty}(P_0)\Big)\widehat{\otimes}_{D^{\infty}(P_0)}(\pi^{\infty})^{\vee} &\cong (\varprojlim_{C_0} D^{\infty}(G_0/P_0)_{C_0/P_0})\widehat{\otimes}_E(\pi^{\infty})^{\vee}\\
\cong \varprojlim_{C_0}(D^{\infty}(G_0/P_0)_{C_0/P_0}\widehat{\otimes}_E(\pi^{\infty})^{\vee}) &\cong \varprojlim_{C_0}\Big((D^{\infty}(G_0/P_0)_{C_0/P_0}\widehat{\otimes}_E D^{\infty}(P_0))\widehat{\otimes}_{D^{\infty}(P_0)}(\pi^{\infty})^{\vee}\Big)\\
&\cong \varprojlim_{C_0}(D^{\infty}(G_0)_{C_0}\widehat{\otimes}_{D^{\infty}(P_0)}(\pi^{\infty})^{\vee})
\end{aligned}$$

where the first and last isomorphisms follow from (4.24), the second and fourth from [27, Proposition 1.1.29] (recall there are countably many C_0) and the third and fifth from [13, Lemme 3.1]. For $r \in \mathcal{I}$ let $(\pi^{\infty})^{\vee}_r \overset{\text{def}}{=} D^{\infty}(P_0)_r \otimes_{D^{\infty}(P_0)} (\pi^{\infty})^{\vee}$. By the same proof as for (i) of Lemma 4.2.7 replacing G_0 by P_0 and G_1^m by $P_1^m = G_1^m \cap P_0$ (see the discussion above Lemma 4.2.5) shows that $D^{\infty}(P_0)_{C,r}$ is finite dimensional. Applying this with $C = P_0$ and since $(\pi^{\infty})^{\vee}$ is in $\mathcal{C}_{D(G)_P}$, it follows that $(\pi^{\infty})^{\vee}_r$ is a finite dimensional E-vector space. Then by Lemma 4.2.12 we have an isomorphism of Fréchet spaces

$$D^{\infty}(G_0)_{C_0}\widehat{\otimes}_{D^{\infty}(P_0)}(\pi^{\infty})^{\vee} \cong \varprojlim_{r}\left(D^{\infty}(G_0)_{C_0,r}\otimes_{D^{\infty}(P_0)_r}(\pi^{\infty})^{\vee}_r\right)$$

Putting everything together we obtain an isomorphism of Fréchet spaces

$$\widehat{D}(G)_X\widehat{\otimes}_{D(G)_P}(\pi^{\infty})^{\vee} \cong \varprojlim_{C_0,}\varprojlim_{r}(D^{\infty}(G_0)_{C_0,r}\otimes_{D^{\infty}(P_0)_r}(\pi^{\infty})^{\vee}_r). \tag{4.44}$$

Given $r \in \mathcal{I}$ let $m \geq 0$, $s = r^{p^m}$ as in Lemma 4.2.7, and define $\pi_r^{\infty} \overset{\text{def}}{=} (\pi^{\infty})^{P_1^m}$ which is a finite dimensional representation of the finite group P_0/P_1^m. As $\{P_1^m\}_{r\in\mathcal{I}}$ is a system of open neighborhoods of 1 inside P_0, we have $\pi^{\infty} = \varinjlim_{r\in\mathcal{I}} \pi_r^{\infty}$. Recall from (i) of Lemma 4.2.7 (with G_0 there replaced with P_0) that

$$D^{\infty}(P_0)_r \cong E[P_0/P_1^m] \cong E \otimes_{D^{\infty}(P_1^m)} D^{\infty}(P_0), \tag{4.45}$$

which implies

$$(\pi^{\infty})^{\vee}_r = D^{\infty}(P_0)_r \otimes_{D^{\infty}(P_0)} (\pi^{\infty})^{\vee} \cong E \otimes_{D^{\infty}(P_1^m)} (\pi^{\infty})^{\vee} \cong (\pi_r^{\infty})^{\vee}. \tag{4.46}$$

For any compact open subset $C_0 \subseteq X_0$ such that $C_0P_0 = C_0$, by (ii) of Lemma 4.2.7 we have $D^{\infty}(G_0)_{C_0,r} \cong E[C_0G_1^m/G_1^m]$, which together with (4.46) and (4.45) implies

$$D^\infty(G_0)_{C_0,r} \otimes_{D^\infty(P_0)_r} (\pi^\infty)_r^\vee \cong E[C_0G_1^m/G_1^m] \otimes_{E[P_0/P_1^m]} (\pi_r^\infty)^\vee \cong \big(\mathrm{ind}_{P_0/P_1^m}^{C_0G_1^m/G_1^m} \pi_r^\infty\big)^\vee \tag{4.47}$$

where $\mathrm{ind}_{P_0/P_1^m}^{C_0G_1^m/G_1^m} \pi_r^\infty$ is the set of functions $\phi : C_0G_1^m/G_1^m \to \pi_r^\infty$ satisfying $\phi(x\overline{h}) = \overline{h^{-1}} \cdot \phi(x)$ for $x \in C_0G_1^m/G_1^m$ and $\overline{h} \in P_0/P_1^m$. If we lift such $\phi : C_0G_1^m/G_1^m \to \pi_r^\infty$ to a locally constant function on $C_0G_1^m$, take its restriction to C_0 and then extend it by zero on the open subset $X_0 \setminus C_0$ of X_0, we obtain a locally constant function $f : X_0 \to \pi_r^\infty \subseteq \pi^\infty$ which is supported on the compact open subset C_0 of X_0, constant on the compact open subsets $yG_1^m \cap C_0 \subseteq C_0$ for $y \in C_0$ and which satisfies $f(yh) = h^{-1} \cdot f(y)$ for $y \in C_0$ and $h \in P_0$. Using $X_0/P_0 \xrightarrow{\sim} X/P$ (see the paragraph before Lemma 4.2.11) there exists a unique extension of $f : X_0 \to \pi^\infty$ to a locally constant function (still denoted) $f : X \to \pi^\infty$ such that $f(yh) = h^{-1} \cdot f(y)$ for $y \in X$ and $h \in P$. This defines an injection

$$\mathrm{ind}_{P_0/P_1^m}^{C_0G_1^m/G_1^m} \pi_r^\infty \hookrightarrow (\mathrm{ind}_P^X \pi^\infty)^\infty \tag{4.48}$$

with image consisting of those $f : X \to \pi^\infty$ such that

- f is supported on C_0P;
- the restriction $f|_{C_0}$ has image in π_r^∞ and is constant on $yG_1^m \cap C_0$ for $y \in C_0$.

As each element of $(\mathrm{ind}_P^X \pi^\infty)^\infty$ obviously satisfies the above two conditions for some compact open subset $C_0 \subseteq X_0$ and some $r \in \mathcal{I}$, we deduce from (4.48) a topological isomorphism with both sides endowed with the finest locally convex topology

$$\varinjlim_{C_0} \varinjlim_{r} \mathrm{ind}_{P_0/P_1^m}^{C_0G_1^m/G_1^m} \pi_r^\infty \xrightarrow{\sim} (\mathrm{ind}_P^X \pi^\infty)^\infty, \tag{4.49}$$

and in particular $(\mathrm{ind}_P^X \pi^\infty)^\infty$ is of compact type. Taking the dual of (4.49), by [65, Proposition 16.10] and (4.47), we obtain topological isomorphisms of Fréchet spaces

$$\begin{aligned}\big((\mathrm{ind}_P^X \pi^\infty)^\infty\big)^\vee &\xrightarrow{\sim} \big(\varinjlim_{C_0} \varinjlim_{r} \mathrm{ind}_{P_0/P_1^m}^{C_0G_1^m/G_1^m} \pi_r^\infty\big)^\vee \cong \varprojlim_{C_0} \varprojlim_{r} \big(\mathrm{ind}_{P_0/P_1^m}^{C_0G_1^m/G_1^m} \pi_r^\infty\big)^\vee \\ &\cong \varprojlim_{C_0} \varprojlim_{r} \big(D^\infty(G_0)_{C_0,r} \otimes_{D^\infty(P_0)_r} (\pi^\infty)_r^\vee\big).\end{aligned}$$

Together with (4.44) this finishes the proof of (4.42). The left $D(Q)$-equivariance of the isomorphisms in (4.42) is easy and left to the reader. □

4.3 Fréchet Completion of Objects of $\mathcal{O}^{\mathfrak{b}}_{\mathrm{alg}}$

We prove several statements on the canonical Fréchet completions of $U(\mathfrak{g})$-modules in $\mathcal{O}^{\mathfrak{b}}_{\mathrm{alg}}$ defined in Sect. 4.1 and use them to give a useful description of the continuous dual of Orlik-Strauch representations (Proposition 4.3.6).

We fix $I \subseteq \Delta$ and use the notation of Sect. 4.2 with $\mathbf{G}_0 \stackrel{\mathrm{def}}{=} \mathrm{GL}_n/\mathcal{O}_K$ and $\mathbf{P}_0 \stackrel{\mathrm{def}}{=} P_I/\mathcal{O}_K$, so that we are back with $P_I = \mathbf{P}_0(K) \subseteq G = \mathbf{G}_0(K) = \mathrm{GL}_n(K)$, $\mathfrak{p}_I = \mathrm{Lie}(P_I)$ and we have $G_0 = \mathrm{GL}_n(\mathcal{O}_K)$, $P_{I,0} = P_I(\mathcal{O}_K)$, etc. We use without further ado that a $U(\mathfrak{p}_I)$-module which is finite dimensional over E is an algebraic representation of P_I ([52, Lemma 3.2]), and hence in particular is a $D(P_I)$-module, and thus also a $D(P_I)_1$-module. We start by recalling Schmidt's result on the canonical Fréchet completion of objects in $\mathcal{O}^{\mathfrak{b}}_{\mathrm{alg}}$ [63]. We define $\widehat{\mathcal{O}}^{\mathfrak{b}}_{\mathrm{alg}} \subseteq \mathcal{C}_{D(G)_1}$ as the full subcategory consisting of those coadmissible left $D(G)_1$-modules D such that $D|_{U(\mathfrak{t})}$ is a small Fréchet $U(\mathfrak{t})$-module in the sense of (ii) of Definition 4.1.7 and the (left) $U(\mathfrak{g})$-module $\bigoplus_{\mu\in\Lambda} D_\mu$ is an object of $\mathcal{O}^{\mathfrak{b}}_{\mathrm{alg}}$ (recall that $D_\mu \subseteq D$ is the eigenspace of D for the weight μ).

Proposition 4.3.1 *(i) The algebra $D(G)_1$ is flat over $U(\mathfrak{g})$.*

(ii) The functor $M \mapsto \mathcal{M} \stackrel{def}{=} D(G)_1 \otimes_{U(\mathfrak{g})} M$ induces an equivalence of (abelian) categories between $\mathcal{O}^{\mathfrak{b}}_{\mathrm{alg}}$ and $\widehat{\mathcal{O}}^{\mathfrak{b}}_{\mathrm{alg}}$, with a quasi-inverse given by $D \mapsto \bigoplus_{\mu\in\Lambda} D_\mu$.

(iii) Let M in $\mathcal{O}^{\mathfrak{p}_I}_{\mathrm{alg}}$ and X a finite dimensional $U(\mathfrak{p}_I)$-module such that one has a surjection $q : U(\mathfrak{g}) \otimes_{U(\mathfrak{p}_I)} X \twoheadrightarrow M$. Then $D(G)_1\ker(q) \subseteq D(G)_1 \otimes_{U(\mathfrak{p}_I)} X$ is a coadmissible (left) $D(G)_1$-submodule of $D(G)_1 \otimes_{U(\mathfrak{p}_I)} X \xrightarrow{\sim} D(G)_1 \otimes_{D(P_I)_1} X$ and we have an isomorphism in $\widehat{\mathcal{O}}^{\mathfrak{b}}_{\mathrm{alg}} \subseteq \mathcal{C}_{D(G)_1}$:

$$\mathcal{M} = D(G)_1 \otimes_{U(\mathfrak{g})} M \cong (D(G)_1 \otimes_{D(P_I)_1} X)/(D(G)_1\ker(q)). \tag{4.50}$$

Proof Part (i) is [63, Theorem 4.3.3] and part (iii) is the case $I = \emptyset$ of [63, Theorem 4.3.1], with the harmless difference that we have the extra condition that the weights of M are integral. We prove (iii). By applying (i) and (ii) to $0 \to \ker(q) \to U(\mathfrak{g}) \otimes_{U(\mathfrak{p}_I)} X \to M \to 0$, we obtain a short exact sequence in $\widehat{\mathcal{O}}^{\mathfrak{b}}_{\mathrm{alg}} \subseteq \mathcal{C}_{D(G)_1}$:

$$0 \longrightarrow D(G)_1 \otimes_{U(\mathfrak{g})} \ker(q) \longrightarrow D(G)_1 \otimes_{U(\mathfrak{p}_I)} X \longrightarrow D(G)_1 \otimes_{U(\mathfrak{g})} M \longrightarrow 0. \tag{4.51}$$

By the density of $U(\mathfrak{g})$ in $D(G)_1$ and 4.51 we know that $D(G)_1 \otimes_{U(\mathfrak{g})} \ker(q)$ is the closure of $\ker(q)$ in $D(G)_1 \otimes_{U(\mathfrak{p}_I)} X$ and by (ii) we have

$$\bigoplus_{\mu\in\Lambda}(D(G)_1 \otimes_{U(\mathfrak{g})} \ker(q))_\mu \cong \ker(q).$$

As $\ker(q)$ is a finitely generated $U(\mathfrak{g})$-submodule of $U(\mathfrak{g}) \otimes_{U(\mathfrak{p}_I)} X$, $D(G)_1\ker(q)$ is a finitely generated $D(G)_1$-submodule of $D(G)_1 \otimes_{U(\mathfrak{p}_I)} X$, hence is coadmissible by [72, Corollary 3.4.iv] and thus closed in $D(G)_1 \otimes_{U(\mathfrak{p}_I)} X$ by [72, Lemma 3.6]. As it contains $\ker(q)$ as dense subspace, we deduce $D(G)_1 \otimes_{U(\mathfrak{g})} \ker(q) \xrightarrow{\sim} D(G)_1\ker(q)$. Comparing (4.50) with (4.51), it remains to show that the natural map $D(G)_1 \otimes_{U(\mathfrak{p}_I)} X \to D(G)_1 \otimes_{D(P_I)_1} X$ is an isomorphism. It is enough to prove that we have an isomorphism

$$X \xrightarrow{\sim} D(P_I)_1 \otimes_{U(\mathfrak{p}_I)} X. \tag{4.52}$$

Since the action of $U(\mathfrak{p}_I)$ on X extends to $D(P_I)_1$, there is a natural surjection $D(P_I)_1 \otimes_{U(\mathfrak{p}_I)} X \twoheadrightarrow X, \mathfrak{d} \otimes x \mapsto \mathfrak{d}x$ which is the identity of X when composed with (4.52). Hence the map (4.52) is injective, and it is enough to prove its surjectivity. But using [72, Corollary 3.4] applied to the Fréchet-Stein algebra $D(P_I)_1$, one can check that $D(P_I)_1 \otimes_{U(\mathfrak{p}_I)} X$ is a coadmissible $D(P_I)_1$-module, hence is a Fréchet space. Since the map $X \to D(P_I)_1 \otimes_{U(\mathfrak{p}_I)} X$ has dense image (as $U(\mathfrak{p}_I)$ is dense in $D(P_I)_1$) with its source finite dimensional, it follows that it is surjective (and that $D(P_I)_1 \otimes_{U(\mathfrak{p}_I)} X$ is finite dimensional). □

Let $I \subseteq \Delta$, following [70, Sect. 3] we say that a smooth representation π^∞ of $P_{I,0}$ over E is strongly admissible if there exists a $P_{I,0}$-equivariant embedding $\pi^\infty \hookrightarrow C^\infty(P_{I,0})^m$ for some $m \geq 0$. Since $P_{I,0}$ is compact, $C^\infty(P_{I,0})$ is a direct sum of (finite dimensional) irreducible smooth representations of $P_{I,0}$ over E. In particular, π^∞ is a direct summand of $C^\infty(P_{I,0})^m$. The following lemma can be extracted from the beginning of [76, Sect. 4.4], we provide a proof for the reader's convenience.

Lemma 4.3.2 *Let $I \subseteq \Delta$, π^∞ a strongly admissible smooth representation of $P_{I,0}$ over E (see [69, Sect. 2]) and X a finite dimensional left $D(P_{I,0})$-module. Then $X \otimes_E (\pi^\infty)^\vee$ admits a resolution by finite free $D(P_{I,0})$-modules. In other words, $X \otimes_E (\pi^\infty)^\vee$ satisfies the condition (FIN) from [73, p. 321].*

Proof Since π^∞ is a strongly admissible smooth representation of $P_{I,0}$ over E, taking duals there exists $m \geq 0$ such that $(\pi^\infty)^\vee$ is a direct summand of $D^\infty(P_{I,0})^m$, and in particular $(\pi^\infty)^\vee$ admits a resolution

$$[\cdots \to Y_i \to \cdots \to Y_0] \xrightarrow{\sim} (\pi^\infty)^\vee$$

with $Y_i = D^\infty(P_{I,0})^m$ for each $i \geq 0$ (writing $D^\infty(P_{I,0})^m \cong D^\infty \oplus (\pi^\infty)^\vee$, the differential maps alternatively project onto $(\pi^\infty)^\vee$ or onto D^∞). Recall from [73, p.307] that $D^\infty(P_{I,0})$ admits the following (standard) resolution by finite free $D(P_{I,0})$-modules

$$D(P_{I,0}) \otimes_E \wedge^\bullet \mathfrak{p}_I \to D^\infty(P_{I,0}).$$

We set $X_{i,j} \stackrel{\text{def}}{=} Y_i \otimes_E \wedge^j \mathfrak{p}_I = (D(P_{I,0}) \otimes_E \wedge^j \mathfrak{p}_I)^m$ for each $i, j \geq 0$, so that $X_{i,\bullet}$ is a resolution of Y_i by finite free $D(P_{I,0})$-modules for $i \geq 0$. Thanks to a construction of C.T.C Wall (see [44, Chapitre V (3.1), (3.2)] and [41, Theorem 6.1]), there exists a map $d_{i,j}^k : X_{i,j} \to X_{i-k,j+k-1}$ for each $i, j, k \geq 0$ such that the data $(X_{i,j}, d_{i,j}^k)_{i,j,k\geq 0}$ altogether form a Wall complex, whose total complex $(X_r, \Delta_r)_{r\geq 0}$ (defined as $X_r \stackrel{\text{def}}{=}$

$\bigoplus_{i+j=r} X_{i,j}$ and $\Delta_r \stackrel{\mathrm{def}}{=} \sum_{i=0}^{r}\sum_{k=0}^{i} d^k_{i,r-i} : X_r \to X_{r-1}$) is a resolution of $(\pi^\infty)^\vee$. Since $X_{i,j}$ is a finite free $D(P_{I,0})$-module which is non-zero only if $i, j \geq 0$, we conclude that X_r is a finite free $D(P_{I,0})$-module for each $r \geq 0$ and thus that $(\pi^\infty)^\vee$ satisfies (FIN). Since X is a finite dimensional left $D(P_{I,0})$-module, we deduce from [76, Lemma 3.5] that $X \otimes_E X_r$ is a finite free $D(P_{I,0})$-module for each $r \geq 0$. We thus obtain the resolution $X \otimes_E X_\bullet$ of $X \otimes_E (\pi^\infty)^\vee$ by finite free $D(P_{I,0})$-modules, namely $X \otimes_E (\pi^\infty)^\vee$ also satisfies the condition (FIN). □

Lemma 4.3.3 *Let M in $\mathcal{O}^{\mathfrak{p}_I}_{\mathrm{alg}}$ and X a finite dimensional $\mathfrak{t}$-semi-simple $U(\mathfrak{p}_I)$-module such that one has a surjection $q : U(\mathfrak{g}) \otimes_{U(\mathfrak{p}_I)} X \twoheadrightarrow M$ in $\mathcal{O}^{\mathfrak{p}_I}_{\mathrm{alg}}$. Then the left $D(G)_1$-action on $\mathcal{M} \stackrel{def}{=} D(G)_1 \otimes_{U(\mathfrak{g})} M \in \mathcal{C}_{D(G)_1}$ (see (ii) of Proposition 4.3.1) naturally extends to a left $D(G)_{P_I}$-action which fits into an isomorphism of coadmissible $D(G)_{P_I}$-modules:*

$$\mathcal{M} \cong (D(G)_{P_I} \otimes_{D(P_I)} X)/(D(G)_{P_I}\ker(q)) \xrightarrow{\sim} (D(G)_{P_I}\widehat{\otimes}_{D(P_I)}X)/(D(G)_{P_I}\ker(q)).$$

Proof First note that, by Lemma 4.3.2, the finite dimensional representation X fits into an exact sequence of the form

$$\cdots \to D^k \to \cdots \to D^0 \to X \to 0 \tag{4.53}$$

where D^k is a finite free $D(P_{I,0})$-modules for $k \leq 0$. Moreover, it follows from the argument in the proof of [73, Lemma 6.3i] with $D(G)_{P_I}$ instead of $D(G)$ that applying $D(G_0)_{P_{I,0}} \otimes_{D(P_{I,0})} (-)$ to (4.53) gives another exact sequence

$$\cdots \to D(G_0)_{P_{I,0}} \otimes_{D(P_{I,0})} D^k \to \cdots \to D(G_0)_{P_{I,0}} \otimes_{D(P_{I,0})} D^0 \to D(G_0)_{P_{I,0}} \otimes_{D(P_{I,0})} X \to 0$$

(using $D(G)_{P_I} \cong D(G_0)_{P_{I,0}} \otimes_{D(P_{I,0})} D(P_I)$). Note that the argument of *loc. cit.* indeed extends since $D(G_0)_{P_{I,0},r}$ is flat over $D(P_{I,0})_r$ for $r \in \mathcal{I}$, which follows from the flatness of $D(G_0)_r$ over $D(P_{I,0})_r$ ([62, Proposition 2.6]) and the fact that $D(G_0)_{P_{I,0},r}$ is a direct summand of $D(G_0)_r$ as $D(P_{I,0})_r$-module by (ii) of Lemma 4.2.7 (applied with $C = G_0$ and $H_0 = P_{I,0}$). This implies that $D(G)_{P_I} \otimes_{D(P_I)} X$ is a $D(G_0)_{P_{I,0}}$-module of finite presentation, hence is coadmissible by [72, Corollary 3.4.v]. In particular we have $D(G)_{P_I} \otimes_{D(P_I)} X \xrightarrow{\sim} D(G)_{P_I}\widehat{\otimes}_{D(P_I)}X$ (compare with [11, Remark 5.1.3(ii)]). Since we have seen in the proof of (iii) of Proposition 4.3.1 that $D(G)_1 \otimes_{D(P_I)_1} X \cong D(G)_1 \otimes_{U(\mathfrak{p}_I)} X$ is a coadmissible, hence complete, $D(G)_1$-module, we also have $D(G_0)_1 \otimes_{D(P_{I,0})_1} X \cong D(G)_1 \otimes_{D(P_I)_1} X \xrightarrow{\sim} D(G)_1\widehat{\otimes}_{D(P_I)_1}X$.

By (iii) of Proposition 4.3.1 we have an isomorphism of coadmissible left $D(G)_1$-modules

$$\mathcal{M} = D(G)_1 \otimes_{U(\mathfrak{g})} M \cong (D(G)_1 \otimes_{D(P_I)_1} X)/(D(G)_1\ker(q)),$$

hence it is enough to prove that the natural map

$$D(G_0)_1 \otimes_{D(P_{I,0})_1} X \longrightarrow D(G_0)_{P_{I,0}} \otimes_{D(P_{I,0})} X \cong D(G)_{P_I} \otimes_{D(P_I)} X$$

is a topological isomorphism which sends $D(G)_1\ker(q)$ to $D(G)_{P_I}\ker(q)$. For the first statement, using Lemma 4.2.12 (noting that X is a finitely generated $D(P_I)_{1,r}$-module or $D(P_{I,0})_r$-module) and the above discussion, it suffices to prove that we have canonical isomorphisms of E-Banach spaces for $r \in \mathcal{I}$

$$D(G)_{1,r} \otimes_{D(P_I)_{1,r}} X \xrightarrow{\sim} D(G_0)_{P_{I,0},r} \otimes_{D(P_{I,0})_r} X.$$

But this follows from

$$D(G)_{1,r}\otimes_{D(P_I)_{1,r}} D(P_{I,0})_r \cong D(G_0)_{P_{I,0},r}, \tag{4.54}$$

which itself follows from (i) of Lemma 4.2.7 (applied with $C = P_{I,0}$ and both G_0 and $P_{I,0}$, noting that $P_{I,0}/P^m_{I,1} \xrightarrow{\sim} P_{I,0}G^m_1/G^m_1$). Thus $D(G)_1 \otimes_{D(P_I)_1} X$ is a coadmissible left $D(G)_{P_I}$-module. As the left $U(\mathfrak{p}_I)$-action on $\ker(q)$ extends to a left $D(P_I)$-action (this is so for any object of $\mathcal{O}^{\mathfrak{p}_I}_{\mathrm{alg}}$, see [52, Sect. 3.4]), we deduce that $D(G)_1\ker(q)$ (which is the closure of $\ker(q)$ in $D(G)_1 \otimes_{D(P_I)_1} X$, see (4.50)) is a closed subspace of $D(G)_{P_I} \otimes_{D(P_I)} X$ which is both $D(P_I)$-stable and $D(G)_1$-stable. As $D(P_I)$ and $D(G)_1$ generate a dense subalgebra of $D(G)_{P_I}$ (see [40, Proposition 1.2.12]), we deduce that $D(G)_1\ker(q)$ is a closed $D(G)_{P_I}$-submodule of $D(G)_{P_I} \otimes_{D(P_I)} X$. In particular $D(G)_1\ker(q) \cong D(G)_{P_I}\ker(q)$ since both spaces are the closure of $\ker(q)$ in $D(G)_{P_I} \otimes_{D(P_I)} X$. This finishes the proof. □

We will need the following result.

Lemma 4.3.4 *Let M in $\mathcal{O}^{\mathfrak{p}_I}_{\mathrm{alg}}$ and $\mathcal{M}$ the canonical Fréchet completion of M as a coadmissible $D(G_0)_{P_{I,0}}$-module (Lemma 4.3.3) with $\mathcal{M} \cong \varprojlim_{r\in\mathcal{I}} \mathcal{M}_r$ where $\mathcal{M}_r \stackrel{def}{=} D(G_0)_{P_{I,0},r} \otimes_{D(G_0)_{P_{I,0}}} \mathcal{M}$. Then there exists a family of standard semi-norms $\{|\cdot|_r\}_{r\in\mathcal{I}}$ on M (see Definition 4.1.3) such that $\mathcal{M}_r$ is the completion of M under $|\cdot|_r$ for $r \in \mathcal{I}$. Moreover the natural map $\mathcal{M} \to \mathcal{M}_r$ is a continuous injection.*

Proof We first prove that the first statement implies the second. Assume that the (continuous) map $\mathcal{M} \to \mathcal{M}_r$ is not injective, then its non-zero kernel is a closed Fréchet $U(\mathfrak{t})$-submodule of $\mathcal{M}$. By Lemma 4.1.10 its intersection with M is non-zero. By Remark 4.1.12 applied to the standard semi-norm $|\cdot|_r$, the completion under $|\cdot|_r$ of this intersection is also non-zero in $\mathcal{M}_r$, which is a contradiction. Hence the map $\mathcal{M} \to \mathcal{M}_r$ is injective.

Let X be a finite dimensional $\mathfrak{t}$-semi-simple $U(\mathfrak{p}_I)$-module such that one has a surjection $q : U(\mathfrak{g}) \otimes_{U(\mathfrak{p}_I)} X \twoheadrightarrow M$ in $\mathcal{O}^{\mathfrak{p}_I}_{\mathrm{alg}}$. We fix $r \in \mathcal{I}$ and let m, s be as before Lemma 4.2.5. The isomorphism $\mathfrak{g} \cong \mathfrak{n}_I^+ \oplus \mathfrak{p}_I$ induces an isomorphism of locally K-analytic manifolds $G^m_1 = (G^m_1 \cap N_I^+) \times (G^m_1 \cap P_I)$ (see the paragraph before Lemma 4.2.5), which together with [61, Propositions 5.9, 6.2] (see also Lemma 4.2.5) implies that

$$D(G)_{1,r} = D(G)_{G_1^m,r} \cong D(G_1^m)_s \cong D(G_1^m \cap N_I^+)_s \widehat{\otimes}_E D(G_1^m \cap P_I)_s$$
$$\cong D(N_I^+)_{G_1^m \cap N_I^+,r} \widehat{\otimes}_E D(P_I)_{G_1^m \cap P_I,r} = D(N_I^+)_{1,r} \widehat{\otimes}_E D(P_I)_{1,r}.$$

This together with (4.54) gives the following isomorphisms

$$D(G_0)_{P_{I,0},r} \otimes_{D(P_{I,0})} X = D(G_0)_{P_{I,0},r} \otimes_{D(P_{I,0})_r} (D(P_{I,0})_r \otimes_{D(P_{I,0})} X)$$
$$\cong D(G_0)_{P_{I,0},r} \otimes_{D(P_{I,0})_r} X \cong (D(G)_{1,r} \otimes_{D(P_I)_{1,r}} D(P_{I,0})_r) \otimes_{D(P_{I,0})_r} X$$
$$\cong D(G)_{1,r} \otimes_{D(P_I)_{1,r}} X \cong D(N_I^+)_{1,r} \otimes_E X,$$

and taking projective limit over $r \in \mathcal{I}$ recovers $D(G_0)_{P_{I,0}} \otimes_{D(P_{I,0})} X \cong D(N_I^+)_1 \otimes_E X$. By replacing $\mathfrak{g}$ in [40, Theorem 1.4.2] with $\mathfrak{n}_I^+$, we know that $U(\mathfrak{n}_I^+)$ admits a standard semi-norm $|\cdot|'_r$ for which its completion is $D(N_I^+)_{1,r}$ (and thus $D(N_I^+)_1 \cong \varprojlim_{r\in\mathcal{I}} D(N_I^+)_{1,r}$ is the completion of $U(\mathfrak{n}_I^+)$ under the Fréchet topology defined by $\{|\cdot|'_r\}_{r\in\mathcal{I}}$). By (ii) of Lemma 4.1.9 we know that $U(\mathfrak{g}) \otimes_{U(\mathfrak{p}_I)} X \cong U(\mathfrak{n}_I^+) \otimes_E X$ admits a standard semi-norm $|\cdot|''_r$ under which its completion is $D(G_0)_{P_{I,0},r} \otimes_{D(P_{I,0})} X \cong D(N_I^+)_{1,r} \otimes_E X$. In particular, still denoting by $|\cdot|''_r$ the induced semi-norm on $D(G_0)_{P_{I,0},r} \otimes_{D(P_{I,0})} X$, the Fréchet topology on the small Fréchet $U(\mathfrak{t})$-module $D(G_0)_{P_{I,0}} \otimes_{D(P_{I,0})} X \cong D(N_I^+)_1 \otimes_E X$ can be defined by the family of semi-norms $\{|\cdot|''_r\}_{r\in\mathcal{I}}$. By Lemma 4.3.3 $\mathcal{M}$ is a quotient of the coadmissible $D(G_0)_{P_{I,0}}$-module $D(G_0)_{P_{I,0}} \otimes_{D(P_{I,0})} X$ and for $r \in \mathcal{I}$ the $D(G_0)_{P_{I,0},r}$-module $\mathcal{M}_r$ is a quotient of $D(G_0)_{P_{I,0},r} \otimes_{D(P_{I,0})} X$. In particular, $\mathcal{M}_r$ is a quotient of $D(G_0)_{P_{I,0},r} \otimes_{D(P_{I,0})} X$ by a (closed) Banach subspace stable under the action of $U(\mathfrak{t})$, so by Remark 4.1.12 the standard semi-norm $|\cdot|''_r$ on $U(\mathfrak{g}) \otimes_{U(\mathfrak{p}_I)} X$ induces a standard semi-norm $|\cdot|_r$ on its quotient M so that $\mathcal{M}_r$ is the completion of M under $|\cdot|_r$. □

For M in $\mathcal{O}^{\mathfrak{p}_I}_{\text{alg}}$, by Lemma 4.3.3 the canonical completion $\mathcal{M}$ of M is a coadmissible left $D(G)_{P_I}$-module. For π^∞ in $\text{Rep}^\infty_{\text{adm}}(L_I)$, recall that $(\pi^\infty)^\vee$ is a coadmissible left $D(G)_{P_I}$-module via the surjections (where the middle isomorphism follows from Lemma 4.2.2)

$$D(G)_{P_I} \twoheadrightarrow D^\infty(G)_{P_I} \cong D^\infty(P_I) \twoheadrightarrow D^\infty(L_I).$$

We will use the following lemma.

Lemma 4.3.5 *Let M in $\mathcal{O}^{\mathfrak{p}_I}_{\text{alg}}$, $\mathcal{M} = D(G)_1 \otimes_{U(\mathfrak{g})} M$ its canonical Fréchet completion, X a finite dimensional $U(\mathfrak{p}_I)$-module with a surjection $q : U(\mathfrak{g}) \otimes_{U(\mathfrak{p}_I)} X \twoheadrightarrow M$. Let π^∞ in $\text{Rep}^\infty_{\text{adm}}(L_I)$ and $D \stackrel{\text{def}}{=} (\pi^\infty)^\vee$.*

(i) The $D(P_I)$-module $X \otimes_E D$ (with the diagonal action) is coadmissible and we have an isomorphism of coadmissible $D(G)_{P_I}$-modules

$$D(G)_{P_I} \widehat{\otimes}_{D(P_I)} (X \otimes_E D) \xrightarrow{\sim} (D(G)_{P_I} \otimes_{D(P_I)} X) \widehat{\otimes}_E D \tag{4.55}$$

with the diagonal action of $D(G)_{P_I}$ on the right hand side.

(ii) The $D(G)_{P_I}$-module $\mathcal{M} \widehat{\otimes}_E D$ (with the diagonal action) is coadmissible and we have a short exact sequence of coadmissible $D(G)_{P_I}$-modules (with the diagonal

action)

$$0 \longrightarrow \big(D(G)_{P_I}\ker(q)\big)\widehat{\otimes}_E D \longrightarrow (D(G)_{P_I} \otimes_{D(P_I)} X)\widehat{\otimes}_E D \longrightarrow \mathcal{M}\widehat{\otimes}_E D \longrightarrow 0. \tag{4.56}$$

Proof Note first that for the diagonal actions, one uses the comultiplication map $D(G)_{P_I} \to D(G)_{P_I}\widehat{\otimes}_E D(G)_{P_I}$ deduced from (4.11) as in [73, Sect. A]. We prove (i). By the proof of [52, Lemma 2.4(i)], the $D(P_I)$-module $X \otimes_E D$ is coadmissible, which implies that $(X \otimes_E D)_r \overset{\text{def}}{=} D(P_{I,0})_r \otimes_{D(P_{I,0})} (X \otimes_E D)$ is a finitely generated $D(P_{I,0})_r$-module for $r \in \mathcal{I}$ (see [52, Sect. 3]). Moreover, using the universal property of the inductive tensor product ([65, Sect. 17.A]), one has a topological isomorphism $(D(G_0)_{P_{I,0}} \otimes_{D(P_{I,0})} D(P_I)) \otimes_{D(P_I)} (X \otimes_E D) \cong D(G_0)_{P_{I,0}} \otimes_{D(P_{I,0})} (X \otimes_E D)$. Then we have topological isomorphisms

$$\begin{aligned} D(G)_{P_I}\widehat{\otimes}_{D(P_I)}(X \otimes_E D) &\cong D(G_0)_{P_{I,0}}\widehat{\otimes}_{D(P_{I,0})}(X \otimes_E D) \\ &\xrightarrow{\sim} \varprojlim_r \big(D(G_0)_{P_{I,0},r} \otimes_{D(P_{I,0})_r} (X \otimes_E D)_r\big) \end{aligned}$$

where the first isomorphism follows from $D(G)_{P_I} \cong D(G_0)_{P_{I,0}} \otimes_{D(P_{I,0})} D(P_I)$ and the fact that $D(G)_{P_I}\widehat{\otimes}_{D(P_I)}(X \otimes_E D)$ is the completion of $D(G)_{P_I} \otimes_{D(P_I)} (X \otimes_E D) \cong D(G_0)_{P_{I,0}} \otimes_{D(P_{I,0})} (X \otimes_E D)$, and the second from Lemma 4.2.12 and the beginning of the proof. Now, since we have for $r \le r'$ in $\mathcal{I}$

$$\begin{aligned} D(G_0)_{P_{I,0},r} &\otimes_{D(G_0)_{P_{I,0},r'}} \big(D(G_0)_{P_{I,0},r'} \otimes_{D(P_{I,0})_{r'}} (X \otimes_E D)_{r'}\big) \\ &\cong D(G_0)_{P_{I,0},r} \otimes_{D(P_{I,0})_{r'}} (X \otimes_E D)_{r'} \\ &\cong D(G_0)_{P_{I,0},r} \otimes_{D(P_{I,0})_r} \big(D(P_{I,0})_r \otimes_{D(P_{I,0})_{r'}} (X \otimes_E D)_{r'}\big) \\ &\cong D(G_0)_{P_{I,0},r} \otimes_{D(P_{I,0})_r} (X \otimes_E D)_r \end{aligned}$$

which is a finitely generated $D(G_0)_{P_{I,0},r}$-module, it follows from the definition of coadmissibility ([52, p. 152]) that $D(G)_{P_I}\widehat{\otimes}_{D(P_I)}(X \otimes_E D)$ is a coadmissible $D(G)_{P_I}$-module. The $D(P_I)$-equivariant embedding $X \hookrightarrow D(G)_{P_I} \otimes_{D(P_I)} X, x \mapsto 1 \otimes x$ induces a continuous $D(P_I)$-equivariant embedding $X \otimes_E D \hookrightarrow (D(G)_{P_I} \otimes_{D(P_I)} X)\widehat{\otimes}_E D$ (with diagonal action of $D(P_I)$ on both sides via [73, Sect. A]), which itself induces a continuous $D(G)_{P_I}$-equivariant map

$$D(G)_{P_I}\widehat{\otimes}_{D(P_I)}(X \otimes_E D) \longrightarrow (D(G)_{P_I} \otimes_{D(P_I)} X)\widehat{\otimes}_E D \tag{4.57}$$

(where we have used that the right hand side is complete thanks to the first paragraph of the proof of Lemma 4.3.3). Hence it suffices to show that (4.57) is a topological isomorphism. Note that we also have a topological isomorphism $X \otimes_E D \xrightarrow{\sim} X \otimes_E (\varprojlim_r D_r) \cong \varprojlim_r (X \otimes_E D_r)$ where $D_r \overset{\text{def}}{=} D(P_{I,0})_r \otimes_{D(P_{I,0})} D \cong D^\infty(P_0)_r \otimes_{D^\infty(P_0)} D$ (a finite dimensional E-vector space). Using Lemma 4.2.12 (noting that both X and $X \otimes_E D_r$ are finitely generated $D(P_{I,0})_r$-modules) it suffices to prove that the natural morphism of E-Banach spaces

$$D(G_0)_{P_{I,0},r} \otimes_{D(P_{I,0})_r} (X \otimes_E D_r) \longrightarrow (D(G_0)_{P_{I,0},r} \otimes_{D(P_{I,0})_r} X) \otimes_E D_r \qquad (4.58)$$

is a topological isomorphism. By (i) of Lemma 4.2.7 we have an isomorphism

$$D(G_0)_{1,r} \otimes_{D(P_{I,0})_{1,r}} D(P_{I,0})_r \xrightarrow{\sim} D(G_0)_{P_{I,0},r} \qquad (4.59)$$

(using *loc. cit.* and the discussion above Lemma 4.2.5 one checks that both are free over $D(G_0)_{1,r}$ with same basis). So (4.58) can be rewritten as

$$D(G_0)_{1,r} \otimes_{D(P_{I,0})_{1,r}} (X \otimes_E D_r) \longrightarrow (D(G_0)_{1,r} \otimes_{D(P_{I,0})_{1,r}} X) \otimes_E D_r. \qquad (4.60)$$

But since $D(P_{I,0})_{1,r}$ acts on D_r via the surjection $D(P_{I,0})_{1,r} \twoheadrightarrow D^{\infty}(P_{I,0})_{1,r} \cong E$, one trivially checks that (4.60) is an isomorphism of E-Banach spaces.

We prove (ii). By Lemma 4.3.3 we have a short exact sequence in $\mathcal{C}_{D(G)_{P_I}}$

$$0 \longrightarrow D(G)_{P_I}\ker(q) \longrightarrow D(G)_{P_I} \otimes_{D(P_I)} X \longrightarrow \mathcal{M} \longrightarrow 0$$

which gives a short exact sequence of Fréchet spaces as in (4.56) by [76, Lemme 4.13]. So it suffices to show that $\mathcal{M}\widehat{\otimes}_E D$ and $(D(G)_{P_I}\ker(q))\widehat{\otimes}_E D$ (with the diagonal $D(G)_{P_I}$-action) are coadmissible $D(G)_{P_I}$-modules. By (i) we know that $(D(G)_{P_I} \otimes_{D(P_I)} X)\widehat{\otimes}_E D$ is a coadmissible $D(G)_{P_I}$-module. Since $(D(G)_{P_I}\ker(q))\widehat{\otimes}_E D$ is closed in $(D(G)_{P_I} \otimes_{D(P_I)} X)\widehat{\otimes}_E D$ (by the above (4.56)) and stable under $D(G)_{P_I}$, it is a coadmissible $D(G)_{P_I}$-module by [72, Lemma 3.6], hence $\mathcal{M}\widehat{\otimes}_E D$ is a coadmissible $D(G)_{P_I}$-module by [72, Lemma 3.4.ii]. □

Given M in $\mathcal{O}^{\mathfrak{p}_I}_{\mathrm{alg}}$ and π^∞ in $\mathrm{Rep}^{\infty}_{\mathrm{adm}}(L_I)$, Orlik and Strauch define a representation $\mathcal{F}^G_{P_I}(M, \pi^\infty)$ in $\mathrm{Rep}^{\mathrm{an}}_{\mathrm{adm}}(G)$, we refer the reader to [52] for details (see also Theorem 4.3.7 below). The following result gives a convenient description of the continuous dual $\mathcal{F}^G_{P_I}(M, \pi^\infty)^\vee$ of $\mathcal{F}^G_{P_I}(M, \pi^\infty)$ using the completion functor $M \mapsto \mathcal{M}$ of Proposition 4.3.1.

Proposition 4.3.6 *For M in $\mathcal{O}^{\mathfrak{p}_I}_{\mathrm{alg}}$, $\mathcal{M} = D(G)_1 \otimes_{U(\mathfrak{g})} M$ and π^∞ a smooth admissible representation of L_I over E we have a canonical isomorphism of coadmissible left $D(G)$-modules*

$$\mathcal{F}^G_{P_I}(M, \pi^\infty)^\vee \cong D(G)\widehat{\otimes}_{D(G)_{P_I}} (\mathcal{M}\widehat{\otimes}_E (\pi^\infty)^\vee). \qquad (4.61)$$

Proof We let X be a finite dimensional left $U(\mathfrak{p}_I)$-module X which is $\mathfrak{t}$-semi-simple and such that we have a surjection $q : U(\mathfrak{g}) \otimes_{U(\mathfrak{p}_I)} X \twoheadrightarrow M$ in $\mathcal{O}^{\mathfrak{p}_I}_{\mathrm{alg}}$, and we define $D_1 \stackrel{\mathrm{def}}{=} D(G)_{P_I}\ker(q)$, $D_2 \stackrel{\mathrm{def}}{=} D(G)_{P_I} \otimes_{D(P_I)} X$ and $D_3 \stackrel{\mathrm{def}}{=} \mathcal{M}$. By (4.56) (applied with $D = (\pi^\infty)^\vee$) we have a short exact sequence in $\mathcal{C}_{D(G)_{P_I}}$

$$0 \longrightarrow D_1\widehat{\otimes}_E(\pi^\infty)^\vee \longrightarrow D_2\widehat{\otimes}_E(\pi^\infty)^\vee \longrightarrow D_3\widehat{\otimes}_E(\pi^\infty)^\vee \longrightarrow 0.$$

Let $R_k \stackrel{\text{def}}{=} D(G)\widehat{\otimes}_{D(G)_{P_I}}(D_k\widehat{\otimes}_E(\pi^\infty)^\vee)$ for $1 \le k \le 3$, then by [76, Lemma 4.27] (the proof of which extends to our setting) we deduce a short exact sequence of (left) $D(G)$-modules

$$0 \longrightarrow R_1 \longrightarrow R_2 \longrightarrow R_3 \longrightarrow 0 \tag{4.62}$$

(one could also again argue using Lemma 4.2.12 and (ii) of Lemma 4.2.7). Let $(\pi^\infty)^\vee_r \stackrel{\text{def}}{=} D(P_0)_r \otimes_{D(P_0)} (\pi^\infty)^\vee \cong D^\infty(P_0)_r \otimes_{D^\infty(P_0)} (\pi^\infty)^\vee$, we have isomorphisms of coadmissible $D(G)_{P_I}$-modules:

$$D_2\widehat{\otimes}_E(\pi^\infty)^\vee \cong D(G)_{P_I}\widehat{\otimes}_{D(P_I)}(X \otimes_E (\pi^\infty)^\vee) \xrightarrow{\sim} \varprojlim_{r\in\mathcal{I}} \big(D(G_0)_{P_{I,0},r} \otimes_{D(P_{I,0})_r} (X \otimes_E (\pi^\infty)^\vee_r)\big)$$

where the first isomorphism follows from (4.55) and the second from the proof of (i) of Lemma 4.3.5. Thus (using Lemma 4.2.12 again with (ii) of Lemma 4.2.7) we have isomorphisms of topological $D(G)$-modules

$$\begin{aligned} R_2 &\cong \varprojlim_{r\in\mathcal{I}} \Big(D(G_0)_r \otimes_{D(G_0)_{P_{I,0},r}} \big(D(G_0)_{P_{I,0},r} \otimes_{D(P_{I,0})_r} (X \otimes_E (\pi^\infty)^\vee_r)\big)\Big) \\ &\cong \varprojlim_{r\in\mathcal{I}} \big(D(G_0)_r \otimes_{D(P_{I,0})_r} (X \otimes_E (\pi^\infty)^\vee_r)\big) \\ &\cong D(G)\widehat{\otimes}_{D(P_I)}(X \otimes_E (\pi^\infty)^\vee) \end{aligned} \tag{4.63}$$

where the last $D(G)$-module is coadmissible by [52, Lemma 2.4(i)] (recall we have an isomorphism of Fréchet spaces $R_2 \cong ((\mathrm{Ind}_{P_I}^G(X^\vee \otimes_E \pi^\infty))^{\mathrm{an}})^\vee$ using e.g. [41, (52), (56), Remark 5.4] with [13, Lemme 3.1]). It then follows from [72, Lemma 3.6] that R_1 and R_3 are also coadmissible $D(G)$-modules.

By [52, (3.2.2), (4.4.1)] we have a natural pairing

$$(U(\mathfrak{g}) \otimes_{U(\mathfrak{p}_I)} X) \otimes_E \big(\mathrm{Ind}_{P_I}^G(X^\vee \otimes_E \pi^\infty)\big)^{\mathrm{an}} \longrightarrow C^{\mathrm{an}}(G,\pi^\infty) \tag{4.64}$$

which sends $(\delta \otimes_E x) \otimes_E f$ (for $\delta \in U(\mathfrak{g})$, $x \in X$ and $f \in (\mathrm{Ind}_{P_I}^G(X^\vee \otimes_E \pi^\infty))^{\mathrm{an}}$) to

$$\big[g \mapsto \delta\big(h \mapsto f(gh)(x)\big)\big] \in C^{\mathrm{an}}(G,\pi^\infty)$$

with $f(gh)(x) \in \pi^\infty$, $[h \mapsto f(gh)(x)] \in C^{\mathrm{an}}(G,\pi^\infty)$ and $\delta(h \mapsto f(gh)(x)) \in \pi^\infty$. Let $g \in G$, composing (4.64) with the evaluation map $C^{\mathrm{an}}(G,\pi^\infty) \to \pi^\infty$, $\varphi \mapsto \varphi(g)$ induces a pairing

$$(U(\mathfrak{g}) \otimes_{U(\mathfrak{p}_I)} X) \otimes_E \big(\mathrm{Ind}_{P_I}^G(X^\vee \otimes_E \pi^\infty)\big)^{\mathrm{an}} \longrightarrow \pi^\infty \tag{4.65}$$

which sends $(\delta \otimes_E x) \otimes_E f$ to $\delta(h \mapsto f(gh)(x))$. By pairing with $(\pi^\infty)^\vee$ on both sides of (4.65), we obtain the pairing

$$\big((U(\mathfrak{g}) \otimes_{U(\mathfrak{p}_I)} X) \otimes_E (\pi^\infty)^\vee\big) \otimes_E \big(\mathrm{Ind}_{P_I}^G(X^\vee \otimes_E \pi^\infty)\big)^{\mathrm{an}} \longrightarrow E \tag{4.66}$$

which sends $((\delta \otimes_E x) \otimes_E \delta^\infty) \otimes_E f$ to $\delta^\infty(\delta(h \mapsto f(gh)(x)))$ for δ, x, f as above and $\delta^\infty \in (\pi^\infty)^\vee$. From the isomorphism $R_2 \cong ((\mathrm{Ind}^G_{P_I}(X^\vee \otimes_E \pi^\infty))^{\mathrm{an}})^\vee$, we have a perfect pairing

$$R_2 \otimes_E (\mathrm{Ind}^G_{P_I} X^\vee \otimes_E \pi^\infty)^{\mathrm{an}} \longrightarrow E. \tag{4.67}$$

Then (4.66) can be reinterpreted as the pairing (recall $\delta_g \in D(G)$ is the Dirac distribution)

$$(\delta_g \cdot ((U(\mathfrak{g}) \otimes_{U(\mathfrak{p}_I)} X) \otimes_E (\pi^\infty)^\vee)) \otimes_E (\mathrm{Ind}^G_{P_I} X^\vee \otimes_E \pi^\infty)^{\mathrm{an}} \longrightarrow E$$

obtained from (4.67) via composition with the natural map

$$\delta_g \cdot ((U(\mathfrak{g}) \otimes_{U(\mathfrak{p}_I)} X) \otimes_E (\pi^\infty)^\vee) \longrightarrow R_2 \cong D(G)\widehat{\otimes}_{D(P_I)}(X \otimes_E (\pi^\infty)^\vee).$$

The representation $\mathcal{F}^G_{P_I}(M, \pi^\infty)$ is defined in [52, (4.4.1)] as the closed (invariant) subspace of $(\mathrm{Ind}^G_{P_I} X^\vee \otimes_E \pi^\infty)^{\mathrm{an}}$ which is the orthogonal of $\ker(q)$ under the pairing (4.64). Since $\varphi \in C^{\mathrm{an}}(G, \pi^\infty)$ is zero if and only if $\varphi(g) = 0$ for all $g \in G$, we deduce that $\mathcal{F}^G_{P_I}(M, \pi^\infty)$ is the closed subspace of $(\mathrm{Ind}^G_{P_I} X^\vee \otimes_E \pi^\infty)^{\mathrm{an}}$ which is the orthogonal of the image of $\sum_{g\in G} \delta_g \cdot (\ker(q) \otimes_E (\pi^\infty)^\vee)$ in R_2 under the pairing (4.67), that is, $\mathcal{F}^G_{P_I}(M, \pi^\infty)^\vee$ is the quotient of R_2 by the closure of

$$\sum_{g\in G} \delta_g \cdot (\ker(q) \otimes_E (\pi^\infty)^\vee). \tag{4.68}$$

As $\ker(q)$ is dense in D_1 (see the last sentence of the proof of Lemma 4.3.3), $\ker(q) \otimes_E (\pi^\infty)^\vee$ is dense in $D_1\widehat{\otimes}_E(\pi^\infty)^\vee$, and as $\{\delta_g \mid g \in G\}$ is dense in $D(G)$, one easily checks that (4.68) is dense inside $R_1 = D(G)\widehat{\otimes}_{D(G)_{P_I}}(D_1\widehat{\otimes}_E(\pi^\infty)^\vee) \subseteq R_2$ (see (4.62)). Hence, we deduce from *loc. cit.* that

$$\mathcal{F}^G_{P_I}(M, \pi^\infty)^\vee \cong R_2/R_1 \cong R_3 = D(G)\widehat{\otimes}_{D(G)_{P_I}}(\mathcal{M}\widehat{\otimes}_E(\pi^\infty)^\vee)$$

which finishes the proof. □

We recall for later convenience the main theorem of [52] in the following form.

Theorem 4.3.7 *(i) The functor $\mathcal{F}^G_{P_I}(-, -)$ is contravariant (resp. covariant) in the first (resp. the second) argument and is exact for both arguments.*

(ii) For $I_1 \subseteq I \subseteq \Delta$, M in $\mathcal{O}^{\mathfrak{p}_I}_{\mathrm{alg}}$ and π_1^∞ in $\mathrm{Rep}^\infty_{\mathrm{adm}}(L_{I_1})$, we have a canonical isomorphism

$$\mathcal{F}^G_{P_{I_1}}(M, \pi_1^\infty) \cong \mathcal{F}^G_{P_I}(M, i^\infty_{I_1,I}(\pi_1^\infty)).$$

(iii) Let $I \subseteq \Delta$, M in $\mathcal{O}^{\mathfrak{p}_I}_{\mathrm{alg}}$ with I maximal for M and π^∞ in $\mathrm{Rep}^\infty_{\mathrm{adm}}(L_I)$. If both M and π^∞ are simple, then so is $\mathcal{F}^G_{P_I}(M, \pi^\infty)$.

(iv) Let $I \subseteq \Delta$, M in $\mathcal{O}^{\mathfrak{p}_I}_{\mathrm{alg}}$ and π^∞ in $\mathrm{Rep}^\infty_{\mathrm{adm}}(L_I)$. Then a Jordan-Hölder factor of $\mathcal{F}^G_{P_I}(M, \pi^\infty)$ has the form $\mathcal{F}^G_{P_{I_1}}(M_1, \pi_1^\infty)$ for some Jordan-Hölder factor M_1 of

M such that $I_1 \supseteq I$ is maximal for $M_1 \in \mathcal{O}_{\mathrm{alg}}^{\mathfrak{p}_{I_1}}$, and some Jordan-Hölder factor π_1^∞ of $i_{I,I_1}^\infty(\pi^\infty)$.

We finish this section with several remarks.

Remark 4.3.8 For $I \subseteq \Delta$, recall the subcategory $\mathcal{O}_{\mathrm{alg}}^{\mathfrak{p}_I,\infty} \subseteq \widetilde{\mathcal{O}}_{\mathrm{alg}}^{\mathfrak{b}} \cap \mathcal{C}_{\mathrm{alg}}^{\mathrm{fin}}$ from the discussion at the beginning of Sect. 3.1. Recall also from [2] that, at least for smooth strongly admissible representations π^∞ of L_I over E, the construction of $\mathcal{F}_{P_I}^G(M, \pi^\infty)$ can be extended to M in $\mathcal{O}_{\mathrm{alg}}^{\mathfrak{p}_I,\infty}$ (modulo a choice of logarithm that we ignore here). For M in $\mathcal{O}_{\mathrm{alg}}^{\mathfrak{p}_I,\infty}$, we can still define its canonical Fréchet completion as $\mathcal{M} \overset{\mathrm{def}}{=} D(G)_1 \otimes_{U(\mathfrak{g})} M$. Let $\widehat{\mathcal{O}}_{\mathrm{alg}}^{\mathfrak{p}_I,\infty} \subseteq \mathcal{C}_{D(G)_1}$ be the abelian full subcategory of coadmissible $D(G)_1$-modules D such that D is a (generalized) small Fréchet $U(\mathfrak{t})$-module in the sense of Remark 4.1.16 and $\bigoplus_{\mu\in\Lambda} D_\mu$ lies in $\mathcal{O}_{\mathrm{alg}}^{\mathfrak{p}_I,\infty}$. Then using Remark 4.1.16 one can check that the functor $M \mapsto \mathcal{M}$ induces an equivalence of categories $\mathcal{O}_{\mathrm{alg}}^{\mathfrak{p}_I,\infty} \xrightarrow{\sim} \widehat{\mathcal{O}}_{\mathrm{alg}}^{\mathfrak{p}_I,\infty}$. Lemma 4.3.3, Proposition 4.3.6 and Theorem 4.3.7 also remain true for M in $\mathcal{O}_{\mathrm{alg}}^{\mathfrak{p}_I,\infty}$.

Remark 4.3.9 If $I \subseteq \Delta$, π^∞ is a smooth strongly admissible representations of L_I over E and X is a finite dimensional left $U(\mathfrak{p}_I)$-module, then by Lemma 4.3.2 $X \otimes_E (\pi^\infty)^\vee$ admits a resolution by finite free $D(P_{I,0})$-modules (we use here the strong admissibility of π^∞). Arguing as in the first paragraph of the proof of Lemma 4.3.3 we then deduce that $D(G)_{P_I} \otimes_{D(P_I)} (X \otimes_E D)$ is a coadmissible $D(G)_{P_I}$-module, in particular we have a topological isomorphism $D(G)_{P_I} \otimes_{D(P_I)} (X \otimes_E (\pi^\infty)^\vee) \xrightarrow{\sim} D(G)_{P_I} \widehat{\otimes}_{D(P_I)} (X \otimes_E (\pi^\infty)^\vee)$ (so no need to complete on the right). By the same kind of arguments we also have $R_2 \cong D(G) \otimes_{D(P_I)} (X \otimes_E (\pi^\infty)^\vee)$ in (4.63). However, it is not clear to us if we can replace $D(G) \widehat{\otimes}_{D(G)_{P_I}} (-)$ by $D(G) \otimes_{D(G)_{P_I}} (-)$ in (4.61).

Remark 4.3.10 Let E' be a finite extension of E. Given two $D(G) = D(G, E)$-modules D_1 and D_2, we may consider the $D(G, E') = D(G) \otimes_E E'$-modules $D_1 \otimes_E E'$ and $D_2 \otimes_E E'$. Note that any injective resolution of $D_1 \otimes_E E'$ in $\mathrm{Mod}_{D(G,E')}$ restricts to an injective resolution of $D_1 \otimes_E E'$ in $\mathrm{Mod}_{D(G)}$, and any injective resolution of D_1 in $\mathrm{Mod}_{D(G)}$ remains by scalar extension to E' an injective resolution of $D_1 \otimes_E E'$ seen in $\mathrm{Mod}_{D(G)}$. We thus have canonical isomorphisms for $k \geq 0$

$$\mathrm{Ext}^k_{D(G)}(D_2, D_1) \otimes_E E' \xrightarrow{\sim} \mathrm{Ext}^k_{D(G)}(D_2, D_1 \otimes_E E') \xrightarrow{\sim} \mathrm{Ext}^k_{D(G,E')}(D_2 \otimes_E E', D_1 \otimes_E E').$$

Moreover, one has the following results.

(i) Let $I \subseteq \Delta$, M in $\mathcal{O}_{\mathrm{alg}}^{\mathfrak{p}_I}$, π^∞ in $\mathrm{Rep}^\infty_{\mathrm{adm}}(L_I)$ and $D \overset{\mathrm{def}}{=} \mathcal{F}_{P_I}^G(M, \pi^\infty)^\vee$, then we have $D \otimes_E E' \cong \mathcal{F}_{P_I}^G(M \otimes_E E', \pi^\infty \otimes_E E')^\vee$ as coadmissible $D(G, E')$-modules. If moreover $\pi^\infty \in \mathcal{B}_\Sigma^I$ for some left $W(L_I)$-coset $\Sigma \subseteq \widehat{T}^\infty$ (see above (2.13)), then it follows from Remark 2.1.22 (using the last statement of Remark 2.1.12) that D is a simple $D(G)$-module if and only if $D \otimes_E E'$ is a simple $D(G, E')$-module.

(ii) Let D be a multiplicity free finite length $D(G)$-module with irreducible constituents all of the form $\mathcal{F}_{P_I}^G(L(\mu), \pi^\infty)^\vee$ for some $I \subseteq \Delta$, $\mu \in \Lambda_I^{\mathrm{dom}}$, $\pi^\infty \in \mathcal{B}_\Sigma^I$ for some left $W(L_I)$-coset $\Sigma \subseteq \widehat{T}^\infty$. Then from (i) the scalar extension $(-) \otimes_E E'$ induces a bijection of partially-ordered sets

$$\mathrm{JH}_{D(G)}(D) \xrightarrow{\sim} \mathrm{JH}_{D(G,E')}(D \otimes_E E').$$

4.4 Orlik-Strauch Representations and the Bruhat Filtration

Using the results of Sect. 4.3 we describe explicitly the duals of the graded pieces of the Bruhat filtration on Orlik-Strauch representations.

For $i = 0, 1$ we fix $I_i \subseteq \Delta$ and recall that W^{I_0,I_1} is the set of minimal length representatives of $W(L_{I_0})\backslash W(G)/W(L_{I_1})$. We have the Bruhat decomposition ([20, Lemma 5.5])

$$G = \bigsqcup_{w \in W^{I_0,I_1}} P_{I_1} w^{-1} P_{I_0}. \tag{4.69}$$

For $w \in W^{I_0,I_1}$ we write $S_w \stackrel{\mathrm{def}}{=} P_{I_1} w^{-1} P_{I_0}$ and $\overline{S_w}$ its closure in G. The following lemma is surely well-known, but we couldn't find a proof.

Lemma 4.4.1 *We have*

$$\overline{S_w} = \bigsqcup_{w,w' \in W^{I_0,I_1},\ w' \leq w} S_{w'}.$$

Proof By [26, Lemma 2.12(1)] $W(L_{I_1})w^{-1}W(L_{I_0})$ is an interval in $W(G)$, hence there is a unique maximal element $w_{\max}$ in $W(L_{I_1})w^{-1}W(L_{I_0})$ for the Bruhat order (and also a unique minimal element which is w^{-1}). It follows that all cosets $Bw'B$ appearing in S_w are such that $w' \leq w_{\max}$ and hence are in the closure $\overline{Bw_{\max}B}$ of $Bw_{\max}B$. This implies $S_w \subseteq \overline{Bw_{\max}B}$ and hence $\overline{S_w} \subseteq \overline{Bw_{\max}B}$. But since $Bw_{\max}B \subseteq S_w$, we have $\overline{S_w} = \overline{Bw_{\max}B} = \bigsqcup_{w' \leq w_{\max}} Bw'B$. Since $P_{I_1}\overline{S_w}P_{I_0} = \overline{S_w}$, it follows that $\overline{S_w} = \bigcup_{w' \leq w_{\max}} P_{I_1}w'P_{I_0}$. Writing $w_{\max} = x_{\max,1}w^{-1}x_{\max,0}$ with $x_{\max,i} \in W(L_{I_i})$ such that $x_{\max,1}w^{-1}$ is minimal in $x_{\max,1}w^{-1}W(L_{I_0})$ ([20, Lemma 5.4(iii)]), by the subword property of the Bruhat order and [6, Proposition 2.4.4] any $w' \leq w_{\max}$ can be written $w' = x_1'w''x_0'$ with $x_i' \leq x_{\max,i}$ in $W(L_{I_i})$ and $w'' \leq w^{-1}$. Thus $\overline{S_w} = \bigcup_{w'' \leq w^{-1}} P_{I_1}w''P_{I_0}$, or equivalently $\overline{S_w} = \bigcup_{w'' \leq w} P_{I_1}w''^{-1}P_{I_0}$. The result follows replacing w'' by its minimal representative in $P_{I_0}w''P_{I_1}$. □

In particular it follows from Lemma 4.4.1 and its proof that S_w is open in its closure $\overline{S_w}$, i.e. S_w is locally closed in G. The closed subsets $(\overline{S_w})_{w \in W^{I_0,I_1}}$ of G induce an exhaustive (by (4.69)) left $D(P_{I_1})$-equivariant and right $D(P_{I_0})$-equivariant filtration

$$\mathrm{Fil}_w(D(G)) \overset{\mathrm{def}}{=} D(G)_{\overline{S_w}} \subseteq D(G)$$

on $D(G)$ indexed by W^{I_0,I_1} (using Lemma 4.4.1). Note that by *loc. cit.* we have $\mathrm{Fil}_1(D(G)) = D(G)_{P_{I_1}P_{I_0}} \subseteq \mathrm{Fil}_w(D(G))$ for any $w \in W^{I_0,I_1}$. The terminology "filtration" comes from $\mathrm{Fil}_{w'}(D(G)) \subseteq \mathrm{Fil}_w(D(G))$ whenever $w' \leq w$. For $w \in W^{I_0,I_1}$ recall that $\widehat{D}(G)_{S_w}$ was defined in (4.27) (using (4.25)), and we have

$$\mathrm{gr}_w(D(G)) \overset{\mathrm{def}}{=} \mathrm{Fil}_w(D(G))/\sum_{w'<w} \mathrm{Fil}_{w'}(D(G)) \cong D(G)_{\overline{S_w}}/D(G)_{\overline{S_w}\setminus S_w} \cong \widehat{D}(G)_{S_w}$$

where the first isomorphism follows from Lemma 4.4.1 and the second from (4.32).

Let M_0 in $\mathcal{O}_{\mathrm{alg}}^{\mathfrak{p}_{I_0}}$, π_0^∞ a smooth admissible representation of L_{I_0} over E and let $V_0 \overset{\mathrm{def}}{=} \mathcal{F}_{P_{I_0}}^G(M_0, \pi_0^\infty)$. Recall from Proposition 4.3.6 that

$$V_0^\vee \cong D(G)\widehat{\otimes}_{D(G)_{P_{I_0}}}(\mathcal{M}_0\widehat{\otimes}_E(\pi_0^\infty)^\vee) \tag{4.70}$$

where $\mathcal{M}_0 = D(G)_1 \otimes_{U(\mathfrak{g})} M_0$ is a coadmissible $D(G)_{P_{I_0}}$-module (Lemma 4.3.3). We define the following filtration indexed by $w \in W^{I_0,I_1}$ on the $D(P_{I_1})$-module $V_0^\vee$:

$$\mathrm{Fil}_w(V_0^\vee) \overset{\mathrm{def}}{=} D(G)_{\overline{S_w}}\widehat{\otimes}_{D(G)_{P_{I_0}}}(\mathcal{M}_0\widehat{\otimes}_E(\pi_0^\infty)^\vee). \tag{4.71}$$

Lemma 4.4.2 *For $w \in W^{I_0,I_1}$ the inclusion $D(G)_{\overline{S_w}} \subseteq D(G)$ induces a closed embedding of left $D(G)_{P_{I_1}}$-modules* $\mathrm{Fil}_w(V_0^\vee) \hookrightarrow V_0^\vee$.

Proof It is clear from (4.70) and (4.71) that the natural map $\mathrm{Fil}_w(V_0^\vee) \to V_0^\vee$ is $D(G)_{P_{I_1}}$-equivariant (as $\overline{S_w} = P_{I_1}\overline{S_w}$). Hence we have to prove that it is a closed embedding. Since $\mathcal{M}_0\widehat{\otimes}_E(\pi_0^\infty)^\vee$ is a coadmissible $D(G)_{P_{I_0}}$-module by (ii) of Lemma 4.3.5, we have

$$\mathcal{M}_0\widehat{\otimes}_E(\pi_0^\infty)^\vee \xrightarrow{\sim} \varprojlim_{r\in\mathcal{I}}(\mathcal{M}_0\widehat{\otimes}_E(\pi_0^\infty)^\vee)_r \tag{4.72}$$

where $(\mathcal{M}_0\widehat{\otimes}_E(\pi_0^\infty)^\vee)_r \overset{\mathrm{def}}{=} D(G_0)_{P_{I_0,0},r} \otimes_{D(G_0)_{P_{I_0,0}}} (\mathcal{M}_0\widehat{\otimes}_E(\pi_0^\infty)^\vee)$ and $P_{I_0,0} \overset{\mathrm{def}}{=} P_{I_0} \cap G_0$. Set $Y \overset{\mathrm{def}}{=} \overline{S_w}$ and $Y_0 \overset{\mathrm{def}}{=} Y \cap G_0$, we have a topological isomorphisms by (4.38)

$$\begin{aligned} D(G)_Y \cong D(G_0)_{Y_0} \otimes_{D(P_{I_0,0})} D(P_{I_0}) &\cong D(G_0)_{Y_0} \otimes_{D(G_0)_{P_{I_0,0}}} (D(G_0)_{P_{I_0,0}} \otimes_{D(P_{I_0,0})} D(P_{I_0})) \\ &\cong D(G_0)_{Y_0} \otimes_{D(G_0)_{P_{I_0,0}}} D(G)_{P_{I_0}}. \end{aligned}$$

From the universal property of the inductive tensor product, we obtain a topological isomorphism

$$D(G)_Y\widehat{\otimes}_{D(G)_{P_{I_0}}}(\mathcal{M}_0\widehat{\otimes}_E(\pi_0^\infty)^\vee) \cong D(G_0)_{Y_0}\widehat{\otimes}_{D(G_0)_{P_{I_0,0}}}(\mathcal{M}_0\widehat{\otimes}_E(\pi_0^\infty)^\vee), \tag{4.73}$$

hence (taking completions) a topological isomorphism

$$\mathrm{Fil}_w(V_0^\vee) \cong D(G_0)_{Y_0} \widehat{\otimes}_{D(G_0)_{P_{I_0,0}}} (\mathcal{M}_0 \widehat{\otimes}_E (\pi_0^\infty)^\vee). \tag{4.74}$$

By (ii) of Lemma 4.2.7 $D(G_0)_r$ and $D(G_0)_{Y_0,r}$ are finite free over $D(G_0)_{P_{I_0,0},r}$. From (4.72), (4.74) and Lemma 4.2.12 we deduce isomorphisms of Fréchet spaces

$$\begin{aligned} \mathrm{Fil}_w(V_0^\vee) &\cong \varprojlim_r \big(D(G_0)_{Y_0,r} \otimes_{D(G_0)_{P_{I_0,0},r}} (\mathcal{M}_0 \widehat{\otimes}_E (\pi_0^\infty)^\vee)_r\big) \qquad (4.75)\\ V_0^\vee &\cong \varprojlim_r \big(D(G_0)_r \otimes_{D(G_0)_{P_{I_0,0},r}} (\mathcal{M}_0 \widehat{\otimes}_E (\pi_0^\infty)^\vee)_r\big). \end{aligned}$$

The result follows from the fact that $D(G_0)_{Y_0,r}$ is a direct summand of $D(G_0)_r$ as $D(G_0)_{P_{I_0,0},r}$-module by (ii) of Lemma 4.2.7 again. □

For $w \in W^{I_0,I_1}$, we define $P_w \overset{\mathrm{def}}{=} w^{-1} P_{I_0} w$ and $L_w \overset{\mathrm{def}}{=} w^{-1} L_{I_0} w$. Recall that $\pi_0^{\infty,w}$ is the representation of L_w with the same underlying vector space as π_0^∞ but where $h \in L_w = w^{-1} L_{I_0} w$ acts by whw^{-1} (by inflation $\pi_0^{\infty,w}$ is also a representation of P_w). Likewise we write $\mathcal{M}_0^w$ for the coadmissible $D(G)_{P_w}$-module with the same underlying (topological) vector space as $\mathcal{M}_0$ but where $\delta \in D(G)_{P_w}$ acts by $w\delta w^{-1} \in D(G)_{P_{I_0}}$.

Proposition 4.4.3 *For $w \in W^{I_0,I_1}$ let $\mathrm{gr}_w(V_0^\vee) \overset{\mathrm{def}}{=} \mathrm{Fil}_w(V_0^\vee)/\sum_{w'<w} \mathrm{Fil}_{w'}(V_0^\vee)$ (via Lemma 4.4.2), we have a canonical (left) $D(P_{I_1})$-equivariant isomorphism of Fréchet spaces*

$$\mathrm{gr}_w(V_0^\vee) \cong \widehat{D}(G)_{P_{I_1} P_w} \widehat{\otimes}_{D(G)_{P_w}} (\mathcal{M}_0^w \widehat{\otimes}_E (\pi_0^{\infty,w})^\vee) \tag{4.76}$$

where $\widehat{D}(G)_{P_{I_1} P_w}$ is defined in (4.27) with its left $D(P_{I_1})$-action as before Lemma 4.2.11.

Proof For $w \in W^{I_0,I_1}$ let $P_{w,0} \overset{\mathrm{def}}{=} P_w \cap G_0$, $S \subseteq G$ a closed subset such that $S = SP_w = P_{I_1} S$ and $S_0 \overset{\mathrm{def}}{=} S \cap G_0$. Arguing as for (4.74) and (4.75) we have $D(G)_{P_{I_1}}$-equivariant isomorphisms of Fréchet spaces (where $(\mathcal{M}_0^w \widehat{\otimes}_E (\pi_0^{\infty,w})^\vee)_r \overset{\mathrm{def}}{=} D(G_0)_{P_{w,0},r} \otimes_{D(G_0)_{P_{w,0}}} (\mathcal{M}_0^w \widehat{\otimes}_E (\pi_0^{\infty,w})^\vee)$)

$$\begin{aligned} D(G)_{S\cdot w^{-1}} \widehat{\otimes}_{D(G)_{P_{I_0}}} (\mathcal{M}_0 \widehat{\otimes}_E (\pi_0^\infty)^\vee) &\cong D(G)_S \widehat{\otimes}_{D(G)_{P_w}} (\mathcal{M}_0^w \widehat{\otimes}_E (\pi_0^{\infty,w})^\vee) \qquad (4.77)\\ &\cong D(G_0)_{S_0} \widehat{\otimes}_{D(G_0)_{P_{w,0}}} (\mathcal{M}_0^w \widehat{\otimes}_E (\pi_0^{\infty,w})^\vee)\\ &\cong \varprojlim_r \big(D(G_0)_{S_0,r} \otimes_{D(G_0)_{P_{w,0},r}} (\mathcal{M}_0^w \widehat{\otimes}_E (\pi_0^{\infty,w})^\vee)_r\big). \end{aligned}$$

Let $X \overset{\mathrm{def}}{=} S_w w = P_{I_1} P_w \subseteq G$, $Y \overset{\mathrm{def}}{=} \overline{X}$ and $Z \overset{\mathrm{def}}{=} \overline{X} \setminus X$, applying these isomorphisms with $S = Y$ and $S = Z$ and using (4.35), (4.36), (4.37) and Lemma 4.2.4 we obtain an exact sequence of Fréchet spaces

$$0 \longrightarrow D(G_0)_{Z_0}\widehat{\otimes}_{D(G_0)_{P_{w,0}}}(\mathcal{M}_0^w\widehat{\otimes}_E(\pi_0^{\infty,w})^\vee) \longrightarrow D(G_0)_{Y_0}\widehat{\otimes}_{D(G_0)_{P_{w,0}}}(\mathcal{M}_0^w\widehat{\otimes}_E(\pi_0^{\infty,w})^\vee)$$
$$\longrightarrow \widehat{D}(G_0)_{X_0}\widehat{\otimes}_{D(G_0)_{P_{w,0}}}(\mathcal{M}_0^w\widehat{\otimes}_E(\pi_0^{\infty,w})^\vee) \longrightarrow 0.$$

Using the definition of $\widehat{D}(G)_X$ (see (4.27)) and arguing as before (4.74), we deduce a $D(P_I)$-equivariant exact sequence of Fréchet spaces

$$0 \longrightarrow D(G)_Z\widehat{\otimes}_{D(G)_{P_w}}(\mathcal{M}_0^w\widehat{\otimes}_E(\pi_0^{\infty,w})^\vee) \longrightarrow D(G)_Y\widehat{\otimes}_{D(G)_{P_w}}(\mathcal{M}_0^w\widehat{\otimes}_E(\pi_0^{\infty,w})^\vee)$$
$$\longrightarrow \widehat{D}(G)_X\widehat{\otimes}_{D(G)_{P_w}}(\mathcal{M}_0^w\widehat{\otimes}_E(\pi_0^{\infty,w})^\vee) \longrightarrow 0.$$

By (4.74) and using (4.77) "backwards" together with Lemma 4.4.1, it remains to prove

$$D(G_0)_{Z_0'}\widehat{\otimes}_{D(G_0)_{P_{I_0,0}}}(\mathcal{M}_0\widehat{\otimes}_E(\pi_0^\infty)^\vee) \cong \sum_{w'<w}\mathrm{Fil}_{w'}(V_0^\vee)$$

where $Z_0' \overset{\text{def}}{=} \cup_{w'<w}(\overline{S_{w'}}\cap G_0)$. Let $D \overset{\text{def}}{=} \mathcal{M}_0\widehat{\otimes}_E(\pi_0^\infty)^\vee$, it is enough to prove that if C_1, C_2 are two closed subsets of G_0 such that $C_i = C_iP_{I_0}$ then we have inside $D(G_0)\widehat{\otimes}_{D(G_0)_{P_{I_0,0}}}D$:

$$D(G_0)_{C_1}\widehat{\otimes}_{D(G_0)_{P_{I_0,0}}}D + D(G_0)_{C_2}\widehat{\otimes}_{D(G_0)_{P_{I_0,0}}}D = D(G_0)_{C_1\cup C_2}\widehat{\otimes}_{D(G_0)_{P_{I_0,0}}}D.$$

Using (ii) of Lemma 4.2.7, for $r \in \mathcal{I}$ we have an isomorphism of finite free $D(G_0)_{P_{I_0,0,r}}$-modules $D(G_0)_{C_1,r} + D(G_0)_{C_2,r} \overset{\sim}{\to} D(G_0)_{C_1\cup C_2,r}$ which gives an isomorphism of Banach spaces where $D_r \overset{\text{def}}{=} D(G_0)_{P_{I_0,0,r}} \otimes_{D(G_0)_{P_{I_0,0}}} D$

$$(D(G_0)_{C_1,r} + D(G_0)_{C_2,r}) \otimes_{D(G_0)_{P_{I_0,0,r}}} D_r \overset{\sim}{\longrightarrow} D(G_0)_{C_1\cup C_2,r} \otimes_{D(G_0)_{P_{I_0,0,r}}} D_r.$$

Using Lemma 4.2.12, it is enough to prove

$$\varprojlim_{r\in\mathcal{I}}(D(G_0)_{C_1,r} + D(G_0)_{C_2,r}) \cong D(G_0)_{C_1} + D(G_0)_{C_2}$$

(inside $\varprojlim_r D(G_0)_r \cong D(G_0)$). The image of $D(G_0)_{C_i}$ is dense in $D(G_0)_{C_i,r}$, hence the image of $D(G_0)_{C_1} + D(G_0)_{C_2}$ is dense in $D(G_0)_{C_1,r} + D(G_0)_{C_2,r}$ (inside $D(G_0)_r$). As $D(G_0)_{C_1,r} + D(G_0)_{C_2,r}$ is closed in $D(G_0)_r$ (see above), it follows that $\varprojlim_r(D(G_0)_{C_1,r} + D(G_0)_{C_2,r})$ is the closure of $D(G_0)_{C_1} + D(G_0)_{C_2}$ inside $D(G_0)$. Hence it is enough to prove $D(G_0)_{C_1} + D(G_0)_{C_2} \cong D(G_0)_{C_1\cap C_2}$ (the latter being closed in $D(G_0)$). Let U_1, U_2 be compact open subsets of G_0 containing respectively C_1, C_2, then we have a short exact sequence of locally convex E-vector spaces of compact type

$$0 \longrightarrow C^{\mathrm{an}}(U_1\cup U_2) \longrightarrow C^{\mathrm{an}}(U_1)\oplus C^{\mathrm{an}}(U_2) \longrightarrow C^{\mathrm{an}}(U_1\cap U_2) \longrightarrow 0$$

where the maps are the restrictions. Taking the colimit over such U_1, U_2, we deduce a short exact sequence of locally convex E-vector spaces of compact type

$$0 \longrightarrow \varinjlim_{U_1,U_2} C^{\mathrm{an}}(U_1 \cup U_2) \longrightarrow \varinjlim_{U_1} C^{\mathrm{an}}(U_1) \oplus \varinjlim_{U_2} C^{\mathrm{an}}(U_2) \longrightarrow \varinjlim_{U_1,U_2} C^{\mathrm{an}}(U_1 \cup U_2) \longrightarrow 0.$$

In particular the injection on the left is a closed embedding. Noting that compact open subsets of G_0 of the form $U_1 \cup U_2$, where U_i are compact open subsets of G_0 containing C_i, are cofinal among compact open subsets of G_0 containing $C_1 \cup C_2$, by [13, (3.3)] and [65, Corollary 9.4] we deduce a surjection of Fréchet spaces $D(G_0)_{C_1} \oplus D(G_0)_{C_2} \twoheadrightarrow D(G_0)_{C_1 \cup C_2}$. In particular we have $D(G_0)_{C_1} + D(G_0)_{C_2} = D(G_0)_{C_1 \cap C_2}$ in $D(G_0)$. □

4.5 Ext Groups of Orlik-Strauch Representations

We prove several results on the Ext groups of Orlik-Strauch representations, in particular that they are finite dimensional when their smooth entries are finite length representations (Theorem 4.5.16). The most important statements are Corollary 4.5.11 (which follows from Theorem 4.5.10) and Corollary 4.5.13.

We keep the notation of Sect. 4.4, in particular we have $I_0, I_1 \subseteq \Delta$, $M_0 \in \mathcal{O}_{\mathrm{alg}}^{\mathfrak{p}_{I_0}}$, π_0^∞ a smooth strongly admissible representation of L_{I_0} over E and $V_0 = \mathcal{F}_{P_{I_0}}^G(M_0, \pi_0^\infty)$. Note that we assume π_0^∞ strongly admissible ([70, Sect. 3]) instead of just admissible and we recall that if π_0^∞ is of finite length then it is strongly admissible ([69, Proposition 2.2]). From now until Corollary 4.5.13 (included), we let $M_1 \overset{\mathrm{def}}{=} M^{I_1}(\mu) \in \mathcal{O}_{\mathrm{alg}}^{\mathfrak{p}_{I_1}}$ for some $\mu \in \Lambda_{I_1}^{\mathrm{dom}}$ (a generalized Verma module, see (3.5) and the lines below it), π_1^∞ a smooth strongly admissible representation of L_{I_1} over E and $V_1 \overset{\mathrm{def}}{=} \mathcal{F}_{P_{I_1}}^G(M_1, \pi_1^\infty)$. Our main aim in this section is to study the E-vector spaces $\mathrm{Ext}^k_{D(G)}(V_1^\vee, V_0^\vee)$ for $k \geq 0$.

Recall from Remark 4.3.9 and (4.63) (noting that $R_1 = 0$ in (4.62)) that we have a $D(G)$-equivariant isomorphism $V_1^\vee \cong D(G) \otimes_{D(P_{I_1})} (L^{I_1}(\mu) \otimes_E (\pi_1^\infty)^\vee)$. By Lemma 4.3.2 and the proof of [73, Lemma 6.3(ii)] (where we use [62, Proposition 2.6] instead of [73, Lemma 6.2] as we are locally K-analytic) we deduce isomorphisms for $k \geq 0$

$$\mathrm{Ext}^k_{D(G)}(V_1^\vee, V_0^\vee) \cong \mathrm{Ext}^k_{D(P_{I_1})}(L^{I_1}(\mu) \otimes_E (\pi_1^\infty)^\vee, V_0^\vee). \qquad (4.78)$$

Thus our main aim is to compute $\mathrm{Ext}^k_{D(P_{I_1})}(L^{I_1}(\mu) \otimes_E (\pi_1^\infty)^\vee, V_0^\vee)$ for $k \geq 0$.

For M in $\mathrm{Mod}_{U(\mathfrak{p}_{I_1})}$, we endow $\mathrm{Hom}_E(L^{I_1}(\mu), M)$ with a structure of $U(\mathfrak{p}_{I_1})$-module by

$$(u \cdot f)(x) = u \cdot f(x) - f(u \cdot x)$$

($u \in \mathfrak{p}_{I_1}$, $f \in \mathrm{Hom}_E(L^{I_1}(\mu), M)$, $x \in L^{I_1}(\mu)$). In particular

$$H^0(\mathfrak{p}_{I_1}, \mathrm{Hom}_E(L^{I_1}(\mu), M)) \cong \mathrm{Hom}_{U(\mathfrak{p}_{I_1})}(L^{I_1}(\mu), M). \tag{4.79}$$

It is easy to check for M' in $\mathrm{Mod}_{U(\mathfrak{p}_{I_1})}$:

$$\mathrm{Hom}_{U(\mathfrak{p}_{I_1})}(M', \mathrm{Hom}_E(L^{I_1}(\mu), M)) \cong \mathrm{Hom}_{U(\mathfrak{p}_{I_1})}(M' \otimes_E L^{I_1}(\mu), M) \tag{4.80}$$

with $U(\mathfrak{p}_{I_1})$ acting diagonally on $M' \otimes_E L^{I_1}(\mu)$. It follows from (4.80) and the exactness of $\mathrm{Hom}_E(L^{I_1}(\mu), -)$ that if $M^\bullet$ is an injective resolution of M in $\mathrm{Mod}_{U(\mathfrak{p}_{I_1})}$, then $\mathrm{Hom}_E(L^{I_1}(\mu), M^\bullet)$ is an injective resolution of $\mathrm{Hom}_E(L^{I_1}(\mu), M)$ in $\mathrm{Mod}_{U(\mathfrak{p}_{I_1})}$. Using (4.79) this implies canonical isomorphisms for $\ell \geq 0$

$$\mathrm{Ext}^{\ell}_{U(\mathfrak{p}_{I_1})}(L^{I_1}(\mu), M) \cong H^{\ell}(\mathfrak{p}_{I_1}, \mathrm{Hom}_E(L^{I_1}(\mu), M)). \tag{4.81}$$

Recall that $H^{\ell}(\mathfrak{p}_{I_1}, \mathrm{Hom}_E(L^{I_1}(\mu), M))$ can also be computed by the Chevalley-Eilenberg complex $C^\bullet$ where for $\ell \geq 0$

$$C^{\ell} \stackrel{\mathrm{def}}{=} \mathrm{Hom}_E(\wedge^{\ell}\mathfrak{p}_{I_1}, \mathrm{Hom}_E(L^{I_1}(\mu), M)) \cong \mathrm{Hom}_E(\wedge^{\ell}\mathfrak{p}_{I_1} \otimes_E L^{I_1}(\mu), M) \tag{4.82}$$

(see also (3.7)). The left (continuous) action of P_I on $L^{I_1}(\mu)$ induces a left action on $\mathrm{Hom}_E(L^{I_1}(\mu), E)$ defined by $(g \cdot f)(x) = f(g^{-1}x)$ ($g \in P_{I_1}$, $f \in \mathrm{Hom}_E(L^{I_1}(\mu), E)$, $x \in L^{I_1}(\mu)$). As $\mathrm{Hom}_E(L^{I_1}(\mu), E)$ is finite dimensional, this action extends to a left $D(P_{I_1})$-action. Let D be any $D(P_{I_1})$-module, we endow $\mathrm{Hom}_E(L^{I_1}(\mu), D) \cong \mathrm{Hom}_E(L^{I_1}(\mu), E) \otimes_E D$ with the diagonal (left) action of $D(P_{I_1})$ (via [73, Sect. A]). It is easy to check that for any $D(P_{I_1})$-module D':

$$\mathrm{Hom}_{D(P_{I_1})}(D', \mathrm{Hom}_E(L^{I_1}(\mu), D)) \cong \mathrm{Hom}_{D(P_{I_1})}(D' \otimes_E L^{I_1}(\mu), D) \tag{4.83}$$

with $D(P_{I_1})$ acting diagonally on $D' \otimes_E L^{I_1}(\mu)$. Hence, if $D^\bullet$ is an injective resolution of D in $\mathrm{Mod}_{D(P_{I_1})}$, then $\mathrm{Hom}_E(L^{I_1}(\mu), D^\bullet)$ is an injective resolution of $\mathrm{Hom}_E(L^{I_1}(\mu), D)$ satisfying

$$\mathrm{Hom}_{D(P_{I_1})}(L^{I_1}(\mu) \otimes_E (\pi_1^{\infty})^\vee, D^k) \cong \mathrm{Hom}_{D(P_{I_1})}((\pi_1^{\infty})^\vee, \mathrm{Hom}_E(L^{I_1}(\mu), D^k))$$

for $k \geq 0$. We thus obtain canonical isomorphisms for $k \geq 0$

$$\mathrm{Ext}^k_{D(P_{I_1})}(L^{I_1}(\mu) \otimes_E (\pi_1^{\infty})^\vee, D) \cong \mathrm{Ext}^k_{D(P_{I_1})}((\pi_1^{\infty})^\vee, \mathrm{Hom}_E(L^{I_1}(\mu), D)). \tag{4.84}$$

It follows from [73, Sect. 3] (more precisely from [73, p. 307 line 5] together with [73, p. 306 line –11] applied with $X = X^\bullet = \mathrm{Hom}_E(L^{I_1}(\mu), D)$ and $Y^\bullet = (\pi_1^{\infty})^\vee$, both in degree 0) that we have a spectral sequence

$$\mathrm{Ext}^k_{D^\infty(P_{I_1})}((\pi_1^{\infty})^\vee, H^{\ell}(\mathfrak{p}_{I_1}, \mathrm{Hom}_E(L^{I_1}(\mu), D))) \implies \mathrm{Ext}^{k+\ell}_{D(P_{I_1})}((\pi_1^{\infty})^\vee, \mathrm{Hom}_E(L^{I_1}(\mu), D)),$$

(in particular $H^\ell(\mathfrak{p}_{I_1}, \mathrm{Hom}_E(L^{I_1}(\mu), D))$ is naturally a $D^\infty(P_{I_1})$-module) which together with (4.84) and (4.81) (applied with $M = D$) gives a spectral sequence

$$\mathrm{Ext}^k_{D^\infty(P_{I_1})}\big((\pi_1^\infty)^\vee, \mathrm{Ext}^\ell_{U(\mathfrak{p}_{I_1})}(L^{I_1}(\mu), D)\big) \implies \mathrm{Ext}^{k+\ell}_{D(P_{I_1})}\big(L^{I_1}(\mu) \otimes_E (\pi_1^\infty)^\vee, D\big). \tag{4.85}$$

The spectral sequence (4.85) applied to graded pieces of $V_0^\vee$ will be our primary means of accessing $\mathrm{Ext}^\bullet_{D(P_{I_1})}(L^{I_1}(\mu) \otimes_E (\pi_1^\infty)^\vee, V_0^\vee)$ (and hence $\mathrm{Ext}^k_{D(G)}(V_1^\vee, V_0^\vee)$ by (4.78)).

From now until Theorem 4.5.10 (included) we fix $w \in W^{I_0,I_1}$. We write $C^\bullet$ (resp. $\mathcal{C}^\bullet$) for the Chevalley-Eilenberg complex (4.82) with $M = M_0^w$ (resp. $M = \mathcal{M}_0^w$) where M_0^w is defined above (3.47) and $\mathcal{M}_0^w$ as in Proposition 4.4.3. Note that as $\wedge^\ell \mathfrak{p}_{I_1} \otimes_E L^{I_1}(\mu)$ is finite dimensional each $\mathcal{C}^\ell$ is a Fréchet space. The algebra $U(\mathfrak{p}_{I_1})$ acts on both $\wedge^\ell \mathfrak{p}_{I_1} \otimes_E L^{I_1}(\mu)$ and M_0^w (resp. $\mathcal{M}_0^w$), and thus acts diagonally on C^ℓ (resp. $\mathcal{C}^\ell$). We write $d^\ell : C^\ell \to C^{\ell+1}$ (resp. $\delta^\ell : \mathcal{C}^\ell \to \mathcal{C}^{\ell+1}$) for the differential maps of the complex $C^\bullet$ (resp. $\mathcal{C}^\bullet$), see [73, p. 305] for instance. The $U(\mathfrak{g})$-equivariant embedding $1 \otimes \mathrm{id} : M_0 \hookrightarrow \mathcal{M}_0 = D(G)_1 \otimes_{U(\mathfrak{g})} M_0$ induces an $U(\mathfrak{g})$-equivariant embedding $M_0^w \hookrightarrow \mathcal{M}_0^w$ which induces a map of complexes $C^\bullet \to \mathcal{C}^\bullet$ which induces for $\ell \geq 0$

$$\kappa^\ell : \mathrm{Ext}^\ell_{U(\mathfrak{p}_{I_1})}(L^{I_1}(\mu), M_0^w) \longrightarrow \mathrm{Ext}^\ell_{U(\mathfrak{p}_{I_1})}(L^{I_1}(\mu), \mathcal{M}_0^w). \tag{4.86}$$

Lemma 4.5.1 *For $\ell \geq 0$ κ^ℓ in (4.86) is an isomorphism of finite dimensional E-vector spaces.*

Proof By dévissage using the short exact sequence $0 \to M_0^w \to \mathcal{M}_0^w \to (\mathcal{M}_0/M_0)^w \to 0$, it suffices to show for $\ell \geq 0$:

$$\mathrm{Ext}^\ell_{U(\mathfrak{p}_{I_1})}(L^{I_1}(\mu), (\mathcal{M}_0/M_0)^w) = 0. \tag{4.87}$$

Let $e \in \mathfrak{t}$ such that $\alpha(e) \in \mathbb{Z}_{>0}$ for all $\alpha \in \Phi^+$. We divide the proof into three steps.

Step 1: We prove that $\mathrm{ad}(w^{-1})(e) - N$ acts bijectively on the $U(\mathfrak{t})$-module $H^{\ell_1}(\mathfrak{u}, (\mathcal{M}_0/M_0)^w)$ for $N \in \mathbb{Z}$ and $\ell_1 \geq 0$.
By (ii) of Proposition 4.3.1 and Lemma 4.1.15 $e - N$ for $N \in \mathbb{Z}$ acts bijectively on $\mathcal{M}_0/M_0$ and thus $\mathrm{ad}(w^{-1})(e) - N$ acts bijectively on $(\mathcal{M}_0/M_0)^w$. As $\wedge^{\ell_1}\mathfrak{u}$ is $\mathfrak{t}$-semi-simple we have $\wedge^{\ell_1}\mathfrak{u} = \bigoplus_{M \in \mathbb{Z}} \wedge^{\ell_1}\mathfrak{u}|_{\mathrm{ad}(w^{-1})(e)=M}$, and we see that $\mathrm{ad}(w^{-1})(e) - N$ acts invertibly on $\mathrm{Hom}_E(\wedge^{\ell_1}\mathfrak{u}, (\mathcal{M}_0/M_0)^w)$, and thus on any of its $U(\mathfrak{t})$-subquotient, in particular on its $U(\mathfrak{t})$-subquotient $H^{\ell_1}(\mathfrak{u}, (\mathcal{M}_0/M_0)^w)$.

Step 2: We prove $\mathrm{Ext}^{\ell_2}_{U(\mathfrak{t})}(\mu_1, H^{\ell_1}(\mathfrak{u}, (\mathcal{M}_0/M_0)^w)) = 0$ for $\ell_1, \ell_2 \geq 0$ and $\mu_1 \in \Lambda$.
Let $\mathfrak{t}' \stackrel{\mathrm{def}}{=} E(\mathrm{ad}(w^{-1})(e)) \subseteq \mathfrak{t}$, for $\mu_1 \in \Lambda$ and any $U(\mathfrak{t})$-module we have a Hochschild-Serre spectral sequence

$$H^{\ell_2''}(\mathfrak{t}/\mathfrak{t}', H^{\ell_2'}(\mathfrak{t}', \mathrm{Hom}_E(\mu_1, D))) \implies H^{\ell_2'+\ell_2''}(\mathfrak{t}, \mathrm{Hom}_E(\mu_1, D)) \cong \mathrm{Ext}_{U(\mathfrak{t})}^{\ell_2'+\ell_2''}(\mu_1, D),$$

(where the last isomorphism is proved as (4.81)). Taking $D = H^{\ell_1}(\mathfrak{u}, (\mathcal{M}_0/M_0)^w))$ it suffices to prove for $\ell_2' \geq 0$:

$$H^{\ell_2'}\big(\mathfrak{t}', \mathrm{Hom}_E(\mu_1, H^{\ell_1}(\mathfrak{u}, (\mathcal{M}_0/M_0)^w))\big) = 0. \tag{4.88}$$

But since $\dim_E \mathfrak{t}' = 1$ the Chevalley-Eilenberg complex that computes (4.88) is just

$$\mathrm{Hom}_E(\mu_1, H^{\ell_1}(\mathfrak{u}, (\mathcal{M}_0/M_0)^w)) \overset{\mathrm{ad}(w^{-1})(e)}{\longrightarrow} \mathrm{Hom}_E(\mu_1, H^{\ell_1}(\mathfrak{u}, (\mathcal{M}_0/M_0)^w))$$

and by Step 1 the unique differential map is an isomorphism, whence the result.

Step 3: We prove for $\ell \geq 0$ and $\mu_1 \in \Lambda$:

$$\mathrm{Ext}^{\ell}_{U(\mathfrak{p}_{I_1})}(L^{I_1}(\mu_1), (\mathcal{M}_0/M_0)^w) = 0. \tag{4.89}$$

Let $\mu_1 \in \Lambda$, by (3.12) applied with $I = \emptyset$ we have the spectral sequence

$$\mathrm{Ext}^{\ell_2}_{U(\mathfrak{t})}(\mu_1, H^{\ell_1}(\mathfrak{u}, (\mathcal{M}_0/M_0)^w)) \implies \mathrm{Ext}^{\ell_1+\ell_2}_{U(\mathfrak{b})}(\mu_1, (\mathcal{M}_0/M_0)^w),$$

which together with Step 2 implies for $\ell \geq 0$

$$\mathrm{Ext}^{\ell}_{U(\mathfrak{p}_{I_1})}(U(\mathfrak{p}_{I_1}) \otimes_{U(\mathfrak{b})} \mu_1, (\mathcal{M}_0/M_0)^w) \cong \mathrm{Ext}^{\ell}_{U(\mathfrak{b})}(\mu_1, (\mathcal{M}_0/M_0)^w) = 0 \tag{4.90}$$

where the isomorphism in (4.90) is Shapiro's lemma. Recall $U(\mathfrak{p}_{I_1}) \otimes_{U(\mathfrak{b})} \mu_1 \cong U(\mathfrak{l}_{I_1}) \otimes_{U(\mathfrak{b}_{I_1})} \mu_1$. Hence if $U(\mathfrak{l}_{I_1}) \otimes_{U(\mathfrak{b}_{I_1})} \mu_1 \cong L^{I_1}(\mu_1)$, we are done. In general, we argue by induction using [33, Theorem 5.1] applied to $\mathfrak{l}_{I_1}$ and a dévissage on the constituents of $U(\mathfrak{l}_{I_1}) \otimes_{U(\mathfrak{b}_{I_1})} \mu_1$. Finally, applying (4.89) with $\mu_1 = \mu$ gives (4.87). □

For $\ell \geq 0$ the map $\delta^{\ell} : \mathcal{C}^{\ell} \to \mathcal{C}^{\ell+1}$ is a continuous map between Fréchet spaces, in particular $\ker(\delta^{\ell}) \subseteq \mathcal{C}^{\ell}$ is a closed subspace. We endow $\ker(\delta^{\ell})$ with the subspace topology of $\mathcal{C}^{\ell}$ and $H^{\ell}(\mathcal{C}^{\bullet}) = \ker(\delta^{\ell})/\mathrm{im}(\delta^{\ell-1})$ with the quotient topology ([65, Sect. 5.B]).

Lemma 4.5.2 *For $\ell \geq 0$ the differential map $\delta^{\ell} : \mathcal{C}^{\ell} \to \mathcal{C}^{\ell+1}$ has closed image and $H^{\ell}(\mathcal{C}^{\bullet})$ is a finite dimensional separated E-vector space (with its natural Banach topology).*

Proof Let $\overline{\mathrm{im}(\delta^{\ell-1})}$ be the closure of $\mathrm{im}(\delta^{\ell-1})$ in $\mathcal{C}^{\ell}$, which is still contained in $\ker(\delta^{\ell})$, then $H^{\ell}(\mathcal{C}^{\bullet})$ is separated if and only if $\mathrm{im}(\delta^{\ell-1}) = \overline{\mathrm{im}(\delta^{\ell-1})}$. The identification $M_0 = \bigoplus_{\mu_1 \in \Lambda} (\mathcal{M}_0)_{\mu_1}$ from (ii) of Proposition 4.3.1 implies $M_0^w = \bigoplus_{\mu_1 \in \Lambda} (\mathcal{M}_0^w)_{\mu_1}$ and thus $\mathcal{C}^{\ell} = \bigoplus_{\mu_1 \in \Lambda} (\mathcal{C}^{\ell})_{\mu_1}$ for $\ell \geq 0$ with each $(\mathcal{C}^{\ell})_{\mu_1}$ being finite dimensional (recall $\wedge^{\ell}\mathfrak{p}_{I_1} \otimes_E L^{I_1}(\mu)$ is finite dimensional and $U(\mathfrak{t})$-semi-simple). In particular, together with the $U(\mathfrak{t})$-equivariance of the differential maps $d^\bullet$ and $\delta^\bullet$ we deduce

$$\mathrm{im}(d^{\ell-1}) = \bigoplus_{\mu_1\in\Lambda} \left(\mathrm{im}(\delta^{\ell-1})\right)_{\mu_1} = \bigoplus_{\mu_1\in\Lambda} \left(\overline{\mathrm{im}(\delta^{\ell-1})}\right)_{\mu_1} \text{ and } \ker(d^{\ell}) = \bigoplus_{\mu_1\in\Lambda} \left(\ker(\delta^{\ell})\right)_{\mu_1}.$$

and from (4.8)

$$\bigoplus_{\mu_1\in\Lambda} \left(\ker(\delta^{\ell})/\overline{\mathrm{im}(\delta^{\ell})}\right)_{\mu_1} \cong \Big(\bigoplus_{\mu_1\in\Lambda} \left(\ker(\delta^{\ell})\right)_{\mu_1}\Big)/\Big(\bigoplus_{\mu_1\in\Lambda} \overline{\mathrm{im}(\delta^{\ell})}\big)_{\mu_1}\Big) \cong \ker(d^{\ell})/\mathrm{im}(d^{\ell}).$$

This forces the composition $\ker(d^{\ell})/\mathrm{im}(d^{\ell}) \xrightarrow{\kappa^{\ell}} \ker(\delta^{\ell})/\mathrm{im}(\delta^{\ell}) \xrightarrow{\theta^{\ell}} \ker(\delta^{\ell})/\overline{\mathrm{im}(\delta^{\ell})}$ to be an injection. But κ^{ℓ} is a bijection by Lemma 4.5.1 and θ^{ℓ} is a surjection by definition, so θ^{ℓ} must also be an isomorphism, which means $\overline{\mathrm{im}(\delta^{\ell})} = \mathrm{im}(\delta^{\ell})$. The rest of the statement follows from Lemma 4.5.1 (and (4.81)). □

We write $D^{\infty} \stackrel{\mathrm{def}}{=} (\pi_0^{\infty,w})^{\vee}$ and $D \stackrel{\mathrm{def}}{=} \mathcal{M}_0^{w}\widehat{\otimes}_E D^{\infty}$ for short. For $g \in G$ we write $\mathcal{M}_0^{wg^{-1}}$, $D^{\infty,g^{-1}}$, $D^{g^{-1}}$ for the $D(G)_{gP_wg^{-1}}$-module with the same underlying space as $\mathcal{M}_0^{w}$, D^{∞}, D (respectively) but with $\delta \in D(G)_{gP_wg^{-1}}$ acting by $\delta_g^{-1}\delta\delta_g$. For $g \in G$ and $h \in gP_wg^{-1}$, the map $v \mapsto \delta_h v$ gives a $D(G)_{gP_wg^{-1}} = D(G)_{hgP_w(hg)^{-1}}$-equivariant topological isomorphism

$$\mathcal{M}_0^{wg^{-1}} \xrightarrow{\sim} \mathcal{M}_0^{w(hg)^{-1}} \tag{4.91}$$

and likewise $D(G)_{gP_wg^{-1}}$-equivariant isomorphisms $D^{\infty,g^{-1}} \xrightarrow{\sim} D^{\infty,(hg)^{-1}}$, $D^{g^{-1}} \xrightarrow{\sim} D^{(hg)^{-1}}$.

For $g \in G$, we consider the Chevalley-Eilenberg complex attached to $\mathcal{M}_0^{wg^{-1}}$:

$$\mathcal{C}_g^{\bullet} \stackrel{\mathrm{def}}{=} \mathrm{Hom}_E(\wedge^{\bullet}\mathfrak{p}_{I_1} \otimes_E L^{I_1}(\mu), \mathcal{M}_0^{wg^{-1}})$$

and we denote by $\delta_g^{\ell} : \mathcal{C}_g^{\ell} \to \mathcal{C}_g^{\ell+1}$ the differential maps (not to be confused with the Dirac distribution δ_g!). For $\ell \geq 0$ we write $\mathcal{D}_g^{\ell} \stackrel{\mathrm{def}}{=} \delta_g^{\ell-1}(\mathcal{C}_g^{\ell-1})$ and

$$\begin{aligned}\mathcal{H}_g^{\ell} \stackrel{\mathrm{def}}{=} H^{\ell}(\mathcal{C}_g^{\bullet}) = \ker(\delta_g^{\ell})/\mathcal{D}_g^{\ell} &= H^{\ell}(\mathfrak{p}_{I_1}, \mathrm{Hom}_E(L^{I_1}(\mu), \mathcal{M}_0^{wg^{-1}})) \\ &\stackrel{(4.81)}{\cong} \mathrm{Ext}^{\ell}_{U(\mathfrak{p}_{I_1})}(L^{I_1}(\mu), \mathcal{M}_0^{wg^{-1}}).\end{aligned}$$

For $\ell \geq 0$, $g \in G$ and $h \in gP_wg^{-1}$, the $D(G)_{gP_wg^{-1}}$-equivariant topological isomorphism (4.91) induces a $U(\mathfrak{p}_{I_1})$-equivariant topological isomorphism

$$\mathcal{C}_g^{\ell} \xrightarrow{\sim} \mathcal{C}_{hg}^{\ell}, \tag{4.92}$$

which further induces topological isomorphisms $\mathcal{D}_g^\ell \xrightarrow{\sim} \mathcal{D}_{hg}^\ell$, $\ker(\delta_g^\ell) \xrightarrow{\sim} \ker(\delta_{hg}^\ell)$ and

$$\mathcal{H}_g^\ell \xrightarrow{\sim} \mathcal{H}_{hg}^\ell. \tag{4.93}$$

It follows from (4.92) and (4.93) that the complex $\mathcal{C}_g^\bullet$ and the cohomology space $\mathcal{H}_g^\ell$ only depend on the coset $(gP_wg^{-1})g = gP_w$ up to natural $U(\mathfrak{p}_{I_1})$-equivariant topological isomorphisms. We tacitly use this in the sequel.

Let $\ell \geq 0$, $g \in G$, $h \in P_{I_1}$. As $\wedge^\ell \mathfrak{p}_{I_1} \otimes_E L^{I_1}(\mu)$ is a finite dimensional $D(P_{I_1})$-module, the map $v \mapsto \delta_h v$ induces an isomorphism of $U(\mathfrak{p}_{I_1})$-modules $\wedge^\bullet \mathfrak{p}_{I_1} \otimes_E L^{I_1}(\mu) \xrightarrow{\sim} (\wedge^\bullet \mathfrak{p}_{I_1} \otimes_E L^{I_1}(\mu))^{h^{-1}}$, which induces a topological isomorphism

$$\begin{aligned}\theta_{g,h}^\ell : \mathcal{C}_g^\ell &= \mathrm{Hom}_E(\wedge^\ell \mathfrak{p}_{I_1} \otimes_E L^{I_1}(\mu), \mathcal{M}_0^{wg^{-1}}) \\ &\cong \mathrm{Hom}_E((\wedge^\ell \mathfrak{p}_{I_1} \otimes_E L^{I_1}(\mu))^{h^{-1}}, \mathcal{M}_0^{wg^{-1}h^{-1}}) \\ &\xrightarrow{\sim} \mathrm{Hom}_E(\wedge^\ell \mathfrak{p}_{I_1} \otimes_E L^{I_1}(\mu), \mathcal{M}_0^{w(hg)^{-1}}) = \mathcal{C}_{hg}^\ell. \end{aligned} \tag{4.94}$$

Moreover under (4.94) the differential map δ_g^ℓ corresponds to δ_{hg}^ℓ, hence we deduce a topological isomorphism for $\ell \geq 0$, $g \in G$ and $h \in P_{I_1}$

$$\omega_{g,h}^\ell : \mathcal{H}_g^\ell = \mathrm{Ext}_{U(\mathfrak{p}_{I_1})}^\ell (L^{I_1}(\mu), \mathcal{M}_0^{wg^{-1}}) \xrightarrow{\sim} \mathrm{Ext}_{U(\mathfrak{p}_{I_1})}^\ell (L^{I_1}(\mu), \mathcal{M}_0^{w(hg)^{-1}}) = \mathcal{H}_{hg}^\ell. \tag{4.95}$$

By a direct check for $g \in G$ and $h, h' \in P_{I_1}$ we have $\theta_{hg,h'}^\ell \circ \theta_{g,h}^\ell = \theta_{g,h'h}^\ell$ and therefore

$$\omega_{hg,h'}^\ell \circ \omega_{g,h}^\ell = \omega_{g,h'h}^\ell. \tag{4.96}$$

Note that $\mathcal{C}_g^\ell$ (for $g \in G$) contains the ($U(\mathfrak{p}_{I_1})$-equivariant) subcomplex

$$C_g^\bullet \stackrel{\text{def}}{=} \mathrm{Hom}_E(\wedge^\bullet \mathfrak{p}_{I_1} \otimes_E L^{I_1}(\mu), M_0^{wg^{-1}})$$

with C_g^ℓ being a dense subspace of $\mathcal{C}_g^\ell$ for each $\ell \geq 0$. The map between complexes $C_g^\bullet \to \mathcal{C}_g^\bullet$ induces a natural map for $\ell \geq 0$

$$H^\ell(C_g^\bullet) \longrightarrow H^\ell(\mathcal{C}_g^\bullet) = \mathcal{H}_g^\ell. \tag{4.97}$$

Lemma 4.5.3 *Assume that $g \in X = P_{I_1}P_w$. Then for $\ell \geq 0$, the morphism (4.97) is an isomorphism, the image $\mathcal{D}_g^\ell$ of the differential map $\delta_g^{\ell-1} : \mathcal{C}_g^{\ell-1} \to \mathcal{C}_g^\ell$ is closed and $\mathcal{H}_g^\ell = H^\ell(\mathcal{C}_g^\bullet) = \ker(\delta_g^\ell)/\mathcal{D}_g^\ell$ is a finite dimensional separated E-vector space (with its natural Banach topology).*

Proof As $g \in X = P_{I_1}P_w$, there exists $h \in gP_wg^{-1}$ such that $hg \in P_{I_1}$. Using the $U(\mathfrak{p}_{I_1})$-equivariant topological isomorphism (4.92) (which restricts to an isomorphism $C_g^\ell \xrightarrow{\sim} C_{hg}^\ell$) and upon replacing g with hg, we can assume $g \in P_{I_1}$. As

$g \in P_{I_1}$, we know that δ_g^ℓ corresponds to $\delta_1^\ell = \delta^\ell$ under (4.94) for $h = g^{-1}$. Consequently, the desired results for $\delta_g^{\ell-1}$, $\mathcal{D}_g^\ell$ and $\mathcal{H}_g^\ell$ follow from those for $\delta_1^{\ell-1} = \delta^{\ell-1}$, $\mathcal{D}_1^\ell$ and $\mathcal{H}_1^\ell$, which are proven in Lemmas 4.5.1 and 4.5.2. In particular, it follows from (4.94) (for $h = g^{-1} \in P_{I_1}$) and (4.96) that $\omega_{g,g^{-1}}^\ell = (\omega_{1,g}^\ell)^{-1} : \mathcal{H}_g^\ell \xrightarrow{\sim} \mathcal{H}_1^\ell$ is a topological isomorphism of finite dimensional E-vector spaces (with their natural Banach topology). □

Lemma 4.5.4 *For $g, h \in P_{I_1}$ such that $h \in g(P_{I_1} \cap P_w)g^{-1}$ we have*

$$\omega_{g,h}^\ell = \mathrm{Id}_{\mathcal{H}_g^\ell}. \tag{4.98}$$

In particular, the map $\omega_{g,h}^\ell$ only depends on the cosets gP_w and hgP_w.

Proof For $\ell \geq 0$ and $g \in P_{I_1}$ by Lemma 4.5.3 the embedding $M_0^{wg^{-1}} \hookrightarrow \mathcal{M}_0^{wg^{-1}}$ induces an isomorphism $\mathrm{Ext}_{U(\mathfrak{p}_{I_1})}^\ell(L^{I_1}(\mu), M_0^{wg^{-1}}) \xrightarrow{\sim} \mathrm{Ext}_{U(\mathfrak{p}_{I_1})}^\ell(L^{I_1}(\mu), \mathcal{M}_0^{wg^{-1}}) \cong \mathcal{H}_g^\ell$. If $g^{-1}hg \in P_{I_1} \cap P_w$, the $D(G)_{gP_wg^{-1}} = D(G)_{hgP_w(hg)^{-1}}$-equivariant isomorphism (4.91) of Fréchet spaces restricts to an isomorphism of $U(\mathfrak{g})$-modules $M_0^{wg^{-1}} \xrightarrow{\sim} M_0^{w(hg)^{-1}}$. In the following we identify $\mathcal{M}_0^{w(hg)^{-1}}$ with $\mathcal{M}_0^{wg^{-1}}$ and $M_0^{w(hg)^{-1}}$ with $M_0^{wg^{-1}}$ via (4.91). Hence (under (4.91)) $h \mapsto \theta_{g,h}^\ell$ gives an action of $g(P_{I_1} \cap P_w)g^{-1}$ on $\mathcal{C}_g^\ell$ which preserves the subspace

$$C_g^\ell = \mathrm{Hom}_E(\wedge^\ell \mathfrak{p}_{I_1} \otimes_E L^{I_1}(\mu), M_0^{wg^{-1}}) \subseteq \mathcal{C}_g^\ell.$$

By the definition of $\theta_{g,h}^\ell$, we see that the action of $g(P_{I_1} \cap P_w)g^{-1}$ on C_g^ℓ is algebraic, and its derivative at $1 \in g(P_{I_1} \cap P_w)g^{-1}$ recovers the natural action of $g(\mathfrak{p}_{I_1} \cap \mathfrak{p}_w)g^{-1}$ on C_g^ℓ. Since $\mathcal{H}_g^\ell \cong \mathrm{Ext}_{U(\mathfrak{p}_{I_1})}^\ell(L^{I_1}(\mu), M_0^{wg^{-1}})$ is a $g(P_{I_1} \cap P_w)g^{-1}$-subquotient of C_g^ℓ on which $g(\mathfrak{p}_{I_1} \cap \mathfrak{p}_w)g^{-1}$ (and even $g\mathfrak{p}_{I_1}g^{-1} = \mathfrak{p}_{I_1}$) acts trivially, we deduce that $g(P_{I_1} \cap P_w)g^{-1}$ also acts trivially on $\mathcal{H}_g^\ell$, i.e. $\omega_{g,h}^\ell = \mathrm{Id}_{\mathcal{H}_g^\ell}$ for $h \in g(P_{I_1} \cap P_w)g^{-1}$. This proves (4.98).

Together with (4.96) this implies that the map $\omega_{g,h}^\ell$ for $g, h \in P_{I_1}$ only depends on the cosets $g(P_{I_1} \cap P_w)$ and $hg(P_{I_1} \cap P_w)g^{-1}$. For the final statement, note that, as $g, h \in P_{I_1}$, the cosets $g(P_{I_1} \cap P_w)$ and $hg(P_{I_1} \cap P_w)g^{-1}$ determine uniquely the cosets gP_w and hgP_w and vice versa. □

Let $P_{w,0} \stackrel{\text{def}}{=} P_w \cap G_0$ and $P_{I_1,0} \stackrel{\text{def}}{=} P_{I_1} \cap G_0$. For $r \in \mathcal{I}$ we let $(\mathcal{M}_0^w)_r \stackrel{\text{def}}{=} D(G_0)_{P_{w,0},r} \otimes_{D(G_0)_{P_{w,0}}} \mathcal{M}_0^w$ and consider the Chevalley-Eilenberg complex

$$\mathcal{C}_r^\ell \stackrel{\text{def}}{=} \mathrm{Hom}_E(\wedge^\ell \mathfrak{p}_{I_1} \otimes_E L^{I_1}(\mu), (\mathcal{M}_0^w)_r)$$

where we denote by $\delta_r^\ell : \mathcal{C}_r^\ell \to \mathcal{C}_r^{\ell+1}$ the differential maps. We let $\mathcal{D}_r^\ell$ be the closure of $\mathcal{D}^\ell \stackrel{\text{def}}{=} \delta^{\ell-1}(\mathcal{C}^{\ell-1}) \subseteq \mathcal{C}^\ell$ in $\mathcal{C}_r^\ell$. Then $\mathcal{D}_r^\ell$ is a closed subspace of $\ker(\delta_r^\ell) \subseteq \mathcal{C}_r^\ell$ and we define $\mathcal{H}_r^\ell \stackrel{\text{def}}{=} \ker(\delta_r^\ell)/\mathcal{D}_r^\ell$ (with the quotient topology). The natural continuous map $\mathcal{M}_0^w \to (\mathcal{M}_0^w)_r$ induces continuous maps $\mathcal{C}^\ell \to \mathcal{C}_r^\ell$, $\mathcal{D}^\ell \to \mathcal{D}_r^\ell$, $\ker(\delta^\ell) \to \ker(\delta_r^\ell)$ and

$$\mathcal{H}^\ell \longrightarrow \mathcal{H}_r^\ell. \tag{4.99}$$

As $\mathcal{M}_0^w \to (\mathcal{M}_0^w)_r$ is a continous injection with dense image (see Lemma 4.3.4), so are $\mathcal{C}^\ell \to \mathcal{C}_r^\ell$ and $\mathcal{D}^\ell \to \mathcal{D}_r^\ell$.

Lemma 4.5.5 *For $r \in \mathcal{I}$ the map $\ker(\delta^\ell) \to \ker(\delta_r^\ell)$ has dense image and the map (4.99) is a topological isomorphism of finite dimensional E-Banach spaces.*

Proof It follows from Lemma 4.3.4 (after conjugation by w) that there exists a family of standard semi-norms $\{|\cdot|_r\}_{r\in\mathcal{I}}$ (see Definition 4.1.3) on M_0^w such that $(\mathcal{M}_0^w)_r$ is the completion of M_0^w under $|\cdot|_r$. By (ii) of Lemma 4.1.9, there exists a norm $|\cdot|$ on the finite dimensional Fréchet $U(\mathfrak{t})$-module $\mathrm{Hom}_E(\wedge^\ell \mathfrak{p}_{I_1} \otimes_E L^{I_1}(\mu), E)$ such that the family of semi-norms $|\cdot| \otimes_E |\cdot|_r$ on $\mathcal{C}^\ell = \mathrm{Hom}_E(\wedge^\ell \mathfrak{p}_{I_1} \otimes_E L^{I_1}(\mu), M_0^w)$ is standard with $\mathcal{C}_r^\ell$ being the completion of $\mathcal{C}^\ell$ under $|\cdot| \otimes_E |\cdot|_r$. By Remark 4.1.12 we know that $\ker(\delta_r^\ell) \cap \mathcal{C}^\ell$ is dense in $\ker(\delta_r^\ell)$. Since $\ker(\delta_r^\ell) \cap \mathcal{C}^\ell = \ker(\delta_r^\ell|_{\mathcal{C}^\ell}) = \ker(\delta^\ell|_{\mathcal{C}^\ell}) \subseteq \ker(\delta^\ell)$, we deduce that $\ker(\delta^\ell)$ is dense in $\ker(\delta_r^\ell)$ (here we use $\mathcal{C}^\ell \hookrightarrow \mathcal{C}_r^\ell$), and thus (4.99) has dense image. By Lemma 4.5.1 the injection $M_0^w \hookrightarrow \mathcal{M}_0^w$ induces an isomorphism $H^\ell(C^\bullet) \xrightarrow{\sim} H^\ell(\mathcal{C}^\bullet) = \mathcal{H}^\ell$ and by Lemma 4.5.2 $\mathcal{H}^\ell$ is a finite dimensional E-Banach space. As $\delta^{\ell-1}$ is continuous, $\delta^{\ell-1}(C^{\ell-1})$ is dense in $\delta^\ell(\mathcal{C}^{\ell-1}) = \mathcal{D}^\ell$ and hence in $\mathcal{D}_r^\ell$, which forces $\delta^{\ell-1}(C^{\ell-1}) = \mathcal{D}_r^\ell \cap C^\ell$ by the bijection statement in Remark 4.1.12. It follows that the map $H^\ell(C^\bullet) \to \mathcal{H}_r^\ell$ is an injection, and hence (with the isomorphism $H^\ell(C^\bullet) \cong \mathcal{H}^\ell$) that (4.99) is a (continuous) injection. Hence (4.99) is a continuous injection with dense image from a finite dimensional E-Banach space to an E-Banach space, it must therefore be a topological isomorphism of finite dimensional E-Banach spaces. □

Let us fix $r \in \mathcal{I}$. For $g \in G_0$ we define $(\mathcal{M}_0^w)_r^{g^{-1}}$ as $D(G_0)_{gP_{w,0}g^{-1}}$-modules. Using $D(G_0)_{P_{w,0},r} = D(G_0)_{P_{w,0}G_1^m,r}$ (see (i) of Lemma 4.2.7 with m defined from r as in *loc. cit.*), for $g \in G_0$ and $h \in gP_{w,0}G_1^m g^{-1}$ the map $v \mapsto \delta_h v$ gives a $D(G_0)_{gP_{w,0}g^{-1},r} = D(G_0)_{hgP_{w,0}(hg)^{-1},r}$-equivariant topological isomorphism

$$(\mathcal{M}_0^w)_r^{g^{-1}} \xrightarrow{\sim} (\mathcal{M}_0^w)_r^{(hg)^{-1}}. \tag{4.100}$$

We consider the following Chevalley-Eilenberg complex

$$\mathcal{C}_{g,r}^\ell \stackrel{\text{def}}{=} \mathrm{Hom}_E(\wedge^\ell \mathfrak{p}_{I_1} \otimes_E L^{I_1}(\mu), (\mathcal{M}_0^w)_r^{g^{-1}})$$

and denote by $\delta_{g,r}^\ell : \mathcal{C}_{g,r}^\ell \to \mathcal{C}_{g,r}^{\ell+1}$ the differential maps. We write $\mathcal{D}_{g,r}^\ell$ for the closure of the image of $\mathcal{D}_g^\ell = \delta_g^{\ell-1}(\mathcal{C}_g^{\ell-1})$ in $\mathcal{C}_{g,r}^\ell$ and define $\mathcal{H}_{g,r}^\ell \stackrel{\text{def}}{=} \ker(\delta_{g,r}^\ell)/\mathcal{D}_{g,r}^\ell$. For $g \in G_0$ and $h \in gP_{w,0}G_1^m g^{-1}$, (4.100) induces a $U(\mathfrak{p}_{I_1})$-equivariant topological isomorphism $\mathcal{C}_{g,r}^\ell \xrightarrow{\sim} \mathcal{C}_{hg,r}^\ell$, which further induces topological isomorphisms $\mathcal{D}_{g,r}^\ell \xrightarrow{\sim} \mathcal{D}_{hg,r}^\ell$,

$\ker(\delta^{\ell}_{g,r}) \xrightarrow{\sim} \ker(\delta^{\ell}_{hg,r})$ and $\mathcal{H}^{\ell}_{g,r} \xrightarrow{\sim} \mathcal{H}^{\ell}_{hg,r}$. Hence $\mathcal{C}^{\ell}_{g,r}$, $\mathcal{D}^{\ell}_{g,r}$, $\ker(\delta^{\ell}_{g,r})$ and $\mathcal{H}^{\ell}_{g,r}$ only depend on the coset $gP_{w,0}G_1^m$, or equivalently on the group $gP_{w,0}G_1^m g^{-1}$ (writing $gh = ghg^{-1}g$ for $h \in P_{w,0}G_1^m$), up to canonical $U(\mathfrak{p}_{I_1})$-equivariant topological isomorphisms.

For $g \in P_{I_1,0}$, using Lemma 4.2.4 and that $\mathcal{D}^{\ell}_g$ is closed in $\mathcal{C}^{\ell}_g$ (Lemma 4.5.3), the projective limit over $r \in \mathcal{I}$ of the strict exact sequence of E-Banach spaces $0 \to \mathcal{D}^{\ell}_{g,r} \to \ker(\delta^{\ell}_{g,r}) \to \mathcal{H}^{\ell}_{g,r} \to 0$ gives back the strict exact sequence of E-Fréchet spaces $0 \to \mathcal{D}^{\ell}_g \to \ker(\delta^{\ell}_g) \to \mathcal{H}^{\ell}_g \to 0$.

Similar to (4.94), for $r \in \mathcal{I}$, $g \in G_0$, $h \in P_{I_1,0}$ and $\ell \geq 0$, we have topological isomorphisms of E-Banach spaces

$$\begin{aligned}\theta^{\ell}_{g,h,r} : \mathcal{C}^{\ell}_{g,r} &= \mathrm{Hom}_E(\wedge^{\ell}\mathfrak{p}_{I_1} \otimes_E L^{I_1}(\mu), (\mathcal{M}_0^w)_r^{g^{-1}}) \\ &\cong \mathrm{Hom}_E((\wedge^{\ell}\mathfrak{p}_{I_1} \otimes_E L^{I_1}(\mu))^{h^{-1}}, (\mathcal{M}_0^w)_r^{g^{-1}h^{-1}}) \\ &\xrightarrow{\sim} \mathrm{Hom}_E(\wedge^{\ell}\mathfrak{p}_{I_1} \otimes_E L^{I_1}(\mu), (\mathcal{M}_0^w)_r^{(hg)^{-1}}) = \mathcal{C}^{\ell}_{hg,r} \end{aligned} \tag{4.101}$$

under which the differential map $\delta^{\ell}_{g,r}$ corresponds to $\delta^{\ell}_{hg,r}$. Hence we deduce a topological isomorphism of E-Banach spaces ($r \in \mathcal{I}$, $g \in G_0$, $h \in P_{I_1,0}$, $\ell \geq 0$)

$$\omega^{\ell}_{g,h,r} : \mathcal{H}^{\ell}_{g,r} \xrightarrow{\sim} \mathcal{H}^{\ell}_{hg,r}. \tag{4.102}$$

By Lemma 4.5.5 $\ker(\delta^{\ell}_1)$ is dense in $\ker(\delta^{\ell}_{1,r})$ and (4.99) is a topological isomorphism of finite dimensional E-Banach spaces. For $g \in P_{I_1,0}$, it then follows from (4.101) (with $1, g$ instead of g, h) that $\ker(\delta^{\ell}_g)$ is dense in $\ker(\delta^{\ell}_{g,r})$, and from (4.102) and (4.95) (also applied with $1, g$) that the natural map

$$\mathcal{H}^{\ell}_g \longrightarrow \mathcal{H}^{\ell}_{g,r} \tag{4.103}$$

is a topological isomorphism of finite dimensional E-Banach space for $r \in \mathcal{I}$.

We write for $r \in \mathcal{I}$

$$\begin{aligned} D^{\infty}_r \overset{\mathrm{def}}{=} D(G_0)_{P_{w,0},r} \otimes_{D(G_0)_{P_{w,0}}} D^{\infty} &\cong \big(D(G_0)_{P_{w,0},r} \otimes_{D(G_0)_{P_{w,0}}} D^{\infty}(G_0)_{P_{w,0}}\big) \otimes_{D^{\infty}(G_0)_{P_{w,0}}} D^{\infty} \\ &\cong D^{\infty}(G_0)_{P_{w,0},r} \otimes_{D^{\infty}(P_{w,0})} D^{\infty} \cong D^{\infty}(P_{w,0})_r \otimes_{D^{\infty}(P_{w,0})} D^{\infty} \end{aligned} \tag{4.104}$$

where we recall that the second isomorphism follows from $D(G_0)_{P_{w,0},r} \otimes_{U(\mathfrak{g})} E \cong D^{\infty}(G_0)_{P_{w,0},r}$ (see Step 2 in the proof of Lemma 4.2.7) and $D^{\infty}(G_0)_{P_{w,0}} \cong D^{\infty}(P_{w,0})$ (see Lemma 4.2.2) and that the last follows using (4.59). Recall also that

$$D = \mathcal{M}_0^w \widehat{\otimes}_E D^{\infty} \cong \varprojlim_{r \in \mathcal{I}} D_r \tag{4.105}$$

where $D_r \stackrel{\text{def}}{=} (\mathcal{M}_0^w)_r \otimes_E D_r^\infty$ with D_r^∞ finite dimensional (use the coadmissibility of $\mathcal{M}_0$ (Lemma 4.3.3) together with Lemma 4.2.12). For $g \in G_0$ and $r \in \mathcal{I}$ we define $D_r^{\infty,g^{-1}}$ and $D_r^{g^{-1}}$ as $D(G_0)_{gP_{w,0}g^{-1}}$-modules.

Recall that $X = P_{I_1}P_w$ and $X_0 = X \cap G_0$. We write $\mathcal{X}_0$ for the set of compact open subsets of X_0 stable under right multiplication by $P_{w,0}$, and $\mathcal{X}$ for the set of open subsets of X which are stable under right multiplication by P_w and which have compact image in X/P_w. We denote by U_0 and U a general element of respectively $\mathcal{X}_0$ and $\mathcal{X}$ (do not confuse U here with the unipotent radical of the Borel subgroup $B \subseteq G$!). Since $X_0/P_{w,0} = X/P_w$, the map $\mathcal{X}_0 \to \mathcal{X} : U_0 \mapsto U_0P_w$ is a bijection with inverse given by $U \mapsto U \cap G_0$.

Recall from (4.25) that we have topological isomorphisms

$$\widehat{D}(G_0)_{X_0} \cong \varprojlim_{U_0 \in \mathcal{X}_0} D(G_0)_{U_0} \cong \varprojlim_{U_0 \in \mathcal{X}_0, r \in \mathcal{I}} D(G_0)_{U_0,r},$$

and from (ii) of Lemma 4.2.7 that $D(G_0)_{U_0,r}$ is a finite free $D(G_0)_{P_{w,0},r}$-module. Applying Lemma 4.2.12 to $V = \widehat{D}(G_0)_{X_0}$, $W = D$, $A = D(G_0)_{P_{w,0}}$ (with the index r replaced by U_0, r) and again to $V = D(G_0)_{U_0}$, $W = D$, $A = D(G_0)_{P_{w,0}}$, we obtain topological isomorphisms (where $\widehat{\otimes}_{D(G_0)_{P_{w,0}}}$ is defined before Lemma 4.2.12)

$$\begin{aligned} \mathrm{gr}_w(V_0^\vee) &\cong \widehat{D}(G_0)_{X_0} \widehat{\otimes}_{D(G_0)_{P_{w,0}}} D \cong \varprojlim_{U_0 \in \mathcal{X}_0, r \in \mathcal{I}} \big(D(G_0)_{U_0,r} \otimes_{D(G_0)_{P_{w,0},r}} D_r\big) \\ &\cong \varprojlim_{U_0 \in \mathcal{X}_0} \big(D(G_0)_{U_0} \widehat{\otimes}_{D(G_0)_{P_{w,0}}} D\big) \cong \varprojlim_{U \in \mathcal{X}} \big(D(G)_U \widehat{\otimes}_{D(G)_{P_w}} D\big). \end{aligned} \tag{4.106}$$

where the first isomorphism follows from Proposition 4.4.3 and its proof, and the last isomorphism follows from

$$D(G_0)_{U_0} \widehat{\otimes}_{D(G_0)_{P_{w,0}}} D \cong D(G)_U \widehat{\otimes}_{D(G)_{P_w}} D \tag{4.107}$$

which is analogous to (4.73).

By (4.82) for $\ell \geq 0$ we have $\mathrm{Ext}^\ell_{U(\mathfrak{p}_{I_1})}(L^{I_1}(\mu), \mathrm{gr}_w(V_0^\vee)) \cong H^\ell(C_w^\bullet)$ where

$$C_w^\ell \stackrel{\text{def}}{=} \mathrm{Hom}_E(\wedge^\ell \mathfrak{p}_{I_1}, \mathrm{Hom}_E(L^{I_1}(\mu), \mathrm{gr}_w(V_0^\vee))) \cong \mathrm{Hom}_E(\wedge^\ell \mathfrak{p}_{I_1} \otimes_E L^{I_1}(\mu), \mathrm{gr}_w(V_0^\vee))$$

and where C_w^ℓ is a left $D(P_{I_1})$-module by the discussion before (4.83) replacing $L^{I_1}(\mu)$ there by $\wedge^\ell \mathfrak{p}_{I_1} \otimes_E L^{I_1}(\mu)$ with the diagonal $D(P_{I_1})$-action (via [73, Sect. A] and with the adjoint action of P_{I_1} on $\wedge^\ell \mathfrak{p}_{I_1}$). We write $\delta_w^\ell : C_w^\ell \to C_w^{\ell+1}$ for the differential maps.

The topological isomorphisms (4.106) induce topological isomorphisms for $\ell \geq 0$

$$\mathcal{C}_w^\ell \cong \varprojlim_{U_0 \in \mathcal{X}_0, r \in \mathcal{I}} \mathcal{C}_{U_0,r}^\ell \cong \varprojlim_{U_0 \in \mathcal{X}_0} \mathcal{C}_{U_0}^\ell \cong \varprojlim_{U \in \mathcal{X}} \mathcal{C}_U^\ell \tag{4.108}$$

where

$$\mathcal{C}_{U_0,r}^\ell \overset{\text{def}}{=} \mathrm{Hom}_E\big(\wedge^\ell \mathfrak{p}_{I_1} \otimes_E L^{I_1}(\mu), D(G_0)_{U_0,r} \otimes_{D(G_0)_{P_{w,0},r}} D_r\big)$$

and similarly with $\mathcal{C}_{U_0}^\ell, \mathcal{C}_U^\ell$. Note that (4.107) gives a natural identification $\mathcal{C}_{U_0}^\ell = \mathcal{C}_U^\ell$. We write $\delta_{U_0,r}^\ell, \delta_{U_0}^\ell$ and δ_U^ℓ for the corresponding differential maps. Under (4.108), we have $\delta_w^\ell = \varprojlim_{U_0,r} \delta_{U_0,r}^\ell = \varprojlim_{U_0} \delta_{U_0}^\ell = \varprojlim_U \delta_U^\ell$. Recall from (ii) of Lemma 4.2.7 that $D(G_0)_{U_0,r}$ is a finite free $D(G_0)_{P_{w,0},r}$-module with a basis given by $\{\delta_g\}_{g \in U_0 G_1^m / P_{w,0} G_1^m}$, so that we have

$$\begin{aligned} D(G_0)_{U_0,r} \otimes_{D(G_0)_{P_{w,0},r}} D_r &\cong \bigoplus_{g \in U_0 G_1^m / P_{w,0} G_1^m} D_r^{g^{-1}} \\ &\cong \bigoplus_{g \in U_0 G_1^m / P_{w,0} G_1^m} \big(((\mathcal{M}_0^w)_r)^{g^{-1}} \otimes_E D_r^{\infty, g^{-1}}\big). \end{aligned} \tag{4.109}$$

This induces $U(\mathfrak{p}_{I_1})$-equivariant isomorphisms for $\ell \geq 0$

$$\mathcal{C}_{U_0,r}^\ell \cong \bigoplus_{g \in U_0 G_1^m / P_{w,0} G_1^m} \mathrm{Hom}_E(\wedge^\ell \mathfrak{p}_{I_1} \otimes_E L^{I_1}(\mu), D_r^{g^{-1}}) \cong \bigoplus_{g \in U_0 G_1^m / P_{w,0} G_1^m} (\mathcal{C}_{g,r}^\ell \otimes_E D_r^{\infty, g^{-1}}) \tag{4.110}$$

and the differential maps satisfy

$$\delta_{U_0,r}^\ell = \bigoplus_{g \in U_0 G_1^m / P_{w,0} G_1^m} (\delta_{g,r}^\ell \otimes_E \mathrm{Id}_{D_r^{\infty, g^{-1}}}). \tag{4.111}$$

We write $\mathcal{D}_w^\ell$ for the closure of $\delta_w^{\ell-1}(\mathcal{C}_w^{\ell-1})$ in $\mathcal{C}_w^\ell$ and define $\mathcal{H}_w^\ell \overset{\text{def}}{=} \ker(\delta_w^\ell)/\mathcal{D}_w^\ell$, so that we have a strict short exact sequence

$$0 \longrightarrow \mathcal{D}_w^\ell \longrightarrow \ker(\delta_w^\ell) \longrightarrow \mathcal{H}_w^\ell \longrightarrow 0. \tag{4.112}$$

For $U_0 \in \mathcal{X}_0$ and $r \in \mathcal{I}$, we similarly define $\mathcal{D}_*^\ell$ and $\mathcal{H}_*^\ell$ with $*$ being U_0, r or U_0 or $U = U_0 P_w$. In particular, we have a short exact sequence of E-Banach spaces

$$0 \longrightarrow \mathcal{D}_{U_0,r}^\ell \longrightarrow \ker(\delta_{U_0,r}^\ell) \longrightarrow \mathcal{H}_{U_0,r}^\ell \longrightarrow 0. \tag{4.113}$$

The projective limit over $r \in \mathcal{I}$ of (4.113) gives a strict exact sequence of E-Fréchet spaces

$$0 \longrightarrow \mathcal{D}_{U_0}^\ell \longrightarrow \ker(\delta_{U_0}^\ell) \longrightarrow \mathcal{H}_{U_0}^\ell \longrightarrow 0, \tag{4.114}$$

and the projective limit of (4.114) over $U_0 \in \mathcal{X}_0$ gives back (4.112) (using (4.108) and Lemma 4.2.4). Moreover we have topological isomorphisms by (4.110) and (4.111)

$$\ker(\delta^{\ell}_{U_0,r}) \cong \bigoplus_{g\in U_0G_1^m/P_{w,0}G_1^m} \left(\ker(\delta^{\ell}_{g,r}) \otimes_E D_r^{\infty,g^{-1}}\right) \text{ and } \mathcal{D}^{\ell}_{U_0,r} \cong \bigoplus_{g\in U_0G_1^m/P_{w,0}G_1^m} \left(\mathcal{D}^{\ell}_{g,r} \otimes_E D_r^{\infty,g^{-1}}\right),$$

which together with (4.114) give a topological isomorphism

$$\mathcal{H}^{\ell}_{U_0,r} \cong \bigoplus_{g\in U_0G_1^m/P_{w,0}G_1^m} \left(\mathcal{H}^{\ell}_{g,r} \otimes_E D_r^{\infty,g^{-1}}\right). \tag{4.115}$$

For $g \in X/P_w = X_0/P_{w,0}$ there is a natural continuous map $\mathcal{C}^{\ell}_g \to \mathcal{C}^{\ell}_w$ induced by the continuous map $D^{g^{-1}} = \delta_g D \to \mathrm{gr}_w(V_0^{\vee})$ (see (4.76)). Taking the direct sum over $g \in X/P_w$, we obtain a map

$$\mathcal{C}^{\ell}_{w,\flat} \overset{\text{def}}{=} \bigoplus_{g\in X/P_w} \mathcal{C}^{\ell}_g \widehat{\otimes}_E (D^{\infty})^{g^{-1}} \longrightarrow \mathcal{C}^{\ell}_w. \tag{4.116}$$

Similarly, for $U \in \mathcal{X}$, we have a map

$$\mathcal{C}^{\ell}_{U,\flat} \overset{\text{def}}{=} \bigoplus_{g\in U/P_w} \mathcal{C}^{\ell}_g \widehat{\otimes}_E (D^{\infty})^{g^{-1}} \longrightarrow \mathcal{C}^{\ell}_U. \tag{4.117}$$

Lemma 4.5.6 *The maps (4.116) and (4.117) are injective with dense image.*

Proof Let $U \in \mathcal{X}$, $U_0 \overset{\text{def}}{=} U \cap G_0 \in \mathcal{X}_0$ and note that $U_0/P_{w,0} = U/P_w$. For $g, g' \in U_0$ and $r \in \mathcal{I}$ one checks from (4.110) that $\mathcal{C}^{\ell}_{g,r} = \mathcal{C}^{\ell}_{g',r} \subseteq \mathcal{C}^{\ell}_{U_0,r}$ if and only if $gP_{w,0}G_1^m = g'P_{w,0}G_1^m$. Let $S \subseteq U_0$ be a finite subset such that the cosets $gP_{w,0}$ are distinct for $g \in S$. (Since $X_0/P_{w,0} = X/P_w$, this is equivalent to gP_w being distinct for $g \in S$.) Then for $r \in \mathcal{I}$ sufficiently close to 1 the cosets $gP_{w,0}G_1^m$ are distinct for $g \in S$ and thus from (4.110) $\mathcal{C}^{\ell}_{U_0,r}$ (for such an r) contains as closed subspace the direct sum $\bigoplus_{g\in S} \mathcal{C}^{\ell}_{g,r} \otimes_E (D_r^{\infty})^{g^{-1}}$. Taking $\varprojlim_{r\in\mathcal{I}}$ we obtain by Lemma 4.2.12 a closed embedding

$$\bigoplus_{g\in S} \mathcal{C}^{\ell}_g \widehat{\otimes}_E (D^{\infty})^{g^{-1}} \hookrightarrow \mathcal{C}^{\ell}_{U_0} \cong \mathcal{C}^{\ell}_U, \tag{4.118}$$

and taking the projective limit over $U_0 \in \mathcal{X}_0$ a closed embedding (using (4.108))

$$\bigoplus_{g\in S} \mathcal{C}^{\ell}_g \widehat{\otimes}_E (D^{\infty})^{g^{-1}} \hookrightarrow \mathcal{C}^{\ell}_w. \tag{4.119}$$

Taking the colimit over S in (4.118) and (4.119) gives the maps (4.117) and (4.116) which are thus injective. As the map (4.117) induces a surjection $\mathcal{C}^{\ell}_{U_0,\flat} \twoheadrightarrow \mathcal{C}^{\ell}_{U_0,r}$ for each $r \in \mathcal{I}$ (see (4.110)), it has dense image. The inclusion $U_0 \subseteq X_0$ induces a projection $\mathcal{C}^{\ell}_{w,\flat} \twoheadrightarrow \mathcal{C}^{\ell}_{U_0,\flat}$ which is compatible with the natural projection $\mathcal{C}^{\ell}_w \twoheadrightarrow \mathcal{C}^{\ell}_{U_0}$. Since $\mathcal{C}^{\ell}_w \cong \varprojlim_{U_0\in\mathcal{X}_0} \mathcal{C}^{\ell}_{U_0}$ and since (4.117) has dense image for each U_0, it follows that (4.116) also has dense image. □

In particular, the natural map $\mathcal{C}_g^\ell \to \mathcal{C}_w^\ell$ for $g \in X$ (resp. $\mathcal{C}_g^\ell \to \mathcal{C}_U^\ell$ for $U \in \mathcal{X}$ and $g \in U$) is injective.

Note that the dense subspace $\mathcal{C}_{w,\flat}^\ell \subseteq \mathcal{C}_w^\ell$ is P_{I_1}-stable with $h \in P_{I_1}$ acting (on the left) via the collection of maps (for $g \in X/P_w$)

$$\theta_{g,h}^\ell \widehat{\otimes}_E \kappa_{g,h} : \mathcal{C}_g^\ell \widehat{\otimes}_E (D^\infty)^{g^{-1}} \longrightarrow \mathcal{C}_{hg}^\ell \widehat{\otimes}_E (D^\infty)^{(hg)^{-1}}$$

where

$$\kappa_{g,h} : (D^\infty)^{g^{-1}} \longrightarrow (D^\infty)^{(hg)^{-1}} \tag{4.120}$$

sends v to $\delta_h v$. As $\delta_{hg}^\ell = \delta_h \delta_g^\ell \delta_{h^{-1}}$ for $\ell \geq 0$, $g \in X/P_w$ and $h \in P_{I_1}$, the action of P_{I_1} on $\mathcal{C}_{w,\flat}^\ell$ stabilizes the two (closed) subspaces

$$\bigoplus_{g \in X/P_w} \mathcal{D}_g^\ell \widehat{\otimes}_E (D^\infty)^{g^{-1}} \subseteq \ker(\delta_w^\ell) \cap \mathcal{C}_{w,\flat}^\ell = \bigoplus_{g \in X/P_w} \ker(\delta_g^\ell) \widehat{\otimes}_E (D^\infty)^{g^{-1}}$$

and thus induces a left action of P_{I_1} on

$$\mathcal{H}_{w,\flat}^\ell \stackrel{\text{def}}{=} \bigoplus_{g \in X/P_w} \mathcal{H}_g^\ell \otimes_E (D^\infty)^{g^{-1}}$$

with $h \in P_{I_1}$ acting via the collection of maps (for $g \in X/P_w$)

$$\omega_{g,h}^\ell \otimes_E \kappa_{g,h} : \mathcal{H}_g^\ell \otimes_E (D^\infty)^{g^{-1}} \longrightarrow \mathcal{H}_{hg}^\ell \otimes_E (D^\infty)^{(hg)^{-1}}. \tag{4.121}$$

The maps (4.116) and (4.117) induce maps

$$\mathcal{H}_{w,\flat}^\ell \longrightarrow \mathcal{H}_w^\ell$$

and for $U \in \mathcal{X}$

$$\mathcal{H}_{U,\flat}^\ell \stackrel{\text{def}}{=} \bigoplus_{g \in U/P_w} \mathcal{H}_g^\ell \otimes_E (D^\infty)^{g^{-1}} \longrightarrow \mathcal{H}_U^\ell. \tag{4.122}$$

Lemma 4.5.7 *For $U \in \mathcal{X}$ such that $U \subseteq P_{I_1,0} P_w$ the map (4.122) is injective with dense image.*

Proof Since $U \subseteq P_{I_1,0} P_w$ and thus $U_0 \subseteq P_{I_1,0} P_{w,0}$, each (representative) g in the decomposition (4.115) can be chosen in $P_{I_1,0}$, in which case we have the topological isomorphism (4.103). This forces the composition $\mathcal{H}_{U_0,\flat}^\ell \to \mathcal{H}_{U_0}^\ell \to \mathcal{H}_{U_0,r}^\ell$ to be a surjection for each $r \in \mathcal{I}$, and thus (4.122) has dense image. By a similar argument as in the proof of Lemma 4.5.6 using (4.115), (4.103), (4.113) and (4.114) the map

$$\bigoplus_{g \in S} \mathcal{H}_g^\ell \otimes_E (D^\infty)^{g^{-1}} \longrightarrow \mathcal{H}_U^\ell = \mathcal{H}_{U_0}^\ell$$

is injective for each finite subset $S \subseteq U_0$ such that the cosets $gP_{w,0}$ for $g \in S$ are distinct. Taking the colimit over such S, it follows that (4.122) is injective. □

By Lemma 4.5.4 applied with $1, g$ ($g \in P_{I_1}$) instead of g, h the map $\omega^\ell_{1,g}$ in (4.95) for $g \in P_{I_1}$ only depends on the coset gP_w. Consequently, we can define the map

$$\zeta^\ell : \mathcal{H}^\ell_1 \otimes_E \Big(\bigoplus_{g \in X/P_w} (D^\infty)^{g^{-1}} \Big) \longrightarrow \mathcal{H}^\ell_{w,\flat} \tag{4.123}$$

by sending $(x, y) \in \mathcal{H}^\ell_1 \otimes_E (D^\infty)^{g^{-1}}$ to $(\omega^\ell_{1,g}(x), y) \in \mathcal{H}^\ell_g \otimes_E (D^\infty)^{g^{-1}}$ for $g \in P_{I_1}$. Note that since $\omega^\ell_{1,g}$ is bijective, the map ζ^ℓ is also bijective. We defined above a left action of P_{I_1} on $\mathcal{H}^\ell_{w,\flat}$. We define one on the left hand side of (4.123) by letting P_{I_1} act trivially on $\mathcal{H}^\ell_1$ and by (4.120) on $\bigoplus_{g \in X/P_w} (D^\infty)^{g^{-1}}$.

Lemma 4.5.8 *The map ζ^ℓ is P_{I_1}-equivariant.*

Proof We have for $h \in P_{I_1}$, $g \in X/P_w$ and $(x, y) \in \mathcal{H}^\ell_1 \otimes_E (D^\infty)^{g^{-1}}$:

$$\zeta^\ell(\delta_h \cdot (x, y)) = \zeta^\ell((x, \kappa_{g,h}(y))) = (\omega^\ell_{1,hg}(x), \kappa_{g,h}(y)) = (\omega^\ell_{g,h}(\omega^\ell_{1,g}(x)), \kappa_{g,h}(y)) = \delta_h \cdot \zeta^\ell(x, y)$$

where the third equality follows from (4.96) and the last from (4.121). □

For $U \in \mathcal{X}$ the map ζ^ℓ in (4.123) restricts to a bijection

$$\zeta^\ell_U : \mathcal{H}^\ell_1 \otimes_E \Big(\bigoplus_{g \in U/P_w} (D^\infty)^{g^{-1}} \Big) \xrightarrow{\sim} \mathcal{H}^\ell_{U,\flat}. \tag{4.124}$$

We recall that $\mathcal{H}^\ell_1$ is finite dimensional by Lemma 4.5.5.

Lemma 4.5.9 *For $U \in \mathcal{X}$ such that $U \subseteq P_{I_1,0}P_w$ the differential map δ^ℓ_U has closed image and the bijection ζ^ℓ_U uniquely extends into an isomorphism of E-Fréchet spaces*

$$\mathcal{H}^\ell_1 \otimes_E \big(D^\infty(G)_U \widehat{\otimes}_{D^\infty(G)_{P_w}} D^\infty \big) \xrightarrow{\sim} \mathcal{H}^\ell_U. \tag{4.125}$$

Proof Note that, although U is not compact, $U/P_w = U_0/P_{w,0}$ is and by (4.28) we have $D^\infty(G)_U \cong \widehat{D}^\infty(G)_U$. Recall that given $U, U' \in \mathcal{X}$ satisfying $U \cap U' = \emptyset$, we have $D(G)_{U \sqcup U'} \cong D(G)_U \oplus D(G)_{U'}$ and $D^\infty(G)_{U \sqcup U'} \cong D^\infty(G)_U \oplus D^\infty(G)_{U'}$, and thus $\mathcal{C}^\ell_{U \sqcup U'} \cong \mathcal{C}^\ell_U \oplus \mathcal{C}^\ell_{U'}$, $\delta^\ell_{U \sqcup U'} = \delta^\ell_U \oplus \delta^\ell_{U'}$, $\mathcal{D}^\ell_{U \sqcup U'} \cong \mathcal{D}^\ell_U \oplus \mathcal{D}^\ell_{U'}$ and $\mathcal{H}^\ell_{U \sqcup U'} \cong \mathcal{H}^\ell_U \oplus \mathcal{H}^\ell_{U'}$. Moreover we have $P_{I_1,0}P_w \setminus U \in \mathcal{X}$ for $U \in \mathcal{X}$ such that $U \subseteq P_{I_1,0}P_w$ (writing $U = U_0 P_w$ where $U_0 = U \cap G_0 \subseteq P_{I_1,0}P_{w,0}$, one checks that $P_{I_1,0}P_w \setminus U = (P_{I_1,0}P_{w,0} \setminus U_0)P_w$). It follows from all this that the statement of the lemma for $U = P_{I_1,0}P_w$ is the "direct sum" of the statement for U and for $P_{I_1,0}P_w \setminus U$. Hence it is enough to prove (4.125) for $U = P_{I_1,0}P_w$ or equivalently $U_0 = P_{I_1,0}P_{w,0}$.

Given a norm $|\cdot|$ on a E-vector space V equipped with a left action by a group H, we say that $|\cdot|$ is H-invariant if $|h \cdot x| = |x|$ for each $h \in H$ and $x \in V$. For

$r \in \mathcal{I}$, let m, s be as before Lemma 4.2.5. By [61, Proposition 5.6] the natural norm on the Banach algebra $D(G_1^m)_s$ is multiplicative, which together with [61, Proposition 6.2] (cf. Lemma 4.2.5) implies that the natural norm on $D(G_0)_{1,r} = D(G_0)_{G_1^m,r}$ is multiplicative and in particular G_1^m-invariant (for the natural left action of G_1^m on $D(G_0)_{G_1^m,r}$). This together with (i) of Lemma 4.2.7 implies that there exists a $P_{w,0}G_1^m$-invariant norm on $D(G_0)_{P_{w,0},r} = D(G_0)_{P_{w,0}G_1^m,r}$ which defines its Banach topology. As $(\mathcal{M}_0^w)_r$ is a finitely generated $D(G_0)_{P_{w,0},r} = D(G_0)_{P_{w,0}G_1^m,r}$-module, the $P_{w,0}G_1^m$-invariant norm on $D(G_0)_{P_{w,0}G_1^m,r}$ induces a $P_{w,0}G_1^m$-invariant norm on $(\mathcal{M}_0^w)_r$ that defines its Banach topology ([72, Proposition 2.1.i]). Similarly, we can choose a $P_{w,0}G_1^m$-invariant norm $|\cdot|_r^\infty$ on D_r^∞ (which defines its Banach topology). We write $|\cdot|_r$ (resp. $|\cdot|_r^\infty$) for the induced semi-norm on $\mathcal{M}_0^w$ (resp. on D^∞) and $|\cdot|_{D,r} \stackrel{\text{def}}{=} |\cdot|_r \otimes_E |\cdot|_r^\infty$ for the induced semi-norm on D (under which the completion of is D_r). (Note that this semi-norm $|\cdot|_r$ on $\mathcal{M}_0^w$ might be different than the one in the proof of Lemma 4.5.5.)

For $r \in \mathcal{I}$ we let $J_r \stackrel{\text{def}}{=} P_{I_1,0}P_{w,0}G_1^m/P_{w,0}G_1^m$ (a finite set). We fix a choice of representatives $\widetilde{J}_r \subseteq P_{I_1,0}$ for J_r, and choose them in a compatible way so that we have a surjection $\widetilde{J}_{r'} \twoheadrightarrow \widetilde{J}_r$ for $r \leq r'$.

Given $g \in P_{I_1,0}$, we write $|\cdot|_r^{g^{-1}}$ (resp. $|\cdot|_r^{\infty,g^{-1}}$, $|\cdot|_{D^{g^{-1}},r}$) for the corresponding semi-norm on $\mathcal{M}_0^{wg^{-1}}$ (resp. $D^{\infty,g^{-1}}$, $D^{g^{-1}}$), and $(\mathcal{M}_0^{wg^{-1}})_r$ (resp. $(D^{\infty,g^{-1}})_r$, $(D^{g^{-1}})_r$) for the corresponding completion. We have obvious identifications $(\mathcal{M}_0^{wg^{-1}})_r = (\mathcal{M}_0^w)_r^{g^{-1}}$, $(D^{\infty,g^{-1}})_r = D_r^{\infty,g^{-1}}$ and $(D^{g^{-1}})_r = D_r^{g^{-1}}$. Since we have $|\delta_h \cdot x|_r = |x|_r$ for $h \in P_{w,0}G_1^m$ and $x \in (\mathcal{M}_0^w)_r$, the norm $|\cdot|_r^{g^{-1}}$ on $(\mathcal{M}_0^{wg^{-1}})_r$ only depends on the coset $gP_{w,0}G_1^m$. Similar facts hold for the norms $|\cdot|_r^{\infty,g^{-1}}$ and $|\cdot|_{D^{g^{-1}},r}$ on respectively $(D^{\infty,g^{-1}})_r$ and $(D^{g^{-1}})_r$.

We fix a norm $|\cdot|_{\mathfrak{p}_{I_1}}$ on $\mathfrak{p}_{I_1}$. For $x \in \mathfrak{p}_{I_1}$ the map $g \mapsto |\mathrm{Ad}(g)(x)|_{\mathfrak{p}_{I_1}}$ is a locally constant function on $P_{I_1,0}$. Hence, replacing $|\cdot|_{\mathfrak{p}_{I_1}}$ by the norm $x \mapsto |\int_{P_{I_1,0}} \mathrm{Ad}(g)(x)|_{\mathfrak{p}_{I_1}} dg$ for some Haar measure dg on $P_{I_1,0}$, we can assume $|\mathrm{Ad}(g)(x)|_{\mathfrak{p}_{I_1}} = |x|_{\mathfrak{p}_{I_1}}$ for $g \in P_{I_1,0}$ and $x \in \mathfrak{p}_{I_1}$. For $\ell \geq 0$, the norm $|\cdot|_{\mathfrak{p}_{I_1}}$ induces a norm on $\otimes_E^\ell \mathfrak{p}_{I_1}$ (the tensor product norm) and then a norm $|\cdot|_{\wedge^\ell \mathfrak{p}_{I_1}}$ on $\wedge^\ell \mathfrak{p}_{I_1}$ (the quotient norm) which satisfies ($g \in P_{I_1,0}$, $x \in \wedge^\ell \mathfrak{p}_{I_1}$)

$$|\mathrm{Ad}(g)(x)|_{\wedge^\ell \mathfrak{p}_{I_1}} = |x|_{\wedge^\ell \mathfrak{p}_{I_1}}. \tag{4.126}$$

For $g \in P_{I_1,0}$, the norm $|\cdot|_{\wedge^\ell \mathfrak{p}_{I_1}}$ on $\wedge^\ell \mathfrak{p}_{I_1}$ with the semi-norm $|\cdot|_r^{g^{-1}}$ on $\mathcal{M}_0^{wg^{-1}}$ induce a semi-norm $|\cdot|_{\mathcal{C}_g^\ell,r}$ on $\mathcal{C}_g^\ell$ such that the corresponding completion is $\mathcal{C}_{g,r}^\ell$ (in particular $\{|\cdot|_{\mathcal{C}_g^\ell,r}\}_{r\in\mathcal{I}}$ defines the Fréchet topology on $\mathcal{C}_g^\ell$). It follows from (4.126) and from (4.94) that for $g, h \in P_{I_1,0}$, $r \in \mathcal{I}$ and $x \in \mathcal{C}_g^\ell$

$$|\theta_{g,h}^\ell(x)|_{\mathcal{C}_{hg}^\ell,r} = |x|_{\mathcal{C}_g^\ell,r}. \tag{4.127}$$

The (induced) norm $|\cdot|_{\mathcal{C}^\ell_g,r}$ on $\mathcal{C}^\ell_{g,r}$ induces a norm on the closed subspaces $\mathcal{D}^\ell_{g,r}$, $\ker(\delta^\ell_{g,r})$ and thus a norm $|\cdot|_{\mathcal{H}^\ell_g,r}$ on $\mathcal{H}^\ell_{g,r}$, which satisfies for $g,h\in P_{I_1,0}$, $r\in\mathcal{I}$ and $x\in\mathcal{H}^\ell_g$ (using (4.95))

$$|\omega^\ell_{g,h}(x)|_{\mathcal{H}^\ell_{hg},r}=|x|_{\mathcal{H}^\ell_g,r}. \tag{4.128}$$

As the norm $|\cdot|^{g^{-1}}_r$ on $(\mathcal{M}_0^{wg^{-1}})_r$ only depends on $gP_{w,0}G_1^m$, so does the norm $|\cdot|_{\mathcal{C}^\ell_g,r}$ on $\mathcal{C}^\ell_{g,r}$ and the norm $|\cdot|_{\mathcal{H}^\ell_g,r}$ on $\mathcal{H}^\ell_{g,r}$.

Using (4.110), we obtain a norm $|\cdot|_{\mathcal{C}^\ell_{U_0},r}$ on $\mathcal{C}^\ell_{U_0,r}$ by taking the maximum of $|\cdot|_{\mathcal{C}^\ell_g,r}\otimes_E(|\cdot|^\infty_r)^{g^{-1}}$ on the direct summands $\mathcal{C}^\ell_{g,r}\otimes_E(D^\infty_r)^{g^{-1}}$ for $g\in\widetilde{J}_r$ (recall $U_0=P_{I_1,0}P_{w,0}$). As both $|\cdot|_{\mathcal{C}^\ell_g,r}$ and $(|\cdot|^\infty_r)^{g^{-1}}$ only depend on $gP_{w,0}G_1^m$, so does $|\cdot|_{\mathcal{C}^\ell_g,r}\otimes_E(|\cdot|^\infty_r)^{g^{-1}}$, and thus $|\cdot|_{\mathcal{C}^\ell_{U_0},r}$ does not depend on the choice of $\widetilde{J}_r$. Since $\mathcal{C}^\ell_{U_0,r}$ is a Banach space with norm $|\cdot|_{\mathcal{C}^\ell_{U_0},r}$ and $\mathcal{C}^\ell_{U_0}=\varprojlim_{r\in\mathcal{I}}\mathcal{C}^\ell_{U_0,r}$, the family of semi-norms $|\cdot|_{\mathcal{C}^\ell_{U_0},r}$ defines the Fréchet topology on $\mathcal{C}^\ell_{U_0}$. Note that for $g\in P_{I_1,0}$ the restriction of $|\cdot|_{\mathcal{C}^\ell_{U_0},r}$ to $\mathcal{C}^\ell_g\widehat{\otimes}_E(D^\infty)^{g^{-1}}$ via (4.117) (and Lemma 4.5.6) is $|\cdot|_{\mathcal{C}^\ell_g,r}\otimes_E(|\cdot|^\infty_r)^{g^{-1}}$. Finally the norm $|\cdot|_{\mathcal{C}^\ell_{U_0},r}$ on $\mathcal{C}^\ell_{U_0,r}$ induces a norm on the closed subspaces $\mathcal{D}^\ell_{U_0,r}$, $\ker(\delta^\ell_{U_0,r})$ and thus a norm on $\mathcal{H}^\ell_{U_0,r}$ which, under (4.115), is explicitly

$$|\cdot|_{\mathcal{H}^\ell_{U_0},r}\overset{\text{def}}{=}\max_{g\in\widetilde{J}_r}\big(|\cdot|_{\mathcal{H}^\ell_g,r}\otimes_E(|\cdot|^\infty_r)^{g^{-1}}\big). \tag{4.129}$$

We now prove that the image of the differential map $\delta^\ell_{U_0}$ is closed for $\ell\geq 0$. Recall that the Fréchet topology on $\mathcal{C}^\ell_{U_0}$ is defined by the family of semi-norms $\{|\cdot|_{\mathcal{C}^\ell_{U_0},r}\}_{r\in\mathcal{I}}$. We write $|\cdot|'_{\mathcal{C}^\ell_{U_0},r}$ for the induced semi-norm on the quotient $\mathcal{C}^\ell_{U_0}/\ker(\delta^\ell_{U_0})$. Since $\delta^\ell_1:\mathcal{C}^\ell_1/\ker(\delta^\ell_1)\xrightarrow{\sim}\mathcal{D}^{\ell+1}_1\subseteq\mathcal{C}^{\ell+1}_1$ is a topological isomorphism by Lemma 4.5.3 (applied to $g=1$), given $r\in\mathcal{I}$, there exists $r'\geq r$ and $A_{r,r'}\in\mathbb{Q}_{>0}$ such that

$$|x|'_{\mathcal{C}^\ell_1,r}\leq A_{r,r'}|\delta^\ell_1(x)|_{\mathcal{C}^{\ell+1}_1,r'} \tag{4.130}$$

for $x\in\mathcal{C}^\ell_1/\ker(\delta^\ell_1)$. Let $g\in P_{I_1,0}$, by (4.94) (applied with $1,g$ instead of h,g), we have a topological isomorphism $\theta^\ell_{1,g}:\mathcal{C}^\ell_1\xrightarrow{\sim}\mathcal{C}^\ell_g$. This together with $\theta^{\ell+1}_{1,g}\circ\delta^\ell_1=\delta^\ell_g\circ\theta^\ell_{1,g}$, (4.130) and (4.127) implies

$$|y|'_{\mathcal{C}^\ell_g,r}\leq A_{r,r'}|\delta^\ell_g(y)|_{\mathcal{C}^{\ell+1}_g,r'} \tag{4.131}$$

for $y\in\mathcal{C}^\ell_g/\ker(\delta^\ell_g)$. By continuity (4.131) holds for any $g\in P_{I_1,0}$ and any $y\in\mathcal{C}^\ell_{g,r'}/\ker(\delta^\ell_{g,r'})$. Using (4.110) (for r and r'), we deduce for $y\in\mathcal{C}^\ell_{U_0,r'}/\ker(\delta^\ell_{U_0,r'})$

$$|y|'_{\mathcal{C}^\ell_{U_0},r}\leq A_{r,r'}|\delta^\ell_{U_0}(y)|_{\mathcal{C}^{\ell+1}_{U_0},r'}. \tag{4.132}$$

As (4.132) *a fortiori* holds for any $y \in \mathcal{C}^{\ell}_{U_0}/\ker(\delta^{\ell}_{U_0})$, we see that $\delta^{\ell}_{U_0} : \mathcal{C}^{\ell}_{U_0}/\ker(\delta^{\ell}_{U_0}) \hookrightarrow \mathcal{C}^{\ell+1}_{U_0}$ is a closed embedding for $\ell \geq 0$ (in particular $\delta^{\ell}_{U_0}(\mathcal{C}^{\ell}_{U_0}) = \mathcal{D}^{\ell+1}_{U_0}$).

We finally prove the isomorphism (4.125). Recall first that by Lemma 4.2.12 and (ii) of Lemma 4.2.7 we have isomorphisms

$$\begin{aligned} D^{\infty}(G_0)_{U_0}\widehat{\otimes}_{D^{\infty}(G_0)_{P_{w,0}}}D^{\infty} &\cong \varprojlim_{r\in\mathcal{I}}\left(D^{\infty}(G_0)_{U_0,r}\otimes_{D^{\infty}(G_0)_{P_{w,0},r}}D^{\infty}_r\right) \\ &\cong \varprojlim_{r\in\mathcal{I}}\Big(\bigoplus_{g\in\widetilde{J}_r}(D^{\infty}_r)^{g^{-1}}\Big). \end{aligned} \tag{4.133}$$

By an argument similar to the proof of the injectivity of (4.117) in Lemma 4.5.6, we see that (4.133) contains as a dense subspace

$$\bigoplus_{g\in U/P_w}(D^{\infty})^{g^{-1}} = \bigoplus_{g\in U_0/P_{w,0}}(D^{\infty})^{g^{-1}}. \tag{4.134}$$

The Fréchet topology on (4.133) is defined by $\{|\cdot|^{\infty}_{U_0,r}\}_{r\in\mathcal{I}}$ where $|\cdot|^{\infty}_{U_0,r} \stackrel{\text{def}}{=} \max_{g\in\widetilde{J}_r}(|\cdot|^{\infty}_r)^{g^{-1}}$), and (4.133) is the completion of (4.134) under this topology. In particular, the left hand side of (4.125) is the completion of the left hand side of (4.124) under the Fréchet topology defined by the family of semi-norms

$$\{|\cdot|_{\mathcal{H}^{\ell}_{1},r}\otimes_E|\cdot|^{\infty}_{U_0,r}\}_{r\in\mathcal{I}}. \tag{4.135}$$

On the other hand the Fréchet topology on $\mathcal{H}^{\ell}_{U_0}$ is defined by the family of semi-norms

$$\{|\cdot|_{\mathcal{H}^{\ell}_{U_0},r}\}_{r\in\mathcal{I}} \tag{4.136}$$

from (4.129). Since the map (4.122) is injective with dense image by Lemma 4.5.7, $\mathcal{H}^{\ell}_{U_0}$ can be identified with the completion of $\mathcal{H}^{\ell}_{U_0,\flat}$ under $\{|\cdot|_{\mathcal{H}^{\ell}_{U_0},r}\}_{r\in\mathcal{I}}$.

By (4.128) (applied with $1, g$ instead of h, g), for $g \in P_{I_1,0}$ the norm $|\cdot|_{\mathcal{H}^{\ell}_{1},r}$ on $\mathcal{H}^{\ell}_{1,r}$ is identical to the norm $|\cdot|_{\mathcal{H}^{\ell}_{g},r}$ on $\mathcal{H}^{\ell}_{g,r}$ under (4.102). Consequently, the two families of semi-norms (4.136) and (4.135) are identical under the isomorphism $\zeta^{\ell}_{U_0}$. It follows that $\zeta^{\ell}_{U_0}$ uniquely extends by completion into a topological isomorphism of E-Fréchet spaces as in (4.125). □

We can finally prove the first important result of this section.

Theorem 4.5.10 *We have a (left) $D^{\infty}(P_{I_1})$-equivariant isomorphism of Fréchet spaces for $\ell \geq 0$*

$$\begin{aligned} &\mathrm{Ext}^{\ell}_{U(\mathfrak{p}_{I_1})}(L^{I_1}(\mu), M^{w}_0)\otimes_E\left(\widehat{D}^{\infty}(G)_{P_{I_1}P_w}\widehat{\otimes}_{D^{\infty}(G)_{P_w}}(\pi^{\infty,w}_0)^{\vee}\right) \\ &\qquad\xrightarrow{\sim}\mathrm{Ext}^{\ell}_{U(\mathfrak{p}_{I_1})}(L^{I_1}(\mu), \mathrm{gr}_w(V^{\vee}_0)) \end{aligned} \tag{4.137}$$

with trivial $D^\infty(P_{I_1})$*-action on the finite dimensional* E*-vector space* $\mathrm{Ext}^\ell_{U(\mathfrak{p}_{I_1})}(L^{I_1}(\mu), M_0^w)$ *and where* $\widehat{D}^\infty(G)_{P_{I_1}P_w}$ *is defined in (4.28).*

Proof Recall that $\mathrm{Ext}^\ell_{U(\mathfrak{p}_{I_1})}(L^{I_1}(\mu), M_0^w) \xrightarrow{\sim} \mathrm{Ext}^\ell_{U(\mathfrak{p}_{I_1})}(L^{I_1}(\mu), \mathcal{M}_0^w)$ is finite dimensional by Lemma 4.5.1. Let $U \in \mathcal{X}$. If there exists $t \in T$ such that $U \subseteq tP_{I_1,0}P_w = tP_{I_1,0}t^{-1}P_w$, then replacing G_0 (resp. $P_{I_1,0}$, $P_{w,0}$) by tG_0t^{-1} (resp. $tP_{I_1,0}t^{-1}$, $tP_{w,0}t^{-1}$), a parallel argument as in the proof of Lemma 4.5.9 shows that δ_U^ℓ is strict and that ζ_U^ℓ uniquely extends into a topological isomorphism as in (4.125). In general, since U/P_w is compact there exists a finite partition $U = \bigsqcup_i U_i$ and elements $t_i \in T$ such that $U_i \in \mathcal{X}$ and $U_i \subseteq t_iP_{I_1,0}P_w = t_iP_{I_1,0}t_i^{-1}P_w$ for each i (use the open covering $(tP_{I_1,0}t^{-1}/(tP_{I_1,0}t^{-1} \cap P_w))_{t\in T}$ of $P_{I_1}/(P_{I_1} \cap P_w)$). Since $\mathcal{C}_U^\ell \cong \bigoplus_i \mathcal{C}_{U_i}^\ell$, $\delta_U^\ell = \bigoplus_i \delta_{U_i}^\ell$ and $\zeta_U^\ell = \bigoplus_i \zeta_{U_i}^\ell$, we deduce that δ_U^ℓ is strict and that ζ_U^ℓ uniquely extends into a topological isomorphism (4.125). Since $\mathcal{C}_w^\ell \cong \varprojlim_{U\in\mathcal{X}} \mathcal{C}_U^\ell$ with $\delta_w^\ell = \varprojlim_{U\in\mathcal{X}} \delta_U^\ell$ (see (4.108)), we deduce that $\delta_w^{\ell-1}$ is strict with closed image $\mathcal{D}_w^\ell$ and thus $H^\ell(\mathcal{C}_w^\bullet) = \mathcal{H}_w^\ell$. Taking $\varprojlim_{U\in\mathcal{X}}$ on (4.125), we obtain a topological isomorphism

$$\varprojlim_{U\in\mathcal{X}} \zeta_U^\ell : \mathcal{H}_1^\ell \otimes_E \left(\widehat{D}^\infty(G)_X \widehat{\otimes}_{D^\infty(G)_{P_w}} D^\infty\right) \xrightarrow{\sim} \mathcal{H}_w^\ell = H^\ell(\mathcal{C}_w^\bullet) \tag{4.138}$$

as the composition of the topological isomorphisms

$$\begin{aligned}\mathcal{H}_1^\ell \otimes_E \left(\widehat{D}^\infty(G)_X \widehat{\otimes}_{D^\infty(G)_{P_w}} D^\infty\right) &\cong \mathcal{H}_1^\ell \otimes_E \left((\varprojlim_{U\in\mathcal{X}} D^\infty(G)_U)\widehat{\otimes}_{D^\infty(G)_{P_w}} D^\infty\right)\\ &\cong \varprojlim_{U\in\mathcal{X}} \left(\mathcal{H}_1^\ell \otimes_E D^\infty(G)_U \widehat{\otimes}_{D^\infty(G)_{P_w}} D^\infty\right) \cong \varprojlim_{U\in\mathcal{X}} \mathcal{H}_U^\ell \cong \mathcal{H}_w^\ell.\end{aligned}$$

Here we have used the topological isomorphisms

$$\begin{aligned}\widehat{D}^\infty(G)_X \widehat{\otimes}_{D^\infty(G)_{P_w}} D^\infty &\cong \widehat{D}^\infty(G_0)_{X_0} \widehat{\otimes}_{D^\infty(G_0)_{P_{w,0}}} D^\infty\\ &\cong \varprojlim_{U_0\in\mathcal{X}_0} \left(D^\infty(G_0)_{U_0} \widehat{\otimes}_{D^\infty(G_0)_{P_{w,0}}} D^\infty\right) \cong \varprojlim_{U\in\mathcal{X}} \left(D^\infty(G)_U \widehat{\otimes}_{D^\infty(G)_{P_w}} D^\infty\right)\end{aligned}$$

which follows by similar argument as for (4.106). The source (resp. target) of (4.138) contains $\mathcal{H}_1^\ell \otimes_E \left(\bigoplus_{g\in X/P_w} (D^\infty)^{g^{-1}}\right)$ (resp. $\mathcal{H}_{w,\mathrm{b}}^\ell$) as a P_{I_1}-stable dense subspace and the isomorphism (4.138) restricts to the isomorphism ζ^ℓ of (4.123) on these dense subspaces. Then Lemma 4.5.8 forces (4.138) to be $D(P_{I_1})$-equivariant (hence $D^\infty(P_{I_1})$-equivariant). Finally, by (4.82) and (4.81) the isomorphism (4.138) is (4.137). □

We write $\Sigma_i \overset{\text{def}}{=} W(L_{I_i}) \cdot \mathcal{J}(\pi_i^\infty) \subseteq \widehat{T}^\infty$ for $i = 0, 1$ where $\mathcal{J}(\pi_i^\infty)$ is in Definition 2.1.4 and the dot action $\cdot$ in (2.11). Recall that the categories $\mathcal{B}_{\Sigma_i}^{I_i}$ are defined above (2.13) and that any π_i^∞ in $\mathcal{B}_{\Sigma_i}^{I_i}$ is of finite length. The following corollary to Theorem 4.5.10 is crucial.

Corollary 4.5.11 *Keep the setting of Theorem 4.5.10 and assume moreover that the smooth representations π_i^∞ of L_{I_i} are of finite length for $i = 0, 1$. We have canonical isomorphisms of finite dimensional E-vector spaces for $w \in W^{I_0,I_1}$ and $k, \ell \geq 0$*

$$\begin{aligned}&\mathrm{Ext}^k_{D^\infty(P_{I_1})}\big((\pi_1^\infty)^\vee, \mathrm{Ext}^\ell_{U(\mathfrak{p}_{I_1})}(L^{I_1}(\mu), \mathrm{gr}_w(V_0^\vee))\big)\\ &\quad\cong \mathrm{Ext}^\ell_{U(\mathfrak{p}_{I_1})}(L^{I_1}(\mu), M_0^w) \otimes_E \mathrm{Ext}^k_{L_{I_1}}(i^\infty_{I_0,I_1,w}(J_{I_0,I_1,w}(\pi_0^\infty)), \pi_1^\infty)^\infty\end{aligned} \quad (4.139)$$

where $i^\infty_{I_0,I_1,w}$ is defined in (2.21) and $J_{I_0,I_1,w}$ in (2.19). If moreover Σ_i is a single G-regular $W(L_{I_i})$-coset, π_i^∞ is in $\mathcal{B}^{I_i}_{\Sigma_i}$ for $i = 0, 1$ and $\Sigma_1 \cap W(G) \cdot \Sigma_0 \neq \emptyset$, then (4.139) is non-zero for at most one $w \in W^{I_0,I_1}$, which is the unique $w \in W^{I_0,I_1}$ such that $\Sigma_1 \cap w^{-1} \cdot \Sigma_0 \neq \emptyset$.

Proof Recall from Lemma 4.2.13 and its proof (applied with $P = P_w$, $X = P_{I_1} P_w$ and $Q = P_{I_1}$) that $(\mathrm{ind}_{P_w}^{P_{I_1} P_w} \pi_0^{\infty,w})^\infty$ (with the finest locally convex topology) is a smooth representation of P_{I_1} on a vector space of compact type and that we have isomorphisms of E-Fréchet spaces with separately continuous $D^\infty(P_{I_1})$-actions

$$\widehat{D}^\infty(G)_{P_{I_1} P_w} \widehat{\otimes}_{D^\infty(G)_{P_w}} (\pi_0^{\infty,w})^\vee \cong \big((\mathrm{ind}_{P_w}^{P_{I_1} P_w} \pi_0^{\infty,w})^\infty\big)^\vee \cong \big((\mathrm{ind}_{P_{I_1}\cap P_w}^{P_{I_1}} \pi_0^{\infty,w})^\infty\big)^\vee. \quad (4.140)$$

From (4.140) and Theorem 4.5.10 we deduce isomorphisms for $k, \ell \geq 0$ (using that $D^\infty(P_{I_1})$ acts trivially on $\mathrm{Ext}^\ell_{U(\mathfrak{p}_{I_1})}(L^{I_1}(\mu), M_0^w)$)

$$\begin{aligned}&\mathrm{Ext}^k_{D^\infty(P_{I_1})}\big((\pi_1^\infty)^\vee, \mathrm{Ext}^\ell_{U(\mathfrak{p}_{I_1})}(L^{I_1}(\mu), \mathrm{gr}_w(V_0^\vee))\big)\\ &\quad\cong \mathrm{Ext}^\ell_{U(\mathfrak{p}_{I_1})}(L^{I_1}(\mu), M_0^w) \otimes_E \mathrm{Ext}^k_{D^\infty(P_{I_1})}\big((\pi_1^\infty)^\vee, ((\mathrm{ind}_{P_{I_1}\cap P_w}^{P_{I_1}} \pi_0^{\infty,w})^\infty)^\vee\big).\end{aligned} \quad (4.141)$$

By Lemma 4.2.3 we have for $k_1 \geq 0$ (writing 1 for the trivial representation of N_{I_1})

$$\mathrm{Ext}^{k_1}_{D^\infty(N_{I_1})}\Big(1, \big((\mathrm{ind}_{P_{I_1}\cap P_w}^{P_{I_1}} \pi_0^{\infty,w})^\infty\big)^\vee\Big) \cong \mathrm{Ext}^{k_1}_{N_{I_1}}\Big(\big((\mathrm{ind}_{P_{I_1}\cap P_w}^{P_{I_1}} \pi_0^{\infty,w})^\infty\big), 1\Big)^\infty. \quad (4.142)$$

Since Jacquet functors are exact on smooth representations, we deduce from (4.142) that we have for $k_1 > 0$

$$\mathrm{Ext}^{k_1}_{D^\infty(N_{I_1})}\Big(1, \big((\mathrm{ind}_{P_{I_1}\cap P_w}^{P_{I_1}} \pi_0^{\infty,w})^\infty\big)^\vee\Big) = 0 \quad (4.143)$$

and a canonical $D^\infty(L_{I_1})$-equivariant isomorphism of E-Fréchet spaces

$$\mathrm{Hom}_{D^\infty(N_{I_1})}\Big(1, \big((\mathrm{ind}_{P_{I_1}\cap P_w}^{P_{I_1}} \pi_0^{\infty,w})^\infty\big)^\vee\Big) \cong \Big(\big((\mathrm{ind}_{P_{I_1}\cap P_w}^{P_{I_1}} \pi_0^{\infty,w})^\infty\big)_{N_{I_1}}\Big)^\vee. \quad (4.144)$$

Combining (4.143) and (4.144) with the (standard "Hochschild-Serre type") spectral sequence

$$\mathrm{Ext}^{k_2}_{D^\infty(L_{I_1})}\big((\pi_1^\infty)^\vee, \mathrm{Ext}^{k_1}_{D^\infty(N_{I_1})}(1, -)\big) \implies \mathrm{Ext}^{k_1+k_2}_{D^\infty(P_{I_1})}((\pi_1^\infty)^\vee, -)$$

we deduce isomorphisms for $k \geq 0$

$$\mathrm{Ext}^k_{D^\infty(P_{I_1})}\Big((\pi_1^\infty)^\vee, \big((\mathrm{ind}_{P_{I_1}\cap P_w}^{P_{I_1}} \pi_0^{\infty,w})^\infty\big)^\vee\Big)$$
$$\cong \mathrm{Ext}^k_{D^\infty(L_{I_1})}\Big((\pi_1^\infty)^\vee, \big(((\mathrm{ind}_{P_{I_1}\cap P_w}^{P_{I_1}} \pi_0^{\infty,w})^\infty)_{N_{I_1}}\big)^\vee\Big). \quad (4.145)$$

Recall from (2.25) the isomorphism of smooth admissible representations of L_{I_1}

$$\big((\mathrm{ind}_{P_{I_1}\cap P_w}^{P_{I_1}} \pi_0^{\infty,w})^\infty\big)_{N_{I_1}} \cong i^\infty_{I_0,I_1,w}(J_{I_0,I_1,w}(\pi_0^\infty)),$$

which together with (4.145), (4.141) and Lemma 4.2.3 gives the isomorphism (4.139). The finite dimensionality of $\mathrm{Ext}^k_{L_{I_1}}(i^\infty_{I_0,I_1,w}(J_{I_0,I_1,w}(\pi_0^\infty)), \pi_1^\infty)^\infty$ comes from the finite length of both representations $i^\infty_{I_0,I_1,w}(J_{I_0,I_1,w}(\pi_0^\infty))$ and π_1^∞ ([68, Sect. 3 Corollary 3] noting that $i^\infty_{I_0,I_1,w}(J_{I_0,I_1,w}(\pi_0^\infty))$ is of finite length as π_0^∞ is by [59, Sect. VI.6.4] and [59, Sect. VI.6.2]). For the last statement, note first that by the definition of regularity (Definition 2.1.4), we have $\Sigma_1 \cap W(G)\cdot\Sigma_0 \neq \emptyset$ if and only if there exists $w \in W(G)$ such that $\Sigma_1 \cap w^{-1}\cdot\Sigma_0 \neq \emptyset$, and we can take w in W^{I_0,I_1} which is then unique. The last statement then follows from the first and (ii) of Lemma 2.1.18 (applied with $L_I = G$). □

Remark 4.5.12 For $i = 0, 1$ and Σ_i as before Corollary 4.5.11, assume that π_i^∞ is in $\mathcal{B}^{I_i}_{\Sigma_i}$. For $w \in W^{I_0,I_1}$, the isomorphism (4.137) is functorial in π_0^∞ and the isomorphism (4.139) is functorial in both π_0^∞ and π_1^∞.

We now use Corollary 4.5.11 to derive several results on the groups $\mathrm{Ext}^\bullet_G(V_0, V_1)$ of (4.14).

Corollary 4.5.13 *Keep the setting of Theorem 4.5.10, let $\Sigma_i = W(L_{I_i})\cdot\mathcal{J}(\pi_i^\infty) \subseteq \widehat{T}^\infty$ and assume moreover that the smooth representations π_i^∞ of L_{I_i} are of finite length ($i = 0, 1$).*

(i) The E-vector space $\mathrm{Ext}^k_G(V_0, V_1)$ is finite dimensional for $k \geq 0$.

(ii) Assume that π_i^∞ is in $\mathcal{B}^{I_i}_{\Sigma_i}$ for $i = 0, 1$. If $\Sigma_1 \cap W(G)\cdot\Sigma_0 = \emptyset$, then $\mathrm{Ext}^k_G(V_0, V_1) = 0$ for $k \geq 0$.

(iii) Assume that π_i^∞ is in $\mathcal{B}^{I_i}_{\Sigma_i}$ for $i = 0, 1$. Let $\xi : Z(\mathfrak{l}_{I_1}) \to E$ be the unique infinitesimal character such that $L^{I_1}(\mu)_\xi \neq 0$. If $\mathrm{Ext}^k_G(V_0, V_1) \neq 0$ for some $k \geq 0$, then there exists $w \in W^{I_0,I_1}$ and $\ell \leq k$ such that $\Sigma_1 \cap w^{-1}\cdot\Sigma_0 \neq \emptyset$ and $H^\ell(\mathfrak{n}_{I_1}, M_0^w)_\xi \neq 0$ (see before Lemma 3.1.3 for the notation).

(iv) Assume that Σ_i is a single G-regular $W(L_{I_i})$-coset and that π_i^∞ is in $\mathcal{B}^{I_i}_{\Sigma_i}$ for $i = 0, 1$. If $\Sigma_1 \cap W(G)\cdot\Sigma_0 \neq \emptyset$, then there exists a unique $w \in W^{I_0,I_1}$ such that $\Sigma_1 \cap w^{-1}\cdot\Sigma_0 \neq \emptyset$, and we have a spectral sequence

$$\mathrm{Ext}^{\ell}_{U(\mathfrak{g})}(M_1, M_0^w) \otimes_E \mathrm{Ext}^k_{L_{I_1}}\big(i^{\infty}_{I_0,I_1,w}(J_{I_0,I_1,w}(\pi_0^{\infty})), \pi_1^{\infty}\big)^{\infty} \implies \mathrm{Ext}^{k+\ell}_G(V_0, V_1). \tag{4.146}$$

Proof We prove (i). By the first statement of Corollary 4.5.11 the E-vector space $\mathrm{Ext}^k_{D^{\infty}(P_{I_1})}((\pi_1^{\infty})^{\vee}, \mathrm{Ext}^{\ell}_{U(\mathfrak{p}_{I_1})}(L^{I_1}(\mu), \mathrm{gr}_w(V_0^{\vee})))$ is finite dimensional for $k \geq 0$, $\ell \geq 0$ and $w \in W^{I_0,I_1}$. By the spectral sequence (4.85) this implies that $\mathrm{Ext}^k_{D(P_{I_1})}(L^{I_1}(\mu) \otimes_E (\pi_1^{\infty})^{\vee}, \mathrm{gr}_w(V_0^{\vee}))$ is finite dimensional for $k \geq 0$ and $w \in W^{I_0,I_1}$. By an obvious dévissage and (4.78) we deduce that $\mathrm{Ext}^k_G(V_0, V_1)$ is finite dimensional for $k \geq 0$.

We prove (ii). As $\Sigma_1 \cap W(G) \cdot \Sigma_0 = \emptyset$ and by (2.31), we have $i^{\infty}_{I_0,I_1,w}(J_{I_0,I_1,w}(\pi_0^{\infty}))_{\mathcal{B}^{I_1}_{\Sigma_1}} = 0$ using the property of the Bernstein block $\mathcal{B}^{I_1}_{\Sigma_1}$ (see the paragraph after Remark 2.1.6). Since $i^{\infty}_{I_0,I_1,w}(J_{I_0,I_1,w}(\pi_0^{\infty}))$ and π_1^{∞} live in different Bernstein blocks, we deduce for $k \geq 0$:

$$\mathrm{Ext}^k_{L_{I_1}}(i^{\infty}_{I_0,I_1,w}(J_{I_0,I_1,w}(\pi_0^{\infty})), \pi_1^{\infty})^{\infty} = 0.$$

Together with (4.85) applied with $D = \mathrm{gr}_w(V_0^{\vee})$, (4.139) and a dévissage with respect to the W^{I_0,I_1}-filtration $\mathrm{Fil}_{\bullet}(V_0^{\vee})$ on $V_0^{\vee}$, we deduce $\mathrm{Ext}^k_{D(P_{I_1})}(L^{I_1}(\mu) \otimes_E (\pi_1^{\infty})^{\vee}, V_0^{\vee}) = 0$ for $k \geq 0$ and hence by (4.78) $\mathrm{Ext}^k_G(V_0, V_1) = 0$ for $k \geq 0$.

We prove (iii). If $\mathrm{Ext}^k_G(V_0, V_1) \neq 0$ we have $\mathrm{Ext}^k_{D(P_{I_1})}(L^{I_1}(\mu) \otimes_E (\pi_1^{\infty})^{\vee}, V_0^{\vee}) \neq 0$ by (4.78), and thus there exists $w \in W^{I_0,I_1}$ such that $\mathrm{Ext}^k_{D(P_{I_1})}(L^{I_1}(\mu) \otimes_E (\pi_1^{\infty})^{\vee}, \mathrm{gr}_w(V_0^{\vee}))$ $\neq 0$. By (4.85) (with $D = \mathrm{gr}_w(V_0^{\vee})$), (4.139) and Lemma 4.5.1 (together with Shapiro's lemma for Lie algebra cohomology) there exist $\ell, k' \leq k$ such that $\mathrm{Ext}^{\ell}_{U(\mathfrak{g})}(M_1, M_0^w) \neq 0$ and $\mathrm{Ext}^{k'}_{L_{I_1}}(i^{\infty}_{I_0,I_1,w}(J_{I_0,I_1,w}(\pi_0^{\infty})), \pi_1^{\infty})^{\infty} \neq 0$. The latter implies $\Sigma_1 \cap w^{-1} \cdot \Sigma_0 \neq \emptyset$ (otherwise $i^{\infty}_{I_0,I_1,w}(J_{I_0,I_1,w}(\pi_0^{\infty}))$ and π_1^{∞} would live in distinct blocks by (2.31) and the paragraph after Remark 2.1.6). By (3.12) the former implies $\mathrm{Ext}^{\ell_1}_{U(\mathfrak{l}_{I_1})}(L^{I_1}(\mu), H^{\ell_2}(\mathfrak{n}_{I_1}, M_0^w)) \neq 0$ for some $\ell_1, \ell_2 \geq 0$ such that $\ell_1 + \ell_2 = \ell$. We thus deduce $H^{\ell_2}(\mathfrak{n}_{I_1}, M_0^w)_{\xi} \neq 0$ from (ii) of Lemma 3.1.3.

We prove (iv). By the last statement in Corollary 4.5.11 we have that w is unique, and together with (4.85) applied with $D = \mathrm{gr}_{w'}(V_0^{\vee})$ for $w' \neq w \in W^{I_0,I_1}$ we deduce $\mathrm{Ext}^k_{D(P_{I_1})}(L^{I_1}(\mu) \otimes_E (\pi_1^{\infty})^{\vee}, \mathrm{gr}_{w'}(V_0^{\vee})) = 0$ for $k \geq 0$ and such w'. Via a dévissage with respect to the W^{I_0,I_1}-filtration $\mathrm{Fil}_{\bullet}(V_0^{\vee})$ on $V_0^{\vee}$, we deduce for $k \geq 0$

$$\mathrm{Ext}^k_{D(P_{I_1})}(L^{I_1}(\mu) \otimes_E (\pi_1^{\infty})^{\vee}, V_0^{\vee}) \cong \mathrm{Ext}^k_{D(P_{I_1})}(L^{I_1}(\mu) \otimes_E (\pi_1^{\infty})^{\vee}, \mathrm{gr}_w(V_0^{\vee})). \tag{4.147}$$

Finally, we apply (4.85) with $D = \mathrm{gr}_w(V_0^{\vee})$, which together with (4.139) and the isomorphisms (4.147), (4.78) give the spectral sequence (4.146). □

By the last statement in (i) of Lemma 2.1.15 and the comment below Lemma 2.1.15, Corollary 4.5.11 and (iv) of Corollary 4.5.13 can in particular be applied to G-basic π_i^{∞}.

Remark 4.5.14 If $I_1 = \Delta$ (with $\mu_1 \in \Lambda^{\mathrm{dom}}$ and $V_1 \cong L(\mu_1)^{\vee} \otimes_E \pi_1^{\infty}$), then $W^{I_0,\Delta} = \{1\}$ and the same argument as at the end of the proof of (iv) of Corollary 4.5.13 (for

$\pi_0^\infty, \pi_1^\infty$ of finite length) gives a spectral sequence of finite dimensional E-vector spaces

$$\mathrm{Ext}^\ell_{U(\mathfrak{g})}(L(\mu_1), M_0) \otimes_E \mathrm{Ext}^k_G(i_{I_0}^\infty(\pi_0^\infty), \pi_1^\infty)^\infty \implies \mathrm{Ext}^{k+\ell}_G(V_0, V_1).$$

If moreover $I_0 = \Delta$ (with $\mu_0 \in \Lambda^{\mathrm{dom}}$ and $V_0 \cong L(\mu_0)^\vee \otimes_E \pi_0^\infty$), we obtain a spectral sequence

$$\mathrm{Ext}^\ell_{U(\mathfrak{g})}(L(\mu_1), L(\mu_0)) \otimes_E \mathrm{Ext}^k_G(\pi_0^\infty, \pi_1^\infty)^\infty \implies \mathrm{Ext}^{k+\ell}_G(V_0, V_1). \tag{4.148}$$

Note that, as $\mu_0, \mu_1 \in \Lambda^{\mathrm{dom}}$, we have $L(\mu_0)_\xi \neq 0 \neq L(\mu_1)_\xi$ for some $\xi : Z(\mathfrak{g}) \to E$ if and only if $\mu_0 = \mu_1$. Hence by (ii) of Lemma 3.1.3 we have $\mathrm{Ext}^\ell_{U(\mathfrak{g})}(L(\mu_1), L(\mu_0)) \neq 0$ for some $\ell \geq 0$ if and only if $\mu_0 = \mu_1$, in which case a translation functor argument gives a canonical isomorphism $\mathrm{Ext}^\ell_{U(\mathfrak{g})}(L(\mu_0), L(\mu_0)) \cong H^\ell(\mathfrak{g}, 1_{\mathfrak{g}})$ for $\ell \geq 0$. Consequently, when $I_0 = I_1 = \Delta$, $\mathrm{Ext}^k_G(V_0, V_1) \neq 0$ for some $k \geq 0$ only if $\mu_0 = \mu_1$, in which case $\mathrm{Ext}^k_G(V_0, V_1) = 0$ for $k < d(\pi_0^\infty, \pi_1^\infty)$ (see (iii) of Definition 2.1.4 for $d(\pi_0^\infty, \pi_1^\infty)$) and we have a canonical isomorphism

$$\mathrm{Ext}_G^{d(\pi_0^\infty, \pi_1^\infty)}(\pi_0^\infty, \pi_1^\infty)^\infty \xrightarrow{\sim} \mathrm{Ext}_G^{d(\pi_0^\infty, \pi_1^\infty)}(V_0, V_1). \tag{4.149}$$

Lemma 4.5.15 *For $i = 0, 1$ let $V_i = \mathcal{F}^G_{P_{I_i}}(M_i, \pi_i^\infty)$ with $I_i \subseteq \Delta$, M_i in $\mathcal{O}^{\mathfrak{p}_{I_i}}_{\mathrm{alg}}$ and π_i^∞ smooth finite length representations of L_{I_i} such that π_i^∞ is in $\mathcal{B}^{I_i}_{\Sigma_i}$ with $\Sigma_i = W(L_{I_i}) \cdot \mathcal{J}(\pi_i^\infty) \subseteq \widehat{T}^\infty$. Assume that*

$$\Sigma_0 \cap W(G) \cdot \Sigma_1 = \emptyset. \tag{4.150}$$

Then we have for $k \geq 0$

$$\mathrm{Ext}^k_G(V_0, V_1) = 0. \tag{4.151}$$

Proof By dévissage we can assume M_1 simple of the form $L(\mu)$ for some $\mu \in \Lambda^{\mathrm{dom}}_{I_1}$. If $M_1 = L(\mu) = M^{I_1}(\mu)$, then (4.151) follows directly from (4.150) and (ii) of Corollary 4.5.13. In general, we assume inductively that

$$\mathrm{Ext}^k_G(V_0, \mathcal{F}^G_{P_{I_1}}(L(\mu'), \pi_1^\infty)) = 0 \tag{4.152}$$

for $\mu' \in \Lambda^{\mathrm{dom}}_{I_1}$ such that $\mu' - \mu \in \mathbb{Z}_{\geq 0}\Phi^+$ and $\mu' \neq \mu$. Recall $M^{I_1}(\mu)$ fits into $0 \to N^{I_1}(\mu) \to M^{I_1}(\mu) \to L(\mu) \to 0$ with all Jordan-Hölder factors of $N^{I_1}(\mu)$ of the form $L(\mu')$ for some $\mu' \in \Lambda^{\mathrm{dom}}_{I_1}$ such that $\mu' - \mu \in \mathbb{Z}_{\geq 0}\Phi^+$ and $\mu' \neq \mu$ (use [33, Theorem 5.1]). This together with (4.152) (and (i) of Theorem 4.3.7) implies $\mathrm{Ext}^k_G(V_0, \mathcal{F}^G_{P_{I_1}}(N^{I_1}(\mu), \pi_1^\infty)) = 0$ for $k \geq 0$. Using $0 \to V_1 \to \mathcal{F}^G_{P_{I_1}}(M^{I_1}(\mu), \pi_1^\infty) \to \mathcal{F}^G_{P_{I_1}}(N^{I_1}(\mu), \pi_1^\infty) \to 0$ we obtain an isomorphism $\mathrm{Ext}^k_G(V_0, V_1) \xrightarrow{\sim} \mathrm{Ext}^k_G(V_0, \mathcal{F}^G_{P_{I_1}}(M^{I_1}(\mu), \pi_1^\infty))$ for $k \geq 0$. We then again deduce (4.151) from (4.150) and (ii) of Corollary 4.5.13, which finishes the proof by induction. □

We isolate the following result because it has its own interest.

Theorem 4.5.16 *For $i = 0, 1$ let $V_i = \mathcal{F}^G_{P_{I_i}}(M_i, \pi_i^\infty)$ with $I_i \subseteq \Delta$, M_i in $\mathcal{O}^{\mathfrak{p}_{I_i}}_{\mathrm{alg}}$ and π_i^∞ smooth finite length representations of L_{I_i} over E. Then the E-vector space $\mathrm{Ext}^k_G(V_0, V_1)$ is finite dimensional for $k \geq 0$.*

Proof By dévissage we can assume M_1 simple of the form $L(\mu)$ for some $\mu \in \Lambda^{\mathrm{dom}}_{I_1}$. If $M_1 = L(\mu) = M^{I_1}(\mu)$, then the result is (i) of Corollary 4.5.13. In general, we argue by induction as in the proof of Lemma 4.5.15 using the short exact sequence $0 \to N^{I_1}(\mu) \to M^{I_1}(\mu) \to L(\mu) \to 0$. □

We can obtain better vanishing results when Σ_i is a single G-regular left $W(L_{I_i})$-coset for $i = 0, 1$.

Lemma 4.5.17 *For $i = 0, 1$ let $V_i = \mathcal{F}^G_{P_{I_i}}(M_i, \pi_i^\infty)$ with $I_i \subseteq \Delta$, M_i in $\mathcal{O}^{\mathfrak{p}_{I_i}}_{\mathrm{alg}}$ and π_i^∞ smooth finite length representations of L_{I_i} such that π_i^∞ is in $\mathcal{B}^{I_i}_{\Sigma_i}$ with $\Sigma_i = W(L_{I_i}) \cdot \mathcal{J}(\pi_i^\infty) \subseteq \widehat{T}^\infty$. Assume that Σ_i is a single G-regular left $W(L_{I_i})$-coset and recall that $d(\pi_0^\infty, \pi_1^\infty)$ is defined in (iii) of Definition 2.1.4.*

(i) We have $\mathrm{Ext}^k_G(V_0, V_1) = 0$ for $k < d(\pi_0^\infty, \pi_1^\infty)$.

(ii) Let $w \in W^{I_0,I_1}$ such that $\Sigma \overset{def}{=} \Sigma_1 \cap w^{-1} \cdot \Sigma_0 \neq \emptyset$ and $I \overset{def}{=} w^{-1}(I_0) \cap I_1$. Then Σ is a single G-regular $W(L_I)$-coset and, if $J_{I_0,I_1,w}(\pi_0^\infty)_{\mathcal{B}^I_\Sigma} = 0$, we have $\mathrm{Ext}^k_G(V_0, V_1) = 0$ for $k \geq 0$.

Proof We prove (i). By (2.7) and (i) of Lemma 2.1.18 (applied with $I = \Delta$) we have for $k \geq 0$

$$\mathrm{Ext}^k_G(i^\infty_{I_0,\Delta}(\pi_0^\infty), i^\infty_{I_1,\Delta}(\pi_1^\infty))^\infty \cong \bigoplus_{w \in W^{I_0,I_1}} \mathrm{Ext}^k_{L_{I_1}}(i^\infty_{I_0,I_1,w}(J_{I_0,I_1,w}(\pi_0^\infty)), \pi_1^\infty)$$

and thus $\mathrm{Ext}^k_{L_{I_1}}(i^\infty_{I_0,I_1,w}(J_{I_0,I_1,w}(\pi_0^\infty)), \pi_1^\infty) = 0$ for $w \in W^{I_0,I_1}$ and $k < d(\pi_0^\infty, \pi_1^\infty)$. This together with (4.146) gives (i) when $M_1 = M^{I_1}(\mu_1)$ for some $\mu_1 \in \Lambda^{\mathrm{dom}}_{I_1}$. The result for general M_1 follows from the same induction as in the proof of Lemma 4.5.15.

We prove (ii). The fact that Σ is a single G-regular $W(L_I)$-coset follows easily from the fact that Σ_i, for $i = 0, 1$, is also a single G-regular $W(L_I)$-coset (last statement in (i) of Lemma 2.1.15). We have for $k \geq 0$ (see (2.6) for $J'_{I_1,I}(\pi_1^\infty)$)

$$\begin{aligned}\mathrm{Ext}^k_{L_{I_1}}(i^\infty_{I_0,I_1,w}(J_{I_0,I_1,w}(\pi_0^\infty)), \pi_1^\infty)^\infty &\cong \mathrm{Ext}^k_{L_I}(J_{I_0,I_1,w}(\pi_0^\infty), J'_{I_1,I}(\pi_1^\infty))^\infty \\ &\cong \mathrm{Ext}^k_{L_I}(J_{I_0,I_1,w}(\pi_0^\infty)_{\mathcal{B}^I_\Sigma}, J'_{I_1,I}(\pi_1^\infty)_{\mathcal{B}^I_\Sigma})^\infty = 0\end{aligned}$$

where the first isomorphism is (2.68), the second follows from Lemma 2.1.29 and Remark 2.1.30 (arguing as above (2.70)) and where the last equality follows from the assumption $J_{I_0,I_1,w}(\pi_0^\infty)_{\mathcal{B}^I_\Sigma} = 0$. Together with (4.146) this implies (ii) when $M_1 = M^{I_1}(\mu_1)$ for some $\mu_1 \in \Lambda^{\mathrm{dom}}_{I_1}$. The result for general M_1 again follows from the same induction as in the proof of Lemma 4.5.15. □

Lemma 4.5.18 *For $i = 0, 1$ let $V_i = \mathcal{F}^G_{P_{I_i}}(M_i, \pi_i^\infty)$ with $I_i \subseteq \Delta$, M_i in $\mathcal{O}^{\mathfrak{p}_{I_i}}_{\mathrm{alg}}$ and π_i^∞ smooth finite length representations of L_{I_i} such that π_i^∞ is in $\mathcal{B}^{I_i}_{\Sigma_i}$ with $\Sigma_i = W(L_{I_i}) \cdot \mathcal{J}(\pi_i^\infty) \subseteq \widehat{T}^\infty$. If $\mathrm{Ext}^k_G(V_0, V_1) \neq 0$ for some $k \geq 0$ then there exists $\xi : Z(\mathfrak{g}) \to E$ such that $M_{0,\xi} \neq 0 \neq M_{1,\xi}$.*

Proof By dévissage we can assume M_1 simple of the form $L(\mu)$ for some $\mu \in \Lambda^{\mathrm{dom}}_{I_1}$. Moreover, we may choose μ so that any $\mu' \in \Lambda^{\mathrm{dom}}_{I_1}$ such that $\mu' \in W(G) \cdot \mu \setminus \{\mu\}$ and $\mu' - \mu \in \mathbb{Z}_{\geq 0}\Phi^+$ must also satisfy for $k \geq 0$

$$\mathrm{Ext}^k_G(V_0, \mathcal{F}^G_{P_{I_1}}(L(\mu'), \pi_1^\infty)) = 0. \tag{4.153}$$

Define $N^{I_1}(\mu)$ as in the proof of Lemma 4.5.15. As the irreducible constituents $L(\mu')$ of $N^{I_1}(\mu)$ satisfy $\mu' \in \Lambda^{\mathrm{dom}}_{I_1}$, $\mu' \in W(G) \cdot \mu \setminus \{\mu\}$ and $\mu' - \mu \in \mathbb{Z}_{\geq 0}\Phi^+$ (use [33, Theorem 5.1]), we have $\mathrm{Ext}^k_G(V_0, \mathcal{F}^G_{P_{I_1}}(N^{I_1}(\mu), \pi_1^\infty)) = 0$ for $k \geq 0$ by (4.153), and thus the surjection $M^{I_1}(\mu) \twoheadrightarrow L(\mu)$ induces an isomorphism for $k \geq 0$

$$\mathrm{Ext}^k_G(V_0, \mathcal{F}^G_{P_{I_1}}(L(\mu), \pi_1^\infty)) \xrightarrow{\sim} \mathrm{Ext}^k_G(V_0, \mathcal{F}^G_{P_{I_1}}(M^{I_1}(\mu), \pi_1^\infty))$$

(which is non-zero for some $k \geq 0$ by assumption). By the proof of (iii) of Corollary 4.5.13 we have $\mathrm{Ext}^\ell_{U(\mathfrak{g})}(M^{I_1}(\mu), M_0^w) \neq 0$ for some $w \in W^{I_0, I_1}$ and some $\ell \geq 0$, which together with (ii) of Lemma 3.1.3 implies $(M_0^w)_\xi \neq 0$ for the unique infinitesimal character $\xi : Z(\mathfrak{g}) \to E$ such that $M^{I_1}(\mu)_\xi \neq 0 \neq L(\mu)_\xi$. As the adjoint action of G, and in particular of $w \in W(G)$, on $Z(\mathfrak{g})$ (inside $U(\mathfrak{g})$) is the identity, this is equivalent to $(M_{0,\xi})^w \neq 0$, i.e. to $M_{0,\xi} \neq 0$. □

Remark 4.5.19 Let V be a finite length object in $\mathrm{Rep}^{\mathrm{an}}_{\mathrm{adm}}(G)$ with each constituent of the form $\mathcal{F}^G_{P_I}(M, \pi^\infty)$ for some $I \subseteq \Delta$, some M in $\mathcal{O}^{\mathfrak{p}_I}_{\mathrm{alg}}$ and some smooth (finite length) representation π^∞ of L_I in $\mathcal{B}^I_{W(L_I)\cdot\mathcal{J}(\pi^\infty)}$. Given $\xi : Z(\mathfrak{g}) \to E$ and a $W(G)$-coset $\Sigma \subseteq \widehat{T}^\infty$, there exists a maximal closed subrepresentation $V_{\xi,\Sigma} \subseteq V$ such that each constituent of $V_{\xi,\Sigma}$ has the form $\mathcal{F}^G_{P_I}(M, \pi^\infty)$ for I, M, π^∞ with $M = M_\xi$ and $\mathcal{J}(\pi^\infty) \subseteq \Sigma$. It follows from Lemmas 4.5.15 and 4.5.18 that V is the direct sum of $V_{\xi,\Sigma}$ over all pairs ξ, Σ. Similarly, we can define V_ξ just taking ξ into account, as well as V_Σ. Taking continuous duals we can define $D_{\xi,\Sigma}$, D_ξ and D_Σ for a finite length coadmissible $D(G)$-module with constituents of the form $\mathcal{F}^G_{P_I}(M, \pi^\infty)^\vee$ (with M, π^∞ as above).

Remark 4.5.20 Let $I_0 \subseteq \Delta$, $x_0 \in W(G)$ with $I_0 = \Delta \setminus D_L(x_0)$, $M_0 \overset{\mathrm{def}}{=} L(x_0) \in \mathcal{O}^{\mathfrak{p}_{I_0}}_{\mathrm{alg}}$ (Lemma 3.1.1), π_0^∞ a smooth strongly admissible representation of L_{I_0} over E and $V_0 \overset{\mathrm{def}}{=} \mathcal{F}^G_{P_{I_0}}(M_0, \pi_0^\infty)$. We have $H^0(\mathfrak{u}, M_0^w) = 0$ for $1 \neq w \in W^{I_0,\emptyset}$ by (ii) of Lemma 3.3.1 (applied with $I' = I_0$ and $I = \emptyset$) and $H^0(\mathfrak{u}, M_0)$ is finite dimensional. Since $H^0(\mathfrak{u}, \mathcal{M}_0^w) \subseteq \mathcal{M}_0^w$ is a closed Fréchet $U(\mathfrak{t})$-submodule, it is small by Lemma 4.1.10, and thus contains $H^0(\mathfrak{u}, M_0^w) = H^0(\mathfrak{u}, \mathcal{M}_0^w) \cap M_0^w$ as a dense subspace. In particular $H^0(\mathfrak{u}, \mathcal{M}_0^w) = 0$ for $1 \neq w \in W^{I_0,\emptyset}$ and the injection $M_0 \hookrightarrow \mathcal{M}_0$ induces a (topological) isomorphism of finite dimensional E-vector spaces $H^0(\mathfrak{u}, M_0) \xrightarrow{\sim} H^0(\mathfrak{u}, \mathcal{M}_0)$. We fix a compact open subgroup $G_0 \subseteq G$ and for $w \in W^{I_0,\emptyset}$, we write $P_w \stackrel{\text{def}}{=} w^{-1}P_{I_0}w$, $P_{w,0} \stackrel{\text{def}}{=} P_w \cap G_0$, and $\mathcal{X}_w$ for the set of compact open subsets of $(BP_w) \cap G_0$ stable under right multiplication by $P_{w,0}$. We write U_w for a general element of $\mathcal{X}_w$. For $r \in \mathcal{I}$, it follows from Lemma 4.3.4 that $(\mathcal{M}_0^w)_r = D(G_0)_{P_{w,0},r} \otimes_{D(G_0)_{P_{w,0}}} \mathcal{M}_0^w$ is the completion of $\mathcal{M}_0^w$ under a standard semi-norm, which by Remark 4.1.12 implies that $H^0(\mathfrak{u}, M_0^w) = H^0(\mathfrak{u}, (\mathcal{M}_0^w)_r) \cap M_0^w$ is dense inside the closed Banach subspace $H^0(\mathfrak{u}, (\mathcal{M}_0^w)_r) \subseteq (\mathcal{M}_0^w)_r$. Consequently, for $1 \neq w \in W^{I_0,\emptyset}$, we have for $g \in B$ and $r \in \mathcal{I}$ (since $H^0(\mathfrak{u}, M_0^w) = 0$)

$$H^0(\mathfrak{u}, ((\mathcal{M}_0^w)_r)^{g^{-1}}) \cong H^0(g\mathfrak{u}g^{-1}, ((\mathcal{M}_0^w)_r)^{g^{-1}}) \cong H^0(\mathfrak{u}, (\mathcal{M}_0^w)_r) = 0,$$

and therefore

$$\begin{aligned}
H^0(\mathfrak{u}, \mathrm{gr}_w(V_0^\vee)) &\cong \big(\varprojlim_{U_w\in\mathcal{X}_w, r\in\mathcal{I}}(D(G_0)_{U_w,r} \otimes_{D(G_0)_{P_{w,0},r}} D_r)\big)[\mathfrak{u}] \\
&\cong \varprojlim_{U_w\in\mathcal{X}_w, r\in\mathcal{I}} \big((D(G_0)_{U_w,r} \otimes_{D(G_0)_{P_{w,0},r}} D_r)[\mathfrak{u}]\big) \\
&\cong \varprojlim_{U_w\in\mathcal{X}_w, r\in\mathcal{I}} \Big(\bigoplus\nolimits_{g\in U_wG_1^m/P_{w,0}G_1^m} \big(((\mathcal{M}_0^w)_r)^{g^{-1}}[\mathfrak{u}] \otimes_E D_r^{\infty,g^{-1}}\big)\Big) \\
&= 0
\end{aligned}$$

where the first isomorphism uses (4.106) and the third isomorphism uses (4.109), and where D_r and D_r^∞ are the same as in *loc. cit.* and depend on the choice of w (see (4.105), (4.104) and the paragraph before (4.91)). For $w = 1$, we have

$$\begin{aligned}
H^0(\mathfrak{u}, \mathrm{gr}_1(V_0^\vee)) &\cong \big(\varprojlim_{r\in\mathcal{I}}((\mathcal{M}_0)_r \otimes_E (\pi_0^\infty)_r^\vee)\big)[\mathfrak{u}] \cong \varprojlim_{r\in\mathcal{I}} \big(((\mathcal{M}_0)_r \otimes_E (\pi_0^\infty)_r^\vee)[\mathfrak{u}]\big) \\
&\cong \varprojlim_{r\in\mathcal{I}}((\mathcal{M}_0)_r[\mathfrak{u}] \otimes_E (\pi_0^\infty)_r^\vee) \quad \cong \varprojlim_{r\in\mathcal{I}}((\mathcal{M}_0)_r[\mathfrak{u}])\widehat{\otimes}_E \varprojlim_{r\in\mathcal{I}}(\pi_0^\infty)_r^\vee) \\
&\cong \mathcal{M}_0[\mathfrak{u}]\widehat{\otimes}_E(\pi_0^\infty)^\vee \\
&\cong \mathcal{M}_0[\mathfrak{u}] \otimes_E (\pi_0^\infty)^\vee
\end{aligned}$$

where the first isomorphism follows from Proposition 4.4.3 (applied with $I_1 = \emptyset$) and from (4.105), the fourth follows from Lemma 4.2.12, and the last follows from the finite dimensionality of $\mathcal{M}_0[\mathfrak{u}]$. We deduce from all this $H^0(U, \mathrm{gr}_w(V_0^\vee)) = 0$ for $1 \neq w \in W^{I_0,\emptyset}$ and $H^0(U, \mathrm{gr}_1(V_0^\vee)) \cong H^0(\mathfrak{u}, M_0) \otimes_E (J_{I_0,\emptyset}(\pi_0^\infty))^\vee$ (cf. (4.144) with $I_1 = \emptyset$ and $w = 1$). By dévissage on the $W^{I_0,\emptyset}$-filtration $\mathrm{Fil}_\bullet(V_0^\vee)$ we finally obtain the following isomorphism of $D(T)$-modules (which will be used in the proof of Theorem 5.4.4 below)

$$H^0(U, V_0^\vee) \cong H^0(\mathfrak{u}, M_0) \otimes_E (J_{I_0,\emptyset}(\pi_0^\infty))^\vee. \tag{4.154}$$

Note that (4.154) is a finite dimensional (coadmissible) $D(T)$-module which is non-zero when π_0^∞ is G-basic (and $M_0 = L(x_0)$).

Chapter 5
Complexes of Locally Analytic Representations

We describe the global sections of the de Rham complex of the Drinfeld space in dimension $n-1$ as a complex of explicit finite length coadmissible $D(G)$-modules, and we describe an explicit section in the derived category (of duals of finite length admissible locally K-analytic representations of G) to the canonical morphism to its H^{n-1} (shifted). The proof works essentially *verbatim* for the complex of holomorphic discrete series of [64].

5.1 Results on Locally Analytic Ext^0, Ext^1 and Ext^2 Groups

We prove various useful results on Hom, Ext^1 and Ext^2 groups between certain Orlik-Strauch representations.

We use the notation of Sects. 2.1 and 2.3, in particular $\mu_0 \in \Lambda^{\mathrm{dom}}$ is a fixed weight, $L(w) = L(w \cdot \mu_0)$ for $w \in W(G)$, $N^I(w) = \ker(M^I(w) \twoheadrightarrow L(w))$, etc. For $x \in W(G)$ we write $I_x \overset{\mathrm{def}}{=} \Delta \setminus D_L(x)$. Throughout this section for $i = 0, 1$, I_i is a subset of Δ, M_i is a $U(\mathfrak{g})$-module in $\mathcal{O}_{\mathrm{alg}}^{\mathfrak{p}_{I_i}}$ (see below (3.4)), Σ_i is a finite subset of $\widehat{T}^\infty$ preserved under the left action (2.11) of $W(L_{I_i})$ and π_i^∞ is a smooth (finite length) representation of L_{I_i} in the category $\mathcal{B}_{\Sigma_i}^{I_i}$ (see above (2.13)). We write $V_i \overset{\mathrm{def}}{=} \mathcal{F}_{P_{I_i}}^G(M_i, \pi_i^\infty)$ for $i = 0, 1$. Depending on the statements, we will add various assumptions on I_i, M_i, Σ_i or π_i^∞ (and thus V_i).

Lemma 5.1.1 *For $i = 0, 1$ assume that $I_i = \Delta \setminus D_L(x_i)$ and $M_i = L(x_i)$ for some $x_i \in W(G)$. Then we have $\mathrm{Hom}_G(V_0, V_1) \neq 0$ only if $x_0 = x_1$ (and $I_0 = I_1$), in which case we have a canonical isomorphism*

$$\mathrm{Hom}_G(V_0, V_1) \cong \mathrm{Hom}_{U(\mathfrak{g})}(M_1, M_0) \otimes_E \mathrm{Hom}_{L_{I_1}}(\pi_0^\infty, \pi_1^\infty). \tag{5.1}$$

C. Breuil and Z. Qian, *Splitting and Making Explicit the de Rham Complex of the Drinfeld Space*, Lecture Notes in Mathematics 2397,
https://doi.org/10.1007/978-3-032-20095-2_5

In particular, if V_0 and V_1 are irreducible (which forces π_i^∞ to be irreducible for $i = 0, 1$), they are isomorphic if and only if $x_0 = x_1$ and $\pi_0^\infty \cong \pi_1^\infty$.

Proof Let $V_2 \overset{\text{def}}{=} \mathcal{F}_{P_{I_1}}^G(M^{I_1}(x_1), \pi_1^\infty)$, the surjection $M^{I_1}(x_1) \twoheadrightarrow L(x_1)$ induces an injection $V_1 \hookrightarrow V_2$ and thus an injection

$$\mathrm{Hom}_G(V_0, V_1) \hookrightarrow \mathrm{Hom}_G(V_0, V_2). \tag{5.2}$$

Recall from (iii) of Lemma 3.3.1 that $L(x_0)^w$ is in $\mathcal{O}_{\mathrm{alg}}^{\mathfrak{b}}$ for some $w \in W^{I_0,I_1}$ if and only if $w = 1$. Hence for $1 \neq w \in W^{I_0,I_1}$ we have (using Shapiro's lemma for the first isomorphism)

$$\mathrm{Hom}_{U(\mathfrak{g})}(M^{I_1}(x_1), L(x_0)^w) \cong \mathrm{Hom}_{U(\mathfrak{p}_{I_1})}(L^{I_1}(x_1), L(x_0)^w) = 0.$$

By (4.78) (applied with $\mathrm{gr}_w(V_0^\vee)$ instead of $V_0^\vee$ there, which doesn't change the argument), (4.85) (applied with $D = \mathrm{gr}_w(V_0^\vee)$ and $k = \ell = 0$) and Corollary 4.5.11 (for $k = \ell = 0$), we deduce $\mathrm{Hom}_{D(G)}(V_2^\vee, \mathrm{gr}_w(V_0^\vee)) = 0$ for $1 \neq w \in W^{I_0,I_1}$ and for $w = 1$:

$$\mathrm{Hom}_{D(G)}(V_2^\vee, \mathrm{gr}_1(V_0^\vee)) \cong \mathrm{Hom}_{U(\mathfrak{g})}(M^{I_1}(x_1), L(x_0)) \otimes_E \mathrm{Hom}_{L_{I_1}}(i_{I_0,I_1,1}^\infty(J_{I_0,I_1,1}(\pi_0^\infty)), \pi_1^\infty).$$

By a dévissage with respect to $(\mathrm{Fil}_w(V_0^\vee))_{w \in W^{I_0,I_1}}$ we obtain

$$\begin{aligned}\mathrm{Hom}_G(V_0, V_2) &\cong \mathrm{Hom}_{D(G)}(V_2^\vee, V_0^\vee)\\ &\cong \mathrm{Hom}_{U(\mathfrak{g})}(M^{I_1}(x_1), L(x_0)) \otimes_E \mathrm{Hom}_{L_{I_1}}(i_{I_0,I_1,1}^\infty(J_{I_0,I_1,1}(\pi_0^\infty)), \pi_1^\infty).\end{aligned} \tag{5.3}$$

If $\mathrm{Hom}_G(V_0, V_1) \neq 0$ then $\mathrm{Hom}_G(V_0, V_2) \neq 0$ by (5.2), thus $\mathrm{Hom}_{U(\mathfrak{g})}(M^{I_1}(x_1), L(x_0)) \neq 0$ by (5.3) and thus

$$\mathrm{Hom}_{U(\mathfrak{g})}(L(x_1), L(x_0)) \xrightarrow{\sim} \mathrm{Hom}_{U(\mathfrak{g})}(M^{I_1}(x_1), L(x_0)) \neq 0 \tag{5.4}$$

(as $M^{I_1}(x_1)$ has irreducible cosocle $L(x_1)$) which forces $x_0 = x_1$ and hence $I_0 = I_1$. As any irreducible constituent $L(x')$ of $N^{I_1}(x_1)$ satisfies $x' > x_1 = x_0$, the argument above with $M^{I_1}(x')$ and $L(x')$ instead of $M^{I_1}(x_1)$ and $L(x_1)$ (note that $I_1 = \Delta \setminus D_L(x')$ by Lemma 3.1.1) shows that $\mathrm{Hom}_G(V_0, \mathcal{F}_{P_{I_1}}^G(L(x'), \pi_1^\infty)) = 0$. As $V_2/V_1 \cong \mathcal{F}_{P_{I_1}}^G(N^{I_1}(x_1), \pi_1^\infty)$ (see (i) of Theorem 4.3.7), we deduce by dévissage $\mathrm{Hom}_G(V_0, V_2/V_1) = 0$ and hence $\mathrm{Hom}_G(V_0, V_1) \xrightarrow{\sim} \mathrm{Hom}_G(V_0, V_2)$. Then (5.1) follows from (5.3) (with $I_0 = I_1$) and (5.4). □

Lemma 5.1.2 *Let $x \in W(G)$, $I \overset{\text{def}}{=} \Delta \setminus D_L(x)$, Σ a finite subset of $\widehat{T}^\infty$ preserved under the left action (2.11) of $W(L_I)$, π^∞ a smooth (finite length) multiplicity free representation of L_I in $\mathcal{B}_\Sigma^I$ and $V \overset{\text{def}}{=} \mathcal{F}_{P_I}^G(L(x), \pi^\infty)$. Then V is of finite length and* *multiplicity free, and the functor $\mathcal{F}_{P_I}^G(L(x), -)$ induces a bijection $\mathrm{JH}_{L_I}(\pi^\infty) \xrightarrow{\sim} \mathrm{JH}_G(V)$ between partially ordered sets (see Sect. 1.4 for the definition of this partial order).*

Proof It follows from Theorem 4.3.7 and Lemma 5.1.1 that V is finite length and multiplicity free with $\mathrm{JH}_G(V) = \{\mathcal{F}_{P_I}^G(L(x), \tau^\infty) \mid \tau^\infty \in \mathrm{JH}_{L_I}(\pi^\infty)\}$. Let $\tau_0^\infty, \tau_1^\infty$ be two distinct irreducible constituents of π^∞ and define σ^∞ as the unique subrepresentation of π^∞ with cosocle τ_0^∞. Recall that, by definition of the partial order on $\mathrm{JH}_{L_I}(\pi^\infty)$, $\tau_1^\infty < \tau_0^\infty$ if and only if $\tau_1^\infty \in \mathrm{JH}_{L_I}(\sigma^\infty)$. Let $W \stackrel{\mathrm{def}}{=} \mathcal{F}_{P_I}^G(L(x), \sigma^\infty)$ and $W_i \stackrel{\mathrm{def}}{=} \mathcal{F}_{P_I}^G(L(x), \tau_i^\infty)$ for $i = 0, 1$. By Theorem 4.3.7 and the last claim in Lemma 5.1.1 $\tau_1^\infty \in \mathrm{JH}_{L_I}(\sigma^\infty)$ if and only if $W_1 \in \mathrm{JH}_G(W)$. By (5.1) applied with $x_0 = x_1 = x$, $\pi_0^\infty = \sigma^\infty$ and π_1^∞ any constituent of σ^∞, we deduce that W has simple cosocle W_0, and thus $W_1 < W_0$ if and only if $W_1 \in \mathrm{JH}_G(W)$. We have shown $\tau_1^\infty < \tau_0^\infty$ if and only if $W_1 < W_0$, which proves the lemma. □

Recall that G-regular and G-basic smooth representations of L_I over E are defined in (ii) of Definition 2.1.4. Recall also that a G-basic representation of L_I is in $\mathcal{B}_\Sigma^I$ for $\Sigma = W(L_I) \cdot \mathcal{J}(\pi^\infty)$ (see below Lemma 2.1.15) and is multiplicity free ((i) of *loc. cit.*), in particular we can apply Lemma 5.1.2 when π^∞ is G-basic.

Lemma 5.1.3 *For $i = 0, 1$ assume that $I \stackrel{def}{=} I_0 = I_1 = \Delta \setminus D_L(x)$ and $M_0 = M_1 = L(x)$ for some $x \in W(G)$. Assume moreover that π_0^∞ and π_1^∞ have no common Jordan-Hölder factor. Then the functor $\mathcal{F}_{P_I}^G(L(x), -)$ induces a canonical isomorphism*

$$\mathrm{Ext}^1_{L_I}(\pi_0^\infty, \pi_1^\infty)^\infty \xrightarrow{\sim} \mathrm{Ext}^1_G(V_0, V_1). \tag{5.5}$$

If moreover π_0^∞ and π_1^∞ are G-basic, then (5.5) is one dimensional if non-zero.

Proof Recall first that both spaces in (5.5) are finite dimensional over E, as follows from Theorem 4.5.16 and (the references in) its proof.

Step 1: We prove that (5.5) is injective.
If $\mathrm{Ext}^1_{L_I}(\pi_0^\infty, \pi_1^\infty)^\infty = 0$, there is nothing to prove. Otherwise, let π^∞ in $\mathrm{Rep}^\infty_{\mathrm{adm}}(L_I)$ that fits into a non-split extension

$$0 \to \pi_1^\infty \to \pi^\infty \to \pi_0^\infty \to 0. \tag{5.6}$$

and note that π^∞ is in $\mathcal{B}_\Sigma^I$ for $\Sigma = \Sigma_0 \cup \Sigma_1$ and is multiplicity free. In particular, there exists $\sigma_i^\infty \in \mathrm{JH}_{L_I}(\pi_i^\infty)$ such that π^∞ admits a length 2 subquotient σ^∞ with socle σ_1^∞ and cosocle σ_0^∞. Applying $\mathcal{F}_{P_I}^G(L(x), -)$ to (5.6) and using Lemma 5.1.2, it follows that $V \stackrel{\mathrm{def}}{=} \mathcal{F}_{P_I}^G(L(x), \pi^\infty)$ admits a length 2 subquotient $\mathcal{F}_{P_I}^G(L(x), \sigma^\infty)$ with socle $\mathcal{F}_{P_I}^G(L(x), \sigma_1^\infty)$ and cosocle $\mathcal{F}_{P_I}^G(L(x), \sigma_0^\infty)$. In particular the short exact sequence $0 \to V_1 \to V \to V_0 \to 0$ is non-split. This proves the injectivity of (5.5).

Step 2: We prove $\dim_E \mathrm{Ext}^1_G(V_0, V_1) \le \dim_E \mathrm{Ext}^1_{L_I}(\pi_0^\infty, \pi_1^\infty)^\infty$.
We can assume $\mathrm{Ext}^1_G(V_0, V_1) \neq 0$. Let $V_2 \stackrel{\mathrm{def}}{=} \mathcal{F}_{P_I}^G(M^I(x), \pi_1^\infty)$, we have $V_1 \hookrightarrow V_2$

with $V_2/V_1 \cong \mathcal{F}_{P_I}^G(N^I(x), \pi_1^\infty)$. Since $L(x) \notin \mathrm{JH}_{U(\mathfrak{g})}(N^I(x))$, by Lemma 5.1.1 (using Lemma 3.1.1) we know that V_2/V_1 and V_0 share no common Jordan-Hölder factor and in particular $\mathrm{Hom}_G(V_0, V_2/V_1) = 0$. A dévissage on $0 \to V_1 \to V_2 \to V_2/V_1 \to 0$ then gives an injection

$$\mathrm{Ext}_G^1(V_0, V_1) \hookrightarrow \mathrm{Ext}_G^1(V_0, V_2) \tag{5.7}$$

and thus $\mathrm{Ext}_G^1(V_0, V_2) \neq 0$ since $\mathrm{Ext}_G^1(V_0, V_1) \neq 0$. By a dévissage on $(\mathrm{Fil}_w(V_0^\vee))_{w \in W^{I,I}}$ there exists $w \in W^{I,I}$ such that $\mathrm{Ext}^1_{D(G)}(V_2^\vee, \mathrm{gr}_w(V_0^\vee)) \neq 0$, which together with (4.78) (applied with $\mathrm{gr}_w(V_0^\vee)$ instead of $V_0^\vee$), (4.85) (applied with $D = \mathrm{gr}_w(V_0^\vee)$) and Corollary 4.5.11 implies

$$\mathrm{Ext}^k_{L_I}(i^\infty_{I,I,w}(J_{I,I,w}(\pi_0^\infty)), \pi_1^\infty)^\infty \otimes_E \mathrm{Ext}^\ell_{U(\mathfrak{g})}(M^I(x), L(x)^w) \neq 0 \tag{5.8}$$

for some $k, \ell \geq 0$ such that $k + \ell = 1$. By (iii) of Lemma 3.3.1 (for $\ell = 0$) and (ii) of Remark 3.3.6 (for $\ell = 1$) we have $\mathrm{Ext}^\ell_{U(\mathfrak{g})}(M^I(x), L(x)^{w'}) = 0$ for $\ell \leq 1$ and $1 \neq w' \in W^{I,I}$. Hence, (5.8) can hold only when $w = 1$. Since π_0^∞ and π_1^∞ have no common Jordan-Hölder factor, we have $\mathrm{Hom}_{L_I}(\pi_0^\infty, \pi_1^\infty) = 0$, so that the only possibly non-zero term in (5.8) is for $k = 1$ and $\ell = 0$. The spectral sequence (4.85) (with Corollary 4.5.11) and (5.7) then yield

$$\begin{aligned} \dim_E \mathrm{Ext}_G^1(V_0, V_1) &\leq \dim_E \mathrm{Ext}_G^1(V_0, V_2) \\ &\leq \dim_E \mathrm{Ext}^1_{L_I}(\pi_0^\infty, \pi_1^\infty)^\infty \dim_E \mathrm{Hom}_{U(\mathfrak{g})}(M^I(x), L(x)). \end{aligned} \tag{5.9}$$

As $\mathrm{Hom}_{U(\mathfrak{g})}(M^I(x), L(x))$ is one dimensional, (5.9) implies the statement.

Step 1 and Step 2 imply that (5.5) is an isomorphism. When π_0^∞ and π_1^∞ are moreover G-basic and (5.5) is non-zero, then $d_I(\pi_0^\infty, \pi_1^\infty) = 1$ (see (iii) of Definition 2.1.4) and Lemma 2.2.4 then implies that (5.5) is one dimensional. □

Lemma 5.1.4 *Let $I \subseteq \Delta$, $L(x)$ in $\mathcal{O}^{\mathfrak{p}_I}_{\mathrm{alg}}$, π^∞ a smooth G-basic representation of L_I and $V \stackrel{\mathrm{def}}{=} \mathcal{F}_{P_I}^G(L(x), \pi^\infty)$. Let V' be a smooth multiplicity free finite length representation of G such that $\mathrm{JH}_G(V') = \mathrm{JH}_G(V)$ as partially ordered sets. Then we have $V \cong V'$.*

Proof Using (ii) of Theorem 4.3.7 we can assume $I = \Delta \setminus D_L(x)$ and by (iii) of Lemma 2.1.15 π^∞ is (finite length) multiplicity free. By Lemma 5.1.2 V is multiplicity free and the functor $\mathcal{F}_{P_I}^G(L(x), -)$ induces a bijection of partially ordered sets $\mathrm{JH}_{L_I}(\pi^\infty) \xrightarrow{\sim} \mathrm{JH}_G(V)$. We prove the statement by induction on the length of π^∞. If π^∞ is irreducible, then V is irreducible by (iii) of Theorem 4.3.7, so $\mathrm{JH}_G(V) = \mathrm{JH}_G(V')$ forces $V \cong V'$ and there is nothing to prove. If π^∞ is reducible, by Corollary 2.1.27 there exist smooth G-basic representations $\pi_0^\infty, \pi_1^\infty$ of L_I and a non-split short exact sequence $0 \to \pi_1^\infty \to \pi^\infty \to \pi_0^\infty \to 0$. Applying $\mathcal{F}_{P_I}^G(L(x), -)$ we obtain a short exact sequence $0 \to V_1 \to V \to V_0 \to 0$ with $V_i \stackrel{\mathrm{def}}{=} \mathcal{F}_{P_I}^G(L(x), \pi_i^\infty)$, which is non-split by Step 1 of the proof of Lemma 5.1.3. Since

$\mathrm{JH}_G(V) = \mathrm{JH}_G(V')$ as partially ordered sets, V' also fits into a non-split short exact sequence $0 \to V_1' \to V' \to V_0' \to 0$ with $\mathrm{JH}_G(V_i) = \mathrm{JH}_G(V_i')$ as partially ordered sets for $i = 0, 1$. As π_i^∞ has strictly smaller length than π^∞, by induction we have $V_i \cong V_i'$ for $i = 0, 1$. As both V and V' fit into non-split short exact sequences $0 \to V_1 \to * \to V_0 \to 0$ and $\dim_E \mathrm{Ext}^1_G(V_0, V_1) = 1$ by Lemma 5.1.3, we deduce $V \cong V'$. □

Lemma 5.1.5 *For $i = 0, 1$ assume that $I_i = \Delta \setminus D_L(x_i)$ and $M_i = L(x_i)$ for some $x_i \in W(G)$. Assume moreover that $x_0 \neq x_1$, that π_0^∞ and π_1^∞ are G-basic and that*

$$\mathrm{Hom}_{L_{I_1}}(i^\infty_{I,I_1}(J_{I_0,I}(\pi_0^\infty)), \pi_1^\infty) \neq 0 \tag{5.10}$$

where $I \overset{\text{def}}{=} I_0 \cap I_1$. Then there exists a unique G-basic representation π^∞ of L_I which is both a subrepresentation of $J'_{I_1,I}(\pi_1^\infty)$ (see (2.6)) and a quotient of $J_{I_0,I}(\pi_0^\infty)$. Moreover, we have a canonical injection

$$\mathrm{Hom}_{L_{I_0}}(\pi_0^\infty, i^\infty_{I,I_0}(\pi^\infty)) \otimes_E \mathrm{Ext}^1_{U(\mathfrak{g})}(M_1, M_0) \otimes_E \mathrm{Hom}_{L_{I_1}}(i^\infty_{I,I_1}(\pi^\infty), \pi_1^\infty) \hookrightarrow \mathrm{Ext}^1_G(V_0, V_1). \tag{5.11}$$

Proof **Step** 1: We construct the desired G-basic π^∞.
By (2.8), (5.10) is equivalent to

$$\mathrm{Hom}_{L_I}(J_{I_0,I}(\pi_0^\infty), J'_{I_1,I}(\pi_1^\infty)) \neq 0. \tag{5.12}$$

For $i = 0, 1$ let $\Sigma_i \overset{\text{def}}{=} W(L_{I_i}) \cdot \mathcal{J}(\pi_i^\infty)$. By (2.12) and (i) of Lemma 2.1.15 $W(L_{I_1}) \cdot \mathcal{J}(J'_{I_1,I}(\pi_1^\infty)) = \Sigma_1$ is a single left $W(L_{I_1})$-coset and Σ_0 is a single left $W(L_{I_0})$-coset, and thus (using the regularity of the characters, see (i) of Lemma 2.1.15)

$$\Sigma \overset{\text{def}}{=} \Sigma_0 \cap W(L_{I_1}) \cdot \mathcal{J}(J'_{I_1,I}(\pi_1^\infty)) = \Sigma_0 \cap \Sigma_1 \tag{5.13}$$

(which is non-empty using (5.12)) is a single left $W(L_I)$-coset. In particular, (5.12) is equivalent to

$$\mathrm{Hom}_{L_I}\left(J_{I_0,I}(\pi_0^\infty)_{\mathcal{B}^I_\Sigma}, J'_{I_1,I}(\pi_1^\infty)_{\mathcal{B}^I_\Sigma}\right) \neq 0$$

with both $J_{I_0,I}(\pi_0^\infty)_{\mathcal{B}^I_\Sigma}$ and $J'_{I_1,I}(\pi_1^\infty)_{\mathcal{B}^I_\Sigma}$ non-zero and thus G-basic by Lemma 2.1.29 and Remark 2.1.30. The desired π^∞ is necessarily the image of a non-zero map $J_{I_0,I}(\pi_0^\infty) \to J'_{I_1,I}(\pi_1^\infty)$, which itself is necessarily the image of the unique (up to scalar) non-zero map $J_{I_0,I}(\pi_0^\infty)_{\mathcal{B}^I_\Sigma} \to J'_{I_1,I}(\pi_1^\infty)_{\mathcal{B}^I_\Sigma}$ (the unicity follows from the fact G-basic representations are multiplicity free with simple socle (and cosocle), see (iv) of Remark 2.1.16). In particular π^∞ is a quotient of $J_{I_0,I}(\pi_0^\infty)_{\mathcal{B}^I_\Sigma}$ with simple socle and cosocle and thus is also G-basic by Corollary 2.1.26. The definition of π^∞ together with (2.7) and (2.8) implies

$$\mathrm{Hom}_{L_{I_0}}(\pi_0^\infty, i^\infty_{I,I_0}(\pi^\infty)) \neq 0 \text{ and } \mathrm{Hom}_{L_{I_1}}(i^\infty_{I,I_1}(\pi^\infty), \pi_1^\infty) \neq 0 \tag{5.14}$$

with both spaces being one dimensional (using (iv) of Remark 2.1.16 as above).

Step 2: We construct the map (5.11).
The exact functor $\mathcal{F}_{P_I}^G(-, \pi^\infty)$ induces a canonical map

$$\mathrm{Ext}^1_{U(\mathfrak{g})}(M_1, M_0) \longrightarrow \mathrm{Ext}^1_G(\mathcal{F}_{P_I}^G(M_0, \pi^\infty), \mathcal{F}_{P_I}^G(M_1, \pi^\infty)). \tag{5.15}$$

From (ii) of Theorem 4.3.7 and Lemma 5.1.1 we have canonical isomorphisms

$$\mathrm{Hom}_{L_{I_0}}(\pi_0^\infty, i_{I,I_0}^\infty(\pi^\infty)) \cong \mathrm{Hom}_G(V_0, \mathcal{F}_{P_I}^G(M_0, \pi^\infty))$$

and

$$\mathrm{Hom}_{L_{I_1}}(i_{I,I_1}^\infty(\pi^\infty), \pi_1^\infty) \cong \mathrm{Hom}_G(\mathcal{F}_{P_I}^G(M_1, \pi^\infty), V_1),$$

which together with (5.15) and obvious functorialities give a canonical map (5.11). In the rest of the proof we assume that the left hand side of (5.11) is non-zero, equivalently $\mathrm{Ext}^1_{U(\mathfrak{g})}(M_1, M_0) \neq 0$ by the end of Step 1 (otherwise (5.11) is obviously an injection).

Step 3: We reduce to the case π_0^∞ is a subobject of $i_{I,I_0}^\infty(\pi^\infty)$ and π_1^∞ a quotient of $i_{I,I_1}^\infty(\pi^\infty)$.

Let $\pi_{0,-}^\infty$ (resp. $\pi_{1,-}^\infty$) be the image of the unique (up to scalar) non-zero map $\pi_0^\infty \to i_{I,I_0}^\infty(\pi^\infty)$ (resp. $i_{I,I_1}^\infty(\pi^\infty) \to \pi_1^\infty$) which is G-basic by Corollary 2.1.26. As π_0^∞ and π_1^∞ are multiplicity free and $x_0 \neq x_1$, by the last statement in Lemma 5.1.1 V_0 and V_1 are multiplicity free with no common Jordan-Hölder factor. In particular

$$\begin{aligned}&\mathrm{Hom}_G\big(V_0, V_1/\mathcal{F}_{P_{I_1}}^G(M_1, \pi_1^\infty/\pi_{1,-}^\infty)\big) = 0\\ &= \mathrm{Hom}_G\big(\mathcal{F}_{P_{I_0}}^G(M_0, \ker(\pi_0^\infty \twoheadrightarrow \pi_{0,-}^\infty)), \mathcal{F}_{P_{I_1}}^G(M_1, \pi_{1,-}^\infty)\big)\end{aligned}$$

hence we have injections (using $V_0 \twoheadrightarrow \mathcal{F}_{P_{I_0}}^G(M_0, \pi_{0,-}^\infty)$ and $\mathcal{F}_{P_{I_1}}^G(M_1, \pi_{1,-}^\infty) \hookrightarrow V_1$)

$$\mathrm{Ext}^1_G(\mathcal{F}_{P_{I_0}}^G(M_0, \pi_{0,-}^\infty), \mathcal{F}_{P_{I_1}}^G(M_1, \pi_{1,-}^\infty)) \hookrightarrow \mathrm{Ext}^1_G(V_0, \mathcal{F}_{P_{I_1}}^G(M_1, \pi_{1,-}^\infty)) \hookrightarrow \mathrm{Ext}^1_G(V_0, V_1).$$

It easily follows that, to prove the injectivity of (5.11), we can assume in the rest of the proof $\pi_0^\infty = \pi_{0,-}^\infty$ and $\pi_1^\infty = \pi_{1,-}^\infty$. We fix an injection $f_0 : \pi_0^\infty \hookrightarrow i_{I,I_0}^\infty(\pi^\infty)$ and a surjection $f_1 : i_{I,I_1}^\infty(\pi^\infty) \twoheadrightarrow \pi_1^\infty$ (unique up to scalar).

Step 4: We reduce to the case π^∞, π_0^∞ and π_1^∞ are all simple.

We choose an arbitrary non-split extension $0 \to M_0 \to M \to M_1 \to 0$, which induces an extension $0 \to R_1 \to R \to R_0 \to 0$ where $R \stackrel{\mathrm{def}}{=} \mathcal{F}_{P_I}^G(M, \pi^\infty)$ and $R_i \stackrel{\mathrm{def}}{=} \mathcal{F}_{P_I}^G(M_i, \pi^\infty) \cong \mathcal{F}_{P_{I_i}}^G(M_i, i_{I,I_i}^\infty(\pi^\infty))$ for $i = 0, 1$. We write V for the extension

$$0 \to V_1 \to V \to V_0 \to 0 \tag{5.16}$$

induced from R by pullback and pushforward along f_0 and f_1. We want to prove that (5.16) is non-split. Let σ^∞ be an arbitrary irreducible (G-regular) constituent of π^∞, then $\mathcal{J}(\sigma^\infty) \subseteq \mathcal{J}(\pi^\infty) \subseteq \mathcal{J}(J_{I_0,I}(\pi_0^\infty)_{\mathcal{B}_\Sigma^I})$ using the first statement in (i) of Lemma 2.1.15 applied to $J_{I_0,I}(\pi_0^\infty)_{\mathcal{B}_\Sigma^I}$. By the last statement in Lemma 2.1.29 the irreducible (G-regular) constituents of $J_{I_0,I}(\pi_0^\infty)_{\mathcal{B}_\Sigma^I}$ are the $J_{I_0,I}(\sigma_0^\infty)_{\mathcal{B}_\Sigma^I}$ for σ_0^∞ an irreducible (G-regular) constituent of π_0^∞. Hence it follows from the last statement in (ii) of Lemma 2.1.15 that there exists a unique irreducible constituent σ_0^∞ of π_0^∞ such that $\sigma^\infty \cong J_{I_0,I}(\sigma_0^\infty)_{\mathcal{B}_\Sigma^I}$. A similar argument using moreover Remark 2.1.30 gives a unique irreducible constituent σ_1^∞ of π_1^∞ such that $\sigma^\infty \cong J'_{I_1,I}(\sigma_1^\infty)_{\mathcal{B}_\Sigma^I}$. By (2.7) and $J_{I_0,I}(\sigma_0^\infty)_{\mathcal{B}_\Sigma^I} \xrightarrow{\sim} \sigma^\infty$ (resp. (2.8) and $\sigma^\infty \xrightarrow{\sim} J'_{I_1,I}(\sigma_1^\infty)_{\mathcal{B}_\Sigma^I}$), we deduce $\sigma_0^\infty \xrightarrow{\sim} \mathrm{soc}_{L_{I_0}}(i_{I,I_0}^\infty(\sigma^\infty))$ (resp. $\mathrm{cosoc}_{L_{I_1}}(i_{I,I_1}^\infty(\sigma^\infty)) \xrightarrow{\sim} \sigma_1^\infty$).

Now, let us assume that $\mathcal{F}_{P_I}^G(M,\sigma^\infty)$ admits a uniserial length 2 subquotient with socle $\mathcal{F}_{P_{I_1}}^G(M_1,\sigma_1^\infty)$ and cosocle $\mathcal{F}_{P_{I_0}}^G(M_0,\sigma_0^\infty)$, then so does $R=\mathcal{F}_{P_I}^G(M,\pi^\infty)$. But such a uniserial length 2 subquotient of R must also be a subquotient of V as R is multiplicity free (using Lemma 5.1.1) and $\mathcal{F}_{P_{I_i}}^G(M_i,\sigma_i^\infty)$ for $i=0,1$ is a constituent of V_i, hence of V. In particular (5.16) is then non-split. Hence to prove the injectivity of (5.11), we see that can we replace π^∞, π_0^∞ and π_1^∞ by σ^∞, σ_0^∞ and σ_1^∞ respectively, and it is enough to prove that $\mathcal{F}_{P_I}^G(M,\pi^\infty)$ admits such a subquotient.

Step 5: We assume π^∞, π_0^∞, π_1^∞ (G-regular) irreducible and prove that $\mathcal{F}_{P_I}^G(M,\pi^\infty)$ has a uniserial length 2 subquotient with socle $\mathcal{F}_{P_{I_1}}^G(M_1,\pi_1^\infty)$ and cosocle $\mathcal{F}_{P_{I_0}}^G(M_0,\pi_0^\infty)$.

Note that we then have $\pi_0^\infty \xrightarrow{\sim} \mathrm{soc}_{L_{I_0}}(i_{I,I_0}^\infty(\pi^\infty))$ and $\mathrm{cosoc}_{L_{I_1}}(i_{I,I_1}^\infty(\pi^\infty)) \xrightarrow{\sim} \pi_1^\infty$. As Lemma 5.1.2 (together with (iv) of Remark 2.1.16) implies that $\mathcal{F}_{P_I}^G(M_i,\pi^\infty)$ has simple socle $\mathcal{F}_{P_{I_i}}^G(M_i,\mathrm{soc}_{L_{I_i}}(i_{I,I_i}^\infty(\pi^\infty)))$ and simple cosocle $\mathcal{F}_{P_{I_i}}^G(M_i,\mathrm{cosoc}_{L_{I_i}}(i_{I,I_i}^\infty(\pi^\infty)))$ for $i=0,1$, we see that the existence of such a subquotient forces $\mathcal{F}_{P_I}^G(M,\pi^\infty)$ to have simple socle $\mathcal{F}_{P_{I_1}}^G(M_1,\mathrm{soc}_{L_{I_1}}(i_{I,I_1}^\infty(\pi^\infty)))$ and simple cosocle $\mathcal{F}_{P_{I_0}}^G(M_0,\mathrm{soc}_{L_{I_0}}(i_{I,I_0}^\infty(\pi^\infty)))$. Also, as $\mathrm{Ext}^1_{U(\mathfrak{g})}(M_1,M_0)\neq 0$ we have either $x_0 \prec x_1$ or $x_1 \prec x_0$ by (ii) of Lemma 3.2.4. We have the following two possibilities.

Case 5.1: Assume $\mathrm{Hom}_{U(\mathfrak{g})}(M^I(x_1),M)=0$. Since $I\subseteq I_1$ $M^{I_1}(x_1)$ is a quotient of $M^I(x_1)$ ([33, Theorem 9.4(c)]) and thus we *a fortiori* have $\mathrm{Hom}_{U(\mathfrak{g})}(M^{I_1}(x_1),M)=0$. Let $V_2 \stackrel{\mathrm{def}}{=} \mathcal{F}_{P_{I_1}}^G(M^{I_1}(x_1),\pi_1^\infty)$, by (4.146) for $k=\ell=0$ and $w=1$ this implies $\mathrm{Hom}_G(R,V_2)=0$. The surjection $M^{I_1}(x_1)\twoheadrightarrow L(x_1)$ induces an injection $V_1 \hookrightarrow V_2$ and thus an injection $\mathrm{Hom}_G(R,V_1)\hookrightarrow \mathrm{Hom}_G(R,V_2)$ which thus implies $\mathrm{Hom}_G(R,V_1)=0$. As $\mathrm{cosoc}_G(R_1)\xrightarrow{\sim} V_1$ by Lemma 5.1.2 and $\mathrm{cosoc}_{L_{I_1}}(i_{I,I_1}^\infty(\pi^\infty))\xrightarrow{\sim}\pi_1^\infty$, there exists an irreducible constituent $V_0'=\mathcal{F}_{P_{I_0}}^G(L(x_0),\tau_0^\infty)$ of R_0 (with τ_0^∞ an irreducible constituent of $i_{I,I_0}^\infty(\pi^\infty)$) such that R admits a subquotient V' with socle V_1 and cosocle V_0'. In particular $\mathrm{Ext}^1_G(V_0',V_1)\neq 0$, which together with (i) of Lemma 4.5.17 implies $d(\tau_0^\infty,\pi_1^\infty)\le 1$. Now by (2.7) we have for $k\ge 0$

$$\mathrm{Ext}^k_G(i^{\infty}_{I_0,\Delta}(\tau_0^{\infty}), i^{\infty}_{I_1,\Delta}(\pi_1^{\infty}))^{\infty} \cong \mathrm{Ext}^k_{L_{I_1}}(J_{\Delta,I_1}(i^{\infty}_{I_0,\Delta}(\tau_0^{\infty})), \pi_1^{\infty})^{\infty}$$

and thus from (i) of Lemma 2.1.18 applied with $I = \Delta$ we deduce

$$\mathrm{Ext}^k_G(i^{\infty}_{I_0,\Delta}(\tau_0^{\infty}), i^{\infty}_{I_1,\Delta}(\pi_1^{\infty}))^{\infty} \cong \bigoplus_{w\in W^{I_0,I_1}} \mathrm{Ext}^k_{L_{I_1}}\big(i^{\infty}_{I_0,I_1,w}(J_{I_0,I_1,w}(\tau_0^{\infty})), \pi_1^{\infty}\big)^{\infty}.$$

But from (2.31) we have $\mathcal{J}(i^{\infty}_{I_0,I_1,w}(J_{I_0,I_1,w}(\pi_0^{\infty}))) \subseteq W(L_{I_1})w^{-1}\Sigma_0$, and from (5.13) and the G-regularity ((i) of Lemma 2.1.15) we have $W(L_{I_1})w^{-1}\Sigma_0 \cap \Sigma_1 = \emptyset$ if $w \neq 1$. It follows that $(i^{\infty}_{I_0,I_1,w}(J_{I_0,I_1,w}(\tau_0^{\infty})))_{\mathcal{B}^{I_1}_{\Sigma_1}} = 0$ if $w \neq 1$ and thus $\mathrm{Ext}^k_G(i^{\infty}_{I_0,\Delta}(\tau_0^{\infty}), i^{\infty}_{I_1,\Delta}(\pi_1^{\infty}))^{\infty} \cong \mathrm{Ext}^k_{L_{I_1}}\big(i^{\infty}_{I,I_1}(J_{I_0,I}(\tau_0^{\infty})), \pi_1^{\infty}\big)^{\infty}$ for $k \geq 0$. Then the last statement of (ii) of Lemma 2.2.11 (applied with $w = 1$, $\sigma_0^{\infty} = \tau_0^{\infty}$ and $\sigma_1^{\infty} = \pi_1^{\infty}$) implies $\tau_0^{\infty} \cong \pi_0^{\infty} \cong \mathrm{soc}_{L_{I_0}}(i^{\infty}_{I,I_0}(\pi^{\infty}))$. In other words, we must have $V_0' \cong V_0$ and V' is the desired V.

Case 5.2: Assume $\mathrm{Hom}_{U(\mathfrak{g})}(M^I(x_1), M) \neq 0$. As $x_0 \neq x_1$ we must have a surjection $M^I(x_1) \twoheadrightarrow M$, which induces an injection $R \hookrightarrow R' \overset{\mathrm{def}}{=} \mathcal{F}^G_{P_I}(M^I(x_1), \pi^{\infty})$ and thus an injection

$$\mathrm{Hom}_G(V_0, R) \hookrightarrow \mathrm{Hom}_G(V_0, R'). \tag{5.17}$$

By (4.146) (for $k = \ell = 0$ and $w = 1$) we have $\mathrm{Hom}_G(V_0, R') = 0$ as $\mathrm{Hom}_{U(\mathfrak{g})}(M^I(x_1), L(x_0)) = 0$, and by (5.17) we deduce $\mathrm{Hom}_G(V_0, R) = 0$. As $V_0 \cong \mathrm{soc}_G(R_0)$ by Lemma 5.1.2, there exists an irreducible constituent $V_1'' = \mathcal{F}^G_{P_{I_1}}(L(x_1), \tau_1^{\infty})$ of R_1 such that R admits a subquotient V'' with socle V_1'' and cosocle V_0. In particular $\mathrm{Ext}^1_G(V_0, V_1'') \neq 0$, which together with (i) of Lemma 4.5.17 implies $d(\pi_0^{\infty}, \tau_1^{\infty}) \leq 1$. By an argument analogous to the one in Case 5.1, (ii) of Lemma 2.2.11 (applied with $w = 1$, $\sigma_0^{\infty} = \pi_0^{\infty}$ and $\sigma_1^{\infty} = \tau_1^{\infty}$) implies $\tau_1^{\infty} \cong \pi_1^{\infty} \cong \mathrm{cosoc}_{L_{I_1}}(i^{\infty}_{I,I_1}(\pi^{\infty}))$. In other words, we must have $V_1'' \cong V_1$ and V'' is the desired V. This finishes the proof. □

Lemma 5.1.6 *For $i = 0, 1$ let $x_i \in W(G)$ such that $x_0 \neq x_1$, $I_i \overset{\mathrm{def}}{=} \Delta \setminus D_L(x_i)$, $I \subseteq I_0 \cap I_1$ and M a uniserial length 2 object in $\mathcal{O}^{\mathfrak{p}_I}_{\mathrm{alg}}$ with socle $L(x_0)$ and cosocle $L(x_1)$. Let Σ be a G-regular finite subset of $\widehat{T}^{\infty}$ preserved under the left action (2.11) of $W(L_I)$, π^{∞} a smooth (finite length) multiplicity free representation of L_I in $\mathcal{B}^I_{\Sigma}$ and $V \overset{\mathrm{def}}{=} \mathcal{F}^G_{P_I}(M, \pi^{\infty})$. Then we have $\mathrm{soc}_G(V) = \mathcal{F}^G_{P_{I_1}}(L(x_1), \mathrm{soc}_{L_{I_1}}(i^{\infty}_{I,I_1}(\pi^{\infty})))$ and $\mathrm{cosoc}_G(V) = \mathcal{F}^G_{P_{I_0}}(L(x_0), \mathrm{cosoc}_{L_{I_0}}(i^{\infty}_{I,I_0}(\pi^{\infty})))$.*

Proof Replacing I by $I_0 \cap I_1$ and π^{∞} by $i^{\infty}_{I,I_0\cap I_1}(\pi^{\infty})$, we can assume $I = I_0 \cap I_1$. Using (2.13) we can assume that Σ is a single (G-regular) $W(L_I)$-coset, i.e. $\Sigma = W(L_I) \cdot \chi$ for some regular χ. Then using Remark 2.1.12 the irreducible constituents of π^{∞} are (some of) the irreducible constituents of $i^{\infty}_{\emptyset,I}(\chi)$, and using (iii) of Lemma 2.1.15 for I_i $(i = 0, 1)$ we deduce that $i^{\infty}_{I,I_i}(\pi^{\infty})$ is still multiplicity free. Hence by Lemma 5.1.2 (and (ii) of Theorem 4.3.7) $V_i \overset{\mathrm{def}}{=} \mathcal{F}^G_{P_I}(L(x_i), \pi^{\infty}) \cong \mathcal{F}^G_{P_{I_i}}(L(x_i), i^{\infty}_{I,I_i}(\pi^{\infty}))$ is multiplicity free with socle $\mathcal{F}^G_{P_{I_i}}(L(x_i), \mathrm{soc}_{L_{I_i}}(i^\infty_{I,I_i}(\pi^\infty)))$ and cosocle $\mathcal{F}^G_{P_{I_i}}(L(x_i), \mathrm{cosoc}_{L_{I_i}}(i^\infty_{I,I_i}(\pi^\infty)))$ for $i = 0, 1$. Fix $\sigma_0^\infty \in \mathrm{JH}_{L_{I_0}}(\mathrm{soc}_{L_{I_0}}(i^\infty_{I,I_0}(\pi^\infty)))$, $\sigma_1^\infty \in \mathrm{JH}_{L_{I_1}}(\mathrm{cosoc}_{L_{I_1}}(i^\infty_{I,I_1}(\pi^\infty)))$ (both irreducible G-regular) and let $W_i \stackrel{\mathrm{def}}{=} \mathcal{F}^G_{P_{I_i}}(L(x_i), \sigma_i^\infty)$ for $i = 0, 1$. It suffices to show that W_0 (resp. W_1) does not show up in the socle (resp. cosocle) of V. By (2.7) and the assumption on π^∞ we have isomorphisms

$$0 \neq \mathrm{Hom}_{L_{I_0}}(\sigma_0^\infty, i^\infty_{I,I_0}(\pi^\infty)) \cong \mathrm{Hom}_{L_I}(J_{I_0,I}(\sigma_0^\infty), \pi^\infty) \cong \mathrm{Hom}_{L_I}(J_{I_0,I}(\sigma_0^\infty)_{\mathcal{B}^I_\Sigma}, \pi^\infty), \tag{5.18}$$

which by the last statement in Lemma 2.1.29 (and our assumption on Σ) implies that $\tau_0^\infty \stackrel{\mathrm{def}}{=} J_{I_0,I}(\sigma_0^\infty)_{\mathcal{B}^I_\Sigma}$ is G-regular irreducible. Another application of (2.7) (with τ_0^∞ instead of π^∞) together with the last statement in (ii) of Lemma 2.1.15 show that $\sigma_0^\infty \xrightarrow{\sim} \mathrm{soc}_{L_{I_0}}(i^\infty_{I,I_0}(\tau_0^\infty))$. Note that any injection $W_0 \hookrightarrow V_0$ determines by (5.1) an injection $\sigma_0^\infty \hookrightarrow i^\infty_{I,I_0}(\pi^\infty)$, then by (5.18) an injection $\tau_0^\infty \hookrightarrow \pi^\infty$, and finally an injection $\widetilde{W}_0 \stackrel{\mathrm{def}}{=} \mathcal{F}^G_{P_I}(M, \tau_0^\infty) \hookrightarrow V$. Moreover $\widetilde{W}_0$ is multiplicity free as V is by Lemma 5.1.1 (and using that the V_i are multiplicity free). It follows from Step 5 of the proof of Lemma 5.1.5 (replacing π^∞ there by τ_0^∞) that $\widetilde{W}_0$ contains a unique length 2 subquotient with socle $\mathcal{F}^G_{P_{I_1}}(L(x_1), \mathrm{cosoc}_{L_{I_1}}(i^\infty_{I,I_1}(\tau_0^\infty)))$ and cosocle W_0, and thus the pullback of $0 \to V_1 \to V \to V_0 \to 0$ along $W_0 \hookrightarrow V_0$ is non-split. In particular, W_0 does not show up in $\mathrm{soc}_G(V)$. Similarly, we can define $\tau_1^\infty \stackrel{\mathrm{def}}{=} J'_{I_1,I}(\sigma_1^\infty)_{\mathcal{B}^I_\Sigma}$ and deduce as above from Remark 2.1.30 and (2.8) (together with (ii) of Lemma 2.1.15 and (2.12)) that τ_1^∞ is G-regular irreducible with $\sigma_1^\infty \cong \mathrm{cosoc}_{L_{I_1}}(i^\infty_{I,I_1}(\tau_1^\infty))$. Parallel to the above argument for W_0, any surjection $V_1 \twoheadrightarrow W_1$ determines a surjection $V \twoheadrightarrow \widetilde{W}_1 \stackrel{\mathrm{def}}{=} \mathcal{F}^G_{P_I}(M, \tau_1^\infty)$ with $\widetilde{W}_1$ admitting a unique length 2 subquotient with socle W_1 and cosocle $\mathcal{F}^G_{P_{I_0}}(L(x_0), \mathrm{soc}_{L_{I_0}}(i^\infty_{I,I_0}(\tau_1^\infty)))$, forcing the pushforward of $0 \to V_1 \to V \to V_0 \to 0$ along $V_1 \twoheadrightarrow W_1$ to be non-split. In particular W_1 does not show up in $\mathrm{cosoc}_G(V)$, which finishes the proof. □

Keeping the notation at the very beginning of this section, we now assume till its end that, for $i = 0, 1$, Σ_i is a single G-regular $W(L_{I_i})$-coset and that the L_{I_i}-representation π_i^∞ (in $\mathcal{B}^{I_i}_{\Sigma_i}$) is multiplicity free. We will add other assumptions, depending on the statements.

Lemma 5.1.7 *Assume that $I_0 = \Delta \setminus D_L(x_0)$ and $M_0 = L(x_0)$ for some $x_0 \in W(G)$.*

(i) If $\mathrm{Hom}_G(V_0, V_1) \neq 0$, then we have $I_1 \subseteq I_0$ and a canonical isomorphism

$$\mathrm{Hom}_G(V_0, V_1) \cong \mathrm{Hom}_{U(\mathfrak{g})}(M_1, M_0) \otimes_E \mathrm{Hom}_{L_{I_0}}(\pi_0^\infty, i^\infty_{I_1,I_0}(\pi_1^\infty)). \tag{5.19}$$

(ii) If $\mathrm{Hom}_G(V_1, V_0) \neq 0$, then we have $I_1 \subseteq I_0$ and a canonical isomorphism

$$\mathrm{Hom}_G(V_1, V_0) \cong \mathrm{Hom}_{U(\mathfrak{g})}(M_0, M_1) \otimes_E \mathrm{Hom}_{L_{I_0}}(i^\infty_{I_1,I_0}(\pi_1^\infty), \pi_0^\infty). \tag{5.20}$$

Proof By Lemma 4.5.18 (and Harish-Chandra's theorem) we can assume that the irreducible constituents of the $U(\mathfrak{g})$-module M_1 are of the form $L(w)$ for some $w \in W(G)$. We only prove (ii) as (i) is symmetric. The assumption $\mathrm{Hom}_G(V_1, V_0) \neq 0$ implies that there is an irreducible constituent $L(x_1)$ of M_1 such that $\mathrm{Hom}_G(\mathcal{F}_{P_{I_1}}^G(L(x_1), \pi_1^\infty), V_0) \neq 0$. By (ii) of Theorem 4.3.7 and Lemma 3.1.1 we can replace I_1 by the larger $\Delta \setminus D_L(x_1)$. Then by the first statement of Lemma 5.1.1 we deduce $x_0 = x_1$, and in particular $I_1 \subseteq I_0$ (note that the smooth induction of π_1^∞ satisfies the condition at the beginning of this section, for instance using (iii) of Lemma 2.1.15 for $I = \Delta \setminus D_L(x_1)$).

Considering each irreducible constituent of $\mathrm{soc}_{U(\mathfrak{g})}(M_1)$ and arguing as above, it follows from (5.1) that we have an isomorphism (where both sides could *a priori* be 0)

$$\begin{aligned}&\mathrm{Hom}_{U(\mathfrak{g})}(M_0, \mathrm{soc}_{U(\mathfrak{g})}(M_1)) \otimes_E \mathrm{Hom}_{L_{I_0}}(i_{I_1,I_0}^\infty(\pi_1^\infty), \pi_0^\infty)\\ &\xrightarrow{\sim} \mathrm{Hom}_G(\mathcal{F}_{P_{I_1}}^G(\mathrm{soc}_{U(\mathfrak{g})}(M_1), \pi_1^\infty), V_0).\end{aligned}$$

Hence, as $\mathrm{Hom}_{U(\mathfrak{g})}(M_0, \mathrm{soc}_{U(\mathfrak{g})}(M_1)) \cong \mathrm{Hom}_{U(\mathfrak{g})}(M_0, M_1)$ (since $M_0 = L(x_0)$), it is enough to show that the map

$$\mathrm{Hom}_G(V_1, V_0) \longrightarrow \mathrm{Hom}_G(\mathcal{F}_{P_{I_1}}^G(M_1/\mathrm{soc}_{U(\mathfrak{g})}(M_1), \pi_1^\infty), V_0) \tag{5.21}$$

induced by the surjection $M_1 \twoheadrightarrow M_1/\mathrm{soc}_{U(\mathfrak{g})}(M_1)$ is zero. Assume on the contrary that (5.21) is non-zero and let M_2 be a quotient of $M_1/\mathrm{soc}_{U(\mathfrak{g})}(M_1)$ of minimal length such that the composition $M_1 \twoheadrightarrow M_1/\mathrm{soc}_{U(\mathfrak{g})}(M_1) \twoheadrightarrow M_2$ induces a non-zero map

$$\mathrm{Hom}_G(V_1, V_0) \longrightarrow \mathrm{Hom}_G(\mathcal{F}_{P_{I_1}}^G(M_2, \pi_1^\infty), V_0).$$

The minimality of M_2 easily implies that $\mathrm{soc}_{U(\mathfrak{g})}(M_2)$ is irreducible, isomorphic to $L(x)$ for some $x \in W(G)$, and that the map $\mathrm{Hom}_G(V_1, V_0) \to \mathrm{Hom}_G(\mathcal{F}_{P_{I_1}}^G(M_2/L(x), \pi_1^\infty), V_0)$ (induced by $M_1 \twoheadrightarrow M_2/L(x)$) is zero. Using the exact sequence

$$\begin{aligned}0 \to \mathrm{Hom}_G\big(\mathcal{F}_{P_{I_1}}^G(\ker(M_1 \to M_2/L(x)), \pi_1^\infty), V_0\big) &\to \mathrm{Hom}_G(V_1, V_0)\\ &\to \mathrm{Hom}_G\big(\mathcal{F}_{P_{I_1}}^G(M_2/L(x), \pi_1^\infty), V_0\big)\end{aligned}$$

we deduce that $\ker(M_1 \to M_2/L(x)) \hookrightarrow M_1$ induces an isomorphism

$$\mathrm{Hom}_G\big(\mathcal{F}_{P_{I_1}}^G(\ker(M_1 \to M_2/L(x)), \pi_1^\infty), V_0\big) \xrightarrow{\sim} \mathrm{Hom}_G(V_1, V_0).$$

Hence we can replace M_1 by $\ker(M_1 \to M_2/L(x))$ and M_2 by $L(x)$. The surjection $M_1 \twoheadrightarrow M_2$ (which is 0 on $\mathrm{soc}_{U(\mathfrak{g})}(M_1)$) then factors through surjections $M_1 \twoheadrightarrow M_1' \twoheadrightarrow M_2$ where M_1' is uniserial of length 2 with cosocle $L(x)$ and socle $L(w)$ for some $w \in W(G)$. Moreover we have $w \neq x$ by (i) of Lemma 3.2.4. Hence $M_1' \twoheadrightarrow M_2$ induces a non-zero map

$$\mathrm{Hom}_G(\mathcal{F}_{P_{I_1}}^G(M_1', \pi_1^\infty), V_0) \longrightarrow \mathrm{Hom}_G(\mathcal{F}_{P_{I_1}}^G(M_2, \pi_1^\infty), V_0). \tag{5.22}$$

The target of (5.22) being non-zero forces $x = x_0$ by Lemma 5.1.1. But by Lemma 5.1.6 (applied with $x_0 = w, x_1 = x, I = I_1, \pi^\infty = \pi_1^\infty$ and $M = M_1'$) $\mathcal{F}_{P_{I_1}}^G(M_1', \pi_1^\infty)$ has cosocle $\mathcal{F}_{P_{I_w}}^G(L(w), \mathrm{cosoc}_{L_{I_w}}(i_{I_1,I_w}^\infty(\pi_1^\infty)))$ where $I_w \stackrel{\mathrm{def}}{=} \Delta \setminus D_L(w) \supseteq I_1$ (by Lemma 3.1.1). Lemma 5.1.1 then forces $\mathrm{Hom}_G(\mathcal{F}_{P_{I_1}}^G(M_1', \pi_1^\infty), V_0) = 0$, a contradiction to (5.22) being non-zero. This finishes the proof. □

Remark 5.1.8 Using Remark 4.5.12, we can check through the proof of Lemma 5.1.1 that the isomorphism (5.1) is functorial in π_0^∞ and π_1^∞. Similarly, the proof of Lemma 5.1.7 shows that both (5.19) and (5.20) are functorial in π_0^∞ and π_1^∞.

Lemma 5.1.9 *Let $I \subseteq \Delta$, Σ a G-regular $W(L_I)$-coset and π^∞ a smooth (finite length) representation of L_I in $\mathcal{B}_\Sigma^I$ such that $i_{I,\Delta}^\infty(\pi^\infty)$ is multiplicity free. Let M, M' in $\mathcal{O}_{\mathrm{alg}}^{\mathfrak{p}_I}$ with all irreducible constituents of the form $L(w)$ for some $w \in W(G)$ and $V \stackrel{def}{=} \mathcal{F}_{P_I}^G(M, \pi^\infty)$, $V' \stackrel{def}{=} \mathcal{F}_{P_I}^G(M', \pi^\infty)$. If V' is a subquotient of V, then there exists a subquotient M'' of M such that $V' \cong \mathcal{F}_{P_I}^G(M'', \pi^\infty)$.*

Proof Note that, although it is possible that $M'' = M'$, we actually do not need that (and we do not prove it below). Replacing M by Q, we can assume that there does not exist a strict subquotient Q of M such that V' is a subquotient of $\mathcal{F}_{P_I}^G(Q, \pi^\infty)$. Let V_-' (resp. V_+') a (closed) subrepresentation (resp. a quotient) of V such that the composition $V_-' \hookrightarrow V \twoheadrightarrow V_+'$ has image V'. It is an easy exercise to check that we can always choose V_-' and V_+' such that the injection $V' \hookrightarrow V_+'$ (resp. the surjection $V_-' \twoheadrightarrow V'$) induces an isomorphism on socles (resp. on cosocles).

Under the first assumption, it is enough to show that $V' = V$, or equivalently $V_-' = V = V_+'$. We consider an arbitrary surjection $q : M \twoheadrightarrow L(x)$, which induces an injection $\mathcal{F}_{P_I}^G(L(x), \pi^\infty) \hookrightarrow V$. The composition

$$\mathcal{F}_{P_I}^G(L(x), \pi^\infty) \hookrightarrow V \twoheadrightarrow V_+' \tag{5.23}$$

must be non-zero, otherwise the surjection $V \twoheadrightarrow V_+'$ factors through $\mathcal{F}_{P_I}^G(\ker(q), \pi^\infty) \twoheadrightarrow V_+'$ which forces V' to be a subquotient of $\mathcal{F}_{P_I}^G(\ker(q), \pi^\infty)$ and thus contradicts our first assumption. Note that $I \subseteq I_x$ by Lemma 3.1.1 (recall $I_x = \Delta \setminus D_L(x)$). As $i_{I_x,\Delta}^\infty(i_{I,I_x}^\infty(\pi^\infty)) \cong i_{I,\Delta}^\infty(\pi^\infty)$ is multiplicity free, so is $i_{I,I_x}^\infty(\pi^\infty)$. By Lemma 5.1.2 (which can be applied since $i_{I,I_x}^\infty(\pi^\infty)$ satisfies the assumption there using the (second statement in) (iii) of Lemma 2.1.15) we deduce that $\mathcal{F}_{P_I}^G(L(x), \pi^\infty) \cong \mathcal{F}_{P_{I_x}}^G(L(x), i_{I,I_x}^\infty(\pi^\infty))$ is multiplicity free with socle $\mathcal{F}_{P_{I_x}}^G(L(x), \sigma_0^\infty)$ where $\sigma_0^\infty \stackrel{\mathrm{def}}{=} \mathrm{soc}_{L_{I_x}}(i_{I,I_x}^\infty(\pi^\infty))$. If the non-zero composition (5.23) is not injective, then $\mathrm{soc}_G(V_+')$ has an irreducible constituent W of the form $\mathcal{F}_{P_{I_x}}^G(L(x), \sigma^\infty)$ for some constituent σ^∞ of $i_{I,I_x}^\infty(\pi^\infty)/\sigma_0^\infty$. Since $\mathrm{soc}_G(V') \xrightarrow{\sim} \mathrm{soc}_G(V_+')$, W is also a constituent of $\mathrm{soc}_G(V')$, in particular $\mathrm{Hom}_G(W, V') \neq 0$. But by (i) of Lemma 5.1.7 (applied with $V_0 = W$

and $V_1 = V'$) $\mathrm{Hom}_G(W, V') \neq 0$ implies $\mathrm{Hom}_{L_{I_x}}(\sigma^\infty, i^\infty_{I,I_x}(\pi^\infty)) \neq 0$, which contradicts $\sigma^\infty \in \mathrm{JH}_{L_{I_x}}(i^\infty_{I,I_x}(\pi^\infty)/\sigma_0^\infty)$ (as $i^\infty_{I,I_x}(\pi^\infty)$ is multiplicity free). Hence, the composition (5.23) must be injective. Since $q : M \twoheadrightarrow L(x)$ was arbitrary, the composition

$$\mathcal{F}^G_{P_I}(\mathrm{cosoc}_{U(\mathfrak{g})}(M), \pi^\infty) \hookrightarrow V \twoheadrightarrow V'_+$$

(induced by $M \twoheadrightarrow \mathrm{cosoc}_{U(\mathfrak{g})}(M)$) has to be injective. Applying Lemma 5.1.6 to $\mathcal{F}^G_{P_I}(N, \pi^\infty)$ where N is any uniserial length 2 subquotient of M (using (i) of Lemma 3.2.4), we obtain (using (iv) of Theorem 4.3.7) that the injection $\mathcal{F}^G_{P_I}(\mathrm{cosoc}_{U(\mathfrak{g})}(M), \pi^\infty) \hookrightarrow V$ contains $\mathrm{soc}_G(V)$. It follows that the composition $\mathrm{soc}_G(V) \hookrightarrow V \twoheadrightarrow V'_+$ is also injective, and thus the surjection $V \twoheadrightarrow V'_+$ is an isomorphism. A symmetric argument (using (ii) of Lemma 5.1.7) shows that the injection $V'_- \hookrightarrow V$ must also be an isomorphism, hence we have $V' = V$. □

Lemma 5.1.10 *For $i = 0, 1$ assume that $I_i = \Delta \setminus D_L(x_i)$, $M_0 = L(x_0)$ and $M_1 = M^{I_1}(x_1)$ for some $x_i \in W(G)$ such that $x_0 \neq x_1$. Then we have a canonical isomorphism*

$$\mathrm{Ext}^1_G(V_0, V_1) \cong \mathrm{Ext}^1_{U(\mathfrak{g})}(M_1, M_0) \otimes_E \mathrm{Hom}_{L_{I_1}}(i^\infty_{I_0\cap I_1, I_1}(J_{I_0, I_0\cap I_1}(\pi_0^\infty)), \pi_1^\infty). \tag{5.24}$$

Proof Assume $\Sigma_1 \cap W(G) \cdot \Sigma_0 = \emptyset$, then $\mathrm{Ext}^1_G(V_0, V_1) = 0$ by (iii) of Corollary 4.5.13 and $\mathrm{Hom}_{L_{I_1}}(i^\infty_{I_0\cap I_1, I_1}(J_{I_0, I_0\cap I_1}(\pi_0^\infty)), \pi_1^\infty) = 0$ by (2.37) for $w = 1$ (using $\mathcal{J}(\pi_0^\infty) \subseteq \Sigma_0$), so (5.24) holds (both sides being 0). Assume $\Sigma_1 \cap W(G) \cdot \Sigma_0 \neq \emptyset$, then by (iv) of Corollary 4.5.13 and as $\mathrm{Hom}_{U(\mathfrak{g})}(M_1, M_0) = 0$ (since $x_0 \neq x_1$), it suffices to show $\mathrm{Ext}^k_{U(\mathfrak{g})}(M_1, M_0^w) = 0$ for $k \leq 1$ and $1 \neq w \in W^{I_0, I_1}$. As $L(x_0)^w$ is not in $\mathcal{O}^{\mathrm{b}}_{\mathrm{alg}}$ for $1 \neq w \in W^{I_0, I_1}$ by (iii) of Lemma 3.3.1, we have $\mathrm{Hom}_{U(\mathfrak{g})}(M_1, M_0^w) = 0$. And we have $\mathrm{Ext}^1_{U(\mathfrak{g})}(M_1, M_0^w) = 0$ for $1 \neq w \in W^{I_0, I_1}$ by (ii) of Remark 3.3.6. □

Lemma 5.1.11 *For $i = 0, 1$ assume that $I_i = \Delta \setminus D_L(x_i)$ and $M_i = L(x_i)$ for some $x_i \in W(G)$ such that $x_0 \neq x_1$. Then we have a canonical injection*

$$\mathrm{Ext}^1_G(V_0, V_1) \hookrightarrow \mathrm{Ext}^1_{U(\mathfrak{g})}(M_1, M_0) \otimes_E \mathrm{Hom}_{L_{I_1}}(i^\infty_{I_0\cap I_1, I_1}(J_{I_0, I_0\cap I_1}(\pi_0^\infty)), \pi_1^\infty). \tag{5.25}$$

In particular $\mathrm{Ext}^1_G(V_0, V_1) \neq 0$ *implies* $\mathrm{Ext}^1_{U(\mathfrak{g})}(M_1, M_0) \neq 0$ *and*

$$\mathrm{Hom}_{L_{I_1}}(i^\infty_{I_0\cap I_1, I_1}(J_{I_0, I_0\cap I_1}(\pi_0^\infty)), \pi_1^\infty) \neq 0. \tag{5.26}$$

Proof It is enough to prove (5.25) when $\mathrm{Ext}^1_G(V_0, V_1) \neq 0$, which we now assume. Let $V_2 \stackrel{\mathrm{def}}{=} \mathcal{F}^G_{P_{I_1}}(M^{I_1}(x_1), \pi_1^\infty)$ and note that there is an injection $V_1 \hookrightarrow V_2$. As $x_0 \neq x_1$, we have $\mathrm{Hom}_{U(\mathfrak{g})}(M^{I_1}(x_1), L(x_0)) = 0$. If $\mathrm{Hom}_G(V_0, V_2) \neq 0$, then $\mathrm{Hom}_{U(\mathfrak{g})}(M^{I_1}(x_1), L(x_0)) \neq 0$ by (i) of Lemma 5.1.7 which is a contradiction, hence $\mathrm{Hom}_G(V_0, V_2) = 0$. This together with the exact sequence $0 \to V_1 \to V_2 \to V_2/V_1 \to 0$ induce an exact sequence

$$0 \to \mathrm{Hom}_G(V_0, V_2/V_1) \to \mathrm{Ext}^1_G(V_0, V_1) \to \mathrm{Ext}^1_G(V_0, V_2). \tag{5.27}$$

Case 1: $\mathrm{Ext}^1_G(V_0, V_2) = 0$.
Then $\mathrm{Hom}_G(V_0, V_2/V_1) \xrightarrow{\sim} \mathrm{Ext}^1_G(V_0, V_1) \neq 0$ by (5.27). Using $V_2/V_1 \cong \mathcal{F}^G_{P_{I_1}}(N^{I_1}(x_1), \pi_1^\infty)$, by (i) of Lemma 5.1.7 we have $I_1 \subseteq I_0$ and an isomorphism

$$\mathrm{Hom}_G(V_0, V_2/V_1) \cong \mathrm{Hom}_{U(\mathfrak{g})}(N^{I_1}(x_1), L(x_0)) \otimes_E \mathrm{Hom}_{L_0}(\pi_0^\infty, i^\infty_{I_1,I_0}(\pi_1^\infty)). \tag{5.28}$$

As $N^{I_1}(x_1)$ is the image of $N(x_1)$ under $M(x_1) \twoheadrightarrow M^{I_1}(x_1)$ and $I_1 \subseteq I_0$, we deduce from [33, Theorem 9.4(c)] and Lemma 3.1.1 that

$$\mathrm{Hom}_{U(\mathfrak{g})}(N^{I_1}(x_1), L(x_0)) \xrightarrow{\sim} \mathrm{Hom}_{U(\mathfrak{g})}(N(x_1), L(x_0)) \neq 0,$$

which by (3.27) and (ii) of Lemma 3.2.2 has dimension $\mu(x_1, x_0)$ (and $x_1 \prec x_0$). As $x_1 < x_0$ and $M^{I_1}(x_1)$ has cosocle $L(x_1)$ we have $\mathrm{Hom}_{U(\mathfrak{g})}(M^{I_1}(x_1), L(x_0)) = 0$ which implies

$$\mathrm{Hom}_{U(\mathfrak{g})}(N^{I_1}(x_1), L(x_0)) \hookrightarrow \mathrm{Ext}^1_{U(\mathfrak{g})}(L(x_1), L(x_0)). \tag{5.29}$$

By (ii) of Lemma 3.2.4 the target of (5.29) also has dimension $\mu(x_1, x_0)$, hence (5.29) is an isomorphism and (5.28) together with $\mathrm{Hom}_G(V_0, V_2/V_1) \xrightarrow{\sim} \mathrm{Ext}^1_G(V_0, V_1)$ give

$$\mathrm{Ext}^1_G(V_0, V_1) \cong \mathrm{Ext}^1_{U(\mathfrak{g})}(L(x_1), L(x_0)) \otimes_E \mathrm{Hom}_{L_0}(\pi_0^\infty, i^\infty_{I_1,I_0}(\pi_1^\infty)). \tag{5.30}$$

The right hand side of (5.30) is exactly the right hand side of (5.25) by (2.7) and $I_0 \cap I_1 = I_1$.

Case 2: $\mathrm{Ext}^1_G(V_0, V_2) \neq 0$.
The isomorphism from Lemma 5.1.10

$$\mathrm{Ext}^1_G(V_0, V_2) \cong \mathrm{Ext}^1_{U(\mathfrak{g})}(M^{I_1}(x_1), L(x_0)) \otimes_E \mathrm{Hom}_{L_{I_1}}(i^\infty_{I_0\cap I_1,I_1}(J_{I_0,I_0\cap I_1}(\pi_0^\infty)), \pi_1^\infty) \tag{5.31}$$

implies $\mathrm{Ext}^1_{U(\mathfrak{g})}(M^{I_1}(x_1), L(x_0)) \neq 0$. Hence the end of Remark 3.2.8 (which can be applied since $\mathrm{Ext}^1_G(V_0, V_1) \neq 0$) implies $\mathrm{Ext}^1_{U(\mathfrak{g})}(L(x_1), L(x_0)) \xrightarrow{\sim} \mathrm{Ext}^1_{U(\mathfrak{g})}(M^{I_1}(x_1), L(x_0))$ and $\mathrm{Hom}_{U(\mathfrak{g})}(N^{I_1}(x_1), L(x_0)) = 0$. In particular (5.31) gives

$$\mathrm{Ext}^1_G(V_0, V_2) \cong \mathrm{Ext}^1_{U(\mathfrak{g})}(L(x_1), L(x_0)) \otimes_E \mathrm{Hom}_{L_{I_1}}(i^\infty_{I_0\cap I_1,I_1}(J_{I_0,I_0\cap I_1}(\pi_0^\infty)), \pi_1^\infty). \tag{5.32}$$

Moreover $\mathrm{Hom}_{U(\mathfrak{g})}(N^{I_1}(x_1), L(x_0)) = 0$ implies $\mathrm{Hom}_G(V_0, V_2/V_1) = 0$ (otherwise use (5.28)) and thus $\mathrm{Ext}^1_G(V_0, V_1) \hookrightarrow \mathrm{Ext}^1_G(V_0, V_2)$ by (5.27). With (5.32) this finishes the proof. □

Remark 5.1.12 As $i^{\infty}_{I_0\cap I_1,I_1}(J_{I_0,I_0\cap I_1}(\pi_0^{\infty}))\cong i^{\infty}_{I_0,I_1,1}(J_{I_0,I_1,1}(\pi_0^{\infty}))$ is a direct summand of $J_{\Delta,I_1}(i^{\infty}_{I_0,\Delta}(\pi_0^{\infty}))$ by (2.36), (5.25) being non-zero forces (using (2.7))

$$0\neq \mathrm{Hom}_{L_{I_1}}(J_{\Delta,I_1}(i^{\infty}_{I_0,\Delta}(\pi_0^{\infty})),\pi_1^{\infty})\cong \mathrm{Hom}_G(i^{\infty}_{I_0,\Delta}(\pi_0^{\infty}),i^{\infty}_{I_1,\Delta}(\pi_1^{\infty})).$$

In particular $\mathrm{Ext}^1_G(V_0,V_1)\neq 0$ and $x_0\neq x_1$ imply $d(\pi_0^{\infty},\pi_1^{\infty})=0$.

Remark 5.1.13 Using Remark 4.5.12, an examination of the proof of Lemma 5.1.10 shows that the isomorphism (5.24) is functorial in π_0^{∞} and π_1^{∞}. Since the proof of Lemma 5.1.11 is based on Lemmas 5.1.10 and 5.1.7, we deduce from Remark 5.1.8 that the injection (5.25) is also functorial in π_0^{∞} and π_1^{∞}.

Proposition 5.1.14 *For $i=0,1$ assume that $I_i=\Delta\setminus D_L(x_i)$ and $M_i=L(x_i)$ for some $x_i\in W(G)$. Assume moreover $x_0\neq x_1$ and that π_0^{∞}, π_1^{∞} are G-basic. Then the injection (5.25) is an isomorphism*

$$\mathrm{Ext}^1_G(V_0,V_1)\xrightarrow{\sim}\mathrm{Ext}^1_{U(\mathfrak{g})}(M_1,M_0)\otimes_E \mathrm{Hom}_{L_{I_1}}(i^{\infty}_{I_0\cap I_1,I_1}(J_{I_0,I_0\cap I_1}(\pi_0^{\infty})),\pi_1^{\infty}). \tag{5.33}$$

Proof It suffices to show that both sides of (5.33) have the same dimension when the right hand side is non-zero. It follows from Lemmas 2.1.29 and 2.1.18 that

$$i^{\infty}_{I_0\cap I_1,I_1}(J_{I_0,I_0\cap I_1}(\pi_0^{\infty}))_{\mathcal{B}^{I_1}_{\Sigma_1}}=i^{\infty}_{I_0,I_1,1}(J_{I_0,I_1,1}(\pi_0^{\infty}))_{\mathcal{B}^{I_1}_{\Sigma_1}}$$

is either zero or G-basic, and by (iv) of Remark 2.1.16 that

$$\mathrm{Hom}_{L_{I_1}}(i^{\infty}_{I_0\cap I_1,I_1}(J_{I_0,I_0\cap I_1}(\pi_0^{\infty})),\pi_1^{\infty})\cong \mathrm{Hom}_{L_{I_1}}(i^{\infty}_{I_0\cap I_1,I_1}(J_{I_0,I_0\cap I_1}(\pi_0^{\infty}))_{\mathcal{B}^{I_1}_{\Sigma_1}},\pi_1^{\infty}) \tag{5.34}$$

is one dimensional if non-zero. Consequently the right hand side of (5.33), if non-zero, has dimension $\dim_E\mathrm{Ext}^1_{U(\mathfrak{g})}(M_1,M_0)$. However, under the assumption that (5.34) is non-zero, it follows from (5.11) and (5.14) that

$$\dim_E\mathrm{Ext}^1_G(V_0,V_1)\geq \dim_E\mathrm{Ext}^1_{U(\mathfrak{g})}(M_1,M_0),$$

which forces (5.33) to be an isomorphism. □

Lemma 5.1.15 *For $i=0,1$ assume that $I_i=\Delta\setminus D_L(x_i)$ and $M_i=L(x_i)$ for some $x_i\in W(G)$. Assume moreover $x_0\neq x_1$ and that π_0^{∞}, π_1^{∞} are G-basic. Let V in $\mathrm{Rep}^{\mathrm{an}}_{\mathrm{adm}}(G)$ which fits into a non-split short exact sequence $0\to V_1\to V\to V_0\to 0$.*

(i) We have $\mathrm{cosoc}_G(V)\cong\mathrm{cosoc}_G(V_0)$ if and only if V has simple cosocle if and only if π_1^{∞} is a quotient of $J_{\Delta,I_1}(i^{\infty}_{I_0,\Delta}(\pi_0^{\infty}))$.
(ii) We have $\mathrm{soc}_G(V)\cong\mathrm{soc}_G(V_1)$ if and only if V has simple socle if and only if π_0^{∞} is a subrepresentation of $J'_{\Delta,I_0}(i^{\infty}_{I_1,\Delta}(\pi_1^{\infty}))$.

Proof Recall that, since Σ_i is a single $W(L_{I_i})$-coset, we have $\Sigma_i = W(L_{I_i}) \cdot \mathcal{J}(\pi_i^\infty)$ for $i = 0, 1$. The assumptions imply that we have (5.26) and by (2.37) (for $w = 1$) we deduce $\Sigma_0 \cap \Sigma_1 \neq \emptyset$. By Lemma 5.1.2 and (the last statement in) (iv) of Remark 2.1.16 V_i has simple socle $\mathcal{F}^G_{P_{I_i}}(M_i, \mathrm{soc}_{L_{I_i}}(\pi_i^\infty))$ and simple cosocle $\mathcal{F}^G_{P_{I_i}}(M_i, \mathrm{cosoc}_{L_{I_i}}(\pi_i^\infty))$ for $i = 0, 1$. Hence V has simple socle (resp. cosocle) if and only if $\mathrm{soc}_G(V) = \mathrm{soc}_G(V_1)$ (resp. $\mathrm{cosoc}_G(V) = \mathrm{cosoc}_G(V_0)$) if and only if the map $\mathrm{Ext}^1_G(V_0, V_1) \to \mathrm{Ext}^1_G(\mathrm{soc}_G(V_0), V_1)$ (resp. $\mathrm{Ext}^1_G(V_0, V_1) \to \mathrm{Ext}^1_G(V_0, \mathrm{cosoc}_G(V_1))$) is non-zero. Since the isomorphism (5.33) is functorial in π_0^∞ and π_1^∞ by Remark 5.1.13, $\mathrm{Ext}^1_G(V_0, V_1) \to \mathrm{Ext}^1_G(\mathrm{soc}_G(V_0), V_1)$ is non-zero if and only if the map

$$\begin{aligned}&\mathrm{Hom}_{L_{I_1}}(i^\infty_{I_0\cap I_1, I_1}(J_{I_0, I_0\cap I_1}(\pi_0^\infty)), \pi_1^\infty)\\ &\qquad\longrightarrow \mathrm{Hom}_{L_{I_1}}(i^\infty_{I_0\cap I_1, I_1}(J_{I_0, I_0\cap I_1}(\mathrm{soc}_{L_{I_0}}(\pi_0^\infty))), \pi_1^\infty) \qquad (5.35)\end{aligned}$$

is non-zero, and $\mathrm{Ext}^1_G(V_0, V_1) \to \mathrm{Ext}^1_G(V_0, \mathrm{cosoc}_G(V_1)))$ is non-zero if and only if the map

$$\begin{aligned}&\mathrm{Hom}_{L_{I_1}}(i^\infty_{I_0\cap I_1, I_1}(J_{I_0, I_0\cap I_1}(\pi_0^\infty)), \pi_1^\infty)\\ &\qquad\longrightarrow \mathrm{Hom}_{L_{I_1}}(i^\infty_{I_0\cap I_1, I_1}(J_{I_0, I_0\cap I_1}(\pi_0^\infty)), \mathrm{cosoc}_{L_{I_1}}(\pi_1^\infty)) \qquad (5.36)\end{aligned}$$

is non-zero. As $\mathrm{cosoc}_{L_{I_1}}(\pi_1^\infty)$ is simple, by (both parts of) Lemma 2.1.18 the map (5.36) is non-zero if and only if π_1^∞ is a quotient of $J_{\Delta, I_1}(i^\infty_{I_0, \Delta}(\pi_0^\infty))_{\mathcal{B}^{I_1}_{\Sigma_1}}$ if and only if π_1^∞ is a quotient of $J_{\Delta, I_1}(i^\infty_{I_0, \Delta}(\pi_0^\infty))$. This proves (i).

By (2.8) followed by (2.7) and by both parts of Lemma 2.1.18 we have functorial isomorphisms in representations $*$ in $\mathrm{Rep}^\infty_{\mathrm{adm}}(L_{I_0})$

$$\begin{aligned}\mathrm{Hom}_{L_{I_0}}(*, J'_{\Delta, I_0}(i^\infty_{I_1, \Delta}(\pi_1^\infty))) &\cong \mathrm{Hom}_{L_{I_1}}(J_{\Delta, I_1}(i^\infty_{I_0, \Delta}(*)), \pi_1^\infty)\\ &\cong \mathrm{Hom}_{L_{I_1}}(i^\infty_{I_0\cap I_1, I_1}(J_{I_0, I_0\cap I_1}(*)), \pi_1^\infty).\end{aligned}$$

Hence (5.35) is the map

$$\mathrm{Hom}_{L_{I_0}}(\pi_0^\infty, J'_{\Delta, I_0}(i^\infty_{I_1, \Delta}(\pi_1^\infty))) \longrightarrow \mathrm{Hom}_{L_{I_0}}(\mathrm{soc}_{L_{I_0}}(\pi_0^\infty), J'_{\Delta, I_0}(i^\infty_{I_1, \Delta}(\pi_1^\infty))). \qquad (5.37)$$

Since $\mathrm{soc}_{L_{I_0}}(\pi_0^\infty)$ is simple, (5.37) is non-zero if and only if π_0^∞ is a subrepresentation of $J'_{\Delta, I_0}(i^\infty_{I_1, \Delta}(\pi_1^\infty))_{\mathcal{B}^{I_1}_{\Sigma_0}}$ if and only if π_0^∞ is a subrepresentation of $J'_{\Delta, I_0}(i^\infty_{I_1, \Delta}(\pi_1^\infty))$. This proves (ii). □

Lemma 5.1.16 *For $i = 0, 1$ assume that $I_i = \Delta \setminus D_L(x_i)$ and $M_i = L(x_i)$ for some $x_i \in W(G)$. Assume moreover that M_0 is not a constituent of $M^{I_1}(x_1)$, that $d(\pi_0^\infty, \pi_1^\infty) \geq 1$ and that $\mathrm{Ext}^2_G(V_0, V_1) \neq 0$. Then we have $\mathrm{Ext}^1_{U(\mathfrak{g})}(M_1, M_0) \neq 0$ and*

$$\mathrm{Ext}^1_{L_{I_1}}(i^\infty_{I_0\cap I_1, I_1}(J_{I_0, I_0\cap I_1}(\pi_0^\infty)), \pi_1^\infty)^\infty \neq 0. \qquad (5.38)$$

Proof Let $V_2 \overset{\text{def}}{=} \mathcal{F}^G_{P_{I_1}}(M^{I_1}(x_1), \pi_1^\infty)$, the surjection $M^{I_1}(x_1) \twoheadrightarrow M_1$ induces an injection $V_1 \hookrightarrow V_2$ with $V_2/V_1 \cong \mathcal{F}^G_{P_{I_1}}(N^{I_1}(x_1), \pi_1^\infty)$. Since M_0 is not a constituent of $M^{I_1}(x_1)$, we have in particular $x_0 \neq x_1'$ for any constituent $L(x_1')$ of $M^{I_1}(x_1)$. We then deduce from $d(\pi_0^\infty, \pi_1^\infty) \geq 1$ and Remark 5.1.12 that $\mathrm{Ext}^1_G(V_0, \mathcal{F}^G_{P_{I_1}}(L(x_1'), \pi_1^\infty)) = 0$ for any constituent $L(x_1')$ of $N^{I_1}(x_1)$ (note that we also implicitly use Lemma 3.1.1 applied to $L(x_1')$ together with (ii) of Theorem 4.3.7). Therefore we have $\mathrm{Ext}^1_G(V_0, V_2/V_1) = 0$ and the exact sequence $0 \to V_1 \to V_2 \to V_2/V_1 \to 0$ induces an embedding

$$0 \neq \mathrm{Ext}^2_G(V_0, V_1) \to \mathrm{Ext}^2_G(V_0, V_2).$$

By a dévissage on $(\mathrm{Fil}_w(V_0^\vee))_{w \in W^{I_0,I_1}}$ there is $w \in W^{I_0,I_1}$ such that $\mathrm{Ext}^2_{D(G)}(V_2^\vee, \mathrm{gr}_w(V_0^\vee)) \neq 0$, which together with (4.78) (applied with $\mathrm{gr}_w(V_0^\vee)$ instead of $V_0^\vee$), (4.85) (applied with $D = \mathrm{gr}_w(V_0^\vee)$) and Corollary 4.5.11 implies

$$\mathrm{Ext}^k_{L_{I_1}}(i^\infty_{I_0,I_1,w}(J_{I_0,I_1,w}(\pi_0^\infty)), \pi_1^\infty)^\infty \otimes_E \mathrm{Ext}^\ell_{U(\mathfrak{g})}(M^{I_1}(x_1), M_0^w) \neq 0 \tag{5.39}$$

for some $k, \ell \geq 0$ such that $k + \ell = 2$. As $\mathrm{Hom}_{U(\mathfrak{g})}(M^{I_1}(x_1), M_0^w) = 0$ using $M_0 \notin \mathrm{JH}_{U(\mathfrak{g})}(M^{I_1}(x_1))$ and (iii) of Lemma 3.3.1, and as $\mathrm{Hom}_{L_{I_1}}(i^\infty_{I_0,I_1,w}(J_{I_0,I_1,w}(\pi_0^\infty)), \pi_1^\infty) = 0$ using $d(\pi_0^\infty, \pi_1^\infty) \geq 1$ and (2.36) for $I = \Delta$ (with (2.7)), we see that (5.39) can hold only when $k = \ell = 1$. As $\mathrm{Ext}^1_{U(\mathfrak{g})}(M^{I_1}(x_1), M_0^w) = 0$ for $1 \neq w \in W^{I_0,I_1}$ by (ii) of Remark 3.3.6, we must have $w = 1$. Then (5.39) implies (5.38) and $\mathrm{Ext}^1_{U(\mathfrak{g})}(M^{I_1}(x_1), M_0) \neq 0$. Finally, it follows from the discussion in the paragraph below (3.41) (applied with w, x being x_1, x_0) that if $\mathrm{Ext}^1_{U(\mathfrak{g})}(M^{I_1}(x_1), M_0) \neq 0$ then $\mathrm{Ext}^1_{U(\mathfrak{g})}(M^{I_1}(x_1), M_0)$ and $\mathrm{Ext}^1_{U(\mathfrak{g})}(M_1, M_0)$ have the same dimension. In particular $\mathrm{Ext}^1_{U(\mathfrak{g})}(M_1, M_0) \neq 0$. □

Lemma 5.1.17 *For $i = 0, 1$ assume that $I_i = \Delta \setminus D_L(x_i)$ and $M_i = L(x_i)$ for some $x_i \in W(G)$. Assume moreover that $\Sigma_0 \cap \Sigma_1 = \emptyset$. Then the following results hold.*

(i) We have $\mathrm{Ext}^1_G(V_0, V_1) = 0$.
(ii) If $d(\pi_0^\infty, \pi_1^\infty) \geq 1$, then we have $\mathrm{Ext}^2_G(V_0, V_1) = 0$.

Proof We let $V_2 \overset{\text{def}}{=} \mathcal{F}^G_{P_{I_1}}(M^{I_1}(x_1), \pi_1^\infty)$, which contains V_1.

We prove (i). Assume $\mathrm{Hom}_G(V_0, V_2/V_1) \neq 0$, then by (i) of Lemma 5.1.7 we have $I_1 \subseteq I_0$ and (using (2.7))

$$\mathrm{Hom}_{L_0}(\pi_0^\infty, i^\infty_{I_1,I_0}(\pi_1^\infty)) \cong \mathrm{Hom}_{L_{I_1}}(i^\infty_{I_0 \cap I_1, I_1}(J_{I_0, I_0 \cap I_1}(\pi_0^\infty)), \pi_1^\infty) \neq 0.$$

But by (2.37) for $w = 1$ this contradicts the assumption $\Sigma_0 \cap \Sigma_1 = \emptyset$. Hence $\mathrm{Hom}_G(V_0, V_2/V_1) = 0$, so the injection $V_1 \hookrightarrow V_2$ induces an injection $\mathrm{Ext}^1_G(V_0, V_1) \hookrightarrow \mathrm{Ext}^1_G(V_0, V_2)$. Assume on the contrary $\mathrm{Ext}^1_G(V_0, V_1) \neq 0$ and thus $\mathrm{Ext}^1_G(V_0, V_2) \neq 0$. By a dévissage on $(\mathrm{Fil}_w(V_0^\vee))_{w \in W^{I_0,I_1}}$ there exists $w \in W^{I_0,I_1}$ such that $\mathrm{Ext}^2_{D(G)}(V_2^\vee, \mathrm{gr}_w(V_0^\vee)) \neq 0$, which as in the previous proof implies

$$\mathrm{Ext}^k_{L_{I_1}}(i^\infty_{I_0,I_1,w}(J_{I_0,I_1,w}(\pi_0^\infty)), \pi_1^\infty)^\infty \otimes_E \mathrm{Ext}^\ell_{U(\mathfrak{g})}(M^{I_1}(x_1), L(x_0)^w) \neq 0 \tag{5.40}$$

for some $k, \ell \geq 0$ such that $k+\ell = 1$. Then (2.37) and $\Sigma_0 \cap \Sigma_1 = \emptyset$ force $w \neq 1$. But then we have $\mathrm{Ext}^{\ell}_{U(\mathfrak{g})}(M^{I_1}(x_1), L(x_0)^w) = 0$ for $\ell \leq 1$ by (iii) of Lemma 3.3.1 ($\ell = 0$) and (ii) of Remark 3.3.6 ($\ell = 1$), which contradicts (5.40).

We prove (ii). By (i), $V_2/V_1 \cong \mathcal{F}^G_{P_{I_1}}(N^{I_1}(x_1), \pi_1^\infty)$ and a dévissage on the constituents of $N^{I_1}(x_1)$, we obtain $\mathrm{Ext}^1_G(V_0, V_2/V_1) = 0$. Assume on the contrary $\mathrm{Ext}^2_G(V_0, V_1) \neq 0$, then we must have $\mathrm{Ext}^2_G(V_0, V_2) \neq 0$, and by the same dévissage as in (i) there exists $w \in W^{I_0,I_1}$ such that (5.40) holds for some $k, \ell \geq 0$ with $k+\ell = 2$. As $\Sigma_0 \cap \Sigma_1 = \emptyset$, we have again $w \neq 1$ and thus $\mathrm{Ext}^{\ell}_{U(\mathfrak{g})}(M^{I_1}(x_1), L(x_0)^w) = 0$ for $\ell \leq 1$. But we also have

$$\mathrm{Hom}_{L_{I_1}}(i^\infty_{I_0,I_1,w}(J_{I_0,I_1,w}(\pi_0^\infty)), \pi_1^\infty) = 0$$

by (2.36) (for $I = \Delta$), (2.7) and the assumption $d(\pi_0^\infty, \pi_1^\infty) \geq 1$. This contradicts (5.40) for all $k, \ell \geq 0$ such that $k+\ell = 2$ and finishes the proof. □

Proposition 5.1.18 *For $i = 0, 1$ assume that $I \stackrel{\text{def}}{=} I_i = \Delta \setminus \{j\}$ and $M_i = L(x_i)$ for some $j \in \Delta$ and some $x_i \in W(G)$ such that $D_L(x_i) = \{j\}$. Assume moreover that $\pi_0^\infty, \pi_1^\infty$ are G-basic, that $\mathrm{Ext}^2_G(V_0, V_1) \neq 0$ and that*

$$i^\infty_{I,I,s_j}(J_{I,I,s_j}(\pi_0^\infty))_{\mathcal{B}^I_{\Sigma_1}} \neq 0. \tag{5.41}$$

Then we have $x_0 = x_1$,

$$\mathrm{Hom}_{L_I}(i^\infty_{I,I,s_j}(J_{I,I,s_j}(\pi_0^\infty)), \pi_1^\infty) \neq 0 \tag{5.42}$$

and

$$\dim_E \mathrm{Ext}^2_G(V_0, V_1) \leq \#S_0 \tag{5.43}$$

where $S_0 = \{x' \mid x' \in W(L_I)x_0,\ \ell(x') = \ell(x_0)+1,\ j \notin D_L(x')\}$.

Proof We have canonical isomorphisms

$$\begin{aligned}\mathrm{Hom}_G(i^\infty_{I,\Delta}(\pi_0^\infty), i^\infty_{I,\Delta}(\pi_1^\infty)) &\cong \mathrm{Hom}_{L_I}(J_{\Delta,I}(i^\infty_{I,\Delta}(\pi_0^\infty)), \pi_1^\infty)\\ &\cong \mathrm{Hom}_{L_I}(i^\infty_{I,I,s_j}(J_{I,I,s_j}(\pi_0^\infty)), \pi_1^\infty) \cong \mathrm{Hom}_{L_I}(i^\infty_{I,I,s_j}(J_{I,I,s_j}(\pi_0^\infty))_{\mathcal{B}^I_{\Sigma_1}}, \pi_1^\infty)\end{aligned}$$

where the first isomorphism is (2.7) and the other two follow from both parts of Lemma 2.1.18 together with (5.41) remembering that $\Sigma_1 = W(L_I)\cdot\mathcal{J}(\pi_1^\infty)$ is a G-regular left $W(L_I)$-coset. In particular, assuming (5.41), (5.42) is equivalent to $d(\pi_0^\infty, \pi_1^\infty) = 0$.

Let $V_2 \stackrel{\text{def}}{=} \mathcal{F}^G_{P_I}(M^I(x_1), \pi_1^\infty)$ and recall $M^I(x_1) \twoheadrightarrow L(x_1)$ induces $V_1 \hookrightarrow V_2$. The exact sequence $0 \to V_1 \to V_2 \to V_2/V_1 \to 0$ induces an exact sequence

$$\mathrm{Ext}^1_G(V_0, V_2/V_1) \to \mathrm{Ext}^2_G(V_0, V_1) \to \mathrm{Ext}^2_G(V_0, V_2). \tag{5.44}$$

Step 1: We prove $\mathrm{Ext}^1_G(V_0, V_2/V_1) = 0$.
We have $V_2/V_1 \cong \mathcal{F}^G_{P_I}(N^I(x_1), \pi_1^\infty)$ which admits a filtration with graded pieces of the form $\mathcal{F}^G_{P_I}(L(x'), \pi_1^\infty)$ for some $x' > x_1$ with $D_L(x') \subseteq D_L(x_1)$ (by Lemma 3.1.1) and hence $D_L(x') = \{j\}$. We deduce from (5.41) and Lemma 2.1.18 that

$$\Sigma_1 = W(L_I) \cdot \mathcal{J}(\pi_1^\infty) \subseteq W(L_I)s_j W(L_I) \cdot \mathcal{J}(\pi_0^\infty), \tag{5.45}$$

which forces

$$\Sigma_1 \cap W(L_I) \cdot \mathcal{J}(\pi_0^\infty) = \Sigma_1 \cap \Sigma_0 = \emptyset. \tag{5.46}$$

Then (5.46) together with (i) of Lemma 5.1.17 applied to each $\mathcal{F}^G_{P_I}(L(x'), \pi_1^\infty)$ and a dévissage imply the statement.

Step 2: We prove the proposition.
Note first that Step 1, (5.44) and the assumption $\mathrm{Ext}^2_G(V_0, V_1) \neq 0$ imply $\mathrm{Ext}^2_G(V_0, V_2) \neq 0$. By (4.146) and (5.45) we have a spectral sequence

$$\mathrm{Ext}^\ell_{U(\mathfrak{g})}(M^I(x_1), L(x_0)^{s_j}) \otimes_E \mathrm{Ext}^k_{L_I}(i^\infty_{I,I,s_j}(J_{I,I,s_j}(\pi_0^\infty)), \pi_1^\infty)^\infty \implies \mathrm{Ext}^{k+\ell}_G(V_0, V_2).$$

Since $\mathrm{Ext}^\ell_{U(\mathfrak{g})}(M^I(x_1), L(x_0)^{s_j}) = 0$ for $\ell \leq 1$ by (iii) of Lemma 3.3.1 ($\ell = 0$) and (ii) of Remark 3.3.6 ($\ell = 1$), we deduce

$$0 \neq \mathrm{Ext}^2_G(V_0, V_2) \cong \mathrm{Hom}_{L_I}(i^\infty_{I,I,s_j}(J_{I,I,s_j}(\pi_0^\infty)), \pi_1^\infty) \otimes_E \mathrm{Ext}^2_{U(\mathfrak{g})}(M^I(x_1), L(x_0)^{s_j}). \tag{5.47}$$

From (5.47) we have (5.42), which is one dimensional by (i) of Lemma 2.2.11, and thus $\dim_E \mathrm{Ext}^2_G(V_0, V_2) = \dim_E \mathrm{Ext}^2_{U(\mathfrak{g})}(M^I(x_1), L(x_0)^{s_j}) \neq 0$, which by Proposition 3.3.9 gives $x_0 = x_1$ and

$$\dim_E \mathrm{Ext}^2_G(V_0, V_2) = \# S_0. \tag{5.48}$$

Finally, we deduce (5.43) from (5.44), Step 1 and (5.48). □

Let $I \subseteq \Delta$, M multiplicity free in $\mathcal{O}^{\mathfrak{p}_I}_{\mathrm{alg}}$ and recall the set $\mathrm{JH}_{U(\mathfrak{g})}(M)$ is equipped with a partial order (see Sect. 1.4). For π^∞ G-basic in $\mathrm{Rep}^\infty_{\mathrm{adm}}(L_I)$ it follows that $\mathcal{F}^G_{P_I}(M, \pi^\infty)$ is also multiplicity free using Lemma 5.1.1 and (iv) of Remark 2.1.16 (and Theorem 4.3.7). As M is multiplicity free, for each $L(x) \in \mathrm{JH}_{U(\mathfrak{g})}(M)$ there is a unique subobject of M with cosocle $L(x)$. This defines an increasing filtration on M indexed by $\mathrm{JH}_{U(\mathfrak{g})}(M)$, where "increasing" means that the inclusions respect the partial order on $\mathrm{JH}_{U(\mathfrak{g})}(M)$. By the exactness of the contravariant functor $\mathcal{F}^G_{P_I}(-, \pi^\infty)$ this in turn defines a decreasing filtration on $\mathcal{F}^G_{P_I}(M, \pi^\infty)$ indexed by $\mathrm{JH}_{U(\mathfrak{g})}(M)$. We call it the $\mathrm{JH}_{U(\mathfrak{g})}(M)$-filtration on $\mathcal{F}^G_{P_I}(M, \pi^\infty)$.

Lemma 5.1.19 *With the above assumptions assume that for each $L(x) \in \mathrm{JH}_{U(\mathfrak{g})}(M)$ there exists a G-basic subquotient σ_x^∞ of $i^\infty_{I,I_x}(\pi^\infty)$ such that $d(\sigma_x^\infty, \sigma_{x'}^\infty) = 0$ when $L(x') \leq L(x)$ in $\mathrm{JH}_{U(\mathfrak{g})}(M)$. Then $\mathcal{F}^G_{P_I}(M, \pi^\infty)$ contains a unique subquotient V such that*

(i) *the* $\mathrm{JH}_{U(\mathfrak{g})}(M)$*-filtration on* $\mathcal{F}_{P_I}^G(M, \pi^\infty)$ *induces a decreasing filtration on* V *indexed by* $\mathrm{JH}_{U(\mathfrak{g})}(M)$ *with* $L(x)$*-graded piece* $V_x \stackrel{def}{=} \mathcal{F}_{P_{I_x}}^G(L(x), \sigma_x^\infty)$ *for* $L(x) \in \mathrm{JH}_{U(\mathfrak{g})}(M)$;

(ii) *for each uniserial length 2 subquotient of* M *with socle* $L(x')$ *and cosocle* $L(x)$, V *contains a (unique) subquotient* $V_{x,x'}$ *which fits into a non-split extension* $0 \to V_x \to V_{x,x'} \to V_{x'} \to 0$;

(iii) *if* σ_x^∞ *is furthermore a subrepresentation (resp. a quotient) of* $i_{I,I_x}^\infty(\pi^\infty)$ *for each* $L(x) \in \mathrm{JH}_{U(\mathfrak{g})}(M)$, *then* V *is a subrepresentation (resp. a quotient) of* $\mathcal{F}_{P_I}^G(M, \pi^\infty)$.

Proof Recall that $i_{I,I_x}^\infty(\pi^\infty)$ and σ_x^∞ are G-basic and thus multiplicity free with simple socle and cosocle ((iv) of Remark 2.1.16). For $L(x) \in \mathrm{JH}_{U(\mathfrak{g})}(M)$, let $\sigma_{x,+}^\infty$ (resp. $\sigma_{x,-}^\infty$) be the unique quotient (resp. subrepresentation) of $i_{I,I_x}^\infty(\pi^\infty)$ with socle $\mathrm{soc}_{L_{I_x}}(\sigma_x^\infty)$ (resp. cosocle $\mathrm{cosoc}_{L_{I_x}}(\sigma_x^\infty)$), which is multiplicity free with simple socle and cosocle. By Corollary 2.1.26 both $\sigma_{x,+}^\infty$ and $\sigma_{x,-}^\infty$ are G-basic. In fact, $\sigma_{x,+}^\infty$ (resp. $\sigma_{x,-}^\infty$) is the unique quotient (resp. subrepresentation) of minimal length of $i_{I,I_x}^\infty(\pi^\infty)$ that admits σ_x^∞ as a subrepresentation (resp. a quotient).

We first prove that (i) implies (iii). It follows from Lemma 5.1.2 and (i) of Lemma 5.1.7 that $\mathcal{F}_{P_I}^G(M, \pi^\infty)$ has socle

$$\bigoplus_{L(x)\in \mathrm{JH}_{U(\mathfrak{g})}(\mathrm{cosoc}_{U(\mathfrak{g})}(M))} \mathcal{F}_{P_{I_x}}^G(L(x), \mathrm{soc}_{L_{I_x}}(i_{I,I_x}^\infty(\pi^\infty))), \tag{5.49}$$

and from Lemma 5.1.2 and (ii) of Lemma 5.1.7 that $\mathcal{F}_{P_I}^G(M, \pi^\infty)$ has cosocle

$$\bigoplus_{L(x)\in \mathrm{JH}_{U(\mathfrak{g})}(\mathrm{soc}_{U(\mathfrak{g})}(M))} \mathcal{F}_{P_{I_x}}^G(L(x), \mathrm{cosoc}_{L_{I_x}}(i_{I,I_x}^\infty(\pi^\infty))). \tag{5.50}$$

If σ_x^∞ is a subrepresentation of $i_{I,I_x}^\infty(\pi^\infty)$ for each $L(x) \in \mathrm{JH}_{U(\mathfrak{g})}(M)$, we have $\mathrm{soc}_{L_{I_x}}(\sigma_x^\infty) = \mathrm{soc}_{L_{I_x}}(i_{I,I_x}^\infty(\pi^\infty))$ for each $L(x) \in \mathrm{JH}_{U(\mathfrak{g})}(M)$ as $i_{I,I_x}^\infty(\pi^\infty)$ is G-basic with simple socle by (iv) of Remark 2.1.16. Hence by Lemma 5.1.2 and (i) we have

$$\mathrm{soc}_G(V_x) \cong \mathcal{F}_{P_{I_x}}^G(L(x), \mathrm{soc}_{L_{I_x}}(\sigma_x^\infty)) = \mathcal{F}_{P_{I_x}}^G(L(x), \mathrm{soc}_{L_{I_x}}(i_{I,I_x}^\infty(\pi^\infty))),$$

which implies by (5.49) and (i) of Lemma 5.1.7

$$\mathrm{soc}_G(V) \supseteq \bigoplus_{L(x)\in \mathrm{JH}_{U(\mathfrak{g})}(\mathrm{cosoc}_{U(\mathfrak{g})}(M))} \mathrm{soc}_G(V_x) = \mathrm{soc}_G(\mathcal{F}_{P_I}^G(M, \pi^\infty)).$$

Since $\mathcal{F}_{P_I}^G(M, \pi^\infty)$ is multiplicity free, this forces V to be a subrepresentation of $\mathcal{F}_{P_I}^G(M, \pi^\infty)$. If σ_x^∞ is a quotient of $i_{I,I_x}^\infty(\pi^\infty)$ for each $L(x) \in \mathrm{JH}_{U(\mathfrak{g})}(M)$, a symmetric argument using (5.50) shows $\mathrm{cosoc}_G(V) \supseteq \mathrm{cosoc}_G(\mathcal{F}_{P_I}^G(M, \pi^\infty))$, which forces V to be a quotient of $\mathcal{F}_{P_I}^G(M, \pi^\infty)$.

We prove (i), (ii) and (iii) by increasing induction on the length $\ell(M) \geq 1$. If $\ell(M) = 1$ with $\mathrm{JH}_{U(\mathfrak{g})}(M) = \{L(x)\}$, then $V = V_x$, which is a subquotient of $\mathcal{F}_{P_I}^G(L(x), \pi^\infty) \cong \mathcal{F}_{P_{I_x}}^G(L(x), i_{I,I_x}^\infty(\pi^\infty))$. Assume from now on $\ell(M) \geq 2$. If $\ell(\mathrm{cosoc}_{U(\mathfrak{g})}(M)) \geq 2$, then there exists $M', M'' \subsetneq M$ such that $M = M' + M''$. In this case, by Theorem 4.3.7 we have

$$\mathcal{F}_{P_I}^G(M, \pi^\infty) \cong \mathcal{F}_{P_I}^G(M', \pi^\infty) \times_{\mathcal{F}_{P_I}^G(M' \cap M'', \pi^\infty)} \mathcal{F}_{P_I}^G(M'', \pi^\infty).$$

By our induction hypothesis there is a subquotient V' (resp. V'') of $\mathcal{F}_{P_I}^G(M', \pi^\infty)$ (resp. of $\mathcal{F}_{P_I}^G(M'', \pi^\infty)$) which satisfies (i) and (ii) (with M replaced by M', M'' respectively). Moreover, still by induction V' and V'' admit a common quotient V''' which is a subquotient of $\mathcal{F}_{P_I}^G(M' \cap M'', \pi^\infty)$ and satisfies (i) and (ii) (with M replaced by $M' \cap M''$). Then we can take $V \stackrel{\mathrm{def}}{=} V' \times_{V'''} V''$.

We assume from now on $\mathrm{cosoc}_{U(\mathfrak{g})}(M) \cong L(x)$ for some $x \in W(G)$ and we write $M' \stackrel{\mathrm{def}}{=} \mathrm{Rad}^1(M) \subsetneq M$. By induction hypothesis there is a subquotient V' of $\mathcal{F}_{P_I}^G(M', \pi^\infty)$ that satisfies (i) and (ii) (with M replaced by M'). By Lemma 5.1.2 (and the definition of $\sigma_{x',-}^\infty$) for $L(x') \in \mathrm{JH}_{U(\mathfrak{g})}(M')$, $\mathcal{F}_{P_{I_{x'}}}^G(L(x'), \sigma_{x',-}^\infty)$ is the unique minimal length subrepresentation of $\mathcal{F}_{P_I}^G(L(x'), \pi^\infty) \cong \mathcal{F}_{P_{I_{x'}}}^G(L(x'), i_{I,I_{x'}}^\infty(\pi^\infty))$ with $V_{x'} = \mathcal{F}_{P_{I_{x'}}}^G(L(x'), \sigma_{x'}^\infty)$ as quotient. As $\sigma_{x',-}^\infty$ is a subrepresentation of $i_{I,I_{x'}}^\infty(\pi^\infty)$ which admits $\sigma_{x'}^\infty$ as quotient, $i_{I_{x'},\Delta}^\infty(\sigma_{x',-}^\infty)$ is a subrepresentation of $i_{I_{x'},\Delta}^\infty(i_{I,I_{x'}}^\infty(\pi^\infty)) \cong i_{I,\Delta}^\infty(\pi^\infty)$ which admits $i_{I_{x'},\Delta}^\infty(\sigma_{x'}^\infty)$ as quotient. Let $L(x'), L(x'') \in \mathrm{JH}_{U(\mathfrak{g})}(M')$ such that $L(x') \leq L(x'')$. The assumption $d(\sigma_{x'}^\infty, \sigma_{x''}^\infty) = 0$, i.e. $\mathrm{Hom}_G(i_{I_{x'},\Delta}^\infty(\sigma_{x'}^\infty), i_{I_{x''},\Delta}^\infty(\sigma_{x''}^\infty)) \neq 0$, implies (with (iv) of Remark 2.1.16)

$$\mathrm{cosoc}_G(i_{I_{x'},\Delta}^\infty(\sigma_{x',-}^\infty)) = \mathrm{cosoc}_G(i_{I_{x'},\Delta}^\infty(\sigma_{x'}^\infty)) \in \mathrm{JH}_G(i_{I_{x''},\Delta}^\infty(\sigma_{x''}^\infty)) \subseteq \mathrm{JH}_G(i_{I_{x''},\Delta}^\infty(\sigma_{x'',-}^\infty)),$$

which forces $i_{I_{x'},\Delta}^\infty(\sigma_{x',-}^\infty) \subseteq i_{I_{x''},\Delta}^\infty(\sigma_{x'',-}^\infty) \subseteq i_{I,\Delta}^\infty(\pi^\infty)$ (using that all these representations are G-basic hence multiplicity free) and thus $d(\sigma_{x',-}^\infty, \sigma_{x'',-}^\infty) = 0$. As $\ell(M') < \ell(M)$, by induction there is subquotient V'_- of $\mathcal{F}_{P_I}^G(M', \pi^\infty)$ with a decreasing filtration indexed by $\mathrm{JH}_{U(\mathfrak{g})}(M')$ such that its $L(x')$-graded piece is $\mathcal{F}_{P_{I_{x'}}}^G(L(x'), \sigma_{x',-}^\infty)$ for each $L(x') \in \mathrm{JH}_{U(\mathfrak{g})}(M')$. By Lemma 5.1.2 again (and the definition of $\sigma_{x',-}^\infty$), for $L(x') \in \mathrm{JH}_{U(\mathfrak{g})}(M')$, $\mathcal{F}_{P_{I_{x'}}}^G(L(x'), \sigma_{x',-}^\infty)$ is the unique minimal length subrepresentation of $\mathcal{F}_{P_I}^G(L(x'), \pi^\infty) \cong \mathcal{F}_{P_{I_{x'}}}^G(L(x'), i_{I,I_{x'}}^\infty(\pi^\infty))$ with $\mathcal{F}_{P_{I_{x'}}}^G(L(x'), \sigma_{x'}^\infty)$ as quotient. Hence by (iii) applied to V'_- (which holds by the induction hypothesis and the beginning of the proof) V'_- is the minimal length subrepresentation of $\mathcal{F}_{P_I}^G(M', \pi^\infty)$ with V' as quotient. As $\mathcal{F}_{P_I}^G(M, \pi^\infty)$ fits into an exact sequence

$$0 \to \mathcal{F}_{P_I}^G(L(x), \pi^\infty) \to \mathcal{F}_{P_I}^G(M, \pi^\infty) \to \mathcal{F}_{P_I}^G(M', \pi^\infty) \to 0,$$

the subrepresentation V'_- of $\mathcal{F}_{P_I}^G(M', \pi^\infty)$ and the quotient $\mathcal{F}_{P_{I_x}}^G(L(x), \sigma_{x,+}^\infty)$ of $\mathcal{F}_{P_I}^G(L(x), \pi^\infty)$ uniquely determine a subquotient W of $\mathcal{F}_{P_I}^G(M, \pi^\infty)$ that fits into

$$0 \to \mathcal{F}_{P_{I_x}}^G(L(x), \sigma_{x,+}^\infty) \to W \to V'_- \to 0. \tag{5.51}$$

We construct the desired V as a subquotient of W through the following steps.

Step 1: We prove that, for each length 2 quotient Q of M (with cosocle $L(x)$ and socle some $L(x')$), the exact sequence

$$0 \to \mathcal{F}_{P_{I_x}}^G(L(x), \sigma_{x,+}^\infty) \to R \to \mathcal{F}_{P_{I_{x'}}}^G(L(x'), \sigma_{x',-}^\infty) \to 0 \tag{5.52}$$

induced from (5.51) is non-split.

As σ_x^∞ is a subrepresentation of $\sigma_{x,+}^\infty$ and $\sigma_{x'}^\infty$ is a quotient of $\sigma_{x',-}^\infty$, we deduce from the assumption $d(\sigma_{x'}^\infty, \sigma_x^\infty) = 0$ that

$$\mathrm{Hom}_G(i_{I_{x'}}^\infty(\sigma_{x',-}^\infty), i_{I_x}^\infty(\sigma_{x,+}^\infty)) \neq 0. \tag{5.53}$$

By Lemma 5.1.6 $\mathcal{F}_{P_I}^G(Q, \pi^\infty)$ has simple socle $\mathcal{F}_{P_{I_x}}^G(L(x), \mathrm{soc}_{L_{I_x}}(i_{I,I_x}^\infty(\pi^\infty)))$ and simple cosocle $\mathcal{F}_{P_{I_{x'}}}^G(L(x'), \mathrm{cosoc}_{L_{I_{x'}}}(i_{I,I_{x'}}^\infty(\pi^\infty)))$. In particular no constituent of $\mathcal{F}_{P_{I_{x'}}}^G(L(x'), \sigma_{x',-}^\infty)$ shows up in the socle of $\mathcal{F}_{P_I}^G(Q, \pi^\infty)$. Hence, if (5.52) splits, an easy diagram chase shows that there must exist a constituent $\mathcal{F}_{P_{I_x}}^G(L(x), \tau^\infty)$ of $\mathcal{F}_{P_I}^G(L(x), \pi^\infty)$ not in $\mathcal{F}_{P_{I_x}}^G(L(x), \sigma_{x,+}^\infty)$ such that $\mathrm{Ext}_G^1(\mathcal{F}_{P_{I_{x'}}}^G(L(x'), \sigma_{x',-}^\infty), \mathcal{F}_{P_{I_x}}^G(L(x), \tau^\infty)) \neq 0$. By Remark 5.1.12 this implies

$$\mathrm{Hom}_G(i_{I_{x'},\Delta}^\infty(\sigma_{x',-}^\infty), i_{I_x,\Delta}^\infty(\tau^\infty)) \neq 0. \tag{5.54}$$

Note that $i_{I_x,\Delta}^\infty(\sigma_{x,+}^\infty)$ and $i_{I_x,\Delta}^\infty(\tau^\infty)$ have no common constituent since $i_{I_x,\Delta}^\infty(i_{I,I_x}^\infty(\pi^\infty)) \cong i_{I,\Delta}^\infty(\pi^\infty)$ is G-basic, thus multiplicity free, and $\tau^\infty \in \mathrm{JH}_{L_{I_x}}(i_{I,I_x}^\infty(\pi^\infty)) \setminus \mathrm{JH}_{L_{I_x}}(\sigma_{x,+}^\infty)$. But (5.53) and (5.54) force both $i_{I_x}^\infty(\sigma_{x,+}^\infty)$ and $i_{I_x}^\infty(\tau^\infty)$ to have the (simple) cosocle of $i_{I_{x'}}^\infty(\sigma_{x',-}^\infty)$ as a Jordan-Hölder factor, which is a contradiction and shows that (5.52) is non-split.

Step 2: We construct V as a subquotient of W.
Since $\sigma_{x,+}^\infty$ is a quotient of $i_{I,I_x}^\infty(\pi^\infty)$ which contains σ_x^∞ as a subrepresentation, $i_{I_x,\Delta}^\infty(\sigma_{x,+}^\infty)$ is a quotient of $i_{I,\Delta}^\infty(\pi^\infty) \cong i_{I_x,\Delta}^\infty(i_{I,I_x}^\infty(\pi^\infty))$ which contains $i_{I_x,\Delta}^\infty(\sigma_x^\infty)$ as a subrepresentation. In fact it is the unique quotient of $i_{I,\Delta}^\infty(\pi^\infty)$ with the same (simple) socle as $i_{I_x,\Delta}^\infty(\sigma_x^\infty)$. Likewise, for $L(y) \in \mathrm{JH}_{U(\mathfrak{g})}(M')$, $i_{I_y,\Delta}^\infty(\sigma_{y,-}^\infty)$ is the unique subrepresentation of $i_{I,\Delta}^\infty(\pi^\infty)$ with the same (simple) cosocle as $i_{I_y,\Delta}^\infty(\sigma_y^\infty)$ (recall all these objects are G-basic and thus multiplicity free with simple socle and cosocle by (iv) of Remark 2.1.16). It follows from $d(\sigma_y^\infty, \sigma_x^\infty) = 0$ that the injection $\sigma_x^\infty \hookrightarrow \sigma_{x,+}^\infty$ and the surjection $\sigma_{y,-}^\infty \twoheadrightarrow \sigma_y^\infty$ induce an isomorphism of 1-dimensional E-vector spaces

$$0 \neq \mathrm{Hom}_G(i_{I_y,\Delta}^\infty(\sigma_y^\infty), i_{I_x,\Delta}^\infty(\sigma_x^\infty)) \xrightarrow{\sim} \mathrm{Hom}_G(i_{I_y,\Delta}^\infty(\sigma_{y,-}^\infty), i_{I_x,\Delta}^\infty(\sigma_{x,+}^\infty)).$$

In particular, the unique (up to scalar) non-zero map $i^{\infty}_{I_y,\Delta}(\sigma^{\infty}_{y,-}) \to i^{\infty}_{I_x,\Delta}(\sigma^{\infty}_{x,+})$ factors through $i^{\infty}_{I_y,\Delta}(\sigma^{\infty}_{y}) \to i^{\infty}_{I_x,\Delta}(\sigma^{\infty}_{x})$ and we have

$$\mathrm{JH}_G(i^{\infty}_{I_y,\Delta}(\sigma^{\infty}_{y,-})) \cap \mathrm{JH}_G(i^{\infty}_{I_x,\Delta}(\sigma^{\infty}_{x,+})) = \mathrm{JH}_G(i^{\infty}_{I_y,\Delta}(\sigma^{\infty}_{y})) \cap \mathrm{JH}_G(i^{\infty}_{I_x,\Delta}(\sigma^{\infty}_{x})) \tag{5.55}$$

(which is the set of constituents of $i^{\infty}_{I,\Delta}(\pi^{\infty})$ "between" the socle of $i^{\infty}_{I_x,\Delta}(\sigma^{\infty}_{x})$ and the cosocle of $i^{\infty}_{I_y,\Delta}(\sigma^{\infty}_{y})$ for the partial order on $\mathrm{JH}_G(i^{\infty}_{I,\Delta}(\pi^{\infty}))$). For $\tau^{\infty}_x \in \mathrm{JH}_{L_{I_x}}(\sigma^{\infty}_{x,+}) \setminus \mathrm{JH}_{L_{I_x}}(\sigma^{\infty}_{x})$ and $\tau^{\infty}_y \in \mathrm{JH}_{L_{I_y,\Delta}}(\sigma^{\infty}_{y,-}) \setminus \mathrm{JH}_{L_{I_y,\Delta}}(\sigma^{\infty}_{y})$ we also have using (5.55)

$$\mathrm{Hom}_G(i^{\infty}_{I_y,\Delta}(\tau^{\infty}_y), i^{\infty}_{I_x,\Delta}(\tau^{\infty}_x)) = 0 \tag{5.56}$$

and

$$\mathrm{Hom}_G(i^{\infty}_{I_y,\Delta}(\tau^{\infty}_y), i^{\infty}_{I_x,\Delta}(\sigma^{\infty}_x)) = 0 = \mathrm{Hom}_G(i^{\infty}_{I_y,\Delta}(\sigma^{\infty}_y), i^{\infty}_{I_x,\Delta}(\tau^{\infty}_x)). \tag{5.57}$$

It follows from (5.56) and (5.57) together with Lemma 5.1.1 (for $\ell = 0$), Remark 5.1.12 and Proposition 5.1.14 (for $\ell = 1$) that for $\ell \leq 1$ and τ_x, τ_y as above

$$\mathrm{Ext}^{\ell}_G(\mathcal{F}^G_{P_{I_y}}(L(y), \tau^{\infty}_y), \mathcal{F}^G_{P_{I_x}}(L(x), \tau^{\infty}_x)) = 0 \tag{5.58}$$

and

$$\mathrm{Ext}^{\ell}_G(\mathcal{F}^G_{P_{I_y}}(L(y), \sigma^{\infty}_y), \mathcal{F}^G_{P_{I_x}}(L(x), \tau^{\infty}_x)) = 0 = \mathrm{Ext}^{\ell}_G(\mathcal{F}^G_{P_{I_y}}(L(y), \tau^{\infty}_y), \mathcal{F}^G_{P_{I_x}}(L(x), \sigma^{\infty}_x)). \tag{5.59}$$

Then by dévissage from (5.58) and (5.59) we deduce an isomorphism for $L(y) \in \mathrm{JH}_{U(\mathfrak{g})}(M')$

$$\mathrm{Ext}^1_G(\mathcal{F}^G_{P_{I_y}}(L(y), \sigma^{\infty}_y), \mathcal{F}^G_{P_{I_x}}(L(x), \sigma^{\infty}_x)) \xrightarrow{\sim} \mathrm{Ext}^1_G(\mathcal{F}^G_{P_{I_y}}(L(y), \sigma^{\infty}_{y,-}), \mathcal{F}^G_{P_{I_x}}(L(x), \sigma^{\infty}_{x,+})). \tag{5.60}$$

As for (5.58) and (5.59) we deduce from (5.56) and (5.57) for $\ell \leq 1$

$$\mathrm{Ext}^{\ell}_G(V', \mathcal{F}^G_{P_{I_x}}(L(x), \sigma^{\infty}_{x,+}/\sigma^{\infty}_x)) = \mathrm{Ext}^{\ell}_G(\ker(V'_- \twoheadrightarrow V'), \mathcal{F}^G_{P_{I_x}}(L(x), \sigma^{\infty}_{x,+})) = 0$$

which implies an isomorphism

$$\mathrm{Ext}^1_G(V', V_x) = \mathrm{Ext}^1_G(V', \mathcal{F}^G_{P_{I_x}}(L(x), \sigma^{\infty}_x)) \xrightarrow{\sim} \mathrm{Ext}^1_G(V'_-, \mathcal{F}^G_{P_{I_x}}(L(x), \sigma^{\infty}_{x,+})).$$

In other words, the object W from (5.51) admits a unique subquotient V that fits into a short exact sequence $0 \to V_x \to V \to V' \to 0$. It is then clear that V satisfies (i).

Step 3: We check (ii) for V as in Step 2.
Let Q' be a subquotient of M which is uniserial of length 2 with socle $L(y)$ and cosocle $L(y')$, we want to show that V in Step 2 admits a (unique) subquotient $V_{y',y}$ which fits into a non-split short exact sequence $0 \to V_{y'} \to V_{y',y} \to V_y \to 0$.

If $y' \neq x$, then Q' is a subquotient of M' and the existence of $V_{y',y}$ follows from the induction hypothesis on V'. If $y' = x$, then it follows from Step 1 that W contains a subrepresentation R which fits into a non-split extension (5.52). But (5.60) ensures that R contains a subquotient $V_{x,y}$ as desired. □

Let $I \subseteq \Delta$, M multiplicity free in $\mathcal{O}_{\mathrm{alg}}^{\mathfrak{p}_I}$ and π^∞ G-basic in $\mathrm{Rep}_{\mathrm{adm}}^\infty(L_I)$. For $L(x) \in \mathrm{JH}_{U(\mathfrak{g})}(M)$ let σ_x^∞ be a G-basic subquotient of $i_{I,I_x}^\infty(\pi^\infty)$ as in Lemma 5.1.19 and let V as in *loc. cit.* Let V_- be the minimal length subrepresentation of $\mathcal{F}_{P_I}^G(M, \pi^\infty)$ with V as quotient. We define $\sigma_{x,-}^\infty$ as in the first paragraph of the proof of Lemma 5.1.19, and as in *loc. cit.* (see the paragraph before (5.51) with M' there replaced by M), V_- has a decreasing filtration (indexed by $\mathrm{JH}_{U(\mathfrak{g})}(M)$) induced by the one on $\mathcal{F}_{P_I}^G(M, \pi^\infty)$ with $L(x)$-graded piece $\mathcal{F}_{P_{I_x}}^G(L(x), \sigma_{x,-}^\infty)$ for $L(x) \in \mathrm{JH}_{U(\mathfrak{g})}(M)$. Now let $x_0 \in W(G)$, $I_0 \stackrel{\mathrm{def}}{=} \Delta \setminus D_L(x_0)$, π_0^∞ a smooth G-basic representation of L_{I_0}, $V_0 \stackrel{\mathrm{def}}{=} \mathcal{F}_{P_{I_0}}^G(L(x_0), \pi_0^\infty)$, $\Sigma_0 \stackrel{\mathrm{def}}{=} W(L_0) \cdot \mathcal{J}(\pi_0^\infty)$ and $\Sigma_{0,x} \stackrel{\mathrm{def}}{=} \Sigma_0 \cap W(L_{I_x}) \cdot \mathcal{J}(\sigma_x^\infty)$ for $L(x) \in \mathrm{JH}_{U(\mathfrak{g})}(M)$. Note that $\Sigma_{0,x}$, if non-empty, is a single left $W(L_{I_0 \cap I_x})$-coset by (i) of Lemma 2.1.15 and G-regularity.

Lemma 5.1.20 *With the above assumptions assume $d(\pi_0^\infty, \sigma_x^\infty) = 0$ for $L(x) \in \mathrm{JH}_{U(\mathfrak{g})}(M)$, $J'_{I_x, I_0 \cap I_x}(\tau_x^\infty)_{\mathcal{B}_{\Sigma_{0,x}}^{I_0 \cap I_x}} = 0$ for $L(x) \in \mathrm{JH}_{U(\mathfrak{g})}(M) \setminus \{L(x_0)\}$ and $\tau_x^\infty \in \mathrm{JH}_{L_{I_x}}(\sigma_{x,-}^\infty) \setminus \mathrm{JH}_{L_{I_x}}(\sigma_x^\infty)$, and $\Sigma_{0,x_0} = \emptyset$ if $L(x_0) \in \mathrm{JH}_{U(\mathfrak{g})}(M)$. Assume moreover that $L(x_0)$ is not a constituent of $M^{I_x}(x)$ for $L(x) \in \mathrm{JH}_{U(\mathfrak{g})}(M)$ with $x \neq x_0$. Then the unique (up to scalar) injection $V_- \hookrightarrow \mathcal{F}_{P_I}^G(M, \pi^\infty)$ and surjection $V_- \twoheadrightarrow V$ induce isomorphisms*

$$\mathrm{Ext}_G^1(V_0, V) \xleftarrow{\sim} \mathrm{Ext}_G^1(V_0, V_-) \xrightarrow{\sim} \mathrm{Ext}_G^1(V_0, \mathcal{F}_{P_I}^G(M, \pi^\infty)). \tag{5.61}$$

Proof Note first that, for any $L(x) \in \mathrm{JH}_{U(\mathfrak{g})}(M)$ and any constituent τ_x^∞ of $i_{I,I_x}^\infty(\pi^\infty)$, it follows from (i) and (iii) of Lemma 2.1.15 that $W(L_{I_x}) \cdot \mathcal{J}(\tau_x^\infty) = W(L_{I_x}) \cdot \mathcal{J}(\sigma_x^\infty) = W(L_{I_x}) \cdot \mathcal{J}(\pi^\infty)$.

Step 1: We prove

$$\mathrm{Ext}_G^1(V_0, \mathcal{F}_{P_{I_x}}^G(L(x), \tau_x^\infty)) = 0 \tag{5.62}$$

for $L(x) \in \mathrm{JH}_{U(\mathfrak{g})}(M)$ and $\tau_x^\infty \in \mathrm{JH}_{L_{I_x}}(i_{I,I_x}^\infty(\pi^\infty)) \setminus \mathrm{JH}_{L_{I_x}}(\sigma_x^\infty)$.
As $d(\pi_0^\infty, \sigma_x^\infty) = 0$, $i_{I_x,\Delta}^\infty(\sigma_x^\infty)$ contains $\mathrm{cosoc}_G(i_{I_0,\Delta}^\infty(\pi_0^\infty))$ as an (irreducible) constituent. As $i_{I_x,\Delta}^\infty(i_{I,I_x}^\infty(\pi^\infty)) \cong i_{I,\Delta}^\infty(\pi^\infty)$ is multiplicity free (since G-basic), for τ_x as above $i_{I_x,\Delta}^\infty(\tau_x^\infty)$ cannot have $\mathrm{cosoc}_G(i_{I_0,\Delta}^\infty(\pi_0^\infty))$ as a constituent, which forces

$$\mathrm{Hom}_G(i_{I_0,\Delta}^\infty(\pi_0^\infty), i_{I_x,\Delta}^\infty(\tau_x^\infty)) = 0. \tag{5.63}$$

If $x \neq x_0$, then (5.63) and Remark 5.1.12 imply (5.62). If $x = x_0$, then $\Sigma_{0,x_0} = \emptyset$ and (i) of Lemma 5.1.17 imply (5.62).

Step 2: We prove

$$\mathrm{Ext}^2_G(V_0, \mathcal{F}^G_{P_{I_x}}(L(x), \tau_x^\infty)) = 0 \tag{5.64}$$

for $L(x) \in \mathrm{JH}_{U(\mathfrak{g})}(M)$ and $\tau_x^\infty \in \mathrm{JH}_{L_{I_x}}(\sigma_{x,-}^\infty) \setminus \mathrm{JH}_{L_{I_x}}(\sigma_x^\infty)$.
It follows from (5.63) that $d(\pi_0^\infty, \tau_x^\infty) \geq 1$. If $x = x_0$, then $\Sigma_{0,x_0} = \emptyset$ together with $d(\pi_0^\infty, \tau_x^\infty) \geq 1$ and (ii) of Lemma 5.1.17 imply (5.64). Assume on the contrary that (5.64) fails for some τ_x^∞ with $x \neq x_0$, then we deduce from $d(\pi_0^\infty, \tau_x^\infty) \geq 1$, Lemma 5.1.16 and (2.8)

$$\mathrm{Ext}^1_{L_{I_0 \cap I_x}}(J_{I_0, I_0 \cap I_x}(\pi_0^\infty), J'_{I_x, I_0 \cap I_x}(\tau_x^\infty))^\infty \cong \mathrm{Ext}^1_{L_{I_x}}(i^\infty_{I_0 \cap I_x, I_x}(J_{I_0, I_0 \cap I_x}(\pi_0^\infty)), \tau_x^\infty)^\infty \neq 0. \tag{5.65}$$

We have $\mathcal{J}(J_{I_0, I_0 \cap I_x}(\pi_0^\infty)) = \mathcal{J}(\pi_0^\infty) \subseteq \Sigma_0$. We also have $\mathcal{J}(J'_{I_x, I_0 \cap I_x}(\tau_x^\infty)) \subseteq W(L_{I_x}) \cdot \mathcal{J}(\tau_x^\infty) = W(L_{I_x}) \cdot \mathcal{J}(\pi^\infty)$ using (2.12) for the first inclusion. Hence we deduce from (5.65)

$$\mathrm{Ext}^1_{L_{I_0 \cap I_x}}(J_{I_0, I_0 \cap I_x}(\pi_0^\infty)_{\mathcal{B}^{I_0 \cap I_x}_{\Sigma_{0,x}}}, J'_{I_x, I_0 \cap I_x}(\tau_x^\infty)_{\mathcal{B}^{I_0 \cap I_x}_{\Sigma_{0,x}}})^\infty \neq 0,$$

which contradicts the assumption $J'_{I_x, I_0 \cap I_x}(\tau_x^\infty)_{\mathcal{B}^{I_0 \cap I_x}_{\Sigma_{0,x}}} = 0$.

Step 3: We prove the isomorphisms in (5.61).
As we have assumed $\Sigma_{0,x_0} = \emptyset$ when $L(x_0) \in \mathrm{JH}_{U(\mathfrak{g})}(M)$, we deduce from Lemma 5.1.1 that V_0 has no constituent in commun with $\mathcal{F}^G_{P_I}(M, \pi^\infty)$, and in particular the injection $V_- \hookrightarrow \mathcal{F}^G_{P_I}(M, \pi^\infty)$ induces an exact sequence

$$0 \to \mathrm{Ext}^1_G(V_0, V_-) \to \mathrm{Ext}^1_G(V_0, \mathcal{F}^G_{P_I}(M, \pi^\infty)) \to \mathrm{Ext}^1_G(V_0, \mathcal{F}^G_{P_I}(M, \pi^\infty)/V_-). \tag{5.66}$$

The surjection $V_- \twoheadrightarrow V$ induces an exact sequence

$$\mathrm{Ext}^1_G(V_0, V_{--}) \to \mathrm{Ext}^1_G(V_0, V_-) \to \mathrm{Ext}^1_G(V_0, V) \to \mathrm{Ext}^2_G(V_0, V_{--}) \tag{5.67}$$

where we write $V_{--} \subseteq V_-$ for the unique subrepresentation such that $V_-/V_{--} \cong V$. By dévissage using (5.62) we have $\mathrm{Ext}^1_G(V_0, \mathcal{F}^G_{P_I}(M, \pi^\infty)/V_-) = 0 = \mathrm{Ext}^1_G(V_0, V_{--})$, and using (5.64) we have $\mathrm{Ext}^2_G(V_0, V_{--}) = 0$. Together with (5.66) and (5.67) we obtain (5.61). □

Remark 5.1.21 There exists a "dual" version of Lemma 5.1.20 with a parallel proof. Let V_+ be the minimal length quotient of $\mathcal{F}^G_{P_I}(M, \pi^\infty)$ which has V as a subrepresentation and define $\sigma_{x,+}^\infty$ as in the proof of Lemma 5.1.19 for $L(x) \in \mathrm{JH}_{U(\mathfrak{g})}(M)$. Similar to V_-, V_+ is equipped with a decreasing filtration indexed by $\mathrm{JH}_{U(\mathfrak{g})}(M)$ which is induced from $\mathcal{F}^G_{P_I}(M, \pi^\infty)$, with $L(x)$-graded piece $\mathcal{F}^G_{P_{I_x}}(L(x), \sigma_{x,+}^\infty)$. Now let x_0, π_0^∞, V_0, Σ_0 and $\Sigma_{0,x}$ (for $L(x) \in \mathrm{JH}_{U(\mathfrak{g})}(M)$) as before Lemma 5.1.20. Assume $d(\sigma_x^\infty, \pi_0^\infty) = 0$ for $L(x) \in \mathrm{JH}_{U(\mathfrak{g})}(M)$, $J_{I_x, I_0 \cap I_x}(\tau_x^\infty)_{\mathcal{B}^{I_0 \cap I_x}_{\Sigma_{0,x}}} = 0$ for $L(x) \in \mathrm{JH}_{U(\mathfrak{g})}(M) \setminus \{L(x_0)\}$ and $\tau_x^\infty \in \mathrm{JH}_{L_{I_x}}(\sigma_{x,+}^\infty) \setminus \mathrm{JH}_{L_{I_x}}(\sigma_x^\infty)$, $\Sigma_{0,x_0} = \emptyset$ if $L(x_0) \in \mathrm{JH}_{U(\mathfrak{g})}(M)$ and $\mathrm{JH}_{U(\mathfrak{g})}(M) \cap \mathrm{JH}_{U(\mathfrak{g})}(M^{I_0}(x_0)) \subseteq \{L(x_0)\}$. Then as in (5.61) the unique (up to scalar) surjection $\mathcal{F}_{P_I}^G(M, \pi^\infty) \twoheadrightarrow V_+$ and injection $V \hookrightarrow V_+$ induce isomorphisms

$$\mathrm{Ext}^1_G(V, V_0) \xleftarrow{\sim} \mathrm{Ext}^1_G(V_+, V_0) \xrightarrow{\sim} \mathrm{Ext}^1_G(\mathcal{F}_{P_I}^G(M, \pi^\infty), V_0).$$

5.2 Ext-Squares and Ext-Cubes of Orlik-Strauch Representations

We use (essentially) all previous results to construct locally analytic representations of G which are either uniserial (Lemma 5.2.33) or "squares" (Propositions 5.2.10 and 5.2.18) and "cubes" (Proposition 5.2.28) of irreducible Orlik-Strauch representations.

We keep the notation of Sect. 5.1. We first define some irreducible locally analytic representations via Orlik-Strauch's construction (Theorem 4.3.7).

For $j, j' \in \Delta$, since $D_L(w_{j,j'}) = \{j\}$ we have $L(w_{j,j'}) \in \mathcal{O}_{\mathrm{alg}}^{\mathfrak{p}_{\widehat{j}}}$ by Lemma 3.1.1, where $\widehat{j} = \Delta \setminus \{j\}$ (see the beginning of Sect. 2.3) and $w_{j,j'}$ is defined in (3.87) (in particular we always have $w_{j,j'} \neq 1$). We set

$$\mathbf{J} = \mathbf{J}_n \stackrel{\mathrm{def}}{=} \{\underline{j} = (j_0, j_1, j_2) \mid 1 \le j_0 \le n-1,\ 1 \le j_1 \le n-1,\ 1 \le j_2 \le n,\ 0 \le j_2 - j_1 \le n-1\}$$

that we equip with the partial order $\underline{j} = (j_0, j_1, j_2) \le \underline{j}' = (j_0', j_1', j_2')$ if and only if $j_0 \le j_0'$, $j_2 \le j_2'$ and $j_2 - j_1 \le j_2' - j_1'$. In particular forgetting j_0 gives a surjection $\mathbf{J} \twoheadrightarrow \mathbf{J}^\infty$ which respects the partial orders (see the beginning of Sect. 2.3 for the set $\mathbf{J}^\infty$). We will see later in Sect. 5.3 that this partial order on $\mathbf{J}$ is motivated by the layer structure of certain admissible finite length multiplicity free locally analytic representations. For $\underline{j}, \underline{j}' \in \mathbf{J}$ we write $d(\underline{j}, \underline{j}') \stackrel{\mathrm{def}}{=} |j_0 - j_0'| + |j_2 - j_2'| + |(j_2 - j_1) - (j_2' - j_1')|$ for the *distance* between $\underline{j}$ and $\underline{j}'$. Finally, for $\underline{j} = (j_0, j_1, j_2) \in \mathbf{J}$, we set

$$C_{\underline{j}} = C_{(j_0, j_1, j_2)} \stackrel{\mathrm{def}}{=} \mathcal{F}_{P_{\widehat{j_1}}}^G(L(w_{j_1, j_0}), \pi^\infty_{j_1, j_2}) \tag{5.68}$$

where $\pi^\infty_{j_1, j_2}$ is the irreducible G-regular smooth representation of $L_{\widehat{j_1}}$ defined in (2.71).

Lemma 5.2.1 *Let $\underline{j}, \underline{j}' \in \mathbf{J}$ such that $\underline{j}' \not\le \underline{j}$. Then $\mathrm{Ext}^1_G(C_{\underline{j}'}, C_{\underline{j}}) \neq 0$ if and only if $d(\underline{j}, \underline{j}') = 1$, in which case it is one dimensional.*

Proof Note first that $\underline{j}' \not\le \underline{j}$ with $j_0 = j_0'$ and $|j_1 - j_1'| \le 1$ force $\underline{j} < \underline{j}'$ in $\mathbf{J}$.

Assume first $w_{j_1, j_0} = w_{j_1', j_0'}$, equivalently $(j_0, j_1) = (j_0', j_1')$. Hence we have $\underline{j} < \underline{j}'$ by the previous sentence, hence $j_2 < j_2'$ since $j_1 = j_1'$, and thus $d(\underline{j}, \underline{j}') \ge 2$. We also have $\Sigma_{j_1, j_2} \cap \Sigma_{j_1', j_2'} = \emptyset$ by (i) of Lemma 2.3.4 (see above (2.73) for the notation), which by (i) of Lemma 5.1.17 implies $\mathrm{Ext}^1_G(C_{\underline{j}'}, C_{\underline{j}}) = 0$. In particular the lemma trivially holds when $w_{j_1,j_0} = w_{j_1',j_0'}$ since both assertions in the statement never happen.

We assume from now on $w_{j_1,j_0} \neq w_{j_1',j_0'}$ and write $I = \Delta \setminus \{j_1, j_1'\}$. It follows from Proposition 5.1.14 that $\mathrm{Ext}^1_G(C_{\underline{j}'}, C_{\underline{j}}) \neq 0$ if and only if $\mathrm{Ext}^1_{U(\mathfrak{g})}(L(w_{j_1,j_0}), L(w_{j_1',j_0'})) \neq 0$ and

$$\mathrm{Hom}_{L_{\widehat{j_1}}}(i^\infty_{I,\widehat{j_1}}(J_{\widehat{j_1},I}(\pi^\infty_{j_1',j_2'})), \pi^\infty_{j_1,j_2}) \neq 0. \tag{5.69}$$

It follows from the last assertion in (ii) of Lemma 3.2.4 that $\mathrm{Ext}^1_{U(\mathfrak{g})}(L(w_{j_1,j_0}), L(w_{j_1',j_0'})) \neq 0$ if and only if either $w_{j_1,j_0} \prec w_{j_1',j_0'}$ or $w_{j_1',j_0'} \prec w_{j_1,j_0}$. By Remark A.10 this is equivalent to $w_{j_1,j_0} < w_{j_1',j_0'}$ with $\ell(w_{j_1',j_0'}) = \ell(w_{j_1,j_0}) + 1$ or $w_{j_1',j_0'} < w_{j_1,j_0}$ with $\ell(w_{j_1,j_0}) = \ell(w_{j_1',j_0'}) + 1$, which is easily checked to be equivalent to $|j_0 - j_0'| + |j_1 - j_1'| = 1$. Hence $\mathrm{Ext}^1_G(C_{\underline{j}'}, C_{\underline{j}}) \neq 0$ if and only if $|j_0 - j_0'| + |j_1 - j_1'| = 1$ and (5.69) holds. In that case, note that $\mathrm{Ext}^1_{U(\mathfrak{g})}(L(w_{j_1,j_0}), L(w_{j_1',j_0'}))$ is one dimensional by the last assertion in (ii) of Lemma 3.2.4.

We distinguish the two cases $j_0 = j_0'$ and $j_0 \neq j_0'$.

Assume $j_0 = j_0'$, then $|j_0 - j_0'| + |j_1 - j_1'| = 1$ is equivalent to $|j_1 - j_1'| = 1$ which implies $\underline{j} < \underline{j}'$ by the first sentence and also $(j_1, j_2) < (j_1', j_2')$ in $\mathbf{J}^\infty$. When $(j_1, j_2) < (j_1', j_2')$ in $\mathbf{J}^\infty$, (5.69) is equivalent to $|j_2 - j_2'| + |(j_2 - j_1) - (j_2' - j_1')| \leq 1$ by (i) of Lemma 2.3.5. It follows that, when $j_0 = j_0'$ (and $\underline{j}' \nleq \underline{j}$, $w_{j_1,j_0} \neq w_{j_1',j_0'}$), $\mathrm{Ext}^1_G(C_{\underline{j}'}, C_{\underline{j}}) \neq 0$ is equivalent to $|j_1 - j_1'| = 1$ and $|j_2 - j_2'| + |(j_2 - j_1) - (j_2' - j_1')| \leq 1$, which is easily checked to be equivalent to $|j_2 - j_2'| + |(j_2 - j_1) - (j_2' - j_1')| = 1$ (using $\underline{j}' \neq \underline{j}$), i.e. $d(\underline{j}, \underline{j}') = 1$.

Assume $j_0 \neq j_0'$, then $|j_0 - j_0'| + |j_1 - j_1'| = 1$ implies $j_1 = j_1'$ and $I = \widehat{j_1}$, and thus (5.69) is equivalent to $j_2 = j_2'$. Thus $\mathrm{Ext}^1_G(C_{\underline{j}'}, C_{\underline{j}}) \neq 0$ is equivalent to $|j_0 - j_0'| = 1$, $j_1 = j_1'$, $j_2 = j_2'$, which is equivalent to $d(\underline{j}, \underline{j}') = 1$ (when $j_0 \neq j_0'$). □

Recall from the paragraph below Lemma 2.3.1 that, for $(j_1, j_2) \in \mathbf{J}^\infty$, we have defined $I^+_{j_1,j_2}, I^-_{j_1,j_2} \subseteq \Delta$ by $\mathrm{soc}_G(i^\infty_{\widehat{j_1},\Delta}(\pi^\infty_{j_1,j_2})) \cong V^\infty_{I^+_{j_1,j_2},\Delta}$ and $\mathrm{cosoc}_G(i^\infty_{\widehat{j_1},\Delta}(\pi^\infty_{j_1,j_2})) \cong V^\infty_{I^-_{j_1,j_2},\Delta}$ (see (2.40) for the G-regular irreducible representation $V^\infty_{I,\Delta}$). Recall also that $L(1) = L(\mu_0)$.

Lemma 5.2.2 *Let* $\underline{j} \in \mathbf{J}$ *and* $I \subseteq \Delta$.

(i) *We have* $\mathrm{Ext}^1_G(L(1)^\vee \otimes_E V^\infty_{I,\Delta}, C_{\underline{j}}) \neq 0$ *if and only if* $j_0 = j_1$ *and* $I = I^+_{j_1,j_2}$, *in which case it is one dimensional.*

(ii) *We have* $\mathrm{Ext}^1_G(C_{\underline{j}}, L(1)^\vee \otimes_E V^\infty_{I,\Delta}) \neq 0$ *if and only if* $j_0 = j_1$ *and* $I = I^-_{j_1,j_2}$, *in which case it is one dimensional.*

Proof We prove (i). As $1 \neq w_{j_1,j_0}$, we deduce from Proposition 5.1.14 that $\mathrm{Ext}^1_G(L(1)^\vee \otimes_E V^\infty_{I,\Delta}, C_{\underline{j}}) \neq 0$ if and only if $\mathrm{Ext}^1_{U(\mathfrak{g})}(L(w_{j_1,j_0}), L(1)) \neq 0$ and $\mathrm{Hom}_G(V^\infty_{I,\Delta}, i^\infty_{\widehat{j_1},\Delta}(\pi^\infty_{j_1,j_2})) \neq 0$, if and only if $\ell(w_{j_1,j_0}) = 1$ i.e. $j_0 = j_1$ (using (ii) of Lemma 3.2.4), and $I = I^+_{j_1,j_2}$ (using (i) of Lemma 2.3.3). The proof of (ii) is completely analogous. □

Lemma 5.2.3 *Let $\underline{j}, \underline{j}' \in \mathbf{J}$ such that $\underline{j} < \underline{j}'$ and $j_0' \le j_0 + 1$.*

(i) If $\mathrm{Ext}^2_G(C_{\underline{j}'}, C_{\underline{j}}) \neq 0$ then $(j_1', j_2') \in \{(j_1+1, j_2+1), (j_1-1, j_2), (j_1, j_2+1), (j_1, j_2)\}$.

(ii) Assume $(j_1', j_2') = (j_1, j_2+1)$. Then $\mathrm{Ext}^2_G(C_{\underline{j}'}, C_{\underline{j}}) \neq 0$ implies $j_0' = j_0$, and $\mathrm{Ext}^2_G(C_{\underline{j}'}, C_{\underline{j}})$ is at most one dimensional except when $2 \le j_0 = j_1 \le n-2$ in which case it is at most two dimensional.

Proof We first prove (i). Assume on the contrary that there exists $\underline{j}, \underline{j}' \in \mathbf{J}$ that satisfies $\underline{j} < \underline{j}'$, $j_0' \le j_0 + 1$, $(j_1', j_2') \notin \{(j_1+1, j_2+1), (j_1-1, j_2), (j_1, j_2+1), (j_1, j_2)\}$ and $\mathrm{Ext}^2_G(C_{\underline{j}'}, C_{\underline{j}}) \neq 0$. The condition $(j_1', j_2') \notin \{(j_1+1, j_2+1), (j_1-1, j_2), (j_1, j_2+1), (j_1, j_2)\}$ together with (i) of Lemma 2.3.2 implies $d(\pi^\infty_{j_1', j_2'}, \pi^\infty_{j_1, j_2}) \ge 1$. We have two cases.

Case 1: $j_1 \neq j_1'$. Then $w_{j_1, j_0} \neq w_{j_1', j_0'}$ and $L(w_{j_1', j_0'})$ is not a constituent of $M^{\widehat{j_1}}(w_{j_1, j_0})$ using Lemma 3.1.1 and [33, Theorem 9.4(c)]. We can then apply Lemma 5.1.16 which gives $\mathrm{Ext}^1_{U(\mathfrak{g})}(L(w_{j_1, j_0}), L(w_{j_1', j_0'})) \neq 0$ and $\mathrm{Ext}^1_{L_{\widehat{j_1}}}(i^\infty_{\widehat{j'}_1 \cap \widehat{j_1}, \widehat{j_1}}(J_{\widehat{j'}_1, \widehat{j'}_1 \cap \widehat{j_1}}(\pi^\infty_{j_1', j_2'})), \pi^\infty_{j_1, j_2})^\infty \neq 0$. The first inequality implies $|j_1 - j_1'| \le 1$ by (ii) of Lemma 3.2.4 and Remark A.10. By (2.7) we have $\mathrm{Ext}^1_G(i^\infty_{\widehat{j'}_1, \Delta}(\pi^\infty_{j_1', j_2'}), i^\infty_{\widehat{j_1}, \Delta}(\pi^\infty_{j_1, j_2}))^\infty = \mathrm{Ext}^1_{L_{\widehat{j_1}}}(J_{\Delta, \widehat{j_1}}(i^\infty_{\widehat{j'}_1, \Delta}(\pi^\infty_{j_1', j_2'})), \pi^\infty_{j_1, j_2})^\infty$, which is non-zero by (2.36) (applied with $I = \Delta$ and taking $w = 1$) and the second inequality. Hence $d(\pi^\infty_{j_1', j_2'}, \pi^\infty_{j_1, j_2}) = 1$, which by (ii) of Lemma 2.3.2 implies $(j_1', j_2') \in \{(j_1+2, j_2+2), (j_1-2, j_2)\}$ and thus $|j_1 - j_1'| = 2$. This contradicts $|j_1 - j_1'| \le 1$.

Case 2: $j_1 = j_1'$. Then $j_2 < j_2'$ and (i) of Lemma 2.3.4 forces $\Sigma_{j_1, j_2} \cap \Sigma_{j_1', j_2'} = \emptyset$, which by (ii) of Lemma 5.1.17 implies $\mathrm{Ext}^2_G(C_{\underline{j}'}, C_{\underline{j}}) = 0$, a contradiction.

We prove (ii). As $(j_1', j_2') = (j_1, j_2+1)$ by (ii) of Lemma 2.3.5 we have

$$\mathrm{Hom}_{L_{\widehat{j_1}}}(i^\infty_{\widehat{j_1}, \widehat{j_1}, s_{j_1}}(J_{\widehat{j_1}, \widehat{j_1}, s_{j_1}}(\pi^\infty_{j_1', j_2'})), \pi^\infty_{j_1, j_2}) \neq 0.$$

By Proposition 5.1.18 if $\mathrm{Ext}^2_G(C_{\underline{j}'}, C_{\underline{j}}) \neq 0$ then $j_0 \neq j_0'$, and when $j_0 = j_0'$ then

$$\dim_E \mathrm{Ext}^2_G(C_{\underline{j}'}, C_{\underline{j}}) \le \# S_0$$

where $S_0 = \{x' \mid x' \in W(L_{\widehat{j_1}}) w_{j_1, j_0},\ \ell(x') = \ell(w_{j_1, j_0}) + 1,\ j_1 \notin D_L(x')\}$. Let $x' \in S_0$ and $j \in D_L(x')$. The condition $j_1 \notin D_L(x')$ forces $j = j_1 + 1$ if $j_1 > j_0$, $j = j_1 - 1$ if $j_1 < j_0$, and $j \in \{j_1 - 1, j_1 + 1\}$ if $j_1 = j_0$. In particular, we always have $S_0 \subseteq \{w_{j_1+1, j_0}, w_{j_1-1, j_0}\}$, with $w_{j_1+1, j_0} \in S_0$ if and only if $j_0 \le j_1 < n-1$, and $w_{j_1-1, j_0} \in S_0$ if and only if $j_0 \ge j_1 > 1$. This finishes the proof. □

Lemma 5.2.4 *Let $\underline{j} \in \mathbf{J}$ and $I \subseteq \Delta$.*

(i) If $I \neq I^+_{j_1, j_2}$ and if $\mathrm{Ext}^2_G(L(1)^\vee \otimes_E V^\infty_{I, \Delta}, C_{\underline{j}}) \neq 0$ then $j_0 = j_1$, $I \notin [I^+_{j_1, j_2}, I^-_{j_1, j_2}]$ and $d(I, I^+_{j_1, j_2}) = 1$ (see (2.50) and (2.49) for the notation).

(ii) If $I \neq I^-_{j_1, j_2}$ and if $\mathrm{Ext}^2_G(C_{\underline{j}}, L(1)^\vee \otimes_E V^\infty_{I, \Delta}) \neq 0$ then $j_0 = j_1$, $I \notin [I^+_{j_1, j_2}, I^-_{j_1, j_2}]$ and $d(I^-_{j_1, j_2}, I) = 1$.

Proof We first prove (i). As $I \neq I^+_{j_1,j_2}$, we have $d(V^\infty_{I,\Delta}, \pi^\infty_{j_1,j_2}) \geq 1$ by (i) of Lemma 2.3.3. By Lemma 5.1.16 (note that $L(1)$ is not a constituent of $M^{\widehat{j_1}}(w_{j_1,j_0})$) $\mathrm{Ext}^2_G(L(1)^\vee \otimes_E V^\infty_{I,\Delta}, C_{\underline{j}}) \neq 0$ implies $\mathrm{Ext}^1_{U(\mathfrak{g})}(L(w_{j_1,j_0}), L(1)) \neq 0$ and $\mathrm{Ext}^1_{L_{\widehat{j_1}}}(i^\infty_{I\cap\widehat{j_1},\widehat{j_1}}(J_{I,I\cap\widehat{j_1}}(V^\infty_{I,\Delta})), \pi^\infty_{j_1,j_2})^\infty \neq 0$. The first inequality is equivalent to $j_0 = j_1$ by (ii) of Lemmas 3.2.4 and 3.2.5. Arguing as in Case 1 of the proof of Lemma 5.2.3 using (2.36), (2.37) and that $\pi^\infty_{j_1,j_2}$ is in a single Bernstein block (as it is irreducible), the second inequality is equivalent to

$$\mathrm{Ext}^1_G(V^\infty_{I,\Delta}, i^\infty_{\widehat{j_1},\Delta}(\pi^\infty_{j_1,j_2}))^\infty \overset{(31)}{\cong} \mathrm{Ext}^1_{L_{\widehat{j_1}}}(J_{\Delta,\widehat{j_1}}(V^\infty_{I,\Delta}), \pi^\infty_{j_1,j_2})^\infty \neq 0,$$

which is equivalent to $d(V^\infty_{I,\Delta}, \pi^\infty_{j_1,j_2}) = 1$ (since $d(V^\infty_{I,\Delta}, \pi^\infty_{j_1,j_2}) \geq 1$), which is equivalent to $I \notin [I^+_{j_1,j_2}, I^-_{j_1,j_2}]$ and $d(I, I^+_{j_1,j_2}) = 1$ by (ii) of Lemma 2.3.3. The proof of (ii) is similar. □

Lemma 5.2.5 *Let $\underline{j} \in \mathbf{J}$.*

(i) If $(j_1, j_2) \neq (1, n)$ and $\underline{j} \neq (2, 2, n)$, then we have $\mathrm{Ext}^2_G(L(1)^\vee \otimes_E \mathrm{St}^\infty_n, C_{\underline{j}}) = 0$.
(ii) If $(j_1, j_2) \neq (1, 1)$ and $\underline{j} \neq (2, 2, 2)$, then we have $\mathrm{Ext}^2_G(C_{\underline{j}}, L(1)^\vee \otimes_E \mathrm{St}^\infty_n) = 0$.

Proof It follows from Lemma 2.3.1 that $I^+_{j_1,j_2} \neq \emptyset$ if and only if $(j_1, j_2) \neq (1, n)$, $I^-_{j_1,j_2} \neq \emptyset$ if and only if $(j_1, j_2) \neq (1, 1)$, and $\emptyset \notin [I^+_{j_1,j_2}, I^-_{j_1,j_2}]$ (equivalently $I^+_{j_1,j_2} \cap I^-_{j_1,j_2} \neq \emptyset$) if and only if $j_1 > 1$.

If $(j_1, j_2) \neq (1, n)$ and $\underline{j} \neq (2, 2, n)$, we have either $j_1 = 1$ in which case $\emptyset \in [I^+_{j_1,j_2}, I^-_{j_1,j_2}]$, or $j_1 > 2$ in which case $d(\emptyset, I^+_{j_1,j_2}) = \#I^+_{j_1,j_2} > 1$, or $j_1 = 2, j_2 < n$ in which case $d(\emptyset, I^+_{j_1,j_2}) > 1$ (again), or $(j_1, j_2) = (2, n)$ in which case $j_0 \neq j_1$ by assumption. In all these cases (i) of Lemma 5.2.4 implies $\mathrm{Ext}^2_G(L(1)^\vee \otimes_E \mathrm{St}^\infty_n, C_{\underline{j}}) = 0$, which gives (i).

If $(j_1, j_2) \neq (1, 1)$ and $\underline{j} \neq (2, 2, 2)$, we have either $j_1 = 1$ in which case $\emptyset \in [I^+_{j_1,j_2}, I^-_{j_1,j_2}]$, or $j_1 > 2$ in which case $d(\emptyset, I^-_{j_1,j_2}) = \#I^-_{j_1,j_2} > 1$, or $j_1 = 2$, $j_2 > 2$ in which case $d(\emptyset, I^-_{j_1,j_2}) > 1$ (again), or $(j_1, j_2) = (2, 2)$ in which case $j_0 \neq j_1$ by assumption. In all these cases (ii) of Lemma 5.2.4 implies $\mathrm{Ext}^2_G(C_{\underline{j}}, L(1)^\vee \otimes_E \mathrm{St}^\infty_n) = 0$, which gives (ii). □

Let Γ_{OS} be the set of pairs (x, π^∞) with $x \in W(G)$ and π^∞ an isomorphism class of G-basic representation in $\mathrm{Rep}^\infty_{\mathrm{adm}}(L_{I_x})$ where $I_x \overset{\mathrm{def}}{=} \Delta \setminus D_L(x)$ ("OS" for "Orlik-Strauch"). For $(x, \pi^\infty) \in \Gamma_{\mathrm{OS}}$ we write $V_{x,\pi^\infty} \overset{\mathrm{def}}{=} \mathcal{F}^G_{P_{I_x}}(L(x), \pi^\infty)$ (which is not assumed to be irreducible). We consider a finite length object V in $\mathrm{Rep}^{\mathrm{an}}_{\mathrm{adm}}(G)$ equipped with a decreasing filtration (for some $d \geq 0$)

$$0 = \mathrm{Fil}^{d+1}(V) \subsetneq \mathrm{Fil}^d(V) \subsetneq \cdots \subsetneq \mathrm{Fil}^1(V) \subsetneq \mathrm{Fil}^0(V) = V, \tag{5.70}$$

and a finite subset $\Gamma(V) \subseteq \Gamma_{\mathrm{OS}}$ equipped with a partition $\Gamma(V) = \bigsqcup_{k=0}^{d} \Gamma_k(V)$, such that there is an isomorphism for $0 \leq k \leq d$

$$0 \neq \mathrm{gr}^k(V) \stackrel{\mathrm{def}}{=} \mathrm{Fil}^k(V)/\mathrm{Fil}^{k+1}(V) \cong \bigoplus_{(x,\pi^\infty)\in\Gamma_k(V)} V_{x,\pi^\infty}. \tag{5.71}$$

We will *always* assume that the pairs $(x, W(L_{I_x}) \cdot \mathcal{J}(\pi^\infty))$ are distinct for different choices of $(x, \pi^\infty) \in \Gamma(V)$. Under this assumption, Lemmas 5.1.1 and 2.1.15 imply that V is multiplicity free and that the set $\Gamma(V)$ is uniquely determined by V (but not necessarily the partition $\bigsqcup_{k=0}^{d} \Gamma_k(V)$ of $\Gamma(V)$ nor the filtration (5.70) of V). In particular, the constituents of the subquotient V_{x,π^∞} of V are the constituents of V of the form $\mathcal{F}_{P_{I_x}}^G(L(x), \sigma^\infty)$ with $\sigma^\infty \in \mathcal{B}^{I_x}_{W(L_{I_x})\cdot\mathcal{J}(\pi^\infty)}$ and two distinct (x, π^∞) in $\Gamma(V)$ lead to two multiplicity free V_{x,π^∞} which have no constituent in commun.

For V fixed as above and V' a subquotient of V, we define $\mathrm{Fil}^k(V') \subseteq V'$ for $k \in \{0, \ldots, d+1\}$ as the maximal (for inclusion) subrepresentation of V' such that its constituents are constituents of $\mathrm{Fil}^k(V)$ (recall that V is multiplicity free of finite length). We say that a subquotient V' of V is a *basic subquotient* of V if there exists $(x, \pi^\infty) \in \Gamma(V)$ such that $V' \cong V_{x,\pi^\infty}$, and is a *good subquotient* of V if there exists $\Gamma_k(V') \subseteq \Gamma_k(V)$ for every $k \in \{0, \ldots, d\}$ such that

$$\mathrm{gr}^k(V') = \mathrm{Fil}^k(V')/\mathrm{Fil}^{k+1}(V') \cong \bigoplus_{(x,\pi^\infty)\in\Gamma_k(V')} V_{x,\pi^\infty}.$$

In particular every subquotient of V is good if the V_{x,π^∞} are irreducible for every $(x, \pi^\infty) \in \Gamma(V)$. (Note that "basic" here is relative to a fixed representation V, and is quite different from "G-basic" as in Definition 2.1.4.)

The partial order on $\mathrm{JH}_G(V)$ induces a partial order on $\Gamma(V)$ as follows. Given two distinct $(x_0, \pi_0^\infty), (x_1, \pi_1^\infty) \in \Gamma(V)$, we write $(x_1, \pi_1^\infty) < (x_0, \pi_0^\infty)$ if there exists $W_i \in \mathrm{JH}_G(V_{x_i,\pi_i^\infty}) \subseteq \mathrm{JH}_G(V)$ for $i = 0, 1$ such that $W_1 < W_0$ in the sense of Sect. 1.4. If $d_i \in \{0, \ldots, d\}$ is such that $(x_i, \pi_i^\infty) \in \Gamma_{d_i}(V)$, we note from (5.70) and (5.71) that $(x_1, \pi_1^\infty) < (x_0, \pi_0^\infty)$ implies $d_0 < d_1$. If there does not exist $(x, \pi^\infty) \in \Gamma(V)$ such that $(x_1, \pi_1^\infty) < (x, \pi^\infty) < (x_0, \pi_0^\infty)$, then there is a good subquotient V' of V that fits into a non-split short exact sequence

$$0 \to V_{x_1,\pi_1^\infty} \to V' \to V_{x_0,\pi_0^\infty} \to 0.$$

The following definition is reminiscent of (though different from) Definition 3.4.1.

Definition 5.2.6 Let d be some integer in $\mathbb{Z}_{\geq 0}$.

(i) A finite length multiplicity free representation V as in (5.70) and (5.71) is an Ext-*hypercube* of rank d if the following properties hold

- the partially ordered set $\Gamma(V)$ admits a unique maximal element which is (necessarily) in $\Gamma_0(V)$ and a unique minimal element which is (necessarily) in $\Gamma_d(V)$;

- for $0 \le d_0, d_1 \le d$ and $(x_i, \pi_i^\infty) \in \Gamma_{d_i}(V)$, $i = 0, 1$, we have

$$\mathrm{Ext}^1_G(V_{x_0,\pi_0^\infty}, V_{x_1,\pi_1^\infty}) \neq 0 \tag{5.72}$$

if and only if $d_1 = d_0 + 1$ and $(x_1, \pi_1^\infty) < (x_0, \pi_0^\infty)$, in which case (5.72) is one dimensional.

(ii) An Ext-hypercube is an Ext-*square* if $d = 2$, and an Ext-*cube* if $d = 3$.
(iii) An Ext-hypercube V is *strict* if it has simple socle and cosocle, in which case $\mathrm{soc}_G(V) = \mathrm{soc}_G(\mathrm{gr}^d(V))$ and $\mathrm{cosoc}_G(V) = \mathrm{cosoc}_G(\mathrm{gr}^0(V))$.
(iv) An Ext-hypercube V of rank d is *minimal* if for any good subquotient V' of V which is an Ext-hypercube of rank $d' \le d$, there does not exist an Ext-hypercube V'' of rank d' such that $\mathrm{gr}^0(V'') \cong \mathrm{gr}^0(V')$, $\mathrm{gr}^{d'}(V'') \cong \mathrm{gr}^{d'}(V')$ and $\mathrm{gr}^k(V'') \subseteq \mathrm{gr}^k(V')$ is a good direct summand of $\mathrm{gr}^k(V')$ for $1 \le k \le d' - 1$ with at least one inclusion being strict.

If V is an Ext-hypercube of rank d the conditions in (i) imply that both $\Gamma_0(V)$, $\Gamma_d(V)$ are singletons. In fact, it is not difficult to see that the filtration (5.70) is uniquely determined by (the isomorphism class of) V, and that the second condition in (i) above implies that, for $0 \le d_0, d_1 \le d$ and $(x_i, \pi_i^\infty) \in \Gamma_{d_i}(V)$ $(i = 0, 1)$ such that $d_1 > d_0 + 1$ and $(x_1, \pi_1^\infty) < (x_0, \pi_0^\infty)$, there exists $(y_k, \sigma_k^\infty) \in \Gamma_k(V)$ for $d_0 \le k \le d_1$ such that $(y_{d_i}, \sigma_{d_i}^\infty) = (x_i, \pi_i^\infty)$ for $i = 0, 1$ and $(y_\ell, \sigma_\ell^\infty) < (y_{\ell+1}, \sigma_{\ell+1}^\infty)$ for $d_0 \le \ell \le d_1 - 1$. Moreover for $(x_0, \pi_0^\infty), (x_1, \pi_1^\infty) \in \Gamma(V)$ such that $(x_1, \pi_1^\infty) < (x_0, \pi_0^\infty)$, one easily checks that V admits a unique good subquotient $V_{x_1,\pi_1^\infty}^{x_0,\pi_0^\infty}$ with $\Gamma(V_{x_1,\pi_1^\infty}^{x_0,\pi_0^\infty})$ consisting exactly of those $(y, \sigma^\infty) \in \Gamma(V)$ such that $(x_1, \pi_1^\infty) \le (y, \sigma^\infty) \le (x_0, \pi_0^\infty)$, and that $V_{x_1,\pi_1^\infty}^{x_0,\pi_0^\infty}$ is an Ext-hypercube. Conversely any good subquotient of V which is an Ext-hypercube has the form $V_{x_1,\pi_1^\infty}^{x_0,\pi_0^\infty}$ for some $(x_1, \pi_1^\infty) \le (x_0, \pi_0^\infty) \in \Gamma(V)$.

Lemma 5.2.7 *Let V be an* Ext*-hypercube of rank d. Assume that there does not exist another* Ext*-hypercube V' of rank d such that $\mathrm{gr}^0(V') \cong \mathrm{gr}^0(V)$, $\mathrm{gr}^d(V') \cong \mathrm{gr}^d(V)$ and $\mathrm{gr}^k(V') \subseteq \mathrm{gr}^k(V)$ is a good direct summand for $1 \le k \le d-1$ with at least one inclusion being strict. Then $\mathrm{Ext}^1_G(\mathrm{gr}^0(V), \mathrm{Fil}^1(V))$ is one dimensional.*

Proof We can assume $d \ge 2$. The existence of V forces $\mathrm{Ext}^1_G(\mathrm{gr}^0(V), \mathrm{Fil}^1(V)) \neq 0$. Assume on the contrary that $\mathrm{Ext}^1_G(\mathrm{gr}^0(V), \mathrm{Fil}^1(V))$ has dimension ≥ 2. Choose a good subrepresentation $V_1 \subseteq \mathrm{Fil}^1(V)$ such that $\mathrm{Fil}^1(V)/V_1$ is a basic (hence nonzero) subquotient of V, then we have $\dim_E \mathrm{Ext}^1_G(\mathrm{gr}^0(V), \mathrm{Fil}^1(V)/V_1) = 1$ by the second condition in (i) of Definition 5.2.6, which by dévissage from $0 \to V_1 \to \mathrm{Fil}^1(V) \to \mathrm{Fil}^1(V)/V_1 \to 0$ forces $\mathrm{Ext}^1_G(\mathrm{gr}^0(V), V_1) \neq 0$. Hence, there exists V'' that fits into a non-split short exact sequence $0 \to V_1 \to V'' \to \mathrm{gr}^0(V) \to 0$. We equip V'' with the filtration $\mathrm{Fil}^0(V'') \overset{\mathrm{def}}{=} V'$ and $\mathrm{Fil}^k(V'') \overset{\mathrm{def}}{=} \mathrm{Fil}^k(V) \cap V_1$ for $k \ge 1$. We now define V' as the minimal length good subrepresentation of V'' such that $\mathrm{gr}^0(V') = \mathrm{gr}^0(V'') = \mathrm{gr}^0(V)$ (note that V' can be strictly smaller than V'' since $\mathrm{gr}^0(V'')$ may have more than one maximal element in $\Gamma(V'')$). Then the conditions in (i) of Definition 5.2.6 for V imply the similar conditions for V', in particular V' is an Ext-hypercube of rank d, which contradicts the minimality of V as $\mathrm{gr}^1(V') \subsetneq \mathrm{gr}^1(V)$. □

The following formal lemma gives a rigidity property of Ext-hypercubes.

Lemma 5.2.8 *Let V be a minimal* Ext-*hypercube and V' a finite length multiplicity free representation in* $\mathrm{Rep}^{\mathrm{an}}_{\mathrm{adm}}(G)$. *If* $\mathrm{JH}_G(V) = \mathrm{JH}_G(V')$ *as partially ordered sets, then* $V \cong V'$.

Proof Let $d \geq 0$ be the rank of V. If $d = 0$, Lemma 5.1.4 implies $V \cong V'$. We assume from now $d \geq 1$ and prove $V \cong V'$ by an increasing induction on $d \geq 0$. The equality $\mathrm{JH}_G(V) = \mathrm{JH}_G(V')$ as partially ordered sets implies that V' admits a unique quotient V_0' such that $\mathrm{JH}_G(V_0') = \mathrm{JH}_G(\mathrm{gr}^0(V))$ as partially ordered sets, which by Lemma 5.1.4 implies $V_0' \cong \mathrm{gr}^0(V)$. For $k \geq 1$ and $(x, \pi^\infty) \in \Gamma_k(V)$, V admits a unique good subrepresentation $\widetilde{V}_{x,\pi^\infty}$ such that $\Gamma(\widetilde{V}_{x,\pi^\infty})$ consists of all pairs $(w, \sigma^\infty) \in \Gamma(V)$ such that $(w, \sigma^\infty) \leq (x, \pi^\infty)$. It is clear from (i) of Definition 5.2.6 that $\widetilde{V}_{x,\pi^\infty}$ is itself an Ext-hypercube of rank $d - k \geq 0$. Let $S \subseteq \Gamma(V) \setminus \Gamma_0(V)$ be a subset such that any $(w, \sigma^\infty) \in \Gamma(V)$ satisfying $(w, \sigma^\infty) \leq (x, \pi^\infty)$ for some $(x, \pi^\infty) \in S$ also satisfies $(w, \sigma^\infty) \in S$ and define the good subrepresentation $\mathrm{Fil}_S(V) \overset{\mathrm{def}}{=} \bigcup_{(x,\pi^\infty)\in S} \widetilde{V}_{x,\pi^\infty}$. The equality $\mathrm{JH}_G(V) = \mathrm{JH}_G(V')$ as partially ordered sets implies that V' admits unique subrepresentations $\widetilde{V}'_{x,\pi^\infty}$ for $(x, \pi^\infty) \in \Gamma(V) \setminus \Gamma_0(V)$ and $\mathrm{Fil}_S(V')$ for $S \subseteq \Gamma(V)$ as above such that $\mathrm{JH}_G(\widetilde{V}_{x,\pi^\infty}) = \mathrm{JH}_G(\widetilde{V}'_{x,\pi^\infty})$ and $\mathrm{JH}_G(\mathrm{Fil}_S(V)) = \mathrm{JH}_G(\mathrm{Fil}_S(V'))$ as partially ordered sets. By induction we have in $\mathrm{Rep}^{\mathrm{an}}_{\mathrm{adm}}(G)$ for $(x, \pi^\infty) \in \Gamma(V) \setminus \Gamma_0(V)$

$$\widetilde{V}_{x,\pi^\infty} \cong \widetilde{V}'_{x,\pi^\infty}. \tag{5.73}$$

We prove $\mathrm{Fil}_S(V) \cong \mathrm{Fil}_S(V')$ by induction on $\#S \geq 1$. Let (x, π^∞) be a maximal element of S, $S' \overset{\mathrm{def}}{=} S \setminus \{(x, \pi^\infty)\}$ and S'' the set of $(w, \sigma^\infty) \in \Gamma(V)$ such that $(w, \sigma^\infty) \leq (x, \pi^\infty)$. If $S'' = S$ (i.e. (x, π^∞) is the unique maximal element of S), then $\mathrm{Fil}_S(V) \cong \widetilde{V}_{x,\pi^\infty} \cong \widetilde{V}'_{x,\pi^\infty} \cong \mathrm{Fil}_S(V')$ by (5.73). Otherwise we have $S', S'' \subsetneq S$ which implies $\mathrm{Fil}_{S'}(V) \cong \mathrm{Fil}_{S'}(V')$, $\mathrm{Fil}_{S''}(V) \cong \mathrm{Fil}_{S''}(V')$ and $\mathrm{Fil}_{S' \cap S''}(V) \cong \mathrm{Fil}_{S' \cap S''}(V')$ by induction on S. It follows from the first condition in (i) of Definition 5.2.6 (applied to V) that $\mathrm{Fil}_{S'}(V)$, $\mathrm{Fil}_{S''}(V)$ and $\mathrm{Fil}_{S' \cap S''}(V)$ are indecomposable representations, and hence that $\mathrm{Fil}_S(V)$ is the amalgamate sum of $\mathrm{Fil}_{S'}(V)$ and $\mathrm{Fil}_{S''}(V)$ over $\mathrm{Fil}_{S' \cap S''}(V)$ (recall all representations are multiplicity free). Likewise $\mathrm{Fil}_{S'}(V')$, $\mathrm{Fil}_{S''}(V')$, $\mathrm{Fil}_{S' \cap S''}(V')$ are indecomposable and $\mathrm{Fil}_S(V')$ is the amalgamate sum of $\mathrm{Fil}_{S'}(V')$ and $\mathrm{Fil}_{S''}(V')$ over $\mathrm{Fil}_{S' \cap S''}(V')$. Hence we deduce $\mathrm{Fil}_S(V) \cong \mathrm{Fil}_S(V')$. Now take $S = \Gamma(V) \setminus \Gamma_0(V)$, so that $\mathrm{Fil}_S(V) = \mathrm{Fil}^1(V)$. From $\mathrm{JH}_G(V) = \mathrm{JH}_G(V')$ and the previous paragraph V' fits into a non-split short exact sequence

$$0 \longrightarrow \mathrm{Fil}^1(V) \cong \mathrm{Fil}_S(V') \longrightarrow V' \longrightarrow V_0' \cong \mathrm{gr}^0(V) \longrightarrow 0. \tag{5.74}$$

The minimality of V and Lemma 5.2.7 imply $\dim_E \mathrm{Ext}^1_G(\mathrm{gr}^0(V), \mathrm{Fil}^1(V)) = 1$, and thus it follows from (5.74) that we must have $V' \cong V$. □

Lemma 5.2.9 *Let V be an* Ext-*square. Assume that there exist good subrepresentations* $\mathrm{gr}^2(V) \subseteq V_2 \subseteq V_1 \subseteq \mathrm{Fil}^1(V)$ *such that V_1/V_2 is a basic subquotient of V and* $\mathrm{Ext}^1_G(V/V_1, V_2) = 0$. *Then V is minimal.*

Proof We fix throughout the proof V_1, V_2 as in the statement. Let V_2' and V_1' be good subrepresentations of V such that $\mathrm{gr}^2(V) \subseteq V_2' \subseteq V_2 \subseteq V_1 \subseteq V_1' \subseteq \mathrm{Fil}^1(V)$. As V is multiplicity free, the injection $V_2' \hookrightarrow V_2$ and the surjection $V_1 \twoheadrightarrow V_1'$ induce an injection $\mathrm{Ext}^1_G(V/V_1', V_2') \hookrightarrow \mathrm{Ext}^1_G(V/V_1, V_2)$ which implies

$$\mathrm{Ext}^1_G(V/V_1', V_2') = 0. \tag{5.75}$$

Step 1: We prove that $\dim_E \mathrm{Ext}^1_G(\mathrm{gr}^0(V), \mathrm{Fil}^1(V)) = 1$.
The existence of the Ext-square V ensures $\mathrm{Ext}^1_G(\mathrm{gr}^0(V), \mathrm{Fil}^1(V)) \neq 0$. As V_1/V_2 is a basic direct summand of $\mathrm{gr}^1(V)$ by (5.71), we have a surjection $q : \mathrm{Fil}^1(V) \twoheadrightarrow V_1/V_2$ which induces an exact sequence

$$\mathrm{Ext}^1_G(\mathrm{gr}^0(V), \ker(q)) \to \mathrm{Ext}^1_G(\mathrm{gr}^0(V), \mathrm{Fil}^1(V)) \to \mathrm{Ext}^1_G(\mathrm{gr}^0(V), V_2/V_1). \tag{5.76}$$

Assume $\dim_E \mathrm{Ext}^1_G(\mathrm{gr}^0(V), \mathrm{Fil}^1(V)) \geq 2$. Then (5.76) and $\dim_E \mathrm{Ext}^1_G(\mathrm{gr}^0(V), V_2/V_1) = 1$ (see the second condition in (i) of Definition 5.2.6) imply $\mathrm{Ext}^1_G(\mathrm{gr}^0(V), \ker(q)) \neq 0$. Similar arguments as in the proof of Lemma 5.2.7 with V_1 there replaced by $\ker(q)$ then show that there exists an Ext-square V' with $\mathrm{gr}^0(V') = \mathrm{gr}^0(V)$, $\mathrm{gr}^2(V') = \mathrm{gr}^2(V)$ and $\mathrm{gr}^1(V')$ a good direct summand of $\ker(q)/\mathrm{gr}^2(V) \cong \mathrm{gr}^1(V)/(V_1/V_2)$. But such a V' necessarily fits into a non-split extension $0 \to V_2' \to V' \to V/V_1' \to 0$ for some good subrepresentations V_1', V_2' of V such that $\mathrm{gr}^2(V) \subseteq V_2' \subseteq V_2 \subseteq V_1 \subseteq V_1' \subseteq \mathrm{Fil}^1(V)$, which contradicts (5.75).

Step 2: We prove that V is minimal.
By (iv) of Definition 5.2.6, it suffices to show that there does not exist an Ext-square V' such that $\mathrm{gr}^0(V') = \mathrm{gr}^0(V)$, $\mathrm{gr}^2(V') = \mathrm{gr}^2(V)$ and $\mathrm{gr}^1(V')$ is a proper good direct summand of $\mathrm{gr}^1(V)$. Assume on the contrary that such V' exists. Let V_0 be a basic subquotient of V which is a direct summand of $\mathrm{gr}^1(V)/\mathrm{gr}^1(V')$ and fix a surjection $q' : \mathrm{Fil}^1(V) \twoheadrightarrow V_0$. Using that V is multiplicity free, the injection $\mathrm{Fil}^1(V') \subseteq \ker(q')$ induces an injection

$$0 \neq \mathrm{Ext}^1_G(\mathrm{gr}^0(V), \mathrm{Fil}^1(V')) \to \mathrm{Ext}^1_G(\mathrm{gr}^0(V), \ker(q')), \tag{5.77}$$

and the short exact sequence $0 \to \ker(q') \to \mathrm{Fil}^1(V) \to V_0 \to 0$ induces an exact sequence

$$0 \to \mathrm{Ext}^1_G(\mathrm{gr}^0(V), \ker(q')) \to \mathrm{Ext}^1_G(\mathrm{gr}^0(V), \mathrm{Fil}^1(V)) \to \mathrm{Ext}^1_G(\mathrm{gr}^0(V), V_0). \tag{5.78}$$

By (i) of Definition 5.2.6 $\dim_E \mathrm{Ext}^1_G(\mathrm{gr}^0(V), V_0) = 1$ and the last map in (5.78) is non-zero. By Step 1 it is an isomorphism, which forces $\mathrm{Ext}^1_G(\mathrm{gr}^0(V), \ker(q')) = 0$, contradicting (5.77). □

Recall that for $\underline{j} \in \mathbf{J}$ the representation $C_{\underline{j}}$ is defined in (5.68).

Proposition 5.2.10 *Let $\underline{j}, \underline{j}' \in \mathbf{J}$ and write $\underline{j} = (j_0, j_1, j_2)$, $\underline{j}' = (j_0', j_1', j_2')$.*

(i) If $\underline{j}' = (j_0 + 1, j_1 - 1, j_2)$ there exists a unique Ext-*square $V_{\underline{j},\underline{j}'}$ such that*

$$\begin{cases} \mathrm{gr}^0(V_{\underline{j},\underline{j}'}) \cong C_{\underline{j}'} \\ \mathrm{gr}^1(V_{\underline{j},\underline{j}'}) \cong C_{(j_0+1,j_1,j_2)} \oplus C_{(j_0,j_1-1,j_2)} \\ \mathrm{gr}^2(V_{\underline{j},\underline{j}'}) \cong C_{\underline{j}}. \end{cases}$$

(ii) If $\underline{j}' = (j_0 + 1, j_1 + 1, j_2 + 1)$ there exists a unique Ext-*square $V_{\underline{j},\underline{j}'}$ such that*

$$\begin{cases} \mathrm{gr}^0(V_{\underline{j},\underline{j}'}) \cong C_{\underline{j}'} \\ \mathrm{gr}^1(V_{\underline{j},\underline{j}'}) \cong C_{(j_0+1,j_1,j_2)} \oplus C_{(j_0,j_1+1,j_2+1)} \oplus L(1)^\vee \otimes_E \pi^\infty_{\underline{j},\underline{j}'} \\ \mathrm{gr}^2(V_{\underline{j},\underline{j}'}) \cong C_{\underline{j}} \end{cases}$$

where $\pi^\infty_{\underline{j},\underline{j}'} \cong V^\infty_{[1,j_1],\Delta}$ if $j_0 = j_1 = j_2$, $\pi^\infty_{\underline{j},\underline{j}'}$ is the unique G-basic length 2 representation of G with socle $V^\infty_{[j_2-j_1+1,j_2],\Delta}$ and cosocle $V^\infty_{[j_2-j_1,j_2],\Delta}$ if $j_0 = j_1 < j_2$ (first statement in Lemma 2.2.5), and $\pi^\infty_{\underline{j},\underline{j}'}$ is zero otherwise.

(iii) If $j_1 = j_0 + 1$ there exists a unique Ext-*square $V_{\underline{j},\infty}$ such that*

$$\begin{cases} \mathrm{gr}^0(V_{\underline{j},\infty}) \cong L(1)^\vee \otimes_E \pi^\infty_{\underline{j},\infty} \\ \mathrm{gr}^1(V_{\underline{j},\infty}) \cong C_{(j_0,j_0,j_2)} \oplus C_{(j_0+1,j_0+1,j_2)} \\ \mathrm{gr}^2(V_{\underline{j},\infty}) \cong C_{\underline{j}} \end{cases}$$

where $\pi^\infty_{\underline{j},\infty}$ is the unique G-basic length two representation of G with socle $V^\infty_{[j_2-j_0+1,j_2]\cap\Delta,\Delta}$ and cosocle $V^\infty_{[j_2-j_0,j_2]\cap\Delta,\Delta}$ (Lemma 2.2.5).

(iv) If $j_1 = j_0 - 1$ there exists a unique Ext-*square $V_{\infty,\underline{j}}$ such that*

$$\begin{cases} \mathrm{gr}^0(V_{\infty,\underline{j}}) \cong C_{\underline{j}} \\ \mathrm{gr}^1(V_{\infty,\underline{j}}) \cong C_{(j_0,j_0,j_2)} \oplus C_{(j_0-1,j_0-1,j_2)} \\ \mathrm{gr}^2(V_{\infty,\underline{j}}) \cong L(1)^\vee \otimes_E \pi^\infty_{\infty,\underline{j}} \end{cases}$$

where $\pi^\infty_{\infty,\underline{j}} \cong V^\infty_{[1,j_2-1],\Delta}$ if $j_1 + 1 = j_0 = j_2$, and $\pi^\infty_{\infty,\underline{j}}$ is the unique G-basic length 2 representation of G with socle $V^\infty_{[j_2-j_0+1,j_2-1],\Delta}$ and cosocle $V^\infty_{[j_2-j_0,j_2-1],\Delta}$ (Lemma 2.2.5) if $j_1 + 1 = j_0 < j_2$.

Moreover, all Ext-*squares above are minimal and strict (see Definition 5.2.6).*

Proof We make crucial use of the Ext-squares of $U(\mathfrak{g})$-modules constructed in Sect. 3.4. We divide the proof into two steps.

Step 1: We construct $I \subseteq \Delta$, Q in $\mathcal{O}^{\mathfrak{p}_I}_{\mathrm{alg}}$ (see Sect. 3.1) and a G-regular irreducible π^∞ in $\mathrm{Rep}^\infty_{\mathrm{adm}}(L_I)$ such that $\mathcal{F}^G_{P_I}(Q, \pi^\infty)$ contains a unique subquotient of the form $V_{\underline{j},\underline{j}'}$ (resp. $V_{\underline{j},\infty}$, $V_{\infty,\underline{j}}$) which is a strict Ext-square as described in (i) and (ii) (resp. (iii), (iv)).

We define $I \overset{\mathrm{def}}{=} \Delta \setminus \{j_1, j_1'\} = \widehat{j_1} \cap \widehat{j_1'}$ in (i) and (ii), and $I \overset{\mathrm{def}}{=} \Delta \setminus \{j_0, j_1\} = \widehat{j_0} \cap \widehat{j_1}$ in (iii) and (iv). In case (i) and (ii) we set $Q \overset{\mathrm{def}}{=} Q_1(w_{j_1,j_0}, w_{j_1',j_0'})$ using Proposition 3.4.9. In case (iii) we set $Q \overset{\mathrm{def}}{=} Q_1(w_{j_1,j_0}, 1)$ and in case (iv) we set $Q \overset{\mathrm{def}}{=} Q_1(1, w_{j_1,j_0})$ using Remark 3.4.10. All these Ext-squares of $U(\mathfrak{g})$-modules are actually in $\mathcal{O}^{\mathfrak{p}_I}_{\mathrm{alg}}$ using Lemma 3.1.1 and [33, Proposition 9.3(c)].

We consider the $W(L_I)$-coset Σ given by $\Sigma_{j_1,j_2} \cap \Sigma_{j_1',j_2'}$ in (i) and (ii), and by $\Sigma_{j_1,j_2} \cap \Sigma_{j_0,j_2}$ in (iii) and (iv). Thanks to (i) of Lemma 2.3.5 and (i) of Lemma 2.2.11, we can define a G-regular irreducible smooth representation π^∞ of L_I in (i) and (ii) as

$$\pi^\infty \overset{\mathrm{def}}{=} J_{\widehat{j_1},I}(\pi^\infty_{j_1',j_2'})_{B^I_\Sigma} \cong J'_{\widehat{j_1},I}(\pi^\infty_{j_1,j_2})_{B^I_\Sigma}, \tag{5.79}$$

and note that $\pi^\infty_{j_1,j_2}$ (resp. $\pi^\infty_{j_1',j_2'}$) is a subquotient of $i^\infty_{I,\widehat{j_1}}(\pi^\infty)$ (resp. $i^\infty_{I,\widehat{j_1'}}(\pi^\infty)$) by (2.8) (resp. by (2.7)). Similarly, using (i) of Lemma 2.3.5 (applied with (j_1, j_1', j_2, j_2') there being (j_1, j_0, j_2, j_2)) and (i) of Lemma 2.2.11) we have in (iii)

$$\pi^\infty \overset{\mathrm{def}}{=} J_{\widehat{j_0},I}(\pi^\infty_{j_0,j_2})_{B^I_\Sigma} \cong J'_{\widehat{j_1},I}(\pi^\infty_{j_1,j_2})_{B^I_\Sigma}$$

and using (i) of Lemma 2.3.5 (applied with (j_1, j_1', j_2, j_2') there being (j_0, j_1, j_2, j_2)) and (i) of Lemma 2.2.11) we have in (iv)

$$\pi^\infty \overset{\mathrm{def}}{=} J_{\widehat{j_1},I}(\pi^\infty_{j_1,j_2})_{B^I_\Sigma} \cong J'_{\widehat{j_0},I}(\pi^\infty_{j_0,j_2})_{B^I_\Sigma}. \tag{5.80}$$

Moreover in (iii) and (iv) $\pi^\infty_{j_1,j_2}$ (resp. $\pi^\infty_{j_0,j_2}$) is a subquotient of $i^\infty_{I,\widehat{j_1}}(\pi^\infty)$ (resp. $i^\infty_{I,\widehat{j_0}}(\pi^\infty)$) using again (2.7) and (2.8). Using Lemma 2.3.1 applied to $\sigma^\infty_{j_1,j_2} = \pi^\infty$ one easily checks that:

- in (ii) $\pi^\infty_{\underline{j},\underline{j}'}$ is a subquotient of $i^\infty_{I,\Delta}(\pi^\infty)$ which is a subrepresentation of $i^\infty_{\widehat{j_1},\Delta}(\pi^\infty_{j_1,j_2})$ and a quotient of $i^\infty_{\widehat{j_1'},\Delta}(\pi^\infty_{j_1',j_2'})$;

- in (iii) $\pi^\infty_{\underline{j},\infty}$ is a subquotient of $i^\infty_{I,\Delta}(\pi^\infty)$ which is a subrepresentation of $i^\infty_{\widehat{j_0},\Delta}(\pi^\infty_{j_0,j_2})$ such that $\mathrm{Hom}_G(\pi^\infty_{\underline{j},\infty}, i^\infty_{\widehat{j_1},\Delta}(\pi^\infty_{j_1,j_2})) \neq 0$;

- in (iv) $\pi^\infty_{\infty,\underline{j}}$ is a subquotient of $i^\infty_{I,\Delta}(\pi^\infty)$ which is a quotient of $i^\infty_{\widehat{j_0},\Delta}(\pi^\infty_{j_0,j_2})$ such that $\mathrm{Hom}_G(i^\infty_{\widehat{j_1},\Delta}(\pi^\infty_{j_1,j_2}), \pi^\infty_{\infty,\underline{j}}) \neq 0$.

We see that all assumptions in Lemma 5.1.19 are satisfied and thus $\mathcal{F}_{P_I}^G(Q, \pi^\infty)$ contains a unique subquotient of the form $V_{\underline{j},\underline{j}'}$ (resp. $V_{\underline{j},\infty}$, $V_{\infty,\underline{j}}$) in (i) and (ii) (resp. in (iii), in (iv)). More precisely the first condition in (i) of Definition 5.2.6 is obvious and the second condition for $V_{\underline{j},\underline{j}'}$, $V_{\underline{j},\infty}$ and $V_{\infty,\underline{j}}$ respectively can be checked using Lemmas 5.2.1, 5.2.2 and (ii) of Lemma 5.1.19. In (i) and (ii), since $\mathrm{gr}^0(V_{\underline{j},\underline{j}'})$ and $\mathrm{gr}^2(V_{\underline{j},\underline{j}'})$ are simple, the Ext-square $V_{\underline{j},\underline{j}'}$ is strict. In (iii), as $\mathrm{gr}^0(V_{\underline{j},\infty})$ is the only reducible basic subquotient of $V_{\underline{j},\infty}$ and $\pi_{\underline{j},\infty}^\infty$ is a subrepresentation of $i_{\widehat{j_0},\Delta}^\infty(\pi_{\widehat{j_0},j_2}^\infty)$, we deduce from (ii) of Lemma 5.1.15 (applied with $I_0 = \Delta$) that $V_{\underline{j},\infty}$ is strict. Similarly, in (iv), as $\mathrm{gr}^2(V_{\infty,\underline{j}})$ is the only reducible basic subquotient of $V_{\infty,\underline{j}}$ and $\pi_{\infty,\underline{j}}^\infty$ is a quotient of $i_{\widehat{j_0},\Delta}^\infty(\pi_{\widehat{j_0},j_2}^\infty)$, we deduce from (i) of Lemma 5.1.15 (applied with $I_1 = \Delta$) that $V_{\infty,\underline{j}}$ is strict.

Step 2: We prove the minimality of the Ext-squares in the statement.
In each case, we choose a subrepresentation $M \subseteq Q$ and a quotient M' of Q/M as follows.

- In (i) and (ii), M' has length 2 with socle $L(w_{j_1,j_0+1})$ and cosocle $L(w_{j_1,j_0})$ if $j_1 \leq j_0$, and $M' \stackrel{\mathrm{def}}{=} L(w_{j_1,j_0})$ if $j_1 \geq j_0 + 1$.
- In (i), M has length 2 with socle $L(w_{j_1-1,j_0+1})$ and cosocle $L(w_{j_1-1,j_0})$ if $j_1 \geq j_0 + 1$, and $M \stackrel{\mathrm{def}}{=} L(w_{j_1-1,j_0+1})$ if $j_1 \leq j_0$. In (ii), M has length 2 with socle $L(w_{j_1+1,j_0+1})$ and cosocle $L(w_{j_1+1,j_0})$ if $j_1 \geq j_0$, and $M \stackrel{\mathrm{def}}{=} L(w_{j_1+1,j_0+1})$ if $j_1 \leq j_0 - 1$.
- In (iii), $M' \stackrel{\mathrm{def}}{=} L(w_{j_1,j_0})$ and M has length 2 with socle $L(1)$ and cosocle $L(s_{j_0})$.
- In (iv), $M' \stackrel{\mathrm{def}}{=} L(1)$ and M has length 2 with socle $L(w_{j_1,j_0})$ and cosocle $L(s_{j_1})$.

Since M' is a proper quotient of Q/M, we have for $k \leq 1$

$$\mathrm{Ext}^k_{U(\mathfrak{g})}(M', M) = 0 \tag{5.81}$$

by minimality of the Ext-square Q, see Proposition 3.4.9 and Remark 3.4.10. By (ii) of Lemma 5.1.19 M uniquely determines a quotient V of $V_{\underline{j},\underline{j}'}$ in (i) and (ii) (resp. $V_{\underline{j},\infty}$ in (iii), $V_{\infty,\underline{j}}$ in (iv)) and M' uniquely determines a subrepresentation V' of $V_{\underline{j},\underline{j}'}$ (resp. $V_{\underline{j},\infty}$, $V_{\infty,\underline{j}}$). We prove the minimality of the Ext-square $V_{\underline{j},\underline{j}'}$, $V_{\underline{j},\infty}$ and $V_{\infty,\underline{j}}$ using Lemma 5.2.9. To check the assumption of Lemma 5.2.9, it suffices to prove in each case above

$$\mathrm{Ext}^1_G(V, V') = 0. \tag{5.82}$$

The idea is to apply (4.146) and use (5.81). But in order to do so, we need to replace V' by a parabolic Verma module as this is part of the assumptions in (4.146).

Case 2.1: In (i) and (ii), we have $V \cong \mathcal{F}_{P_{\widehat{j_1'}}}^G(M, \pi_{j_1',j_2'}^\infty)$ and $V' \cong \mathcal{F}_{P_{\widehat{j_1}}}^G(M', \pi_{j_1,j_2}^\infty)$. We first note that M' is a quotient of $M(w_{j_1,j_0})$ (using for instance (3.27)), hence of $M^{\widehat{j_1}}(w_{j_1,j_0})$ by [33, Theorem 9.4(c)]. By *loc. cit.* and Lemma 3.1.1, $M^{\widehat{j_1}}(w_{j_1,j_0})$ and M do not share any constituent, and thus $\mathcal{F}_{P_{\widehat{j_1}}}^G(M^{\widehat{j_1}}(w_{j_1,j_0}), \pi_{j_1,j_2}^\infty)$ and V do not share any constituent by Lemma 5.1.1. Hence the surjection $M^{\widehat{j_1}}(w_{j_1,j_0}) \twoheadrightarrow M'$ induces an embedding

$$\mathrm{Ext}^1_G(V, V') \hookrightarrow \mathrm{Ext}^1_G\big(V, \mathcal{F}^G_{P_{\widehat{j_1}}}(M^{\widehat{j_1}}(w_{j_1,j_0}), \pi^\infty_{j_1,j_2})\big). \tag{5.83}$$

Let $M'' \subseteq M^{\widehat{j_1}}(w_{j_1,j_0})$ such that $M^{\widehat{j_1}}(w_{j_1,j_0})/M'' \cong M'$ and let $L(x) \in \mathrm{JH}_{U(\mathfrak{g})}(M'')$. We have $x \geq w_{j_1,j_0}$ (from the structure of $M(w_{j_1,j_0})$), $D_L(x) = \{j_1\}$ (from Lemma 3.1.1) and $x \notin \{w_{j_1,j_0}, w_{j_1,j_0+1}\}$ (from $L(x) \notin \mathrm{JH}_{U(\mathfrak{g})}(M')$ and the fact both $L(w_{j_1,j_0})$ and $L(w_{j_1,j_0+1})$ have multiplicity ≤ 1 in $M(w_{j_1,j_0})$ by (i) of Lemma A.12 for instance). Since each $L(y) \in \mathrm{JH}_{U(\mathfrak{g})}(M)$ satisfies $y \in \{w_{j_1',j_0}, w_{j_1',j_0+1}\}$ we deduce $\mathrm{Ext}^1_{U(\mathfrak{g})}(L(x), L(y)) = 0$ from (iii) of Lemma A.11. Using this and $L(x) \notin \mathrm{JH}_{U(\mathfrak{g})}(M)$ we deduce by an obvious dévissage on $0 \to M'' \to M^{\widehat{j_1}}(w_{j_1,j_0}) \to M' \to 0$:

$$\mathrm{Ext}^k_{U(\mathfrak{g})}(M', M) \xrightarrow{\sim} \mathrm{Ext}^k_{U(\mathfrak{g})}(M^{\widehat{j_1}}(w_{j_1,j_0}), M) \text{ for } k \leq 1,$$

which together with (5.81) implies $\mathrm{Ext}^k_{U(\mathfrak{g})}(M^{\widehat{j_1}}(w_{j_1,j_0}), M) = 0$ for $k \leq 1$. As $\Sigma_{j_1,j_2} \cap \Sigma_{j_1',j_2'} \neq \emptyset$ ((i) of Lemma 2.3.4), we can apply (4.146) with $w = 1$ and obtain

$$\mathrm{Ext}^1_G(V, \mathcal{F}^G_{P_{\widehat{j_1}}}(M^{\widehat{j_1}}(w_{j_1,j_0}), \pi^\infty_{j_1,j_2})) = 0.$$

By (5.83) this implies (5.82).

Case 2.2: In (iii), we have $V' \cong C_{\underline{j}}$ and V is a subrepresentation of $\mathcal{F}^G_{P_{\widehat{j_0}}}(M, \pi^\infty_{j_0,j_2})$ as $\pi^\infty_{\underline{j},\infty}$ is a subrepresentation of $i^\infty_{\widehat{j_0},\Delta}(\pi^\infty_{j_0,j_2})$. As $(j_0, j_2) = (j_1 - 1, j_2)$, we have by (i) of Lemma 2.3.5

$$\mathrm{Hom}_{L_{\widehat{j_1}}}(i^\infty_{\widehat{j_0}\cap\widehat{j_1}}(J_{\widehat{j_0},\widehat{j_0}\cap\widehat{j_1}}(\pi^\infty_{j_0,j_2})), \pi^\infty_{j_1,j_2}) \neq 0,$$

which together with (ii) of Lemma 2.2.11 implies $J_{\widehat{j_1}}(\tau^\infty)_{\mathcal{B}^{\widehat{j_1}}_{\Sigma_{j_1,j_2}}} = 0$ for any constituent τ^∞ of $i^\infty_{\widehat{j_0},\Delta}(\pi^\infty_{j_0,j_2})/\pi^\infty_{\underline{j},\infty}$. Note that $\mathrm{JH}_{U(\mathfrak{g})}(M) \cap \mathrm{JH}_{U(\mathfrak{g})}(M^{\widehat{j_1}}(w_{j_1,j_0})) = \emptyset$ (using $j_1 > j_0$). Hence, we deduce from Remark 5.1.21 (applied when $V_+ = V$) an isomorphism

$$\mathrm{Ext}^1_G(\mathcal{F}^G_{P_{\widehat{j_0}}}(M, \pi^\infty_{j_0,j_2}), V') \xrightarrow{\sim} \mathrm{Ext}^1_G(V, V'). \tag{5.84}$$

Note that $\mathcal{F}^G_{P_{\widehat{j_0}}}(M, \pi^\infty_{j_0,j_2})$ and $\mathcal{F}^G_{P_{\widehat{j_1}}}(M^{\widehat{j_1}}(w_{j_1,j_0}), \pi^\infty_{j_1,j_2})$ also have no common constituent by Lemma 5.1.1. So the surjection $M^{\widehat{j_1}}(w_{j_1,j_0}) \twoheadrightarrow L(w_{j_1,j_0})$ induces an embedding

$$\mathrm{Ext}^1_G(\mathcal{F}^G_{P_{\widehat{j_0}}}(M, \pi^\infty_{j_0,j_2}), V') \hookrightarrow \mathrm{Ext}^1_G(\mathcal{F}^G_{P_{\widehat{j_0}}}(M, \pi^\infty_{j_0,j_2}), \mathcal{F}^G_{P_{\widehat{j_1}}}(M^{\widehat{j_1}}(w_{j_1,j_0}), \pi^\infty_{j_1,j_2})). \tag{5.85}$$

By a similar argument as in Case 2.1 using (iii) of Lemma A.11 and (5.81), one shows $\mathrm{Ext}^k_{U(\mathfrak{g})}(M^{\widehat{j_1}}(w_{j_1,j_0}), M) = 0$ for $k \leq 1$. As $\Sigma_{j_1,j_2} \cap \Sigma_{j_0,j_2} \neq \emptyset$ (from (i) of Lemma 2.3.4 and $j_0 = j_1 - 1$ here) we can apply (4.146) with $w = 1$ and deduce

$$\mathrm{Ext}^1_G(\mathcal{F}^G_{P_{\widehat{j_0}}}(M, \pi^\infty_{j_0,j_2}), \mathcal{F}^G_{P_{\widehat{j_1}}}(M^{\widehat{j_1}}(w_{j_1,j_0}), \pi^\infty_{j_1,j_2})) = 0.$$

Then (5.85) and (5.84) imply (5.82).

Case 2.3: In (iv), we have $V \cong \mathcal{F}^G_{P_{\widehat{j_1}}}(M, \pi^\infty_{j_1,j_2})$ and $V' \cong L(1)^\vee \otimes_E \pi^\infty_{\infty,\underline{j}}$, and the assumption $\Sigma_{j_1,j_2} \cap W(G) \cdot \mathcal{J}(\pi^\infty_{\infty,\underline{j}}) \neq \emptyset$ is satisfied since both cosets contain $\mathcal{J}(\pi^\infty)$ (see case (iv) below (5.80)). Thus we can apply directly (4.146) (with $w = 1$) and obtain (5.82) from (5.81). □

A $D(G)$-module is said to be $Z(\mathfrak{g})$-finite if every element is killed by an ideal of finite codimension in $Z(\mathfrak{g})$ (recall that $Z(\mathfrak{g})$ lies in the center of $D(G)$ by [70, Prop. 3.7]). For $\lambda, \mu \in \Lambda \times \Lambda$, [35, (2)] defines (by the formula (1.22)) an exact endofunctor $\mathcal{T}^\mu_\lambda$ on the abelian category of $Z(\mathfrak{g})$-finite $D(G)$-modules such that for $I \subseteq \Delta$, M in $\mathcal{O}^{\mathfrak{p}_I}_{\mathrm{alg}}$ and π^∞ a strongly admissible smooth representation of L_I over E we have an isomorphism of $D(G)$-modules (see [35, Theorem 2])

$$\mathcal{T}^\mu_\lambda(\mathcal{F}^G_{P_I}(M, \pi^\infty)^\vee) \cong \mathcal{F}^G_{P_I}(T^\mu_\lambda(M), \pi^\infty)^\vee \tag{5.86}$$

where T^μ_λ is defined in (3.84) (note that $T^\mu_\lambda(M)$ remains in $\mathcal{O}^{\mathfrak{p}_I}_{\mathrm{alg}}$ when M is in $\mathcal{O}^{\mathfrak{p}_I}_{\mathrm{alg}}$ using the argument in the proof of [33, Theorem 1.1(d)]). Recall from Sect. 4.2 that $\mathcal{C}_{D(G)}$ is the abelian category of coadmissible $D(G)$-modules over E. If V is an admissible locally analytic representation of G over E such that $V^\vee$ is $Z(\mathfrak{g})$-finite, it follows from the discussion below [35, Definition 2.4.5] that $\mathcal{T}^\mu_\lambda(V^\vee)$ is again in $\mathcal{C}_{D(G)}$, hence we can define an admissible locally analytic representation by $\mathcal{T}^\mu_\lambda(V) \stackrel{\mathrm{def}}{=} \mathcal{T}^\mu_\lambda(V^\vee)^\vee$. Note also that if V has a central character, then (using that an irreducible algebraic representation of G always has a central character), we deduce from [35, (2)] that $\mathcal{T}^\mu_\lambda(V)$ also has a central character.

Let $j \in \{1, \ldots, n-1\}$ and $\mu \in \Lambda$ such that $\langle \mu + \rho, \alpha^\vee \rangle \geq 0$ for $\alpha \in \Phi^+$ and the stabilizer of μ in $W(G)$ for the dot action is $\{1, s_j\}$. By the same formula as below (3.84) we define a wall-crossing functor $\Theta_\mu \stackrel{\mathrm{def}}{=} \mathcal{T}^{w_0 \cdot \mu_0}_\mu \circ \mathcal{T}^\mu_{w_0 \cdot \mu_0}$ which is an exact endofunctor on the abelian category of $Z(\mathfrak{g})$-finite $D(G)$-modules. If V is an admissible locally analytic representation of G over E such that $V^\vee$ is $Z(\mathfrak{g})$-finite, we define $\Theta_\mu(V) \stackrel{\mathrm{def}}{=} \Theta_\mu(V^\vee)^\vee$. By (5.86) for $I \subseteq \Delta$, M in $\mathcal{O}^{\mathfrak{p}_I}_{\mathrm{alg}}$ and π^∞ a strongly admissible smooth representation of L_I over E we have

$$\Theta_\mu(\mathcal{F}^G_{P_I}(M, \pi^\infty)^\vee) \cong \mathcal{F}^G_{P_I}(\Theta_{s_j}(M), \pi^\infty)^\vee. \tag{5.87}$$

As $\Theta_\mu(\mathcal{F}^G_{P_I}(M, \pi^\infty)^\vee)$ only depends on s_j (and on μ_0) by (5.87), we write $\Theta_{s_j}(\mathcal{F}^G_{P_I}(M, \pi^\infty)^\vee)$ in that case.

Remark 5.2.11 Though we do not need it, it is possible that the endofunctor $\Theta_\mu = \mathcal{T}^{w_0 \cdot \mu_0}_\mu \circ \mathcal{T}^\mu_{w_0 \cdot \mu_0}$ only depends on s_j (and on μ_0) up to isomorphism. Let $\mu' \in \Lambda$ such that $\langle \mu' + \rho, \alpha^\vee \rangle \geq 0$ for $\alpha \in \Phi^+$ and the stabilizer of μ' in $W(G)$ for the dot action is $\{1, s_j\}$. By [5, Proposition 5.0.8(c)] we have isomorphisms of functors $T_\mu^{\mu'} \circ T_{w_0\cdot\mu_0}^{\mu} \xrightarrow{\sim} T_{w_0\cdot\mu_0}^{\mu'}$ and $T_\mu^{w_0\cdot\mu_0} \circ T_{\mu'}^{\mu} \xrightarrow{\sim} T_{\mu'}^{w_0\cdot\mu_0}$. It would be enough to prove the analogous statements with $\mathcal{T}_\mu^{\mu'} \circ \mathcal{T}_{w_0\cdot\mu_0}^{\mu}$ and $\mathcal{T}_\mu^{w_0\cdot\mu_0} \circ \mathcal{T}_{\mu'}^{\mu}$. Indeed, we would then have isomorphisms of endofunctors

$$\begin{aligned}\mathcal{T}_{\mu'}^{w_0\cdot\mu_0} \circ \mathcal{T}_{w_0\cdot\mu_0}^{\mu'} &\cong (\mathcal{T}_\mu^{w_0\cdot\mu_0} \circ \mathcal{T}_{\mu'}^{\mu}) \circ (\mathcal{T}_\mu^{\mu'} \circ \mathcal{T}_{w_0\cdot\mu_0}^{\mu}) \\ &\cong \mathcal{T}_\mu^{w_0\cdot\mu_0} \circ (\mathcal{T}_{\mu'}^{\mu} \circ \mathcal{T}_\mu^{\mu'}) \circ \mathcal{T}_{w_0\cdot\mu_0}^{\mu} \cong \mathcal{T}_\mu^{w_0\cdot\mu_0} \circ \mathcal{T}_{w_0\cdot\mu_0}^{\mu}\end{aligned}$$

where the last isomorphism follows from $\mathcal{T}_{\mu'}^{\mu} \circ \mathcal{T}_\mu^{\mu'} \cong \mathrm{id}$ (see [35, Theorem 3.2.1]). If M is a $Z(\mathfrak{g})$-finite $D(G)$-module, seeing M as a $U(\mathfrak{g})$-module we have an isomorphism of $U(\mathfrak{g})$-modules $(T_\mu^{\mu'} \circ T_{w_0\cdot\mu_0}^{\mu})(M) \xrightarrow{\sim} T_{w_0\cdot\mu_0}^{\mu'}(M)$, and since the functors $\mathcal{T}_\lambda^{\mu}$ in [35, §1] are just T_λ^{μ} on the underlying $U(\mathfrak{g})$-modules, it would be enough to prove that this isomorphism is $D(G)$-equivariant. An examination of the proof of [35, Theorem 3.2.1] shows that it would even be enough to prove this $D(G)$-equivariance for M the form $D(G) \otimes_{U(\mathfrak{g})} N$ where N is in $\mathrm{Mod}_{U(\mathfrak{g})}$. However, we couldn't find a quick argument for this $D(G)$-equivariance (if true).

Lemma 5.2.12 *Let $x \in W(G)$ and π^∞ an irreducible G-regular smooth representation of L_{I_x} such that $V = \mathcal{F}_{P_{I_x}}^G(L(x), \pi^\infty)$ is irreducible (see (iii) of Theorem 4.3.7).*

(i) If $j \notin D_R(x)$ then $\Theta_{w_0 s_j w_0}(V) = 0$.
(ii) If $j \in D_R(x)$ then $\Theta_{w_0 s_j w_0}(V)$ has V as both socle and cosocle.

Proof By Proposition 3.4.5 $\Theta_{w_0 s_j w_0}(L(x))$ is zero if $j \notin D_R(x)$, and has Loewy length 3 with both socle and cosocle isomorphic to $L(x)$ (and middle layer not containing $L(x)$) if $j \in D_R(x)$. Then (i) follows from (5.87). Assume $j \in D_R(x)$. Let W be an irreducible constituent of $\Theta_{w_0 s_j w_0}(V)$, which we can write $W = \mathcal{F}_{I_w}^G(L(w), \sigma^\infty)$ for some $w \in W(G)$ such that $I_x \subseteq I_w$ and some irreducible G-regular smooth representation σ^∞ which is a subquotient of $i_{I_x, I_w}^\infty(\pi^\infty)$ (using (ii) of Theorem 4.3.7 and (i), (iii) of Lemma 2.1.15). By the first sentence of this proof together with (i) of Lemma 5.1.7 applied with $I_0 = I_w$, $I_1 = I_x$, $M_0 = L(w)$, $M_1 = \Theta_{w_0 s_j w_0}(L(x))$, $\pi_0^\infty = \sigma^\infty$ and $\pi_1^\infty = \pi^\infty$, we have $\mathrm{Hom}_G(W, \Theta_{w_0 s_j w_0}(V)) \neq 0$ if and only if $w = x$ and $\sigma^\infty = \pi^\infty$ if and only if $W = V$ (using Lemma 5.1.1 for the last equivalence). Moreover the space of homomorphisms is then 1-dimensional (still by (i) of Lemma 5.1.7). It follows that V is the socle of $\Theta_{w_0 s_j w_0}(V)$. An analogous argument using (ii) of Lemma 5.1.7 gives that V is also the cosocle of $\Theta_{w_0 s_j w_0}(V)$. □

Lemma 5.2.13 *Let $\underline{j} = (j_0, j_1, j_2) \in \mathbf{J}$ and $\mu \in \Lambda$ such that $\langle \mu + \rho, \alpha^\vee \rangle \geq 0$ for $\alpha \in \Phi^+$ and the stabilizer of μ in $W(G)$ for the dot action is $\{1, w_0 s_{j_0} w_0\}$.*

(i) The representation $\Theta_{w_0 s_{j_0} w_0}(C_{\underline{j}})$ of G has simple socle and cosocle $C_{\underline{j}}$ and middle layer

$$\mathrm{rad}_G(\Theta_{w_0 s_{j_0} w_0}(C_{\underline{j}}))/\mathrm{soc}_G(\Theta_{w_0 s_{j_0} w_0}(C_{\underline{j}})) \cong C_{(j_0-1, j_1, j_2)} \oplus C_{(j_0+1, j_1, j_2)} \oplus L(1)^\vee \otimes_E \pi^\infty$$

where π^∞ is non-zero if and only if $j_0 = j_1$, in which case $\pi^\infty \cong i^\infty_{\widehat{j_0},\Delta}(\pi^\infty_{j_0,j_2})$, and where we omit $C_{(j_0-1,j_1,j_2)}$ when $j_0 = 1$ and $C_{(j_0+1,j_1,j_2)}$ when $j_0 = n-1$.

(ii) Assume $j_0 < n-1$ and let $\underline{j}' \in \{(j_0, j_1+1, j_2+1), (j_0, j_1-1, j_2)\}$, $\underline{j}'' \stackrel{def}{=} (j_0+1, j_1', j_2')$ and $V_{\underline{j},\underline{j}''}$ the minimal Ext*-square with socle $C_{\underline{j}}$ and cosocle $C_{\underline{j}''}$ constructed in (i) or (ii) of Proposition 5.2.10. Let V be the unique length 2 representation of G with socle $C_{\underline{j}}$ and cosocle $C_{\underline{j}'}$ defined from Lemma 5.2.1. Then $\Theta_\mu(V)$ admits a unique subquotient isomorphic to $V_{\underline{j},\underline{j}''}$ which is moreover a subrepresentation.*

Proof We prove (i). It follows from Lemma 5.2.12 that $\Theta_{w_0 s_{j_0} w_0}(C_{\underline{j}})$ has simple socle and cosocle $C_{\underline{j}}$. By Proposition 3.4.5 and Remark 3.4.6 $\Theta_{w_0 s_{j_0} w_0}(L(w_{j_1,j_0}))$ has Loewy length 3 with both socle and cosocle $L(w_{j_1,j_0})$ and with middle layer $\mathrm{rad}_1(\Theta_{w_0 s_{j_0} w_0}(L(w_{j_1,j_0})))$ semi-simple and multiplicity free. More precisely, from Remark 3.4.6 $L(x)$ is a constituent of $\mathrm{rad}_1(\Theta_{w_0 s_{j_0} w_0}(L(w_{j_1,j_0})))$ if and only if either $j_0 = j_1$ and $x = 1$, or $j_0 > 1$ and $x = w_{j_1,j_0-1}$, or $j_0 < n-1$ and $x = w_{j_1,j_0+1}$. If $x = w_{j_1,j}$ for $j \in \{j_0-1, j_0+1\}$, we have $\mathcal{F}^G_{P_{\widehat{j_1}}}(L(x), \pi^\infty_{j_1,j_2}) = C_{(j,j_1,j_2)}$ by definition (see (5.68)). If $x = 1$ (with $j_0 = j_1$), we have $\mathcal{F}^G_{P_{\widehat{j_1}}}(L(1), \pi^\infty_{j_1,j_2}) = L(1)^\vee \otimes_E i^\infty_{\widehat{j_0},\Delta}(\pi^\infty_{j_0,j_2})$ using (ii) of Theorem 4.3.7. By (5.87) this finishes the proof of (i).

We prove (ii). Let $I \stackrel{\text{def}}{=} \Delta \setminus \{j_1, j_1'\} = \widehat{j_1} \cap \widehat{j_1'}$ and M_0 the (unique) length 2 object in $\mathcal{O}^{\mathfrak{p}_I}_{\mathrm{alg}}$ with socle $L(w_{j_1',j_0})$ and cosocle $L(w_{j_1,j_0})$. Let $\Sigma \stackrel{\text{def}}{=} \Sigma_{j_1,j_2} \cap \Sigma_{j_1',j_2'}$ and π^∞ as in (5.79). Then by Step 5 in the proof of Lemma 5.1.5 we know that V (as in the statement) is a subquotient of $\mathcal{F}^G_{P_I}(M_0, \pi^\infty)$. By Lemma 3.4.7 $L(w_{j_1',j_0+1})$ appears with multiplicity one in $\Theta_{w_0 s_{j_0} w_0}(M_0)$ and the unique quotient of $\Theta_{w_0 s_{j_0} w_0}(M_0)$ with socle $L(w_{j_1',j_0+1})$ is isomorphic to $Q_1(w_{j_1,j_0}, w_{j_1',j_0+1})$ (see Proposition 3.4.9). Moreover $\pi^\infty_{j_1'',j_2''} = \pi^\infty_{j_1',j_2'}$ appears with multiplicity one in the G-basic, hence multiplicity free, representation $i^\infty_{I,\widehat{j_1}}(\pi^\infty)$ (see below (5.79)). Using Theorem 4.3.7 and Lemma 5.1.1 this implies that $C_{\underline{j}''}$ appears with multiplicity one in $\Theta_{w_0 s_{j_0} w_0}(\mathcal{F}^G_{P_I}(M_0, \pi^\infty)) \cong \mathcal{F}^G_{P_I}(\Theta_{w_0 s_{j_0} w_0}(M_0), \pi^\infty)$. Moreover the unique subrepresentation of $\Theta_{w_0 s_{j_0} w_0}(\mathcal{F}^G_{P_I}(M_0, \pi^\infty))$ with cosocle $C_{\underline{j}''}$ is a subrepresentation of $\mathcal{F}^G_{P_I}(Q_1(w_{j_1,j_0}, w_{j_1',j_0+1}), \pi^\infty)$ and thus is multiplicity free (using Lemma 5.1.1 and the facts that $Q_1(w_{j_1,j_0}, w_{j_1',j_0+1})$ is multiplicity free and π^∞ irreducible G-regular). By the last paragraph in Step 1 in the proof of Proposition 5.2.10, we know that $V_{\underline{j},\underline{j}''}$ is the unique subquotient of $\mathcal{F}^G_{P_I}(Q_1(w_{j_1,j_0}, w_{j_1',j_0+1}), \pi^\infty)$ with socle $C_{\underline{j}}$ and cosocle $C_{\underline{j}''}$. Since any subquotient of $\Theta_{w_0 s_{j_0} w_0}(\mathcal{F}^G_{P_I}(M_0, \pi^\infty))$ with cosocle $C_{\underline{j}''}$ is a subquotient of $\mathcal{F}^G_{P_I}(Q_1(w_{j_1,j_0}, w_{j_1',j_0+1}), \pi^\infty)$ by the previous discussion, $V_{\underline{j},\underline{j}''}$ is also the unique subquotient of $\Theta_{w_0 s_{j_0} w_0}(\mathcal{F}^G_{P_I}(M_0, \pi^\infty))$ with socle $C_{\underline{j}}$ and cosocle $C_{\underline{j}''}$. As $\Theta_\mu(V)$ is a subquotient of $\Theta_{w_0 s_{j_0} w_0}(\mathcal{F}^G_{P_I}(M_0, \pi^\infty))$ and $C_{\underline{j}}, C_{\underline{j}''} \in \mathrm{JH}_G(\Theta_\mu(V))$ (by (i) applied to $C_{\underline{j}}$ and $C_{\underline{j}'}$), $V_{\underline{j},\underline{j}''}$ is the unique subquotient of $\Theta_\mu(V)$ with socle $C_{\underline{j}}$ and cosocle $C_{\underline{j}''}$. By the explicit description of $V_{\underline{j},\underline{j}''}$ in (i) or (ii) of Proposition 5.2.10 we know that V injects into $V_{\underline{j},\underline{j}''}$. By the first statement in Proposition 3.4.5 and (5.87) (and the exactness of Θ_μ) we have $\Theta_\mu(V_{\underline{j},\underline{j}''}/V) = 0$ and thus the injection $V \hookrightarrow V_{\underline{j},\underline{j}''}$ induces an isomorphism $\Theta_\mu(V) \xrightarrow{\sim} \Theta_\mu(V_{\underline{j},\underline{j}''})$. Moreover, the canonical maps $C_{\underline{j}} \to \Theta_\mu(C_{\underline{j}})$ and $C_{\underline{j}'} \to \Theta_\mu(C_{\underline{j}'})$ are injective (since non-zero), hence so is the canonical map $V \to \Theta_\mu(V) \cong \Theta_\mu(V_{\underline{j},\underline{j}''})$. We deduce that the restriction of $V_{\underline{j},\underline{j}''} \to \Theta_\mu(V_{\underline{j},\underline{j}''})$ to V is injective. Since V and $V_{\underline{j},\underline{j}''}$ have same socle $C_{\underline{j}}$, it follows that $V_{\underline{j},\underline{j}''}$ injects into $\Theta_\mu(V_{\underline{j},\underline{j}''}) \cong \Theta_\mu(V)$. □

Remark 5.2.14 Similarly, we can prove that $\Theta_{w_0 s_{j_0+1} w_0}(C_{(j_0+1,j_1,j_2)})$ has Loewy length 3 with both socle and cosocle $C_{(j_0+1,j_1,j_2)}$, and with middle layer not containing $C_{(j_0+1,j_1,j_2)}$ but containing $C_{(j_0,j_1,j_2)}$ with multiplicity one. Keeping the notation and assumption in (ii) of Lemma 5.2.13, let V' be the unique length 2 representation of G with socle $C_{(j_0+1,j_1,j_2)}$ and cosocle $C_{\underline{j}''}$ (using Lemma 5.2.1), then $\Theta_\mu(V')$ also uniquely admits $V_{\underline{j},\underline{j}''}$ as a subquotient (with $V_{\underline{j},\underline{j}''}$ constructed in (i) or (ii) of Proposition 5.2.10), which is moreover a quotient.

Remark 5.2.15 Using similar arguments as in the proof of (ii) of Lemma 5.2.13, we can also prove the following results (where we choose $\mu \in \Lambda$ such that $\langle \mu + \rho, \alpha^\vee \rangle \geq 0$ for $\alpha \in \Phi^+$ and the stabilizer of μ in $W(G)$ for the dot action is $\{1, w_0 s_j w_0\}$).

(i) Let V be the (unique) length 2 representation of G with socle $C_{(j_0,j_0+1,j_2)}$ and cosocle $C_{(j_0,j_0,j_2)}$ for some $1 \leq j_0 < j_2 \leq n$ (Lemma 5.2.1), then $V_{(j_0,j_0+1,j_2),\infty}$ in (iii) of Proposition 5.2.10 is isomorphic to the unique subrepresentation of $\Theta_\mu(V)$ with cosocle $L(1)^\vee \otimes_E V^\infty_{[j_2-j_0,j_2]\cap\Delta,\Delta}$.

(ii) Let V' be the (unique) length 2 representation of G with socle $C_{(j_0,j_0,j_2)}$ and cosocle $C_{(j_0,j_0-1,j_2)}$ for some $1 < j_0 \leq j_2 \leq n$ (Lemma 5.2.1), then $V_{\infty,(j_0,j_0-1,j_2)}$ in (iv) of Proposition 5.2.10 is isomorphic to the unique quotient of $\Theta_\mu(V')$ with socle $L(1)^\vee \otimes_E V^\infty_{[j_2-j_0+1,j_2-1],\Delta}$.

We now introduce some technical but useful notation, which will be used in Lemmas 5.2.16 and 5.2.17 below. For a fixed $\underline{j} \in \mathbf{J}$ with $(j_0, j_1, j_2+1) \in \mathbf{J}$ we define

$$S_{j_0,j_1} \stackrel{\text{def}}{=} \{x \in W(G) \mid w_{j_1,j_0} \prec x,\ j_1 \notin D_L(x)\}. \tag{5.88}$$

It follows from (ii) of Lemma A.11 that $S_{j_0,j_1} = \{w_{j_1+1,j_0}\}$ if $j_1 > j_0$ and $j_1+1 \in \Delta$, $S_{j_0,j_1} = \{w_{j_1-1,j_0}\}$ if $j_1 < j_0$ and $j_1 - 1 \in \Delta$, $S_{j_0,j_1} = \{w_{j,j_0} \mid j \in \{j_1-1, j_1+1\} \cap \Delta\}$ if $j_1 = j_0$, and $S_{j_0,j_1} = \emptyset$ otherwise. We define

$$\begin{cases} A_{\underline{j}} \stackrel{\text{def}}{=} C_{(j_0,j_1-1,j_2)} & \text{if } j_1 > j_0 \\ A_{\underline{j}} \stackrel{\text{def}}{=} C_{(j_0,j_1+1,j_2+1)} & \text{if } j_1 < j_0 \\ A_{\underline{j}} \stackrel{\text{def}}{=} L(1)^\vee \otimes_E V^\infty_{[j_2-j_0+1,j_2],\Delta} & \text{if } j_1 = j_0. \end{cases}$$

We also define

$$B_{\underline{j}} \stackrel{\text{def}}{=} \oplus_{\underline{j}'} C_{\underline{j}'},$$

the direct sum being over those $\underline{j}' \in \mathbf{J}$ such that $\underline{j} \leq \underline{j}'$, $j_0 = j_0'$, $d(\underline{j}, \underline{j}') = |j_2' - j_2| + |(j_2' - j_1') - (j_2 - j_1)| = 1$ and $w_{j_1',j_0} \in S_{j_0,j_1}$. An easy check shows that there is in fact a bijection between S_{j_0,j_1} and $\mathrm{JH}_G(B_{\underline{j}})$ given by $w_{j_1+1,j_0} \mapsto C_{(j_0,j_1+1,j_2+1)}$, $w_{j_1-1,j_0} \mapsto C_{(j_0,j_1-1,j_2)}$. For $S \subseteq S_{j_0,j_1}$ we define $B_{\underline{j},S}$ as the direct summand of $B_{\underline{j}}$ corresponding to S under this bijection (with $B_{\underline{j},\emptyset} \stackrel{\mathrm{def}}{=} 0$). We finally define $W_{\underline{j}}$ as the unique (up to isomorphism) representation with socle $C_{\underline{j}}$ which fits into an exact sequence $0 \to C_{\underline{j}} \to W_{\underline{j}} \to B_{\underline{j}} \to 0$. Note that the existence and unicity of $W_{\underline{j}}$ follows from Lemma 5.2.1.

Lemma 5.2.16 *Let $\underline{j} \in \mathbf{J}$ with $(j_0, j_1, j_2+1) \in \mathbf{J}$. Then we have*

$$\mathrm{Ext}^1_G(C_{(j_0,j_1,j_2+1)}, W_{\underline{j}}) = 0. \tag{5.89}$$

Proof If $S_{j_0,j_1} = \emptyset$, i.e. $B_{\underline{j}} = 0$, we have $W_{\underline{j}} = C_{\underline{j}}$ and (5.89) follows from Lemma 5.2.1 (note that $d(\underline{j}, (j_0, j_1, j_2+1)) = 2$). We assume from now on $S_{j_0,j_1} \neq \emptyset$ and write M for the unique $U(\mathfrak{g})$-module with cosocle $L(w_{j_1,j_0})$ which fits into an exact sequence $0 \to L_{S_{j_0,j_1}} \to M \to L(w_{j_1,j_0}) \to 0$ (recall that for $S \subseteq W(G)$ we define $L_S = \oplus_{x \in S} L(x)$). We let $I \subseteq \Delta$ be the maximal subset such that M is in $\mathcal{O}^{\mathfrak{p}_I}_{\mathrm{alg}}$. In the following, we only prove (5.89) when $j_0 = j_1$ and $S_{j_1,j_0} = \{j_1 - 1, j_1 + 1\}$, the other cases being simpler and left to the (interested) reader. In particular we have $2 \leq j_0 = j_1 \leq n-2$ and $I = \Delta \setminus \{j_1 - 1, j_1, j_1 + 1\}$. We recall the notation (above) Lemma 2.3.6: $\Sigma_\pm = \Sigma_{j_1,j_2} \cap s_{j_1} \cdot \Sigma_{j_1,j_2+1}$ (a left $W(L_I)$-coset) and

$$\pi^\infty_\pm = J_{\widehat{j_1},\widehat{j_1},s_{j_1}}(\pi^\infty_{j_1,j_2+1})_{\mathcal{B}^I_{\Sigma_\pm}} \cong J'_{\widehat{j_1},I}(\pi^\infty_{j_1,j_2})_{\mathcal{B}^I_{\Sigma_\pm}}$$

which is an irreducible G-regular representation of L_I over E (note that $I_\pm = \Delta \setminus \{j_1 - 1, j_1, j_1 + 1\} = I$). In particular $\mathcal{F}^G_{P_I}(M, \pi^\infty_\pm)$ is multiplicity free by Lemma 5.1.1.

Step 1: We prove that $W_{\underline{j}}$ is a subquotient of $\mathcal{F}^G_{P_I}(M, \pi^\infty_\pm)$.
By (2.8) applied to (v) above Lemma 2.3.6, $\pi^\infty_{j_1,j_2}$ is in the cosocle of $i^\infty_{I,\widehat{j_1}}(\pi^\infty_\pm)$. By (ii) of Lemma 2.3.6 (and the definition of $\pi^\infty_{+,1}$, $\pi^\infty_{-,1}$ above *loc. cit.*) $\pi^\infty_{*,1} \cong \mathrm{cosoc}_{L_{I_*}}(i^\infty_{I,I_*}(\pi^\infty_\pm))$ for $* \in \{+,-\}$. By (2.7) applied to (iii) and (iv) above Lemma 2.3.6, $\pi^\infty_{j_1+1,j_2+1}, \pi^\infty_{j_1-1,j_2}$ is in the socle of $i^\infty_{I_+,\Delta\setminus\{j_1+1\}}(\pi^\infty_{+,1}), i^\infty_{I_-,\Delta\setminus\{j_1-1\}}(\pi^\infty_{-,1})$ respectively. Thus $\pi^\infty_{j_1+1,j_2+1}, \pi^\infty_{j_1-1,j_2}$ is a subquotient of $i^\infty_{I,\Delta\setminus\{j_1+1\}}(\pi^\infty_\pm), i^\infty_{I,\Delta\setminus\{j_1-1\}}(\pi^\infty_\pm)$ respectively. Moreover we have $d(\pi^\infty_{j_1+1,j_2+1}, \pi^\infty_{j_1,j_2}) = d(\pi^\infty_{j_1-1,j_2}, \pi^\infty_{j_1,j_2}) = 0$ by (i) of Lemma 2.3.2. We can thus apply Lemma 5.1.19 to M as above and $\pi^\infty_\pm$, which gives that $W_{\underline{j}}$ is a subquotient of $\mathcal{F}^G_{P_I}(M, \pi^\infty_\pm)$. We define $W_{\underline{j},-}$ as the (unique) minimal length subrepresentation of $\mathcal{F}^G_{P_I}(M, \pi^\infty_\pm)$ which admits $W_{\underline{j}}$ as a quotient.

Step 2: We prove that the injection $W_{\underline{j},-} \hookrightarrow \mathcal{F}^G_{P_I}(M, \pi^\infty_\pm)$ and surjection $W_{\underline{j},-} \twoheadrightarrow W_{\underline{j}}$ induce isomorphisms

$$\mathrm{Ext}^1_G(C_{(j_0,j_1,j_2+1)}, W_{\underline{j}}) \xleftarrow{\sim} \mathrm{Ext}^1_G(C_{(j_0,j_1,j_2+1)}, W_{\underline{j},-}) \xrightarrow{\sim} \mathrm{Ext}^1_G(C_{(j_0,j_1,j_2+1)}, \mathcal{F}^G_{P_I}(M, \pi^\infty_\pm)). \tag{5.90}$$

We wish to apply Lemma 5.1.20 to $V_0 = C_{(j_0,j_1,j_2+1)}$, $\pi^\infty = \pi^\infty_\pm$ and $V = W_{\underline{j}}$, so we check that all assumptions there are satisfied (using the notation of *loc. cit.*). We recall that $j_0 = j_1$, so $w_{j_1,j_0} = s_{j_1}$. First, for $L(x) \in \mathrm{JH}_{U(\mathfrak{g})}(M)$ with $x \neq w_{j_1,j_0} = s_{j_1}$ we have $j_1 \notin D_L(x)$, and hence $L(s_{j_1})$ is not a constituent of $M^{I_x}(x)$ by [33, Theorem 9.4(c)] and Lemma 3.1.1. Secondly we have $d(\pi^\infty_{j_1,j_2+1}, \pi^\infty_{j_1,j_2}) = d(\pi^\infty_{j_1,j_2+1}, \pi^\infty_{j_1+1,j_2+1}) = d(\pi^\infty_{j_1,j_2+1}, \pi^\infty_{j_1-1,j_2}) = 0$ by (i) of Lemma 2.3.2. Thirdly, writing $\Sigma_{0,x} = \Sigma_{j_1,j_2+1} \cap W(L_{I_x}) \cdot \mathcal{J}(\sigma_x^\infty)$ for $L(x) \in \mathrm{JH}_{U(\mathfrak{g})}(M)$, we have $\Sigma_{0,s_{j_1}} = \Sigma_{j_1,j_2+1} \cap \Sigma_{j_1,j_2} = \emptyset$ by (i) of Lemma 2.3.4. Finally we need to check that for $L(x) \in \mathrm{JH}_{U(\mathfrak{g})}(M) \setminus \{L(s_{j_1})\}$ and $\tau_x^\infty \in \mathrm{JH}_{L_{I_x}}(\sigma_{x,-}^\infty) \setminus \{\sigma_x^\infty\}$ we have

$$J'_{I_x, I_x \cap \widehat{j_1}}(\tau_x^\infty)_{\mathcal{B}^{I_x \cap \widehat{j_1}}_{\Sigma_{0,x}}} = 0$$

where $\sigma_x^\infty \stackrel{\mathrm{def}}{=} \pi^\infty_{j_1+1,j_2+1}$ if $x = w_{j_1+1,j_0}$ and $\sigma_x^\infty \stackrel{\mathrm{def}}{=} \pi^\infty_{j_1-1,j_2}$ if $x = w_{j_1-1,j_0}$. But this follows from (iii) of Lemma 2.3.6, noting that the constituents in $\mathrm{JH}_{L_{I_x}}(\sigma_{x,-}^\infty) \setminus \{\sigma_x^\infty\}$ are exactly the constituents τ_x^∞ in $i^\infty_{I,I_x}(\pi_\pm^\infty)$ such that $\tau_x^\infty < \sigma_x^\infty$.

Step 3: We prove (for $j_0 = j_1$)

$$\mathrm{Ext}^1_G(C_{(j_0,j_1,j_2+1)}, \mathcal{F}^G_{P_I}(M, \pi_\pm^\infty)) = 0. \tag{5.91}$$

As M has cosocle $L(s_{j_1})$ and socle $L(w_{j_1-1,j_1}) \oplus L(w_{j_1+1,j_1})$, it is a quotient of $M(s_{j_1})$ (using (3.27)), which by [33, Theorem 9.4(c)] implies that M is a quotient of $M^I(s_{j_1})$. Let $W \stackrel{\mathrm{def}}{=} \mathcal{F}^G_{P_I}(M^I(s_{j_1}), \pi_\pm^\infty)$, then $W/\mathcal{F}^G_{P_I}(M, \pi_\pm^\infty) \cong \mathcal{F}^G_{P_I}(Q, \pi_\pm^\infty)$ where $Q \stackrel{\mathrm{def}}{=} \ker(M^I(s_{j_1}) \twoheadrightarrow M)$ ((i) of Theorem 4.3.7). It is clear that $L(s_{j_1}) \notin \mathrm{JH}_{U(\mathfrak{g})}(Q)$ and thus we have by Lemma 5.1.1

$$\mathrm{Hom}_G(C_{(j_0,j_1,j_2+1)}, W/\mathcal{F}^G_{P_I}(M, \pi_\pm^\infty)) = 0.$$

Hence, the injection $\mathcal{F}^G_{P_I}(M, \pi_\pm^\infty) \hookrightarrow W$ induces an embedding

$$\mathrm{Ext}^1_G(C_{(j_0,j_1,j_2+1)}, \mathcal{F}^G_{P_I}(M, \pi_\pm^\infty)) \hookrightarrow \mathrm{Ext}^1_G(C_{(j_0,j_1,j_2+1)}, W). \tag{5.92}$$

By (ii) of Lemma 3.3.1 applied with $I' = \widehat{j_1}$ and $j_1 \notin I$ we have $H^0(\mathfrak{n}_I, L(s_{j_1})^{s_{j_1}}) = 0$. By (i) of Lemma 3.3.4 applied with $I' = \widehat{j_1}$ we also have $\mathrm{Hom}_{U(\mathfrak{l}_I)}(L^I(s_{j_1}), H^1(\mathfrak{n}_I, L(s_{j_1})^{s_{j_1}})) = 0$. By (3.12) applied with $M_I = L^I(s_{j_1})$ and $M = L(s_{j_1})^{s_{j_1}}$ we deduce for $\ell \leq 1$

$$\mathrm{Ext}^\ell_{U(\mathfrak{g})}(M^I(s_{j_1}), L(s_{j_1})^{s_{j_1}}) = 0. \tag{5.93}$$

Since $\Sigma_\pm \cap s_{j_1} \cdot \Sigma_{j_1,j_2+1} = \Sigma_\pm \neq \emptyset$ we can apply (4.146) with $w = s_{j_1}$ which gives using (5.93)

$$\mathrm{Ext}^1_G(C_{(j_0,j_1,j_2+1)}, W) = 0.$$

This together with (5.92) gives (5.91). Then (5.89) follows from (5.91) and (5.90). □

For $S \subseteq S_{j_0,j_1}$ we define $\widetilde{B}_{\underline{j},S}$ as the unique (up to isomorphism) representation with cosocle $C_{(j_0,j_1,j_2+1)}$ which fits into an exact sequence $0 \to A_{\underline{j}} \oplus B_{\underline{j},S} \to \widetilde{B}_{\underline{j},S} \to C_{(j_0,j_1,j_2+1)} \to 0$. Note that the existence and unicity of $\widetilde{B}_{\underline{j},S}$ follows from Lemma 5.2.1 and (ii) of Lemma 5.2.2.

Lemma 5.2.17 *Let $\underline{j} \in \mathbf{J}$ with $(j_0, j_1, j_2+1) \in \mathbf{J}$. Then we have for $S \subsetneq S_{j_0,j_1}$*

$$\mathrm{Ext}^1_G(\widetilde{B}_{\underline{j},S}, C_{\underline{j}}) = 0. \tag{5.94}$$

Proof Let $S \subsetneq S_{j_0,j_1}$. By (iii) of Lemma 3.2.2 we have $\dim_E \mathrm{Ext}^1_{\mathcal{O}^{\mathfrak{b}}_{\mathrm{alg}}}(L(s_{j_1} w_{j_1,j_0}), L(w_{j_1,j_0})) = 1$ (recall $j_1 \in D_L(w_{j_1,j_0})$). By (i) of Lemma 3.2.4 and the explicit description of S_{j_0,j_1} below (5.88) we have $\dim_E \mathrm{Ext}^1_{\mathcal{O}^{\mathfrak{b}}_{\mathrm{alg}}}(L(x), L(w_{j_1,j_0})) = 1$ for $x \in S_{j_0,j_1}$. Hence there is a unique M in $\mathcal{O}^{\mathfrak{b}}_{\mathrm{alg}}$ with socle $L(w_{j_1,j_0})$ which fits into an exact sequence (where $L_S = \oplus_{x\in S} L(x)$)

$$0 \longrightarrow L(w_{j_1,j_0}) \longrightarrow M \longrightarrow L(s_{j_1} w_{j_1,j_0}) \oplus L_S \longrightarrow 0.$$

Let $I \subseteq \Delta$ be the maximal subset such that M is in $\mathcal{O}^{\mathfrak{p}_I}_{\mathrm{alg}}$. An explicit check using Lemma 3.1.1 and the explicit description of S_{j_0,j_1} below (5.88) shows that we have the following cases (with the notation in (2.84) and using that S is *strictly* smaller than S_{j_0,j_1}):

- $I = I_-$ if either $j_1 > j_0$ with $S = \emptyset$, or $j_1 = j_0$ with $S = \{w_{j_1-1,j_0}\}$;
- $I = I_+$ if either $j_1 < j_0$ with $S = \emptyset$, or $j_1 = j_0$ with $S = \{w_{j_1+1,j_0}\}$;
- $I = \widehat{j_1}$ if $j_1 = j_0$ with $S = \emptyset$.

We let $\Sigma \overset{\mathrm{def}}{=} \Sigma_{*,0}$ if $I = I_*$ with $* \in \{+,-\}$ and $\Sigma \overset{\mathrm{def}}{=} \Sigma_{j_1,j_2+1}$ if $I = \widehat{j_1}$ (see (2.85)). From (2.86) and (2.85) we have

$$\Sigma_{j_1,j_2} \cap s_{j_1} \cdot \Sigma = \Sigma_{j_1,j_2} \cap s_{j_1} \cdot \Sigma_{j_1,j_2+1} = \Sigma_{\pm} \neq \emptyset. \tag{5.95}$$

We define $\pi^\infty \overset{\mathrm{def}}{=} J_{\widehat{j_1},I}(\pi^\infty_{\widehat{j_1},j_2+1})_{B^I_\Sigma}$, an irreducible G-regular representation of L_I (so $\pi^\infty = \pi^\infty_{j_1,j_2+1}$ if $I = \widehat{j_1}$ and $\pi^\infty = \pi^\infty_{*,0}$ if $I = I_*$ for $* \in \{+,-\}$ with the notation above Lemma 2.3.6). By (2.7) and using (iv) of Remark 2.1.16 $\pi^\infty_{\widehat{j_1},j_2+1}$ is the socle of $i^\infty_{I,\widehat{j_1}}(\pi^\infty)$, and by (2.8) with the isomorphism in (i) (resp. (ii)) above Lemma 2.3.6 $\pi^\infty_{\widehat{j_1+1},j_2+1}$ (resp. $\pi^\infty_{\widehat{j_1-1},j_2}$) is the cosocle of $i^\infty_{I,\Delta\setminus\{j_1+1\}}(\pi^\infty)$ if $I = I_+$ (resp. of $i^\infty_{I,\Delta\setminus\{j_1-1\}}(\pi^\infty)$ if $I = I_-$). By Lemma 2.3.1 (and $j_1 \leq j_2 < j_2+1$) we have

$$\mathrm{cosoc}_G(i^\infty_{\widehat{j_1},\Delta}(\pi^\infty_{\widehat{j_1},j_2+1})) \cong V^\infty_{[j_2-j_1+1,j_2],\Delta} \cong \mathrm{soc}_G(i^\infty_{\widehat{j_1},\Delta}(\pi^\infty_{\widehat{j_1},j_2})), \tag{5.96}$$

hence $V^\infty_{[j_2-j_1+1,j_2],\Delta}$ is a subquotient of $i^\infty_{I,\Delta}(\pi^\infty)$. By (i) of Lemma 2.3.2 and (i) Lemma 2.3.3 we have

$$d(\pi^\infty_{\widehat{j_1},j_2+1}, \pi^\infty_{\widehat{j_1-1},j_2}) = d(\pi^\infty_{\widehat{j_1},j_2+1}, \pi^\infty_{\widehat{j_1+1},j_2+1}) = d(\pi^\infty_{\widehat{j_1},j_2+1}, V^\infty_{[j_2-j_1+1,j_2],\Delta}) = 0.$$

Hence we can apply Lemma 5.1.19 with $\sigma_x^\infty \overset{\text{def}}{=} \pi^\infty_{\widehat{j_1},j_2+1}$ if $x = w_{j_1,j_0}$, $\sigma_x^\infty \overset{\text{def}}{=} \pi^\infty_{\widehat{j_1-1},j_2}$ if $x = w_{j_1-1,j_0}$, $\sigma_x^\infty \overset{\text{def}}{=} \pi^\infty_{\widehat{j_1+1},j_2+1}$ if $x = w_{j_1+1,j_0}$ and $\sigma_x^\infty \cong V^\infty_{[j_2-j_1+1,j_2],\Delta}$ if $j_0 = j_1$ and $x = 1$. This gives that $\widetilde{B}_{\underline{j},S}$ is a subquotient of $\mathcal{F}^G_{P_I}(M,\pi^\infty)$. Since M is multiplicity free and $i^\infty_{I,I_x}(\pi^\infty)$ is multiplicity free for any constituent $L(x)$ of M (see (iv) of Remark 2.1.16), $\mathcal{F}^G_{P_I}(M,\pi^\infty)$ is also multiplicity free by Lemmas 5.1.2 and 5.1.1. We can thus define $\widetilde{B}^+_{\underline{j},S}$ as the minimal length quotient of $\mathcal{F}^G_{P_I}(M,\pi^\infty)$ which admits $\widetilde{B}_{\underline{j},S}$ as a subrepresentation. By a similar argument as in the paragraph before Step 1 of the proof of Lemma 5.1.19, $\widetilde{B}^+_{\underline{j},S}$ admits a decreasing filtration indexed by $\mathrm{JH}_{U(\mathfrak{g})}(M)$ with $L(x)$-graded piece given by $\mathcal{F}^G_{P_{I_x}}(L(x),\sigma^\infty_{x,+})$, where $\sigma^\infty_{x,+}$ is the minimal length quotient of $i^\infty_{I,I_x}(\pi^\infty)$ which admits σ^∞_x as a subrepresentation. For $L(x) \in \mathrm{JH}_{U(\mathfrak{g})}(M)$, we have from the previous discussion

- if $x = w_{j_1-1,j_0}$, $I = I_-$ and $\sigma^\infty_x \cong \pi^\infty_{\widehat{j_1-1},j_2} \cong \mathrm{cosoc}(i^\infty_{I,\Delta\setminus\{j_1-1\}}(\pi^\infty))$, thus $\sigma^\infty_{x,+} = \sigma^\infty_x$;
- if $x = w_{j_1+1,j_0}$, $I = I_+$ and $\sigma^\infty_x \cong \pi^\infty_{\widehat{j_1+1},j_2+1} \cong \mathrm{cosoc}(i^\infty_{I,\Delta\setminus\{j_1+1\}}(\pi^\infty))$, thus $\sigma^\infty_{x,+} = \sigma^\infty_x$;
- if $x = w_{j_1,j_0}$, $\sigma^\infty_x \cong \pi^\infty_{\widehat{j_1},j_2+1} \cong \mathrm{soc}_{L_{\widehat{j_1}}}(i^\infty_{I,\widehat{j_1}}(\pi^\infty))$, thus $\sigma^\infty_{x,+} = i^\infty_{I,\widehat{j_1}}(\pi^\infty)$;
- If $x = 1$, $\sigma^\infty_{x,+}$ is the unique quotient of $i^\infty_{I,\Delta}(\pi^\infty)$ with socle $\sigma^\infty_x \cong V^\infty_{[j_2-j_1+1,j_2],\Delta}$.

We need to make $\sigma^\infty_{x,+}$ a bit more explicit when $x = 1$. We have isomorphisms

$$\begin{aligned}&\mathrm{Hom}_G(i^\infty_{I,\Delta}(\pi^\infty), i^\infty_{\widehat{j_1},\Delta}(\pi^\infty_{\widehat{j_1},j_2})) \cong \mathrm{Hom}_{L_{\widehat{j_1}}}(J_{\Delta,\widehat{j_1}}(i^\infty_{I,\Delta}(\pi^\infty)), \pi^\infty_{\widehat{j_1},j_2})\\ &\cong \mathrm{Hom}_{L_{\widehat{j_1}}}(i^\infty_{I,\widehat{j_1},s_{j_1}}(J_{I,\widehat{j_1},s_{j_1}}(\pi^\infty)), \pi^\infty_{\widehat{j_1},j_2}) \cong \mathrm{Hom}_{L_I}(J_{I,\widehat{j_1},s_{j_1}}(\pi^\infty), J'_{\widehat{j_1},I_\pm}(\pi^\infty_{\widehat{j_1},j_2}))\end{aligned}$$

where the first isomorphism follows from (2.7), the second from (i) and (ii) of Lemma 2.1.18 and from $\Sigma_{j_1,j_2} \subseteq W(L_{\widehat{j_1}})s_{j_1}W(L_I)\cdot\mathcal{J}(\pi^\infty)$ (which follows from (5.95) and $\Sigma = W(L_I)\cdot\mathcal{J}(\pi^\infty)$), and the last from $s_{j_1}(I)\cap\widehat{j_1} = s_{j_1}(\widehat{j_1})\cap\widehat{j_1} = I_\pm$ (see (2.21) and (2.84)) followed by (2.8). Moreover these spaces are all non-zero since we have

$$J'_{\widehat{j_1},I_\pm}(\pi^\infty_{\widehat{j_1},j_2})_{\mathcal{B}^I_{\Sigma_\pm}} \cong J_{\widehat{j_1},\widehat{j_1},s_{j_1}}(\pi^\infty_{\widehat{j_1},j_2+1})_{\mathcal{B}^I_{\Sigma_\pm}} \cong J_{I_*,\widehat{j_1},s_{j_1}}(\pi^\infty_{*,0})_{\mathcal{B}^I_{\Sigma_\pm}} \neq 0$$

by (v) above Lemma 2.3.6 and (ii) of Lemma 2.3.6. Since $\sigma^\infty_x \cong V^\infty_{[j_2-j_1+1,j_2],\Delta}$ (when $x = 1$) is the socle of $i^\infty_{\widehat{j_1},\Delta}(\pi^\infty_{\widehat{j_1},j_2})$ by (5.96) (and since all representations are multiplicity free by (iv) of Remark 2.1.16), we finally deduce that $\sigma^\infty_{x,+}$ injects into $i^\infty_{\widehat{j_1},\Delta}(\pi^\infty_{\widehat{j_1},j_2})$.

We divide the rest of the proof into two steps.

Step 1: We prove that

$$\mathrm{Ext}^1_G(\mathcal{F}^G_{P_I}(M,\pi^\infty), C_{\underline{j}}) = 0. \tag{5.97}$$

We first claim that $\mathcal{F}_{P_I}^G(M,\pi^\infty)$ has no constituent isomorphic to $\mathcal{F}_{P_{\widehat{j_1}}}^G(L(x),\pi_{j_1,j_2}^\infty)$ for any x such that $D_L(x)=\{j_1\}$. Assume on the contrary that W is such a constituent, then by Lemma 5.1.1 (and (ii) of Theorem 4.3.7) W must be a constituent of $\mathcal{F}_{P_{\widehat{j_1}}}^G(L(x), i_{I,\widehat{j_1}}^\infty(\pi^\infty))$, and thus π_{j_1,j_2}^∞ is a constituent of $i_{I,\widehat{j_1}}^\infty(\pi^\infty)$ ((iv) of Theorem 4.3.7). This implies $\Sigma_{j_1,j_2}=W(L_{\widehat{j_1}})\cdot\mathcal{J}(i_{I,\widehat{j_1}}^\infty(\pi^\infty))$ since both are single $W(L_{\widehat{j_1}})$-cosets by the last statement in (i) of Lemma 2.1.15. By (2.28) this implies $\Sigma=W(L_I)\cdot\mathcal{J}(\pi^\infty)\subseteq\Sigma_{j_1,j_2}$. But by (2.85) we have $\Sigma\subseteq\Sigma_{j_1,j_2+1}$, and since $\Sigma_{j_1,j_2+1}\cap\Sigma_{j_1,j_2}=\emptyset$ by (i) of Lemma 2.3.4, we derive a contradiction. Now, since any constituent $L(x)$ of $N^{\widehat{j_1}}(w_{j_1,j_0})$ is such that $D_L(x)=\{j_1\}$ by Lemma 3.1.1, we deduce in particular

$$\mathrm{Hom}_G\big(\mathcal{F}_{P_I}^G(M,\pi^\infty),\mathcal{F}_{P_{\widehat{j_1}}}^G(N^{\widehat{j_1}}(w_{j_1,j_0}),\pi_{j_1,j_2}^\infty)\big)=0,$$

which together with $0\to N^{\widehat{j_1}}(w_{j_1,j_0})\to M^{\widehat{j_1}}(w_{j_1,j_0})\to L(w_{j_1,j_0})\to 0$ gives an embedding

$$\mathrm{Ext}_G^1(\mathcal{F}_{P_I}^G(M,\pi^\infty),C_{\underline{j}})\hookrightarrow\mathrm{Ext}_G^1\big(\mathcal{F}_{P_I}^G(M,\pi^\infty),\mathcal{F}_{P_{\widehat{j_1}}}^G(M^{\widehat{j_1}}(w_{j_1,j_0}),\pi_{j_1,j_2}^\infty)\big). \tag{5.98}$$

By Remark 3.4.12 (with j, w, S_0 there being j_1, w_{j_1,j_0}, S_{j_0,j_1}) we have for $\ell\leq 1$

$$\mathrm{Ext}_{U(\mathfrak{g})}^\ell(M^{\widehat{j_1}}(w_{j_1,j_0}),M^{s_{j_1}})=0. \tag{5.99}$$

By (5.95) we can apply (4.146), which gives using (5.99)

$$\mathrm{Ext}_G^1\big(\mathcal{F}_{P_I}^G(M,\pi^\infty),\mathcal{F}_{P_{\widehat{j_1}}}^G(M^{\widehat{j_1}}(w_{j_1,j_0}),\pi_{j_1,j_2}^\infty)\big)=0.$$

By (5.98) this gives (5.97).

Step 2: We prove that the injection $\widetilde{B}_{\underline{j},S}\hookrightarrow\widetilde{B}_{\underline{j},S}^+$ and the surjection $\mathcal{F}_{P_I}^G(M,\pi^\infty)\twoheadrightarrow\widetilde{B}_{\underline{j},S}^+$ induce isomorphisms

$$\mathrm{Ext}_G^1(\widetilde{B}_{\underline{j},S},C_{\underline{j}})\xrightarrow{\sim}\mathrm{Ext}_G^1(\widetilde{B}_{\underline{j},S}^+,C_{\underline{j}})\xleftarrow{\sim}\mathrm{Ext}_G^1(\mathcal{F}_{P_I}^G(M,\pi^\infty),C_{\underline{j}}). \tag{5.100}$$

We wish to apply Remark 5.1.21 to $V_0=C_{\underline{j}}$ and $V=\widetilde{B}_{\underline{j},S}$, so we check that all assumptions there are satisfied (using the notation of *loc. cit.*). From the definition of M, Lemma 3.1.1 and [33, Theorem 9.4(c)], we have $\mathrm{JH}_{U(\mathfrak{g})}(M)\cap\mathrm{JH}_{U(\mathfrak{g})}(M^{\widehat{j_1}}(w_{j_1,j_0}))=\{L(w_{j_1,j_0})\}$. By (i) of Lemma 2.3.2 and (i) Lemma 2.3.3 we have $d(\sigma_x^\infty,\pi_{j_1,j_0}^\infty)=0$ for all constituents $L(x)$ of M. By (i) of Lemma 2.3.4 we have $\Sigma_{j_1,j_2+1}\cap\Sigma_{j_1,j_2}=\emptyset$. Hence it remains to check $J_{\widehat{j_1}}(\tau_x^\infty)_{\mathcal{B}_{\Sigma_{j_1,j_2}}^{\widehat{j_1}}}=0$ for each $L(x)\in\mathrm{JH}_{U(\mathfrak{g})}(M)$ with $x\neq w_{j_1,j_0}$ and each $\tau_x^\infty\in\mathrm{JH}_{L_{I_x}}(\sigma_{x,+}^\infty)\setminus\{\sigma_x^\infty\}$. Since $\sigma_{x,+}^\infty=\sigma_x^\infty$ when $x\in\{w_{j_1-1,j_0},w_{j_1+1,j_0}\}$ by the explicit description of $\sigma_{x,+}^\infty$ before Step 1, such a τ_x^∞ only exists when $x=1$. In this case $\sigma_x^\infty\cong V_{[j_2-j_1+1,j_2],\Delta}^\infty$ and

$\tau_x^\infty \in \mathrm{JH}_G(i^\infty_{\widehat{j_1},\Delta}(\pi^\infty_{j_1,j_2})) \setminus \{V^\infty_{[j_2-j_1+1,j_2],\Delta}\}$ by the discussion just before Step 1. We claim that we have isomorphisms

$$J_{\Delta,\widehat{j_1}}(V^\infty_{[j_2-j_1+1,j_2],\Delta})_{\mathcal{B}^{\widehat{j_1}}_{\Sigma_{j_1,j_2}}} \cong \pi^\infty_{j_1,j_2} \cong J_{\Delta,\widehat{j_1}}(i^\infty_{\widehat{j_1},\Delta}(\pi^\infty_{j_1,j_2}))_{\mathcal{B}^{\widehat{j_1}}_{\Sigma_{j_1,j_2}}}. \tag{5.101}$$

Applying (2.7) to $V^\infty_{[j_2-j_1+1,j_2],\Delta} \hookrightarrow i^\infty_{\widehat{j_1},\Delta}(\pi^\infty_{j_1,j_2})$ and using the last assertion of Lemma 2.1.29 gives the first isomorphism. The second isomorphism is the very last isomorphism in (ii) of Lemma 2.1.18 applied with $I, I_1, I_0, w, \pi_0^\infty$ there being $\Delta, \widehat{j_1}, \widehat{j_1}, 1, \pi^\infty_{j_1,j_2}$. By exactness of $J_{\Delta,\widehat{j_1}}$ we deduce from (5.101) $J_{\Delta,\widehat{j_1}}(\tau_x^\infty)_{\mathcal{B}^{\widehat{j_1}}_{\Sigma_{j_1,j_2}}} = 0$ for τ_x as above (and $x = 1$). By Remark 5.1.21 we thus have (5.100). Finally, (5.94) follows from (5.100) and (5.97). □

Proposition 5.2.18 *Let* $\underline{j} \in \mathbf{J}$ *such that* $\underline{j}' = (j_0', j_1', j_2') \overset{def}{=} (j_0, j_1, j_2+1)$ *is still in* $\mathbf{J}$.

(i) If $j_0 \neq j_1$ *there exists a unique minimal* Ext*-square* $V_{\underline{j},\underline{j}'}$ *such that*

$$\begin{cases} \mathrm{gr}^0(V_{\underline{j},\infty}) \cong C_{(j_0,j_1,j_2+1)} \\ \mathrm{gr}^1(V_{\underline{j},\infty}) \cong C_{(j_0,j_1-1,j_2)} \oplus C_{(j_0,j_1+1,j_2+1)} \\ \mathrm{gr}^2(V_{\underline{j},\infty}) \cong C_{(j_0,j_1,j_2)}. \end{cases}$$

(ii) If $j_0 = j_1$ *there exists a unique minimal* Ext*-square* $V_{\underline{j},\underline{j}'}$ *such that*

$$\begin{cases} \mathrm{gr}^0(V_{\underline{j},\infty}) \cong C_{(j_0,j_1,j_2+1)} \\ \mathrm{gr}^1(V_{\underline{j},\infty}) \cong C_{(j_0,j_1-1,j_2)} \oplus C_{(j_0,j_1+1,j_2+1)} \oplus L(1)^\vee \otimes_E V^\infty_{[j_2-j_0+1,j_2],\Delta} \\ \mathrm{gr}^2(V_{\underline{j},\infty}) \cong C_{(j_0,j_1,j_2)}. \end{cases}$$

In both (i) and (ii) we omit $C_{(j_0,j_1+1,j_2+1)}$ *if* $j_1 = j_2 = n-1$ *and* $C_{(j_0,j_1-1,j_2)}$ *if* $j_1 = 1$.

Proof Note that, with the notation before Lemma 5.2.16, we have $\mathrm{gr}^1(V_{\underline{j},\underline{j}'}) \cong A_{\underline{j}} \oplus B_{\underline{j}}$, and we prove (i) and (ii) simultaneously. We define $W_{\underline{j}}^+$ as the unique (up to isomorphism) representation with socle $C_{\underline{j}}$ which fits into an exact sequence $0 \to C_{\underline{j}} \to W_{\underline{j}}^+ \to A_{\underline{j}} \oplus B_{\underline{j}} \to 0$. Note that the existence and unicity of $W_{\underline{j}}^+$ follows from Lemma 5.2.1 and (i) of Lemma 5.2.2, and that we have an exact sequence $0 \to W_{\underline{j}} \to W_{\underline{j}}^+ \to A_{\underline{j}} \to 0$ (see before Lemma 5.2.16 for $W_{\underline{j}}$). An obvious dévissage gives

$$\dim_E \mathrm{Ext}^1_G(C_{\underline{j}'}, W_{\underline{j}}^+) \leq \dim_E \mathrm{Ext}^1_G(C_{\underline{j}'}, W_{\underline{j}}) + \dim_E \mathrm{Ext}^1_G(C_{\underline{j}'}, A_{\underline{j}}) = 0 + 1 = 1 \tag{5.102}$$

where the last equality follows from (5.89), Lemma 5.2.1 and (ii) of Lemma 5.2.2. The short exact sequence $0 \to C_{\underline{j}} \to W_{\underline{j}}^+ \to A_{\underline{j}} \oplus B_{\underline{j}} \to 0$ yields an exact sequence

$$\mathrm{Ext}^1_G(C_{\underline{j}'}, C_{\underline{j}}) \to \mathrm{Ext}^1_G(C_{\underline{j}'}, W^+_{\underline{j}}) \to \mathrm{Ext}^1_G(C_{\underline{j}'}, A_{\underline{j}} \oplus B_{\underline{j}}) \to \mathrm{Ext}^2_G(C_{\underline{j}'}, C_{\underline{j}}). \tag{5.103}$$

We have by Lemma 5.2.1

$$\mathrm{Ext}^1_G(C_{\underline{j}'}, C_{\underline{j}}) = 0 \tag{5.104}$$

and by Lemma 5.2.1 and (ii) of Lemma 5.2.2

$$\dim_E \mathrm{Ext}^1_G(C_{\underline{j}'}, A_{\underline{j}} \oplus B_{\underline{j}}) = 1 + \#S_{j_0,j_1}. \tag{5.105}$$

By (ii) of Lemma 5.2.3 and the explicit description of S_{j_0,j_1} below (5.88) we have

$$\dim_E \mathrm{Ext}^2_G(C_{\underline{j}'}, C_{\underline{j}}) \leq \#S_{j_0,j_1},$$

which together with (5.104), (5.103) and (5.105) implies

$$\dim_E \mathrm{Ext}^1_G(C_{\underline{j}'}, W^+_{\underline{j}}) \geq 1 + \#S_{j_0,j_1} - \#S_{j_0,j_1} = 1.$$

With (5.102), we deduce $\dim_E \mathrm{Ext}^1_G(C_{\underline{j}'}, W^+_{\underline{j}}) = 1$. The unique (up to scalar) non-zero class in $\mathrm{Ext}^1_G(C_{\underline{j}'}, W^+_{\underline{j}})$ determines a unique (up to isomorphism) representation of G over E, which is moreover multiplicity free using Lemma 5.1.1. We define $V_{\underline{j},\underline{j}'}$ as its unique subrepresentation with cosocle $C_{\underline{j}'}$. It is clear from its definition and from Lemmas 5.2.1 and 5.2.2 that $V_{\underline{j},\underline{j}'}$ is an Ext-square. Moreover we have by (5.89) (and since $A_{\underline{j}}$ is irreducible)

$$A_{\underline{j}} \subseteq \mathrm{gr}^1(V_{\underline{j},\underline{j}'}) \subseteq A_{\underline{j}} \oplus B_{\underline{j}}. \tag{5.106}$$

If the second inclusion in (5.106) is strict, there exists $S \subsetneq S_{j_0,j_1}$ such that $\mathrm{gr}^1(V_{\underline{j},\underline{j}'}) \cong A_{\underline{j}} \oplus B_{\underline{j},S}$, and the existence of $V_{\underline{j},\underline{j}'}$ forces $\mathrm{Ext}^1_G(\widetilde{B}_{\underline{j},S}, C_{\underline{j}}) \neq 0$, which contradicts (5.94). Hence we have $W_{\underline{j}} \hookrightarrow V_{\underline{j},\underline{j}'}$. Finally, the minimality of $V_{\underline{j},\underline{j}'}$ follows from Lemma 5.2.9 (applied with $V_1 = \mathrm{Fil}^1(V_{\underline{j},\underline{j}'})$, $V_2 = W_{\underline{j}}$) and Lemma 5.2.16. □

Until the end of the proof of Proposition 5.2.28, we fix $\underline{j} \in \mathbf{J}$ such that $\underline{j}' = (j_0', j_1', j_2') \stackrel{\mathrm{def}}{=} (j_0 + 1, j_1, j_2 + 1)$ is still in $\mathbf{J}$. Note that we have $1 \leq j_0 \leq n-2$, $1 \leq j_1 \leq j_2 \leq n-1$ and $j_2 - j_1 \leq n - 2 - j_0$. The following notation is convenient

$$\begin{array}{lll}
C^{0,0,0} \stackrel{\mathrm{def}}{=} C_{(j_0,j_1,j_2)} & & \\
C^{1,0,0} \stackrel{\mathrm{def}}{=} C_{(j_0+1,j_1,j_2)} & C^{0,1,0} \stackrel{\mathrm{def}}{=} C_{(j_0,j_1+1,j_2+1)} & C^{0,0,1} \stackrel{\mathrm{def}}{=} C_{(j_0,j_1-1,j_2)} \\
C^{1,1,0} \stackrel{\mathrm{def}}{=} C_{(j_0+1,j_1+1,j_2+1)} & C^{0,1,1} \stackrel{\mathrm{def}}{=} C_{(j_0,j_1,j_2+1)} & C^{1,0,1} \stackrel{\mathrm{def}}{=} C_{(j_0+1,j_1-1,j_2)} \\
C^{1,1,1} \stackrel{\mathrm{def}}{=} C_{(j_0+1,j_1,j_2+1)} & &
\end{array}$$

(so $C^{0,0,0} = C_{\underline{j}}$ and $C^{1,1,1} = C_{\underline{j}'}$). By Proposition 5.2.18 there exists a minimal Ext-square V_0 with socle $C^{0,0,0}$, cosocle $C^{0,1,1}$ and middle layer containing $C^{0,1,0} \oplus$

$C^{0,0,1}$. By *loc. cit.* applied with $\underline{j}, \underline{j}'$ there being $(j_0+1, j_1, j_2), (j_0+1, j_1, j_2+1)$, there exists a minimal Ext-square V_1 with socle $C^{1,0,0}$, cosocle $C^{1,1,1}$ and middle layer containing $C^{1,1,0} \oplus C^{1,0,1}$. By *loc.cit.* V_0 (resp. V_1) is multiplicity free and admits a locally algebraic constituent if and only if $j_1 = j_0$ (resp. $j_1 = j_0 + 1$).

Until Proposition 5.2.28 we also fix $\mu \in \Lambda$ such that $\langle \mu + \rho, \alpha^\vee \rangle \geq 0$ for all $\alpha \in \Phi^+$ and the stabilizer of μ in $W(G)$ for the dot action is $\{1, w_0 s_{j_0} w_0\}$. The following lemma gives technical results on $\Theta_{w_0 s_{j_0} w_0}(C^{0,b,c})$ for $b, c \in \{0,1\}$ which will be used in various proofs, in particular the proof of Proposition 5.2.28.

Lemma 5.2.19 *For each* $(b, c) \in \{0,1\}^2$ *we have:*

(i) $C^{a',b',c'}$ *is a constituent of* $\Theta_{w_0 s_{j_0} w_0}(C^{0,b,c})$ *for some* $a', b', c' \in \{0,1\}$ *if and only if* $b' = b$ *and* $c' = c$;
(ii) $C^{1,b,c}$ *(resp.* $C^{0,b,c}$*) appears with multiplicity* 1 *(resp.* 2*) in* $\Theta_{w_0 s_{j_0} w_0}(C^{0,b,c})$ *and in* $\Theta_\mu(V_0)$;
(iii) $\Theta_{w_0 s_{j_0} w_0}(C^{0,b,c})$ *has simple socle and cosocle* $C^{0,b,c}$, *and admits a unique length* 2 *subrepresentation (resp. quotient) with socle* $C^{0,b,c}$ *and cosocle* $C^{1,b,c}$ *(resp. with socle* $C^{1,b,c}$ *and cosocle* $C^{0,b,c}$*).*

Proof We fix $b, c \in \{0,1\}$ and write $C^{0,b,c} = \mathcal{F}_{P_{I_x}}^G(L(x), \sigma^\infty)$ for some $x \in W(G)$ and some irreducible G-regular smooth σ^∞, which can be made explicit using (5.68). In particular by Proposition 3.4.5 we have $\Theta_{w_0 s_{j_0} w_0}(L(x)) \neq 0$.

We prove (i). For $a', b', c' \in \{0,1\}$ we write $C^{a',b',c'} = \mathcal{F}_{P_{I_w}}^G(L(w), \pi^\infty)$ for some $w \in W(G)$ and some irreducible G-regular smooth π^∞. Note that $\#D_L(x) = \#D_L(w) = 1$, and thus $I_w \supseteq I_x$ if and only if $I_x = I_w$. Hence by (5.87) and Lemma 5.1.1 $C^{a',b',c'}$ is a constituent of $\Theta_{w_0 s_{j_0} w_0}(C^{0,b,c})$ if and only if $L(w)$ is a constituent of $\Theta_{w_0 s_{j_0} w_0}(L(x))$, $I_x = I_w$ and $\sigma^\infty \cong \pi^\infty$. On one hand, $I_x = I_w$ and $\sigma^\infty \cong \pi^\infty$ force $b = b'$ and $c = c'$ (by definition of $C^{0,b,c}$ and $C^{a',b',c'}$). On the other hand, if $b = b'$ and $c = c'$, then $\sigma^\infty \cong \pi^\infty$ and a case by case check from the definition of $C^{0,b,c}$, $C^{a',b',c'}$ gives that $x = w_{j,j_0}$ for some $j \in \{j_1 - 1, j_1, j_1 + 1\} \cap \Delta$ and either $w = x$ or $w = w_{j,j_0+1}$. In all these cases $C^{a',b',c'}$ is a constituent of $\Theta_{w_0 s_{j_0} w_0}(C^{0,b,c})$ by Proposition 3.4.5 (with Remark 3.4.6).

We prove (ii). By the first statement in Proposition 3.4.5 and (5.87) we have $\Theta_{w_0 s_{j_0} w_0}(W) = 0$ if W is a locally algebraic constituent of V_0. As V_0 is multiplicity free with non locally algebraic constituents exactly given by $C^{0,b,c}$ for $b, c \in \{0,1\}$ (see Proposition 5.2.18), we deduce from the exactness of Θ_μ that $\Theta_\mu(V_0)$ admits a filtration with graded pieces given by $\Theta_{w_0 s_{j_0} w_0}(C^{0,b,c})$ for $b, c \in \{0,1\}$. This together with (i) implies that $C^{1,b,c}$ (resp. $C^{0,b,c}$) appears in $\Theta_{w_0 s_{j_0} w_0}(C^{0,b,c})$ and $\Theta_\mu(V_0)$ with the same multiplicity. It then follows from the (second half of) the first sentence in (i) of Lemma 5.2.13 that $C^{1,b,c}$ (resp. $C^{0,b,c}$) appears with multiplicity one (resp. two) in $\Theta_{w_0 s_{j_0} w_0}(C^{0,b,c})$ (we apply *loc. cit.* with $\underline{j}$ there being $(j_0, j_1, j_2), (j_0, j_1+1, j_2+1)$, (j_0, j_1-1, j_2) and (j_0, j_1, j_2+1) which corresponds to (b, c) being $(0,0)$, $(1,0)$, $(0,1)$ and $(1,1)$).

Finally, (iii) follows from the rest of the statement in (i) of Lemma 5.2.13. □

For $\lambda, \lambda' \in \Lambda$ and C, D any $Z(\mathfrak{g})$-finite $D(G)$-modules, by [35, Theorem 2.4.7] we have canonical isomorphisms

$$\mathrm{Hom}_{D(G)}(T_{\lambda}^{\lambda'}(C), D) \cong \mathrm{Hom}_{D(G)}(C, T_{\lambda'}^{\lambda}(D)), \tag{5.107}$$

from which we (formally) obtain canonical functorial adjunction maps $\Theta_\lambda(D) \to D$, $D \to \Theta_\lambda(D)$ for D any $Z(\mathfrak{g})$-finite $D(G)$-module and $\lambda \in \Lambda$ such that $\langle \lambda + \rho, \alpha^\vee \rangle \geq 0$ for $\alpha \in \Phi^+$ and the stabilizer of λ in $W(G)$ for the dot action is $\{1, s_j\}$ for some $j \in \{0, \ldots, n-1\}$. It is clear that these two adjunction maps are non-zero when both D and $\Theta_\lambda(D)$ are non-zero. If V is an admissible locally analytic representation of G over E such that $V^\vee$ is $Z(\mathfrak{g})$-finite, we therefore have canonical functorial adjunctions maps $V \to \Theta_\lambda(V)$ and $\Theta_\lambda(V) \to V$. Note that, when $D = \mathcal{F}_{P_I}^G(M, \pi^\infty)^\vee$ for $I \subseteq \Delta$, M in $\mathcal{O}_{\mathrm{alg}}^{\mathfrak{p}_I}$ and π^∞ a strongly admissible smooth representation of L_I, an examination of the proof of (5.87) in [35, Theorem 4.1.12] shows that the adjunction maps $\Theta_\lambda(D) = \Theta_{s_j}(D) \to D$ coming from (i) the above argument and (ii) the adjunction map $\Theta_{s_j}(M) \to M$ (see below (3.84)) by functoriality of the Orlik-Strauch functor are the same. Likewise with the adjunction maps $D \to \Theta_{s_j}(D)$.

Going back to our previous running notation, since $C^{1,1,1}$ occurs with multiplicity 1 in $\Theta_\mu(V_0)$ by (ii) of Lemma 5.2.19, we can (and do) define $V_{\underline{j},\underline{j}'}$ as the unique subrepresentation of $\Theta_\mu(V_0)$ with cosocle $C^{1,1,1}$.

Lemma 5.2.20 *The representations $\Theta_\mu(V_0)$ and $V_{\underline{j},\underline{j}'}$ have socle $C^{0,0,0}$. Moreover, $V_{\underline{j},\underline{j}'}$ satisfies the following properties:*

(i) V_0 injects into $V_{\underline{j},\underline{j}'}$;

(ii) for each (b, c), $(b', c') \in \{0, 1\}^2$ such that $b \leq b'$, $c \leq c'$ and $b' + c' = b + c + 1$, $\Theta_\mu(V_0)$ admits a subquotient which is the minimal Ext-*square (with socle $C^{0,b,c}$, cosocle $C^{1,b',c'}$ and middle layer containing $C^{0,b',c'} \oplus C^{1,b,c}$) constructed in (i) or (ii) of Proposition 5.2.10;*

(iii) for each $(a, b, c) \in \{0, 1\}^3$, $C^{a,b,c}$ appears in $V_{\underline{j},\underline{j}'}$ with multiplicity 1;

(iv) for each (a, b, c), $(a', b', c') \in \{0, 1\}^3$ such that $a \leq a'$, $b \leq b'$, $c \leq c'$ and $a' + b' + c' = a + b + c + 1$, $V_{\underline{j},\underline{j}'}$ admits a unique length 2 subquotient with socle $C^{a,b,c}$ and cosocle $C^{a',b',c'}$.

Proof By the first statement in (i) of Lemma 5.2.13, for $b, c \in \{0, 1\}$ we have (up to non-zero scalars) a canonical injection $C^{0,b,c} \hookrightarrow \Theta_{w_0 s_{j_0} w_0}(C^{0,b,c})$ and a canonical surjection $\Theta_{w_0 s_{j_0} w_0}(C^{0,b,c}) \twoheadrightarrow C^{0,b,c}$. Since V_0 has socle $C^{0,0,0}$ and cosocle $C^{0,1,1}$, it follows from the exactness and functoriality of Θ_μ that the adjunction map $V_0 \to \Theta_\mu(V_0)$ is injective and the adjunction map $\Theta_\mu(V_0) \to V_0$ is surjective. As the composition $C^{0,b,c} \hookrightarrow \Theta_{w_0 s_{j_0} w_0}(C^{0,b,c}) \twoheadrightarrow C^{0,b,c}$ is (obviously) zero for $b, c \in \{0, 1\}$ and as Θ_μ kills any locally algebraic constituent of V_0 (by the first statement in Proposition 3.4.5 and (5.87)), by functoriality and exactness again of Θ_μ the composition $V_0 \to \Theta_\mu(V_0) \to V_0$ is also zero. As $C^{1,1,1}$ is not a constituent of V_0 and as $C^{0,b,c}$ appears in $\Theta_\mu(V_0)$ (resp. V_0) with multiplicity 2 (resp. 1) for each $(b, c) \in \{0, 1\}^2$

by (ii) of Lemma 5.2.19, we deduce that $V_{\underline{j},\underline{j}'}$ injects into the kernel of the surjection $\Theta_\mu(V_0) \twoheadrightarrow V_0$ and that each $C^{0,b,c}$ appears in $V_{\underline{j},\underline{j}'}$ with multiplicity at most one.

We prove that $\Theta_\mu(V_0)$ has socle $C^{0,0,0}$ (and thus its subrepresentation $V_{\underline{j},\underline{j}'}$ also has socle $C^{0,0,0}$). We choose an arbitrary irreducible $W \hookrightarrow \mathrm{soc}_G(\Theta_\mu(V_0))$. By the discussion before this lemma, we have canonical isomorphisms

$$0 \neq \mathrm{Hom}_G(W, \Theta_\mu(V_0)) \cong \mathrm{Hom}_G(T^{\mu}_{w_0\cdot\mu_0}(W), T^{\mu}_{w_0\cdot\mu_0}(V_0)) \cong \mathrm{Hom}_G(\Theta_{w_0 s_{j_0} w_0}(W), V_0). \tag{5.108}$$

In particular, we have $\Theta_{w_0 s_{j_0} w_0}(W) \neq 0$ and W is not locally algebraic by Theorem 4.3.7 and (5.87). By Lemma 5.2.12 $\Theta_{w_0 s_{j_0} w_0}(W)$ has simple cosocle W, which together with (5.108) forces W to be a non locally algebraic factor of V_0, and thus $W = C^{0,b,c}$ for some $b, c \in \{0, 1\}$. But by (i) of Lemma 5.2.19 we know that the socle $C^{0,0,0}$ of V_0 is a constituant of $\Theta_{w_0 s_{j_0} w_0}(C^{0,b,c})$ if and only if $b = c = 0$. Hence, a non-zero map $\Theta_{w_0 s_{j_0} w_0}(W) \to V_0$ exists only if $W = C^{0,0,0}$, in which case this map must factor through $\Theta_{w_0 s_{j_0} w_0}(W) \twoheadrightarrow C^{0,0,0} = \mathrm{soc}_G(V_0)$ and (5.108) is one dimensional by (iii) of Lemma 5.2.19. It follows from this discussion that $\Theta_\mu(V_0)$ has simple socle $C^{0,0,0}$.

We prove (i). The surjection $V_0 \twoheadrightarrow C^{0,1,1}$ induces a surjection $\Theta_\mu(V_0) \twoheadrightarrow \Theta_{w_0 s_{j_0} w_0}(C^{0,1,1})$ such that (by functoriality of the adjunction map) the composition $V_0 \hookrightarrow \Theta_\mu(V_0) \twoheadrightarrow \Theta_{w_0 s_{j_0} w_0}(C^{0,1,1})$ coincides with the composition

$$V_0 \twoheadrightarrow C^{0,1,1} \cong \mathrm{soc}_G(\Theta_{w_0 s_{j_0} w_0}(C^{0,1,1}) \hookrightarrow \Theta_{w_0 s_{j_0} w_0}(C^{0,1,1})$$

(see (iii) of Lemma 5.2.19 for the above isomorphism). By *loc. cit.* $\Theta_{w_0 s_{j_0} w_0}(C^{0,1,1})$ admits a unique length 2 subrepresentation with socle $C^{0,1,1}$ and cosocle $C^{1,1,1}$, and we denote by $\widetilde{V}$ its inverse image in $\Theta_\mu(V_0)$ via $\Theta_\mu(V_0) \twoheadrightarrow \Theta_{w_0 s_{j_0} w_0}(C^{0,1,1})$. Then $C^{0,1,1}$ has multiplicity 1 in $\widetilde{V}$ since it has multiplicity 2 in $\Theta_\mu(V_0)$ and $\Theta_{w_0 s_{j_0} w_0}(C^{0,1,1})$ (by (ii) of Lemma 5.2.19). As V_0 has cosocle $C^{0,1,1}$ and $V_0 \subseteq \widetilde{V}$ by the above discussion, it follows that V_0 is the unique subrepresentation of $\widetilde{V}$ with cosocle $C^{0,1,1}$. Moreover any subrepresentation of $\widetilde{V}$ which contains $C^{0,1,1}$ as a constituent also contains the subrepresentation V_0. Since $V_{\underline{j},\underline{j}'}$ has cosocle $C^{1,1,1}$ (which appears with multiplicity 1 in $\Theta_\mu(V_0)$), its image in $\Theta_{w_0 s_{j_0} w_0}(C^{0,1,1})$ is the unique subrepresentation of cosocle $C^{1,1,1}$, in particular $V_{\underline{j},\underline{j}'} \subseteq \widetilde{V}$, and thus $V_0 \subseteq V_{\underline{j},\underline{j}'}$. Hence, by the end of the last sentence of the first paragraph of the proof, we also deduce that each $C^{0,b,c}$ for $b, c \in \{0, 1\}$ appears in $V_{\underline{j},\underline{j}'}$ with multiplicity exactly one.

We prove (ii). Let $(b, c), (b', c') \in \{0, 1\}^2$ such that $b \leq b'$, $c \leq c'$ and $b' + c' = b + c + 1$, then V_0 admits a unique length 2 subquotient V with socle $C^{0,b,c}$ and cosocle $C^{0,b',c'}$. It follows from (ii) of Lemma 5.2.13 that $\Theta_\mu(V)$ admits a unique subquotient which is the Ext-square constructed in (i) or (ii) of Proposition 5.2.10 with socle $C^{0,b,c}$, cosocle $C^{1,b',c'}$ and middle layer containing $C^{1,b,c} \oplus C^{0,b',c'}$. Since $\Theta_\mu(V)$ is a subquotient of $\Theta_\mu(V_0)$, this proves (ii). In particular $\Theta_\mu(V_0)$ admits a length 2 subquotient with socle $C^{1,b,c}$ and cosocle $C^{1,b',c'}$.

We prove (iii). Let $(b, c), (b', c') \in \{0, 1\}^2$ such that $b \le b'$, $c \le c'$, and recall from (ii) of Lemma 5.2.19 that $C^{1,b,c}$ and $C^{1,b',c'}$ have multiplicity one in $\Theta_\mu(V_0)$. Hence, the end of the previous paragraph implies that $\Theta_\mu(V_0)$ admits a unique length 2 subquotient with socle $C^{1,b,c}$ and cosocle $C^{1,b',c'}$ when $b' + c' = b + c + 1$. Since $C^{1,1,1}$ appears in $V_{\underline{j},\underline{j}'}$, we deduce that $C^{1,0,1}$ and $C^{1,1,0}$, and then $C^{1,0,0}$, must all appear in $V_{\underline{j},\underline{j}'}$. So $V_{\underline{j},\underline{j}'}$ admits a unique length 2 subquotient with socle $C^{1,b,c}$ and cosocle $C^{1,b',c'}$ when $b' + c' = b + c + 1$. Together with the last sentence of the proof of (i), we deduce (iii).

We prove (iv). As V_0 injects into $V_{\underline{j},\underline{j}'}$ by (i) and V_0 admits a unique length 2 subquotient with socle $C^{0,b,c}$ and cosocle $C^{0,b',c'}$ when $b' + c' = b + c + 1$, such a subquotient uniquely appears in $V_{\underline{j},\underline{j}'}$ (unicity uses multiplicity 1 in (iii)). As $C^{1,b,c}$ has multiplicity one in $\Theta_\mu(V_0)$ and $V_{\underline{j},\underline{j}'}$ by the previous paragraph, $V_{\underline{j},\underline{j}'}$ contains the unique subrepresentation of $\Theta_\mu(V_0)$ with cosocle $C^{1,b,c}$. As $\Theta_{w_0 s_{j_0} w_0}(C^{0,b,c})$ admits a length 2 subrepresentation with socle $C^{0,b,c}$ and cosocle $C^{1,b,c}$ by (iii) of Lemma 5.2.19, this subrepresentation of $\Theta_\mu(V_0)$, and hence $V_{\underline{j},\underline{j}'}$, also admit a length 2 subquotient with socle $C^{0,b,c}$ and cosocle $C^{1,b,c}$. Moreover, for $V_{\underline{j},\underline{j}'}$ this subquotient is unique again by multiplicity 1 in (iii). Together with the one but last sentence of the proof of (iii), this gives all cases of (iv). □

Remark 5.2.21 Let $q : V_0 \to \Theta_\mu(V_0)$ be any non-zero map, which is necessarily injective as V_0 is multiplicity free with socle $C^{0,0,0}$ and $\Theta_\mu(V_0)$ has socle $C^{0,0,0}$ by (the first statement of) Lemma 5.2.20, and denote by $p : \Theta_\mu(V_0) \twoheadrightarrow V_0$ the canonical surjection (cf. the first paragraph of the proof of Lemma 5.2.20). If the composition $p \circ q : V_0 \to V_0$ is non-zero, then it has to be an isomorphism as V_0 is multiplicity free with socle $C^{0,0,0}$. By exactness of Θ_μ and (iii) of Lemma 5.2.19, the restriction $(p \circ q)|_{\mathrm{soc}_G(V_0)}$ is the composition $C^{0,0,0} \hookrightarrow \Theta_\mu(C^{0,0,0}) \twoheadrightarrow C^{0,0,0}$ which is zero, a contradiction to $p \circ q$ being an isomorphism. Hence, we have $p \circ q = 0$ and thus $\mathrm{im}(q) \subseteq \ker(p)$. But from (ii) of Lemma 5.2.19 $C^{0,1,1}$ has multiplicity one in $\ker(p)$, so $\mathrm{im}(q)$ has to be the unique subrepresentation of $\ker(p)$ with cosocle $C^{0,1,1}$. It follows that q is unique up to a scalar, or equivalently $\dim_E \mathrm{Hom}_G(V_0, \Theta_\mu(V_0)) = 1$.

For each $(b, c) \in \{0, 1\}^2$, we write $V_{b,c}$ for the unique subrepresentation of V_0 with cosocle $C^{0,b,c}$. By exactness of Θ_μ and since Θ_μ kills any locally algebraic constituent of V_0, $\Theta_\mu(V_0)$ admits a natural increasing filtration $\mathrm{Fil}_{b,c}(\Theta_\mu(V_0)) \stackrel{\text{def}}{=} \Theta_\mu(V_{b,c})$ with graded piece

$$\mathrm{gr}_{b,c}(\Theta_\mu(V_0)) \stackrel{\text{def}}{=} \mathrm{Fil}_{b,c}(\Theta_\mu(V_0)) \Big/ \sum_{b'+c'<b+c} \mathrm{Fil}_{b',c'}(\Theta_\mu(V_0)) \cong \Theta_{w_0 s_{j_0} w_0}(C^{0,b,c}).$$

The filtration $\{\mathrm{Fil}_{b,c}(\Theta_\mu(V_0))\}_{0 \le b,c \le 1}$ induces a filtration $\{\mathrm{Fil}_{b,c}(V_{\underline{j},\underline{j}'})\}_{0 \le b,c \le 1}$ on $V_{\underline{j},\underline{j}'} \subseteq \Theta_\mu(V_0)$ with graded piece $\mathrm{gr}_{b,c}(V_{\underline{j},\underline{j}'}) \subseteq \Theta_{w_0 s_{j_0} w_0}(C^{0,b,c})$. We deduce from (ii) and (iii) of Lemma 5.2.19 combined with (iii) of Lemma 5.2.20 that we have inclusions for $(b, c) \in \{0, 1\}^2$

$$\mathrm{gr}_{b,c}(V_{\underline{j},\underline{j}'}) \subseteq \mathrm{rad}_G(\Theta_{w_0 s_{j_0} w_0}(C^{0,b,c})) = \ker(\Theta_{w_0 s_{j_0} w_0}(C^{0,b,c}) \twoheadrightarrow C^{0,b,c}). \quad (5.109)$$

Lemma 5.2.22 *Assume $j_1 \neq j_0 + 1$. Then $V_{\underline{j},\underline{j}'}$ admits a locally algebraic constituent if and only if $j_0 = j_1$. Moreover, if $j_0 = j_1$ then $\mathrm{gr}_{b,c}(V_{\underline{j},\underline{j}'})$ (for $(b, c) \in \{0, 1\}^2$) admits a locally algebraic constituent if and only if $b = c = 0$.*

Proof Assume that $V_{\underline{j},\underline{j}'}$ admits a locally algebraic constituent W, then there exists $(b, c) \in \{0, 1\}^2$ such that W occurs in $\mathrm{gr}_{b,c}(V_{\underline{j},\underline{j}'}) \subseteq \Theta_{w_0 s_{j_0} w_0}(C^{0,b,c})$. Writing $C^{0,b,c} = \mathcal{F}^G_{P_{I_x}}(L(x), \pi^\infty)$ with $x \in W(G)$ and π^∞ irreducible G-regular, it follows from (5.87) and Lemma 5.1.1 (and (ii) of Theorem 4.3.7) that $L(1)$ is a constituent of $\Theta_{w_0 s_{j_0} w_0}(L(x))$, which by Proposition 3.4.5 forces $x = s_{j_0}$. By checking the definition of $C^{0,b,c}$ for $b, c \in \{0, 1\}$ (see above Lemma 5.2.19 with (5.68)) and using $j_1 \neq j_0 + 1$, we have either $b = c$ with $j_1 = j_0$, or $b = 1$, $c = 0$ with $j_0 = j_1 + 1$.

Case 1: Assume W occurs in $\mathrm{gr}_{1,1}(V_{\underline{j},\underline{j}'})$. By (iii) of Lemma 5.2.19 $\Theta_{w_0 s_{j_0} w_0}(C^{0,1,1})$ admits a unique length 2 subrepresentation V with socle $C^{0,1,1}$ and cosocle $C^{1,1,1}$, and by the proof of (i) of Lemma 5.2.20, V is the image of $V_{\underline{j},\underline{j}'}$ via $\Theta_\mu(V_0) \twoheadrightarrow \Theta_{w_0 s_{j_0} w_0}(C^{0,1,1})$, i.e. $\mathrm{gr}_{1,1}(V_{\underline{j},\underline{j}'}) \cong V$. In particular, $\mathrm{gr}_{1,1}(V_{\underline{j},\underline{j}'})$ has no locally algebraic constituent, a contradiction.

Case 2: Assume W occurs in $\mathrm{gr}_{1,0}(V_{\underline{j},\underline{j}'})$ (and $j_0 = j_1 + 1$). Then by Proposition 3.4.5 and (5.109), W occurs in the cosocle of $\mathrm{gr}_{1,0}(V_{\underline{j},\underline{j}'})$. As $V_{\underline{j},\underline{j}'}$ has cosocle $C^{1,1,1}$, there must exist $W' \in \mathrm{JH}_G(\mathrm{gr}_{1,1}(V_{\underline{j},\underline{j}'})) = \{C^{0,1,1}, C^{1,1,1}\}$ (see **Case** 1 for the equality) such that $V_{\underline{j},\underline{j}'}$ admits a length 2 subquotient V' with socle W and cosocle W'. Recall that V_0 is the unique subrepresentation of $V_{\underline{j},\underline{j}'}$ with cosocle $C^{0,1,1}$ (use (i) and (iii) of Lemma 5.2.20). Thus, if $W' = C^{0,1,1}$ then W must occur in V_0. But V_0 admits a locally algebraic factor if and only if $j_0 = j_1$ (see Proposition 5.2.18), which contradicts $j_0 = j_1 + 1$. Hence we must have $W' = C^{1,1,1}$. Since $\mathrm{Ext}^1_{U(\mathfrak{g})}(L(1), L(w_{j_1,j_0+1})) = 0$ when $j_0 = j_1 + 1$ by (ii) of Lemma 3.2.4, Proposition 5.1.14 forces $\mathrm{Ext}^1_G(C^{1,1,1}, W) = 0$, which contradicts the existence of V' above.

So the only remaining case is $b = c = 0$ with $j_1 = j_0$. This proves the "only if" of the two statements of the lemma. For the "if", we can use that V_0 has a locally algebraic constituent when $j_1 = j_0$ by (ii) of Proposition 5.2.18, hence so does $V_{\underline{j},\underline{j}'}$ by (i) of Lemma 5.2.20. □

Lemma 5.2.23 *Assume $j_1 \neq j_0 + 1$. The non-locally algebraic constituents of $V_{\underline{j},\underline{j}'}$ are the $C^{a,b,c}$ for $(a, b, c) \in \{0, 1\}^3$.*

Proof Assume on the contrary that there exists a non-locally algebraic constituent W_0 of $V_{\underline{j},\underline{j}'}$ such that $W_0 \neq C^{a,b,c}$ for any $(a, b, c) \in \{0, 1\}^3$. Then there exists $(b, c) \in \{0, 1\}^2$ such that W_0 occurs in $\mathrm{gr}_{b,c}(V_{\underline{j},\underline{j}'}) \subseteq \Theta_{w_0 s_{j_0} w_0}(C^{0,b,c})$. We take (b, c) such that $b + c$ is maximal among those $(b, c) \in \{0, 1\}^2$ such that $\mathrm{gr}_{b,c}(V_{\underline{j},\underline{j}'})$ has a non-locally algebraic constituent distinct from the $C^{a,b,c}$. By definition we can write $C^{0,b,c} = \mathcal{F}^G_{P_{\hat{j}}}(L(w_{j,j_0}), \pi^\infty)$ for some $j \in \{j_1 - 1, j_1, j_1 + 1\} \cap \Delta$ and some irreducible G-regular π^∞ (see above Lemma 5.2.19), and note that, by definition again, we have $C^{1,b,c} = \mathcal{F}^G_{P_{\hat{j}}}(L(w_{j,j_0+1}), \pi^\infty)$. It then follows from Proposition 3.4.5, Remark 3.4.6,

(5.87) and Lemma 5.1.1 (and the above assumptions on W_0) that we must have $W_0 \cong \mathcal{F}_{P_{\hat{j}}}^G(L(w_{j,j_0-1}), \pi^\infty)$, and using moreover (5.109) and the first statement of (iii) of Lemma 5.2.19 that W_0 must be in the cosocle of $\mathrm{gr}_{b,c}(V_{\underline{j},\underline{j}'})$. Since $V_{\underline{j},\underline{j}'}$ has cosocle $C^{1,1,1} \neq W_0$ there must exist a constituent W_1 of $\mathrm{gr}_{b',c'}(V_{\underline{j},\underline{j}'})$ for some $b', c' \in \{0, 1\}$ with $b' + c' > b + c$ such that $V_{\underline{j},\underline{j}'}$ admits a length 2 subquotient V with socle W_0 and cosocle W_1. As $b' + c' > b + c \geq 0$, it follows from Lemma 5.2.22 that W_1 is not locally algebraic. By maximality of $b + c$, we deduce $W_1 = C^{a',b',c'}$ for some $(a', b', c') \in \{0, 1\}^3$. If $a' = 0$, then W_1 occurs in V_0 by (i) and (iii) of Lemma 5.2.20, but W_0 does not, a contradiction. Assume $a' = 1$, and thus $W_1 = C^{1,b',c'} = \mathcal{F}_{P_{\hat{j}'}}^G(L(w_{j',j_0+1}), \sigma^\infty)$ for some $j' \in \{j_1 - 1, j_1, j_1 + 1\} \cap \Delta$ and some irreducible G-regular σ^∞ (see above Lemma 5.2.19). By (ii) of Lemma 3.2.4 and (the very last statement in) Remark A.10, we have $\mathrm{Ext}^1_{U(\mathfrak{g})}(L(w_{j,j_0-1}), L(w_{j',j_0+1})) = 0$. By Proposition 5.1.14 this forces $\mathrm{Ext}^1_G(W_1, W_0) = 0$, a contradiction to the existence of V. □

Lemma 5.2.24 *Assume $j_1 \neq j_0 + 1$. Then $V_{\underline{j},\underline{j}'}$ admits a unique quotient isomorphic to V_1 (see above Lemma 5.2.19 for V_1).*

Proof Using (iii) of Lemma 5.2.20 we define V_1' as the unique quotient of $V_{\underline{j},\underline{j}'}$ with socle $C^{1,0,0}$. We let $\mathrm{Fil}_{b,c}(V_1')$ be the image of $\mathrm{Fil}_{b,c}(V_{\underline{j},\underline{j}'})$ via $V_{\underline{j},\underline{j}'} \twoheadrightarrow V_1'$. It follows from (ii) of Lemma 5.2.19 and from (5.109) that $C^{1,0,0}$ occurs in the cosocle of $\mathrm{gr}_{0,0}(V_{\underline{j},\underline{j}'})$, which forces $\mathrm{gr}_{0,0}(V_1') \cong C^{1,0,0}$. Since $\mathrm{gr}_{b,c}(V_{\underline{j},\underline{j}'})$, and hence $\mathrm{gr}_{b,c}(V_1')$, have no locally algebraic constituent when $b + c > 0$ by Lemma 5.2.22, it follows that V_1' has no locally algebraic constituent. By Lemma 5.2.23, all constituents of V_1' are therefore of the form $C^{a,b,c}$ for some $(a, b, c) \in \{0, 1\}^3$. In particular V_1' is multiplicity free by (iii) of Lemma 5.2.20 with socle $C^{1,0,0}$ and cosocle $C^{1,1,1}$. Since V_0 maps to 0 via $V_0 \hookrightarrow V_{\underline{j},\underline{j}'} \twoheadrightarrow V_1'$ (using (i) of Lemma 5.2.20 and the fact V_1' has cosocle $C^{1,0,0}$ which doesn't occur in V_0), it follows from (iii) of Lemma 5.2.20 that the constituents of V_1' are of the form $C^{1,b',c'}$ for some $(b', c') \in \{0, 1\}^2$. Since $C^{1,b,c}$ is a constituent of $\mathrm{gr}_{b',c'}(V_{\underline{j},\underline{j}'})$ if and only if $b = b'$ and $c = c'$ by (i) of Lemma 5.2.19, we deduce with (iv) of Lemma 5.2.20 that V_1' has Loewy length 3 with socle $C^{1,0,0}$, cosocle $C^{1,1,1}$ and middle layer $C^{1,0,1} \oplus C^{1,1,0}$. By Lemma 5.2.8 and the definition of V_1, this implies $V_1' \cong V_1$. □

For $I \subseteq \Delta$ we write (see also (2.40))

$$V_{I,\Delta}^{\mathrm{alg}} \stackrel{\mathrm{def}}{=} L(1)^\vee \otimes_E V_{I,\Delta}^\infty = L(\mu_0)^\vee \otimes_E V_{I,\Delta}^\infty. \tag{5.110}$$

Lemma 5.2.25 *Assume $j_1 \neq j_0 + 1$. Then $V_{\underline{j},\underline{j}'}$ is multiplicity free. Moreover, when $j_1 = j_0$, $V_{\underline{j},\underline{j}'}$ admits a unique quotient isomorphic to $V_{\infty,\underline{j}'}$ (see (iv) of Proposition 5.2.10) which contains all its locally algebraic constituents.*

Proof If $j_1 \neq j_0$ (and $j_1 \neq j_0 + 1$), then by Lemma 5.2.22 $V_{\underline{j},\underline{j}'}$ has no locally algebraic constituent, which by Lemma 5.2.23 and (iii) of Lemma 5.2.20 implies that $V_{\underline{j},\underline{j}'}$ is multiplicity free. We assume $j_1 = j_0$ in the rest of the proof. Note that, by *loc. cit.* each non locally algebraic constituent of $V_{\underline{j},\underline{j}'}$ occurs with multiplicity 1. Recall from Proposition 3.4.5 that $L(1)$ has multiplicity 1 in $\Theta_{w_0 s_{j_0} w_0}(L(s_{j_0}))$. It then follows from (5.87), (ii) of Theorem 4.3.7 and Lemma 5.1.1 that $L(1)^\vee \otimes_E i^\infty_{\widehat{j_0},\Delta}(\pi^\infty_{\widehat{j_0},j_2})$ contains all the locally algebraic constituents of $\Theta_{w_0 s_{j_0} w_0}(C^{0,0,0})$, and therefore of $V_{\underline{j},\underline{j}'}$ by Lemma 5.2.22. As $i^\infty_{\widehat{j_0},\Delta}(\pi^\infty_{\widehat{j_0},j_2})$ is multiplicity free by (iv) of Remark 2.1.16, we deduce that $V_{\underline{j},\underline{j}'}$ is multiplicity free.

It remains to prove that $V_{\underline{j},\underline{j}'}$ admits a (unique) quotient isomorphic to $V_{\infty,\underline{j}'}$ which contains all its locally algebraic constituents. As $\mathrm{gr}_{0,0}(V_{\underline{j},\underline{j}'}) \subseteq \Theta_{w_0 s_{j_0} w_0}(C^{0,0,0})$, and the subquotient $L(1)^\vee \otimes_E i^\infty_{\widehat{j_0},\Delta}(\pi^\infty_{\widehat{j_0},j_2})$ of $\Theta_{w_0 s_{j_0} w_0}(C^{0,0,0})$ contains all its locally algebraic constituents, there exists a (unique) subrepresentation $\pi^\infty \subseteq i^\infty_{\widehat{j_0},\Delta}(\pi^\infty_{\widehat{j_0},j_2})$ such that $L(1)^\vee \otimes_E \pi^\infty$ is a subquotient of $V_{\underline{j},\underline{j}'}$ which contains its locally algebraic constituents. Moreover by (5.109), Proposition 3.4.5 and Lemma 5.1.1 we know that $L(1)^\vee \otimes_E \pi^\infty$ is a quotient of $\mathrm{gr}_{0,0}(V_{\underline{j},\underline{j}'})$.

Now we study all possible length 2 subquotients V of $V_{\underline{j},\underline{j}'}$ with socle W_0 and cosocle W_1 such that W_0 is locally algebraic and W_1 is not. By Lemma 5.2.23 $W_1 = C^{a,b,c}$ for some $(a,b,c) \in \{0,1\}^3$ and by (i) of Lemma 5.2.19 W_1 is a constituent of $\mathrm{gr}_{b,c}(V_{\underline{j},\underline{j}'})$. As W_0 is necessarily a constituent of $L(1)^\vee \otimes_E \pi^\infty$ which is a quotient of $\mathrm{gr}_{0,0}(V_{\underline{j},\underline{j}'})$, we must have $b+c>0$. If $a=0$, then W_1 occurs in V_0 (using (i) and (iii) of Lemma 5.2.20), which together with the layer structure of V_0 (see (ii) of Proposition 5.2.18) forces $W_1 = C^{0,1,1}$ and $W_0 = L(1)^\vee \otimes_E V^\infty_{[j_2-j_0+1,j_2],\Delta}$. Assume $a=1$ and thus $W_1 = \mathcal{F}^G_{P_{\widehat{j}}}(L(w_{j,j_0+1}),\tau^\infty)$ for some $j \in \{j_1-1, j_1, j_1+1\} \cap \Delta$ and some irreducible G-regular τ^∞ (see above Lemma 5.2.19). The existence of V implies $\mathrm{Ext}^1_G(W_1,W_0) \neq 0$, which together with Proposition 5.1.14 forces $\mathrm{Ext}^1_{U(\mathfrak{g})}(L(1), L(w_{j,j_0+1})) \neq 0$. By (ii) of Lemma 3.2.4 and our assumption $j_0 = j_1$, we deduce $j = j_0+1 = j_1+1$ and thus $W_1 = C^{1,1,0}$ (with $w_{j,j_0+1} = s_{j_0+1}$ and $\tau^\infty = \pi^\infty_{\widehat{j_0+1},j_2+1}$). We can write $W_0 = L(1)^\vee \otimes_E \sigma^\infty$ for some irreducible G-regular constituent σ^∞ of π^∞. Then $\mathrm{Ext}^1_G(W_1,W_0) \neq 0$ and the last statement in Remark 5.1.12 imply $\mathrm{Hom}_G(\sigma^\infty, i^\infty_{\Delta\setminus\{j_0+1\},\Delta}(\pi^\infty_{\widehat{j_0+1},j_2+1})) \neq 0$. By Lemma 2.3.1 we deduce $\sigma^\infty = V^\infty_{[j_2-j_0,j_2],\Delta}$ if $j_0 = j_1 < j_2$ and $\sigma^\infty = V^\infty_{[j_2-j_0+1,j_2],\Delta}$ if $j_0 = j_1 = j_2$. We have thus shown that all possible pairs (W_0, W_1) are $(V^{\mathrm{alg}}_{[j_2-j_0+1,j_2],\Delta}, C^{0,1,1})$, $(V^{\mathrm{alg}}_{[j_2-j_0,j_2],\Delta}, C^{1,1,0})$ when $j_0 = j_1 < j_2$, and $(V^{\mathrm{alg}}_{[j_2-j_0+1,j_2],\Delta}, C^{1,1,0})$ when $j_0 = j_1 = j_2$.

Let σ^∞ be a constituent of $\mathrm{cosoc}_G(\pi^\infty)$, then since $L(1)^\vee \otimes_E \pi^\infty$ is a quotient of $\mathrm{gr}_{0,0}(V_{\underline{j},\underline{j}'})$ there must exist (W_0,W_1) in the above list with $W_0 = L(1)^\vee \otimes_E \sigma^\infty$. An examination of Lemma 2.3.1 shows that $\pi^\infty \subseteq i^\infty_{\widehat{j_0},\Delta}(\pi^\infty_{\widehat{j_0},j_2})$ must be the length 2 subrepresentation with socle $V^\infty_{[j_2-j_0+1,j_2],\Delta}$ and cosocle $V^\infty_{[j_2-j_0,j_2],\Delta}$ when $j_0 = j_1 < j_2$, and the irreducible representation $V^\infty_{[j_2-j_0+1,j_2],\Delta}$ when $j_0 = j_1 = j_2$. From the description of the possible pairs above, from Lemma 5.2.23 and (iii), (iv) of Lemma 5.2.20, and from the fact $V_{\underline{j},\underline{j}'}$ has (irreducible) cosocle $C^{1,1,1}$, we can deduce that $V_{\underline{j},\underline{j}'}$ admits a unique (multiplicity free) quotient with constituents

$$\{V^{\mathrm{alg}}_{[j_2-j_0+1,j_2],\Delta},\ V^{\mathrm{alg}}_{[j_2-j_0,j_2],\Delta},\ C^{0,1,1},\ C^{1,1,0},\ C^{1,1,1}\}$$

with partial order (in the sense of Sect. 1.4) which can only be: $V^{\rm alg}_{[j_2-j_0+1,j_2],\Delta} \leq V^{\rm alg}_{[j_2-j_0,j_2],\Delta}$, $V^{\rm alg}_{[j_2-j_0+1,j_2],\Delta} \leq C^{0,1,1}$, $V^{\rm alg}_{[j_2-j_0,j_2],\Delta} \leq C^{1,1,0}$, $C^{0,1,1} \leq C^{1,1,1}$ and $C^{1,1,0} \leq C^{1,1,1}$ (with $V^{\rm alg}_{[j_2-j_0,j_2],\Delta}$ omitted when $j_0 = j_1 = j_2$). It then follows from (iv) of Proposition 5.2.10 (with $\underline{j}$ there replaced by $\underline{j}'$ here) together with the minimality of the Ext-square $V_{\infty,\underline{j}'}$ in *loc. cit.* (last statement of Proposition 5.2.10) and with Lemma 5.2.8 that this quotient must be $V_{\infty,\underline{j}'}$. □

Lemma 5.2.26 *Assume $j_1 \neq j_0 + 1$. Then $V_{\underline{j},\underline{j}'}$ is an* Ext-*cube such that*

(i) $\mathrm{gr}^0(V_{\underline{j},\underline{j}'}) \cong C_{\underline{j}'}$ *and* $\mathrm{gr}^3(V_{\underline{j},\underline{j}'}) \cong C_{\underline{j}}$;

(ii) $V_{\underline{j},\underline{j}'}$ *contains a unique subquotient of the form $V_{\underline{j}'',\underline{j}'''}$ for each pair $(\underline{j}'', \underline{j}''') \in \mathbf{J}^2$ satisfying $\underline{j} \leq \underline{j}'' \leq \underline{j}''' \leq \underline{j}'$ and $d(\underline{j}'', \underline{j}''') = 2$ (these $V_{\underline{j}'',\underline{j}'''}$ are defined in (i), (ii) of Proposition 5.2.10 and (i), (ii) of Proposition 5.2.18).*

Proof We first prove (ii). Note that $\underline{j}'' \in \mathbf{J}$ satisfying $\underline{j} \leq \underline{j}'' \leq \underline{j}'$ is equivalent to the choice of $(a, b, c) \in \{0, 1\}^3$ (writing $\underline{j}'' = (j_0 + a, j_1 + c - b, j_2 + c)$). In particular, the choice of $\underline{j}'', \underline{j}''' \in \mathbf{J}$ such that $\underline{j} \leq \underline{j}'' \leq \underline{j}''' \leq \underline{j}'$ and $d(\underline{j}'', \underline{j}''') = 2$ is equivalent to the choice of $(a, b, c), (a', b', c') \in \{0, 1\}^3$ such that $a \leq a'$, $b \leq b'$, $c \leq c'$ and $a' + b' + c' = a + b + c + 2$. It follows from (iii) of Lemma 5.2.20 that if $V_{\underline{j}'',\underline{j}'''}$ is a subquotient of $V_{\underline{j},\underline{j}'}$, then it is necessarily the unique subquotient of $V_{\underline{j},\underline{j}'}$ with socle $C_{\underline{j}''}$ and cosocle $C_{\underline{j}'''}$. If $b = b'$ or $c = c'$, then the existence of a subquotient $V_{\underline{j}'',\underline{j}'''}$ of $V_{\underline{j},\underline{j}'}$ follows from (ii), (iii) of Lemma 5.2.20 and (ii) of Lemma 5.2.19. If $a = a' = 0$, then $V_{\underline{j}'',\underline{j}'''} = V_0$ and the existence of a subquotient V_0 of $V_{\underline{j},\underline{j}'}$ follows from (i) of Lemma 5.2.20. If $a = a' = 1$, then $V_{\underline{j}'',\underline{j}'''} = V_1$ and the existence of a subquotient V_1 of $V_{\underline{j},\underline{j}'}$ follows from Lemma 5.2.24. This proves (ii).

Now we construct a filtration on $V_{\underline{j},\underline{j}'}$ that makes it an Ext-cube (see Definition 5.2.6). Recall that $V_{\underline{j},\underline{j}'}$ has socle $C^{0,0,0}$ and cosocle $C^{1,1,1}$. It follows from (ii) above, from (iii) and (iv) of Lemma 5.2.20, from Lemma 5.2.23 and from Lemma 5.2.25 (with (iv) and (ii) of Proposition 5.2.10) that $V_{\underline{j},\underline{j}'}$ admits a unique decreasing filtration $(\mathrm{Fil}^k(V_{\underline{j},\underline{j}'}))_{0\leq k\leq 3}$ such that for $0 \leq k \leq 3$

$$\mathrm{gr}^k(V_{\underline{j},\underline{j}'}) \cong C_{\infty,k} \bigoplus \Big(\bigoplus_{a+b+c=3-k} C^{a,b,c} \Big), \tag{5.111}$$

where $C_{\infty,k} \stackrel{\rm def}{=} \mathrm{gr}^2(V_{\infty,\underline{j}'})$ when $k = 2$ and $j_1 = j_0$, and $C_{\infty,k} \stackrel{\rm def}{=} 0$ otherwise. In particular, (i) holds for the filtration $(\mathrm{Fil}^k(V_{\underline{j},\underline{j}'}))_k$. It remains to check that $V_{\underline{j},\underline{j}'}$ is an Ext-cube for this filtration. But the first condition in (i) of Definition 5.2.6 holds by definition of $V_{\underline{j},\underline{j}'}$ and of the filtration $(\mathrm{Fil}^k(V_{\underline{j},\underline{j}'}))_k$, and the second condition in (i) of Definition 5.2.6 holds by Lemmas 5.2.1 and 5.2.2. □

Lemma 5.2.27 *Assume $j_1 \neq j_0 + 1$. Then the* Ext-*cube $V_{\underline{j},\underline{j}'}$ is strict and minimal (see (iii) and (iv) of Definition 5.2.6).*

Proof The Ext-cube $V_{\underline{j},\underline{j}'}$ is strict as it has simple socle ($C^{0,0,0} = C_{\underline{j}}$) and simple cosocle ($C^{1,1,1} = C_{\underline{j}'}$). By (i), (ii) of Propositions 5.2.10 and 5.2.18, $V_{\underline{j}'',\underline{j}'''}$ is minimal for each $(\underline{j}'', \underline{j}''') \in \mathbf{J}^2$ such that $\underline{j} \le \underline{j}'' \le \underline{j}''' \le \underline{j}'$ and $d(\underline{j}'', \underline{j}''') = 2$, and by (iv) of Proposition 5.2.10 $V_{\infty,\underline{j}'}$ is minimal when $j_1 = j_0$. Hence to check the minimality of $V_{\underline{j},\underline{j}'}$, it suffices to show that any Ext-cube V' such that $\mathrm{gr}^0(V') = C_{\underline{j}'}$, $\mathrm{gr}^3(V') = C_{\underline{j}}$ and $\mathrm{gr}^k(V')$ is a good direct summand of $\mathrm{gr}^k(V_{\underline{j},\underline{j}'})$ for $k = 1, 2$ must satisfy $\mathrm{gr}^k(V') = \mathrm{gr}^k(V_{\underline{j},\underline{j}'})$ for $k = 1, 2$. Let V' be such an Ext-cube, then $\mathrm{gr}^k(V') \neq 0$ for $k = 1, 2$. Thus by (5.111) $\mathrm{gr}^1(V')$ contains a constituent $C_{\underline{j}''}$ for some $\underline{j} < \underline{j}'' < \underline{j}'$ such that $d(\underline{j}, \underline{j}'') = 2$. It follows from Lemma 5.2.26 and its proof that the unique subrepresentation V'' of V' with cosocle $C_{\underline{j}''}$ is an Ext-square such that $\mathrm{gr}^0(V'') = C_{\underline{j}''}$, $\mathrm{gr}^2(V'') = C_{\underline{j}}$ and $\mathrm{gr}^1(V'')$ is a good direct summand of $\mathrm{gr}^1(V_{\underline{j},\underline{j}''})$. The minimality of $V_{\underline{j},\underline{j}''}$ forces $\mathrm{gr}^1(V'') \cong \mathrm{gr}^1(V_{\underline{j},\underline{j}''})$, and thus for each $\underline{j}'''$ such that $\underline{j} < \underline{j}''' < \underline{j}''$ the constituent $C_{\underline{j}'''}$ shows up in $\mathrm{gr}^1(V'')$, and therefore in $\mathrm{gr}^2(V')$. A similar argument using the minimality of $V_{\underline{j}''',\underline{j}'}$ for $\underline{j}'''$ such that $\underline{j} < \underline{j}''' < \underline{j}''$ gives $\mathrm{gr}^1(V') \cong \mathrm{gr}^1(V_{\underline{j},\underline{j}'})$. In particular $C_{\underline{j}''}$ occurs in $\mathrm{gr}^1(V')$ for *each* $\underline{j} < \underline{j}'' < \underline{j}'$ with $d(\underline{j}, \underline{j}'') = 2$, and we can repeat the previous argument with the subrepresentation V'' of V' with cosocle $C_{\underline{j}''}$. We now deduce that $C_{\underline{j}'''}$ occurs in $\mathrm{gr}^1(V'')$, and therefore in $\mathrm{gr}^2(V')$, for *any* $\underline{j}'''$ such that $\underline{j} < \underline{j}''' < \underline{j}'$ and $d(\underline{j}, \underline{j}''') = 1$. If $j_0 \neq j_1$, there are no constituents left and we have proven $\mathrm{gr}^k(V') = \mathrm{gr}^k(V_{\underline{j},\underline{j}'})$ for $k = 1, 2$. We now assume $j_1 = j_0$. We already have $\mathrm{gr}^1(V') \cong \mathrm{gr}^1(V_{\underline{j},\underline{j}'})$ but it remains to prove $\mathrm{gr}^2(V') \cong \mathrm{gr}^2(V_{\underline{j},\underline{j}'})$, and for that we have to prove that the locally algebraic constituents of $V_{\underline{j},\underline{j}'}$, equivalently of $\mathrm{gr}^2(V_{\underline{j},\underline{j}'})$ by Lemma 5.2.25, are also in $\mathrm{gr}^2(V')$. Let $\underline{j}'' = (j_0 + 1, j_1 + 1, j_2 + 1)$, from the minimality of $V_{\underline{j},\underline{j}''}$ in (ii) of Proposition 5.2.10 and the fact $C_{\underline{j}''}$ occurs in $\mathrm{gr}^1(V')$, we deduce $\mathrm{gr}^2(V_{\infty,\underline{j}'}) \subseteq \mathrm{gr}^2(V')$. This finally implies $\mathrm{gr}^2(V') \cong \mathrm{gr}^2(V_{\underline{j},\underline{j}'})$ and finishes the proof. □

We now note that Lemma 5.2.19 has a symmetric statement replacing V_0 by V_1, $\Theta_{w_0 s_{j_0} w_0}$ (resp. Θ_μ) by $\Theta_{w_0 s_{j_0+1} w_0}$ (resp. by Θ_μ where $\mu \in \Lambda$ is such that $\langle \mu + \rho, \alpha^\vee \rangle \ge 0$ for $\alpha \in \Phi^+$ and the stabilizer of μ in $W(G)$ for the dot action is $\{1, w_0 s_{j_0+1} w_0\}$) and switching $C^{0,b,c}$ and $C^{1,b,c}$ everywhere. Then we can define $\widetilde{V}_{\underline{j},\underline{j}'}$ as the unique quotient of $\Theta_\mu(V_1)$ with socle $C_{\underline{j}}$. Then Lemma 5.2.20 also has a symmetric version ($\Theta_\mu(V_1)$ and $\widetilde{V}_{\underline{j},\underline{j}'}$ have cosocle $C^{1,1,1}$, $\widetilde{V}_{\underline{j},\underline{j}'}$ surjects onto V_1, etc.) and likewise all statements from Lemma 5.2.22 to Lemma 5.2.27 have symmetric versions replacing the assumption $j_1 \neq j_0 + 1$ by the assumption $j_1 \neq j_0$, V_0 by V_1 and the case $j_1 = j_0$ by the case $j_1 = j_0 + 1$. We let the reader work out by himself the symmetric statements. For instance the symmetric Lemma 5.2.25 is: assume $j_1 \neq j_0$, then $V_{\underline{j},\underline{j}'}$ is multiplicity free, and moreover when $j_1 = j_0 + 1$, $\widetilde{V}_{\underline{j},\underline{j}'}$ admits a unique subrepresentation isomorphic to $V_{\underline{j},\infty}$ which contains all its locally algebraic constituents. Note that all minimal Ext-squares that are used in the (symmetric) proofs are still provided by Propositions 5.2.10 and 5.2.18. It also follows from Lemma 5.2.26 and its proof, from the minimality in Lemma 5.2.27, from their symmetric versions, and from Lemma 5.2.8 that $V_{\underline{j},\underline{j}'} \cong \widetilde{V}_{\underline{j},\underline{j}'}$ when $j_1 \notin \{j_0, j_0+1\}$.

In order to avoid too much notation, we will now denote by $V_{\underline{j},\underline{j}'}$ the Ext-cube previously (also) denoted $V_{\underline{j},\underline{j}'}$ when $j_1 \neq j_0+1$, and by $V_{\underline{j},\underline{j}'}$ the Ext-cube previously denoted $\widetilde{V}_{\underline{j},\underline{j}'}$ when $j_1 \neq j_0$. Note that $V_{\underline{j},\underline{j}'}$ is well defined by the above isomorphism. The following proposition sums up some of the previous results on $V_{\underline{j},\underline{j}'}$ which are proven in Lemmas 5.2.27, 5.2.23, 5.2.26, 5.2.25 and in their symmetric versions.

Proposition 5.2.28 *Let $\underline{j} \in \mathbf{J}$ such that $\underline{j}' = (j_0', j_1', j_2') \overset{def}{=} (j_0+1, j_1, j_2+1)$ is still in $\mathbf{J}$. There exists a minimal* Ext*-cube $V_{\underline{j},\underline{j}'}$ such that*

(i) the non-locally algebraic constituents of $V_{\underline{j},\underline{j}'}$ are the $C_{\underline{j}''}$ for $\underline{j} \leq \underline{j}'' \leq \underline{j}'$;

(ii) $\mathrm{gr}^0(V_{\underline{j},\underline{j}'}) \cong C_{(j_0+1,j_1,j_2+1)}$ and $\mathrm{gr}^3(V_{\underline{j},\underline{j}'}) \cong C_{(j_0,j_1,j_2)}$;

(iii) $V_{\underline{j},\underline{j}'}$ contains a unique subquotient of the form $V_{\underline{j}'',\underline{j}'''}$ for each pair $(\underline{j}'', \underline{j}''') \in \mathbf{J}^2$ satisfying $\underline{j} \leq \underline{j}'' \leq \underline{j}''' \leq \underline{j}'$ and $d(\underline{j}'', \underline{j}''') = 2$ ((i), (ii) of Proposition 5.2.10 and (i), (ii) of Proposition 5.2.18);

(iv) $V_{\underline{j},\underline{j}'}$ admits a locally algebraic constituent if and only if one of the following holds: either $j_1 = j_0$ and $V_{\underline{j},\underline{j}'}$ admits a unique quotient isomorphic to $V_{\infty,\underline{j}'}$ ((iv) of Proposition 5.2.10) which contains all its locally algebraic constituents, or $j_1 = j_0+1$ and $V_{\underline{j},\underline{j}'}$ admits a unique subrepresentation isomorphic to $V_{\underline{j},\infty}$ ((iii) of Proposition 5.2.10) which contains all its locally algebraic constituents.

Remark 5.2.29 In fact one can prove that the representations $V_{\underline{j},\underline{j}'}$ and $\widetilde{V}_{\underline{j},\underline{j}'}$ just below Lemma 5.2.27 are *always* isomorphic (even if $j_1 \in \{j_0, j_0+1\}$). But we won't need that result.

We end up this section with three lemmas, two of which construct other finite length representations $V_{\underline{j},\infty}^+$ and $V_{\underline{j}}$ of G in $\mathrm{Rep}^{\mathrm{an}}_{\mathrm{adm}}(G)$ which will be important to define key representations in Sect. 5.3.

Lemma 5.2.30 *Let $1 \leq j_0 = j_1 \leq j_2 < n-1$ and let $\underline{j} \overset{def}{=} (j_0, j_0, j_2)$, $\underline{j}' \overset{def}{=} (j_0, j_0, j_2+1)$, $\underline{j}'' \overset{def}{=} (j_0+1, j_0+1, j_2+1)$. Then there exists a unique multiplicity free finite length representation $V_{\underline{j},\infty}^+$ in* $\mathrm{Rep}^{\mathrm{an}}_{\mathrm{adm}}(G)$ *such that (see (5.110) for the notation)*

(i) $V_{\underline{j},\infty}^+$ has socle $C_{\underline{j}}$ and cosocle $V^{\mathrm{alg}}_{[j_2-j_0+1,j_2+1],\Delta}$;

(ii) both $V_{\underline{j},\underline{j}'}$ and $V_{\underline{j},\underline{j}''}$ inject into $V_{\underline{j},\infty}^+$ (see (ii) of Proposition 5.2.18 and (ii) of Proposition 5.2.10 respectively);

(iii) the quotient $V_{\underline{j},\infty}^+/(V_{\underline{j},\underline{j}'} + V_{\underline{j},\underline{j}''})$ is uniserial of length 2 with socle $V^{\mathrm{alg}}_{[j_2-j_0+2,j_2+1],\Delta}$ and cosocle $V^{\mathrm{alg}}_{[j_2-j_0+1,j_2+1],\Delta}$.

Proof We first prove unicity of $V_{\underline{j},\infty}^+$. We fix $\mu \in \Lambda$ such that $\langle \mu+\rho, \alpha^\vee\rangle \geq 0$ for $\alpha \in \Phi^+$ and the stabilizer of μ in $W(G)$ for the dot action is $\{1, w_0 s_{j_0} w_0\}$. Condition (iii) together with Proposition 3.4.5 and (5.87) imply $\Theta_\mu(V_{\underline{j},\infty}^+/V_{\underline{j},\underline{j}'}) = 0$.

The exactness of Θ_μ then imply $\Theta_\mu(V_{\underline{j},\underline{j}'}) \xrightarrow{\sim} \Theta_\mu(V^+_{\underline{j},\infty})$. Since $V^+_{\underline{j},\infty}$ has socle $C_{\underline{j}}$ by condition (ii), it follows from the exactness and functoriality of Θ_μ that the adjunction map $V^+_{\underline{j},\infty} \to \Theta_\mu(V^+_{\underline{j},\infty})$ is injective. We thus deduce an injection $V^+_{\underline{j},\infty} \hookrightarrow \Theta_\mu(V_{\underline{j},\underline{j}'})$. It follows from Proposition 3.4.5 that $L(1)$ is a constituent of $\Theta_{w_0 s_{j_0} w_0}(L(x))$ for some $x \in W(G)$ if and only if $x = s_{j_0}$, in which case $L(1)$ occurs with multiplicity 1. Together with (ii) of Proposition 5.2.18, (5.87) and the fact Θ_μ kills locally algebraic constituents of $V_{\underline{j},\underline{j}'}$ (Proposition 3.4.5), we see that if W is a constituent of $V_{\underline{j},\underline{j}'}$, then $\Theta_\mu(W)$ admits locally algebraic constituents if and only if $W \in \{C_{\underline{j}}, C_{\underline{j}'}\}$, and together with (ii) of Theorem 4.3.7 that $\Theta_\mu(C_{\underline{j}})$ (resp. $\Theta_\mu(C_{\underline{j}'})$) admits a subquotient $L(1)^\vee \otimes_E i^\infty_{\widehat{j_0},\Delta}(\pi^\infty_{j_0,j_2})$ (resp. $L(1)^\vee \otimes_E i^\infty_{\widehat{j_0},\Delta}(\pi^\infty_{j_0,j_2+1})$) which contains all its locally algebraic constituents. By Lemma 2.3.1 $i^\infty_{\widehat{j_0},\Delta}(\pi^\infty_{j_0,j_2})$ and $i^\infty_{\widehat{j_0},\Delta}(\pi^\infty_{j_0,j_2+1})$ are multiplicity free, and $V^\infty_{[j_2-j_0+1,j_2+1],\Delta}, V^\infty_{[j_2-j_0+2,j_2+1],\Delta} \in \mathrm{JH}_G(i^\infty_{\widehat{j_0},\Delta}(\pi^\infty_{j_0,j_2+1})) \setminus \mathrm{JH}_G(i^\infty_{\widehat{j_0},\Delta}(\pi^\infty_{j_0,j_2}))$. This implies that $V^{\mathrm{alg}}_{[j_2-j_0+1,j_2+1],\Delta}, V^{\mathrm{alg}}_{[j_2-j_0+2,j_2+1],\Delta} \in \mathrm{JH}_G(\Theta_\mu(C_{\underline{j}'})) \setminus \mathrm{JH}_G(\Theta_\mu(C_{\underline{j}}))$, that $V^{\mathrm{alg}}_{[j_2-j_0+1,j_2+1],\Delta}, V^{\mathrm{alg}}_{[j_2-j_0+2,j_2+1],\Delta}$ both appear with multiplicity 1 in $\Theta_\mu(V_{\underline{j},\underline{j}'})$, and that $\Theta_\mu(V_{\underline{j},\underline{j}'})$ contains as a subquotient the unique non-split extension of $V^{\mathrm{alg}}_{[j_2-j_0+1,j_2+1],\Delta}$ by $V^{\mathrm{alg}}_{[j_2-j_0+2,j_2+1],\Delta}$ (see Lemmas 2.3.1 and 5.1.3). In particular the second statement in condition (i) and the beginning of the proof force $V^+_{\underline{j},\infty}$, if it exists, to be the unique subrepresentation of $\Theta_\mu(V_{\underline{j},\underline{j}'})$ with cosocle $V^{\mathrm{alg}}_{[j_2-j_0+1,j_2+1],\Delta}$. This proves unicity of $V^+_{\underline{j},\infty}$.

We now prove that $V^+_{\underline{j},\infty}$, defined as the unique subrepresentation of $\Theta_\mu(V_{\underline{j},\underline{j}'})$ with cosocle $V^{\mathrm{alg}}_{[j_2-j_0+1,j_2+1],\Delta}$, is multiplicity free and satisfies (i), (ii), (iii). We already note that (i) follows from the fact that $\Theta_\mu(V_{\underline{j},\underline{j}'})$ has socle $C_{\underline{j}}$ (see the first statement of Lemma 5.2.20 and note that $V_{\underline{j},\underline{j}'}$ is the representation denoted V_0 there in the case $j_1 = j_0$).

Step 1: We prove that $V_{\underline{j},\underline{j}'}$ injects into $V^+_{\underline{j},\infty}$.
Let π^∞ be the unique smooth length 2 representation of G with socle $V^\infty_{[j_2-j_0+2,j_2+1],\Delta}$ and cosocle $V^\infty_{[j_2-j_0+1,j_2+1],\Delta}$, which is G-basic by Lemma 2.2.5. By (i) of Lemma 2.3.1 (applied with j_1, j_2 there being j_0, j_2+1) π^∞ injects into $i^\infty_{\widehat{j_0},\Delta}(\pi^\infty_{j_0,j_2+1})$, and thus $d(\pi^\infty, \pi^\infty_{j_0,j_2+1}) = 0$. Let M be the unique length 2 $U(\mathfrak{g})$-module with socle $L(1)$ and cosocle $L(s_{j_0})$ ((ii) of Lemma 3.2.4), then (ii) of Lemma 5.1.19 implies that $\mathcal{F}^G_{P_{\widehat{j_0}}}(M, \pi^\infty_{j_0,j_2+1})$ contains a unique subrepresentation V which fits into a non-split extension $0 \to C_{\underline{j}'} \to V \to L(1)^\vee \otimes_E \pi^\infty \to 0$. Moreover, as π^∞ injects into $i^\infty_{\widehat{j_0},\Delta}(\pi^\infty_{j_0,j_2+1})$, we deduce from (ii) of Lemma 5.1.15 (applied with $V_1 = C_{\underline{j}'}$ and $V_0 = L(1)^\vee \otimes_E \pi^\infty$) that V is uniserial of length 3, with socle $C_{\underline{j}'}$, cosocle $V^{\mathrm{alg}}_{[j_2-j_0+1,j_2+1],\Delta}$ and middle layer $V^{\mathrm{alg}}_{[j_2-j_0+2,j_2+1],\Delta}$. As M is a quotient of $\Theta_{w_0 s_{j_0} w_0}(L(s_{j_0}))$ by Proposition 3.4.5, we deduce from (5.87) that $\mathcal{F}^G_{P_{\widehat{j_0}}}(M, \pi^\infty_{j_0,j_2+1})$ is a subrepresentation of $\Theta_\mu(C_{\underline{j}'}) \cong \mathcal{F}^G_{P_{\widehat{j_0}}}(\Theta_{w_0 s_{j_0} w_0}(L(s_{j_0})), \pi^\infty_{j_0,j_2+1})$. Hence V is also a subrepresentation of $\Theta_\mu(C_{\underline{j}'})$, and is necessarily the unique subrepresentation

with cosocle $V^{\mathrm{alg}}_{[j_2-j_0+1,j_2+1],\Delta}$ (note that $V^{\mathrm{alg}}_{[j_2-j_0+1,j_2+1],\Delta}$ occurs with multiplicity 1 in $\Theta_\mu(C_{\underline{j}'})$ arguing as in the first paragraph of the proof of Lemma 5.2.25). Let us consider the surjection

$$\Theta_\mu(V_{\underline{j},\underline{j}'}) \twoheadrightarrow \Theta_\mu(C_{\underline{j}'}) \tag{5.112}$$

induced by $V_{\underline{j},\underline{j}'} \twoheadrightarrow C_{\underline{j}'}$. As $V^{\mathrm{alg}}_{[j_2-j_0+1,j_2+1],\Delta}$ occurs with multiplicity 1 in both $\Theta_\mu(V_{\underline{j},\underline{j}'})$ and $\Theta_\mu(C_{\underline{j}'})$, it follows from the definition of $V^+_{\underline{j},\infty}$ that V is the image of $V^+_{\underline{j},\infty}$ under (5.112). In particular $V^+_{\underline{j},\infty}$ contains the constituent $C_{\underline{j}'}$. As $C_{\underline{j}'}$ does not occur in the kernel of (5.112) by (i) of Lemma 5.2.19, it follows from (ii) of Lemma 5.2.19 that $C_{\underline{j}'}$ has multiplicity 1 in the inverse image $\widetilde{V}$ of V via (5.112) (which contains $V^+_{\underline{j},\infty}$). Note that, by the same proof as for $V^+_{\underline{j},\infty}$, the adjunction map $V_{\underline{j},\underline{j}'} \to \Theta_\mu(V_{\underline{j},\underline{j}'})$ is injective. By functoriality of Θ_μ the composition of (5.112) with the injection $V_{\underline{j},\underline{j}'} \hookrightarrow \Theta_\mu(V_{\underline{j},\underline{j}'})$ factors through $V_{\underline{j},\underline{j}'} \twoheadrightarrow C_{\underline{j}'} \hookrightarrow \Theta_\mu(C_{\underline{j}'})$. Hence the image of $V_{\underline{j},\underline{j}'}$ under (5.112) is just its cosocle $C_{\underline{j}'}$, and we deduce that $V_{\underline{j},\underline{j}'}$ is the unique subrepresentation of $\widetilde{V}$ with cosocle $C_{\underline{j}'}$. Since $V^+_{\underline{j},\infty}$ is a subrepresentation of $\widetilde{V}$ which contains $C_{\underline{j}'}$, it follows that $V_{\underline{j},\underline{j}'}$ is also the unique subrepresentation of $V^+_{\underline{j},\infty}$ with cosocle $C_{\underline{j}'}$. In particular $V_{\underline{j},\underline{j}'}$ injects into $V^+_{\underline{j},\infty}$.

Step 2: We prove that $V_{\underline{j},\underline{j}''}$ injects into $V^+_{\underline{j},\infty}$.
Let V' be the unique length 2 representation of G with socle $C_{(j_0,j_0+1,j_2+1)}$ and cosocle $C_{(j_0,j_0,j_2+1)}$ (Lemma 5.2.1). Note that we have a canonical surjection $V_{\underline{j},\underline{j}'} \twoheadrightarrow V'$ by (ii) of Proposition 5.2.18. By (i) of Remark 5.2.15 (applied with j_0, j_2 there being j_0, j_2+1), the representation $V_{(j_0,j_0+1,j_2+1),\infty}$ in (iii) of Proposition 5.2.10 is isomorphic to the unique subrepresentation of $\Theta_\mu(V')$ with cosocle $V^{\mathrm{alg}}_{[j_2-j_0+1,j_2+1],\Delta}$. Let us consider the surjection

$$\Theta_\mu(V_{\underline{j},\underline{j}'}) \twoheadrightarrow \Theta_\mu(V') \tag{5.113}$$

induced from $V_{\underline{j},\underline{j}'} \twoheadrightarrow V'$. As $V^{\mathrm{alg}}_{[j_2-j_0+1,j_2+1],\Delta}$ occurs with multiplicity 1 in both $\Theta_\mu(V_{\underline{j},\underline{j}'})$ and $\Theta_\mu(V')$, it follows from the definition of $V^+_{\underline{j},\infty}$ that $V_{(j_0,j_0+1,j_2+1),\infty}$ is the image of $V^+_{\underline{j},\infty}$ under (5.113). As $C_{\underline{j}''}$ occurs in $V_{(j_0,j_0+1,j_2+1),\infty}$ and $\Theta_\mu(V_{\underline{j},\underline{j}'})$ with multiplicity 1 (by (iii) of Proposition 5.2.10 and (ii) of Lemma 5.2.19 respectively), we deduce that $C_{\underline{j}''}$ occurs in $V^+_{\underline{j},\infty}$ with multiplicity 1. Now let V'' be the unique length 2 representation of G with socle $C_{(j_0,j_0,j_2)}$ and cosocle $C_{(j_0,j_0+1,j_2+1)}$ (Lemma 5.2.1). Note that we have a canonical injection $V'' \hookrightarrow V_{\underline{j},\underline{j}'}$ by (ii) of Proposition 5.2.18. By (ii) of Lemma 5.2.13 $V_{\underline{j},\underline{j}''}$ is isomorphic to the unique subrepresentation of $\Theta_\mu(V'')$ with cosocle $C_{\underline{j}''}$. The injection $V'' \hookrightarrow V_{\underline{j},\underline{j}'}$ induces an injection $\Theta_\mu(V'') \hookrightarrow \Theta_\mu(V_{\underline{j},\underline{j}'})$, which therefore allows us to identify $V_{\underline{j},\underline{j}''}$ with the unique subrepresentation of $\Theta_\mu(V_{\underline{j},\underline{j}'})$ with cosocle $C_{\underline{j}''}$. Since $V^+_{\underline{j},\infty}$ is a subrepresentation of $\Theta_\mu(V_{\underline{j},\underline{j}'})$ which contains $C_{\underline{j}''}$, it follows that $V_{\underline{j},\underline{j}''}$ is also the unique subrepresentation of $V^+_{\underline{j},\infty}$ with cosocle $C_{\underline{j}''}$. In particular $V_{\underline{j},\underline{j}''}$ injects into $V^+_{\underline{j},\infty}$.

Step 3: We prove that, when $j_0 > 1$, there does not exist $W \in \mathrm{JH}_G(\Theta_\mu(C_{(j_0,j_0-1,j_2)}))$ such that $\mathrm{Ext}^1_G(V_I^{\mathrm{alg}}, W) \neq 0$ for some $I \in \{[j_2 - j_0 + 2, j_2 + 1], [j_2 - j_0 + 1, j_2 + 1]\}$.
By (5.87) and Lemma 5.1.1 a constituent of $\Theta_\mu(C_{(j_0,j_0-1,j_2)})$ has the form $W = \mathcal{F}^G_{P_{\Delta\setminus\{j_0-1\}}}(L(x), \pi^\infty_{j_0-1,j_2})$ where $L(x)$ is a constituent of $\Theta_{w_0 s_{j_0} w_0}(L(w_{j_0-1,j_0}))$, and by Proposition 3.4.5 and Remark 3.4.6 we have $D_L(x) = \{j_0 - 1\}$. Assume on the contrary that $\mathrm{Ext}^1_G(V_I^{\mathrm{alg}}, W) \neq 0$ for some $I \in \{[j_2 - j_0 + 2, j_2 + 1], [j_2 - j_0 + 1, j_2 + 1]\}$, then we have $d(V^\infty_{I,\Delta}, \pi^\infty_{j_0-1,j_2}) = 0$ by the last statement in Remark 5.1.12, which implies $I = [j_2 - j_0 + 2, j_2]$ by (i) of Lemma 2.3.1, a contradiction.

Step 4: We prove that the constituents in the following list

$$\left(\mathrm{JH}_G(V_{\underline{j},\underline{j}'}) \cup \mathrm{JH}_G(V_{\underline{j},\underline{j}''})\right) \sqcup \left\{V^{\mathrm{alg}}_{[j_2-j_0+2,j_2+1],\Delta}, V^{\mathrm{alg}}_{[j_2-j_0+1,j_2+1],\Delta}\right\} \tag{5.114}$$

occur with multiplicity 1 in $V^+_{\underline{j},\infty}$.
Note first that, from Step 1, Step 2 and the definition of $V^+_{\underline{j},\infty}$, all these constituents occur in $V^+_{\underline{j},\infty}$. By the end of the first paragraph of the proof of Lemma 5.2.20, we have a surjective adjunction map

$$\Theta_\mu(V_{\underline{j},\underline{j}'}) \twoheadrightarrow V_{\underline{j},\underline{j}'}. \tag{5.115}$$

Since the cosocle $V^{\mathrm{alg}}_{[j_2-j_0+1,j_2+1],\Delta}$ of $V^+_{\underline{j},\infty}$ doesn't occur in $V_{\underline{j},\underline{j}'}$ by (ii) of Proposition 5.2.18, $V^+_{\underline{j},\infty}$ is contained in the kernel of (5.115). Hence it is enough to prove that all constituents in (5.114) occur with multiplicity ≤ 1 in this kernel. For $V^{\mathrm{alg}}_{[j_2-j_0+2,j_2+1],\Delta}$ and $V^{\mathrm{alg}}_{[j_2-j_0+1,j_2+1],\Delta}$, this holds because they occur with multiplicity 1 in $\Theta_\mu(V_{\underline{j},\underline{j}'})$ (see above Step 1). For the non-locally algebraic constituents in $\mathrm{JH}_G(V_{\underline{j},\underline{j}'}) \cup \mathrm{JH}_G(V_{\underline{j},\underline{j}''})$, this also holds using (ii) of Lemma 5.2.19. The discussion in the first paragraph of the proof shows that only $\Theta_\mu(C_{\underline{j}})$ and $\Theta_\mu(C_{\underline{j}'})$ have locally algebraic constituents given by the constituents of $L(1)^\vee \otimes_E i^\infty_{\hat{j}_0,\Delta}(\pi^\infty_{j_0,j_2})$ and $L(1)^\vee \otimes_E i^\infty_{\hat{j}_0,\Delta}(\pi^\infty_{j_0,j_2+1})$ respectively. By Lemma 2.3.1 the only common constituent of these two locally algebraic representations is $V^{\mathrm{alg}}_{[j_2-j_0+1,j_2],\Delta}$, which occurs in $V_{\underline{j},\underline{j}'}$ by (ii) of Proposition 5.2.18. Hence all these locally algebraic constituents finally occur with multiplicity 1 in the kernel of (5.115).

Step 5: We prove that $V^+_{\underline{j},\infty}$ is multiplicity free and satisfies (iii).
For $V^+_{\underline{j},\infty}$ multiplicity free, by Step 4 it is enough to prove

$$\mathrm{JH}_G(V^+_{\underline{j},\infty}) = \left(\mathrm{JH}_G(V_{\underline{j},\underline{j}'}) \cup \mathrm{JH}_G(V_{\underline{j},\underline{j}''})\right) \sqcup \left\{V^{\mathrm{alg}}_{[j_2-j_0+2,j_2+1],\Delta}, V^{\mathrm{alg}}_{[j_2-j_0+1,j_2+1],\Delta}\right\}. \tag{5.116}$$

By Step 4 the right hand side of (5.116) is contained in $\mathrm{JH}_G(V^+_{\underline{j},\infty})$. Assume on the contrary that (5.116) is a strict inclusion. Then $V^+_{\underline{j},\infty}/(V_{\underline{j},\underline{j}'} + V_{\underline{j},\underline{j}''})$ admits a quotient containing a constituent W which is not in the list (5.116). Taking such a quotient of *minimal* length and using Step 4, we can assume that it contains a length 2 subrepresentation with socle W and cosocle V_I^{alg} for some $I \in \{[j_2 - j_0 + 2, j_2 + 1], [j_2 - j_0 + 1, j_2 + 1]\}$ (since any quotient of $V^+_{\underline{j},\infty}/(V_{\underline{j},\underline{j}'} + V_{\underline{j},\underline{j}''})$ has cosocle $V^{\mathrm{alg}}_{[j_2-j_0+1,j_2+1],\Delta}$ and since V_I^{alg} for $I \in \{[j_2 - j_0 + 2, j_2 + 1], [j_2 - j_0 + 1, j_2 + 1]\}$ are the only "remaining" constituents in (5.116)). This forces $\mathrm{Ext}^1_G(V_I^{\mathrm{alg}}, W) \neq 0$. Define V''' as $C_{\underline{j}}$ if $j_0 = 1$, and as the unique length 2 representation of G with socle $C_{\underline{j}}$ and cosocle $C_{(j_0,j_0-1,j_2)}$ if $j_0 > 1$ (using Lemma 5.2.1). By (ii) of Proposition 5.2.18 $\ker(V_{\underline{j},\underline{j}'} \to V')/V'''$ is locally algebraic and thus $\Theta_\mu(\ker(V_{\underline{j},\underline{j}'} \to V')/V''') = 0$ by Proposition 3.4.5 and (5.87). This together with the exactness of Θ_μ allows us to identify $\Theta_\mu(V''')$ with the kernel of the surjection (5.113). We have seen in Step 2 that the image of $V^+_{\underline{j},\infty}$ under (5.113) is isomorphic to $V_{(j_0,j_0+1,j_2+1),\infty}$. By (iii) of Proposition 5.2.10 (with j_2 there being $j_2 + 1$) $\mathrm{JH}_G(V_{(j_0,j_0+1,j_2+1),\infty})$ is contained in the right and side of (5.116). Since $\Theta_\mu(V''')$ is the kernel of (5.113), we deduce that $W \in \mathrm{JH}_G(\Theta_\mu(V''')) \setminus (\mathrm{JH}_G(V_{\underline{j},\underline{j}'}) \cup \mathrm{JH}_G(V_{\underline{j},\underline{j}''}))$. Moreover by Step 3 $W \notin \mathrm{JH}_G(\Theta_\mu(C_{(j_0,j_0-1,j_2)}))$ when $j_0 > 1$. Therefore, by definition of V''' (and the exactness of Θ_μ), we must have

$$W \in \mathrm{JH}_G\big(\Theta_\mu(C_{\underline{j}})\big) \setminus \big(\mathrm{JH}_G(V_{\underline{j},\underline{j}'}) \cup \mathrm{JH}_G(V_{\underline{j},\underline{j}''})\big).$$

Since W is a constituent of $\Theta_\mu(C_{\underline{j}})$, by (5.87), (ii) of Theorem 4.3.7 and Lemma 5.1.1 it has the form $W = \mathcal{F}^G_{P_x}(L(x), \sigma^\infty)$ where $L(x)$ is a constituent of $\Theta_{w_0 s_{j_0} w_0}(L(s_{j_0}))$ and σ^∞ a constituent of $i^\infty_{j_0,I_x}(\pi^\infty_{j_0,j_2})$. If $x \neq 1$, then $\mathrm{Ext}^1_G(V_I^{\mathrm{alg}}, W) \neq 0$ together Proposition 5.1.14 and Remark 5.1.12 force $\mathrm{Ext}^1_{U(\mathfrak{g})}(L(x), L(1)) \neq 0$ and $d(V^\infty_{I,\Delta}, \sigma^\infty) = 0$. By Proposition 3.4.5 and (ii) of Lemma 3.2.4 we deduce $x = s_{j_0}$, which forces $W = C_{\underline{j}}$ and $\sigma^\infty = \pi^\infty_{j_0,j_2}$, a contradiction as $d(V^\infty_{I,\Delta}, \pi^\infty_{j_0,j_2}) = 0$ if and only if $I = [j_2 - j_0 + 1, j_2]$ (Lemma 2.3.1) but we have $I \in \{[j_2 - j_0 + 2, j_2 + 1], [j_2 - j_0 + 1, j_2 + 1]\}$. Hence we have $x = 1$, i.e. $W = L(1)^\vee \otimes_E \sigma^\infty$. Since W is not a constituent of $V_{\underline{j},\underline{j}''}$ and since $L(1)^\vee \otimes_E \pi^\infty_{\underline{j},\underline{j}''}$ is a subquotient of $V_{\underline{j},\underline{j}''}$ by (ii) of Proposition 5.2.10, it follows that

$$\sigma^\infty \in \mathrm{JH}_G(i^\infty_{j_0,\Delta}(\pi^\infty_{j_0,j_2})) \setminus \mathrm{JH}_G(\pi^\infty_{\underline{j},\underline{j}''}) = \{V^\infty_{[j_2-j_0+1,j_2-1],\Delta}, V^\infty_{[j_2-j_0,j_2-1],\Delta}\}$$

with $V^\infty_{[j_2-j_0,j_2-1],\Delta}$ omitted when $j_0 = j_2$ (the equality follows from Lemma 2.3.1). Write $\sigma^\infty = V^\infty_{I',\Delta}$, then one can check for each $I \in \{[j_2 - j_0 + 2, j_2 + 1], [j_2 - j_0 + 1, j_2 + 1]\}$ that $d(V^\infty_{I,\Delta}, V^\infty_{I',\Delta}) > 1$ (for instance using [48]) and thus $\mathrm{Ext}^1_G(V_I^{\mathrm{alg}}, W) = 0$ by Lemma 5.1.3, another contradiction. We conclude that W cannot exist, and thus (5.116) holds. Hence $V^+_{\underline{j},\infty}$ is multiplicity free. The statement (iii) comes from the fact $\Theta_\mu(V_{\underline{j},\underline{j}'})$ contains as a subquotient the unique non-split extension of $V^{\mathrm{alg}}_{[j_2-j_0+1,j_2+1],\Delta}$ by $V^{\mathrm{alg}}_{[j_2-j_0+2,j_2+1],\Delta}$ (see above Step 1). □

Remark 5.2.31 An inspection of Proposition 5.2.18 and (ii) of Proposition 5.2.10 shows that the (multiplicity free) subrepresentation $V_{\underline{j},\underline{j}'} + V_{\underline{j},\underline{j}''}$ of $V^+_{\underline{j},\infty}$ must be the amalgamate sum of $V_{\underline{j},\underline{j}'}$ and $V_{\underline{j},\underline{j}''}$ over the length 3 subrepresentation of $V^+_{\underline{j},\infty}$ with socle $C_{\underline{j}}$ and cosocle $C_{(j_0,j_0+1,j_2+1)} \oplus V^{\mathrm{alg}}_{[j_2-j_0+1,j_2],\Delta}$. This determines the partially ordered set $\mathrm{JH}_G(V_{\underline{j},\underline{j}'}+V_{\underline{j},\underline{j}''})$ completely. The partial order on $\mathrm{JH}_G(V^+_{\underline{j},\infty})$ is then determined by the one on $\mathrm{JH}_G(V_{\underline{j},\underline{j}'}+V_{\underline{j},\underline{j}''})$ and the relations $C_{\underline{j}'} \leq V^{\mathrm{alg}}_{[j_2-j_0+2,j_2+1],\Delta} \leq V^{\mathrm{alg}}_{[j_2-j_0+1,j_2+1],\Delta}$ and $C_{\underline{j}''} \leq V^{\mathrm{alg}}_{[j_2-j_0+1,j_2+1],\Delta}$, using that the representation $V_{(j_0,j_0+1,j_2+1),\infty}$ of (iii) of Proposition 5.2.10 is a quotient of $V^+_{\underline{j},\infty}$ (see Step 2 of the proof of Lemma 5.2.30) and using $\mathrm{Ext}^1_G(V^{\mathrm{alg}}_{[j_2-j_0+2,j_2+1],\Delta}, C_{\underline{j}''})=0$ which follows from (i) of Lemma 5.2.2.

Recall from (i) of Lemma 5.2.30 that $V^+_{\underline{j},\infty}$ has cosocle $V^{\mathrm{alg}}_{[j_2-j_0+1,j_2+1],\Delta}$ and socle $C_{\underline{j}}$, and thus fits into a non-split extension $0 \to \mathrm{rad}_G(V^+_{\underline{j},\infty}) \to V^+_{\underline{j},\infty} \to V^{\mathrm{alg}}_{[j_2-j_0+1,j_2+1],\Delta} \to 0$. In particular

$$\mathrm{Ext}^1_G(V^{\mathrm{alg}}_{[j_2-j_0+1,j_2+1],\Delta}, \mathrm{rad}_G(V^+_{\underline{j},\infty})) \neq 0. \tag{5.117}$$

Lemma 5.2.32 *The vector space in (5.117) is one dimensional, and any multiplicity free finite length $V^\flat$ in $\mathrm{Rep}^{\mathrm{an}}_{\mathrm{adm}}(G)$ which satisfies $\mathrm{JH}_G(V^\flat)=\mathrm{JH}_G(\mathrm{rad}_G(V^+_{\underline{j},\infty}))$ as partially ordered sets must satisfy $V^\flat \cong \mathrm{rad}_G(V^+_{\underline{j},\infty})$.*

Proof We borrow all notation from the proof of Lemma 5.2.30.

We prove that the vector space in (5.117) is one dimensional, and it suffices to show that it has dimension at most one. Using Remark 5.2.31, we check that there is a unique increasing filtration on $\mathrm{rad}_G(V^+_{\underline{j},\infty})$ whose only reducible graded pieces are the length 2 quotient U_0 of $\mathrm{rad}_G(V^+_{\underline{j},\infty})$ with socle $C_{(j_0,j_0+1,j_2+1)}$ and cosocle $C_{\underline{j}''}$, and the length 2 subrepresentation U_1 of $\mathrm{rad}_G(V^+_{\underline{j},\infty})$ with socle $C_{\underline{j}}$ and cosocle $V^{\mathrm{alg}}_{[j_2-j_0+1,j_2],\Delta}$. By Lemmas 5.1.3 and 2.2.3 we have $\dim_E \mathrm{Ext}^1_G(V^{\mathrm{alg}}_{[j_2-j_0+1,j_2+1],\Delta}, V^{\mathrm{alg}}_{[j_2-j_0+2,j_2+1],\Delta})=1$, and by (i) of Lemma 5.2.2 (and Remark 5.2.31) we have $\mathrm{Ext}^1_G(V^{\mathrm{alg}}_{[j_2-j_0+1,j_2+1],\Delta}, W)=0$ for $W \in \mathrm{JH}_G(\mathrm{rad}_G(V^+_{\underline{j},\infty})) \setminus (\mathrm{JH}_G(U_0) \sqcup \mathrm{JH}_G(U_1) \sqcup \{V^{\mathrm{alg}}_{[j_2-j_0+2,j_2+1],\Delta}\})$. Hence by an easy dévissage it is enough to prove

$$\mathrm{Ext}^1_G(V^{\mathrm{alg}}_{[j_2-j_0+1,j_2+1],\Delta}, U_i)=0 \ \text{ for } i=0,1. \tag{5.118}$$

The surjection $\mathrm{gr}^0(V_{(j_0,j_0+1,j_2+1),\infty}) \twoheadrightarrow V^{\mathrm{alg}}_{[j_2-j_0+1,j_2+1],\Delta}$ induces an embedding

$$\mathrm{Ext}^1_G(V^{\mathrm{alg}}_{[j_2-j_0+1,j_2+1],\Delta}, U_0) \hookrightarrow \mathrm{Ext}^1_G(\mathrm{gr}^0(V_{(j_0,j_0+1,j_2+1),\infty}), U_0).$$

But the minimality of $V_{(j_0,j_0+1,j_2+1),\infty}$ in (iii) of Proposition 5.2.10 (applied with j_2 there being j_2+1) implies $\mathrm{Ext}^1_G(\mathrm{gr}^0(V_{(j_0,j_0+1,j_2+1),\infty}), U_0)=0$. Hence we deduce (5.118) for $i=0$. Assume on the contrary that (5.118) fails for $i=$

1. As $\mathrm{Ext}^1_G(V^{\mathrm{alg}}_{[j_2-j_0+1,j_2+1],\Delta}, C_{\underline{j}}) = 0$ by (i) of Lemma 5.2.2, there must exist a uniserial length 3 representation $\widetilde{U}_1$ containing U_1 with socle $C_{\underline{j}}$, cosocle $V^{\mathrm{alg}}_{[j_2-j_0+1,j_2+1],\Delta}$ and middle layer $V^{\mathrm{alg}}_{[j_2-j_0+1,j_2],\Delta}$. By Lemmas 5.1.3 and 2.2.3 there exists a unique length 2 representation $L(1) \otimes_E \tau^\infty$ with τ^∞ of length 2 with socle $V^\infty_{[j_2-j_0+1,j_2],\Delta}$ and cosocle $V^\infty_{[j_2-j_0+1,j_2+1],\Delta}$ (and thus G-basic by Lemma 2.2.5). By Lemma 2.3.1 $\mathrm{Hom}_G(\tau^\infty, i^\infty_{\widehat{j_0},\Delta}(\pi^\infty_{\widehat{j_0},j_2})) = 0$, which together with the last statement in Remark 5.1.12 forces $\mathrm{Ext}^1_G(L(1)^\vee \otimes_E \tau^\infty, C_{\underline{j}}) = 0$, contradicting the existence of $\widetilde{U}_1$. Hence (5.118) holds and thus the vector space in (5.117) has dimension 1.

Now we let $V^\flat$ in $\mathrm{Rep}^{\mathrm{an}}_{\mathrm{adm}}(G)$ be a multiplicity free finite length representation such that $\mathrm{JH}_G(V^\flat) = \mathrm{JH}_G(\mathrm{rad}_G(V^+_{\underline{j},\infty}))$ as partially ordered sets. By Remark 5.2.21 (recall $V_{\underline{j},\underline{j}'}$ is the representation V_0 there for $j_0 = j_1$) $\Theta_\mu(V_{\underline{j},\underline{j}'})$ contains a unique subrepresentation isomorphic to $V_{\underline{j},\underline{j}'}$. By (ii) of Lemma 5.2.19 and (ii) of Lemma 5.2.20 (and the fact $\Theta_\mu(V_{\underline{j},\underline{j}'})$ has socle $C_{\underline{j}}$) we see that $\Theta_\mu(V_{\underline{j},\underline{j}'})$ also contains a unique subrepresentation isomorphic to $V_{\underline{j},\underline{j}''}$. Finally, by the first paragraph of the proof of Lemma 5.2.30, recall that $V^{\mathrm{alg}}_{[j_2-j_0+1,j_2+1],\Delta}$ has multiplicity one in $\Theta_\mu(V_{\underline{j},\underline{j}'})$.

Given $V^\flat$ as above, the equality of partially ordered sets $\mathrm{JH}_G(V^\flat) = \mathrm{JH}_G(\mathrm{rad}_G(V^+_{\underline{j},\infty}))$ forces $V^\flat$ to have socle $C_{\underline{j}}$. Lemma 5.2.8 then forces $V^\flat$ to contain both $V_{\underline{j},\underline{j}'}$ and $V_{\underline{j},\underline{j}''}$. Using the first statement of Proposition 3.4.5, (5.87) and the explicit description of $\mathrm{JH}_G(V^\flat/V_{\underline{j},\underline{j}'}) = \mathrm{JH}_G(\mathrm{rad}_G(V^+_{\underline{j},\infty})/V_{\underline{j},\underline{j}'})$ (cf. Remark 5.2.31), we have $\Theta_\mu(V^\flat/V_{\underline{j},\underline{j}'}) = 0$ and thus $\Theta_\mu(V_{\underline{j},\underline{j}'}) \xrightarrow{\sim} \Theta_\mu(V^\flat)$. As $V^\flat$ and $\Theta_\mu(V_{\underline{j},\underline{j}'})$ have socle $C_{\underline{j}}$, the canonical (non-zero) map $V^\flat \to \Theta_\mu(V^\flat) \cong \Theta_\mu(V_{\underline{j},\underline{j}'})$ is injective, hence $V^\flat$ is a subrepresentation of $\Theta_\mu(V_{\underline{j},\underline{j}'})$. Finally the equality $\mathrm{JH}_G(V^\flat) = \mathrm{JH}_G(\mathrm{rad}_G(V^+_{\underline{j},\infty}))$ and the above unicity statements in $\Theta_\mu(V_{\underline{j},\underline{j}'})$ force $V^\flat \cong \mathrm{rad}_G(V^+_{\underline{j},\infty})$. □

Lemma 5.2.33 *Let $\underline{j} \in \mathbf{J}$ with $1 \le j_0 = j_1 < j_2 < n$. Then there exists a unique uniserial length 3 representation $V_{\underline{j}}$ in $\mathrm{Rep}^{\mathrm{an}}_{\mathrm{adm}}(G)$ with socle $L(1)^\vee \otimes_E V^\infty_{[j_2-j_1,j_2-1],\Delta}$, cosocle $L(1)^\vee \otimes_E V^\infty_{[j_2-j_1+1,j_2],\Delta}$ and middle layer $C_{\underline{j}}$.*

Proof We write $I \overset{\mathrm{def}}{=} \Delta \setminus \{j_1\}$. By Lemma 3.4.14 there exists a $\mathfrak{z}$-semi-simple uniserial $U(\mathfrak{g})$-module M of length 3 with both socle and cosocle $L(1)$ and middle layer $L(s_{j_1})$. Let M^- be the unique length 2 $U(\mathfrak{g})$-module with socle $L(s_{j_1})$ and cosocle $L(1)$ (using (ii) of Lemma 3.2.4), then M^- is isomorphic to the unique length 2 quotient of M, and the unique length 2 $U(\mathfrak{g})$-submodule of M is isomorphic to $(M^-)^\tau$ (see (3.2 for the notation) which is the unique length 2 $U(\mathfrak{g})$-module with socle $L(1)$ and cosocle $L(s_{j_1})$ (again by (ii) of Lemma 3.2.4).

We define $W \overset{\mathrm{def}}{=} \mathcal{F}^G_{P_I}(M, \pi^\infty_{j_1,j_2})$, which is well defined by Remark 4.3.8 and comes with an injection $q_1 : \mathcal{F}^G_{P_I}(M^-, \pi^\infty_{j_1,j_2}) \hookrightarrow W$ and a surjection $q_2 : W \twoheadrightarrow \mathcal{F}^G_{P_I}((M^-)^\tau, \pi^\infty_{j_1,j_2})$ such that $q_2 \circ q_1$ has (simple) image $C_{\underline{j}} \cong \mathcal{F}^G_{P_I}(L(s_{j_1}), \pi^\infty_{j_1,j_2})$. As $j_1 < j_2 < n$, recall that $V^\infty_{[j_2-j_1,j_2-1],\Delta} \cong \mathrm{cosoc}_G(i^\infty_I(\pi^\infty_{j_1,j_2}))$ and $V^\infty_{[j_2-j_1+1,j_2],\Delta} \cong \mathrm{soc}_G(i^\infty_I(\pi^\infty_{j_1,j_2}))$ by (i) of Lemma 2.3.1. Now by Lemma 5.1.19 $\mathcal{F}^G_{P_I}(M^-, \pi^\infty_{j_1,j_2})$

(resp. $\mathcal{F}_{P_I}^G((M^-)^\tau, \pi_{j_1,j_2}^\infty)$) admits a quotient W_1 (resp. a subrepresentation W_2) which is uniserial of length 2 with socle $L(1)^\vee \otimes_E \mathrm{cosoc}_G(i_I^\infty(\pi_{j_1,j_2}^\infty)) \cong V_{[j_2-j_1,j_2-1],\Delta}^{\mathrm{alg}}$ and cosocle $C_{\underline{j}}$ (resp. with socle $C_{\underline{j}}$ and cosocle $L(1)^\vee \otimes_E \mathrm{soc}_G(i_I^\infty(\pi_{j_1,j_2}^\infty)) \cong V_{[j_2-j_1+1,j_2],\Delta}^{\mathrm{alg}}$). Taking the pushforward of W along $\mathcal{F}_{P_I}^G(M^-, \pi_{j_1,j_2}^\infty) \twoheadrightarrow W_1$, and then the pullback along $W_2 \hookrightarrow \mathcal{F}_{P_I}^G((M^-)^\tau, \pi_{j_1,j_2}^\infty)$ gives a length 3 subquotient $V_{\underline{j}}$ of W which admits W_1 as a subrepresentation and W_2 as a quotient. As we have $d(V_{[j_2-j_1+1,j_2],\Delta}^\infty, V_{[j_2-j_1,j_2-1],\Delta}^\infty) = 2 > 1$ by [48], we have by (the sentence before) (4.149)

$$\mathrm{Ext}_G^1(V_{[j_2-j_1+1,j_2],\Delta}^{\mathrm{alg}}, V_{[j_2-j_1,j_2-1],\Delta}^{\mathrm{alg}}) = 0.$$

We also have by Lemma 5.2.2

$$\dim_E \mathrm{Ext}_G^1(V_{[j_2-j_1+1,j_2],\Delta}^{\mathrm{alg}}, C_{\underline{j}}) = 1 = \dim_E \mathrm{Ext}_G^1(C_{\underline{j}}, V_{[j_2-j_1,j_2-1],\Delta}^{\mathrm{alg}}).$$

All this implies (by an easy dévissage) that the representation $V_{\underline{j}}$ is uniserial and unique. □

5.3 Hooking All Constituents Together

We hook together the various Ext-squares, Ext-cubes and uniserial representations constructed in Sect. 5.2 to define two complexes of explicit finite length coadmissible $D(G)$-modules $\mathbf{D}^\bullet$, $\widetilde{\mathbf{D}}^\bullet$ and we use $\widetilde{\mathbf{D}}^\bullet$ to construct an explicit section in the derived category to $\mathbf{D}^\bullet \twoheadrightarrow H^{n-1}(\mathbf{D}^\bullet)[-(n-1)]$ (Theorem 5.3.13).

From now on we tacitly identify the set $\mathbf{J}$ of Sect. 5.2 with the set $\{C_{\underline{j}} \mid \underline{j} \in \mathbf{J}\}$ of irreducible admissible representations of G over E defined in (5.68), and we recall that $\mathbf{J}$ is equipped with the partial order $\underline{j} \le \underline{j}'$ if and only if $j_0 \le j_0'$, $j_2 \le j_2'$ and $j_2 - j_1 \le j_2' - j_1'$. In order to take into account locally algebraic constituents, it is convenient to enlarge $\mathbf{J}$ and define

$$\widetilde{\mathbf{J}} \overset{\mathrm{def}}{=} \mathbf{J} \sqcup \{V_{[j,j'],\Delta}^{\mathrm{alg}} \mid 1 \le j \le j' \le n-1\}$$

(see (5.110)) equipped with the (unique) weakest partial order such that

- the partial order on $\widetilde{\mathbf{J}}$ restricts to the one on $\mathbf{J}$;
- $$\begin{cases} C_{(j_0,j_0,j_2)} < V_{[j_2-j_0+1,j_2],\Delta}^{\mathrm{alg}} < C_{(j_0,j_0,j_2+1)} & \text{for } 1 \le j_0 \le j_2 \le n-1 \\ V_{[1,j_0],\Delta}^{\mathrm{alg}} < C_{j_0+1,j_0+1,j_0+1} & \text{for } 1 \le j_0 \le n-2 \\ V_{[j_2-j_0+1,j_2],\Delta}^{\mathrm{alg}} < V_{[j_2-j_0,j_2],\Delta}^{\mathrm{alg}} & \text{for } 1 \le j_0 < j_2 \le n-1. \end{cases}$$

In particular we have $V_{[j_2-j_0+1,j_2],\Delta}^{\mathrm{alg}} \le C_{(j_0,j_0,j_2+1)} \le V_{[j_2-j_0+2,j_2+1],\Delta}^{\mathrm{alg}}$, and an easy induction shows that if $C_{(j_0,j_1,j_2)} \le V_{[j_2'-j_0'+1,j_2'],\Delta}^{\mathrm{alg}}$ then $j_0 \le j_0'$. For $V, W \in \widetilde{\mathbf{J}}$ such that $V \le W$, we then define

$$[V, W] \overset{\text{def}}{=} \{V' \in \widetilde{\mathbf{J}} \mid V \le V' \le W\} \subseteq \widetilde{\mathbf{J}}, \tag{5.119}$$

and for $1 \le j_0 \le n-1$ we define

$$\widetilde{\mathbf{J}}_{j_0} \overset{\text{def}}{=} [C_{(j_0,1,1)}, C_{(j_0,1,n)}] \subseteq \widetilde{\mathbf{J}}, \tag{5.120}$$

which can be more explicitly described as

$$\widetilde{\mathbf{J}}_{j_0} = \{C_{(j_0,j_1,j_2)} \mid (j_0, j_1, j_2) \in \mathbf{J}\} \sqcup \{V^{\text{alg}}_{[j_2-j_0+1,j_2],\Delta} \mid j_0 \le j_2 \le n-1\}. \tag{5.121}$$

Let $\underline{j}, \underline{j}' \in \mathbf{J}$ such that $\underline{j} \le \underline{j}'$ and

$$\max\{j_0' - j_0, j_2' - j_2, (j_2' - j_1') - (j_2 - j_1)\} \le 1, \tag{5.122}$$

and note that the corresponding $C_{\underline{j}'}$ are exactly the 8 constituents above Lemma 5.2.19. When $d(\underline{j}, \underline{j}') = |j_0 - j_0'| + |j_2 - j_2'| + |(j_2 - j_1) - (j_2' - j_1')| \ge 2$ we have defined in (i), (ii) of Propositions 5.2.10, 5.2.18 and 5.2.28 a minimal Ext-hypercube $V_{\underline{j},\underline{j}'}$ of socle $C_{\underline{j}}$ and cosocle $C_{\underline{j}'}$. When $d(\underline{j}, \underline{j}') = 1$ we let $V_{\underline{j},\underline{j}'}$ be the unique non-split extension of $C_{\underline{j}'}$ by $C_{\underline{j}}$ in Lemma 5.2.1, and when $d(\underline{j}, \underline{j}') = 0$ we set $V_{\underline{j},\underline{j}'} \overset{\text{def}}{=} C_{\underline{j}}$. From the definition of the partial order on $\widetilde{\mathbf{J}}$, we immediately check:

Lemma 5.3.1 *Let $\underline{j}, \underline{j}' \in \mathbf{J}$ satisfying $\underline{j} \le \underline{j}'$ and the bound (5.122), we have $\mathrm{JH}_G(V_{\underline{j},\underline{j}'}) = [C_{\underline{j}}, C_{\underline{j}'}]$ as partially ordered sets.*

For $\underline{j}' \in \mathbf{J}$, we set $\widetilde{\mathbf{J}}(C_{\underline{j}'}) \overset{\text{def}}{=} [C_{\underline{j}}, C_{\underline{j}'}] \subseteq \widetilde{\mathbf{J}}$ where $\underline{j} \in \mathbf{J}$ is the (unique) minimal element that satisfies $\underline{j} \le \underline{j}'$ and (5.122) (for instance if $j_0', j_2' \ge 2$ we have $\underline{j} = (j_0' - 1, j_1', j_2' - 1)$, the remaining cases are easily worked out). For $2 \le j_0 \le j_2 \le n-1$, we set

$$\widetilde{\mathbf{J}}(V^{\text{alg}}_{[j_2-j_0+1,j_2],\Delta}) \overset{\text{def}}{=} [C_{(j_0-1,j_0-1,j_2-1)}, V^{\text{alg}}_{[j_2-j_0+1,j_2],\Delta}] = \mathrm{JH}_G(V^{+}_{(j_0-1,j_0-1,j_2-1),\infty})$$

where the last identification (as partially ordered sets) follows from Lemma 5.2.30 and Remark 5.2.31. For $2 \le j_2 \le n-1$ (and $j_0 = 1$) we set $\widetilde{\mathbf{J}}(V^{\text{alg}}_{\{j_2\},\Delta}) \overset{\text{def}}{=} [V^{\text{alg}}_{\{j_2-1\},\Delta}, V^{\text{alg}}_{\{j_2\},\Delta}]$, and (when $j_2 = j_0 = 1$) $\widetilde{\mathbf{J}}(V^{\text{alg}}_{\{1\},\Delta}) \overset{\text{def}}{=} [C_{(1,1,1)}, V^{\text{alg}}_{\{1\},\Delta}]$ (in fact, in each case we have $\widetilde{\mathbf{J}}(V^{\text{alg}}_{[j_2-j_0+1,j_2],\Delta}) = \mathrm{JH}_G(V^{+}_{(j_0-1,j_0-1,j_2-1),\infty})$ where we keep the constituents of $V^{+}_{(j_0-1,j_0-1,j_2-1),\infty}$ which "remain").

Lemma 5.3.2 *Let $\underline{j} \in \mathbf{J}$, $j_0' \in \{1, \dots, n-1\}$ and $W \in \widetilde{\mathbf{J}}_{j_0'}$ such that $C_{\underline{j}} \le W$.*

(i) There exists a unique multiplicity free finite length representation $V_{\underline{j},W}$ in $\mathrm{Rep}^{\text{an}}_{\text{adm}}(G)$ such that $\mathrm{JH}_G(V_{\underline{j},W}) = [C_{\underline{j}}, W] \cap \widetilde{\mathbf{J}}(W)$ as partially ordered sets.

(ii) Let $V_{\underline{j},<W}$ be the unique subrepresentation of $V_{\underline{j},W}$ such that $V_{\underline{j},W}/V_{\underline{j},<W} \cong W$ and assume $V_{\underline{j},<W} \neq 0$, then $V_{\underline{j},<W}$ is the unique multiplicity free finite length

representation in $\mathrm{Rep}^{\mathrm{an}}_{\mathrm{adm}}(G)$ *such that* $\mathrm{JH}_G(V_{\underline{j},<W}) = \mathrm{JH}_G(V_{\underline{j},W}) \setminus \{W\}$ *as partially ordered sets. Moreover we have*

$$\dim_E \mathrm{Ext}^1_G(W, V_{\underline{j},<W}) = 1. \tag{5.123}$$

Proof Assume first $W = C_{\underline{j}'}$ for some $\underline{j}' \in \widetilde{\mathbf{J}}_{j'_0}$ (so we have $\underline{j} \le \underline{j}'$). Then $[C_{\underline{j}}, W] \cap \widetilde{\mathbf{J}}(W) = [C_{\underline{j}''}, W]$ for the (unique) minimal $\underline{j}'' = (j''_0, j''_1, j''_2) \in \mathbf{J}$ such that $\underline{j} \le \underline{j}'' \le \underline{j}'$ and $\max\{j'_0 - j''_0, j'_2 - j''_2, (j'_2 - j'_1) - (j''_2 - j''_1)\} \le 1$. It follows from the definition of the partial order on $\mathbf{J}$ that $\underline{j}''$ is the unique element in $\mathbf{J}$ such that

$$j'_0 - j''_0 = \min\{1, j'_0 - j_0\},\ j'_2 - j''_2 = \min\{1, j'_2 - j_2\},$$
$$(j'_2 - j'_1) - (j''_2 - j''_1) = \min\{1, (j'_2 - j'_1) - (j_2 - j_1)\}.$$

In particular $d(\underline{j}'', \underline{j}') \in \{0, 1, 2, 3\}$. We have $d(\underline{j}'', \underline{j}') = 0$ if and only if $\underline{j}'' = \underline{j}'$ if and only if $\underline{j} = \underline{j}'$, in which case $V_{\underline{j},W} \cong C_{\underline{j}}$. If $d(\underline{j}'', \underline{j}') = 1$ then $V_{\underline{j},W}$ is the unique length 2 representation with socle $C_{\underline{j}''}$ and cosocle $C_{\underline{j}'}$ (Lemma 5.2.1). If $d(\underline{j}'', \underline{j}') = 2$ then $V_{\underline{j},W}$ is the unique minimal Ext-square $V_{\underline{j}'',\underline{j}'}$ constructed in (i), (ii) of Propositions 5.2.10 and 5.2.18. Finally if $d(\underline{j}'', \underline{j}') = 3$ then $V_{\underline{j},W}$ is the unique minimal Ext-cube $V_{\underline{j}'',\underline{j}'}$ constructed in Proposition 5.2.28). This proves (i) in that case. Assume now $V_{\underline{j},<W} \ne 0$, i.e. $\underline{j}'' < \underline{j}'$. The unicity of $V_{\underline{j},<W}$ is obvious when $d(\underline{j}'', \underline{j}') = 1$ (as $V_{\underline{j},<W} \cong C_{\underline{j}''}$), and follows from the minimality of $V_{\underline{j}'',\underline{j}'}$ and Lemma 5.2.8 when $d(\underline{j}'', \underline{j}') \ge 2$. Finally (5.123) follows from Lemma 5.2.1 when $d(\underline{j}'', \underline{j}') = 1$, and from Lemma 5.2.7 (together with the minimality of $V_{\underline{j}'',\underline{j}'}$) when $d(\underline{j}'', \underline{j}') \ge 2$. This proves (ii) in the case $W = C_{\underline{j}'}$.

Assume now $W = V^{\mathrm{alg}}_{[j'_2-j'_0+1,j'_2],\Delta}$ for some $j'_0 \le j'_2 \le n-1$. If $j_0 = j'_0 = 1$, then $[C_{\underline{j}}, W] \cap \widetilde{\mathbf{J}}(W)$ is either $\widetilde{\mathbf{J}}(W)$ or $[C_{(1,1,j'_2)}, W]$, in which case the existence and unicity of $V_{\underline{j},W}$ follows from Lemma 5.2.33 and (i) of Lemma 5.2.2 respectively. Assume from now $j'_0 > 1$ so that $\widetilde{\mathbf{J}}(W) = [C_{(j'_0-1,j'_0-1,j'_2-1)}, W] = \mathrm{JH}_G(V^+_{(j'_0-1,j'_0-1,j'_2-1),\infty})$. From the definition of the partial order on $\widetilde{\mathbf{J}}$ one checks that $C_{\underline{j}} \le V^{\mathrm{alg}}_{[j'_2-j'_0+1,j'_2],\Delta}$ if and only if either $\underline{j} \le (j'_0, j'_0, j'_2 - 1)$ or $\underline{j} \le (j'_0 - 1, j'_0 - 1, j'_2 - 1)$. Hence the intersection $[C_{\underline{j}}, W] \cap \widetilde{\mathbf{J}}(W) = [C_{\underline{j}}, W] \cap [C_{(j'_0-1,j'_0-1,j'_2-1)}, W]$ is equal to $[C_{\underline{j}''}, W]$ for some $\underline{j}'' \in \mathbf{J}$ such that $\underline{j} \le \underline{j}''$, $(j'_0 - 1, j'_0 - 1, j'_2 - 1) \le \underline{j}''$ and either $\underline{j}'' \le (j'_0, j'_0, j'_2 - 1)$ or $\underline{j}'' \le (j'_0 - 1, j'_0 - 1, j'_2 - 1)$. We see that we necessarily have:

$$\underline{j}'' \in \{(j'_0 - 1, j'_0 - 1, j'_2 - 1), (j'_0 - 1, j'_0, j'_2), (j'_0 - 1, j'_0 - 2, j'_2 - 1), (j'_0 - 1, j'_0 - 1, j'_2),$$
$$(j'_0, j'_0 - 1, j'_2 - 1), (j'_0, j'_0, j'_2)\}.$$

But from Remark 5.2.31 any constituent $C_{\underline{j}''}$ for $\underline{j}''$ in the above set is a constituent of $V^+_{(j'_0-1,j'_0-1,j'_2-1),\infty}$, hence it follows that $[C_{\underline{j}}, W] \cap \widetilde{\mathbf{J}}(W) =$

$[C_{\underline{j}''}, W] = \mathrm{JH}_G(Q_{\underline{j}''})$ (as partially ordered sets) where $Q_{\underline{j}''}$ is the unique quotient of $V^+_{(j_0'-1,j_0'-1,j_2'-1),\infty}$ with socle $C_{\underline{j}''}$ (and cosocle $W = V^{\mathrm{alg}}_{[j_2'-j_0'+1,j_2'],\Delta}$). We thus take $V_{\underline{j},W} = Q_{\underline{j}''}$ and $V_{\underline{j},<W} = \mathrm{rad}_G(Q_{\underline{j}''})$. It remains to prove (ii) (which implies the unicity in (i) once we know the existence in (i)). If $\underline{j}'' = (j_0'-1, j_0'-1, j_2'-1)$ then $\mathrm{JH}_G(Q_{\underline{j}''}) = \mathrm{JH}_G(V^+_{(j_0'-1,j_0'-1,j_2'-1),\infty})$, and (ii) follows from Lemma 5.2.32. If $\underline{j}'' = (j_0'-1, j_0', j_2')$ then $\mathrm{JH}_G(Q_{\underline{j}''}) = \mathrm{JH}_G(V_{(j_0'-1,j_0',j_2'),\infty})$ (see Step 2 of the proof of Lemma 5.2.30 and recall that $V_{(j_0'-1,j_0',j_2'),\infty}$ is defined in (iii) of Proposition 5.2.10), and (ii) follows from the existence and minimality of the Ext-square $V_{(j_0'-1,j_0',j_2'),\infty}$ ((iii) of Proposition 5.2.10) and from Lemma 5.2.8. In the remaining 4 cases for $\underline{j}''$, $Q_{\underline{j}''}$ and $\mathrm{rad}_G(Q_{\underline{j}''})$ are uniserial, and (ii) follows by dévissage from Lemma 5.2.1, (i) of Lemmas 5.2.2, 5.1.3 and 2.2.3. □

It follows from (5.121) that $\widetilde{\mathbf{J}}_1 = \mathbf{J} \sqcup \{V^{\mathrm{alg}}_{\{j\},\Delta} \mid 1 \le j \le n-1\}$.

Lemma 5.3.3 *Let $j_0, j_0' \in \{1, \dots, n-1\}$ such that $j_0 \le j_0' \le j_0 + 1$ and $V \in \widetilde{\mathbf{J}}_{j_0}$, $V' \in \widetilde{\mathbf{J}}_{j_0'}$ such that $V < V'$.*

(i) If $V \notin \widetilde{\mathbf{J}}(V')$, then we have $\mathrm{Ext}^1_G(V', V) = 0$.
(ii) If $V \notin \widetilde{\mathbf{J}}(V')$ and $\mathrm{Ext}^2_G(V', V) \neq 0$, then either $j_0 = j_0' = 1$, or $V' \cong C_{(j_0+1,j_0+1,j_2+2)}$ and $V \cong V^{\mathrm{alg}}_{[j_2-j_0+1,j_2],\Delta}$ for some $1 \le j_0 \le j_2 \le n-2$.

Proof Assume first that V and V' are both not locally algebraic. Then $V = C_{\underline{j}}$ and $V' = C_{\underline{j}'}$ with $\underline{j} < \underline{j}'$ in $\mathbf{J}$. The assumption $V \notin \widetilde{\mathbf{J}}(V')$ then implies that (5.122) fails, and it follows from Lemma 5.2.1 and (i) of Lemma 5.2.3 that

$$\mathrm{Ext}^1_G(V', V) = \mathrm{Ext}^2_G(V', V) = 0. \tag{5.124}$$

Assume $V = C_{\underline{j}} = C_{(j_0,j_1,j_2)}$ and V' is locally algebraic. Then there exists $j_2' \in \{j_0', \dots, n-1\}$ such that $V' \cong V^{\mathrm{alg}}_{[j_2'-j_0'+1,j_2'],\Delta}$. An examination of the partial order on $\widetilde{\mathbf{J}}$ shows that $V < V'$ implies $j_2 \le j_2' \le n-1$, and thus $I^+_{j_1,j_2} = [j_2 - j_1 + 1, j_2]$ by Lemma 2.3.1 (see (2.76) for $I^+_{j_1,j_2}$). If $I^+_{j_1,j_2} = [j_2' - j_0' + 1, j_2']$ then $j_2 = j_2'$ and $j_1 = j_0'$, hence (using $j_0' \in \{j_0, j_0+1\}$) $\underline{j} \in \{(j_0', j_0', j_2'), (j_0'-1, j_0', j_2')\}$ which implies $V = C_{\underline{j}} \in \widetilde{\mathbf{J}}(V')$ and contradicts $V \notin \widetilde{\mathbf{J}}(V')$. Hence $I^+_{j_1,j_2} \neq [j_2' - j_0' + 1, j_2']$ and thus $\mathrm{Ext}^1_G(V', V) = 0$ by (i) of Lemma 5.2.2. If $\mathrm{Ext}^2_G(V', V) \neq 0$, then by (i) of Lemma 5.2.4 we have $j_0 = j_1$ and $d([j_2' - j_0' + 1, j_2'], [j_2 - j_1 + 1, j_2]) = d([j_2' - j_0' + 1, j_2'], [j_2 - j_0 + 1, j_2]) = 1$. Since $j_2 \le j_2'$ and $j_0' \in \{j_0, j_0+1\}$, the latter implies either $j_2' = j_2$ and $j_0' = j_0 + 1$, or $j_2' = j_2 + 1$ and $j_0' = j_0 + 1$. In the first case we have $V = C_{(j_0'-1,j_0'-1,j_2')} \in \widetilde{\mathbf{J}}(V')$ and in the second $V = C_{(j_0'-1,j_0'-1,j_2'-1)} \in \widetilde{\mathbf{J}}(V')$. Hence both cases contradict $V \notin \widetilde{\mathbf{J}}(V')$, and we must have $\mathrm{Ext}^2_G(V', V) = 0$.

Assume V is locally algebraic, i.e. $V \cong V^{\mathrm{alg}}_{[j_2-j_0+1,j_2],\Delta}$ for some $j_2 \in \{j_0, \dots, n-1\}$, and $V' = C_{\underline{j}'} = C_{(j_0',j_1',j_2')}$. Then one checks that $V < V'$ implies $j_2' > j_1'$ and $j_2' \ge j_2 + 1$, and thus $I^-_{j_1',j_2'} = [j_2' - j_1', j_2' - 1]$ by Lemma 2.3.1 (see (2.76) for $I^-_{j_1',j_2'}$). If $I^-_{j_1',j_2'} = [j_2 - j_0 + 1, j_2]$ then $j_2 = j_2' - 1$ and $j_1' = j_0$, hence (using

$j'_0 \in \{j_0, j_0+1\}$) $\underline{j}' \in \{(j_0, j_0, j_2+1), (j_0+1, j_0, j_2+1)\}$ which implies $V = C_{\underline{j}} \in \widetilde{\mathbf{J}}(V')$ and contradicts $V \notin \widetilde{\mathbf{J}}(V')$. Hence $I^-_{j'_1, j'_2} \neq [j_2 - j_0 + 1, j_2]$ and thus $\mathrm{Ext}^1_G(V', V) = 0$ by (ii) of Lemma 5.2.2. If $\mathrm{Ext}^2_G(V', V) \neq 0$, then by (ii) of Lemma 5.2.4 we have $j'_0 = j'_1$ and $d([j'_2 - j'_1, j'_2 - 1], [j_2 - j_0 + 1, j_2]) = d([j'_2 - j'_0, j'_2 - 1], [j_2 - j_0 + 1, j_2]) = 1$. Since $j_2 + 1 \leq j'_2$ and $j'_0 \in \{j_0, j_0+1\}$, the latter implies either $j'_2 = j_2 + 1$ and $j'_0 = j_0 + 1$, or $j'_2 = j_2 + 2$ and $j'_0 = j_0 + 1$. In the first case we have $V' \cong C_{(j_0+1, j_0+1, j_2+1)}$ and thus $V \cong V^{\mathrm{alg}}_{[j_2-j_0+1, j_2], \Delta} \in \widetilde{\mathbf{J}}(V')$ which contradicts $V \notin \widetilde{\mathbf{J}}(V')$. So the only possible case is $V \cong V^{\mathrm{alg}}_{[j_2-j_0+1, j_2], \Delta}$ and $V' \cong C_{(j_0+1, j_0+1, j_2+2)}$.

We now finally assume that V and V' are both locally algebraic, i.e. $V \cong V^{\mathrm{alg}}_{[j_2-j_0+1, j_2], \Delta}$ and $V' \cong V^{\mathrm{alg}}_{[j'_2-j'_0+1, j'_2], \Delta}$ for some $j_2 \in \{j_0, \dots, n-1\}$ and some $j'_2 \in \{j'_0, \dots, n-1\}$. The assumption $V < V'$ forces $j_0 \leq j'_0$ and $j_2 \leq j'_2$ by an examination of the partial order on $\widetilde{\mathbf{J}}$. If $j'_0 = 1$ (and hence $j_0 = 1$), then we have $V \cong V^{\mathrm{alg}}_{\{j_2\}, \Delta}$ and $V' \cong V^{\mathrm{alg}}_{\{j'_2\}, \Delta}$ with $j_2 < j'_2$, in which case we have

$$\mathrm{Ext}^1_G(V', V) = 0 \neq \mathrm{Ext}^2_G(V', V)$$

by Lemma 2.2.3 and the sentence before (4.149). In particular this finishes the proof of (i). Assume from now $j'_0 > 1$, which forces $\widetilde{\mathbf{J}}(V') = \mathrm{JH}_G(V^+_{(j'_0-1, j'_0-1, j'_2-1), \infty})$. If $j_2 \in \{j'_2, j'_2 - 1\}$ then the assumption $j_0 \leq j'_0 \leq j_0 + 1$ implies

$$V \in \{V^{\mathrm{alg}}_{[j'_2-j'_0+1, j'_2], \Delta}, V^{\mathrm{alg}}_{[j'_2-j'_0+2, j'_2], \Delta}, V^{\mathrm{alg}}_{[j'_2-j'_0, j'_2-1], \Delta}, V^{\mathrm{alg}}_{[j'_2-j'_0+1, j'_2-1], \Delta}\}$$

which are precisely the 4 locally algebraic constituents of $\widetilde{\mathbf{J}}(V') = \mathrm{JH}_G(V^+_{(j'_0-1, j'_0-1, j'_2-1), \infty})$ by Remark 5.2.31, a contradiction with $V \notin \widetilde{\mathbf{J}}(V')$. Hence we have $j_2 < j'_2 - 1$. The bounds $j'_0 \leq j_0 + 1$ and $j_2 < j'_2 - 1$ imply $j'_2 - j'_0 + 1 \nleq j_2 - j_0 + 1$, hence in particular $j_2 - j_0 + 1 \in [j_2 - j_0 + 1, j_2] \setminus [j'_2 - j'_0 + 1, j'_2]$, and the bounds $j'_0 > 1$, $j_2 < j'_2 - 1$ imply $j'_2, j'_2 - 1 \in [j'_2 - j'_0 + 1, j'_2] \setminus [j_2 - j_0 + 1, j_2]$. Thus $d([j'_2 - j'_0 + 1, j'_2], [j_2 - j_0 + 1, j_2]) \geq 3$ by (2.49). Hence by Lemma 2.2.3 we have $d(V^{\infty}_{[j'_2-j'_0+1, j'_2], \Delta}, V^{\infty}_{[j_2-j_0+1, j_2], \Delta}) \geq 3$, which by the sentence before (4.149) implies (5.124) in this case. By summarizing all the above cases, we have (ii). □

Lemma 5.3.4 *Let* $\underline{j} = (j_0, j_1, j_2) \in \mathbf{J}$, $j'_0 \in \{1, \dots, n-1\}$ *such that* $j_0 \leq j'_0 \leq j_0 + 1$ *and* $W \in \widetilde{\mathbf{J}}_{j'_0}$ *such that* $C_{\underline{j}} \leq W$. *There exists at most one multiplicity free finite length representation* V *in* $\mathrm{Rep}^{\mathrm{an}}_{\mathrm{adm}}(G)$ *such that* $\mathrm{JH}_G(V) = [C_{\underline{j}}, W]$ *as partially ordered sets. Moreover, if* $j_0 = j'_0 > 1$ *then such a* V *exists.*

Proof We fix $\underline{j} \in \mathbf{J}$ and let $\mathcal{C}_{\underline{j}} \subseteq \widetilde{\mathbf{J}}$ be the subset of constituents W such that $C_{\underline{j}} \leq W$ and there exists $j'_0 \in \{j_0, j_0+1\}$ with $j'_0 \leq n-1$ such that $W \in \widetilde{\mathbf{J}}_{j'_0}$. As $W \in \widetilde{\mathbf{J}}_{j'_0}$, using (5.120) we have $C_{(j_0,1,1)} \leq C_{\underline{j}} \leq W \leq C_{(j'_0,1,n)}$. By (5.120) we have $[C_{(j_0,1,1)}, C_{(j'_0,1,n)}] = [C_{(j_0,1,1)}, C_{(j_0,1,n)}] = \widetilde{\mathbf{J}}_{j_0}$ when $j_0 = j'_0$, and an examination of the partial order on $\widetilde{\mathbf{J}}$ gives $[C_{(j_0,1,1)}, C_{(j'_0,1,n)}] = [C_{(j_0,1,1)}, C_{(j_0,1,n)}] \sqcup$

$[C_{(j_0',1,1)}, C_{(j_0',1,n)}] = \widetilde{\mathbf{J}}_{j_0} \sqcup \widetilde{\mathbf{J}}_{j_0'}$ when $j_0' = j_0 + 1$. Hence, we always have $[C_{\underline{j}}, W] \subseteq [C_{j_0,1,1}, C_{j_0',1,n}] = \widetilde{\mathbf{J}}_{j_0} \cup \widetilde{\mathbf{J}}_{j_0'}$ for each $W \in \widetilde{\mathbf{J}}_{j_0'}$ such that $C_{\underline{j}} \leq W$. Note that we have $\mathcal{C}_{\underline{j}} = [C_{\underline{j}}, C_{j_0+1,1,n}]$ when $j_0 < n-1$, and $\mathcal{C}_{\underline{j}} = [C_{\underline{j}}, C_{n-1,1,n}]$ when $j_0 = n-1$.

We now prove the statement by an increasing induction on $W \in \mathcal{C}_{\underline{j}}$ (for the partial order on $\mathcal{C}_{\underline{j}}$ induced by $\widetilde{\mathbf{J}}$). If $W = C_{\underline{j}}$, we have $V = C_{\underline{j}}$ and there is nothing to prove. We assume from now on $C_{\underline{j}} < W$. By induction, for each $W' \in [C_{\underline{j}}, W] \setminus \{W\}$ there exists at most one multiplicity free finite length representation V' in $\mathrm{Rep}^{\mathrm{an}}_{\mathrm{adm}}(G)$ such that $\mathrm{JH}_G(V') = [C_{\underline{j}}, W']$ as partially ordered sets. Taking the amalgamate sum of all such V' for $W' \in [C_{\underline{j}}, W] \setminus \{W\}$ (noting that we need here the unicity in the induction hypothesis) we obtain a representation $\widetilde{V}_{\underline{j},<W}$ in $\mathrm{Rep}^{\mathrm{an}}_{\mathrm{adm}}(G)$ which is the unique (if it exists) multiplicity free finite length representation such that $\mathrm{JH}_G(\widetilde{V}_{\underline{j},<W}) = [C_{\underline{j}}, W] \setminus \{W\}$ (as partially ordered sets). Replacing $[C_{\underline{j}}, W] \setminus \{W\}$ by $[C_{\underline{j}}, W] \setminus \widetilde{\mathbf{J}}(W)$ and noting that any $W'' \in [C_j, W]$ such that $W'' \leq W'$ for some $W' \in [C_j, W] \setminus \widetilde{\mathbf{J}}(W)$ is still in $[C_j, W] \setminus \widetilde{\mathbf{J}}(W)$, we also obtain a unique (if it exists) multiplicity free finite length representation $\widetilde{V}'_{\underline{j},<W}$ in $\mathrm{Rep}^{\mathrm{an}}_{\mathrm{adm}}(G)$ such that $\mathrm{JH}_G(\widetilde{V}'_{\underline{j},<W}) = [C_{\underline{j}}, W] \setminus \widetilde{\mathbf{J}}(W)$ and such that $\widetilde{V}'_{\underline{j},<W}$ is a subrepresentation of $\widetilde{V}_{\underline{j},<W}$. It then follows from (ii) of Lemma 5.3.2 that we have a short exact sequence (if $\widetilde{V}_{\underline{j},<W}$ exists)

$$0 \longrightarrow \widetilde{V}'_{\underline{j},<W} \longrightarrow \widetilde{V}_{\underline{j},<W} \longrightarrow V_{\underline{j},<W} \to 0,$$

which in turn induces a long exact sequence

$$\mathrm{Ext}^1_G(W, \widetilde{V}'_{\underline{j},<W}) \to \mathrm{Ext}^1_G(W, \widetilde{V}_{\underline{j},<W}) \xrightarrow{q} \mathrm{Ext}^1_G(W, V_{\underline{j},<W}) \to \mathrm{Ext}^2_G(W, \widetilde{V}'_{\underline{j},<W}). \tag{5.125}$$

It follows from (i) of Lemma 5.3.3 that $\mathrm{Ext}^1_G(W, W') = 0$ for each constituent W' of $\widetilde{V}'_{\underline{j},<W}$, which by dévissage implies $\mathrm{Ext}^1_G(W, \widetilde{V}'_{\underline{j},<W}) = 0$. Then (5.125) and (5.123) imply $\dim_E \mathrm{Ext}^1_G(W, \widetilde{V}_{\underline{j},<W}) \leq 1$, which shows that V as in the statement is unique if it exists. Note that if $j_0 = j_0' > 1$, then (ii) of Lemma 5.3.3 implies $\mathrm{Ext}^2_G(W, \widetilde{V}'_{\underline{j},<W}) = 0$ by an analogous dévissage and thus q is an isomorphism. Going back through the induction above, we see that V exists if $j_0 = j_0' > 1$. □

Proposition 5.3.5 *Let $1 \leq j_0 \leq n-1$ and $\mu \in \Lambda$ such that $\langle \mu + \rho, \alpha^\vee \rangle \geq 0$ for $\alpha \in \Phi^+$ and the stabilizer of μ in $W(G)$ for the dot action is $\{1, w_0 s_{j_0} w_0\}$.*

(i) For each representation $V \in \widetilde{\mathbf{J}} \setminus (\widetilde{\mathbf{J}}_{j_0} \cap \mathbf{J})$, we have $\Theta_{w_0 s_{j_0} w_0}(V) = 0$.

(ii) Assume $j_0 > 1$ and set $j_0^+ \overset{def}{=} \min\{j_0 + 1, n-1\}$. Let $(j_1, j_2) \in \{(n-1, n-1), (1, n)\}$ and V_0 be the unique multiplicity free finite length representation in $\mathrm{Rep}^{\mathrm{an}}_{\mathrm{adm}}(G)$ such that $\mathrm{JH}_G(V_0) = [C_{(j_0,1,1)}, C_{(j_0,j_1,j_2)}]$ as partially ordered sets (see Lemma 5.3.4). Then we have a short exact sequence

$$0 \longrightarrow V_0^+ \longrightarrow \Theta_\mu(V_0) \longrightarrow V_0^- \longrightarrow 0,$$

where V_0^- *(resp.* V_0^+*) is the unique multiplicity free finite length representation in* $\mathrm{Rep}^{\mathrm{an}}_{\mathrm{adm}}(G)$ *such that* $\mathrm{JH}_G(V_0^-) = [C_{(j_0-1,1,1)}, C_{(j_0,j_1,j_2)}]$ *(resp. such that* $\mathrm{JH}_G(V_0^+) = [C_{(n-1,1,1)}, L(1)^\vee]$ *when* $j_0 = j_1 = j_2 = n-1$*, and* $\mathrm{JH}_G(V_0^+) = [C_{(j_0,1,1)}, C_{(j_0^+,j_1,j_2)}]$ *otherwise) as partially ordered sets.*

Proof (i) follows from (i) of Lemma 5.2.12 and the fact that each $V \in \widetilde{\mathbf{J}} \setminus (\widetilde{\mathbf{J}}_{j_0} \cap \mathbf{J})$ is either locally algebraic or of the form $\mathcal{F}^G_{P_{\widehat{j_1}}}(L(w_{j_1,j_0'}), \pi^\infty_{j_1,j_2})$ with $D_R(w_{j_1,j_0'}) = \{j_0'\}$ and $j_0' \neq j_0$.

We prove (ii). It follows from Lemma 5.3.4 that V_0 exists (using $j_0 > 1$) and is unique, that V_0^- and V_0^+ are unique if they exist, and that V_0^+ exists if $j_0 = n-1 = j_0^+$. By a decreasing induction on $1 < j_0 \le n-1$, it suffices to assume the existence of V_0^+, construct an embedding $V_0^+ \hookrightarrow \Theta_\mu(V_0)$, and then prove that $\Theta_\mu(V_0)/V_0^+$ is a multiplicity free finite length representation that satisfies $\mathrm{JH}_G(\Theta_\mu(V_0)/V_0^+) = [C_{(j_0-1,1,1)}, C_{(j_0,j_1,j_2)}]$ as partially ordered sets (for instance, when $j_0 = n-1$, then $\Theta_\mu(V_0)/V_0^+$ is actually the V_0^+ for $j_0 = n-2$ and the induction goes on). Assume from now that V_0^+ exists. As $C_{(j_0,1,1)} \le C_{(j_0,j_1,j_2)} \le C_{(j_0^+,j_1,j_2)}$, V_0 is the unique subrepresentation of V_0^+ with cosocle $C_{(j_0,j_1,j_2)}$. We divide the rest of the proof of (ii) into the following steps.

Step 1: We prove that the injection $V_0 \hookrightarrow V_0^+$ induces an isomorphism $\Theta_\mu(V_0) \xrightarrow{\sim} \Theta_\mu(V_0^+)$, and that the adjunction map $V_0^+ \to \Theta_\mu(V_0^+)$ is injective. Note that we have

$$\begin{aligned} V_0^+/V_0 &\cong L(1)^\vee \quad \text{if } j_0 = j_0^+ = n-1 \text{ and } (j_1, j_2) = (n-1, n-1) \\ V_0^+ &= V_0 \qquad\;\;\; \text{if } j_0 = j_0^+ = n-1 \text{ and } (j_1, j_2) = (1, n), \end{aligned}$$

and when $j_0^+ = j_0 + 1 \le n-1$:

$$\mathrm{JH}_G(V_0^+/V_0) = [C_{(j_0,1,1)}, C_{(j_0^+,j_1,j_2)}] \setminus [C_{(j_0,1,1)}, C_{(j_0,j_1,j_2)}] = [C_{(j_0^+,1,1)}, C_{(j_0^+,j_1,j_2)}] \subseteq \widetilde{\mathbf{J}}_{j_0'}.$$

Since in the third case $\widetilde{\mathbf{J}}_{j_0^+} \cap \widetilde{\mathbf{J}}_{j_0} = \emptyset$, we have $\mathrm{JH}_G(V_0^+/V_0) \subseteq \widetilde{\mathbf{J}} \setminus \widetilde{\mathbf{J}}_{j_0}$, and *a fortiori* $\mathrm{JH}_G(V_0^+/V_0) \subseteq \widetilde{\mathbf{J}} \setminus (\widetilde{\mathbf{J}}_{j_0} \cap \mathbf{J})$. The first two cases also clearly satisfy $\mathrm{JH}_G(V_0^+/V_0) \subseteq \widetilde{\mathbf{J}} \setminus (\widetilde{\mathbf{J}}_{j_0} \cap \mathbf{J})$. Hence we deduce from (i) that $\Theta_\mu(V_0^+/V_0) = 0$ and thus that the injection $V_0 \hookrightarrow V_0^+$ induces an isomorphism $\Theta_\mu(V_0) \xrightarrow{\sim} \Theta_\mu(V_0^+)$. We define q^+ as the composition $V_0^+ \to \Theta_\mu(V_0^+) \cong \Theta_\mu(V_0)$ (the first map being the adjunction map). As q^+ restricts to an injection $C_{(j_0,1,1)} \hookrightarrow \Theta_{w_0 s_{j_0} w_0}(C_{(j_0,1,1)})$ by (the first statement in) (i) of Lemma 5.2.13 and V_0^+ has socle $C_{(j_0,1,1)}$, the map q^+ is injective.

Step 2: We prove that $\Theta_\mu(V_0)/V_0^+$ is a multiplicity free finite length representation such that there is an equality of *sets*

$$\mathrm{JH}_G(\Theta_\mu(V_0)/V_0^+) = [C_{(j_0-1,1,1)}, C_{(j_0,j_1,j_2)}]. \tag{5.126}$$

It suffices to show that we have an equality in the *Grothendieck group* of finite length admissible representations of G

$$[\Theta_\mu(V_0)] = [C_{(j_0,1,1)}, C_{(j_0^+,j_1,j_2)}] + [C_{(j_0-1,1,1)}, C_{(j_0,j_1,j_2)}] \tag{5.127}$$

(where now $[C_{(j_0,1,1)}, C_{(j_0^+,j_1,j_2)}]$, $[C_{(j_0-1,1,1)}, C_{(j_0,j_1,j_2)}]$ mean the direct sums of the corresponding constituents in this Grothendieck group, but this does not lead to confusions). By (i) of Lemma 5.2.12 we have

$$[\Theta_\mu(V_0)] = \sum_{(j_1',j_2')\leq(j_1,j_2)} [\Theta_{w_0 s_{j_0} w_0}(C_{(j_0,j_1',j_2')})],$$

and we can compute $[\Theta_{w_0 s_{j_0} w_0}(C_{(j_0,j_1',j_2')})]$ using (i) of Lemma 5.2.13. For each $(j_1', j_2') \leq (j_1, j_2)$, we check that $C_{(j_0,j_1',j_2')}$ (resp. $C_{(j_0-1,j_1',j_2')}$) appears with multiplicity 2 (resp. 1) on both sides of (5.127), and that $C_{(j_0+1,j_1',j_2')}$ appears with multiplicity 1 on both sides of (5.127) when $j_0' = j_0 + 1 \leq n - 1$. We now treat the locally algebraic constituents. Assume first $(j_1, j_2) = (1, n)$. Using (i) of Lemmas 5.2.13 and 2.3.1, we check that the contribution of locally algebraic constituents on the left hand side of (5.127) is

$$\sum_{j_2'=j_0}^{n} [L(1)^\vee \otimes_E i^\infty_{j_0,\Delta}(\pi^\infty_{j_0,j_2'})] = 2\sum_{j_2'=j_0}^{n-1} [V^{\mathrm{alg}}_{[j_2'-j_0+1,j_2'],\Delta}] + \sum_{j_2'=j_0+1}^{n-1} [V^{\mathrm{alg}}_{[j_2'-j_0,j_2'],\Delta}]$$
$$+ \sum_{j_2'=j_0}^{n} [V^{\mathrm{alg}}_{[j_2'-j_0+1,j_2'-1],\Delta}]. \tag{5.128}$$

Using the relations

$$\begin{cases} C_{(j_0,j_0,j_2')} \leq V^{\mathrm{alg}}_{[j_2'-j_0+1,j_2'],\Delta} \leq C_{(j_0,j_0,j_2'+1)} & \text{for } j_0 \leq j_2' \leq n-1; \\ C_{(j_0+1,j_0+1,j_2')} \leq V^{\mathrm{alg}}_{[j_2'-j_0,j_2'],\Delta} \leq C_{(j_0+1,j_0+1,j_2'+1)} & \text{for } j_0+1 \leq j_2' \leq n-1; \\ C_{j_0-1,j_0-1,j_2'-1} \leq V^{\mathrm{alg}}_{[j_2'-j_0+1,j_2'-1],\Delta} \leq C_{j_0-1,j_0-1,j_2'} & \text{for } j_0 \leq j_2' \leq n, \end{cases}$$

we check that the contribution of the locally algebraic constituents in $[V_0^+] + [V_0^-]$ is given by exactly the same formula (5.128). Finally, when $(j_1, j_2) = (n-1, n-1)$, using again (i) of Lemmas 5.2.13, 2.3.1 and the above relations, one checks that the contribution of locally algebraic constituents on both sides of (5.127) is $[L(1) \otimes_E i^\infty_{j_0,\Delta}(\pi^\infty_{j_0,j_0})] = [V^{\mathrm{alg}}_{[1,j_0],\Delta}] + [V^{\mathrm{alg}}_{[1,j_0-1],\Delta}]$.

Step 3: We assume $(j_1, j_2) = (1, n)$ and prove that the partial order on $\mathrm{JH}_{\widetilde{G}}(\Theta_\mu(V_0)/V_0^+)$ is at least as strong as the one on $[C_{(j_0-1,1,1)}, C_{(j_0,1,n)}]$ induced by $\widetilde{\mathbf{J}}$.

Note that the underlying set of (5.126) is explicitly given by

$$\{C_{\underline{j}'} \mid (j_0-1,1,1) \leq \underline{j}' \leq (j_0,1,n)\} \sqcup \{V^{\mathrm{alg}}_{[j_2'-j_0+2,j_2'],\Delta} \mid 1 \leq j_2' \leq n-1\}$$
$$\sqcup \{V^{\mathrm{alg}}_{[j_2'-j_0+1,j_2'],\Delta} \mid 2 \leq j_2' \leq n-1\}.$$

Recall from (i) of Lemma 5.2.13 that the adjunction map $\Theta_\mu(C_{(j_0,1,n)}) \to C_{(j_0,1,n)}$ is surjective, and the composition $C_{(j_0,1,n)} \to \Theta_\mu(C_{(j_0,1,n)}) \to C_{(j_0,1,n)}$ with the other adjunction map is zero. Since V_0 has cosocle $C_{(j_0,1,n)}$ and the adjunction maps $V_0 \to \Theta_\mu(V_0) \to V_0$ are compatible with the adjunction maps $C_{(j_0,1,n)} \to \Theta_\mu(C_{(j_0,1,n)}) \to C_{(j_0,1,n)}$ under the surjection $V_0 \twoheadrightarrow C_{(j_0,1,n)}$, we deduce that the adjunction map $\Theta_\mu(V_0) \to V_0$ is surjective and the composition of $V_0 \to \Theta_\mu(V_0) \to V_0$ is zero. If $j_0 = j_0^+ = n-1$, then $V_0 = V_0^+$ and we obtain a surjection $\Theta_\mu(V_0)/V_0^+ \twoheadrightarrow V_0$. If $j_0 < n-1$, then as the cosocle $C_{(j_0+1,1,n)}$ of V_0^+ does not occur in V_0, this forces the composition $V_0^+ \hookrightarrow \Theta_\mu(V_0) \twoheadrightarrow V_0$ to be zero and we also obtain a surjection $\Theta_\mu(V_0)/V_0^+ \twoheadrightarrow V_0$. It follows that the partial order on $\mathrm{JH}_G(\Theta_\mu(V_0)/V_0^+)$ restricts to the one on $\mathrm{JH}_G(V_0) = [C_{(j_0,1,1)}, C_{(j_0,1,n)}]$. In particular, inside $\mathrm{JH}_G(\Theta_\mu(V_0)/V_0^+)$ we have $C_{(j_0,j_1',j_2')} < C_{(j_0,j_1'',j_2'')}$ for $(j_1', j_2') < (j_1'', j_2'') \in \mathbf{J}^\infty$, and $C_{j_0,j_0,j_2'} < V^{\mathrm{alg}}_{[j_2'-j_0+1,j_2'],\Delta} < C_{j_0,j_0,j_2'+1}$ for $2 \le j_2' \le n-1$ (as these relations occur in $[C_{(j_0,1,1)}, C_{(j_0,1,n)}]$). Now, let $(j_1', j_2'), (j_1'', j_2'') \in \mathbf{J}^\infty$ with $(j_1'', j_2'') \in \{(j_1'+1, j_2'+1), (j_1'-1, j_2')\}$, then V_0 admits a unique length 2 subquotient V_1 with socle $C_{(j_0,j_1',j_2')}$ and cosocle $C_{(j_0,j_1'',j_2'')}$ (Lemma 5.2.1). By the last statement in Remark 5.2.14 applied to j_0 (there) being $j_0 - 1$ (here), $\Theta_\mu(V_1)$ admits a unique subquotient isomorphic to $V_{(j_0-1,j_1',j_2'),(j_0,j_1'',j_2'')}$ as in (i) or (ii) of Proposition 5.2.10. Since $\mathrm{soc}_G(V_{(j_0-1,j_1',j_2'),(j_0,j_1'',j_2'')}) \cong C_{(j_0-1,j_1',j_2')} \notin \mathrm{JH}_G(V_0^+)$, it follows that $V_{(j_0-1,j_1',j_2'),(j_0,j_1'',j_2'')}$ is still a subquotient of $\Theta_\mu(V_0)/V_0^+$. In particular, from the structure of $V_{(j_0-1,j_1',j_2'),(j_0,j_1'',j_2'')}$ in *loc. cit.*, inside $\mathrm{JH}_G(\Theta_\mu(V_0)/V_0^+)$ we have $C_{(j_0-1,j_1',j_2')} < C_{(j_0-1,j_1'',j_2'')} < C_{(j_0,j_1'',j_2'')}$, $C_{(j_0-1,j_1',j_2')} < C_{(j_0,j_1',j_2')}$, and

$$\begin{cases} C_{(j_0-1,j_0-1,j_2')} < V^{\mathrm{alg}}_{[j_2'-j_0+2,j_2'],\Delta} < V^{\mathrm{alg}}_{[j_2'-j_0+1,j_2'],\Delta} < C_{(j_0,j_0,j_2'+1)} & \text{if} \quad j_1=1 \quad \text{and } j_2''=j_2'+1>j_0 \\ C_{(j_0-1,j_0-1,j_0-1)} < V^{\mathrm{alg}}_{[1,j_0-1],\Delta} < C_{(j_0,j_0,j_0)} & \text{if } j_1 = 1\, {}_{j_1=1} \text{ and } j_2''=j_2'+1=j_0. \end{cases} \tag{5.129}$$

In particular the partial order on $\mathrm{JH}_G(\Theta_\mu(V_0)/V_0^+)$ and on $[C_{(j_0-1,1,1)}, C_{(j_0,1,n)}]$ have the same restriction to $[C_{(j_0-1,1,1)}, C_{(j_0,1,n)}] \cap \mathbf{J}$ (i.e. non locally algebraic constituents). We now deal with locally algebraic constituents. We prove by an increasing induction on $1 \le j_2' \le n-1$ that, inside $\mathrm{JH}_G(\Theta_\mu(V_0)/V_0^+)$, $V^{\mathrm{alg}}_{[j_2'-j_0+2,j_2'],\Delta}$ is the only locally algebraic constituent V such that:

$$C_{(j_0-1,j_0-1,j_2')} < V < C_{(j_0-1,j_0-1,j_2'+1)}. \tag{5.130}$$

Note first that all constituents of the form $V^{\mathrm{alg}}_{[j_2'-j_0+1,j_2'],\Delta}$ occur in V_0 and hence can't lie below $C_{(j_0-1,j_0-1,j_2'+1)}$ in $\mathrm{JH}_G(\Theta_\mu(V_0)/V_0^+)$ (since $C_{(j_0-1,j_0-1,j_2'+1)} \in \mathrm{JH}_G(\Theta_\mu(V_0)/V_0^+)$ maps to 0 in the quotient V_0 of $\Theta_\mu(V_0)/V_0^+$, the same holds for any constituent below $C_{(j_0-1,j_0-1,j_2'+1)}$ in $\mathrm{JH}_G(\Theta_\mu(V_0)/V_0^+)$). It follows from (5.129) that $C_{(j_0-1,j_0-1,j_2'+1)} \le C_{(j_0-1,j_0-1,j)} < V^{\mathrm{alg}}_{[j-j_0+2,j],\Delta}$ for $j > j_2'$. Our induction hypothesis implies $V^{\mathrm{alg}}_{[j-j_0+2,j],\Delta} < C_{(j_0-1,j_0-1,j+1)} \le C_{(j_0-1,j_0-1,j_2')}$ for $j < j_2'$ (and is empty when $j_2' = j_0 - 1$). Hence, the only locally algebraic constituent V that could satisfy (5.130) is $V = V^{\mathrm{alg}}_{[j_2'-j_0+2,j_2'],\Delta}$. To prove that $V^{\mathrm{alg}}_{[j_2'-j_0+2,j_2'],\Delta}$ is indeed there (in $\mathrm{JH}_G(\Theta_\mu(V_0)/V_0^+)$), one has to check that the unique subquotient V_2 of

$\Theta_\mu(V_0)/V_0^+$ with socle $C_{(j_0-1,j_0-1,j_2')}$ and cosocle $C_{(j_0-1,j_0-1,j_2'+1)}$ has Loewy length 3 with middle layer either $C_{(j_0-1,j_0,j_2'+1)}$ or $C_{(j_0-1,j_0,j_2'+1)} \oplus V^{\mathrm{alg}}_{[j_2'-j_0+2,j_2'],\Delta}$. But by minimality of the Ext-square $V_{(j_0-1,j_0-1,j_2'),(j_0-1,j_0-1,j_2'+1)}$ ((ii) of Proposition 5.2.18 applied with j_0, j_1 there both being j_0-1) we see that $V^{\mathrm{alg}}_{[j_2'-j_0+2,j_2'],\Delta}$ must appear in the middle layer of V_2 and thus we indeed have (5.130) for $V = V^{\mathrm{alg}}_{[j_2'-j_0+2,j_2'],\Delta}$. Since the partial order on $[C_{(j_0-1,1,1)}, C_{(j_0,1,n)}]$ is generated by the relations (5.129), (5.130) and $C_{\underline{j}} < C_{\underline{j}'}$ for $\underline{j} < \underline{j}'$, we have shown that the partial order on $\mathrm{JH}_G(\Theta_\mu(V_0)/V_0^+)$ is at least as strong as the one on $[C_{(j_0-1,1,1)}, C_{(j_0,1,n)}]$ induced by $\widetilde{\mathbf{J}}$.

Step 4: We assume $(j_1, j_2) = (1, n)$ and prove that the equality of sets (5.126) holds as *partially ordered sets*.

We consider an arbitrary length 2 subquotient V_3 of $\Theta_\mu(V_0)/V_0^+$ with socle V and cosocle V'. We have $\mathrm{Ext}^1_G(V', V) \neq 0$ and V is right below V' for the partial order on $\mathrm{JH}_G(\Theta_\mu(V_0)/V_0^+)$. Let us prove $V < V'$ for the partial order on $\widetilde{\mathbf{J}}$. If V and V' are not both locally algebraic, then this follows from Lemmas 5.2.1 and 5.2.2. Assume that V and V' are both locally algebraic. Since $\mathrm{JH}_G(V_0) = [C_{(j_0,1,1)}, C_{(j_0,1,n)}]$ (as partially ordered sets), one easily checks that V_0 does not contain any locally algebraic length 2 subquotient. Hence V is not a constituent of V_0, and we therefore have $V = V^{\mathrm{alg}}_{[j_2'-j_0+2,j_2'],\Delta}$ for some $1 \le j_2' \le n-1$. Then by Lemmas 5.1.3 and 2.2.3 we have $V' = V^{\mathrm{alg}}_{[j_2''-j_0+1,j_2''],\Delta}$ with $d([j_2'-j_0+2, j_2'], [j_2''-j_0+1, j_2'']) = 1$. If $V' = V^{\mathrm{alg}}_{[j_2'-j_0+1,j_2'],\Delta}$ with $j_2' > 1$, then V lies right below V' in $\widetilde{\mathbf{J}}$. Otherwise $V' = V^{\mathrm{alg}}_{[j_2'-j_0+2,j_2'+1],\Delta}$ and we have for the partial order on $\mathrm{JH}_G(\Theta_\mu(V_0)/V_0^+)$

$$V = V^{\mathrm{alg}}_{[j_2'-j_0+2,j_2'],\Delta} < C_{(j_0-1,j_0-1,j_2'+1)} < V^{\mathrm{alg}}_{[j_2'-j_0+3,j_2'+1],\Delta} < V' = V^{\mathrm{alg}}_{[j_2'-j_0+2,j_2'+1],\Delta}$$

(as these relations hold in $\widetilde{\mathbf{J}}$ and the partial order on $\mathrm{JH}_G(\Theta_\mu(V_0)/V_0^+)$ is at least as strong). But this contradicts the existence of V_3. We conclude that (5.126) is an equality of partially ordered sets.

Step 5: We finish the proof of (ii).

If $(j_1, j_2) = (1, n)$, we conclude (ii) from Step 4 together with a decreasing induction on $1 < j_0 \le n-1$ (as explained above Step 1). In particular, for each $1 < j_0 \le n-1$ there exists a unique multiplicity free finite length representation V in $\mathrm{Rep}^{\mathrm{an}}_{\mathrm{adm}}(G)$ such that $\mathrm{JH}_G(V) = [C_{(j_0-1,1,1)}, C_{(j_0,1,n)}]$ as partially ordered sets. Then its unique subrepresentation V' with cosocle $C_{(j_0,n-1,n-1)}$ satisfies $\mathrm{JH}_G(V') = [C_{(j_0-1,1,1)}, C_{(j_0,n-1,n-1)}]$ as partially ordered sets. Assume $(j_1, j_2) = (n-1, n-1)$ and let V_0, V_0^- as in (ii) (both of which exist now). We have a surjection $V_0 \twoheadrightarrow V_0^-$. By a symmetric argument as for V_0^+ in Step 1, this induces an isomorphism $\Theta_\mu(V_0^-) \cong \Theta_\mu(V_0)$ and the adjunction map $\Theta_\mu(V_0^-) \to V_0^-$ is surjective. We define q^- as the composition $\Theta_\mu(V_0) \cong \Theta_\mu(V_0^-) \twoheadrightarrow V_0^-$. As $\mathrm{soc}_G(V_0^-) = C_{(j_0-1,1,1)}$ is not a constituent of V_0^+, the composition $q^- \circ q^+$ (see Step 1 for q^+) is necessarily zero, hence we have a complex $[V_0^+ \xrightarrow{q^+} \Theta_\mu(V_0) \xrightarrow{q^-} V_0^-]$ which is exact on the left by

Step 1 and on the right by above. Then Step 2 (applied with $(j_1, j_2) = (n-1, n-1)$) gives exactness in the middle, and also that (5.126) is again an equality of partially ordered sets. This finishes the proof of (ii). □

Theorem 5.3.6 *For each $\underline{j} = (j_0, j_1, j_2) \in \mathbf{J}$, $j_0' \in \{1, \dots, n-1\}$ such that $j_0 \le j_0' \le j_0 + 1$ and $W \in \widetilde{\mathbf{J}}_{j_0'}$ such that $C_{\underline{j}} \le W$, there exists a unique multiplicity free finite length representation V in $\mathrm{Rep}^{\mathrm{an}}_{\mathrm{adm}}(G)$ such that $\mathrm{JH}_G(V) = [C_{\underline{j}}, W]$ as partially ordered sets.*

Proof The unicity of V follows from Lemma 5.3.4. By (ii) of Proposition 5.3.5 there exists a unique multiplicity free finite length representation V_0 in $\mathrm{Rep}^{\mathrm{an}}_{\mathrm{adm}}(G)$ such that $\mathrm{JH}_G(V_0) = [C_{(j_0,1,1)}, C_{(j_0',1,n)}]$ as partially ordered sets. Since we have $C_{\underline{j}}, W \in [C_{(j_0,1,1)}, C_{(j_0',1,n)}]$ with $C_{\underline{j}} \le W$, V_0 admits a unique subquotient with socle $C_{\underline{j}}$ and cosocle W, which gives the existence of V. □

Remark 5.3.7 It follows from Theorem 5.3.6 that there exists a unique multiplicity free finite length representation V_0 in $\mathrm{Rep}^{\mathrm{an}}_{\mathrm{adm}}(G)$ such that $\mathrm{JH}_G(V_0) = [C_{(1,1,1)}, C_{(1,1,n)}]$ as partially ordered sets. When $n = 2$ the representation V_0 is well-known, it is (up to unramified twist) the representation denoted $\Pi^{1,\sigma}(\underline{k}_\sigma, \underline{D})$ at the bottom of [11, p. 638]. When $n = 3$ and $K = \mathbb{Q}_p$, V_0 is the representation denoted $\Pi^1(\underline{k}, D)$ in [11, Theorem 1.2] (the value of $\underline{k} = (k_1, k_2, k_3)$ being determined by μ_0).

Recall that $\mathrm{St}_n^{\mathrm{alg}} = V^{\mathrm{alg}}_{\emptyset,\Delta} = L(1)^\vee \otimes_E V^\infty_{\emptyset,\Delta} = L(\mu_0)^\vee \otimes_E \mathrm{St}_n^\infty$ (with St_n^∞ being the smooth Steinberg representation of G). Let $Z(G)$ be the center of G and $\chi : Z(G) \to E^\times$ the central character of $\mathrm{St}_n^{\mathrm{alg}}$ (which depends on μ_0). We consider the extension groups $\mathrm{Ext}^\bullet_{D(G),\chi^\vee}$ computed in the full subcategory of $\mathrm{Mod}_{D(G)}$ of $D(G)$-modules where $D(Z(G))$ acts by $\chi^\vee : D(Z(G)) \to E^\times$ (see e.g. [11, Remark 5.1.3(i)]). It follows from the analogue of the spectral sequence (4.148) (with (4.14)) where we fix central characters everywhere and from $\mathrm{Ext}^\ell_{U(\mathfrak{g}),\chi^\vee}(L(1), L(1)) = 0$ for $\ell = 1, 2$ (Whitehead's lemma) that we have isomorphisms for $\ell \le 2$ and $I, I' \subseteq \Delta$

$$\mathrm{Ext}^\ell_{G,1}(V^\infty_{I,\Delta}, V^\infty_{I',\Delta})^\infty \xrightarrow{\sim} \mathrm{Ext}^\ell_{D(G),\chi^\vee}((V^{\mathrm{alg}}_{I',\Delta})^\vee, (V^{\mathrm{alg}}_{I,\Delta})^\vee) \tag{5.131}$$

(where $\mathrm{Ext}^\ell_{G,1}(-,-)^\infty$ means smooth extensions with trivial central character). By [48, Theorem 2] and (5.131) we deduce

$$\mathrm{Ext}^1_{D(G),\chi^\vee}((\mathrm{St}_n^{\mathrm{alg}})^\vee, (\mathrm{St}_n^{\mathrm{alg}})^\vee) = 0. \tag{5.132}$$

Lemma 5.3.8 *Let $j_0^+ = \min\{2, n-1\}$, $(j_1, j_2) \in \{(n-1, n-1), (1, n)\}$ and V_0 (resp. V_0^+) be the unique multiplicity free finite length representation in $\mathrm{Rep}^{\mathrm{an}}_{\mathrm{adm}}(G)$ such that $\mathrm{JH}_G(V_0) = [C_{(1,1,1)}, C_{(1,j_1,j_2)}]$ (resp. $\mathrm{JH}_G(V_0^+) = [C_{(1,1,1)}, L(1)^\vee]$ if $n = 2$ with $(j_1, j_2) = (1, 1)$, and $\mathrm{JH}_G(V_0^+) = [C_{(1,1,1)}, C_{(j_0^+,j_1,j_2)}]$ otherwise) as partially ordered sets (see Theorem 5.3.6). Let $\mu \in \Lambda$ such that $\langle \mu + \rho, \alpha^\vee \rangle \ge 0$ for $\alpha \in \Phi^+$ and the stabilizer of μ in $W(G)$ for the dot action is $\{1, w_0 s_1 w_0\}$. We set $m = 1$ if $(j_1, j_2) = (n-1, n-1)$ and $m = n$ if $(j_1, j_2) = (1, n)$.*

(i) The injection $V_0 \hookrightarrow V_0^+$ *induces an isomorphism* $\Theta_\mu(V_0) \xrightarrow{\sim} \Theta_\mu(V_0^+)$, *and the canonical adjunction map* $V_0^+ \to \Theta_\mu(V_0^+)$ *is an injection.*

(ii) The representation $\Theta_\mu(V_0)/V_0^+$ *admits a central character, has socle* $(\mathrm{St}_n^{\mathrm{alg}})^{\oplus n}$ *and fits into a short exact sequence*

$$0 \longrightarrow (\mathrm{St}_n^{\mathrm{alg}})^{\oplus m} \longrightarrow \Theta_\mu(V_0)/V_0^+ \longrightarrow V_0 \longrightarrow 0. \tag{5.133}$$

Proof The proof of (i) is similar to Step 1 of the proof of (ii) of Proposition 5.3.5. We prove (ii). As V_0 is multiplicity free, it admits a central character, and so do $\Theta_\mu(V_0)$ and $\Theta_\mu(V_0)/V_0^+$. By Step 3 and Step 5 of the proof of (ii) of Proposition 5.3.5, the adjunction $\Theta_\mu(V_0) \to V_0$ induces a surjection $q : \Theta_\mu(V_0)/V_0^+ \twoheadrightarrow V_0$. Similar to Step 2 of the proof of (ii) of Proposition 5.3.5, we can check using (i) of Lemma 5.2.12 and (i) of Lemma 5.2.13 that we have in the Grothendieck group of finite length admissible representations of G

$$[\Theta_\mu(V_0)] - [V_0] - [V_0^+] = m[\mathrm{St}_n^{\mathrm{alg}}], \tag{5.134}$$

which implies $[\ker(q)] = m[\mathrm{St}_n^{\mathrm{alg}}]$ by the previous sentence. Since $\Theta_\mu(V_0)/V_0^+$ has a central character, so does its subrepresentation $\ker(q)$, which together with (5.132) forces $\ker(q) \cong (\mathrm{St}_n^{\mathrm{alg}})^{\oplus m}$ and gives (5.133). It remains to show that $\Theta_\mu(V_0)/V_0^+$ has socle $(\mathrm{St}_n^{\mathrm{alg}})^{\oplus m}$. Since V_0 has socle $C_{(1,1,1)}$, by (5.133) it suffices to show that $C_{(1,1,1)}$ does not appear in the socle of $\Theta_\mu(V_0)/V_0^+$. It follows from (i) of Lemma 5.2.13 that $\Theta_\mu(C_{(1,1,1)})$ admits a quotient V with socle $\mathrm{St}_n^{\mathrm{alg}}$ and cosocle $C_{(1,1,1)}$. Since $\mathrm{St}_n^{\mathrm{alg}}$ is not constituent of V_0^+, the subquotient V of $\Theta_\mu(C_{(1,1,1)})$ and thus of $\Theta_\mu(V_0)$ must be a subquotient of $\Theta_\mu(V_0)/V_0^+$. Since $C_{(1,1,1)}$ has multiplicity 1 in $\Theta_\mu(V_0)/V_0^+$ by (5.134), we deduce that $C_{(1,1,1)}$ cannot show up in the socle of $\Theta_\mu(V_0)/V_0^+$. □

We now define several important finite length multiplicity free coadmissible $D(G)$-modules. By Theorem 5.3.6 we only need to specify the corresponding partially ordered set of irreducible constituents of $\widetilde{\mathbf{J}}$.

We set $X_0 = Y_0 = Z_0 = 0$, and for $1 \le k \le n-1$, we define (with the notation (5.119))

$$\begin{cases} X_k \text{ such that } \mathrm{JH}_G(X_k^\vee) = [C_{(n-k,1,1)}, C_{(n-k,n-1,n-1)}] \\ Y_k \text{ such that } \mathrm{JH}_G(Y_k^\vee) = [C_{(n-k,1,2)}, C_{(n-k,1,n)}] \\ Z_k \text{ such that } \mathrm{JH}_G(Z_k^\vee) = [C_{(n-k,1,1)}, C_{(n-k,1,n)}]. \end{cases} \tag{5.135}$$

We can check that $Y_1^\vee$ has no locally algebraic constituent but that for $k \ge 2$ the locally algebraic constituents of $Y_k^\vee$ are

$$V^{\mathrm{alg}}_{[2,n-k+1],\Delta}, V^{\mathrm{alg}}_{[3,n-k+2],\Delta}, \dots, V^{\mathrm{alg}}_{[k,n-1],\Delta},$$

and that for $k \ge 1$ the locally algebraic constituents of $Z_k^\vee$ are

$$V^{\mathrm{alg}}_{[1,n-k],\Delta}, V^{\mathrm{alg}}_{[2,n-k+1],\Delta}, V^{\mathrm{alg}}_{[3,n-k+2],\Delta}, \dots, V^{\mathrm{alg}}_{[k,n-1],\Delta}.$$

We define

$$\begin{cases} D_0 \text{ such that } \mathrm{JH}_G(D_0^\vee) = [C_{(n-1,1,1)}, V^{\mathrm{alg}}_{\Delta,\Delta}] = [C_{(n-1,1,1)}, L(1)^\vee] \\ \widetilde{D}_0 \text{ such that } \mathrm{JH}_G(\widetilde{D}_0^\vee) = [C_{(n-1,1,1)}, C_{(n-1,1,n)}], \end{cases} \tag{5.136}$$

(note that $\widetilde{D}_0 = Z_1$) and for $1 \leq k \leq n-2$

$$\begin{cases} D_k \text{ such that } \mathrm{JH}_G(D_k^\vee) = [C_{(n-k-1,1,1)}, C_{(n-k,n-1,n-1)}] \\ \widetilde{D}_k \text{ such that } \mathrm{JH}_G(\widetilde{D}_k^\vee) = [C_{(n-k-1,1,1)}, C_{(n-k,1,n)}]. \end{cases} \tag{5.137}$$

From their definition (and the definition of $\widetilde{\mathbf{J}}$), we see that all coadmissible $D(G)$-modules X_k, Y_k, Z_k, D_k and $\widetilde{D}_k$ are indecomposable multiplicity free with irreducible socle and cosocle.

The following remark will be useful.

Remark 5.3.9 Let V_0 be a multiplicity free finite length representation in $\mathrm{Rep}^{\mathrm{an}}_{\mathrm{adm}}(G)$, and let $S_1, S_2 \subseteq \mathrm{JH}_G(V_0)$ be subsets such that $S_1 \cap S_2 = \emptyset$. The partial order on $\mathrm{JH}_G(V_0)$ restricts to a partial order on S_i for $i = 1, 2$. Assume that the following conditions hold for each S_i:

- the partially ordered set S_i admits a unique minimal element V_i' and a unique maximal element V_i'';
- each $V \in \mathrm{JH}_G(V_0)$ such that $V_i' \leq V \leq V_i''$ is in S_i.

Then the following results are easily checked.

(i) For $i = 1, 2$ V_0 admits a unique subquotient V_i such that $\mathrm{JH}_G(V_i) = S_i$.
(ii) If $V_2' \nleq V_1''$ in $\mathrm{JH}_G(V_0)$ and $\mathrm{JH}_G(V_0) = S_1 \sqcup S_2$ as sets, then V_0 fits into a (possibly split) short exact sequence $0 \to V_1 \to V_0 \to V_2 \to 0$.
(iii) If $V_2' \nleq V_1''$ and $V_1' \nleq V_2''$ in $\mathrm{JH}_G(V_0)$, then V_0 admits a unique subquotient isomorphic to $V_1 \oplus V_2$.

We now sum up the main properties of the above coadmissible $D(G)$-modules.

Theorem 5.3.10

(i) The finite length coadmissible $D(G)$-modules X_k, Y_k, Z_k, D_k and $\widetilde{D}_k$ are multiplicity free with simple socle and cosocle and are uniquely determined (up to isomorphism) by their set of constituents endowed with the partial order of Sect. 1.4.
(ii) For $1 \leq k \leq n-1$, the coadmissible $D(G)$-module Z_k admits a unique increasing 3-stage filtration by (closed) $D(G)$-submodules with subrepresentation Y_k, middle graded piece $(V^{\mathrm{alg}}_{[1,n-k],\Delta})^\vee$ and quotient X_k.

(iii) *For $0 \leq k \leq n-1$, the coadmissible $D(G)$-module D_k admits a unique increasing 3-stage filtration by (closed) $D(G)$-submodules with subrepresentation X_k, middle graded piece $(V^{\mathrm{alg}}_{[1,n-k-1],\Delta})^\vee$ and quotient X_{k+1} (with $X_n \overset{def}{=} 0$ if $k = n-1$).*

(iv) *For $0 \leq k \leq n-2$ we have a short exact sequence $0 \to Z_k \to \widetilde{D}_k \to Z_{k+1} \to 0$ and a surjection $\widetilde{D}_k \twoheadrightarrow D_k$. More precisely, for $1 \leq k \leq n-2$, the coadmissible $D(G)$-module $\widetilde{D}_k$ admits a unique increasing 5-stage filtration by (closed) $D(G)$-submodules with subrepresentation Y_k, second graded piece $(V^{\mathrm{alg}}_{[1,n-k],\Delta})^\vee$, third graded piece $X_k \oplus Y_{k+1}$, fourth graded piece $(V^{\mathrm{alg}}_{[1,n-k-1],\Delta})^\vee$ and quotient X_{k+1}.*

Proof (i) follows directly from Theorem 5.3.6. (ii), (iii) and (iv) follow from Remark 5.3.9 and corresponding decompositions inside $\widetilde{\mathbf{J}}$ of the respective partially ordered sets of constituents. Let us prove (iv) and leave the other (easier) cases to the reader. For $1 \leq k \leq n-2$ we have

$$\begin{aligned}\mathrm{JH}_G(\widetilde{D}_k^\vee) = [C_{(n-k-1,1,1)}, C_{(n-k,1,n)}] &= [C_{(n-k-1,1,1)}, C_{(n-k-1,1,n)}] \sqcup [C_{(n-k,1,1)}, C_{(n-k,1,n)}] \\ &= \mathrm{JH}_G(Z_{k+1}^\vee) \sqcup \mathrm{JH}_G(Z_k^\vee),\end{aligned}$$

which by (ii) of Remark 5.3.9 and the fact $C_{(n-k,1,1)} \not\leq C_{(n-k-1,1,n)}$ in $[C_{(n-k-1,1,1)}, C_{(n-k,1,n)}]$ gives the first exact sequence in (iv). We also have

$$\begin{aligned}\mathrm{JH}_G(\widetilde{D}_k^\vee) &= [C_{(n-k-1,1,1)}, C_{(n-k,n-1,n-1)}] \sqcup [C_{(n-k-1,1,2)}, C_{(n-k,1,n)}] \\ &= \mathrm{JH}_G(D_k^\vee) \sqcup [C_{(n-k-1,1,2)}, C_{(n-k,1,n)}],\end{aligned}$$

which by (ii) of Remark 5.3.9 and the fact that $C_{(n-k-1,1,2)} \not\leq C_{(n-k,n-1,n-1)}$ in $[C_{(n-k-1,1,1)}, C_{(n-k,1,n)}]$ gives an injection $D_k^\vee \hookrightarrow \widetilde{D}_k^\vee$, and thus a surjection $\widetilde{D}_k \twoheadrightarrow D_k$. Finally, by (iii) of Remark 5.3.9 and the fact $C_{(n-k-1,1,2)} \not\leq C_{(n-k,n-1,n-1)}$ and $C_{(n-k,1,1)} \not\leq C_{(n-k-1,1,n)}$ (with $\mathrm{JH}_G(X_k^\vee) = [C_{(n-k,1,1)}, C_{(n-k,n-1,n-1)}]$ and $\mathrm{JH}_G(Y_{k+1}^\vee) = [C_{(n-k-1,1,2)}, C_{(n-k-1,1,n)}]$), we deduce that $X_k \oplus Y_{k+1}$ is a subquotient of $\widetilde{D}_k$. □

Theorem 5.3.11

(i) *The E-vector space $\mathrm{Ext}^1_{D(G)}((\mathrm{St}_n^{\mathrm{alg}})^\vee, X_{n-1})$ has dimension 1.*

(ii) *The E-vector spaces $\mathrm{Ext}^1_{D(G)}((\mathrm{St}_n^{\mathrm{alg}})^\vee, Z_{n-1})$ and $\mathrm{Ext}^1_{D(G)}(Z_{n-1}, (\mathrm{St}_n^{\mathrm{alg}})^\vee)$ have dimension n.*

(iii) *For $1 \leq k \leq n-2$ we have $\mathrm{Ext}^1_{D(G)}(D_k, (V^{\mathrm{alg}}_{[1,n-k],\Delta})^\vee) = 0$.*

(iv) *For $1 \leq k \leq n-1$, the E-vector space $\mathrm{Ext}^1_{D(G)}(X_k, (V^{\mathrm{alg}}_{[1,n-k],\Delta})^\vee)$ has dimension 1 and the E-vector space $\mathrm{Ext}^1_{D(G)}((V^{\mathrm{alg}}_{[1,n-k-1],\Delta})^\vee, \widetilde{X}_k)$ has dimension 2 where $\widetilde{X}_k$ is the unique non-split extension of X_k by $(V^{\mathrm{alg}}_{[1,n-k],\Delta})^\vee$.*

Proof We prove (i). By (ii) of Lemma 5.2.2 applied with $I = \emptyset$ and by (ii) of Lemma 5.2.5 we have $\dim_E \mathrm{Ext}^1_G(C_{(1,1,1)}, \mathrm{St}_n^{\mathrm{alg}}) = 1$ and $\mathrm{Ext}^i_G(V, \mathrm{St}_n^{\mathrm{alg}}) = 0$ for

$i = \{1, 2\}$ and $V = C_{(1,j_1,j_1)} \in \mathrm{JH}_G(X_{n-1}^\vee) = [C_{(1,1,1)}, C_{(1,n-1,n-1)}]$ with $2 \leq j_1 \leq n-1$ (note that $I_{j_1,j_1}^- = \emptyset$ if and only if $j_1 = 1$ by Lemma 2.3.1 and (2.76)). The statement then follows by dévissage.

We prove (ii). Recall the functors $\mathrm{Ext}^\bullet_{D(G),\chi^\vee}$ from the discussion above Lemma 5.3.8. Note first that, since $\mathrm{St}_n^{\mathrm{alg}}$ is not a constituent of $Z_{n-1}^\vee$, we have isomorphisms $\mathrm{Ext}^1_{D(G),\chi}((\mathrm{St}_n^{\mathrm{alg}})^\vee, Z_{n-1}) \xrightarrow{\sim} \mathrm{Ext}^1_{D(G)}((\mathrm{St}_n^{\mathrm{alg}})^\vee, Z_{n-1})$ and $\mathrm{Ext}^1_{D(G),\chi^\vee}(Z_{n-1}, (\mathrm{St}_n^{\mathrm{alg}})^\vee) \xrightarrow{\sim} \mathrm{Ext}^1_{D(G)}(Z_{n-1}, (\mathrm{St}_n^{\mathrm{alg}})^\vee)$. By [48, Theorem 2] and (5.131) we deduce for $j \in \{1, \ldots, n-1\}$

$$\dim_E \mathrm{Ext}^1_{D(G),\chi^\vee}((\mathrm{St}_n^{\mathrm{alg}})^\vee, (V_{\{j\}}^{\mathrm{alg}})^\vee) = 1,\ \mathrm{Ext}^2_{D(G),\chi^\vee}((\mathrm{St}_n^{\mathrm{alg}})^\vee, (V_{\{j\}}^{\mathrm{alg}})^\vee) = 0. \tag{5.138}$$

Moreover, by (ii) of Lemma 5.2.2 applied with $I = \emptyset$ (where we recall that $\mathrm{Ext}^1_{D(G),\chi^\vee} = \mathrm{Ext}^1_{D(G)}$ there) and by the analogue of (ii) of Lemma 5.2.5 with $\mathrm{Ext}^2_{D(G),\chi^\vee}$ instead of $\mathrm{Ext}^2_{D(G)}$ (the proof of which is the same), we deduce for $\ell \in \{1, 2\}$ and $V = C_{(1,j_1,j_2)} \in \mathrm{JH}_G(Z_{n-1}^\vee) \setminus \{C_{(1,1,1)}\} = [C_{(1,1,1)}, C_{(1,1,n)}] \setminus \{C_{(1,1,1)}\}$

$$\mathrm{Ext}^\ell_{D(G),\chi^\vee}((\mathrm{St}_n^{\mathrm{alg}})^\vee, V^\vee) = 0. \tag{5.139}$$

Then combining $\dim_E \mathrm{Ext}^1_{D(G),\chi^\vee}((\mathrm{St}_n^{\mathrm{alg}})^\vee, C_{(1,1,1)}^\vee) = 1$ in (i) with (5.138), (5.139) and a dévissage, we obtain the first statement in (ii). Similarly, by [48, Theorem 2] and (5.131) we deduce for $j \in \{1, \ldots, n-1\}$

$$\dim_E \mathrm{Ext}^1_{D(G),\chi^\vee}((V_{\{j\}}^{\mathrm{alg}})^\vee, (\mathrm{St}_n^{\mathrm{alg}})^\vee) = 1,\ \mathrm{Ext}^2_{D(G),\chi^\vee}((V_{\{j\}}^{\mathrm{alg}})^\vee, (\mathrm{St}_n^{\mathrm{alg}})^\vee) = 0. \tag{5.140}$$

Recall from Lemma 2.3.1 and (2.76) (see also the proof of Lemma 5.2.5) that $I_{j_1,j_2}^+ = \emptyset$ if and only if $(j_1, j_2) = (1, n)$. Then by (i) of Lemma 5.2.2 applied with $I = \emptyset$ (noting that $\mathrm{Ext}^1_{D(G),\chi^\vee} = \mathrm{Ext}^1_{D(G)}$ there) and by the analogue of (i) of Lemma 5.2.5 with $\mathrm{Ext}^2_{D(G),\chi^\vee}$ instead of $\mathrm{Ext}^2_{D(G)}$ (the proof of which is the same), we deduce for $\ell \in \{1, 2\}$ and $V = C_{(1,j_1,j_2)} \in \mathrm{JH}_G(Z_{n-1}^\vee) \setminus \{C_{(1,1,n)}\} = [C_{(1,1,1)}, C_{(1,1,n)}] \setminus \{C_{(1,1,n)}\}$

$$\mathrm{Ext}^\ell_{D(G),\chi^\vee}(V^\vee, (\mathrm{St}_n^{\mathrm{alg}})^\vee) = 0. \tag{5.141}$$

Then combining $\dim_E \mathrm{Ext}^1_{D(G),\chi^\vee}(C_{(1,1,n)}^\vee, (\mathrm{St}_n^{\mathrm{alg}})^\vee) = 1$ from (i) of Lemma 5.2.2 with (5.140), (5.141) and a dévissage, we obtain the second statement in (ii).

We prove (iii). By construction D_k admits a unique subquotient D with $\mathrm{JH}_G((D)^\vee) = [C_{(n-k-1,n-k-1,n-k-1)}, C_{(n-k,n-k,n-k)}]$ as partially ordered sets. More precisely $D^\vee$ is the Ext-square constructed in (ii) of Proposition 5.2.10 with socle $C_{(n-k-1,n-k-1,n-k-1)}$, cosocle $C_{(n-k,n-k,n-k)}$, and middle layer $C_{(n-k-1,n-k,n-k)} \oplus C_{(n-k,n-k-1,n-k-1)} \oplus V_{[1,n-k-1],\Delta}^{\mathrm{alg}}$. By (i) of Lemmas 5.2.2 and 5.1.3 together with Lemma 2.2.3 we know that, for $V \in \mathrm{JH}_G(D_k^\vee)$, we have $\mathrm{Ext}^1_{D(G)}(V^\vee, (V_{[1,n-k],\Delta}^{\mathrm{alg}})^\vee) \neq 0$ if and only if $V \in \{C_{n-k,n-k,n-k}, V_{[1,n-k-1],\Delta}^{\mathrm{alg}}\}$. So by dévissage it suffices to show

$$\mathrm{Ext}^1_G(V^{\mathrm{alg}}_{[1,n-k],\Delta}, D^\vee) = 0. \tag{5.142}$$

Assume (5.142) does not hold. Then there exists a non-split extension $0 \to D^\vee \to W \to V^{\mathrm{alg}}_{[1,n-k],\Delta} \to 0$ in $\mathrm{Rep}^{\mathrm{an}}_{\mathrm{adm}}(G)$. As $D^\vee$ has socle $V_1 \stackrel{\mathrm{def}}{=} C_{(n-k-1,n-k-1,n-k-1)}$, so does W. Let W^- be the unique length 2 subrepresentation of $D^\vee \subseteq W$ with socle V_1 and cosocle $V_0 \stackrel{\mathrm{def}}{=} C_{(n-k-1,n-k,n-k)}$. By (i) of Lemma 5.2.12 we have $\Theta_{s_k}(V) = 0$ for each constituent V of W/W^-, and thus the injection $W^- \hookrightarrow W$ induces an isomorphism $\Theta_{s_k}(W^-) \xrightarrow{\sim} \Theta_{s_k}(W)$. As the canonical adjunction map $W \to \Theta_{s_k}(W)$ restricts to an injection $V_1 \hookrightarrow \Theta_{s_k}(V_1)$ (as V_1 is irreducible and the adjunction map is non-zero), and as W has socle V_1, we obtain an injection $W \hookrightarrow \Theta_{s_k}(W) \cong \Theta_{s_k}(W^-)$. But from (i) of Lemma 5.2.13 we deduce $V^{\mathrm{alg}}_{[1,n-k],\Delta} \notin \mathrm{JH}_G(\Theta_{s_k}(W^-)) = \mathrm{JH}_G(\Theta_{s_k}(V_1)) \cup \mathrm{JH}_G(\Theta_{s_k}(V_0))$, which is a contradiction.

We prove (iv). Let $1 \le k \le n-1$ and recall that $\mathrm{JH}_G(X_k^\vee) = [C_{(n-k,1,1)}, C_{(n-k,n-1,n-1)}] = \{C_{(n-k,j,j)} \mid 1 \le j \le n-1\}$. We have $d([1,n-k-1],[1,n-k]) = 1$ (with $[1,n-k-1] = \emptyset$ when $k = n-1$), which by (5.131) and [48, Theorem 1] gives

$$\begin{aligned}&\mathrm{Ext}^2_{D(G),\chi^\vee}((V^{\mathrm{alg}}_{[1,n-k-1],\Delta})^\vee, (V^{\mathrm{alg}}_{[1,n-k],\Delta})^\vee) = 0 \text{ and}\\ &\qquad\dim_E \mathrm{Ext}^1_{D(G),\chi^\vee}((V^{\mathrm{alg}}_{[1,n-k-1],\Delta})^\vee, (V^{\mathrm{alg}}_{[1,n-k],\Delta})^\vee) = 1.\end{aligned} \tag{5.143}$$

By (i) of Lemma 5.2.2 (with (2.76)), we have

$$\mathrm{Ext}^1_{D(G),\chi^\vee}(C^\vee_{(n-k,j,j)}, (V^{\mathrm{alg}}_{[1,n-k],\Delta})^\vee) \neq 0 \text{ if and only if } j = n-k \tag{5.144}$$

and $\dim_E \mathrm{Ext}^1_{D(G),\chi^\vee}(C^\vee_{(n-k,n-k,n-k)}, (V^{\mathrm{alg}}_{[1,n-k],\Delta})^\vee) = 1$. By (ii) of Lemma 5.2.2, we have

$$\mathrm{Ext}^1_{D(G),\chi^\vee}((V^{\mathrm{alg}}_{[1,n-k-1],\Delta})^\vee, C^\vee_{(n-k,j,j)}) \neq 0 \text{ if and only if } j = n-k \tag{5.145}$$

and $\dim_E \mathrm{Ext}^1_{D(G),\chi^\vee}((V^{\mathrm{alg}}_{[1,n-k-1],\Delta})^\vee, C^\vee_{(n-k,n-k,n-k)}) = 1$. By the analogue of (i) of Lemma 5.2.4 (see also the second paragraph in the proof of Lemma 5.3.3) with $\mathrm{Ext}^2_{D(G),\chi^\vee}$ instead of $\mathrm{Ext}^2_{D(G)}$ (the proof of which is the same), we have for $j < n-k$

$$\mathrm{Ext}^2_{D(G),\chi^\vee}(C^\vee_{(n-k,j,j)}, (V^{\mathrm{alg}}_{[1,n-k],\Delta})^\vee) = 0. \tag{5.146}$$

By the analogue of (ii) of Lemma 5.2.4 (see also the third paragraph in the proof of Lemma 5.3.3) with $\mathrm{Ext}^2_{D(G),\chi^\vee}$ instead of $\mathrm{Ext}^2_{D(G)}$, we have for $j > n-k$

$$\mathrm{Ext}^2_{D(G),\chi^\vee}((V^{\mathrm{alg}}_{[1,n-k-1],\Delta})^\vee, C^\vee_{(n-k,j,j)}) = 0. \tag{5.147}$$

Let V_0 (resp. V_1) be the unique quotient of $X_k^\vee$ with socle $C_{(n-k,n-k,n-k)}$ (resp. $C_{(n-k,n-k+1,n-k+1)}$) with $V_1 = 0$ when $k = 1$. By a dévissage using (5.144) and (5.146), we deduce $\mathrm{Ext}^1_{D(G),\chi^\vee}(V_1^\vee, (V^{\mathrm{alg}}_{[1,n-k],\Delta})^\vee) = 0$ and $\mathrm{Ext}^\ell_{D(G),\chi^\vee}(X_k/V_0^\vee, (V^{\mathrm{alg}}_{[1,n-k],\Delta})^\vee) = 0$ for $\ell = 1, 2$. Hence, the surjection $X_k \twoheadrightarrow X_k/V_1^\vee$ and the injection $C^\vee_{(n-k,n-k,n-k)} \hookrightarrow X_k/V_1^\vee$ (with cokernel $X_k/V_0^\vee$) induce isomorphisms between 1-dimensional E-vector spaces

$$\mathrm{Ext}^1_{D(G),\chi^\vee}(X_k, (V^{\mathrm{alg}}_{[1,n-k],\Delta})^\vee) \xleftarrow{\sim} \mathrm{Ext}^1_{D(G),\chi^\vee}(X_k/V_1^\vee, (V^{\mathrm{alg}}_{[1,n-k],\Delta})^\vee)$$
$$\xrightarrow{\sim} \mathrm{Ext}^1_{D(G),\chi^\vee}(C^\vee_{(n-k,n-k,n-k)}, (V^{\mathrm{alg}}_{[1,n-k],\Delta})^\vee).$$

In particular there exists a unique (up to isomorphism) $D(G)$-module that fits into a non-split extension $0 \to (V^{\mathrm{alg}}_{[1,n-k],\Delta})^\vee \to \widetilde{X}_k \to X_k \to 0$. A symmetric argument gives

$$\dim_E \mathrm{Ext}^1_{D(G),\chi^\vee}((V^{\mathrm{alg}}_{[1,n-k-1],\Delta})^\vee, X_k/V_1^\vee) = 1.$$

By (5.143) we have a short exact sequence

$$0 \to \mathrm{Ext}^1_{D(G),\chi^\vee}((V^{\mathrm{alg}}_{[1,n-k-1],\Delta})^\vee, (V^{\mathrm{alg}}_{[1,n-k],\Delta})^\vee) \to \mathrm{Ext}^1_{D(G),\chi^\vee}((V^{\mathrm{alg}}_{[1,n-k-1],\Delta})^\vee, \widetilde{X}_k/V_1^\vee)$$
$$\to \mathrm{Ext}^1_{D(G),\chi^\vee}((V^{\mathrm{alg}}_{[1,n-k-1],\Delta})^\vee, X_k/V_1^\vee) \to 0$$

where the left hand side and the right hand side both have dimension 1. By another dévissage using (5.145) and (5.147) we deduce $\mathrm{Ext}^1_{D(G),\chi^\vee}((V^{\mathrm{alg}}_{[1,n-k-1],\Delta})^\vee, X_k/V_0^\vee) = 0$ and $\mathrm{Ext}^\ell_{D(G),\chi^\vee}((V^{\mathrm{alg}}_{[1,n-k-1],\Delta})^\vee, V_1^\vee) = 0$ for $\ell = 1, 2$, which gives an isomorphism

$$\mathrm{Ext}^1_{D(G),\chi^\vee}((V^{\mathrm{alg}}_{[1,n-k-1],\Delta})^\vee, \widetilde{X}_k) \xrightarrow{\sim} \mathrm{Ext}^1_{D(G),\chi^\vee}((V^{\mathrm{alg}}_{[1,n-k-1],\Delta})^\vee, \widetilde{X}_k/V_1^\vee).$$

In particular, we conclude that $\dim_E \mathrm{Ext}^1_{D(G),\chi^\vee}((V^{\mathrm{alg}}_{[1,n-k-1],\Delta})^\vee, \widetilde{X}_k) = 2$. Finally, since all irreducible constituents in the various $\mathrm{Ext}^1_{D(G),\chi^\vee}$ are actually distinct, we can replace $\mathrm{Ext}^1_{D(G),\chi^\vee}$ by $\mathrm{Ext}^1_{D(G)}$. □

Remark 5.3.12 Note that, when $n = 2$, the first statement in (ii) of Theorem 5.3.11 proves [11, Conjecture 3.2.1].

For $1 \leq j \leq n-1$, define $Z_{n-1,\geq j}$ (resp. $Z_{n-1,\leq j}$) as the $D(G)$-module which is the unique (closed) subspace (resp. (topological) quotient) of Z_{n-1} such that $\mathrm{JH}_G(Z^\vee_{n-1,\geq j}) = [V^{\mathrm{alg}}_{\{j\},\Delta}, C_{(1,1,n)}]$ (resp. $\mathrm{JH}_G(Z^\vee_{n-1,\leq j}) = [C_{(1,1,1)}, V^{\mathrm{alg}}_{\{j\},\Delta}]$). The same argument as in the proof of (ii) of Theorem 5.3.11 shows that the injections $Z_{n-1,\geq n-1} \hookrightarrow \cdots \hookrightarrow Z_{n-1,\geq 1} \hookrightarrow Z_{n-1}$ induce a decreasing filtration of subspaces of $\mathrm{Ext}^1_{D(G)}((\mathrm{St}^{\mathrm{alg}}_n)^\vee, Z_{n-1})$

$$\left\{\mathrm{Ext}^1_{D(G)}((\mathrm{St}^{\mathrm{alg}}_n)^\vee, Z_{n-1,\geq j})\right\}_{1\leq j\leq n-1} \tag{5.148}$$

where $\dim_E \mathrm{Ext}^1_{D(G)}((\mathrm{St}^{\mathrm{alg}}_n)^\vee, Z_{n-1,\geq j}) = n - j$. Similarly, the surjections $Z_{n-1} \twoheadrightarrow Z_{n-1,\leq n-1} \twoheadrightarrow \cdots \twoheadrightarrow Z_{n-1,\leq 1}$ induce an increasing filtration of subspaces of

$\mathrm{Ext}^1_{D(G)}(Z_{n-1}, (\mathrm{St}_n^{\mathrm{alg}})^\vee)$

$$\left\{\mathrm{Ext}^1_{D(G)}(Z_{n-1,\le j}, (\mathrm{St}_n^{\mathrm{alg}})^\vee)\right\}_{1\le j\le n-1}$$

where $\dim_E \mathrm{Ext}^1_{D(G)}(Z_{n-1,\le j}, (\mathrm{St}_n^{\mathrm{alg}})^\vee) = j$.

By (iv) of Theorem 5.3.10, for $0 \le k \le n-2$, $\widetilde{D}_k$ has the form

$$\begin{array}{ccccccccc} & & & & Y_{k+1} & & & & \\ & & & \diagup & & \diagdown & & & \\ \widetilde{D}_k \cong Y_k & \text{———} & (V^{\mathrm{alg}}_{[1,n-k],\Delta})^\vee & & & & (V^{\mathrm{alg}}_{[1,n-k-1],\Delta})^\vee & \text{———} & X_{k+1} \\ & & & \diagdown & & \diagup & & & \\ & & & & X_k & & & & \end{array} \tag{5.149}$$

where we write subrepresentations on the left, quotients on the right, where lines represent non-split extensions, and where $Y_k = (V^{\mathrm{alg}}_{[1,n-k],\Delta})^\vee = X_k = 0$ when $k = 0$. Let D_{n-1} be the unique (up to isomorphism) non-split extension in (i) of Theorem 5.3.11:

$$0 \longrightarrow X_{n-1} \longrightarrow D_{n-1} \longrightarrow (\mathrm{St}_n^{\mathrm{alg}})^\vee \longrightarrow 0.$$

Define $\widetilde{D}_{n-1}$ as *any* non-split extension *with cosocle* $(\mathrm{St}_n^{\mathrm{alg}})^\vee$:

$$0 \longrightarrow Z_{n-1} \longrightarrow \widetilde{D}_{n-1} \longrightarrow (\mathrm{St}_n^{\mathrm{alg}})^\vee \longrightarrow 0.$$

Note that, by (ii) of Theorem 5.3.11 and the discussion around (5.148), the isomorphism class of $\widetilde{D}_{n-1}$ depends on $n-1$ "parameters". More precisely, the set of isomorphism classes of coadmissible $D(G)$-modules $\widetilde{D}_{n-1}$ is in natural bijection with the set

$$\left(\mathrm{Ext}^1_{D(G)}((\mathrm{St}_n^{\mathrm{alg}})^\vee, Z_{n-1}) \setminus \mathrm{Ext}^1_{D(G)}((\mathrm{St}_n^{\mathrm{alg}})^\vee, Z_{n-1,\ge 1})\right)/E^\times$$

which is in non-canonical bijection with $\mathbb{A}^{n-1}(E)$. By (ii) of Theorem 5.3.10 (for $k = n-1$) $\widetilde{D}_{n-1}$ has the form

$$\widetilde{D}_{n-1} \cong Y_{n-1} \text{ ——— } (V^{\mathrm{alg}}_{\{1\},\Delta})^\vee \text{ ——— } X_{n-1} \text{ ——— } (\mathrm{St}_n^{\mathrm{alg}})^\vee. \tag{5.150}$$

It follows from (i), (iii) of Theorem 5.3.10 and the above definition of D_{n-1} that, for $0 \le k \le n-2$, there exists a unique (up to scalar) non-zero map $d^k_{\mathbf{D}} : D_k \to D_{k+1}$ whose image is X_{k+1}. In particular we can consider the complex of finite length coadmissible $D(G)$-modules (with D_k in degree k)

$$\mathbf{D}^\bullet \stackrel{\mathrm{def}}{=} [D_0 \xrightarrow{d^0_{\mathbf{D}}} \cdots \xrightarrow{d^{k-1}_{\mathbf{D}}} D_k \xrightarrow{d^k_{\mathbf{D}}} D_{k+1} \xrightarrow{d^{k+1}_{\mathbf{D}}} \cdots \xrightarrow{d^{n-2}_{\mathbf{D}}} D_{n-1}]. \tag{5.151}$$

For $0 \leq k \leq n-1$ we define $H^k(\mathbf{D}^\bullet) \stackrel{\text{def}}{=} \ker(d_{\mathbf{D}}^k)/\mathrm{im}(d_{\mathbf{D}}^{k-1})$ (with the convention $d_{\mathbf{D}}^{n-1} = 0$). By (iii) of Theorem 5.3.10 and the above definition of D_{n-1} we have for $0 \leq k \leq n-1$

$$H^k(\mathbf{D}^\bullet) \cong (V^{\mathrm{alg}}_{[1,n-k-1],\Delta})^\vee.$$

Similarly it follows from (i), (iv) of Theorem 5.3.10 and the above definition of $\widetilde{D}_{n-1}$ that, for $0 \leq k \leq n-2$, there exists a unique (up to scalar) non-zero map $d^k_{\widetilde{\mathbf{D}}} : \widetilde{D}_k \to \widetilde{D}_{k+1}$ whose image is Z_{k+1}, and we obtain a complex of finite length coadmissible $D(G)$-modules (with $\widetilde{D}_k$ in degree k)

$$\widetilde{\mathbf{D}}^\bullet \stackrel{\text{def}}{=} [\widetilde{D}_0 \xrightarrow{d^0_{\widetilde{\mathbf{D}}}} \cdots \xrightarrow{d^{k-1}_{\widetilde{\mathbf{D}}}} \widetilde{D}_k \xrightarrow{d^k_{\widetilde{\mathbf{D}}}} \widetilde{D}_{k+1} \xrightarrow{d^{k+1}_{\widetilde{\mathbf{D}}}} \cdots \xrightarrow{d^{n-2}_{\widetilde{\mathbf{D}}}} \widetilde{D}_{n-1}]. \tag{5.152}$$

Recall that $\widetilde{\mathbf{D}}^\bullet$ is not unique because $\widetilde{D}_{n-1}$ depends non-canonically on some element in $\mathbb{A}^{n-1}(E)$. For $0 \leq k \leq n-1$ we define $H^k(\widetilde{\mathbf{D}}^\bullet) \stackrel{\text{def}}{=} \ker(d^k_{\widetilde{\mathbf{D}}})/\mathrm{im}(d^{k-1}_{\widetilde{\mathbf{D}}})$ (with the convention $d^{n-1}_{\widetilde{\mathbf{D}}} = 0$). By (iv) of Theorem 5.3.10 and the above definition of $\widetilde{D}_{n-1}$ we have $H^k(\widetilde{\mathbf{D}}^\bullet) = 0$ for $0 \leq k \leq n-2$ and $H^{n-1}(\widetilde{\mathbf{D}}^\bullet) \cong (\mathrm{St}_n^{\mathrm{alg}})^\vee$. In particular the canonical morphism of complexes $\widetilde{\mathbf{D}}^\bullet \twoheadrightarrow H^{n-1}(\widetilde{\mathbf{D}}^\bullet)[-(n-1)] \cong (\mathrm{St}_n^{\mathrm{alg}})^\vee[-(n-1)]$ is a quasi-isomorphism.

For $0 \leq k \leq n-2$, by (iii), (iv) of Theorem 5.3.10 we have a surjection $\widetilde{D}_k \twoheadrightarrow D_k$ which is unique up to scalar (as $\widetilde{D}_k$ is multiplicity free with simple cosocle by (i) of Theorem 5.3.10). It follows from (5.150) and the definition of D_{n-1} that there is a unique (up to scalar) surjection $\widetilde{D}_{n-1} \twoheadrightarrow D_{n-1}$. Consequently, we see from the definition of the complexes (5.151) and (5.152) that there is a natural morphism of complexes of $D(G)$-modules

$$\widetilde{\mathbf{D}}^\bullet \longrightarrow \mathbf{D}^\bullet$$

which is an isomorphism on H^{n-1}. We thus have proven the following theorem.

Theorem 5.3.13 *The canonical morphism of complexes*

$$\mathbf{D}^\bullet \twoheadrightarrow H^{n-1}(\mathbf{D}^\bullet)[-(n-1)] \cong (\mathrm{St}_n^{\mathrm{alg}})^\vee[-(n-1)]$$

admits an explicit section in the derived category of finite length coadmissible $D(G)$-modules with Orlik-Strauch constituents (Theorem 4.3.7) given by

$$(\mathrm{St}_n^{\mathrm{alg}})^\vee[-(n-1)] \longleftarrow \widetilde{\mathbf{D}}^\bullet \longrightarrow \mathbf{D}^\bullet.$$

Recall that $\widetilde{\mathbf{D}}^\bullet$ (and hence the corresponding section) depends on a parameter in $\mathbb{A}^{n-1}(E)$.

At this point, one can ask the following question. For $0 \leq \ell \leq n-1$, define the usual truncated subcomplex $\tau_{\leq \ell}\mathbf{D}^\bullet$ of $\mathbf{D}^\bullet$ (with $\tau_{\leq n-1}\mathbf{D}^\bullet = \mathbf{D}^\bullet$)

$$\tau_{\leq \ell}\mathbf{D}^\bullet \stackrel{\text{def}}{=} [D_0 \longrightarrow \cdots \longrightarrow \cdots \longrightarrow D_{\ell-1} \longrightarrow \ker(d^\ell_{\mathbf{D}})].$$

Then we again have a canonical morphism of complexes

$$\tau_{\leq \ell}\mathbf{D}^\bullet \twoheadrightarrow H^\ell(\mathbf{D}^\bullet)[-\ell] \cong (V^{\mathrm{alg}}_{[1,n-\ell-1],\Delta})^\vee[-\ell]. \tag{5.153}$$

In view of Theorem 5.3.13, it is natural to ask if, for $0 \leq \ell \leq n-2$ there also exists a section to this morphism in the derived category of finite length coadmissible $D(G)$-modules with Orlik-Strauch constituents. This is obvious when $\ell = 0$ (since $\tau_{\leq 0}\mathbf{D}^\bullet \cong H^0(\mathbf{D}^\bullet)[0]$), and not too complicated when $\ell = 1$:

Proposition 5.3.14 *The canonical morphism of complexes*

$$\tau_{\leq 1}\mathbf{D}^\bullet \twoheadrightarrow H^1(\mathbf{D}^\bullet)[-1] \cong (V^{\mathrm{alg}}_{[1,n-2],\Delta})^\vee[-1]$$

admits an explicit section in the derived category of finite length coadmissible $D(G)$-modules with Orlik-Strauch constituents given by

$$(V^{\mathrm{alg}}_{[1,n-2],\Delta})^\vee[-1] \longleftarrow [D_0 \to D_0 \text{ --- } (V^{\mathrm{alg}}_{[1,n-2],\Delta})^\vee] \longrightarrow \tau_{\leq 1}\mathbf{D}^\bullet$$

where $D_0 \text{ --- } (V^{\mathrm{alg}}_{[1,n-2],\Delta})^\vee \cong (V^{\mathrm{alg}}_{[1,n-1],\Delta})^\vee \text{ --- } X_1 \text{ --- } (V^{\mathrm{alg}}_{[1,n-2],\Delta})^\vee$ *is one of the representations in (iv) of Theorem 5.3.11 (applied with $k = 1$) depending on a parameter in $\mathbb{A}^1(E)$, and where the morphisms of complexes are the obvious ones.*

One can prove that the parameter in $\mathbb{A}^1(E)$ corresponds to a choice of a p-adic logarithm $\log : K^\times \to E$. Note that, when $n = 2$, $D_0 \text{ --- } (V^{\mathrm{alg}}_{[1,n-2],\Delta})^\vee \cong \widetilde{D}_1/Y_1$. Theorem 5.3.13 and Proposition 5.3.14 have the following consequence.

Corollary 5.3.15 *For $n = 3$ there exists an explicit splitting in the derived category of finite length coadmissible $D(G)$-modules with Orlik-Strauch constituents $\mathbf{D}^\bullet \cong \oplus_{\ell=0}^2 H^\ell(\mathbf{D}^\bullet)[-\ell]$.*

When $2 \leq \ell \leq n-2$, finding a "nice" explicit section to (5.153) becomes more complicated. For instance when $n = 4$ and $\ell = 2$, we can use a variant of the complex $\widetilde{\mathbf{D}}^\bullet$ of (5.152) for $\mathrm{GL}_3(K)$ combined with parabolic induction to $\mathrm{GL}_4(K)$ to build a complex of finite length coadmissible $D(G)$-modules with Orlik-Strauch constituents which is exact in degrees 0, 1 and maps to $\tau_{\leq 2}\mathbf{D}^\bullet$ with an isomorphism on H^2, hence which gives an explicit section. But this complex is not nice (contrary to $\widetilde{\mathbf{D}}^\bullet$). It just gives us enough confidence to state the following conjecture.

Conjecture 5.3.16 *For $2 \leq \ell \leq n-2$ the morphism of complexes (5.153) admits a section in the derived category of finite length coadmissible $D(G)$-modules with Orlik-Strauch constituents.*

Recall we know that a section exists in the derived category of *all* (abstract) $D(G)$-modules by [76, Theorem 6.1] and [18, Corollary A.1.3].

5.4 Application to De Rham Complex of the Drinfeld Space

We show that, for $E = K$ and $\mu_0 = (0, \cdots, 0)$, the complex of coadmissible $D(G)$-modules $\mathbf{D}^\bullet$ in (5.151) is isomorphic to the global sections of the de Rham complex of the rigid analytic Drinfeld space over K of dimension $n-1$, and for arbitrary μ_0 to the complex of holomorphic discrete series of [64].

Throughout this section, we use the notation $I \stackrel{\text{def}}{=} \widehat{1}$ and $J \stackrel{\text{def}}{=} \widehat{n-1}$. Recall from Sect. 1.4 that w_0 (resp. w_I) is the longest element of $W(G)$ (resp. of $W(L_I)$). We check that $w_0 = w_{n-1,1}w_I = w_I w_{1,n-1}$. We also keep the notation of Sect. 5.3 (in particular we have fixed a weight μ_0 in Λ^{dom}).

We start with two more results on coadmissible $D(G)$-modules which will be used in Theorem 5.4.16 below. The first statement shows that the $D(G)$-modules D_k and $\widetilde{D}_k$ of (5.136), (5.137) have a nice behaviour with respect to wall-crossing functors.

Theorem 5.4.1 *Let $1 \leq k \leq n-1$ and $\mu \in \Lambda$ such that $\langle \mu + \rho, \alpha^\vee \rangle \geq 0$ for $\alpha \in \Phi^+$ and the stabilizer of μ in $W(G)$ for the dot action is $\{1, s_k\}$.*

(i) The morphisms $D_{k-1} \twoheadrightarrow X_k \hookrightarrow D_k$ from (iii) of Theorem 5.3.10 induce isomorphisms $\Theta_\mu(D_{k-1}) \xrightarrow{\sim} \Theta_\mu(X_k) \xrightarrow{\sim} \Theta_\mu(D_k)$.
(ii) The morphisms $\widetilde{D}_{k-1} \twoheadrightarrow Z_k \hookrightarrow \widetilde{D}_k$ from (iv) of Theorem 5.3.10 induce isomorphisms $\Theta_\mu(\widetilde{D}_{k-1}) \xrightarrow{\sim} \Theta_\mu(Z_k) \xrightarrow{\sim} \Theta_\mu(\widetilde{D}_k)$.
(iii) We have non-split short exact sequences $0 \to D_k \to \Theta_\mu(D_k) \to D_{k-1} \to 0$ where $D_k \to \Theta_\mu(D_k)$ is the canonical adjunction map.
(iv) If $1 \leq k \leq n-2$ we have non-split short exact sequences $0 \to \widetilde{D}_k \to \Theta_\mu(\widetilde{D}_k) \to \widetilde{D}_{k-1} \to 0$ where $\widetilde{D}_k \to \Theta_\mu(\widetilde{D}_k)$ is the canonical adjunction map.
(v) If $k = n-1$ the composition $\Theta_\mu(\widetilde{D}_{n-1}) \cong \Theta_\mu(\widetilde{D}_{n-2}) \to \widetilde{D}_{n-2}$ (where the first isomorphism follows from (ii) and the second map is the adjunction map) is surjective and its kernel is the "universal" extension

$$0 \longrightarrow Z_{n-1} \longrightarrow * \longrightarrow \left((\mathrm{St}_n^{\mathrm{alg}})^\vee\right)^{\oplus n} \longrightarrow 0 \tag{5.154}$$

deduced from (ii) of Theorem 5.3.11.

Proof We prove (i) and (ii). One checks that any $V \in \mathrm{JH}_G(\ker(\widetilde{D}_k^\vee \twoheadrightarrow Z_k^\vee))$ (resp. any $V \in \mathrm{JH}_G(\widetilde{D}_{k-1}^\vee / Z_k^\vee))$ either is locally algebraic or is in $\widetilde{\mathbf{J}}_{n-k-1}$ if $k < n-1$ (resp. in $\widetilde{\mathbf{J}}_{n-k+1}$ if $1 < k$) (see (5.120)). Hence we deduce from (i) of Lemma 5.2.12 that $\Theta_\mu(V) = 0$. The exactness of $\Theta_\mu(-)$ then gives the isomorphisms in (ii). The proof of (i) is completely analogous.

We prove (iii) and (iv). When $1 \leq k \leq n-2$, (iii) (resp. (iv)) follows from the definition of D_k (resp. of $\widetilde{D}_k$) in (5.137) and (ii) of Proposition 5.3.5 applied with $V_0^- = D_k^\vee$ (i.e. $j_0 = n-k$) and $(j_1, j_2) = (n-1, n-1)$ (resp. and $(j_1, j_2) = (1, n)$), noting that in the proof of *loc. cit.* we have $\Theta_\mu(V_0) \cong \Theta_\mu(V_0^-)$. When $k = n-1$, (iii) follows from the $(j_1, j_2) = (n-1, n-1)$ case of (ii) of Lemma 5.3.8 (where $V = X_{n-1}^\vee$ and $V^+ = D_{n-2}^\vee$), the fact that D_{n-1} is the unique $D(G)$-module that fits into a non-split extension $0 \to X_{n-1} \to D_{n-1} \to (\mathrm{St}_n^{\mathrm{alg}})^\vee \to 0$ ((i) of Theorem 5.3.11), and $\Theta_\mu(X_{n-1}) \xrightarrow{\sim} \Theta_\mu(D_{n-1})$ since $\Theta_\mu((\mathrm{St}_n^{\mathrm{alg}})^\vee) = 0$ by (i) of Lemma 5.2.12. The non-splitness easily follows from (ii) of Lemma 5.2.12.

Finally (v) follows from the $(j_1, j_2) = (1, n)$ case of (ii) of Lemma 5.3.8 with $V_0 = Z_{n-1}^\vee$ and $V_0^+ = \widetilde{D}_{n-2}^\vee$ in *loc. cit.* (and noting that $\Theta_\mu(V_0) = \Theta_\mu(V_0^+)$). □

Remark 5.4.2 With the notation of Theorem 5.4.1, for $1 \leq k \leq n-1$ the composition $\Theta_\mu(D_k) \twoheadrightarrow D_{k-1} \to D_k$ where the last map is the differential map $d_{\mathbf{D}}^{k-1}$ (see (5.151)) is nothing else than the (non-zero) canonical adjunction map $\Theta_\mu(D_k) \to D_k$. Indeed, by functoriality of the adjunction maps and since $\Theta_\mu(X_k) \xrightarrow{\sim} \Theta_\mu(D_k)$ (see (i) of Theorem 5.4.1), this adjunction map factors as $\Theta_\mu(D_k) \to X_k \hookrightarrow D_k$. One easily checks from (5.135) and (5.137) that the (irreducible) cosocle of D_k does not appear in X_k. It then follows from (iii) of Theorem 5.4.1 that the adjunction map $\Theta_\mu(D_k) \to D_k$ factors through a non-zero map $D_{k-1} \to D_k$, which must be $d_{\mathbf{D}}^{k-1}$ (up to a non-zero scalar) by unicity of $d_{\mathbf{D}}^{k-1}$ (see the references above (5.151)). A similar proof replacing X_k, X_{k-1} by Z_k, Z_{k-1} and using (ii), (iv) and (v) of Theorem 5.4.1 gives that, for $1 \leq k \leq n-1$, the composition $\Theta_\mu(\widetilde{D}_k) \twoheadrightarrow \widetilde{D}_{k-1} \to \widetilde{D}_k$ (where the last map is $d_{\widetilde{\mathbf{D}}}^{k-1}$, see (5.152)) is the (non-zero) adjunction map $\Theta_\mu(\widetilde{D}_k) \to \widetilde{D}_k$ (for $k = n-1$ one has to use (5.154)).

Using Theorem 5.4.1 we can prove the following unicity statements which strengthen Theorem 5.3.10.

Corollary 5.4.3 *We have the following unicity results.*

(i) For $1 \leq k \leq n-1$, Z_k is the unique coadmissible $D(G)$-module of the form

$$Y_k \longrightarrow (V_{[1,n-k],\Delta}^{\mathrm{alg}})^\vee \longrightarrow X_k .$$

(ii) For $0 \leq k \leq n-1$, D_k is the unique coadmissible $D(G)$-module of the form

$$X_k \longrightarrow (V_{[1,n-k-1],\Delta}^{\mathrm{alg}})^\vee \longrightarrow X_{k+1} .$$

(iii) For $0 \leq k \leq n-2$, $\widetilde{D}_k$ is the unique coadmissible $D(G)$-module of the form

$$Z_k \longrightarrow Z_{k+1} .$$

Proof We prove (i). It follows from (i) of Lemma 5.2.2 that for $j \in \{1, \ldots, n-1\}$ we have $\mathrm{Ext}^1_{D(G)}(C_{n-k,j,j}^\vee, (V_{[1,n-k],\Delta}^{\mathrm{alg}})^\vee) \neq 0$ if and only if $j = n-k$, in which case it is one dimensional. It follows from (i) of Lemma 5.2.4 that $\mathrm{Ext}^2_{D(G)}(C_{n-k,j,j}^\vee, (V_{[1,n-k],\Delta}^{\mathrm{alg}})^\vee) = 0$ for $1 \leq j < n-k$. By dévissage and (5.135) this implies

$$\dim_E \mathrm{Ext}^1_{D(G)}(X_k, (V_{[1,n-k],\Delta}^{\mathrm{alg}})^\vee) = 1, \tag{5.155}$$

in particular it is enough to prove $\mathrm{Ext}^1_{D(G)}(X_k, Y_k) = 0$.

Assume the contrary and let M be a non-split extension of X_k by Y_k. Using again (i) of Lemma 5.2.2, for $j \in \{1, \dots, n-1\}$ and $i \in \{1, \dots, k-1\}$ we have

$$\mathrm{Ext}^1_{D(G)}(C^\vee_{n-k,j,j}, (V^{\mathrm{alg}}_{[1+i,n-k+i],\Delta})^\vee) = 0,$$

and hence $\mathrm{Ext}^1_{D(G)}(X_k, (V^{\mathrm{alg}}_{[1+i,n-k+i],\Delta})^\vee) = 0$ for $i \in \{1, \dots, k-1\}$ (see (5.135)). It follows that there exist $j \in \{1, \dots, n-1\}$ and $(n-k, j_1, j_2) \in \mathbf{J}$ with $j_1 < j_2$ such that M has a length 2 subquotient with socle $C^\vee_{n-k,j_1,j_2}$ in Y_k and cosocle $C^\vee_{n-k,j,j}$ in X_k (see (5.135) and the lines that follow). Moreover by Lemma 5.2.1 we must have $|j_2 - j| + |j_2 - j_1| = 1$, i.e. $j_2 = j$ and $j_1 = j_2 - 1 = j - 1$ (which implies $j \geq 2$). In other terms $C^\vee_{n-k,j-1,j}$ is the only constituent of Y_k that can lie "just below" the constituent $C^\vee_{n-k,j,j}$ of X_k.

Using this result we now prove that, for *any* $j' \in \{1, \dots, n-1\}$, M must contain as a subquotient the unique non-split extension of $C^\vee_{n-k,j',j'}$ by $C^\vee_{n-k,j'-1,j'}$ (using again Lemma 5.2.1). We have just seen this holds for $j' = j$. Assume this fails for some $j' > j$ and take the minimal such j'. Then, by definition of j', M contains as a subquotient the unique non-split extension of $C \overset{\mathrm{def}}{=} C^\vee_{n-k,j'-1,j'-1}$ by $B \overset{\mathrm{def}}{=} C^\vee_{n-k,j'-2,j'-1}$. Note that, from (5.135) and the definition of the partial order on $\mathbf{J}$, Y_k and hence M also contain as a subquotient the unique non-split extension of B by $A \overset{\mathrm{def}}{=} C^\vee_{n-k,j'-1,j'}$. Moreover there cannot exist a fourth constituent B' of M distinct from B such that $A < B' < C$ for the partial order on $\mathrm{JH}_{D(G)}(M)$. Indeed, if such a B' exists, one can take it such that $\mathrm{Ext}^1_{D(G)}(B', A) \neq 0$. If $B' \in \mathrm{JH}_{D(G)}(Y_k)$, then the partial order on Y_k and (5.135) force $B' = B$, a contradiction. Hence $B' \in \mathrm{JH}_{D(G)}(X_k)$, but then arguing as at the end of the previous paragraph we must have $B' = C^\vee_{n-k,j',j'}$, contradicting the hypothesis on j'. It follows that M contains as a subquotient a uniserial $D(G)$-module of the form $A \text{ --- } B \text{ --- } C$, contradicting the minimality of the Ext-square $V_{(n-k,j'-1,j'-1),(n-k,j'-1,j')}$ in Proposition 5.2.18. A symmetric argument when $j' < j$ also yields a contradiction.

Finally, applying the previous result with $j' = n-k$ and $j' = n-k+1$ (and using the structures of X_k and Y_k from (5.135)), we deduce that M contains as a subquotient the dual of the minimal Ext-square $V_{(n-k,n-k,n-k),(n-k,n-k,n-k+1)}$ in (ii) of Proposition 5.2.18, in particular M contains $(V^{\mathrm{alg}}_{[1,n-k],\Delta})^\vee$ which contradicts the lines below (5.135). Hence $\mathrm{Ext}^1_{D(G)}(X_k, Y_k) = 0$, which proves (i).

We prove (ii). The case $k = 0$ follows from (5.155) for $k = 1$, while the case $k = n-1$ is (i) of Theorem 5.3.11. We assume $k \in \{1, \dots, n-2\}$ and let M be any coadmissible $D(G)$-module as in (ii). By the same proof as for (i) of Theorem 5.4.1 we have $\Theta_\mu(X_k) \xrightarrow{\sim} \Theta_\mu(M)$ (with μ as in *loc. cit.*), and thus by adjunction (for M) a non-zero map $M \to \Theta_\mu(X_k)$. Since $C^\vee_{(n-k-1,1,1)}$ is the (irreducible) cosocle of X_{k+1}, it also appears in the cosocle of M. Assume first that $C^\vee_{(n-k-1,1,1)}$ maps to 0 in $\Theta_\mu(X_k)$. Then from the form of M it follows that a non-zero strict quotient of X_k embeds into $\Theta_\mu(X_k)$, and thus (from (5.135)) there exists $1 \leq j \leq n-2$ such that $C^\vee_{(n-k,j,j)}$ appears in the socle of $\Theta_\mu(X_k)$. By (i) and (iii) of Theorem 5.4.1, this implies that $C^\vee_{(n-k,j,j)}$ embeds into D_k or into D_{k-1}, which is impossible by (5.137). Hence $C^\vee_{(n-k-1,1,1)}$ still occurs in the (cosocle of the) image of M in $\Theta_\mu(X_k)$. By (i), (iii) of Theorem 5.4.1 and (5.137), $C^\vee_{(n-k-1,1,1)}$ has multiplicity 1 in $\Theta_\mu(X_k)$. Hence the image of M contains the unique (closed) $D(G)$-submodule of $\Theta_\mu(X_k)$ of cosocle $C^\vee_{(n-k-1,1,1)}$. But this submodule is D_k using *loc. cit.* again. Since M and D_k are multiplicity free with the same irreducible constituents, if follows that $M \xrightarrow{\sim} D_k$.

The proof of (iii) is analogous to (ii) replacing X_k, X_{k+1} by Z_k, Z_{k+1} and (i), (iii) of Theorem 5.4.1 by (ii), (iv) of Theorem 5.4.1. □

We now prove another unicity theorem which will play a key role in the comparison of the complex $\mathbf{D}^\bullet$ when $\mu_0 = (0, \cdots, 0)$ with the (global sections of) the de Rham complex of the Drinfeld space (Theorem 5.4.16).

Theorem 5.4.4 *Let C be a finite length coadmissible $D(G)$-module equipped with a decreasing filtration*

$$C = \mathrm{Fil}^0(C) \supseteq \mathrm{Fil}^1(C) \supseteq \cdots \supseteq \mathrm{Fil}^{n-1}(C) = 0$$

which satisfies the following conditions:

(i) we have $H^0(N_J, C) \cong L^J(w_{n-1,1})$;

(ii) for $0 \le \ell \le n-2$, there exists a $U(\mathfrak{g})$-module M_ℓ in $\mathcal{O}_{\mathrm{alg}}^{\mathfrak{p}_{\widehat{\ell+1}}}$ (see the beginning of Sect. 3.1) such that $\mathrm{gr}^0(C) = C/\mathrm{Fil}^1(C)$ fits into a short exact sequence of coadmissible $D(G)$-modules

$$0 \to \mathcal{F}^G_{P_I}(M_0, \pi^\infty_{1,1})^\vee \to \mathrm{gr}^0(C) \to (\mathrm{St}^{\mathrm{alg}}_n)^\vee \to 0,$$

and $\mathrm{gr}^\ell(C) \cong \mathcal{F}^G_{P_{\widehat{\ell+1}}}(M_\ell, \pi^\infty_{\ell+1,\ell+1})^\vee$ if $\ell \neq 0$ (see (2.71) for $\pi^\infty_{\ell+1,\ell+1}$).

Then we have $C \cong D_{n-1}$ (and $M_\ell = L(w_{\ell+1,1})$ for $0 \le \ell \le n-2$).

Proof Note that in (ii) we do not specify M_ℓ neither whether the short exact sequence is split or not. But Condition (i) (which was inspired by [76, Proposition 6.3] when $G = \mathrm{GL}_3(\mathbb{Q}_p)$) "rigidifies" everything.

As $\mathcal{F}^G_{P_{\widehat{\ell+1}}}(M_\ell, \pi^\infty_{\ell+1,\ell+1})^\vee$ injects into $\mathrm{gr}^\ell(C)$ by condition (ii), we define $\mathrm{Fil}^\ell(C)'$ for $0 \le \ell \le n-2$ as the unique (closed) $D(G)$-submodule of $\mathrm{Fil}^\ell(C)$ which fits into a short exact sequence

$$0 \to \mathrm{Fil}^{\ell+1}(C) \to \mathrm{Fil}^\ell(C)' \to \mathcal{F}^G_{P_{\widehat{\ell+1}}}(M_\ell, \pi^\infty_{\ell+1,\ell+1})^\vee \to 0 \tag{5.156}$$

(in particular $\mathrm{Fil}^\ell(C)' = \mathrm{Fil}^\ell(C)$ for $1 \le \ell \le n-2$). For $\ell \ge 1$, we define $\mathrm{Fil}^\ell(D_{n-1})' = \mathrm{Fil}^\ell(D_{n-1})$ as the unique (closed) $D(G)$-submodule of D_{n-1} such that $\mathrm{JH}_G(\mathrm{Fil}^\ell(D_{n-1})^\vee) = [C_{(1,\ell+1,\ell+1)}, C_{(1,n-1,n-1)}]$. For $\ell = 0$ we define $\mathrm{Fil}^0(D_{n-1}) \stackrel{\mathrm{def}}{=} D_{n-1}$ and $\mathrm{Fil}^0(D_{n-1})' \stackrel{\mathrm{def}}{=} X_{n-1} \subseteq D_{n-1}$. We check from the definitions of D_{n-1} and X_{n-1} that $\mathrm{gr}^0(D_{n-1})$ fits into a non-split extension

$$0 \to \mathcal{F}_{P_I}^G(L(s_1), \pi_{1,1}^\infty)^\vee \to \mathrm{gr}^0(D_{n-1}) \to (\mathrm{St}_n^{\mathrm{alg}})^\vee \to 0,$$

and that $\mathrm{gr}^\ell(D_{n-1}) \cong C_{(1,\ell+1,\ell+1)}^\vee = \mathcal{F}_{P_{\widehat{\ell+1}}}^G(L(w_{\ell+1,1}), \pi_{\ell+1,\ell+1}^\infty)^\vee$ for $1 \leq \ell \leq n-2$ (see (5.68)).

Step 1: We prove $M_{n-2} \cong L(w_{n-1,1})$.
From (i) we deduce

$$H^0(U, C) \cong H^0(U_J, H^0(N_J, C)) \cong w_{n-1,1} \cdot \mu_0. \tag{5.157}$$

Note that $\mathcal{F}_{P_J}^G(M_{n-2}, \pi_{n-1,n-1}^\infty)^\vee \cong \mathrm{gr}^{n-2}(C)$ injects into C by (ii). Since $H^0(U, -)$ is left exact and since we have by (4.154)

$$H^0(U, \mathcal{F}_{P_J}^G(M_{n-2}, \pi_{n-1,n-1}^\infty)^\vee) \cong H^0(\mathfrak{u}, M_{n-2}) \otimes_E (J_{J,\emptyset}(\pi_{n-1,n-1}^\infty))^\vee \neq 0,$$

we deduce the following isomorphism of $D(T)$-modules

$$H^0(\mathfrak{u}, M_{n-2}) \otimes_E (J_{J,\emptyset}(\pi_{n-1,n-1}^\infty))^\vee \cong H^0(U, C) \cong w_{n-1,1} \cdot \mu_0.$$

In particular we have $J_{J,\emptyset}(\pi_{n-1,n-1}^\infty) = J_{J,\emptyset}(1_{L_J}) = 1_T$ and an isomorphism of $U(\mathfrak{t})$-modules

$$H^0(\mathfrak{u}, M_{n-2}) \cong w_{n-1,1} \cdot \mu_0. \tag{5.158}$$

Left exactness of $H^0(\mathfrak{u}, -)$ with (ii) of Lemma 3.1.8 force $\mathrm{soc}_{U(\mathfrak{g})}(M_{n-2}) \cong L(w_{n-1,1})$. If $M_{n-2} \ncong L(w_{n-1,1})$, then M_{n-2} contains a length 2 $U(\mathfrak{g})$-submodule M'_{n-2} with cosocle $L(x)$ such that $D_L(x) \subseteq \{n-1\}$ (using $M_{n-2} \in \mathcal{O}_{\mathrm{alg}}^{\mathfrak{p}_J}$ and Lemma 3.1.1), or equivalently $x \in W^{J,\emptyset}$ (see §1.4 for $W^{J,\emptyset}$). Moreover $x \neq w_{n-1,1}$ by (i) of Lemma 3.2.4, and hence $x < w_{n-1,1}$ since $x \in W^{J,\emptyset}$ and $w_{n-1,1}$ is the maximal element in $W^{J,\emptyset}$. But $x < w_{n-1,1}$ together with (3.27) imply that M'_{n-2} is a quotient of $M(x)$, which by (3.13) (applied with $I = \emptyset$) implies

$$0 \neq \mathrm{Hom}_{U(\mathfrak{g})}(M(x), M'_{n-2}) \hookrightarrow \mathrm{Hom}_{U(\mathfrak{g})}(M(x), M_{n-2}) \cong \mathrm{Hom}_{U(\mathfrak{t})}(x \cdot \mu_0, H^0(\mathfrak{u}, M_{n-2})),$$

contradicting (5.158). Hence, we have $M_{n-2} \cong L(w_{n-1,1})$. Note that by (4.154) we have $H^0(U, V^\vee) \neq 0$ for $V = \mathcal{F}_{P_I}^G(M, \pi^\infty)$ with $M \in \mathcal{O}_{\mathrm{alg}}^{\mathfrak{p}_I}$ and π^∞ G-basic ((ii) of Definition 2.1.4). By left exactness of $H^0(U, -)$ we deduce using (5.157) and $M_{n-2} \cong L(w_{n-1,1})$

$$\mathrm{soc}_{D(G)}(C) \cong \mathrm{gr}^{n-2}(C) \cong C_{(1,n-1,n-1)}^\vee. \tag{5.159}$$

Note that, by Remark 4.5.19, (5.159) and condition (ii) imply $(M_\ell)_\xi = M_\ell$ for $0 \leq \ell \leq n-2$ where $\xi : Z(\mathfrak{g}) \to E$ is the unique infinitesimal character such that $L(1)_\xi \neq 0$. In particular any irreducible constituent of M_ℓ (for $0 \leq \ell \leq n-2$) is of the form $L(x)$ for $x \in W(G)$.

Step 2: We prove that $L(1) \notin \mathrm{JH}_{U(\mathfrak{g})}(M_\ell)$ for $1 \le \ell \le n-2$.
We fix $\mu \in \Lambda$ such that $\langle \mu + \rho, \alpha^\vee \rangle \ge 0$ for $\alpha \in \Phi^+$ and the stabilizer of μ in $W(G)$ for the dot action is $\{1, s_{n-1}\}$. Note that $\mathrm{Fil}^1(C) \subseteq C$ has socle $C^\vee_{(1,n-1,n-1)}$ by (5.159) (for $n \ge 3$), and that the canonical adjunction map $\mathrm{Fil}^1(C) \to \Theta_\mu(\mathrm{Fil}^1(C))$ is injective as it restricts to the canonical injection $C^\vee_{(1,n-1,n-1)} \hookrightarrow \Theta_{s_{n-1}}(C^\vee_{(1,n-1,n-1)})$ (see (ii) of Lemma 5.2.12). The decreasing filtration $\mathrm{Fil}^\ell(C)$ on $\mathrm{Fil}^1(C)$ for $1 \le \ell \le n-2$ induces a decreasing filtration $\Theta_\mu(\mathrm{Fil}^\ell(C))$ on $\Theta_\mu(\mathrm{Fil}^1(C))$ with graded pieces $\Theta_{s_{n-1}}(\mathrm{gr}^\ell(C)) \cong \mathcal{F}^G_{P_{\widehat{\ell+1}}}(\Theta_{s_{n-1}}(M_\ell), \pi^\infty_{\ell+1,\ell+1})^\vee$ by condition (ii) and (5.87). It follows from Proposition 3.4.5 that $L(1) \in \mathrm{JH}_{U(\mathfrak{g})}(\Theta_{s_{n-1}}(L(x)))$ for some $x \in W(G)$ if and only if $x = s_1$. Since any $L(x) \in \mathrm{JH}_{U(\mathfrak{g})}(M_\ell)$ satisfies $D_L(x) \subseteq \{\ell+1\}$ by Lemma 3.1.1, we deduce $x \ne s_1$, and thus $L(1) \notin \mathrm{JH}_{U(\mathfrak{g})}(\Theta_{s_{n-1}}(L(x)))$, for any $L(x) \in \mathrm{JH}_{U(\mathfrak{g})}(M_\ell)$ and any $1 \le \ell \le n-2$. By the above discussion we deduce that $\Theta_\mu(\mathrm{Fil}^1(C))$, and hence its $D(G)$-submodule $\mathrm{Fil}^1(C)$, do not admit constituents which are duals of locally algebraic representations, or equivalently $L(1) \notin \mathrm{JH}_{U(\mathfrak{g})}(M_\ell)$ for $1 \le \ell \le n-2$ by condition (ii) (and Lemma 5.1.1).

Step 3: We prove that $M_\ell \ne 0$ for $0 \le \ell \le n-2$.
From Step 1 we have $M_\ell \ne 0$ for $\ell = n-2$. Assume first $M_0 = 0$. As $(\mathrm{St}^{\mathrm{alg}}_n)^\vee$ shows up in C by (ii) but not in the socle of C by (5.159), from condition (ii) there exists $1 \le \ell \le n-2$ and $L(x) \in \mathrm{JH}_{U(\mathfrak{g})}(M_\ell)$ such that

$$\mathrm{Ext}^1_{D(G)}((\mathrm{St}^{\mathrm{alg}}_n)^\vee, \mathcal{F}^G_{P_{\widehat{\ell+1}}}(L(x), \pi^\infty_{\ell+1,\ell+1})^\vee) \ne 0.$$

As $x \ne 1$ by Step 2, the last statement in Remark 5.1.12 implies $d(\pi^\infty_{\ell+1,\ell+1}, \mathrm{St}^\infty_n) = 0$. But by (ii) of Lemma 2.3.1, for $1 \le \ell \le n-2$ the representation $i^\infty_{\widehat{\ell+1},\Delta}(\pi^\infty_{\ell+1,\ell+1})$ has length 2 with socle $V^\infty_{[1,\ell+1],\Delta}$ and cosocle $V^\infty_{[1,\ell],\Delta}$, and thus we can't have $d(\pi^\infty_{\ell+1,\ell+1}, \mathrm{St}^\infty_n) = 0$. Hence we have $M_0 \ne 0$. Assume now $M_\ell = 0$ for some $0 < \ell < n-2$. Since $M_0 \ne 0$, we may choose ℓ such that $M_{\ell-1} \ne 0 = M_\ell$. Let $L(y) \subseteq M_{\ell-1}$ be an arbitrary irreducible $U(\mathfrak{g})$-submodule and recall that $D_L(y) \subseteq \{\ell\}$ by Lemma 3.1.1. As $M_\ell = 0$ and $\mathcal{F}^G_{P_{\widehat{\ell}}}(L(y), \pi^\infty_{\ell,\ell})^\vee$ has no common constituent with $\mathrm{soc}_{D(G)}(C)$ by (5.159) and $D_L(y) \subseteq \{\ell\}$ (using Lemma 5.1.1), there exist $\ell < \ell' \le n-2$ and $L(z) \in \mathrm{JH}_{U(\mathfrak{g})}(M_{\ell'})$ such that

$$\mathrm{Ext}^1_{D(G)}\big(\mathcal{F}^G_{P_{\widehat{\ell}}}(L(y), \pi^\infty_{\ell,\ell})^\vee, \mathcal{F}^G_{P_{\widehat{\ell'+1}}}(L(z), \pi^\infty_{\ell'+1,\ell'+1})^\vee\big) \ne 0.$$

Note that $z \ne 1$ by Step 2 (as $\ell' \ge 1$) and hence $D_L(z) = \{\ell'+1\}$, in particular $z \ne y$ (as $D_L(y) \subseteq \{\ell\}$). But then, the last statement of Remark 5.1.12 again implies $d(\pi^\infty_{\ell'+1,\ell'+1}, \pi^\infty_{\ell,\ell}) = 0$, which contradicts (i) of Lemma 2.3.2 since $\ell < \ell'$. This finishes the proof of Step 3.

Step 4: Let $0 \le \ell < \ell' \le n-2$ and $L(x) \in \mathrm{JH}_{U(\mathfrak{g})}(M_\ell)$, we prove that if

$$\mathrm{Ext}^1_{D(G)}(\mathcal{F}^G_{P_{\widehat{\ell+1}}}(L(x), \pi^\infty_{\ell+1,\ell+1})^\vee, C^\vee_{(1,\ell'+1,\ell'+1)}) \ne 0 \tag{5.160}$$

then $\ell' = \ell+1$ and $x = w_{\ell+1,1}$, in which case (5.160) has dimension 1.

As $D_L(w_{\ell'+1,1}) = \{\ell'+1\}$ and $D_L(x) \subseteq \{\ell+1\}$ (by Lemma 3.1.1), we have $x \neq w_{\ell'+1,1}$. By Proposition 5.1.14 we have $\mathrm{Ext}^1_{U(\mathfrak{g})}(L(x), L(w_{\ell'+1,1})) \neq 0$, or equivalently $\mathrm{Ext}^1_{U(\mathfrak{g})}(L(w_{\ell'+1,1}), L(x)) \neq 0$ by (3.3) and (3.4). As $\ell(w_{\ell'+1,1}) \geq 2$, we have $\mathrm{Ext}^1_{U(\mathfrak{g})}(L(w_{\ell'+1,1}), L(1)) = 0$ by (ii) of Lemmas 3.2.4 and 3.2.9, thus $x \neq 1$ and $D_L(x) = \{\ell+1\}$. By (iii) of Lemma A.11 and $\ell < \ell'$, we obtain $\ell' = \ell+1$ and $x = w_{\ell+1,1}$. Hence $\mathcal{F}^G_{P_{\widehat{\ell+1}}}(L(x), \pi^\infty_{\ell+1,\ell+1})^\vee = C^\vee_{(1,\ell+1,\ell+1)}$ and (5.160) is one dimensional by Lemma 5.2.1.

Step 5: We prove $\mathrm{Fil}^\ell(C)' \cong \mathrm{Fil}^\ell(D_{n-1})'$ for $0 \leq \ell \leq n-2$.

We proceed by a decreasing induction on $0 \leq \ell \leq n-2$. The case $\ell = n-2$ holds by Step 1. We now assume $0 \leq \ell \leq n-3$ and $\mathrm{Fil}^{\ell'}(C)' \cong \mathrm{Fil}^{\ell'}(D_{n-1})'$ for all $\ell' > \ell$, or equivalently $\mathrm{Fil}^{\ell'}(C) \cong \mathrm{Fil}^{\ell'}(D_{n-1})$ (since $\ell' > 0$). Let $L(x) \subseteq M_\ell$ be an arbitrary (irreducible) $U(\mathfrak{g})$-submodule, which induces a $D(G)$-submodule $\mathcal{F}^G_{P_{\widehat{\ell+1}}}(L(x), \pi^\infty_{\ell+1,\ell+1})^\vee \subseteq \mathcal{F}^G_{P_{\widehat{\ell+1}}}(M_\ell, \pi^\infty_{\ell+1,\ell+1})^\vee$. By pullback and the induction hypothesis, (5.156) gives a short exact sequence

$$0 \to \mathrm{Fil}^{\ell+1}(D_{n-1}) \to * \to \mathcal{F}^G_{P_{\widehat{\ell+1}}}(L(x), \pi^\infty_{\ell+1,\ell+1})^\vee \to 0. \tag{5.161}$$

It follows from (5.159) (and $\ell \leq n-3$) that (5.161) is non-split and thus

$$\mathrm{Ext}^1_{D(G)}(\mathcal{F}^G_{P_{\widehat{\ell+1}}}(L(x), \pi^\infty_{\ell+1,\ell+1})^\vee, \mathrm{Fil}^{\ell+1}(D_{n-1})) \neq 0. \tag{5.162}$$

By Step 4 and a dévissage on $\mathrm{Fil}^{\ell+1}(D_{n-1})$, we deduce $x = w_{\ell+1,1}$ (hence $\mathcal{F}^G_{P_{\widehat{\ell+1}}}(L(x), \pi^\infty_{\ell+1,\ell+1})^\vee = C^\vee_{(1,\ell+1,\ell+1)}$) and that the surjection $\mathrm{Fil}^{\ell+1}(D_{n-1}) \twoheadrightarrow C^\vee_{(1,\ell+2,\ell+2)}$ induces an isomorphism between (5.162) and the 1-dimensional vector space $\mathrm{Ext}^1_{D(G)}(C^\vee_{(1,\ell+1,\ell+1)}, C^\vee_{(1,\ell+2,\ell+2)})$. In particular, $\mathrm{soc}_{U(\mathfrak{g})}(M_\ell) \cong L(w_{\ell+1,1})$ and the pushout of (5.161) along $\mathrm{Fil}^{\ell+1}(D_{n-1}) \twoheadrightarrow C^\vee_{(1,\ell+2,\ell+2)}$ is a non-split extension

$$0 \to C^\vee_{(1,\ell+2,\ell+2)} \to * \to C^\vee_{(1,\ell+1,\ell+1)} \to 0. \tag{5.163}$$

If $M_\ell \cong L(w_{\ell+1,1})$, then both $\mathrm{Fil}^\ell(C)'$ and $\mathrm{Fil}^\ell(D_{n-1})'$ fit into a non-split extension

$$0 \to \mathrm{Fil}^{\ell+1}(D_{n-1}) \to * \to C^\vee_{(1,\ell+1,\ell+1)} \to 0,$$

which forces $\mathrm{Fil}^\ell(C)' \cong \mathrm{Fil}^\ell(D_{n-1})'$ as (5.162) is 1-dimensional for $x = w_{\ell+1,1}$. We now prove that we must have $M_\ell \cong L(w_{\ell+1,1})$. Assume on the contrary that M_ℓ is not $L(w_{\ell+1,1})$, then it contains a length 2 $U(\mathfrak{g})$-submodule M'_ℓ with socle $L(w_{\ell+1,1})$ and cosocle some $L(x')$ with $D_L(x') \subseteq \{\ell+1\}$ (so $x' \neq w_{\ell+2,1}$) and $\mathrm{Ext}^1_{\mathcal{O}^{\mathfrak{b}}_{\mathrm{alg}}}(L(x'), L(w_{\ell+1,1})) \neq 0$. By (i) of Lemma 3.2.4 this implies $x' \neq w_{\ell+1,1}$ and $|\ell(x') - \ell(w_{\ell+1,1})|$ odd. The pullback of (5.156) along $\mathcal{F}^G_{P_{\widehat{\ell+1}}}(M'_\ell, \pi^\infty_{\ell+1,\ell+1})^\vee \hookrightarrow \mathcal{F}^G_{P_{\widehat{\ell+1}}}(M_\ell, \pi^\infty_{\ell+1,\ell+1})^\vee$ followed by the pushout along $\mathrm{Fil}^{\ell+1}(D_{n-1}) \twoheadrightarrow C^\vee_{(1,\ell+2,\ell+2)}$ gives a short exact sequence

$$0 \to C^{\vee}_{(1,\ell+2,\ell+2)} \to * \to \mathcal{F}^G_{P_{\widehat{\ell+1}}}(M'_\ell, \pi^\infty_{\ell+1,\ell+1})^\vee \to 0. \qquad (5.164)$$

Note that (5.164) is non-split as it restricts to (5.163) which is non-split. In particular, we have

$$\mathrm{Ext}^1_{D(G)}(\mathcal{F}^G_{P_{\widehat{\ell+1}}}(M'_\ell, \pi^\infty_{\ell+1,\ell+1})^\vee, C^\vee_{(1,\ell+2,\ell+2)}) \neq 0. \qquad (5.165)$$

Case 2.1: If $x' = 1$, then Step 2 forces $\ell = 0$. Note that M'_0 is a quotient of $M^I(1)$ (using (3.27) and [33, Theorem 9.4(c)]) and that $L(w_{2,1}) \notin \mathrm{JH}_{U(\mathfrak{g})}(M^I(1))$ (by Lemma 3.1.1). Thus $C^\vee_{(1,2,2)}$ is not a constituent of $\mathcal{F}^G_{P_I}(M^I(1), \pi^\infty_{1,1})^\vee$ (by Lemma 5.1.1). Hence, the surjection $M^I(1) \twoheadrightarrow M'_0$ induces an injection

$$\mathrm{Ext}^1_{D(G)}(\mathcal{F}^G_{P_I}(M'_0, \pi^\infty_{1,1})^\vee, C^\vee_{(1,2,2)}) \hookrightarrow \mathrm{Ext}^1_{D(G)}(\mathcal{F}^G_{P_I}(M^I(1), \pi^\infty_{1,1})^\vee, C^\vee_{(1,2,2)}). \qquad (5.166)$$

Let $\xi_I : Z(\mathfrak{l}_I) \to E$ be the unique infinitesimal character such that $L^I(1)_{\xi_I} \neq 0$. Since $w_{2,1} \notin W(L_I)$, by exactly the same argument as in the paragraph following (3.100) we have $H^k(\mathfrak{n}_I, L(w_{2,1}))_{\xi_I} = 0$ for $k \geq 0$. By (3.12) and Lemma 3.1.3 this implies $\mathrm{Ext}^k_{U(\mathfrak{g})}(M^I(1), L(w_{2,1})) = 0$ for $k \geq 0$. By (4.146) (applied with $w = 1$, note that the assumption there is satisfed using (i) of Lemma 2.3.4) this forces the right hand side of (5.166) to be zero, which contradicts (5.165) (when $\ell = 0$).

Case 2.2: If $x' \neq 1$, then $D_L(x') = \{\ell + 1\}$. It follows from Lemma A.5 and Remark A.6 that $D_R(x') = \{1\}$ and $D_L(x') = \{\ell+1\}$ imply $x' = w_{\ell+1,1}$. Since we have $x' \neq w_{\ell+1}$ (and $x' \neq 1$), we deduce $D_R(x') \nsubseteq \{1\}$. Choose $j \in D_R(x') \setminus \{1\} \subseteq \{2, \dots, n-1\}$. As $|\ell(x') - \ell(w_{\ell+1,1})|$ is odd, it follows that $|\ell(x') - \ell(w_{\ell+2,1})|$ is even. Since $x' \neq w_{\ell+2,1}$, by (i), (ii) of Lemma 3.2.4 we obtain $\mathrm{Ext}^1_{U(\mathfrak{g})}(L(x'), L(w_{\ell+2,1})) = 0$ and thus by Proposition 5.1.14

$$\mathrm{Ext}^1_{D(G)}(\mathcal{F}^G_{P_{\widehat{\ell+1}}}(L(x'), \pi^\infty_{\ell+1,\ell+1})^\vee, C^\vee_{(1,\ell+2,\ell+2)}) = 0.$$

Let $W \stackrel{\mathrm{def}}{=} \mathcal{F}^G_{P_{\widehat{\ell+1}}}(L(x'), \pi^\infty_{\ell+1,\ell+1})$, then (5.165) forces the existence of a uniserial coadmissible $D(G)$-module D of length 3 with socle $C^\vee_{(1,\ell+2,\ell+2)}$, cosocle $W^\vee$, and middle layer $C^\vee_{(1,\ell+1,\ell+1)}$. Fix $\mu' \in \Lambda$ such that $\langle \mu' + \rho, \alpha^\vee \rangle \geq 0$ for $\alpha \in \Phi^+$ and the stabilizer of μ' in $W(G)$ for the dot action is $\{1, w_0 s_j w_0\}$. As $j \neq 1$, by (i) of Lemma 5.2.12 we have

$$\Theta_{w_0 s_j w_0}(C^\vee_{(1,\ell+2,\ell+2)}) = 0 = \Theta_{w_0 s_j w_0}(C^\vee_{(1,\ell+1,\ell+1)})$$

and by (ii) of Lemma 5.2.12 the canonical map $\Theta_{w_0 s_j w_0}(W^\vee) \to W^\vee$ is surjective. Then the exactness of $\Theta_{\mu'}$ forces the canonical map $\Theta_{\mu'}(D) \to D$ to be a surjection, and we have $\Theta_{\mu'}(D) \cong \Theta_{\mu'}(W^\vee)$. As $x' \neq w_{\ell+2,1}$ and $|\ell(x') - \ell(w_{\ell+2,1})|$ is even, it follows from Proposition 3.4.5 with (i) of Lemma 3.2.4 and (5.87) that $C^\vee_{(1,\ell+2,\ell+2)} \notin \mathrm{JH}_{D(G)}(\Theta_{\mu'}(W^\vee)) = \mathrm{JH}_{D(G)}(\Theta_{\mu'}(D))$, a contradiction to the surjection $\Theta_{\mu'}(D) \twoheadrightarrow D$.

From Step 5 and condition (ii), we deduce that both C and D_{n-1} fit into a short exact sequence $0 \to X_{n-1} = \mathrm{Fil}^0(D_{n-1})' \to * \to (\mathrm{St}_n^{\mathrm{alg}})^\vee \to 0$. Moreover D_{n-1} is the unique such non-split extension by (i) of Theorem 5.3.11 (and the definition of D_{n-1}). But (5.159) implies that C is also a non-split extension, hence we finally deduce $C \cong D_{n-1}$. □

From now on we assume $E = K$. We consider the Drinfeld space $\mathbb{H}_{/K}$ of dimension $n-1$ defined as in (1.1), i.e. $\mathbb{H} \stackrel{\mathrm{def}}{=} \mathbb{P}^{n-1}_{\mathrm{rig}} \setminus \bigcup_{\mathcal{H}} \mathcal{H}$ where $\mathcal{H}$ runs through the K-rational hyperplanes inside $\mathbb{P}^{n-1}_{\mathrm{rig}}$. Recall from [67, Proposition 1.4] that $\mathbb{H}$ is quasi-Stein and that for any coherent sheaf $\mathcal{F}$ on $\mathbb{H}$ we have $H^k(\mathcal{F}) = 0$ for $k > 0$. We follow the convention of [49, 71] and endow $\mathbb{H}$ with the left action of $G = \mathrm{GL}_n(K)$ coming from the left action of G on $\mathbb{P}^{n-1}_{\mathrm{rig}}$ given by $(z_0, \dots, z_{n-1}) \mapsto (z_0, \dots, z_{n-1})g^{-1}$ (matrix product) for $g \in G$. This action in turn induces an action of G (which we will always take on the left) on the global sections $\mathcal{F}(\mathbb{H})$ for any G-equivariant vector bundle $\mathcal{F}$ on $\mathbb{H}$. Moreover if $\mathcal{F}$ is the restriction to $\mathbb{H}$ of a G-equivariant vector bundle on $\mathbb{P}^{n-1}_{\mathrm{rig}}$, then by the argument in [49, p. 593] $\mathcal{F}(\mathbb{H})$ is naturally a coadmissible $D(G)$-module. For instance this applies to the sheaf of differential k-forms on $\mathbb{H}$.

We consider the *global sections* of the de Rham complex of $\mathbb{H}$, which we denote (see (1.1))

$$\Omega^\bullet = [\Omega^0 \longrightarrow \Omega^1 \longrightarrow \cdots \longrightarrow \Omega^{n-1}]. \tag{5.167}$$

By the above discussion this is a complex of coadmissible $D(G)$-modules. The action of G induces an action on each cohomology group $H^k(\Omega^\bullet)$ for $0 \le k \le n-1$. We recall the following seminal result of Schneider-Stuhler [67, Theorem 3.1, Lemma 4.1] (see also Sect. 1.1).

Theorem 5.4.5 *We have a G-equivariant isomorphism $H^k(\Omega^\bullet) \cong (V^\infty_{[1,n-k-1],\Delta})^\vee$ for $0 \le k \le n-1$.*

Given a finite dimensional $U(\mathfrak{p}_I)$-module X over K, we can lift it to a finite dimensional algebraic representation of P_I ([52, Lemma 3.2]), and then consider the algebraic G-equivariant vector bundle over $G/P_I \cong \mathbb{P}^{n-1}_{/K}$

$$\mathcal{F}_X \stackrel{\mathrm{def}}{=} (G \times X)/P_I$$

where $p_I \in P_I$ acts on $(g, x) \in G \times X$ by $p_I(g, x) = (gp_I, p_I^{-1}x)$ (here we view G and P_I as algebraic groups over K rather than their K-points). We also denote by $\mathcal{F}_X$ the analytification of $\mathcal{F}_X$, which is a G-equivariant vector bundle over $\mathbb{P}^{n-1}_{\mathrm{rig}}$, hence by restriction over $\mathbb{H}$, and we denote its global sections (a coadmissible $D(G)$-module) by

$$D_X \stackrel{\mathrm{def}}{=} \mathcal{F}_X(\mathbb{H}).$$

Note that, given a short exact sequence $0 \to X_1 \to X_2 \to X_3 \to 0$ of finite dimensional $U(\mathfrak{p}_I)$-modules, we have a short exact sequence $0 \to \mathcal{F}_{X_1} \to \mathcal{F}_{X_2} \to \mathcal{F}_{X_3} \to$

0 of G-equivariant vector bundles over $\mathbb{H}$. As $\mathbb{H}$ is quasi-Stein, by taking global sections we obtain a short exact sequence of coadmissible $D(G)$-modules

$$0 \to D_{X_1} \to D_{X_2} \to D_{X_3} \to 0. \tag{5.168}$$

Now we recall Schneider's holomorphic discrete series [64]. For $1 \leq k \leq n-1$ we define $\mu_k \stackrel{\text{def}}{=} w_{n-1,n-k} \cdot \mu_0 \in \Lambda_J^{\text{dom}}$ where we recall that $w_{n-1,n-k} = s_{n-1}s_{n-2}\cdots s_{n-k}$ (see (3.87)). For $\mu \in \Lambda_J^{\text{dom}}$, we have $w_{1,n-1}(\mu) \in \Lambda_I^{\text{dom}}$ (where $w_{1,n-1} = s_1 s_2 \cdots s_{n-1}$ and we see $L^I(w_{1,n-1}(\mu))$ as a finite dimensional algebraic representation of P_I over K via the inflation $P_I \twoheadrightarrow L_I$. For $\mu \in \Lambda_J^{\text{dom}}$ we define the coadmissible $D(G)$-module

$$D_\mu \stackrel{\text{def}}{=} D_{L^I(w_{1,n-1}(\mu))}. \tag{5.169}$$

In [64, Sect. 3], Schneider defines a complex denoted there (see [64, Lemma 9] and its proof):

$$[D_{\lambda(0)} \to D_{\lambda(1)} \to \cdots \to D_{\lambda(n-1)}]. \tag{5.170}$$

Unraveling the various conventions between [64, Sect. 3] and this work, we can check that for $0 \leq k \leq n-1$ we have $D_{\lambda(k)} = D_{L^I(w_{1,n-1}(\mu_k))}$. Hence by (5.169), (5.170) is a complex of coadmissible $D(G)$-modules

$$\mathbf{D}^\bullet_{\mu_0} \stackrel{\text{def}}{=} [D_{\mu_0} \longrightarrow D_{\mu_1} \longrightarrow \cdots \longrightarrow D_{\mu_{n-1}}]. \tag{5.171}$$

We recall the following results from [64, Sect. 3].

Proposition 5.4.6

(i) *For $0 \leq k \leq n-1$ we have an isomorphism of coadmissible $D(G)$-modules $\Omega^k \cong D_{w_{n-1,n-k}\cdot 0}$ (with the convention $w_{n-1,n} = 1$).*
(ii) *For any finite dimensional $U(\mathfrak{p}_I)$-module X over K and any $\nu \in \Lambda^{\text{dom}}$, we see $X \otimes_K L(\nu)$ as a (finite dimensional) $U(\mathfrak{p}_I)$-module via the diagonal action of $U(\mathfrak{p}_I)$, then we have an isomorphism of coadmissible $D(G)$-modules*

$$D_{X \otimes_K L(\nu)} \cong D_X \otimes_K L(\nu).$$

Proof We check that (i) (resp. (ii)) is the translation of [64, Proposition 1] (resp. [64, Lemma 5]) under our convention. Recall that the "diagonal" $D(G)$-action on $D_X \otimes_K L(\nu)$ is not completely straightforward, see [35, Sect. 2.3.1]. □

Lemma 5.4.7 *For $\mu \in \Lambda_J^{\text{dom}}$, $Z(\mathfrak{g})$ acts on D_μ by ξ_μ where $\xi_\mu : Z(\mathfrak{g}) \to E$ is the unique infinitesimal character such that $L(\mu)_{\xi_\mu} \neq 0$.*

Proof We use results from [76], where ρ there is denoted $-\delta$ and ξ_μ is denoted $\chi_{\mu-\delta} = \chi_{\mu+\rho}$. By the argument of [76, Lemma 6.4] (which essentially follows from [39, Proposition 8.22]) this implies that $Z(\mathfrak{g})$ acts on D_μ (denoted $D_{w_0(\mu)}$ in *loc. cit.*) by $\chi_{w_0(\mu)+\delta} = \xi_{w_0(\mu)-2\rho} = \xi_{w_0\cdot\mu} = \xi_\mu$. □

We will need the two following (somewhat technical) lemmas.

Lemma 5.4.8 *Let L be a finite dimensional $U(\mathfrak{g})$-module. Then there exist a* decreasing *exhaustive separated filtration* $(\mathrm{Fil}^\ell(L))_{\ell\in\mathbb{Z}}$ *of $U(\mathfrak{p}_I)$-submodules on L, and an* increasing *exhaustive separated filtration* $(\mathrm{Fil}_\ell(L))_{\ell\in\mathbb{Z}}$ *of $U(\mathfrak{p}_I^+)$-submodules on L, such that for each $\ell \in \mathbb{Z}$ the $U(\mathfrak{l}_I)$-modules $\mathrm{gr}^\ell(L)$ and $\mathrm{gr}_\ell(L)$ are either both zero or both simple and isomorphic.*

Proof We consider the decomposition $L|_{U(\mathfrak{l}_I)} \cong \bigoplus_{\nu_1\in\Lambda_I^{\mathrm{dom}}} L_{\nu_1}$ where L_{ν_1} is the $L^I(\nu_1)$-isotypic component of $L|_{U(\mathfrak{l}_I)}$. We fix an arbitrary total order $\le$ on Λ_I^{dom} such that $\nu_1', \nu_1'' \in \Lambda_I^{\mathrm{dom}}$ satisfy $\nu_1' \le \nu_1''$ only if $\nu_1' - \nu_1'' \in \mathbb{Z}_{\ge 0}\Phi^+$. For $\nu_1 \in \Lambda_I^{\mathrm{dom}}$ we define

$$\mathrm{Fil}^{\nu_1}(L) \stackrel{\mathrm{def}}{=} \bigoplus_{\nu_1'\ge\nu_1} L_{\nu_1'} \quad \text{and} \quad \mathrm{Fil}_{\nu_1}(L) \stackrel{\mathrm{def}}{=} \bigoplus_{\nu_1'\le\nu_1} L_{\nu_1'}$$

which is a decreasing, respectively increasing, exhaustive separated filtration of $L|_{U(\mathfrak{l}_I)}$ (by $U(\mathfrak{l}_I)$-submodules) indexed by the totally ordered set Λ_I^{dom}. The key observation (which is easy to check) is that $\mathrm{Fil}^{\nu_1}(L)$ (resp. $\mathrm{Fil}_{\nu_1}(L)$) is $U(\mathfrak{p}_I)$-stable (resp. $U(\mathfrak{p}_I^+)$)-stable in L for each $\nu_1 \in \Lambda_I^{\mathrm{dom}}$. Then we fix an arbitrary filtration on L_{ν_1} for each $\nu_1 \in \Lambda_I^{\mathrm{dom}}$ with graded pieces being either simple or zero, and we further refine and reindex $(\mathrm{Fil}^{\nu_1}(L))_{\nu_1\in\Lambda_I^{\mathrm{dom}}}$ (resp. $(\mathrm{Fil}_{\nu_1}(L))_{\nu_1\in\Lambda_I^{\mathrm{dom}}}$) to get a decreasing filtration $(\mathrm{Fil}^\ell(L))_{\ell\in\mathbb{Z}}$ by $U(\mathfrak{p}_I)$-submodules (resp. an increasing filtration $(\mathrm{Fil}_\ell(L))_{\ell\in\mathbb{Z}}$ by $U(\mathfrak{p}_I^+)$-submodules) as in the statement. □

Recall that, for $\lambda, \mu \in \Lambda$, T_λ^μ is the translation functor on the category of $Z(\mathfrak{g})$-finite $D(G)$-modules, see above (5.86).

Lemma 5.4.9 *Let $\nu \in \Lambda_J^{\mathrm{dom}}$ and $\lambda, \mu \in \Lambda$. Then there exist:*

- *a decreasing exhaustive separated filtration* $(\mathrm{Fil}^i(T_\lambda^\mu(D_\nu)))_{i\in\mathbb{Z}}$ *on $T_\lambda^\mu(D_\nu)$ by (closed) $D(G)$-submodules*
- *an increasing exhaustive separated filtration* $(\mathrm{Fil}_i(T_\lambda^\mu(M^J(\nu))))_{i\in\mathbb{Z}}$ *on $T_\lambda^\mu(M^J(\nu))$ by $U(\mathfrak{g})$-submodules*

such that for $i \in \mathbb{Z}$ we have either $\mathrm{gr}^i(T_\lambda^\mu(D_\nu)) = \mathrm{gr}_i(T_\lambda^\mu(M^J(\nu))) = 0$, or $\mathrm{gr}^i(T_\lambda^\mu(D_\nu)) \cong D_{\kappa_i}$ and $\mathrm{gr}_i(T_\lambda^\mu(M^J(\nu))) \cong M^J(\kappa_i)$ for some $\kappa_i \in \Lambda_J^{\mathrm{dom}}$. Moreover $\mathrm{gr}^i(T_\lambda^\mu(D_\nu))$ and $\mathrm{gr}_i(T_\lambda^\mu(M^J(\nu)))$ are non-zero only for finitely many $i \in \mathbb{Z}$.

Proof By Lemma 5.4.7 we can assume $\xi_\nu = \xi_\lambda$ otherwise we have $T_\lambda^\mu(D_\nu) = T_\lambda^\mu(M^J(\nu)) = 0$ by definition of T_λ^μ and there is nothing to prove. Let L be the unique finite dimensional simple $U(\mathfrak{g})$-module with highest weight in the $W(G)$-orbit of $\mu-\lambda$ (for the naive action of $W(G)$). By definition of $\mathcal{T}_\lambda^\mu$ in [35, (2)] and using (ii) of Proposition 5.4.6 we have

$$\mathcal{T}_\lambda^\mu(D_\nu)=(D_{L^I(w_{1,n-1}(\nu))}\otimes_K L)_{\xi_\mu}\cong(D_{L^I(w_{1,n-1}(\nu))\otimes_K L})_{\xi_\mu}.$$

By Lemma 5.4.8 we fix on L a decreasing exhaustive separated filtration $(\mathrm{Fil}^\ell(L))_{\ell\in\mathbb{Z}}$ by $U(\mathfrak{p}_I)$-submodules, and an increasing exhaustive separated filtration $(\mathrm{Fil}_\ell(L))_{\ell\in\mathbb{Z}}$ by $U(\mathfrak{p}_I^+)$-submodules.

Step 1: We construct the decreasing filtration $(\mathrm{Fil}^i(\mathcal{T}_\lambda^\mu(D_\nu)))_{i\in\mathbb{Z}}$ on $\mathcal{T}_\lambda^\mu(D_\nu)$. We first choose an arbitrary decreasing exhaustive separated filtration $(\mathrm{Fil}^i(L^I(w_{1,n-1}(\nu))\otimes_K L))_{i\in\mathbb{Z}}$ of $U(\mathfrak{p}_I)$-submodules on $L^I(w_{1,n-1}(\nu))\otimes_K L$ which refines the filtration $(L^I(w_{1,n-1}(\nu))\otimes_K \mathrm{Fil}^\ell(L))_{\ell\in\mathbb{Z}}$ and with each graded piece $\mathrm{gr}^i(L^I(w_{1,n-1}(\nu))\otimes_K L)$ either simple (of the form $L^I(w_{1,n-1}(\kappa_i))$ for some $\kappa_i\in\Lambda_J^{\mathrm{dom}}$) or zero. By (5.168) the filtration $\mathrm{Fil}^i(L^I(w_{1,n-1}(\nu))\otimes_K L)$ induces a filtration $\mathrm{Fil}^i(D_{L^I(w_{1,n-1}(\nu))\otimes_K L})\overset{\mathrm{def}}{=}D_{\mathrm{Fil}^i(L^I(w_{1,n-1}(\nu))\otimes_K L)}$ by closed $D(G)$-submodules on $D_{L^I(w_{1,n-1}(\nu))\otimes_K L}$, which further induces a filtration $\mathrm{Fil}^i((D_{L^I(w_{1,n-1}(\nu))\otimes_K L})_{\xi_\mu})\overset{\mathrm{def}}{=}\mathrm{Fil}^i(D_{L^I(w_{1,n-1}(\nu))\otimes_K L})_{\xi_\mu}$ on $(D_{L^I(w_{1,n-1}(\nu))\otimes_K L})_{\xi_\mu}$. In particular, for $i\in\mathbb{Z}$, we have

$$\mathrm{gr}^i((D_{L^I(w_{1,n-1}(\nu))\otimes_K L})_{\xi_\mu})=(D_{\mathrm{gr}^i(L^I(w_{1,n-1}(\nu))\otimes_K L)})_{\xi_\mu},$$

which is non-zero if and only if $\mathrm{gr}^i(L^I(w_{1,n-1}(\nu))\otimes_K L)\cong L^I(w_{1,n-1}(\kappa_i))$ for some $\kappa_i\in\Lambda_J^{\mathrm{dom}}$ such that $L(\kappa_i)_{\xi_\mu}\neq 0$ (using (5.169) and Lemma 5.4.7).

Step 2: We construct the increasing filtration $(\mathrm{Fil}_i(T_\lambda^\mu(M^J(\nu))))_{i\in\mathbb{Z}}$ on $T_\lambda^\mu(M^J(\nu))$.

We first note that $L^I(w_{1,n-1}(\nu))\otimes_K \mathrm{gr}^\ell(L)$ (resp. $L^I(w_{1,n-1}(\nu))\otimes_K \mathrm{gr}_\ell(L)$) is a semi-simple $U(\mathfrak{p}_I)$-module (resp. a semi-simple $U(\mathfrak{p}_I^+)$-module) and they are isomorphic as semi-simple $U(\mathfrak{l}_I)$-modules. Hence, it is possible to choose an increasing filtration $(\mathrm{Fil}_i(L^I(w_{1,n-1}(\nu))\otimes_K L))_{i\in\mathbb{Z}}$ of $U(\mathfrak{p}_I^+)$-modules on $L^I(w_{1,n-1}(\nu))\otimes_K L$ such that for each $i\in\mathbb{Z}$ we have an isomorphism of $U(\mathfrak{l}_I)$-modules $\mathrm{gr}_i(L^I(w_{1,n-1}(\nu))\otimes_K L)\cong\mathrm{gr}^i(L^I(w_{1,n-1}(\nu))\otimes_K L)$ (which by Step 1 is either 0 or $L^I(w_{1,n-1}(\kappa_i))$). For $\mu'\in\Lambda$ and $w\in W(G)$, we write $L^I(\mu')^w$ for the finite dimensional simple $U(w^{-1}\mathfrak{l}_I w)$-module $L^I(\mu')$ where $x\in U(w^{-1}\mathfrak{l}_I w)$ acts by wxw^{-1}. Using $\mathfrak{u}_J=w_{1,n-1}^{-1}\mathfrak{u}_I w_{1,n-1}$ and $\mathfrak{t}=w_{1,n-1}^{-1}\mathfrak{t}w_{1,n-1}$, we have an isomorphism of $U(\mathfrak{t})$-modules $((L^I(w_{1,n-1}(\nu))^{w_{1,n-1}})[\mathfrak{u}_J])_\nu=((L^I(w_{1,n-1}(\nu))[\mathfrak{u}_I])_{w_{1,n-1}(\nu)})^{w_{1,n-1}}\neq 0$. As $L^I(w_{1,n-1}(\nu))^{w_{1,n-1}}$ is a finite dimensional simple $U(w_{1,n-1}^{-1}\mathfrak{l}_I w_{1,n-1})=U(\mathfrak{l}_J)$-modules, this implies

$$L^I(w_{1,n-1}(\nu))^{w_{1,n-1}}\cong L^J(\nu)\tag{5.172}$$

as $U(\mathfrak{l}_J)$-modules. Writing similarly $(L^I(w_{1,n-1}(\nu))\otimes_K L)^{w_{1,n-1}}$ for the $U(w_{1,n-1}^{-1}\mathfrak{p}_I^+ w_{1,n-1})=U(\mathfrak{p}_J)$-module on which $x\in U(\mathfrak{p}_J)$ acts by $w_{1,n-1}xw_{1,n-1}^{-1}$, we obtain the following isomorphisms of $U(w_{1,n-1}^{-1}\mathfrak{p}_I^+ w_{1,n-1}) = U(\mathfrak{p}_J)$-modules

$$(L^I(w_{1,n-1}(\nu)) \otimes_K L)^{w_{1,n-1}} \cong L^I(w_{1,n-1}(\nu))^{w_{1,n-1}} \otimes_K L^{w_{1,n-1}} \cong L^J(\nu) \otimes_K L.$$

In particular we deduce from $(\mathrm{Fil}_i(L^I(w_{1,n-1}(\nu)) \otimes_K L))_{i\in\mathbb{Z}}$ an increasing filtration $(\mathrm{Fil}_i(L^J(\nu) \otimes_K L))_{i\in\mathbb{Z}}$ of $U(\mathfrak{p}_J)$-modules on $L^J(\nu) \otimes_K L$ such that $\mathrm{gr}_i(L^J(\nu) \otimes_K L)$ is either 0 or $L^J(\kappa_i)$. Using the tensor identity (cf. [33, Sect. 3.6])

$$U(\mathfrak{g}) \otimes_{U(\mathfrak{p}_J)} (L^J(\nu) \otimes_K L) \cong (U(\mathfrak{g}) \otimes_{U(\mathfrak{p}_J)} L^J(\nu)) \otimes_K L = M^J(\nu) \otimes_K L$$

we obtain an increasing filtration of $U(\mathfrak{g})$-submodules on $M^J(\nu) \otimes_K L$

$$(\mathrm{Fil}_i(M^J(\nu) \otimes_K L))_{i\in\mathbb{Z}} \overset{\text{def}}{=} (U(\mathfrak{g}) \otimes_{U(\mathfrak{p}_J)} \mathrm{Fil}_i(L^J(\nu) \otimes_K L))_{i\in\mathbb{Z}}$$

such that $\mathrm{gr}_i(M^J(\nu) \otimes_K L)$ is either 0 of $M^J(\kappa_i)$. By exactness of the translation functor T_λ^μ (see (3.84)), we deduce an increasing filtration $(\mathrm{Fil}_i(T_\lambda^\mu(M^J(\nu))))_{i\in\mathbb{Z}} \overset{\text{def}}{=} (\mathrm{Fil}_i(M^J(\nu) \otimes_K L))_{\xi_\mu})_{i\in\mathbb{Z}}$ on $T_\lambda^\mu(M^J(\nu)) = (M^J(\nu) \otimes_K L)_{\xi_\mu}$ which is non-zero if and only if $\mathrm{gr}_i(T_\lambda^\mu(M^J(\nu))) \cong M^J(\kappa_i)$ for some $\kappa_i \in \Lambda_J^{\mathrm{dom}}$ such that $(M^J(\kappa_i))_{\xi_\mu} \neq 0$, or equivalently $L(\kappa_i)_{\xi_\mu} \neq 0$. Comparing with the end of Step 1, we obtain the statement.

The last statement of the lemma follows from the fact $L^I(w_{1,n-1}(\nu)) \otimes_K L$ is finite dimensional. □

Remark 5.4.10 Lemma 5.4.9 is applied in Lemmas 5.4.11 and 5.4.12 below to study certain $\mathcal{T}_\lambda^\mu(D_\nu)$. Although more direct proofs probably exist without using Verma modules (since ultimately it is a matter of decomposing $L^I(w_{1,n-1}(\nu)) \otimes_K L$, with the notation of the proof of Lemma 5.4.9), the present proofs are more convenient for us as we can use standard results on the translation or wall-crossing of Verma modules.

Lemma 5.4.11 *For $0 \leq k \leq n-1$ we have a canonical isomorphism of coadmissible $D(G)$-modules*

$$\mathcal{T}_{w_0\cdot 0}^{w_0\cdot\mu_0}(\Omega^k) \cong D_{\mu_k}. \tag{5.173}$$

Proof Let ξ_0 (resp. ξ_{μ_0}) be the unique infinitesimal character such that $L(0)_{\xi_0} \neq 0$ (resp. $L(\mu_0)_{\xi_{\mu_0}} \neq 0$). By [33, Sect. 7.8] the pair of functors $T_{w_0\cdot 0}^{w_0\cdot\mu_0}$ and $T_{w_0\cdot\mu_0}^{w_0\cdot 0}$ (see (3.84)) are both left and right adjoint of each other, and gives an equivalence of categories $(\mathcal{O}_{\mathrm{alg}}^{\mathfrak{b}})_{\xi_0} \cong (\mathcal{O}_{\mathrm{alg}}^{\mathfrak{b}})_{\xi_{\mu_0}}$ (where $(\mathcal{O}_{\mathrm{alg}}^{\mathfrak{b}})_\xi$ for an infinitesimal character ξ is the subcategory of $\mathcal{O}_{\mathrm{alg}}^{\mathfrak{b}}$ of M such that $M = M_\xi$ (see above Lemma 3.1.3 and recall that [33, Sect. 7.8] uses antidominant weights, whence this w_0 everywhere). By [33, Theorem 7.6] for $w \in W(G)$ we have $T_{w_0\cdot 0}^{w_0\cdot\mu_0}(M(w\cdot 0)) \cong M(w\cdot\mu_0)$ and $T_{w_0\cdot\mu_0}^{w_0\cdot 0}(M(w\cdot\mu_0)) \cong M(w\cdot 0)$. Since $\mathcal{O}_{\mathrm{alg}}^{\mathfrak{p}_J}$ is stable under subobjects and tensoring by finite dimensional $U(\mathfrak{g})$-modules, $T_{w_0\cdot 0}^{w_0\cdot\mu_0}$ and $T_{w_0\cdot\mu_0}^{w_0\cdot 0}$ preserve $\mathcal{O}_{\mathrm{alg}}^{\mathfrak{p}_J}$, and thus induce an equivalence of categories $(\mathcal{O}_{\mathrm{alg}}^{\mathfrak{p}_J})_{\xi_0} \cong (\mathcal{O}_{\mathrm{alg}}^{\mathfrak{p}_J})_{\xi_{\mu_0}}$ (with obvious notation). Let

$k \in \{0\dots, n-1\}$, since $M^J(w_{n-1,n-k}\cdot 0)$ (resp. $M^J(w_{n-1,n-k}\cdot \mu_0)$) is the maximal quotient of $M(w_{n-1,n-k}\cdot 0)$ (resp. $M(w_{n-1,n-k}\cdot \mu_0)$) in $\mathcal{O}^{\mathfrak{p}_J}_{\mathrm{alg}}$ by [33, Theorem 9.4(c)], the isomorphism $T^{w_0\cdot\mu_0}_{w_0\cdot 0}(M(w_{n-1,n-k}\cdot 0)) \cong M(w_{n-1,n-k}\cdot \mu_0)$ necessarily induces an isomorphism $T^{w_0\cdot\mu_0}_{w_0\cdot 0}(M^J(w_{n-1,n-k}\cdot 0)) \cong M^J(w_{n-1,n-k}\cdot \mu_0)$. Since $M^J(\mu)$ for $\mu \in \Lambda_J^{\mathrm{dom}}$ has cosocle $L(\mu)$, any quotient of $M^J(w_{n-1,n-k}\cdot \mu_0)$ of the form $M^J(\mu)$ for some $\mu \in \Lambda_J^{\mathrm{dom}}$ is necessarily $M^J(w_{n-1,n-k}\cdot \mu_0)$ itself. In particular any increasing filtration on $T^{w_0\cdot\mu_0}_{w_0\cdot 0}(M^J(w_{n-1,n-k}\cdot 0))$ as in Lemma 5.4.9 has only one non-zero graded piece, which is $M^J(w_{n-1,n-k}\cdot \mu_0)$. Then Lemma 5.4.9 implies $\mathcal{T}^{w_0\cdot\mu_0}_{w_0\cdot 0}(D_{w_{n-1,n-k}\cdot 0}) \cong D_{\mu_k}$. Thus (5.173) follows from (i) of Proposition 5.4.6. □

Applying $\mathcal{T}^{w_0\cdot\mu_0}_{w_0\cdot 0}$ to the complex (5.167) we deduce from Lemma 5.4.11 an isomorphism of complexes of coadmissible $D(G)$-modules:

$$\mathcal{T}^{w_0\cdot\mu_0}_{w_0\cdot 0}(\Omega^\bullet) \cong \mathbf{D}^\bullet_{\mu_0}.$$

This is essentially equivalent to what Schneider does in [64, Theorem 3]: though there is no mention of infinitesimal characters there, in [64, p.643] he first defines a decreasing filtration on the complex $\Omega^\bullet \otimes_K L(1)$ (note that $L(1) = L(\mu_0)$ is the unique finite dimensional simple $U(\mathfrak{g})$-module with highest weight in the $W(G)$-orbit of $w_0\cdot\mu_0 - w_0\cdot 0 = \mu_0$) and then in the proof of [64, Thm. 3] projects onto the unique graded piece with infinitesimal character ξ_{μ_0}. As $\mathcal{T}^{w_0\cdot\mu_0}_{w_0\cdot 0}$ is exact, we deduce from Theorem 5.4.5 for $0 \le k \le n-1$ (with the notation of (5.110)

$$H^k(\mathbf{D}^\bullet_{\mu_0}) \cong T^{w_0\cdot\mu_0}_{w_0\cdot 0}(H^k(\Omega^\bullet)) \cong L(1)\otimes_K (V^\infty_{[1,n-k-1],\Delta})^\vee = (V^{\mathrm{alg}}_{[1,n-k-1],\Delta})^\vee. \tag{5.174}$$

We need the following result on the wall-crossing of holomorphic discrete series.

Lemma 5.4.12 *Let $1 \le k \le n-1$ and $\mu \in \Lambda$ such that $\langle \mu+\rho, \alpha^\vee\rangle \ge 0$ for $\alpha \in \Phi^+$ and the stabilizer of μ in $W(G)$ for the dot action is $\{1, s_k\}$. We have a short exact sequence of coadmissible $D(G)$-modules*

$$0 \longrightarrow D_{\mu_k} \longrightarrow \Theta_\mu(D_{\mu_k}) \longrightarrow D_{\mu_{k-1}} \longrightarrow 0. \tag{5.175}$$

Proof Recall from above (5.87) that $\Theta_\mu(D_{\mu_k}) = \mathcal{T}^{w_0\cdot\mu_0}_{\mu}(\mathcal{T}^{\mu}_{w_0\cdot\mu_0}(D_{\mu_k}))$. By Lemma 5.4.9 applied with $\nu = \mu_k$ and $\lambda = w_0\cdot\mu_0$, we have a decreasing exhaustive separated filtration $(\mathrm{Fil}^i(D))_{i\in\mathbb{Z}}$ on $D \overset{\mathrm{def}}{=} \mathcal{T}^{\mu}_{w_0\cdot\mu_0}(D_{\mu_k})$, and an increasing exhaustive separated filtration $(\mathrm{Fil}_i(M))_{i\in\mathbb{Z}}$ on $M \overset{\mathrm{def}}{=} T^{\mu}_{w_0\cdot\mu_0}(M^J(\mu_k))$ such that for $i \in \mathbb{Z}$ we have either $\mathrm{gr}^i(D) = \mathrm{gr}_i(M) = 0$, or $\mathrm{gr}^i(D) \cong D_{\kappa_i}$ and $\mathrm{gr}_i(M) \cong M^J(\kappa_i)$ for some $\kappa_i \in \Lambda_J^{\mathrm{dom}}$. Then we apply Lemma 5.4.9 again to $\mathcal{T}^{w_0\cdot\mu_0}_{\mu}(\mathrm{gr}^i(D))$ (i.e. with $\nu = \kappa_i$, $\lambda = \mu$, $\mu = w_0\cdot\mu_0$) for $i \in \mathbb{Z}$ such that $0 \ne \mathrm{gr}^i(D) \cong D_{\kappa_i}$, and obtain a decreasing exhaustive separated filtration $(\mathrm{Fil}^j(\Theta_\mu(D_{\mu_k})))_{j\in\mathbb{Z}}$ on $\Theta_\mu(D_{\mu_k}) = \mathcal{T}^{w_0\cdot\mu_0}_{\mu}(D)$ which refines the decreasing filtration $(\mathcal{T}^{w_0\cdot\mu_0}_{\mu}(\mathrm{Fil}^i(D)))_{i\in\mathbb{Z}}$, and an increasing exhaustive separated filtration $(\mathrm{Fil}_j(\Theta_{s_k}(M^J(\mu_k))))_{j\in\mathbb{Z}}$ on $\Theta_{s_k}(M^J(\mu_k)) = T^{w_0\cdot\mu_0}_{\mu}(M)$

which refines the increasing filtration $(T_{\mu}^{w_0\cdot\mu_0}(\mathrm{Fil}_i(M)))_{i\in\mathbb{Z}})$ such that for each $j\in\mathbb{Z}$ we have either $\mathrm{gr}^j(\Theta_\mu(D_{\mu_k}))=\mathrm{gr}_j(\Theta_{s_k}(M^J(\mu_k)))=0$, or $\mathrm{gr}^j(\Theta_\mu(D_{\mu_k}))\cong D_{\theta_j}$ and $\mathrm{gr}_j(\Theta_{s_k}(M^J(\mu_k)))\cong M^J(\theta_j)$ for some $\theta_j\in\Lambda_J^{\mathrm{dom}}$. However, by (ii) of Lemma 3.4.15 and up to some reindexation, the only such filtration on $\Theta_{s_k}(M^J(\mu_k))$ is

$$0=\mathrm{Fil}_0(\Theta_{s_k}(M^J(\mu_k)))\subsetneq \mathrm{Fil}_1(\Theta_{s_k}(M^J(\mu_k)))=M^J(\mu_{k-1})\subsetneq \mathrm{Fil}_2(\Theta_{s_k}(M^J(\mu_k)))=\Theta_{s_k}(M^J(\mu_k))$$

with $\mathrm{gr}_1(\Theta_{s_k}(M^J(\mu_k)))\cong M^J(\mu_{k-1})$ and $\mathrm{gr}_2(\Theta_{s_k}(M^J(\mu_k)))\cong M^J(\mu_k)$. This forces $\Theta_\mu(D_{\mu_k})$ to admit a filtration

$$\Theta_\mu(D_{\mu_k})=\mathrm{Fil}^1(\Theta_\mu(D_{\mu_k}))\supsetneq \mathrm{Fil}^2(\Theta_\mu(D_{\mu_k}))=D_{\mu_k}\supsetneq \mathrm{Fil}^3(\Theta_\mu(D_{\mu_k}))=0$$

with $\mathrm{gr}^1(\Theta_\mu(D_{\mu_k}))\cong D_{\mu_{k-1}}$ and $\mathrm{gr}^2(\Theta_\mu(D_{\mu_k}))\cong D_{\mu_k}$. Hence we have (5.175). □

Remark 5.4.13 We will see in the proof of Theorem 5.4.16 below that the injection $D_{\mu_k}\to\Theta_\mu(D_{\mu_k})$ in Lemma 5.4.12 is the canonical adjunction map.

We now slightly reformulate (a weak form of) [50, Theorem 2.2].

Theorem 5.4.14 *Let $\mu\in\Lambda_J^{\mathrm{dom}}$. Then D_μ in (5.169) is a finite length coadmissible $D(G)$-module which admits a decreasing filtration*

$$D_\mu=\mathrm{Fil}^0(D_\mu)\supseteq\mathrm{Fil}^1(D_\mu)\supseteq\cdots\supseteq\mathrm{Fil}^{n-1}(D_\mu)$$

satisfying the following conditions:

(i) $\mathrm{gr}^{n-1}(D_\mu)\neq 0$ if and only if $\mu\in\Lambda^{\mathrm{dom}}$, in which case $\mathrm{gr}^{n-1}(D_\mu)\cong L(\mu)$;
(ii) for $0\leq\ell\leq n-2$, there exists M_ℓ in $\mathcal{O}_{\mathrm{alg}}^{\mathfrak{p}_{\widehat{\ell+1}}}$ such that $\mathrm{gr}^\ell(D_\mu)$ fits into a short exact sequence of coadmissible $D(G)$-modules if $w_{\ell+1,n-1}\cdot\mu\in\Lambda^{\mathrm{dom}}$

$$0\to\mathcal{F}_{P_{\widehat{\ell+1}}}^G(M_\ell,\pi_{\ell+1,\ell+1}^\infty)^\vee\to\mathrm{gr}^\ell(D_\mu)\to L(w_{\ell+1,n-1}\cdot\mu)\otimes_K(V_{[1,\ell],\Delta}^\infty)^\vee\to 0,$$

and $\mathrm{gr}^\ell(D_\mu)\cong\mathcal{F}_{P_{\widehat{\ell+1}}}^G(M_\ell,\pi_{\ell+1,\ell+1}^\infty)^\vee$ otherwise.

Proof Recall from (5.169) that $D_\mu=\mathcal{F}_\mu(\mathbb{H})$ where $\mathcal{F}_\mu$ is the analytification on $\mathbb{P}_{\mathrm{rig}}^{n-1}$ of the G-equivariant vector bundle $\mathcal{F}_{L^I(w_{1,n-1}(\mu))}$ on $\mathbb{P}^{n-1}$. It follows from [30, Theorem 6.1] (see also [42, Theorem 6.4]) and from the definition of Λ^{dom} (using $w_0=w_Iw_{1,n-1}$) that $H^{n-1-\ell}(\mathbb{P}^{n-1},\mathcal{F}_{L^I(w_{1,n-1}(\mu))})\neq 0$ if and only if $w_{\ell+1,n-1}\cdot\mu\in\Lambda^{\mathrm{dom}}$, in which case $H^{n-1-\ell}(\mathbb{P}^{n-1},\mathcal{F}_{L^I(w_{1,n-1}(\mu))})\cong L(w_{\ell+1,n-1}\cdot\mu)$. Also recall that we have for $0\leq\ell\leq n-2$ (see the line below (2.71))

$$\pi_{\ell+1,\ell+1}^\infty\cong 1_{\ell+1}\boxtimes_E\mathrm{St}_{n-1-\ell}^\infty=V_{[1,\ell],\widehat{\ell+1}}^\infty.$$

Define for $0\leq\ell\leq n-2$ (see [50, (7)] and the notation there)

$$M_\ell \stackrel{\text{def}}{=} \ker\left(H^{n-1-\ell}_{\mathbb{P}^\ell}(\mathbb{P}^{n-1}, \mathcal{F}_{L^I(w_{1,n-1}(\mu))}) \longrightarrow H^{n-1-\ell}(\mathbb{P}^{n-1}, \mathcal{F}_{L^I(w_{1,n-1}(\mu))})\right),$$

then (i) and (ii) follow from [50, Lemma 2.1], [50, Theorem 2.2] (based on [49, Theorem 1]) and the discussion that follows *loc. cit.* □

Recall that $J = \widehat{n-1}$ and $\mu_k = w_{n-1,n-k} \cdot \mu_0$ for $0 \leq k \leq n-1$. We will need the following technical result.

Lemma 5.4.15 *For $0 \leq k \leq n-1$ (and $\mu_0 \in \Lambda^{\text{dom}}$) we have a $D(L_J)$-equivariant isomorphism*

$$H^0(N_J, D_{\mu_k}) \cong L^J(\mu_k) = L^J(w_{n-1,n-k}). \tag{5.176}$$

Proof Note first that we have an equality of closed vector subspaces of D_{μ_k}

$$H^0(N_J, D_{\mu_k}) = H^0(N_J, H^0(\mathfrak{n}_J, D_{\mu_k})). \tag{5.177}$$

Recall that $\mathbb{H}$ is by definition contained in the rigid analytic space associated with the Zariski open *affine* subscheme $P_J w_0 P_I / P_I$ of $G/P_I = \mathbb{P}^{n-1}_{/K}$. Using $P_J = N_J L_J$, $w_0 L_J w_0 = L_I$ and $w_0 = w_{n-1,1} w_I$, we have (as affine schemes)

$$\begin{aligned} P_J w_0 P_I = N_J L_J w_0 P_I = N_J w_0 (w_0 L_J w_0) P_I = N_J w_0 P_I \\ = N_J w_{n-1,1} P_I \cong N_J \times (w_{n-1,1} P_I). \end{aligned} \tag{5.178}$$

Let $L^I \stackrel{\text{def}}{=} L^I(w_{1,n-1}(\mu_k))$, the restriction $\mathcal{F}_{L^I(w_{1,n-1}(\mu_k))}|_{P_J w_0 P_I / P_I} = \mathcal{F}_{L^I}|_{P_J w_0 P_I / P_I}$ is by definition the quotient $(P_J w_0 P_I \times L^I)/P_I$ where P_I acts on $P_J w_0 P_I \times L^I$ by

$$(P_J w_0 P_I \times L^I) \times P_I \to P_J w_0 P_I \times L^I, \ ((x, v), h) \mapsto (xh, h^{-1} \cdot v).$$

Using (5.178), we write each $x \in P_J w_0 P_I$ as $x = a w_{n-1,1} b$ for some $a \in N_J$ and $b \in P_I$ (uniquely determined by x). We consider the following isomorphism

$$P_J w_0 P_I \times L^I \xrightarrow{\sim} P_J w_0 P_I \times L^I, \ (a w_{n-1,1} b, v) \mapsto (a w_{n-1,1} b, b \cdot v)$$

which descends to an isomorphism

$$(P_J w_0 P_I \times L^I)/P_I \xrightarrow{\sim} (P_J w_0 P_I)/P_I \times L^I, \ (a w_{n-1,1} b, v) P_I \mapsto (a w_{n-1,1} b P_I, b \cdot v). \tag{5.179}$$

We have a (left) P_J-action on $(P_J w_0 P_I \times L^I)/P_I$ given by

$$P_J \times ((P_J w_0 P_I \times L^I)/P_I) \to (P_J w_0 P_I \times L^I)/P_I, \ (g, (x, v) P_I) \mapsto (gx, v) P_I. \tag{5.180}$$

For $g = g' g'' \in N_J L_J = P_J$ and $x = a w_{n-1,1} b \in N_J w_{n-1,1} P_I = P_J w_0 P_I$, we have

$$gx = g'(g'' a (g'')^{-1}) w_{n-1,1} (w_{1,n-1} g'' w_{1,n-1}^{-1}) b$$

with $g'(g''a(g'')^{-1}) \in N_J$ and $(w_{1,n-1}g''w_{1,n-1}^{-1}) \in L_I$. Hence, the P_J-action (5.180) translates via (5.179) into the P_J-action (with the above notation)

$$P_J \times ((P_J w_0 P_I)/P_I \times L^I) \to (P_J w_0 P_I)/P_I \times L^I, \ (g, (xP_I, v)) \\ \mapsto (gxP_I, (w_{1,n-1}g''w_{1,n-1}^{-1}) \cdot v). \quad (5.181)$$

Recall from the notation in Sect. 1.4 that $(L^I)^{w_{1,n-1}}$ is the $L_J = w_{1,n-1}^{-1} L_I w_{1,n-1}$-representation with same underlying space as L^I and $g'' \in L_J$ acting as $w_{1,n-1}g''w_{1,n-1}^{-1}$ on L^I. It follows from the above discussion that (5.179) gives a P_J-equivariant isomorphism of P_J-equivariant algebraic vector bundles on $P_J w_0 P_I/P_I$ (with the P_J-action (5.180) on the left and the P_J-action (5.181) on the right):

$$\mathcal{F}_{L^I}|_{P_J w_0 P_I/P_I} \cong \mathcal{O}_{P_J w_0 P_I/P_I} \times (L^I)^{w_{1,n-1}}.$$

As in (5.172), we have $(L^I)^{w_{1,n-1}} \cong L^J(\mu_k)$ as L_J-representations. Taking rigid analytic sections over $\mathbb{H}$, we obtain a P_J-equivariant topological isomorphism of Fréchet spaces

$$D_{\mu_k} \overset{(5.169)}{=} \mathcal{F}_{L^I}(\mathbb{H}) \cong \mathcal{O}(\mathbb{H}) \otimes_K L^J(\mu_k) = \Omega^0 \otimes_K L^J(\mu_k). \quad (5.182)$$

The N_J-equivariant isomorphism of schemes $P_J w_0 P_I/P_I = N_J w_0 P_I/P_I \xrightarrow{\sim} N_J$ induces a N_J-equivariant isomorphism between their algebraic de Rham complexes $\Omega^\bullet_{P_J w_0 P_I/P_I} \cong \Omega^\bullet_{N_J}$. As N_J is the affine space $\mathbb{A}^{\dim N_J}_{/K}$, $\Omega^\bullet_{N_J}$ is naturally identified with the Chevalley-Eilenberg complex $\mathrm{Hom}_K(\wedge^\bullet \mathfrak{n}_J, \mathcal{O}_{N_J}) \cong \mathrm{Hom}_K(\wedge^\bullet \mathfrak{n}_J, \mathcal{O}_{P_J w_0 P_I/P_I})$. Taking rigid analytic sections over $\mathbb{H}$, we can identify the de Rham complex (5.167) with $\mathrm{Hom}_K(\wedge^\bullet \mathfrak{n}_J, \Omega^0)$, which is the Chevalley-Eilenberg complex of Ω^0 (see also [64, p. 635] for the isomorphism $\Omega^j \cong \mathrm{Hom}_K(\wedge^j \mathfrak{n}_J, \Omega^0)$ for $0 \le j \le n-1$). In particular, we obtain an equality of closed vector subspaces of Ω^0

$$H^0(\mathfrak{n}_J, \Omega^0) = H^0(\Omega^\bullet) \cong 1_G^\vee$$

(where the second isomorphism follows from Theorem 5.4.5). With (5.182) (and as $\mathfrak{n}_J$ acts trivially on $(L^I)^{w_{1,n-1}} \cong L^J(\mu_k)$) this gives a P_J-equivariant isomorphism of Fréchet spaces

$$H^0(\mathfrak{n}_J, D_{\mu_k}) \cong H^0(\mathfrak{n}_J, \Omega^0) \otimes_K L^J(\mu_k) = L^J(\mu_k)$$

with N_J acting trivially on $L^J(\mu_k)$. This together with (5.177) gives (5.176). □

We actually only use Theorem 5.4.14 for $\mu = \mu_{n-1}$, and likewise Lemma 5.4.15 for $k = n-1$, in order to prove the key result that follows.

Theorem 5.4.16 *For $0 \le k \le n-1$ we have an isomorphism of finite length coadmissible $D(G)$-modules*

$$D_k \cong D_{\mu_k}. \tag{5.183}$$

Moreover, for $0 \leq k \leq n-2$, *the differential map* $D_{\mu_k} \to D_{\mu_{k+1}}$ *is the unique (up to scalar) non-zero map* $D_k \to D_{k+1}$ *with image* X_{k+1}.

Proof We prove the first statement. Note first that $D_{\mu_{n-1}} = D_{w_{n-1,1}\cdot\mu_0}$ is a finite length coadmissible $D(G)$-module by the first statement of Theorem 5.4.14. Condition (i) of Theorem 5.4.4 holds for $C = D_{\mu_{n-1}}$ by Lemma 5.4.15. Condition (ii) of Theorem 5.4.4 holds for $C = D_{\mu_{n-1}}$ by Theorem 5.4.14, noting that $w_{\ell+1,n-1} \cdot \mu_{n-1} \in \Lambda^{\mathrm{dom}}$ if and only if $\ell = 0$ (in which case it is μ_0). Hence $D_{n-1} \cong D_{\mu_{n-1}}$ by Theorem 5.4.4. One easily checks from (5.135) and (5.137) that the (irreducible) cosocle of D_{n-1} does not appear in D_{n-2}, hence $\mathrm{Hom}_G(D_{n-1}, D_{n-2}) = 0$. Then it follows from (iii) of Theorem 5.4.1 (for $k = n-1$) that there is (up to a non-zero scalar) a unique injection $D_{n-1} \hookrightarrow \Theta_\mu(D_{n-1})$ (with the notation of *loc. cit.*). In particular this must be the injection in (5.175) for $k = n-1$. Then it follows from (iii) of Theorem 5.4.1 and (5.175) again (both applied with $k = n-1$) that we have $D_{n-2} \cong D_{\mu_{n-2}}$. Since we again have $\mathrm{Hom}_G(D_{n-2}, D_{n-3}) = 0$, the same argument for $k = n-2$ instead of $k = n-1$ gives $D_{n-3} \cong D_{\mu_{n-3}}$. By descending induction we deduce (5.183) for all k.

We prove the second statement. For $0 \leq k \leq n-2$, as $H^{k+1}(\mathbf{D}^\bullet_{\mu_0}) \cong (V^{\mathrm{alg}}_{[1,n-k-2],\Delta})^\vee$ (see (5.174)) and $(V^{\mathrm{alg}}_{[1,n-k-2],\Delta})^\vee$ does not show up in $\mathrm{soc}_{D(G)}(D_{\mu_{k+1}}) \cong \mathrm{soc}_{D(G)}(D_{k+1})$ (using (5.137)), we deduce that the differential map $D_{\mu_k} \to D_{\mu_{k+1}}$ is non-zero. Therefore it must be the unique (up to scalar) non-zero map $D_k \to D_{k+1}$ with image X_{k+1} (see the discussion above (5.151)). □

Remark 5.4.17 By the definition of D_{n-1} below (5.149), we have $\mathrm{cosoc}_{D(G)}(D_{n-1}) = (\mathrm{St}_n^{\mathrm{alg}})^\vee$. When $\mu_0 = (0, \cdots, 0)$, we have $D_{n-1} \cong \Omega^{n-1}$ by Theorem 5.4.16 and (i) of Proposition 5.4.6. Thus we obtain $\mathrm{cosoc}_{D(G)}(\Omega^{n-1}) = (\mathrm{St}_n^\infty)^\vee$. Then, by an argument parallel to [76, Corollary 6.11] using [34], one can expect to deduce from this that Ω^{n-1} is a quotient of the dual of the locally K-analytic Steinberg $(\mathrm{St}_n^{\mathrm{an}})^\vee$, and thus a subquotient of $\mathcal{F}_B^G(U(\mathfrak{g}) \otimes_{U(\mathfrak{b})} 0, 1_T)^\vee$. Using Theorem 5.4.16, Lemma 5.4.11 and (5.86), it would follow that $D_{n-1} \cong D_{\mu_{n-1}} \cong \mathcal{T}_{w\cdot 0}^{w\cdot\mu_0}(\Omega^{n-1})$ is a subquotient of $\mathcal{T}_{w\cdot 0}^{w\cdot\mu_0}(\mathcal{F}_B^G(U(\mathfrak{g}) \otimes_{U(\mathfrak{b})} 0, 1_T)^\vee) \cong \mathcal{F}_B^G(U(\mathfrak{g}) \otimes_{U(\mathfrak{b})} \mu_0, 1_T)^\vee$, which by Lemma 5.4.12 (or (iii) of Theorem 5.4.1) would imply that $D_k \cong D_{\mu_k}$ is a subquotient of

$$\Theta_{s_{k+1}} \cdots \Theta_{s_{n-1}}(\mathcal{F}_B^G(U(\mathfrak{g}) \otimes_{U(\mathfrak{b})} \mu_0, 1_T)^\vee) \cong \mathcal{F}_B^G(Q_k, 1_T)$$

where $Q_k \overset{\mathrm{def}}{=} \Theta_{s_{k+1}} \cdots \Theta_{s_{n-1}}(U(\mathfrak{g}) \otimes_{U(\mathfrak{b})} \mu_0) \in \mathcal{O}^{\mathfrak{b}}_{\mathrm{alg}}$.

Recall that we have defined an explicit complex of finite length coadmissible $D(G)$-modules with Orlik-Strauch constituents $\widetilde{\mathbf{D}}^\bullet$ in (5.152).

Corollary 5.4.18 *For* $\mu_0 \in \Lambda^{\mathrm{dom}}$ *we have two morphisms of complexes of finite length coadmissible* $D(G)$*-modules with Orlik-Strauch constituents*

$$H^{n-1}(\mathbf{D}^\bullet_{\mu_0})[-(n-1)] \longleftarrow \widetilde{\mathbf{D}}^\bullet \longrightarrow \mathbf{D}^\bullet_{\mu_0}$$

which give an explicit section to the morphism of complexes $\mathbf{D}^\bullet_{\mu_0} \twoheadrightarrow H^{n-1}(\mathbf{D}^\bullet_{\mu_0})[-(n-1)]$ *in the derived category of finite length coadmissible* $D(G)$*-modules with Orlik-Strauch constituents.*

Proof This follows directly from (5.151), Theorem 5.3.13, (5.171) and Theorem 5.4.16. □

One also has statements analogous to Proposition 5.3.14 and Corollary 5.3.15 with $\mathbf{D}^\bullet_{\mu_0}$ instead of $\mathbf{D}^\bullet$. In particular, if we define $\widetilde{\Omega}^\bullet \stackrel{\text{def}}{=} \widetilde{\mathbf{D}}^\bullet$ when $\mu_0 = (0, \cdots, 0)$, we at last obtain one of our main results.

Corollary 5.4.19 *We have two morphisms of complexes of finite length coadmissible* $D(G)$*-modules with Orlik-Strauch constituents.*

$$H^{n-1}(\Omega^\bullet)[-(n-1)] \longleftarrow \widetilde{\Omega}^\bullet \longrightarrow \Omega^\bullet$$

which give an explicit section to the morphism of complexes $\Omega^\bullet \twoheadrightarrow H^{n-1}(\Omega^\bullet)[-(n-1)]$ *in the derived category of finite length coadmissible* $D(G)$*-modules with Orlik-Strauch constituents.*

Remark 5.4.20 Assume $\mu_0 = (0, \cdots, 0)$ and, changing notation, denote by $\widetilde{\Omega}^{n-1}$ the unique coadmissible $D(G)$-module with cosocle $((\mathrm{St}_n^\infty)^{\oplus n})^\vee$ which sits in a short exact sequence

$$0 \longrightarrow Z_{n-1} \longrightarrow \widetilde{\Omega}^{n-1} \longrightarrow \big((\mathrm{St}_n^\infty)^{\oplus n}\big)^\vee \longrightarrow 0$$

(see (5.154)). Then the complex $[\widetilde{\Omega}^0 \to \widetilde{\Omega}^1 \to \cdots \to \widetilde{\Omega}^{n-2} \to \widetilde{\Omega}^{n-1}]$ (the same as in (5.152) except that we have modified the term in degree $n-1$) is canonical, exact in degrees $< n-1$ and its H^{n-1} is $((\mathrm{St}_n^\infty)^{\oplus n})^\vee$. That is, we have "made exact" the de Rham complex $\Omega^\bullet$ in degrees $< n-1$ at the expense of replacing $H^{n-1}(\Omega^\bullet) = (\mathrm{St}_n^\infty)^\vee$ by $((\mathrm{St}_n^\infty)^{\oplus n})^\vee$. These properties look similar to the properties of the complexes obtained as χ-isotypic direct factors of the global sections of the de Rham complex of the first covering of $\mathbb{H}$, where χ is a smooth character of $\mathcal{O}_D^\times$ and D is the division algebra over K of invariant $1/n$, see for instance Junger's work [36]. Finally, note that, by (v) of Theorem 5.4.1, with this modified complex (iv) of Theorem 5.4.1 now also holds when $k = n-1$.

Remark 5.4.21 When $n = 2$ and $K = \mathbb{Q}_p$, the study of the global sections of the de Rham complex of Drinfeld's coverings Σ_n $(n \geq 1)$ of $\mathbb{H}$ is very important for the p-adic Langlands program for $\mathrm{GL}_2(\mathbb{Q}_p)$, see for instance the deep results of [23] or [53]. Moreover, when $n = 2$ and K is arbitrary, at least for the first covering Σ_1 we now know from [1, Corollary B], [54, Theorem 7.2.1, Remark 7.2.4] that $H^0(\Sigma_1, \mathcal{O}_{\Sigma_1})$ and $H^0(\Sigma_1, \Omega^1_{\Sigma_1})$ are finite length coadmissible $D(G)$-modules (or rather $D(G^0)$-modules here, where $G^0 \subset G$ is the subgroup of matrices with determinant in $\mathcal{O}_K^\times$).

It is therefore a natural question to ask for the internal structure of these $D(G^0)$-modules, and more generally of all the $H^0(\Sigma_n, \Omega^j_{\Sigma_n})$, $n \geq 1$, $j \geq 0$. However, it is unlikely that one can apply (at least directly) the techniques of this work which are adapted to Orlik-Strauch representations, i.e. subquotients of locally anaytic parabolic inductions of locally algebraic representations.

Appendix A
Combinatorics in $W(G)$

We prove several technical combinatorial lemmas involving specific elements of $W(G)$ that are used throughout this work.

We equip the set $\mathbb{Z} \times \mathbb{Z}$ with the partial order $(a, b) \leq (a', b')$ if and only if $a \leq a'$ and $b \leq b'$. Let $\Sigma \subseteq \mathbb{Z} \times \mathbb{Z}$ be a finite subset such that $1 \leq a - b \leq n - 1$ for each $(a, b) \in \Sigma$. For $k \in \mathbb{Z}$, we set

$$x_{\Sigma,k} \stackrel{\text{def}}{=} \prod_{(a,b)\in\Sigma, a+b=k} s_{a-b}. \tag{A.1}$$

One easily checks that $s_{a-b}s_{a'-b'} = s_{a'-b'}s_{a-b}$ for $(a, b), (a', b') \in \Sigma$ satisfying $a + b = a' + b' = k$, so the definition of $x_{\Sigma,k}$ is independent of the order of the s_{a-b} in (A.1). Note that

$$\ell(x_{\Sigma,k}) = \#\{(a, b) \in \Sigma \mid a + b = k\}.$$

As Σ is a finite set, we have $x_{\Sigma,k} \neq 1$ only for finitely many $k \in \mathbb{Z}$, so we can define

$$x_\Sigma \stackrel{\text{def}}{=} \cdots x_{\Sigma,k} \cdot x_{\Sigma,k-1} \cdots \in W(G).$$

Definition A.1 Let $\Sigma \subseteq \mathbb{Z} \times \mathbb{Z}$ be a finite subset such that $1 \leq a - b \leq n - 1$ for $(a, b) \in \Sigma$.

(i) We say that Σ is an *expansion* of x_Σ if $\ell(x_\Sigma) = \sum_{k\in\mathbb{Z}} \ell(x_{\Sigma,k})$ $(= \#\Sigma)$.
(ii) We say that Σ is *saturated* if the following extra condition holds: for $(a, b), (a', b') \in \Sigma$ and $(a'', b'') \in \mathbb{Z} \times \mathbb{Z}$ such that $(a, b) \leq (a'', b'') \leq (a', b')$, we have $(a'', b'') \in \Sigma$.
(iii) We say that a saturated Σ is *connected* if for each pair of saturated subsets Σ_1 and Σ_2 satisfying $\Sigma_1 \cap \Sigma_2 = \emptyset$ and $\Sigma = \Sigma_1 \sqcup \Sigma_2$, there exists $(a_1, b_1) \in \Sigma_1$ and $(a_2, b_2) \in \Sigma_2$ such that $|(a_1 - b_1) - (a_2 - b_2)| \leq 1$.
(iv) We say that $x \in W(G)$ is *saturated* if $x = x_\Sigma$ for a saturated Σ.
(v) We say a saturated x_Σ is *connected* if Σ can be moreover taken connected.

C. Breuil and Z. Qian, *Splitting and Making Explicit the de Rham Complex of the Drinfeld Space*, Lecture Notes in Mathematics 2397,
https://doi.org/10.1007/978-3-032-20095-2

We define the *saturated closure* of any subset $\Sigma \subseteq \mathbb{Z} \times \mathbb{Z}$ to be the minimal saturated subset of $\mathbb{Z} \times \mathbb{Z}$ which contains Σ, if it exists. Note that the saturated closure can contain elements (a'', b'') such that $a'' - b'' \notin \{1, \ldots, n-1\}$ (see Example A.3 below). Note also that there is an obvious way to decompose an arbitrary saturated subset $\Sigma \subseteq \mathbb{Z} \times \mathbb{Z}$ into its connected components.

For $x \in W(G)$ we let $\mathrm{Supp}(x) \subseteq \Delta$ be the set of $j \in \{1, \ldots, n-1\}$ such that s_j appears in one (and thus all) reduced decomposition of x. Recall that $x \in W(G)$ is called Coxeter if it is a product of all distinct simple reflections (each with multiplicity 1). We say that $x \in W(G)$ is *partial-Coxeter* if there exists a reduced decomposition of x which is multiplicity free, i.e. s_j shows up at most once for each $j \in \Delta$. One easily checks from the braid relations that *any* reduced decomposition of a partial-Coxeter element is multiplicity free.

Lemma A.2 *Let $\Sigma \subseteq \mathbb{Z} \times \mathbb{Z}$ be a saturated subset.*

(i) We have $\ell(x_\Sigma) = \#\Sigma$, i.e. Σ is an expansion of x_Σ. Moreover, there exists a natural bijection between the set of reduced decompositions of x_Σ and the set of total orders on Σ refining the partial order on Σ induced by the one on $\mathbb{Z} \times \mathbb{Z}$.

(ii) We have $j \in D_L(x_\Sigma)$ (resp. $j \in D_R(x_\Sigma)$) if and only if there exists a maximal (resp. minimal) element $(a, b) \in \Sigma$ such that $j = a - b$.

(iii) If Σ is connected and $\Sigma' \subseteq \mathbb{Z} \times \mathbb{Z}$ is another saturated subset such that $x_\Sigma = x_{\Sigma'}$, there exists $c \in \mathbb{Z}$ such that $\Sigma' = \{(a-c, b-c) \mid (a, b) \in \Sigma\}$.

(iv) The element x_Σ is partial-Coxeter if and only if the map

$$\Sigma \longrightarrow \mathrm{Supp}(x_\Sigma),\ (a, b) \longmapsto a - b \tag{A.2}$$

is bijective. Moreover, all partial-Coxeter elements have the form x_Σ for a saturated Σ.

Proof We start with several preliminaries. We first fix an arbitrary finite $\Sigma \subseteq \mathbb{Z} \times \mathbb{Z}$ such that $1 \le a - b \le n - 1$ for $(a, b) \in \Sigma$. We write $\le^*$ for a total order on Σ which refines the fixed partial order $\le$.

For each total order $\le^*$ on Σ that refines $\le$, we consider the word

$$x_\Sigma(\le^*) \overset{\text{def}}{=} \prod_{(a,b)\in\Sigma,\le^*} s_{a,b} \tag{A.3}$$

where $s_{a,b}$ shows up at the *right* of $s_{a',b'}$ whenever $(a, b) <^* (a', b')$. Here we use the symbol $s_{a,b}$ for the copy of s_{a-b} corresponding to (a, b) inside the word (A.3).

There exists a total order $\le^*$ on Σ that refines $\le$ such that x_Σ is the image of the word $x_\Sigma(\le^*)$ in $W(G)$. This is obvious as we can fix an arbitrary total order on $\{(a, b) \in \Sigma \mid a + b = k\}$ for each $k \in \mathbb{Z}$, and then glue them to a total order $\le^*$ on Σ by requiring that $(a, b) \le^* (a', b')$ whenever $a + b \le a' + b'$.

For two different total orders $\le_1^*$ and $\le_2^*$ on Σ which refine $\le$, we say that $\le_1^*$ and $\le_2^*$ are *adjacent* if there exists exactly one pair $(a, b), (a', b') \in \Sigma$ such that

$(a, b) <_1^* (a', b')$ and $(a', b') <_2^* (a, b)$. If both words (A.3) have the same image in $W(G)$, we see that $x_\Sigma(\leq_1^*)$ is reduced if and only if $x_\Sigma(\leq_2^*)$ is reduced (since then both have length $\#\Sigma$).

Given two total orders $\leq_1^*$ and $\leq_2^*$ on Σ which refine $\leq$, we say that $\leq_1^*$ and $\leq_2^*$ are *equivalent* if there exist an integer $t \geq 0$ and total orders $\leq_{1,t'}^*$ on Σ for $0 \leq t' \leq t$ which refine $\leq$ such that $\leq_{1,0}^* = \leq_1^*$, $\leq_{1,t}^* = \leq_2^*$ and $\leq_{1,t'}^*$ is adjacent to $\leq_{1,t'-1}^*$ for $1 \leq t' \leq t$.

We now prove that all total orders on Σ which refine $\leq$ are equivalent. It suffices to prove that any total order $\leq^*$ is equivalent to the total order $\leq_0^*$ defined by $(a, b) \leq_0^* (a', b')$ if and only if $a + b \leq a' + b'$ and $a \leq a'$ (which refines $\leq$). As a total order on Σ corresponds to a bijection between Σ and the set $\{1, \ldots, \#\Sigma\}$, $\leq^*$ differs from $\leq_0^*$ by a permutation, and we write $\ell(\leq^*)$ for the length of this permutation (with $\ell(\leq_0^*) = 0$). We prove that $\leq^*$ is equivalent to $\leq_0^*$ by induction on $\ell(\leq^*)$. If $\ell(\leq^*) = 0$, there is nothing to prove. If $\ell(\leq^*) > 0$, then the subset

$$\{(a, b) \in \Sigma \text{ such that } \exists\, (a', b') \in \Sigma \text{ with } (a', b') <_0^* (a, b) \text{ and } (a, b) <^* (a', b')\}$$

has a maximal element under $\leq_0^*$. We choose such a maximal element (a, b), and we choose (a', b') in the above subset to be minimal under $\leq^*$. Assume that there exists $(a'', b'') \in \Sigma$ such that $(a, b) <^* (a'', b'') <^* (a', b')$. We can assume (a'', b'') to be the minimal such element under $\leq^*$. The minimality of our choice of (a', b') forces $(a, b) <_0^* (a'', b'')$, which in turn contradicts the maximality of (a, b) (as $(a'', b'') <^* (a', b')$ and $(a', b') <_0^* (a, b) <_0^* (a'', b'')$). Hence, there is no element between (a, b) and (a', b') under $\leq^*$, and we can define $\leq_1^*$ by interchanging (a, b) and (a', b') in the total order $\leq^*$. Then $\ell(\leq_1^*) = \ell(\leq^*) - 1$ by [6, Proposition 1.5.8], so $\leq_1^*$ is equivalent to $\leq_0^*$ by induction. As $\leq_1^*$ is adjacent to $\leq^*$, it follows that $\leq^*$ is also equivalent to $\leq_0^*$.

We now assume that Σ is saturated. We prove that if we modify the word $x_\Sigma(\leq^*)$ using a braid relation, the new word is $x_\Sigma(\leq'^*)$ for another total order $\leq'^*$ on Σ (refining $\leq$) which is *adjacent* to $\leq^*$. As the braid relations are $s_j s_{j'} = s_{j'} s_j$ for some $|j - j'| \geq 2$ or $s_j s_{j-1} s_j = s_{j-1} s_j s_{j-1}$ for some $2 \leq j \leq n-1$, from the definition of adjacent it is enough to rule out the second possibility. Assume on the contrary that there exist $(a, b) <^* (a', b') <^* (a'', b'')$ in Σ such that the only element in Σ between (a, b) and (a'', b'') (under $\leq^*$) is (a', b') and such that $a - b = j = a'' - b'' = j$ and $a' - b' = j - 1$ for some $2 \leq j \leq n - 1$. Since $a - b = j = a'' - b'' = j$ and $a' - b' = j - 1$, and since $\leq^*$ refines $\leq$, it is easy to check that we must have $(a, b) < (a', b') < (a'', b'')$. Then it is also easy to see that the saturated condition on Σ forces the existence of at least another $(a''', b''') \in \Sigma$ such that $(a, b) < (a''', b''') < (a'', b'')$ and $a''' - b''' = j + 1$. But this contradicts the fact that (a', b') is the only element between (a, b) and (a'', b'') under $\leq^*$. An analogous argument shows that we cannot have elements $(a, b) <^* (a', b') <^* (a'', b'')$ in Σ such that the only element between (a, b) and (a'', b'') is (a', b') and such that $a - b = j - 1 = a'' - b'' = j - 1$ and $a' - b' = j$ for some $2 \leq j \leq n - 1$. Hence, the only braid relations that can be used to modify $x_\Sigma(\leq^*)$ are $s_j s_{j'} = s_{j'} s_j$ for some

$|j - j'| \geq 2$. This implies that the new word has the form $x_\Sigma(\leq'^*)$ for some $\leq'^*$ which is adjacent to $\leq^*$.

We now prove (i). From the previous statements, we only have to prove that $x_\Sigma(\leq^*)$ is a reduced word for one (equivalently any) total order $\leq^*$ on Σ refining $\leq$. Note first that if two copies of s_j occur in $x_\Sigma(\leq^*)$, they come from two distinct elements $(a_1, b_1), (a_2, b_2) \in \Sigma$ such that $a_1 - b_1 = a_2 - b_2 = j$. We can assume $a_1 < a_2$, and we see that $(a_1 + 1, b_1)$ also belongs to Σ as Σ is saturated. We deduce that s_{j+1} occurs in $x_\Sigma(\leq^*)$ between these two s_j. In particular, the two s_j cannot cancel after applying the braid relations. It follows that $x_\Sigma(\leq^*)$ is reduced.

We prove (ii). An element $(a, b) \in \Sigma$ is the unique maximal (resp. minimal) element under some total order $\leq^*$ if and only if (a, b) is a maximal (resp. minimal) element under the partial order $\leq$. Then the statement follows from $\ell(x_\Sigma) = \#\Sigma$ and [6, Corollary 1.4.6].

We prove (iii). Let $\Sigma' \subseteq \mathbb{Z} \times \mathbb{Z}$ be another saturated subset such that $x_\Sigma = x_{\Sigma'}$, we argue by induction on $\#\Sigma$. If $\#\Sigma = 1$, the claim is obvious. If $\#\Sigma > 1$, there exists $(a_0, b_0) \in \Sigma$ which is either minimal or maximal (for $\leq$), such that $\Sigma \setminus \{(a_0, b_0)\}$ is still saturated and connected. Let $j \stackrel{\text{def}}{=} a_0 - b_0$. If (a_0, b_0) is minimal (resp. maximal) in Σ, we have $j \in D_R(x_\Sigma) = D_R(x_{\Sigma'})$ (resp. $j \in D_L(x_\Sigma) = D_L(x_{\Sigma'})$) (using (ii)), which gives a unique minimal (resp. maximal) element $(a'_0, b'_0) \in \Sigma'$ satisfying $a'_0 - b'_0 = j$. By our induction assumption, there exists $c \in \mathbb{Z}$ such that

$$\Sigma' \setminus \{(a'_0, b'_0)\} = \{(a - c, b - c) \mid (a, b) \in \Sigma \setminus \{(a_0, b_0)\}\}.$$

Thus we only need to prove $(a'_0, b'_0) = (a_0 - c, b_0 - c)$. Assume that (a_0, b_0) is maximal in Σ. Then as Σ is connected, either $(a_0 - 1, b_0)$ or $(a_0, b_0 - 1)$ is in $\Sigma \setminus \{(a_0, b_0)\}$ but none of $(a_0 + 1, b_0)$, $(a_0, b_0 + 1)$. Thus $\Sigma' \setminus \{(a'_0, b'_0)\}$ contains either $(a_0 - c - 1, b_0 - c)$ or $(a_0 - c, b_0 - c - 1)$ but not $(a_0 - c + 1, b_0 - c)$, $(a_0 - c, b_0 - c + 1)$. As (a'_0, b'_0) is maximal in Σ' and Σ' is saturated, this forces $(a'_0, b'_0) = (a_0 - c, b_0 - c)$. The proof for (a_0, b_0) minimal is analogous.

We prove (iv). As $x_\Sigma(\leq^*_0)$ is a reduced word, the map (A.2) takes values in $\mathrm{Supp}(x_\Sigma) \subseteq \{1, \dots, n-1\}$ and is clearly surjective. Hence, (A.2) is a bijection if and only if it is injective if and only if any reduced decomposition of x_Σ is multiplicity free, which is equivalent to x_Σ being partial-Coxeter by definition. We now prove that any partial-Coxeter element $x \in W(G)$ has the form x_Σ for some saturated Σ by induction on $\ell(x)$. The case $\ell(x) = 0$ is trivial. Let $1 \neq x \in W(G)$ be partial-Coxeter and choose $j \in \mathrm{Supp}(x)$ such that $j' \notin \mathrm{Supp}(x)$ for $j' > j$. We set $x' \stackrel{\text{def}}{=} s_j x$ (resp. $x' \stackrel{\text{def}}{=} x s_j$) if $j \in D_L(x)$ (resp. if $j \in D_R(x)$). We have $j \in D_L(x) \cap D_R(x)$ if $j - 1 \notin \mathrm{Supp}(x)$, $j \in D_R(x) \setminus D_L(x)$ if $s_{j-1}s_j \leq x$, $j \in D_L(x) \setminus D_R(x)$ if $s_j s_{j-1} \leq x$, and $\ell(x') = \ell(x) - 1$ in all cases. By our induction assumption, there is a saturated subset Σ' such that $x' = x_{\Sigma'}$. When $j - 1 \in \mathrm{Supp}(x')$ let $(a', b') \in \Sigma'$ be the unique element such that $a' - b' = j - 1$. Then we claim that $x = x_\Sigma$ for $\Sigma \stackrel{\text{def}}{=} \Sigma' \sqcup \{(a, b)\}$ with (a, b) described as follows:

- if $j - 1 \notin \mathrm{Supp}(x)$, then $x = s_j x' = x' s_j$ and we take (a, b) to be any element of $\mathbb{Z} \times \mathbb{Z}$ satisfying $a - b = j$;

- if $s_{j-1}s_j \le x$, then $x = x's_j$ and we take $(a, b) \stackrel{\text{def}}{=} (a', b'-1) \le (a', b')$;
- if $s_j s_{j-1} \le x$, then $x = s_j x'$ and we take $(a, b) \stackrel{\text{def}}{=} (a'+1, b') \ge (a', b')$. □

Example A.3 We take $\mathfrak{g} = \mathfrak{gl}_n$. For $n = 4$, the partial-Coxeter elements of length 3 are $s_3s_2s_1$, $s_1s_2s_3$, $s_1s_3s_2$, $s_2s_1s_3$. The element $s_2s_1s_3s_2$ is saturated and connected (take for instance $\Sigma = \{(2, 0), (2, 1), (3, 0), (3, 1)\}$) but not partial-Coxeter (and is in fact the only such element for $\mathfrak{gl}_4$). The elements $s_1s_2s_1$ and $s_1s_2s_3s_2s_1$ are not saturated (for the first one, one could think of $\Sigma = \{(2, 0), (2, 1), (3, 2)\}$ or $\Sigma = \{(2, 1), (3, 1), (3, 2)\}$ or $\Sigma = \{(2, 0), (2, 1), (3, 1)\}$ but their saturated closure contains the extra element $(2, 2)$ for the first two, the extra element $(3, 0)$ for the third). More generally, for n sufficiently large, $s_js_{j-1}s_{j+1}s_js_{j+2}s_{j+1}s_{j+3}s_{j+2}$ and its inverse are examples of saturated connected elements.

Lemma A.4 *Let $\Sigma \subseteq \mathbb{Z} \times \mathbb{Z}$ be a saturated subset with a unique minimal element (a_1, b_1) and a unique maximal element (a_2, b_2). Assume there is $j \in \mathrm{Supp}(x_\Sigma)$ such that $|j - (a_2 - b_2)| = 1$. Then we have*

$$\{a_1 - b_1\} \subsetneq D_R(s_j x_\Sigma). \tag{A.4}$$

Proof It follows from (ii) of Lemma A.2 that $D_R(x_\Sigma) = \{a_1 - b_1\}$ and $D_L(x_\Sigma) = \{a_2 - b_2\}$. This forces $s_j x_\Sigma > x_\Sigma > x_\Sigma s_{a_1-b_1}$ (recall from the proof of Proposition A.2 that the braid relations in x_Σ are only of the type $s_js_{j'} = s_{j'}s_j$), which implies $\ell(s_jx_\Sigma s_{a_1-b_1}) \le \ell(x_\Sigma s_{a_1-b_1}) + 1 = \ell(s_jx_\Sigma) - 1$ and thus $s_jx_\Sigma > s_jx_\Sigma s_{a_1-b_1}$. Hence $a_1 - b_1 \in D_R(s_jx_\Sigma)$. Now we prove that (A.4) is a strict inclusion by increasing induction on $\#\Sigma$. If $\#\Sigma = 1$, then $\mathrm{Supp}(x_\Sigma) = \{a_2 - b_2\}$ and there is nothing to prove. We can assume $j = a_2 - b_2 + 1$ as the proof for $j = a_2 - b_2 - 1$ is symmetric. We have the following two possibilities.

Case 1: If $a_2 - b_2 - 1 \notin \mathrm{Supp}(x_\Sigma)$, then we must have $a_1 = a_2$ and $\Sigma = \{(a_1, b) \mid b_1 \le b \le b_2\}$. Hence, we have

$$s_jx_\Sigma = s_{a_2-b_2+1}s_{a_2-b_2}s_{a_2-b_2+1}x_{\Sigma'} = s_{a_2-b_2}s_{a_2-b_2+1}s_{a_2-b_2}x_{\Sigma'} = s_{a_2-b_2}s_{a_2-b_2+1}x_{\Sigma'}s_{a_2-b_2}$$

with $\Sigma' \stackrel{\text{def}}{=} \Sigma \setminus \{(a_2, b_2), (a_2, b_2 - 1)\}$. This implies $\{a_1 - b_1\} \subsetneq \{a_2 - b_2, a_1 - b_1\} \subseteq D_R(s_jx_\Sigma)$.

Case 2: If $a_2 - b_2 - 1 \in \mathrm{Supp}(x_\Sigma)$, then we must have $(a_1, b_1) \le (a_2 - 1, b_2) \in \Sigma$. Let Σ' be the saturated closure of $\{(a_1, b_1), (a_2 - 1, b_2)\}$, which is a (saturated) subset of Σ. Moreover, $\Sigma \setminus \Sigma' = \{(a_2, b) \mid b_1 \le b \le b_2\}$ is also a non-empty saturated subset of Σ. It is not difficult to check that $x_\Sigma = x_{\Sigma\setminus\Sigma'}x_{\Sigma'}$. As $j = a_2 - b_2 + 1 \in \mathrm{Supp}(x_\Sigma)$, we also have $(a_1, b_1) \le (a_2, b_2 - 1) \in \Sigma$, and we set $\Sigma'' \stackrel{\text{def}}{=} (\Sigma \setminus \Sigma') \setminus \{(a_2, b_2), (a_2, b_2 - 1)\}$, which is yet another saturated subset of Σ. We observe that

$$\begin{aligned} s_jx_\Sigma &= s_{a_2-b_2+1}s_{a_2-b_2}s_{a_2-b_2+1}x_{\Sigma''}x_{\Sigma'} = s_{a_2-b_2}s_{a_2-b_2+1}s_{a_2-b_2}x_{\Sigma''}x_{\Sigma'} \\ &= s_{a_2-b_2}s_{a_2-b_2+1}x_{\Sigma''}s_{a_2-b_2}x_{\Sigma'} \end{aligned} \tag{A.5}$$

where each decomposition in A.5 is reduced. As $(a_1, b_1) \leq (a_2 - 1, b_2)$ and $(a_1, b_1) \leq (a_2, b_2 - 1)$, we have $(a_1, b_1) \leq (a_2 - 1, b_2 - 1) \leq (a_2 - 1, b_2)$, which implies $(a_2 - 1, b_2 - 1) \in \Sigma'$ and $a_2 - b_2 = (a_2 - 1) - (b_2 - 1) \in \mathrm{Supp}(x_{\Sigma'})$. By our induction assumption applied to Σ' and $a_2 - b_2$, we have $\{a_1 - b_1\} \subsetneq D_R(s_{a_2-b_2}x_{\Sigma'})$, which together with (A.5) finishes the proof. □

Lemma A.5 *Let $x \in W(G)$ and $j \in \Delta$ with $D_R(x) = \{j\}$. Then there exists a saturated connected subset $\Sigma \subseteq \mathbb{Z} \times \mathbb{Z}$ such that $x = x_\Sigma$. Moreover, Σ contains a unique minimal element (a, b) that satisfies $a - b = j$.*

Proof Connectedness is clear otherwise we would have $\#D_R(x) > 1$. We prove the statement by induction on $\ell(x)$. If $\ell(x) = 1$, then we deduce from $D_R(x) = \{j\}$ that $x = s_j$. So we can take $\Sigma = \{(a, b)\}$ for an arbitrary $(a, b) \in \mathbb{Z} \times \mathbb{Z}$ satisfying $a - b = j$. We assume from now on $\ell(x) \geq 2$. We choose an arbitrary $j' \in D_L(x)$ and set $x' \stackrel{\text{def}}{=} s_{j'}x$ which satisfies $\ell(x') = \ell(x) - 1 \geq 1$. For each $j'' \in D_R(x')$, we have $x = s_{j'}x' > x' > x's_{j''}$ which implies $\ell(s_{j'}x's_{j''}) \leq \ell(x's_{j''}) + 1 = \ell(s_{j'}x') - 1$ and thus $x = s_{j'}x' > s_{j'}x's_{j''} = xs_{j''}$, equivalently $j'' \in D_R(x)$. As $D_R(x) = \{j\}$, we conclude that $D_R(x') = \{j\}$, which together with our induction assumption gives a saturated subset $\Sigma' \subseteq \mathbb{Z} \times \mathbb{Z}$ such that $x' = x_{\Sigma'}$. Our induction assumption also says that Σ' contains a unique minimal element (a_0, b_0) that satisfies $a_0 - b_0 = j$. We now construct Σ from Σ', according to the following two possibilities.

Assume $\mathrm{Supp}(x') \cap \{j' + 1, j' - 1\} = \emptyset$. Then $x = s_{j'}x' = x's_{j'}$ and thus $j' \in D_R(x)$, which forces $j = j'$. Thus $x = x's_j$ which contradicts $D_R(x) = D_R(x') = \{j\}$.

We thus have $\mathrm{Supp}(x') \cap \{j' + 1, j' - 1\} \neq \emptyset$. Since Σ' contains a unique minimal element (a_0, b_0), it is easy to see that, for each saturated subset $\Sigma'' \subseteq \Sigma'$ containing (a_0, b_0), $\Sigma' \setminus \Sigma''$ is another saturated subset of Σ' and that $x_{\Sigma'} = x_{\Sigma' \setminus \Sigma''}x_{\Sigma''}$. We choose such Σ'' to be the saturated closure of the following subset

$$\{(a_0, b_0)\} \sqcup \{(a, b) \in \Sigma' \mid |(a - b) - j'| \leq 1\} \subseteq \Sigma'.$$

As $s_{j'}s_{a'-b'} = s_{a'-b'}s_{j'}$ for each $(a', b') \in \Sigma' \setminus \Sigma''$, we have

$$x = s_{j'}x_{\Sigma'} = x_{\Sigma' \setminus \Sigma''}s_{j'}x_{\Sigma''} \tag{A.6}$$

with both sides being reduced (using (i) of Lemma A.2). As $s_{j'}x_{\Sigma'} > x_{\Sigma'}$, we have $s_{j'}x_{\Sigma''} > x_{\Sigma''}$ and thus a maximal element $(a, b) \in \Sigma''$ cannot satisfy $a - b = j'$. As $\mathrm{Supp}(x') \cap \{j' + 1, j' - 1\} \neq \emptyset$, any maximal element $(a, b) \in \Sigma''$ must satisfy $|a - b - j'| = 1$. We now have the following 3 cases.

- Assume that Σ'' has two different maximal elements (a_1, b_1), (a_2, b_2), and we can assume $a_1 \geq a_2$. If $a_1 - b_1 \leq a_2 - b_2$, then $a_1 \geq a_2$ forces $(a_1, b_1) \geq (a_2, b_2)$ which contradicts the fact that both (a_1, b_1) and (a_2, b_2) are maximal in Σ''. So we must have $a_1 - b_1 = j' + 1$ and $a_2 - b_2 = j' - 1$. If $a_1 \geq a_2 + 2$, this implies $b_1 \geq b_2$ and thus $(a_1, b_1) > (a_2, b_2)$, another contradiction. If $a_1 = a_2$, then $b_1 = b_2 - 2$ and thus $(a_1, b_1) < (a_2, b_2)$, also a contradiction. So the only possibility is

$a_1 = a_2 + 1$ and $b_1 = b_2 - 1$, in which case we simply take $\Sigma \overset{\text{def}}{=} \Sigma' \sqcup \{(a_1, b_2)\}$ which is easily seen to be saturated and satisfies $x = s_{j'}x' = x_\Sigma$.

- Assume that Σ'' has a unique maximal element (a_1, b_1) and $j' \notin \mathrm{Supp}(x')$. We can assume $a_1 - b_1 = j' + 1$ as the case $a_1 - b_1 = j' - 1$ is similar. Then Σ'' is the saturated closure of $\{(a_0, b_0), (a_1, b_1)\}$. Since $j' \notin \mathrm{Supp}(x')$, we must have $a_0 = a_1$ and $\Sigma'' = \{(a_1, b) \mid b_0 \leq b \leq b_1\}$. Hence, it is clear that $\Sigma'' \sqcup \{(a_1, b_1 + 1)\}$ is saturated. Using that Σ'' contains the unique minimal element of Σ', we also deduce that $\Sigma \overset{\text{def}}{=} \Sigma' \sqcup \{(a_1, b_1 + 1)\}$ is saturated. As $(a_1, b_1 + 1)$ must also be maximal in Σ and $a_1 - (b_1 + 1) = j'$, we conclude that $x_\Sigma = s_{j'}x_{\Sigma'} = x$.
- Assume that Σ'' has a unique maximal element (a_1, b_1) and $j' \in \mathrm{Supp}(x')$. Here again we can assume $a_1 - b_1 = j' + 1$. As Σ'' is the saturated closure of $\{(a_0, b_0), (a_1, b_1)\}$ and $j' \in \mathrm{Supp}(x')$, we must have $j' \in \mathrm{Supp}(x_{\Sigma''})$ and $a_0 < a_1$ which forces $(a_0, b_0) \leq (a_1 - 1, b_1) \leq (a_1, b_1)$ and $(a_1 - 1, b_1) \in \Sigma''$. But then it follows from Lemma A.4 that $\{a_0 - b_0\} \subsetneq D_R(s_{j'}x_{\Sigma''})$, which together with (A.6) implies that $\{a_0 - b_0\} = \{j\} \subsetneq D_R(s_{j'}x') = D_R(x) = \{j\}$, a contradiction.

We have constructed the desired Σ from Σ' in all possible cases, which finishes the proof. □

Remark A.6 It is an easy check that if a (non-empty) saturated subset $\Sigma \subseteq \mathbb{Z} \times \mathbb{Z}$ admits a unique minimal element (a_0, b_0) and moreover $a_0 - b_0 = 1$, then Σ must have the form $\{(a_0 + c, b_0) \mid 0 \leq c \leq c_0\}$ for some $0 \leq c_0 \leq n - 2$. Similarly, if a (non-empty) saturated subset $\Sigma \subseteq \mathbb{Z} \times \mathbb{Z}$ admits a unique maximal element (a_1, b_1) and moreover $a_1 - b_1 = n - 1$, then Σ must have the form $\{(a_1 - c, b_1) \mid 0 \leq c \leq c_1\}$ for some $0 \leq c_1 \leq n - 2$.

In the lemmas below we use without comment notation related to the Kazhdan-Lusztig polynomials, see the beginning of Sect. 3.2.

Lemma A.7 *Let $j \in \Delta$ and $x, w \in W(G)$ such that $D_L(x) = D_L(w) = \{j\}$. Then there does not exist x' such that $x \prec x' \prec w$ and $j \notin D_L(x')$.*

Proof Assume on the contrary that such a triple x, w, x' exist. By applying Lemma A.5 to x^{-1} and w^{-1} (with $D_R(x^{-1}) = D_R(w^{-1}) = \{j\}$) we know that both x and w are saturated, i.e. there exist saturated subsets $\Sigma_1, \Sigma_2 \subseteq \mathbb{Z} \times \mathbb{Z}$ such that $x = x_{\Sigma_1}$ and $w = x_{\Sigma_2}$. As $j \in D_L(w) \setminus D_L(x')$ and $w \prec x'$, by Lemma 3.2.5 we have $w = s_j x' > x'$ and in particular x' is saturated with $x' = x_{\Sigma_2 \setminus \{(a,b)\}}$ where (a, b) is the maximal element of Σ_2 satisfying $a - b = j$. Note that $D_L(w) = \{j\}$ forces $D_L(x') \subseteq \{j - 1, j + 1\} \cap \Delta$. As $D_L(x) = \{j\}$ and $j \notin D_L(x') \neq \emptyset$, by Lemma 3.2.5 we have $x' = s_{j'}x > x$ for some $j' \in D_L(x') \subseteq \{j - 1, j + 1\} \cap \Delta$. If we write (a', b') for the maximal element of $\Sigma_2 \setminus \{(a, b)\}$ that satisfies $a' - b' = j'$, then we can choose $\Sigma_1 = \Sigma_2 \setminus \{(a, b), (a', b')\}$. Let (a'', b'') be the unique maximal element of Σ_1 such that $a'' - b'' = j$ (using $D_L(x) = \{j\}$ and (ii) of Lemma A.2). As $(a'', b'') \in \Sigma_1$, we have $(a'', b'') = (a - c, b - c)$ for some $c \geq 1$, and in particular $(a - 1, b), (a, b - 1) \in \Sigma_2 \setminus \{(a, b)\}$ as Σ_2 is saturated. Hence, either $(a - 1, b)$ or $(a, b - 1)$ remains in $\Sigma_1 = (\Sigma_2 \setminus \{(a, b), (a', b')\}$, contradicting the maximality of (a'', b'') in Σ_1. □

For $j_1, j_2 \in \Delta$ with $|j_1 - j_2| = 1$ and $x \in W(G)$ with $\#(D_L(x) \cap \{j_1, j_2\}) = 1$, we define $\theta_{j_1,j_2}(x)$ as the unique element in $\{s_{j_1}x, s_{j_2}x\}$ such that $\#(D_L(\theta_{j_1,j_2}(x)) \cap \{j_1, j_2\}) = 1$. It is clear that $j_i \in D_L(x)$ for $i \in \{1, 2\}$ if and only if $j_{3-i} \in D_L(\theta_{j_1,j_2}(x))$.

Lemma A.8 *Let $x, w \in W(G)$ and $j_1, j_2 \in \Delta$ with $|j_1 - j_2| = 1$. Assume $w \notin W(L_{\{j_1,j_2\}})x$ and $\#(D_L(\theta_{j_1,j_2}(x)) \cap \{j_1, j_2\}) = 1 = \#(D_L(\theta_{j_1,j_2}(w)) \cap \{j_1, j_2\})$. Then we have $x \prec w$ if and only if $\theta_{j_1,j_2}(x) \prec \theta_{j_1,j_2}(w)$, in which case $\mu(x, w) = \mu(\theta_{j_1,j_2}(x), \theta_{j_1,j_2}(w))$.*

Proof This is [37, Theorem 4.2(i)]. □

For $j, j' \in \Delta$, recall that $w_{j,j'} \in W(G)$ is defined in (3.87).

Lemma A.9 *Let $x, w \in W(G)$ with $x \prec w$. If $x = 1$, then $w = s_j$ for some $j \in \Delta$. If $x = w_{j,j'}$, then $\ell(w) > \ell(x) + 1$ if and only if $x = s_j$ and $w = s_j s_{j+1} s_{j-1} s_j$ for some $2 \leq j \leq n-2$, in which case we have $\mu(x, w) = 1$.*

Proof The claim when $x = 1$ is clear. Assume from now that $x = w_{j,j'}$ and w satisfies $x \prec w$ and $\ell(w) > \ell(x) + 1$. We can assume $j \geq j'$ (the case $j \leq j'$ is similar). By Lemma 3.2.5 we must have $D_L(w) = D_L(x) = \{j\}$ and $D_R(w) = D_R(x) = \{j'\}$. By Lemma A.5 we know that w is saturated, i.e. $w = x_\Sigma$ for some saturated subset $\Sigma \subseteq \mathbb{Z} \times \mathbb{Z}$. By (ii) of Lemma A.2 the set Σ admits a unique maximal element (a, b) with $a - b = j$ and a unique minimal element (a', b') with $a' - b' = j'$. If either $a = a'$ or $b = b'$, then it is easy to check that $w = w_{j,j'}$ which contradicts $\ell(w) > \ell(x) + 1$. Hence, we must have $a' < a$ and $b' < b$ and in particular $(a-1, b-1), (a-1, b), (a, b-1) \in \Sigma$ (and $j-1, j+1 \in \Delta$). One checks that the only $x' \in W(L_{\{j,j+1\}})x$ satisfying $D_L(x') = \{j\}$ and $D_R(x') = \{j'\}$ is $x' = x$, and in particular $w \notin W(L_{\{j,j+1\}})x$. This together with $x \prec w$, $D_L(x) = D_L(w) = \{j\}$ and Lemma A.8 implies $\theta_{j,j+1}(x) \prec \theta_{j,j+1}(w)$. We check that $\theta_{j,j+1}(x) = s_{j+1}x = w_{j+1,j'}$ with $D_L(\theta_{j,j+1}(x)) = \{j+1\}$ and $\theta_{j,j+1}(w) = s_j w < w$ with $D_L(\theta_{j,j+1}(w)) = \{j-1, j+1\}$ (as $\Sigma \setminus \{(a, b)\}$ has maximal elements $(a-1, b)$ and $(a, b-1)$), which together with Lemma 3.2.5 forces $\theta_{j,j+1}(w) = s_{j-1}\theta_{j,j+1}(x) = s_{j-1}s_{j+1}w_{j,j'}$. Equivalently, we have $x = w_{j,j'} = x_{\Sigma'}$ with $\Sigma' \stackrel{\text{def}}{=} \Sigma \setminus \{(a, b), (a-1, b), (a, b-1)\}$ (which admits $(a-1, b-1)$ as the unique maximal element). This together with $j \geq j'$ forces $b' = b - 1$ and $\Sigma' = \{(a'', b-1) \mid a' \leq a'' \leq a-1\}$. As $\Sigma = \Sigma' \sqcup \{(a, b), (a-1, b), (a, b-1)\}$ is saturated, we must also have $a' = a - 1$, i.e. $x = s_j$ and $w = s_j s_{j-1} s_{j+1} s_j$ (for some $2 \leq j \leq n-2$). Finally, note that $\mu(x, w) = \mu(\theta_{j,j+1}(x), \theta_{j,j+1}(w)) = 1$ by Lemma A.8 (and $\ell(\theta_{j,j+1}(w)) = \ell(\theta_{j,j+1}(x)) + 1$). □

Remark A.10 Let $j_0, j_1, j_0', j_1' \in \Delta$ and consider $w_{j_1,j_0}, w_{j_1',j_0'} \in W(G)$. It is easy to check (using [6, Theorem 2.2.2]) that $w_{j_1,j_0} \leq w_{j_1',j_0'}$ if and only if either $j_1' \geq j_1 \geq j_0 \geq j_0'$ or $j_1' \leq j_1 \leq j_0 \leq j_0'$. Combined with Lemma A.9, we observe that $w_{j_1,j_0} \prec w_{j_1',j_0'}$ if and only if $w_{j_1,j_0} < w_{j_1',j_0'}$ and $\ell(w_{j_1',j_0'}) = \ell(w_{j_1,j_0}) + 1$, in which case $|j_0 - j_0'| + |j_1 - j_1'| = 1$.

Lemma A.11 *Let* $j, j' \in \Delta$ *and* $x \in W(G)$ *such that* $D_L(x) = \{j_1\}$ *for some* $j_1 \neq j$.

(i) *We have* $x \prec w_{j,j'}$ *if and only if* $x = w_{j+1,j'}$ *with* $j' > j$, *or* $x = w_{j-1,j'}$ *with* $j' < j$.
(ii) *We have* $w_{j,j'} \prec x$ *if and only if* $x = w_{j+1,j'}$ *with* $j' \leq j$, *or* $x = w_{j-1,j'}$ *with* $j' \geq j$.
(iii) *We have* $\mathrm{Ext}^1_{U(\mathfrak{g})}(L(w_{j,j'}), L(x)) \neq 0$ *if and only if* $x = w_{j_1,j'}$ *with* $j_1 \in \{j-1, j+1\} \cap \Delta$, *in which case* $\dim_E \mathrm{Ext}^1_{U(\mathfrak{g})}(L(w_{j,j'}), L(x)) = 1$.
(iv) *We have* $w_{j,j'} \prec w$ *with* $j' \notin D_R(w)$ *if and only if* $w = w_{j,j'+1}$ *with* $j \leq j'$, *or* $w = w_{j,j'-1}$ *with* $j \geq j'$.

Proof We prove (i). As $x \prec w_{j,j'}$ and $j \in D_L(w_{j,j'}) \setminus D_L(x)$, by Lemma 3.2.5 we have $w_{j,j'} = s_j x$. As $D_L(x) \neq \emptyset$, we have $x \neq 1$ and thus $j \neq j'$. So either we have $j' > j$ and $x = w_{j+1,j'}$, or we have $j' < j$ and $x = w_{j-1,j'}$. The other direction is obvious.

We prove (ii). As $w_{j,j'} \prec x$ and $j_1 \in D_L(x) \setminus D_L(w_{j,j'})$, by Lemma 3.2.5 we have $x = s_{j_1} w_{j,j'}$. As $D_L(x) = \{j_1\}$, by Lemma A.5 (applied to x^{-1}) we know that $x = x_\Sigma$ for some saturated subset $\Sigma \subseteq \mathbb{Z} \times \mathbb{Z}$. By (ii) of Lemma A.2 we know that Σ admits a unique maximal element (a_1, b_1) with $a_1 - b_1 = j_1$, and that $x_{\Sigma \setminus \{(a_1,b_1)\}} = w_{j,j'}$. In particular, $\Sigma \setminus \{(a_1, b_1)\}$ admits a unique maximal element (a, b) with $a - b = j$ and a unique minimal element (a', b') with $a' - b' = j'$, with $a = a'$ and $b \geq b'$ (resp. $a \geq a'$ and $b = b'$) if $j \geq j'$ (resp. if $j \leq j'$). As Σ is saturated and (a, b) is the unique maximal element of $\Sigma \setminus \{(a_1, b_1)\}$, either we have $(a, b) = (a_1 - 1, b_1)$ with $(a_1, b_1 - 1) \notin \Sigma$, or we have $(a, b) = (a_1, b_1 - 1)$ with $(a_1 - 1, b_1) \notin \Sigma$. In other words, either we have $(a, b) = (a_1 - 1, b_1)$, $a \geq a'$ and $b = b'$ (with $j_1 = j + 1$ and $j \geq j'$), or we have $(a, b) = (a_1, b_1 - 1)$, $a = a'$ and $b \geq b'$ (with $j_1 = j - 1$ and $j \leq j'$). This finishes the proof of (ii) as the other direction is trivial.

(iii) follows directly from (i), (ii) and from (ii) of Lemma 3.2.4.

We prove (iv). First $D_R(w_{j,j'}) = \{j'\}$ and $j' \notin D_R(w)$ together with Lemma 3.2.5 force $w = w_{j,j'} s_{j''}$ for some $j'' \neq j'$. If $|j'' - j'| \neq 1$, then $s_{j'} s_{j''} = s_{j''} s_{j'}$ and thus $j' \in D_R(w)$, a contradiction. Hence $j'' \in \{j' - 1, j' + 1\}$. If $j \leq j'$ (resp. $j \geq j'$), then we have $D_R(w_{j,j'} s_{j'+1}) = \{j' + 1\}$ (resp. $D_R(w_{j,j'} s_{j'-1}) = \{j' - 1\}$), which gives the two cases of the statement. If $j < j'$ and $j'' = j' - 1$ (resp. $j > j'$ and $j'' = j' + 1$), then we have $w_{j,j'} s_{j'-1} = s_{j'} w_{j,j'-1} s_{j'}$ (resp. $w_{j,j'} s_{j'+1} = s_{j'} w_{j,j'+1} s_{j'}$) and thus $j' \in D_R(w_{j,j'} s_{j'-1})$ (resp. $j' \in D_R(w_{j,j'} s_{j'+1})$), a contradiction. Here again the other direction is obvious. □

Lemma A.12 *Let* $x, w \in W(G)$ *with* $x < w$.

(i) *If* w *is partial-Coxeter, then* $L(w)$ *has multiplicity* 1 *in* $M(x)$.
(ii) *If* $w = s_j w_{n-1,1}$ *and* $x = w_{j,1}$ *for some* $j < n - 1$, *then* $L(w)$ *has multiplicity* 1 *in* $M(x)$.

Proof It is enough to prove $P_{w_0 x w_0, w_0 w w_0} = 1$ for both (i) and (ii) (cf. [33, Sect. 8.4]). We write $w' \stackrel{\text{def}}{=} w_0 w w_0$ and $x' \stackrel{\text{def}}{=} w_0 x w_0$ and note that $x' < w'$.

We prove $P_{x',w'} = 1$ for (i). As w is partial-Coxeter, so are w' and x'. Let $j \in D_L(w')$, since w' is partial-Coxeter we have $s_j z > z$ for $z \prec s_j w' < w'$. We deduce from [37, (2.2.c)] the following two possibilities. If $j \in \mathrm{Supp}(x')$, then $j \in D_L(x')$ since $x' < w'$ (and w' is partial-Coxeter), and we have $P_{x',w'} = P_{s_j x', s_j w'}$ in this case. If $j \notin \mathrm{Supp}(x')$, then $x' \leq s_j w'$, $s_j x' > x'$ and we have $P_{x',w'} = P_{x', s_j w'}$ in this case. A simple induction on $\ell(w')$ thus shows $P_{x',w'} = 1$.

We prove $P_{x',w'} = 1$ for (ii). As $w = s_j w_{n-1,1}$ and $x = w_{j,1}$, we have $w' = s_{j'} w_{1,n-1}$ and $x' = w_{j',n-1}$ with $j' \overset{\text{def}}{=} n - j > 1$. We consider any z satisfying $x' \leq z \prec s_{j'} w' = w_{1,n-1}$ and $s_{j'} z < z$. As $w_{1,n-1}$ is partial-Coxeter, the proof of (i) shows that $P_{z, s_{j'} w'} = 1$, which together with $z \prec s_{j'} w' = w_{1,n-1}$ forces $\ell(z) = \ell(w_{n-1,1}) - 1$. Using [10, Theorem 2.2.2], it is easy to check that the only z satisfying $z < w_{n-1,1}$, $\ell(z) = \ell(w_{n-1,1}) - 1$ and $s_{j'} z < z$ is $z = w_{n-1,j'+1} w_{j',1}$. By the proof of (i) we see that $P_{x', s_{j'} w'} = P_{s_{j'} x', s_{j'} w'} = P_{x',z} = 1$, which together with [37, (2.2.c)] gives $P_{x',w'} = P_{s_{j'} x', s_{j'} w'} + q P_{x', s_{j'} w'} - q P_{x',z} = 1$. □

Lemma A.13 *Let $j_0, j_1 \in \Delta$ with $j_1 \geq j_0$.*

(i) If $x \prec w_{j_1,j_0}$ and $x \prec w_{j_1+1,j_0+1}$, then we have $x = w_{j_1,j_0} s_{j_0} = s_{j_1+1} w_{j_1+1,j_0+1}$.
(ii) If $w_{j_1,j_0} \prec x$ and $w_{j_1+1,j_0+1} \prec x$, then we have $x = w_{j_1+1,j_0}$ when $j_1 > j_0$, and $x \in \{w_{j_0+1,j_0}, w_{j_0,j_0+1}\}$ when $j_1 = j_0$.

Proof Note that w_{j_1,j_0} and w_{j_1+1,j_0+1} are both partial-Coxeter and satisfy $\ell(w_{j_1,j_0}) = \ell(w_{j_1+1,j_0+1}) \geq 1$.

We prove (i). By the proof of (i) of Lemma A.12 we have $P_{x,w_{j_1,j_0}} = P_{x,w_{j_1+1,j_0+1}} = 1$, which together with $x \prec w_{j_1,j_0}$ and $x \prec w_{j_1+1,j_0+1}$ forces $\ell(x) = \ell(w_{j_1,j_0}) - 1 = \ell(w_{j_1+1,j_0+1}) - 1$. As $\mathrm{Supp}(x) \subseteq \mathrm{Supp}(w_{j_1,j_0}) \cap \mathrm{Supp}(w_{j_1+1,j_0+1})$, we deduce from [10, Theorem 2.2.2] that $x = w_{j_1,j_0} s_{j_0} = s_{j_1+1} w_{j_1+1,j_0+1}$.

We prove (ii). If $\ell(x) > \ell(w_{j_1,j_0}) + 1 = \ell(w_{j_1+1,j_0+1}) + 1$, then by Lemma 3.2.5 we have $D_R(x) \subseteq \{j_0\} \cap \{j_0 + 1\} = \emptyset$, a contradiction. So we have $\ell(x) = \ell(w_{j_1,j_0}) + 1 = \ell(w_{j_1+1,j_0+1}) + 1$ and $\mathrm{Supp}(x) \supseteq \mathrm{Supp}(w_{j_1,j_0}) \cup \mathrm{Supp}(w_{j_1+1,j_0+1}) = [j_1 + 1, j_0]$. If $j_1 > j_0$, then $x > w_{j_1,j_0}$, $\ell(x) = \ell(w_{j_1,j_0}) + 1$ and $\mathrm{Supp}(x) \supseteq [j_1 + 1, j_0]$ together with [10, Theorem 2.2.2] imply $x \in \{w_{j_1+1,j_0}, w_{j_1,j_0} s_{j_1+1}\}$. Together with $x > w_{j_1+1,j_0+1}$ (and [10, Theorem 2.2.2]) this implies $x = w_{j_1+1,j_0}$. If $j_1 = j_0$, a similar argument shows that $x \in \{w_{j_0+1,j_0}, w_{j_0,j_0+1}\}$. □

Recall that the partially-ordered set $\mathbf{J}^\infty$ is defined at the beginning of Sect. 2.3, and that, for each $(j_1, j_2) \in \mathbf{J}^\infty$, x_{j_1,j_2} is the element of maximal length in the set (2.73).

Lemma A.14 *Let $(j_1, j_2) \in \mathbf{J}^\infty$. The element x_{j_1,j_2} is saturated. More precisely, for each $(a_1, b_1) \in \mathbb{Z} \times \mathbb{Z}$ satisfying $a_1 - b_1 = j_1$, we have $x_{j_1,j_2} = x_\Sigma$ with*

$$\Sigma = \{(a, b) \mid a_1 - j_1 + 1 \leq a \leq a_1, b_1 - j_2 + j_1 + 1 \leq b \leq b_1\} \subseteq \mathbb{Z} \times \mathbb{Z}. \tag{A.7}$$

Proof As $D_L(x_{j_1,j_2}) = \{j_1\}$ by definition, we have $D_R(x_{j_1,j_2}^{-1}) = \{j_1\}$, so x_{j_1,j_2}^{-1} is saturated by Lemma A.5, and thus x_{j_1,j_2} is saturated. We write $x_{j_1,j_2} = x_\Sigma$ and let (a_1, b_1) be the unique maximal element in Σ (see (ii) of Lemma A.2). It remains to check the equality (A.7). We observe that Σ as in (A.7) is the maximal possible saturated subset of $\mathbb{Z} \times \mathbb{Z}$ which admits (a_1, b_1) as its unique maximal element and satisfies $a - b \leq j_2 - 1$ for each $(a, b) \in \Sigma$. In particular, if $(a, b) \leq (a_1, b_1)$ satisfies $1 \leq a' - b' \leq j_2 - 1$ for each $(a, b) \leq (a', b') \leq (a_1, b_1)$, then we have $(a, b) \geq (a_1 - j_1 + 1, b_1 - j_2 + j_1 + 1)$. $\square$

Remark A.15 Let $j' \leq j \leq j'' \in \Delta$. Then by a straightforward generalization of Lemma A.14, the element of maximal length x satisfying $D_L(x) = \{j\}$ and $\mathrm{Supp}(x) \subseteq [j', j'']$ is saturated, and we have $x = x_\Sigma$ with $\Sigma = \{(a', b') \mid a - j + j' \leq a' \leq a, b - j'' + j \leq b' \leq b\}$, for any (a, b) satisfying $a - b = j$.

Recall that w_I for $I \subseteq \Delta$ is the element of $W(L_I)$ of maximal length (see Sect. 1.1).

Lemma A.16 *Let* $(j_1, j_2) < (j_1', j_2')$ *in* $\mathbf{J}^\infty$.

(i) If $j_1' > j_1$ *and* $j_2' - j_1' > j_2 - j_1$*, then* $x_{j_1', j_2'} \not\leq x x_{j_1, j_2}$ *where* x *is the element of maximal length in* $W^{\widehat{j_1} \cap \widehat{j_1'}, \emptyset}(L_{\widehat{j_1}})$ *(see above (2.15) for the notation).*

(ii) If $j_1 = j_1'$*, then* $w_{[1, j_2'-1]} w_{[1, j_2-1]}^{-1} \in W(L_{\widehat{j_1}}) s_{j_1} W(L_{\widehat{j_1}})$ *if and only if either* $j_1 = 1$ *or* $j_2' = j_2 + 1$.

Proof We prove (i). Let $(a_1, b_1) \in \mathbb{Z} \times \mathbb{Z}$ with $a_1 - b_1 = j_1$. We define

$$\Sigma_1 \overset{\text{def}}{=} \{(a, b) \mid a_1 - j_1 + 1 \leq a \leq a_1, b_1 - j_2 + j_1 + 1 \leq b \leq b_1\}$$
$$\Sigma_2 \overset{\text{def}}{=} \{(a, b) \mid a_1 - j_1 + 1 \leq a \leq a_1 + j_1' - j_1, b_1 - j_2' + j_1' + 1 \leq b \leq b_1\}$$
$$\Sigma_3 \overset{\text{def}}{=} \{(a, b) \mid a_1 + 1 \leq a \leq a_1 + j_1' - j_1, b_1 - n + j_1' + 1 \leq b \leq b_1\}.$$

By Lemma A.14 (and Remark A.15 applied to x) we see that $x_{j_1, j_2} = x_{\Sigma_1}$, $x_{j_1', j_2'} = x_{\Sigma_2}$ and $x = x_{\Sigma_3}$. It is easy to check that $\Sigma_1 \sqcup \Sigma_3$ is still saturated with a unique maximal element $(a_1 + j_1' - j_1, b_1)$. Since Σ_2 also admits $(a_1 + j_1' - j_1, b_1)$ as its unique maximal element, we see that $x_{\Sigma_2} = x_{j_1', j_2'} \leq x x_{j_1, j_2} = x_{\Sigma_3} x_{\Sigma_1} = x_{\Sigma_1 \sqcup \Sigma_3}$ if and only if $\Sigma_2 \subseteq \Sigma_1 \sqcup \Sigma_3$. But $j_2' - j_1' > j_2 - j_1$ implies that the minimal element $(a_1 - j_1 + 1, b_1 - j_2' + j_1' + 1)$ is not inside $\Sigma_1 \sqcup \Sigma_3$. So we must have $x_{j_1', j_2'} \not\leq x x_{j_1, j_2}$.

We prove (ii). Let y be the element of maximal length such that $D_R(y) = \{j_2\}$ and $\mathrm{Supp}(y) \subseteq [1, j_2' - 1]$, which satisfies $w_{[1, j_2'-1]} = y w_{[1, j_2-1]} w_{[j_2+1, j_2'-1]}$ with $\ell(w_{[1, j_2'-1]}) = \ell(y) + \ell(w_{[1, j_2-1]}) + \ell(w_{[j_2+1, j_2'-1]})$ and in particular $w_{[1, j_2'-1]} w_{[1, j_2-1]}^{-1} \geq y$. If $j_1 = 1$, then $W^{\widehat{j_1}, \widehat{j_1}} = \{1, s_{j_1}\}$. As $1 = j_1 \in \mathrm{Supp}(y)$, we have $y \in W(L_{\widehat{j_1}}) s_{j_1} W(L_{\widehat{j_1}})$ which together with $w_{[1, j_2'-1]} w_{[1, j_2-1]}^{-1} \geq y$ (and [6, Proposition 2.5.1]) implies $w_{[1, j_2'-1]} w_{[1, j_2-1]}^{-1} \in W(L_{\widehat{j_1}}) s_{j_1} W(L_{\widehat{j_1}})$. If $j_2' = j_2 + 1$, then we have $w_{[1, j_2'-1]} w_{[1, j_2-1]}^{-1} = y = w_{1, j_2} \in W(L_{\widehat{j_1}}) s_{j_1} W(L_{\widehat{j_1}})$ as $j_1 \leq j_2$. We assume from now $j_1 > 1$ and $j_2' > j_2 + 1$. Let $(a, b) \in \mathbb{Z} \times \mathbb{Z}$ with $a - b = j_2$. Applying Remark A.15 to y^{-1}, we see that y^{-1} is saturated and $y^{-1} = x_\Sigma$ for $\Sigma = \{(a', b') \mid a - j_2 + 1 \leq a' \leq a, b - j_2' + j_2 + 1 \leq b' \leq b\}$. Since we have $j_2 - 1 > j_2 - j_1$ and $j_2' > j_2 + 1$, we observe that Σ contains the saturated subset $\Sigma' = \{(a', b') \mid a - j_2 + j_1 - 1 \leq a' \leq a - j_2 + j_1, b - 1 \leq b' \leq b\}$, which implies $y^{-1} \geq x_{\Sigma'}$ and moreover $w_{[1,j_2'-1]}w_{[1,j_2-1]}^{-1} \geq y \geq x_{\Sigma'}^{-1} = x_{\Sigma'} \in W^{\widehat{j_1},\widehat{j_1}}$ (using (ii) of Lemma A.2). By [6, Proposition 2.5.1], we know that $w_{[1,j_2'-1]}w_{[1,j_2-1]}^{-1} \in W(L_{\widehat{j_1}})wW(L_{\widehat{j_1}})$ for some $w \in W^{\widehat{j_1},\widehat{j_1}}$ such that $w \geq x_{\Sigma'} > s_{j_1}$, and in particular $w_{[1,j_2'-1]}w_{[1,j_2-1]}^{-1} \notin W(L_{\widehat{j_1}})s_{j_1}W(L_{\widehat{j_1}})$. □

Appendix B
Some Figures for GL_4 and GL_5

For $G = \mathrm{GL}_4(K)$ or $G = \mathrm{GL}_5(K)$ we draw many of the previous finite length $D(G)$-modules.

In the drawings, as usual a bullet $\bullet$ means an irreducible constituent (the socle being at the bottom of the page and the cosocle being at the top) and a line between two irreducible constituents, that is between two $\bullet$, means a *non-split* extension as subquotient (whether it is dotted or not). For clarity in the three-dimensional drawings, we draw the non-split extensions which are "behind" as dotted lines.

We first draw the complex of $D(G)$-modules $\mathbf{D}^\bullet$ of (5.151) in Fig. B.1 ($\mathrm{GL}_4(K)$) and Fig. B.7 ($\mathrm{GL}_5(K)$).

Then we draw the $D(G)$-modules Y_k —— $(V^{\mathrm{alg}}_{[1,4-k],\Delta})^\vee$ for $1 \le k \le 3$ in Fig. B.2 ($G = \mathrm{GL}_4(K)$) and the $D(G)$-modules Y_k —— $(V^{\mathrm{alg}}_{[1,5-k],\Delta})^\vee$ for $1 \le k \le 4$ in Fig. B.8 ($G = \mathrm{GL}_5(K)$). Recall that Y_k is defined in (5.135) and $V^{\mathrm{alg}}_{[1,4-k],\Delta}$, $V^{\mathrm{alg}}_{[1,5-k],\Delta}$ in (5.110). Moreover the red bullets are the duals of locally algebraic constituents.

We draw the $D(G)$-module $\widetilde{D}_3$ of (5.150) for $G = \mathrm{GL}_4(K)$ in Fig. B.3 and the $D(G)$-module $\widetilde{D}_4$ of (5.150) for $G = \mathrm{GL}_5(K)$ in Fig. B.9. Finally for $G = \mathrm{GL}_4(K)$ we draw the $D(G)$-modules $\widetilde{D}_2$ (Fig. B.4), $\widetilde{D}_1$ (Fig. B.5) and $\widetilde{D}_0$ (Fig. B.6) of (5.149), and for $G = \mathrm{GL}_5(K)$ the $D(G)$-modules $\widetilde{D}_3$ (Fig. B.10), $\widetilde{D}_2$ (Fig. B.11), $\widetilde{D}_1$ (Fig. B.12) and $\widetilde{D}_0$ (Fig. B.13) of (5.149). Note that $\widetilde{D}_0$ is just the "top" of $\widetilde{D}_1$ (keeping in mind that the drawings are 3-dimensional). Note also that, to save place, we sometimes write $V^{\mathrm{alg}}_{[1,4-k-1]}$, $V^{\mathrm{alg}}_{[1,5-k-1]}$ instead of (respectively) $V^{\mathrm{alg}}_{[1,4-k-1],\Delta}$, $V^{\mathrm{alg}}_{[1,5-k-1],\Delta}$.

The reader can have (a bit of) fun guessing on the drawings what are the morphisms $d^k_{\mathbf{D}} : D_k \to D_{k+1}$, the surjections $\widetilde{D}_k \twoheadrightarrow D_k$ and the morphisms $d^k_{\widetilde{\mathbf{D}}} : \widetilde{D}_k \to \widetilde{D}_{k+1}$.

C. Breuil and Z. Qian, *Splitting and Making Explicit the de Rham Complex of the Drinfeld Space*, Lecture Notes in Mathematics 2397,
https://doi.org/10.1007/978-3-032-20095-2

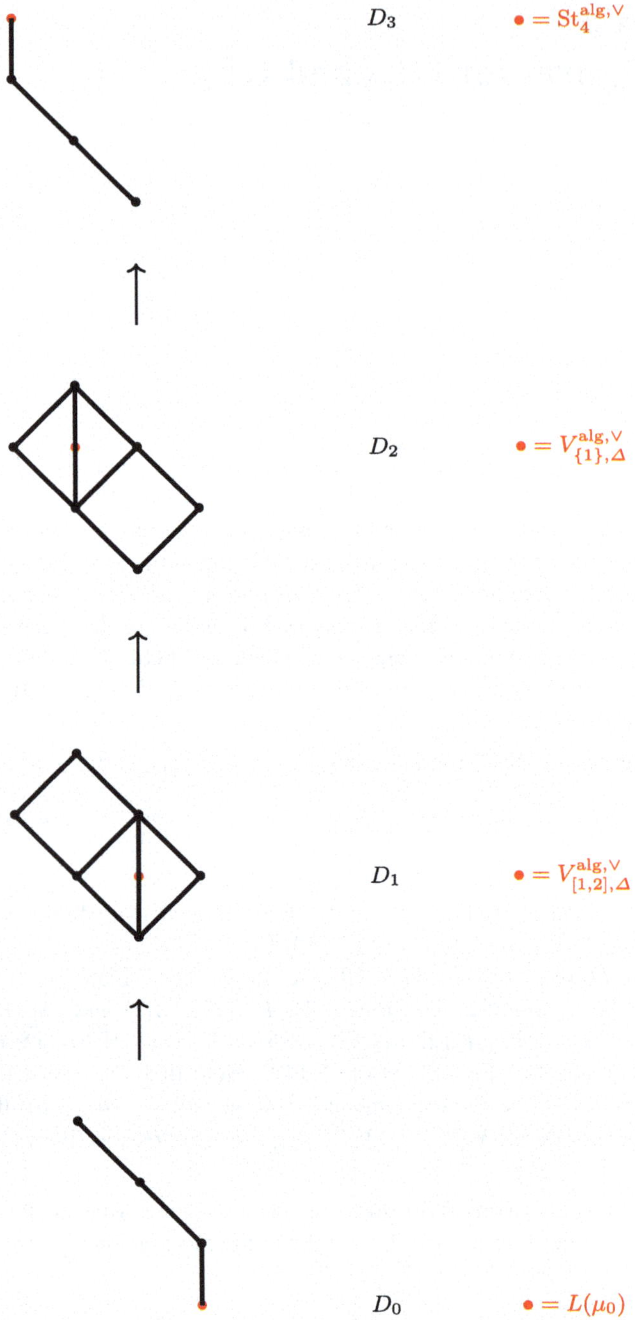

Fig. B.1 $\mathbf{D}^\bullet$ for GL_4

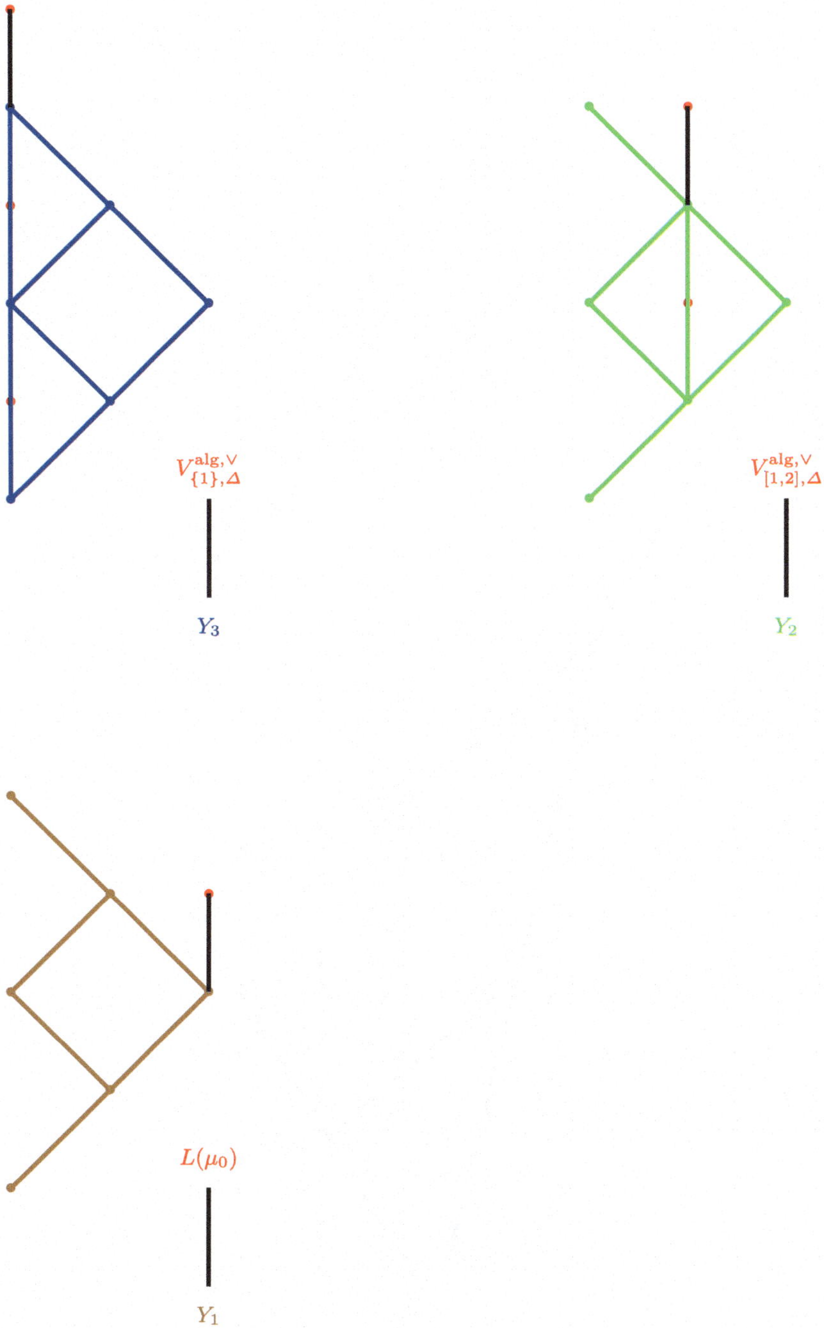

Fig. B.2 $Y_k - V^{alg,\vee}_{[1,4-k],\Delta}$ for GL_4 and $1 \leq k \leq 3$

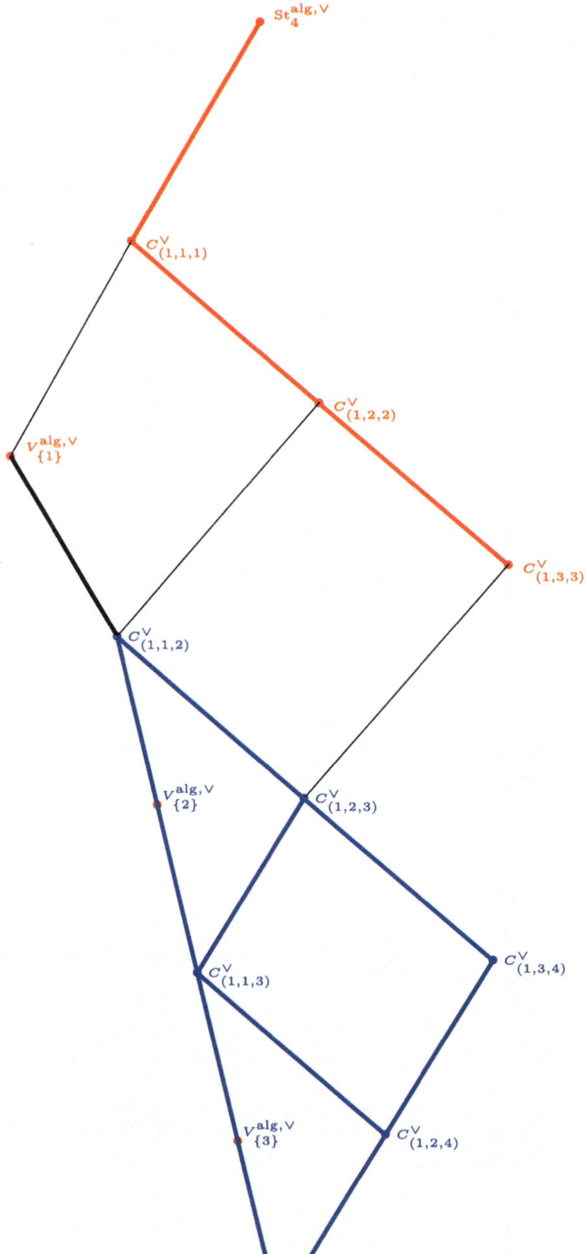

Fig. B.3 $\widetilde{D}_3$ for GL_4

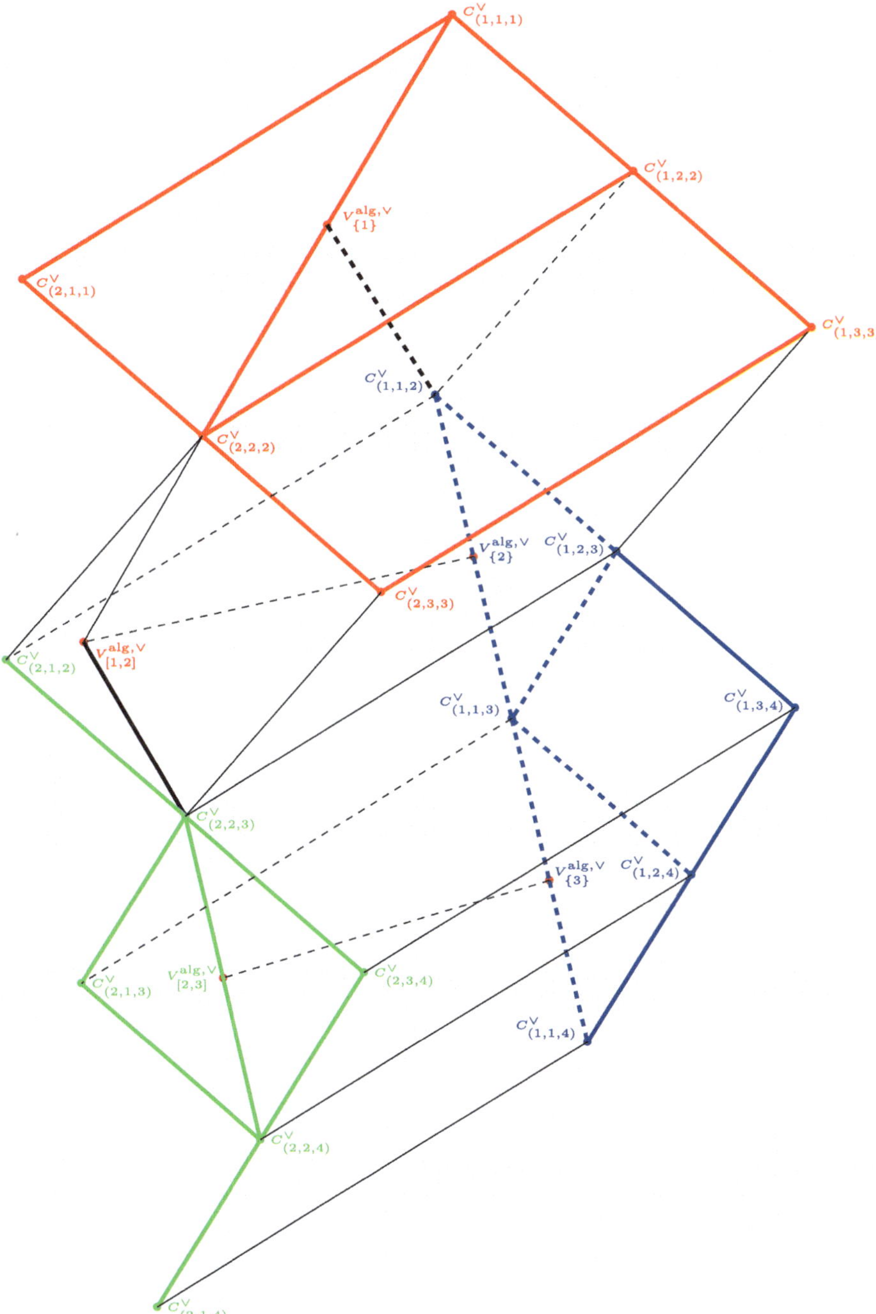

Fig. B.4 $\widetilde{D}_2$ for GL_4

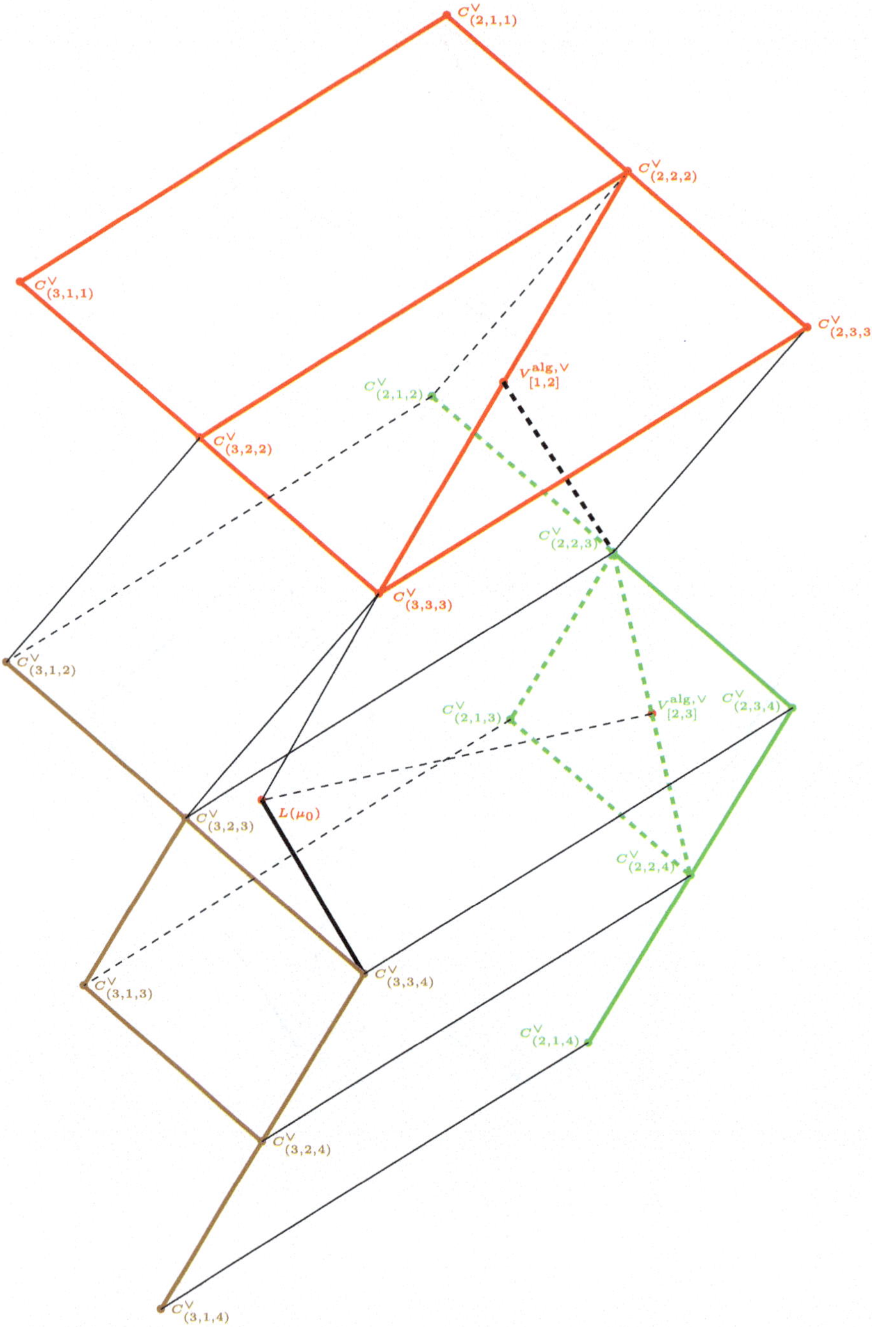

Fig. B.5 $\widetilde{D}_1$ for GL_4

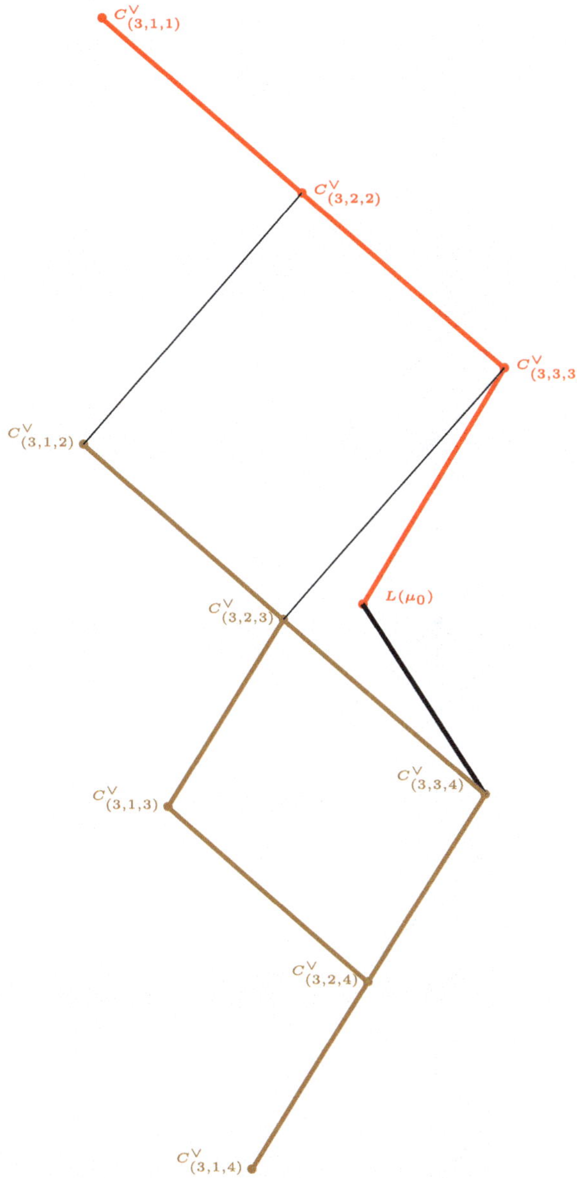

Fig. B.6 $\widetilde{D}_0$ for GL_4

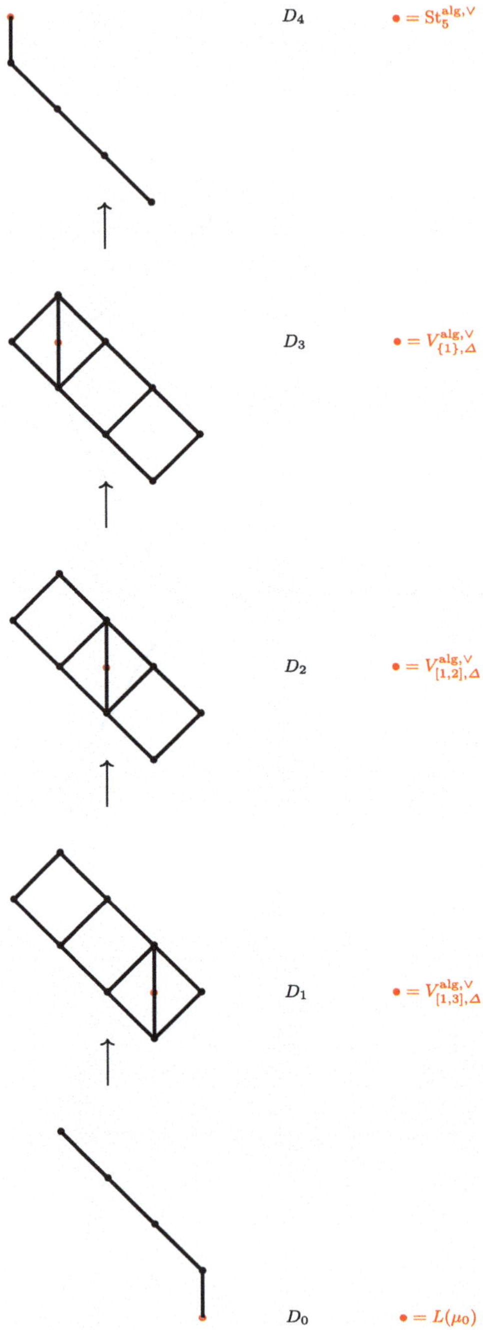

Fig. B.7 $\mathbf{D}^{\bullet}$ for GL_5

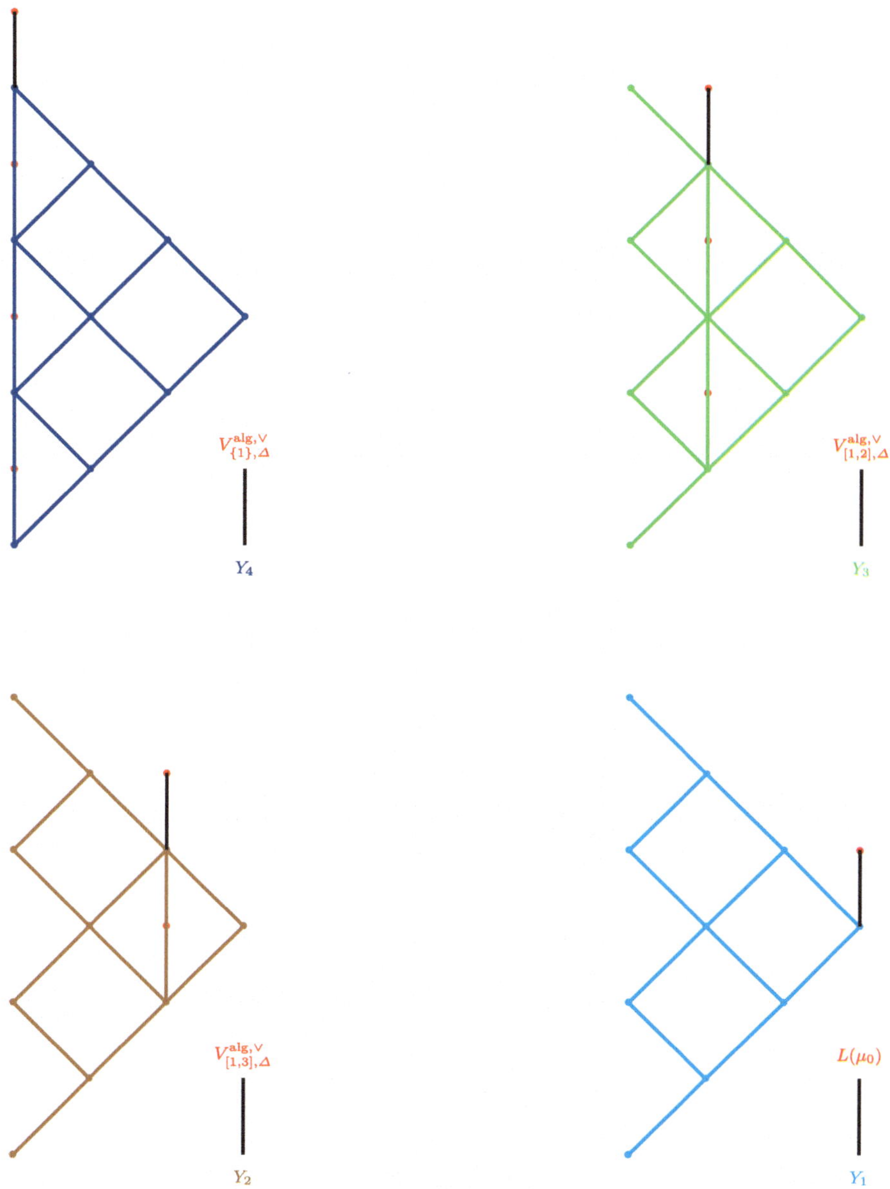

Fig. B.8 $Y_k - V^{\mathrm{alg},\vee}_{[1,5-k],\Delta}$ for GL_5 and $1 \le k \le 4$

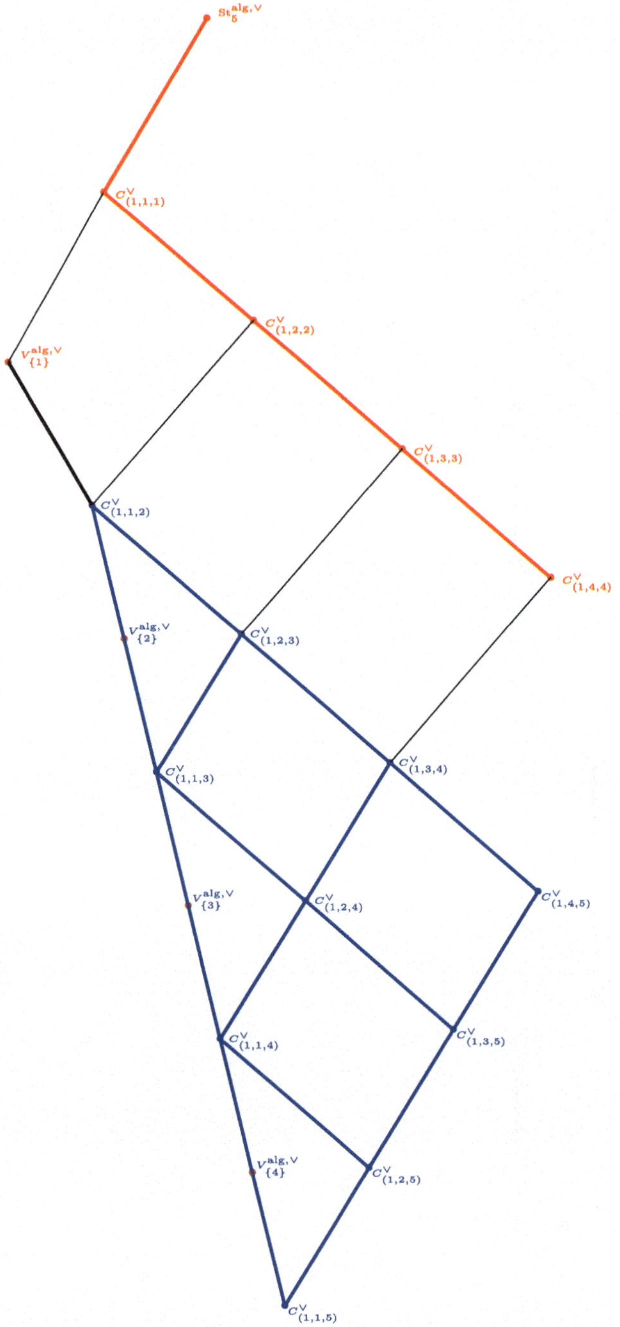

Fig. B.9 $\widetilde{D}_4$ for GL_5

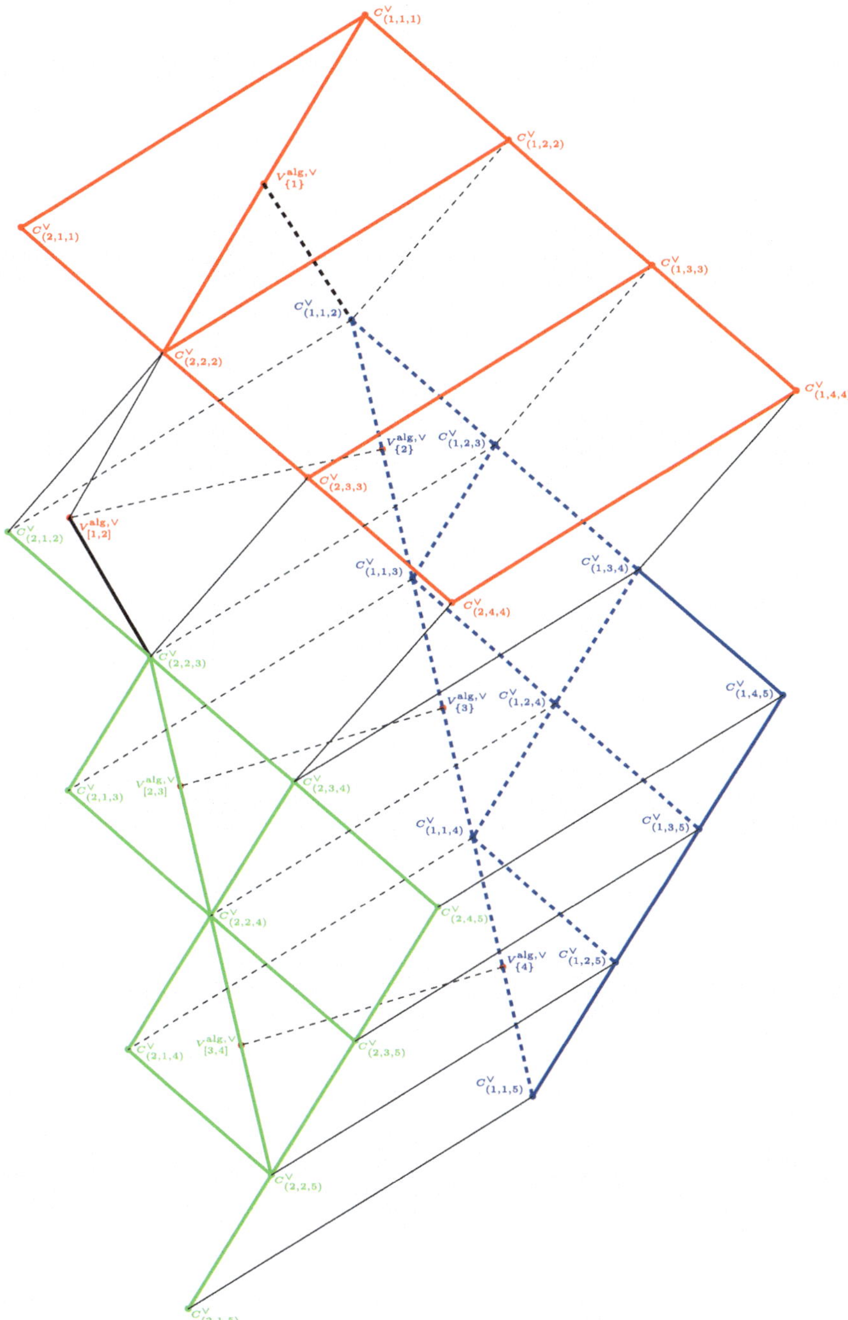

Fig. B.10 $\widetilde{D}_3$ for GL_5

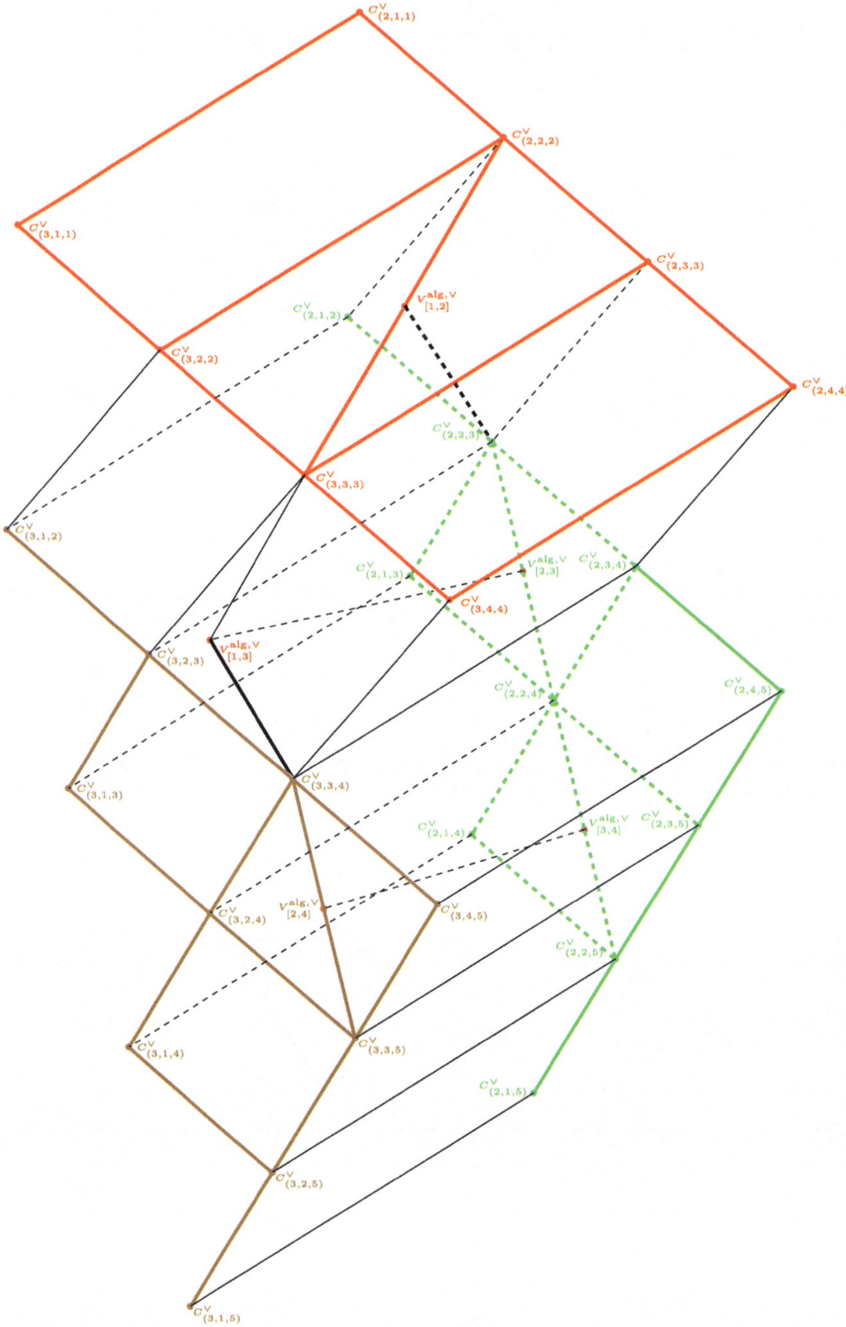

Fig. B.11 $\widetilde{D}_2$ for GL_5

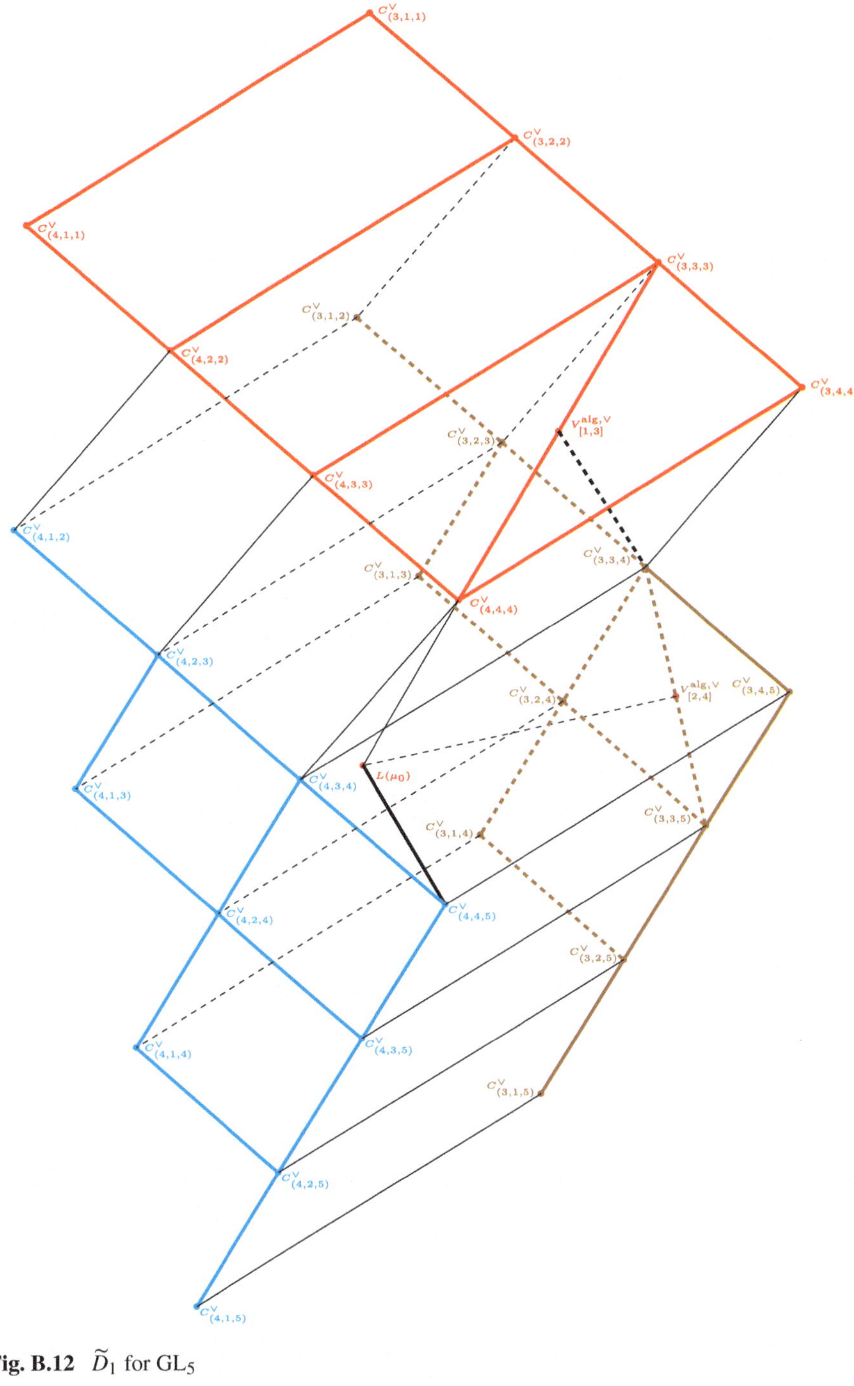

Fig. B.12 $\widetilde{D}_1$ for GL_5

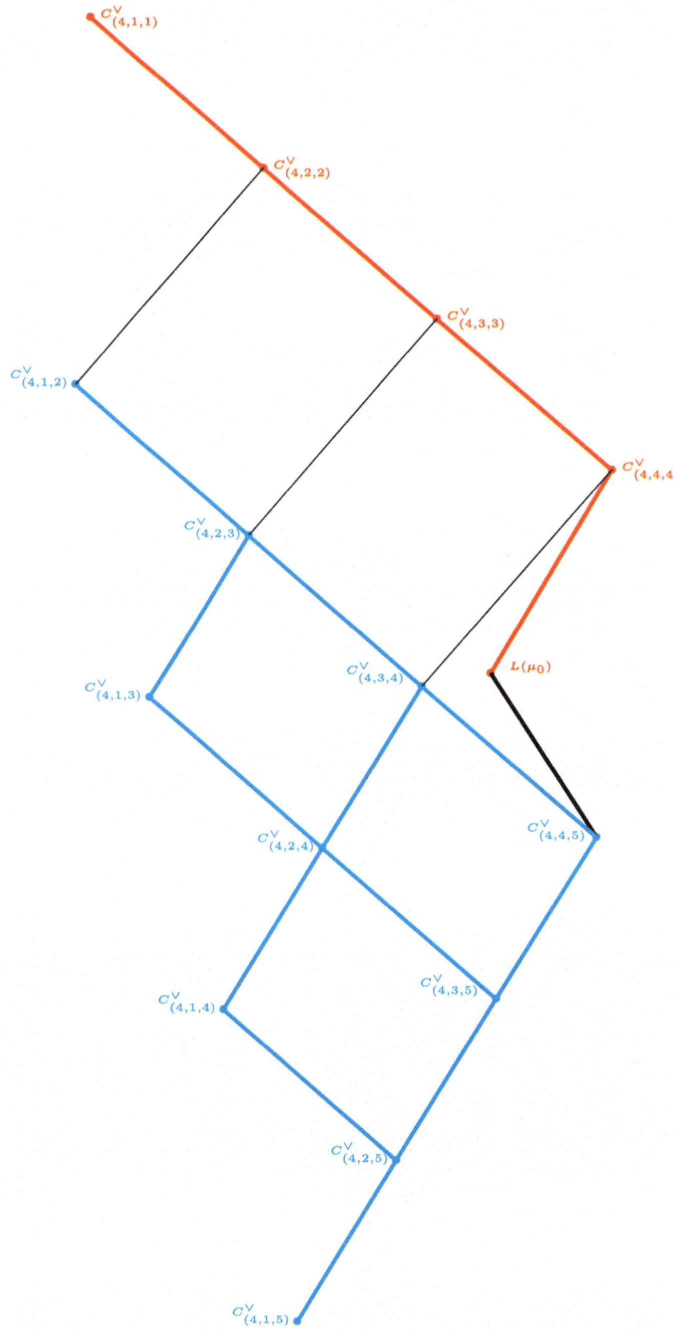

Fig. B.13 $\widetilde{D}_0$ for GL_5

References

1. K. Ardakov, S. Wadsley, Global sections of equivariant line bundles on the p-adic upper half plane. https://arxiv.org/abs/2312.12395
2. S. Agrawal, M. Strauch, From category$\mathcal{O}^{\infty}$ to locally analytic representations. J. Algebra **592**, 169–232 (2022)
3. I.N. Bernstein, Draft of: Representations of p-adic Groups. Lectures at Harvard (1992). https://www.math.tau.ac.il/~bernstei/Unpublished_texts/unpublished_texts/Bernstein93new-harv.lect.from-chic.pdf
4. I.N. Bernstein, A.V. Zelevinsky, Induced representations of reductive p-adic groups I. Ann. Sci. École Norm. Sup. **10**, 441–472 (1977)
5. R. Bezrukavnikov, Canonical basis and representation categories, course at MIT in 2013. https://math.mit.edu/~bezrukav/old/Course_RT.pdf
6. A. Björner, F. Brenti, Combinatorics of Coxeter Groups. Graduate Texts in Mathematics, vol. 231 (2005)
7. A. Borel, N. Wallach, *Continuous Cohomology, Discrete Subgroups, and Representations of Reductive Groups*. Mathematical Surveys and Monographs, vol. 67, 2nd edn (2000)
8. N. Bourbaki, *General Topology: Chapters 1–4* (Springer, 2013)
9. G. Boxer, F. Calegari, T. Gee, V. Pilloni, Modularity theorems for abelian surfaces. https://arxiv.org/abs/2502.20645
10. F. Brenti, Kazhdan-Lusztig polynomials: History Problems, and Combinatorial Invariance. Séminaire Lotharingien de Combinatoire, vol. 49 (2003)
11. C. Breuil, Ext^1 localement analytique et compatibilité local-global. Am. J. Math. **141**, 611–703 (2019)
12. C. Breuil, Y. Ding, Higher $\mathcal{L}$ -invariants for $\mathrm{GL}_3(\mathbb{Q}_p)$ and local-global compatibility. Cambridge J. Math. **8**, 775–951 (2020)
13. C. Breuil, Y. Ding, Sur un problème de compatibilité local-global localement analytique. Mem. Am. Math. Soc. **290**(1442), (2023)
14. C. Bushnell, Induced representations of locally profinite groups. J. Algebra **134**, 104–114 (1990)
15. W. Casselman, M. Osborne, The $\mathfrak{n}$ -cohomology of representations with an infinitesimal character. Compos. Math. **31**, 219–227 (1975)
16. P. Colmez, G. Dospinescu, V. Paškūnas, The p-adic local Langlands correspondence for $\mathrm{GL}_2(\mathbb{Q}_p)$. Cambridge J. Math. **2**, 1–47 (2014)

C. Breuil and Z. Qian, *Splitting and Making Explicit the de Rham Complex of the Drinfeld Space*, Lecture Notes in Mathematics 2397,
https://doi.org/10.1007/978-3-032-20095-2

17. P. Colmez, G. Dospinescu, W. Niziol, Cohomology of p-adic Stein spaces. Invent. Math. **219**, 873–985 (2020)
18. J.-F. Dat, Espaces symétriques de Drinfeld et correspondance de Langlands locale. Ann. Sci. École Norm. Sup. **39**, 1–74 (2006)
19. P. Deligne, Théorèmes de Lefschetz et critères de dégénérescence de suites spectrales. Publ. Math. IHES **35**, 259–278 (1968)
20. F. Digne, J. Michel, *Representations of Finite Groups of Lie Type*. London Mathematical Society Student Texts, vol. 21 (1991)
21. Y. Ding, Towards the wall-crossing of locally $\mathbb{Q}_p$ -analytic representations of $\mathrm{GL}_n(K)$ for a p-adic field K. https://arxiv.org/abs/2404.06315
22. J.D. Dixon, M.P.F. du Sautoy, A. Mann, D. Segal, *Analytic pro-p Groups*. Cambridge Studies in Advanced Mathematics, vol. 61 (1999)
23. G. Dospinescu, A.-C. Le Bras, Revêtements du demi-plan de Drinfeld et correspondance de Langlands p -adique. Ann. Math. **186**, 321–411 (2017)
24. V. Drinfeld, Elliptic modules. Math. USSR Sbornik **23**, 561–592 (1974)
25. V. Drinfeld, Coverings of p-adic symmetric regions. Funct. Anal. Appl. **10**, 107–115 (1976)
26. B. Elias, H. Ko, A single Coxeter presentation. Proc. Lond. Math. Soc. **126**, 837–1062 (2023)
27. M. Emerton, Locally analytic vectors in representations of locally analytic p-adic groups. Mem. Am. Math. Soc. **248**(1175), (2017)
28. J.-M. Fontaine, Représentations p -adiques semi-stables. Astérisque **223**, 113–184 (1994)
29. H. Frommer, The locally analytic principal series of split reductive groups, preprint SFB 478/265 (2003)
30. P. Griffiths, W. Schmid, Locally homogeneous complex manifolds. Acta Math. **123**, 253–302 (1969)
31. A. Grothendieck, J. Dieudonné, EGA III. Étude cohomologique des faisceaux cohérents, Première partie. Publ. Math. IHES, vol. 11, 5–167 (1961)
32. F. Herzig, A Satake isomorphism in characteristic p. Compos. Math. **147**, 263–283 (2011)
33. J.E. Humphreys, *Representations of Semisimple Lie Algebras in the BGG Category* $\mathcal{O}$. Graduate Studies in Mathematics, vol. 94 (2008)
34. A. Iovita, M. Spiess, Logarithmic differential forms on p-adic symmetric spaces. Duke Math. J. **110**, 253–278 (2001)
35. A. Jena, A. Lahiri, M. Strauch, Translation functors for locally analytic representations. Int. J. Number Theory **20**, 2611–2640 (2024)
36. D. Junger, Cohomologie de de Rham du revêtement modéré de l'espace de Drinfeld. J. Inst. Math. Jussieu **23**, 2733–2775 (2024)
37. D. Kazhdan, G. Lusztig, Representations of Coxeter groups and Hecke algebras. Invent. Math. **53**, 165–184 (1979)
38. R. Kiehl, Theorem A und Theorem B in der nichtarchimedischen Funktionentheorie. Invent. Math. **2**, 256–273 (1967)
39. W. Knapp, *Representation Theory of Semisimple Groups*. Princeton Landmarks in Mathematics and Physics (Princeton University Press, 2001)
40. J. Kohlhaase, Invariant distribution on p-adic analytic groups. Duke Math. J. **137**, 19–62 (2007)
41. J. Kohlhaase, The cohomology of locally analytic representations. J. Reine Angew. Math. **651**, 187–240 (2011)
42. B. Kostant, Lie algebra cohomology and the generalized Borel-Weil theorem. Ann. Math. **74**, 329–387 (1961)
43. T. Féaux de Lacroix, Einige Resultate über die topologischen Darstellungen p-adischer Liegruppen auf unendlich dimensionalen Vektorräumen über einem p-adischen Körper. Schriftenreihe Math. Inst. Univ. Münster, vol. 23 (1999)
44. M. Lazard, Groupes analytiques p-adiques. Publ. Math. I.H.E.S **26**, (1965)
45. Y. Morita, A. Murase, Analytic representations of SL_2 over a p-adic number field. J. Fac. Sci. Univ. Tokyo **28**, 891–905 (1982)
46. Y. Morita, Analytic representations of SL_2 over a p-adic number field II. Prog. Math. **46**, 282–297 (1984)

47. Y. Morita, Analytic representations of SL_2 over a p-adic number field III. Adv. Stud. Pure Math. **7**, 185–222 (1985)
48. S. Orlik, On extensions of generalized Steinberg representations. J. Algebra **293**, 611–630 (2005)
49. S. Orlik, Equivariant vector bundles on Drinfeld's upper half space. Invent. Math. **172**, 585–656 (2008)
50. S. Orlik, The de Rham cohomology of Drinfeld's half space. Münster J. Math. **8**, 169–179 (2015)
51. S. Orlik, M. Strauch, On the irreducibility of locally analytic principal series representations. Represent. Theory **14**, 713–746 (2010)
52. S. Orlik, M. Strauch, On Jordan-Hölder series of some locally analytic representations. J. Am. Math. Soc. **28**, 99–157 (2015)
53. L. Pan, On locally analytic vectors of the completed cohomology of modular curves II. Ann. Math. https://arxiv.org/abs/2209.06366
54. D. Patel, T. Schmidt, M. Strauch, Locally analytic representations of $GL(2, L)$ via semistable models of $\mathbb{P}^1$. J. Inst. Math. Jussieu **18**, 125–187 (2019)
55. K. Pohlkamp, *Randwerte holomorpher Funktionen auf p-adischen symmetrischen Raïnen*. Diplomarbeit Universität Münster, 2004
56. Z. Qian, On $\text{Ext}^\bullet$ between locally analytic generalized Steinberg with applications. https://arxiv.org/abs/2512.24279
57. M. Rapoport, E. Viehmann, Towards a theory of local Shimura varieties. Münster J. Math. **7**, 273–326 (2014)
58. M. Rapoport, T. Zink, Period spaces for p -divisible groups. Ann. Math. Stud. **141**, (1996)
59. D. Renard, *Représentations des groupes réductifs p-adiques*, vol. 17. Cours spécialisés. Société Mathématique de France (2010)
60. J. Rodrigues Jacinto, J.E. Rodríguez Camargo, Solid locally analytic representations of p-adic Lie groups. Represent. Theory **25**, 962–1024 (2022)
61. T. Schmidt, Auslander regularity of p-adic distribution algebras. Represent. Theory **12**, 37–57 (2008)
62. T. Schmidt, Analytic vectors in continuous p-adic representations. Compos. Math. **145**, 247–270 (2009)
63. T. Schmidt, Verma modules over p-adic Arens-Michael envelopes of reductive Lie algebras. J. Algebra **390**, 160–180 (2013)
64. P. Schneider, The cohomology of local systems on p-adically uniformized varieties. Math. Ann. **293**, 623–650 (1992)
65. P. Schneider, *Nonarchimedean Functional Analysis*. Springer Monographs in Mathematics (2002)
66. P. Schneider, *p-adic Lie Groups*. Grundlehren der Mathematischen Wissenschaften, vol. 344 (Springer, 2011)
67. P. Schneider, U. Stuhler, The cohomology of p-adic symmetric spaces. Invent. Math. **105**, 47–122 (1991)
68. P. Schneider, U. Stuhler, Resolutions for smooth representations of the general linear group of a local field. J. Reine Angew. Math. **436**, 19–32 (1993)
69. P. Schneider, J. Teitelbaum, $U(\mathfrak{g})$-finite locally analytic representations. Represent. Theory **5**, 111–128 (2001)
70. P. Schneider, J. Teitelbaum, Locally analytic distributions and p-adic representation theory, with applications to GL_2. J. Am. Math. Soc. **15**, 443–468 (2002)
71. P. Schneider, J. Teitelbaum, p-adic boundary values. Astérisque **278**, 51–125 (2002)
72. P. Schneider, J. Teitelbaum, Algebras of p-adic distributions and admissible representations. Invent. Math. **153**, 145–196 (2003)
73. P. Schneider, J. Teitelbaum, Duality for admissible locally analytic representations. Represent. Theory **9**, 297–326 (2005)
74. P. Scholze, J. Weinstein, Berkeley lectures on p -adic geometry. Ann. Math. Stud. **207**, (2020)

75. B. Schraen, Représentations p -adiques de $\mathrm{GL}_2(L)$ et catégories dérivées. Israel J. Math. **176**, 307–361 (2010)
76. B. Schraen, Représentation localement analytiques de $\mathrm{GL}_3(\mathbb{Q}_p)$. Ann. Sci. École Norm. Sup. **44**, 43–145 (2011)
77. M.-F. Vignéras, Extensions between irreducible representations of a p -adic GL_n. Pacific J. Math. **181**, 349–357 (1997)
78. A. Weibel, *An Introduction to Homological Algebra*. Cambridge Studies in Advanced Mathematics, vol. 38 (Cambridge University Press, 1994)
79. A.V. Zelevinsky, Induced representations of reductive p -adic groups. II. On irreducible representations of $\mathrm{GL}(n)$. Ann. Sci. École Norm. Sup. **13**, 165–210 (1980)

Index

C. Breuil and Z. Qian, *Splitting and Making Explicit the de Rham Complex of the Drinfeld Space*, Lecture Notes in Mathematics 2397,
https://doi.org/10.1007/978-3-032-20095-2

LECTURE NOTES IN MATHEMATICS Springer

Editors in Chief: J.-M. Morel, B. Teissier

Editorial Policy

1. Lecture Notes aim to report new developments in all areas of mathematics and their applications – quickly, informally and at a high level. Mathematical texts analysing new developments in modelling and numerical simulation are welcome. See Lecture Notes in Mathematics for the existing online volumes of LNM.

   Manuscripts should be reasonably self-contained and rounded off. Thus, they may, and often will, present not only results of the author but also related work by other people. They may be based on specialised lecture courses. Furthermore, the manuscripts should provide sufficient motivation, examples and applications. This clearly distinguishes Lecture Notes from journal articles or technical reports which normally are very concise. Articles intended for a journal but too long to be accepted by most journals, usually do not have this "lecture notes" character. For similar reasons it is unusual for doctoral theses to be accepted for the Lecture Notes series, though habilitation theses may be appropriate.

2. Besides monographs, multi-author manuscripts resulting from SUMMER SCHOOLS or similar INTENSIVE COURSES are welcome, provided their objective was held to present an active mathematical topic to an audience at the beginning or intermediate graduate level (a list of participants should be provided).

   The resulting manuscript should not be just a collection of course notes but should require advance planning and coordination among the main lecturers. The subject matter should dictate the structure of the book. This structure should be motivated and explained in a scientific introduction, and the notation, references, index and formulation of results should be, if possible, unified by the editors. Each contribution should have an abstract and an introduction referring to the other contributions. In other words, more preparatory work must go into a multi-authored volume than simply assembling a disparate collection of papers, communicated at the event.

3. Manuscripts should be submitted online either at www.editorialmanager.com/lnm to Springer's mathematics editorial in Heidelberg, or electronically to one of the series editors. Authors should be aware that incomplete or insufficiently close-to-final manuscripts almost always result in longer refereeing times and nevertheless unclear referees' recommendations, making further refereeing of a final draft necessary. The strict minimum amount of material that will be considered should include a detailed outline describing the planned contents of each chapter, a bibliography and several sample chapters. Parallel submission of a manuscript to another publisher while under consideration for LNM is not acceptable and can lead to rejection.

4. The type of material considered for publication includes:

   - Research monographs
   - Lectures on a new field or presentations of a new angle in a classical field
   - Summer schools and intensive courses on topics of current research

In general, **monographs** will be sent out to at least 2 external referees for evaluation.

A final decision to publish can be made only on the basis of the complete manuscript, however a refereeing process leading to a preliminary decision can be based on a pre-final or incomplete manuscript.

Volume Editors of **multi-author works** are expected to arrange for the refereeing, to the usual scientific standards, of the individual contributions. If the resulting reports can be forwarded to the LNM Editorial Board, this is very helpful. If no reports are forwarded or if other questions remain unclear in respect of homogeneity etc., the series editors may wish to consult external referees for an overall evaluation of the volume.

5. Manuscripts should in general be submitted in English. Final manuscripts should contain at least 100 pages of mathematical text and should always include:
   - A table of contents
   - An informative introduction, with adequate motivation and perhaps some historical remarks: it should be accessible to a reader not intimately familiar with the topic treated
   - A subject index: as a rule, this is genuinely helpful for the reader
   - For evaluation purposes, manuscripts should be submitted as pdf files.
6. Careful preparation of the manuscripts will help keep production time short besides ensuring satisfactory appearance of the finished book in print and online. After acceptance of the manuscript authors will be asked to prepare the final LaTeX source files plus the corresponding pdf- or zipped ps-file (Preparing your Manuscript). You can also download our LaTeX package for monographs. The LaTeX source files are essential for producing the full-text online version of the book. For more detailed information, please visit the Book Manuscript Guidelines and Key Style Points. The technical production of a Lecture Notes volume takes approximately 12 weeks. Additional instructions, if necessary, are available on request from lnm@springer.com.
7. Authors receive a total of 30 free copies of their volume and free access to their book on SpringerLink, but no royalties. They are entitled to a discount of 40 % on the price of Springer books purchased for their personal use, if ordering directly from Springer.
8. Commitment to publish is made by a *Publishing Agreement*; contributing authors of multi-author books are requested to sign a *Consent to Publish form*. Springer-Verlag registers the copyright for each volume. Authors are free to reuse material contained in their LNM volumes in later publications: a brief written (or e-mail) request for formal permission is sufficient.

Addresses:

Professor Jean-Michel Morel, City University Hongkong, Kowloon Tong, Hong Kong
E-mail: moreljlmichel@gmail.com

Professor Bernard Teissier, Equipe Géométrie et Dynamique,
Institut de Mathématiques de Jussieu – Paris Rive Gauche, Paris, France
E-mail: bernard.teissier@imj-prg.fr

Springer: Ute McCrory, Editor, Mathematics, Heidelberg, Germany.
E-mail: lnm@springer.com

The manufacturer's authorised representative in the EU is Springer Nature Customer Service Centre GmbH, Europaplatz 3, 69115 Heidelberg, Germany. If you have any concerns regarding our products, please contact ProductSafety@springernature.com

Printed and bound by CPI Group (UK) Ltd, Croydon, CR0 4YY
25/09/2026
02222850-0004